Geomicrobiology of Río Tinto

Geomicrobiology of Río Tinto

A Terrestrial Mars Analogue

Edited by
Ricardo Amils
Universidad Autónoma de Madrid, Spain

ELSEVIER

ACADEMIC PRESS
An imprint of Elsevier

Academic Press is an imprint of Elsevier
125 London Wall, London EC2Y 5AS, United Kingdom
50 Hampshire Street, 5th Floor, Cambridge, MA 02139, United States

ISBN: 978-0-443-32948-7

For information on all Academic Press publications visit our website at https://www.elsevier.com/books-and-journals

Publisher: Mara Conner
Acquisitions Editor: Jessica Mack
Editorial Project Manager: Manvi Sharma
Production Project Manager: M. Balakrishnan
Cover Designer: Greg Harris

Typeset by TNQ Tech

Contents

Contributors

José P. Abad
Department of Molecular Biology, Universidad Autónoma de Madrid, Madrid, Spain

Sergey Abramov
Stuttgart University, Environmental Microbiology, Stuttgart, Germany

Ángeles Aguilera
Departmento de Evolución Molecular, Centro de Astrobiología (CAB), INTA-CSIC, Torrejón de Ardoz, Madrid, Spain

Linda Amaral-Zettler
Department of Marine Microbiology and Biogeochemistry, NIOZ Royal Netherlands Institute for Sea Research, Texel, Netherlands; Department of Freshwater and Marine Ecology, Institute for Biodiversity and Ecosystem Dynamics, University of Amsterdam, Amsterdam, Netherlands

Ricardo Amils
Universidad Autónoma de Madrid, Centro de Biología Molecular Severo Ochoa, Department of Molecular Biology, Madrid, Spain; Centro de Astrobiología, Department of Planetology and Habitability, Madrid, Spain

Miguel Arribas-Tiemblo
Departmento de Planetología y Habitabilidad, Centro de Astrobiología (INTA-CSIC), Torrejón de Ardoz, Madrid, Spain; Universidad de Alcalá de Henares, Alcalá de Henares, Madrid, Spain

Carmen Conde
Instituto de Geociencias (CSIC-UCM), Madrid, Spain

Ignacio Carrasco-Ropero
Departmento de Planetología y Habitabilidad, Centro de Astrobiología (INTA-CSIC), Torrejón de Ardoz, Madrid, Spain; Centro de Biología Molecular Severo Ochoa, Universidad Autónoma de Madrid, Cantoblanco, Madrid, Spain; Centro de Biología Molecular Severo Ochoa (CSIC-UAM), Interactions with the Environment, Madrid, Spain

Graciela de Diego-Castilla
Departmento de Evolución Molecular, Centro de Astrobiología (CAB), INTA-CSIC, Torrejón de Ardoz, Madrid, Spain

Vicenta de la Fuente
Universidad Autónoma de Madrid, Campus Cantoblanco, Madrid, Spain

Cristina Escudero
Eberhard-Karls-University Tuebingen, Department of Geosciences, Tübingen, Germany

Angelo Farci
Atalaya Mining Plc, Huelva, Spain

David C. Fernández-Remolar
Macau University of Science and Technology, SKL Lunar and Planetary Sciences, Taipa, Macau; CNSA Macau Center for Space Exploration and Science, Macau, China

Antonio García-Moyano
NORCE Research, Bergen, Norway

Javier Gómez-Elvira
Department of Instrumentation and Space Exploration, Centro de Astrobiologia (CAB), INTA-CSIC, Madrid, Spain; Instituto Nacional de Técnica Aeroespacial, Department of Payloads, Madrid, Spain

David Gómez-Ortiz
Universidad Rey Juan Carlos, Departamento de Geologia, Fisica y Quimica inorganica, Mostoles, Spain; Research Group 'Geofísica y Geoquímica Ambiental', Universidad Rey Juan Carlos, Mostoles, Spain

Felipe Gómez
Departmento de Planetología y Habitabilidad, Centro de Astrobiología (INTA-CSIC), Torrejón de Ardoz, Madrid, Spain

Elena González-Toril
Centro de Astrobiología (CAB), INTA-CSIC, Torrejón de Ardoz, Madrid, Spain; Instituto de Técnica Aeroespacial Esteban Terradas (INTA), Torrejón de Ardoz, Madrid, Spain

Ting Huang
Macau University of Science and Technology, SKL Lunar and Planetary Sciences, Taipa, Macau; CNSA Macau Center for Space Exploration and Science, Macau, China

Andrew H. Knoll
Harvard University, Department of Organismic and Evolutionary Biology, Cambridge, MA, United States

Laurence Lemelle
LGL-TPE, ENS de Lyon, Univ. de Lyon, CNRS, Lyon, France

Guillermo Lopez-Reyes
Research Group ERICA and LaDIS, Universidad de Valladolid, Edificio Alfonso VIII, Valladolid, Spain

Lara Maldanis
Vrije Universiteit Amsterdam, Earth Science Department, Science Faculty, Amsterdam, the Netherlands

Irma Marín
Department of Molecular Biology, Universidad Autónoma de Madrid, Madrid, Spain

Adrián Martínez-Bonilla
Centro de Biología Molecular Severo Ochoa (CSIC-UAM), Interactions with the Environment, Madrid, Spain; Universidad Autónoma de Madrid, Departamento de Biología Molecular, Madrid, Spain; Centro de Biología Molecular Severo Ochoa (CSIC-UAM), Universidad Autónoma de Madrid, Madrid, Spain

Germán Martínez
Department of Instrumentation and Space Exploration, Centro de Astrobiologia (CAB), INTA-CSIC, Madrid, Spain

Daniel Sánchez Mata
Department of Plant Biology II, Faculty of Pharmacy, Universidad Complutense, Ciudad Universitaria, Madrid, Spain; Department of Organismic and Evolutionary Biology, Harvard University Herbaria, Cambridge, MA, United States; Real Colegio Complutense at Harvard University, Cambridge, MA, United States

Guillermo Mateos
Universidad Autónoma de Madrid, Departamento de Biología Molecular, Madrid, Spain; Centro de Biología Molecular Severo Ochoa (CSIC-UAM), Universidad Autónoma de Madrid, Madrid, Spain

Catalina del Moral
Centro de Biología Molecular Severo-Ochoa, Madrid, Spain

Mercedes Moreno-Paz
Centro de Astrobiología (CAB), INTA-CSIC, Torrejón de Ardoz, Madrid, Spain

Sara Navarro
Department of Instrumentation and Space Exploration, Centro de Astrobiologia (CAB), INTA-CSIC, Madrid, Spain

Monike Oggerin de Orube
Centro de Biología Molecular Severo-Ochoa, Madrid, Spain

Victor Parro
Department of Molecular Evolution, Centro de Astrobiologia (CAB), INTA-CSIC, Torrejón de Ardoz, Madrid, Spain

Carlos Pernas-Pleite
Department of Molecular Biology, Universidad Autónoma de Madrid, Madrid, Spain

Fernando Puente-Sánchez
Department of Aquatic Sciences and Assessment, Swedish University of Agricultural Sciences, Uppsala, Sweden

Esteban Ramírez
Real Colegio Complutense at Harvard University, Cambridge, MA, United States; Department of Organismic and Evolutionary Biology, Harvard University Herbaria, Cambridge, MA, United States

Jose A. Rodriguez-Manfredi
Department of Instrumentation and Space Exploration, Centro de Astrobiologia (CAB), INTA-CSIC, Madrid, Spain

Nuria Rodríguez
Centro de Astrobiología (INTA-CSIC), Madrid, Spain

Lourdes Rufo
Instituto de Investigaciones Biosanitarias, Facultad de Ciencias Experimentales, Universidad Francisco de Vitoria, Madrid, Spain

Fernando Rull
Research Group ERICA and LaDIS, Universidad de Valladolid, Edificio Alfonso VIII, Valladolid, Spain

Irene Sánchez-Andrea
Science & Technology School, IE University, Segovia, Spain

Irene Sánchez-Gavilán
Real Colegio Complutense at Harvard University, Cambridge, MA, United States

Jorge Sanjurjo-Sánchez
Universidade da Coruña, Instituto Universitario de Xeología, Campus de Elviña, A Coruña, Spain

Jose Luis Sanz
Universidad Autónoma de Madrid, Madrid, Spain

Aurelio Sanz
Research Group ERICA and LaDIS, Universidad de Valladolid, Edificio Alfonso VIII, Valladolid, Spain

Eduardo Sebastián
Department of Instrumentation and Space Exploration, Centro de Astrobiologia (CAB), INTA-CSIC, Madrid, Spain

Alexandre Simionovici
ISTerre, Univ. Grenoble Alpes, Univ. Savoie Mont Blanc, CNRS, IRD, IFSTTAR, Grenoble, France

Virginia Souza-Egipsy
Instituto Estructura de la Materia, CSIC, Madrid, Spain

Fernando Tornos
Instituto de Geociencias (CSIC-UCM), Madrid, Spain

Josefina Torres
Instituto Nacional de Técnica Aeroespacial, Department of Payloads, Madrid, Spain

Jorge L. Vago
European Space Agency-ESTEC, Noordwijk, Netherlands

Raquel Vega
Atalaya Mining Plc, Huelva, Spain

Erik Zettler
Department of Marine Microbiology and Biogeochemistry, NIOZ Royal Netherlands Institute for Sea Research, Horntje, the Netherlands

Preface

Our scientific adventure on Río Tinto, the Red River, started 40 years ago when our group at the Department of Molecular Biology at the Universidad Autónoma de Madrid became involved in a European project studying the use of Río Tinto thermophilic microorganisms in bioleaching operations. Although the project was not technologically successful, we were fascinated by the microbial ecology of this peculiar extreme acidic ecosystem, Río Tinto. At almost a hundred kilometers in length, it was considered the product of 5000 years of mining activities in the area.

Early observations of the microbial ecology of the river disagreed with the then prevalent idea of an ecosystem produced by mining contamination. But when the oldest terraces of the river were dated and found to be several million years old and to have a mineralogy that required the same geochemistry as that of today, the hypothesis of mining activity as the origin of the river was disproved. After joining the recently inaugurated Centro de Astrobiología (CAB), an associated member of the NASA Astrobiology Institute (NAI), and the incorporation of Río Tinto as one of the model systems at CAB, collaborative study of Río Tinto with NAI members was initiated, giving astrobiological significance to our work, and giving rise to the river being considered one of the best geochemical and mineralogical analogues of Mars. Many researchers have focused their work on diverse aspects of Río Tinto, from its geological and paleontological aspects to the microbial ecology of the system, the development of technologies to be applied on Earth and in current and future Mars exploration missions and, of course, its debatable origin. This book aims to present the work done by twenty research groups over the last 30 years, with diverse perspectives on Río Tinto and its origin, the Iberian Pyrite Belt. We believe that this broad compendium encompasses most of the current knowledge in an array of complementary areas, thus facilitating the transdisciplinary approach required for its use as an astrobiological model.

The first section of the book is devoted to the geology, hydrology and paleontology of Río Tinto. Chapter 1 explains in detail the origin of the Iberian Pyrite Belt and the geological characteristics that allowed so many mining activities to be operational in the area. Chapter 2 gives a historical perspective of the Río Tinto ecosystem centered on its sedimentology and geobiology, which were essential for its consideration as one of the best geochemical and mineralogical analogues of Mars. This chapter also covers the paleontology of the system, fundamental for the design of exploration missions to discover signs of life on the Red Planet. Chapter 3 explains the hydrological and hydrogeochemical characteristics detected at the source of Río Tinto.

The most extensive part of the book is dedicated to the microbial and eukaryotic diversity associated with the Río Tinto basin. Chapter 4 covers the prokaryotic diversity identified along the course of the river, especially the chemolithotrophy that creates the extreme conditions detected in the river. Chapter 5 is centered on the microbial ecology of the anoxic sediments existing along the course of Río Tinto and its possible applications. Chapter 6 covers the characterization of the cyanobacteria associated with the most extreme conditions detected in the river. Chapter 7 is devoted to the examination of two distinct microbial communities detected in close proximity in the deep subsurface of the Iberian Pyrite Belt. The identification of actively growing cyanobacteria in the so-called dark biosphere resolved a long-standing conundrum in this area of research. Chapter 8 covers the predominant, mainly photosynthetic, eukaryotic microbial diversity, existing in the Tinto basin. Chapter 9 is devoted to the fungal diversity identified along the course of the river, its relationship with the high concentration of toxic metals existing in the ecosystem, and their possible biotechnological applications. Chapter 10 is a compendium of the vegetation that is found associated with the Río Tinto basin, which has been ignored until recently due to the view that Río Tinto was a contaminated river. Chapter 11 covers the peculiar observed biomineralizations as well as the metal content detected in various tissues of the plants growing in the Tinto ecosystem.

The third part of the book covers research carried out in preparation for current and future Mars exploration missions as a consequence of Río Tinto's status as a good Mars terrestrial analogue. Chapter 12 covers the main spectroscopic methodologies, especially Raman, designed to explore the mineralogical characteristics of Mars and tested in Río Tinto as a Mars analog. Chapter 13 is devoted to the characterization of methanogenic archaea, producers of methane in the Tinto basin, and the evaluation of the possible biogenicity of the methane detected on Mars. Chapter 14 evaluates the possible metabolisms operating on Mars using the experimental results obtained in Río Tinto and the detection of possible biosignatures. Chapter 15 offers an interesting outlook on the technology developed for the surface and subsurface characterization of the Iberian Pyrite Belt and its possible use in future Mars exploration missions.

The last part of the book is devoted to fundamentals and applications of Río Tinto geomicrobiology. Chapter 16 covers the research done in our attempt to understand the anaerobic oxidation of iron in the deep subsurface of the Iberian Pyrite Belt, which gave rise to the extreme conditions detected in the river. Chapter 17 is devoted to the characterization of the biogeochemical cycles operating in the deep subsurface of the Iberian Pyrite Belt and their relation to the extreme conditions existing in the Tinto basin. The use of metatranscriptomics to decipher the microbial metabolisms operating under the polyextreme conditions existing in Río Tinto is described in Chapter 18. Chapter 19 is devoted to the diverse biotechnological applications that have been investigated using organisms isolated from the Río Tinto basin. And finally, Chapter 20 explores the production of vitamin B12, a critical component required for an operative Iberian Pyrite Belt deep subsurface.

The authors and the editor would like to thank Juan Pérez Mercader, the former Director of the Centro de Astrobiología, for his outstanding support for the work done in Río Tinto.

SECTION

1

Geology, hydrology, and paleontology

CHAPTER

The geological framework of the giant Riotinto massive sulfide district

1

Carmen Conde[1], Fernando Tornos[1], Angelo Farci[2], Raquel Vega[2]
[1]*Instituto de Geociencias (CSIC-UCM), Madrid, Spain;* [2]*Atalaya Mining Plc, Huelva, Spain*

Introduction

The Iberian Pyrite Belt (IPB) is widely regarded as the most outstanding massive sulfide province in the world. It hosts one of the largest concentrations of sulfides on Earth (Laznicka, 1999). The IPB contains estimated total resources originally in place ranging from approximately 1600 Mt to over 2500 Mt of massive sulfides, along with significant resources in the underlying stockwork or feeder zones (estimated between ~250 Mt and >2000 Mt). The IPB includes about 10 “giant” and “world-class” massive sulfide deposits, typically defined as containing over 50 Mt or 100 Mt of ore, respectively, about 80 smaller orebodies with more than 1 Mt, and hundreds of smaller prospects.

Traditionally, the IPB deposits have been compared to the Kuroko type of volcanic-hosted massive sulfide (VMS) deposits (Franklin et al., 1981; Sawkins, 1990). However, their geological setting within a putative transpressive continental basin infilled by dominantly felsic volcanic rocks interbedded with black shale and formed above a siliciclastic sequence deposited in a stable platform, imposes unique features that differentiate the IPB from other VMS districts. These characteristics include extremely high pyrite contents, a large proportion of biogenic sulfur, significant amounts of shale-hosted massive sulfides, and local unusually high Sn grades contents (Tornos, 2006).

The Riotinto mining district is the largest and most emblematic of the giant VMS deposits in the IPB. It is also one of the best-known mining sites globally, recognized for the dimensions of the orebody and its intense mining history, which make it one of the most charismatic mine districts.

Riotinto: Mining history and current operations

The Riotinto district is renowned as one of the world’s oldest mining districts, boasting a history of activity spanning over 4500–5000 years (Pinedo, 1963; Avery, 1974; Rothenberg and Blanco-Freijeiro, 1981; Rothenberg and Pérez, 1987). The oldest direct evidence of systematic mining exploitation dates back to at least 1000 BC,

Geomicrobiology of Río Tinto. https://doi.org/10.1016/B978-0-443-32948-7.00009-6

in the Tartessian times. Later on, numerous cultures engaged in mining in this area, including the Phoenicians, Greeks and Romans (Fig. 1.1a). The Roman period was characterized by large-scale, industrial-level mining, marked by the introduction of new technologies that helped to mine below the water level and in large underground works (Rothenberg and Pérez, 1987). During this time, the primary focus was on exploiting the base of the gossans and the supergene cementation zones to recover copper, gold, and silver. Enduring testimonies to this extensive activity include approximately 16 Mt of Roman slag, as well as historical adits, shafts, and the ruins of ancient smelters and settlements.

FIGURE 1.1 Representative Photos of the Long Mining Tradition of the Riotinto Region.

(a) Archeological remains of a Roman smelter that treated ore from the Corta Lago, a large Roman open pit, during the early centuries of the Roman Empire (Corta Lago); (b) historic steam locomotive preserved at Corta Atalaya, illustrating the industrial heritage associated with 20th-century mining activities.

Following the decline of the Roman Empire in the fourth century AD, mining activity decreased dramatically. Mining became sporadic and remained so until the 19th century. The mining potential of the Iberian Pyrite Belt was rediscovered, leading to the mine's sale in 1873 to the "Rio Tinto Company Ltd." (Avery, 1974), a British entity that operated the deposit until 1953. Initially, the company concentrated on extracting high-grade copper ores in both the cementation zone and primary ores. Subsequently, attention shifted toward the pyrite-rich massive sulfides, which were primarily utilized for the production of sulfuric acid. During its operational tenure, the Rio Tinto Company became the largest industry in Spain, employing up to 40,000 miners. Between 1875 and 1976, a total of 128 Mt of ore was extracted. This period is often described as the "golden age" of mining in the region, serving as a significant driver for Spanish and European industry and a focal point of industrialization (Fig. 1.1b). The gradual decline in the use of pyrite for sulfuric acid production and the grade depletion of the copper ores led to a reduction in mining activities.

From 1954 onwards, Spanish companies, including Compañía Española de Minas de Rio Tinto S.A. and Riotinto Minera SA, took over the operations. A crucial development during this phase was the discovery in 1964 of the Cerro Colorado copper deposit, a concealed stockwork mineralization below a gold-rich gossan. This discovery spurred the construction and expansion of a copper concentrator from 1969 and a gold leach plant starting in 1971. Mining during this period encompassed stockwork ores from 1973 to 1986 and gold/silver gossans from 1974 to 2002.

Production at Riotinto ceased in 1987 due to low copper prices and stopped entirely in 1990. In 1995, Minas de Riotinto S.A.L. (MRT), a workers' cooperative, acquired the mining rights. MRT operated intermittently, focusing on the extraction of remaining gossans and high-grade stockwork zones. The mine closed again in 2001. Subsequently, the mineral rights were acquired by Mantesur Andévalo SL (MSA) and later by EMED Tartessus in 2004, initiating plans for the restart of operations.

The restart of operations occurred in 2015 under the ownership of Atalaya Mining, and with the name of the Riotinto Copper Project. Atalaya commenced resource drilling in 2014. Current mining activities are concentrated in the Cerro Colorado and Corta Atalaya areas, with ongoing preparatory works supporting the future exploitation of the San Dionisio deposit. The operational infrastructure integrates modern mining facilities with historical infrastructure. The current operation also takes into account aspects of cultural heritage preservation and environmental management, including a policy of progressive rehabilitation and regular reviews of closure plans, aiming to equate environmental compliance with safety and production. The gradual resurgence in metal prices since 2004 has facilitated a renewal of exploration across the Iberian Pyrite Belt, and Riotinto stands as one of the four mines in Spain (with Aguas Teñidas, Sotiel and Magdalena) currently in operation. The property benefits from good road connections via a recently renovated national road system, located 75 km from the port and industrial city of Huelva and 88 km from Seville. Copper concentrate is transported by road to the port of Huelva for ocean shipping.

It is estimated that the Riotinto district originally contained total resources exceeding 2500 Mt of stockwork and massive sulfides, covering an area of approximately 4 km^2. Specifically, an estimated 500 Mt of massive sulfides were originally in place, with estimated grades of 45% sulfur, 0.9% copper, 2.1% zinc, 0.8% lead, 0.5 g/t gold, and 26 g/t silver. The stockwork zone in the Cerro Colorado area alone is estimated to hold over 2000 Mt of altered volcanic rocks having irregular copper grades.

Regional geological setting of the Iberian Pyrite

The Iberian Pyrite Belt is situated within the South Portuguese Zone (SPZ), which represents the southernmost domain of the Iberian Variscan Massif. The SPZ is interpreted as an allochthonous or exotic terrain relative to Iberia, which collided with the Iberian Autochthonous Terrain during the Variscan Orogeny (Silva et al., 1990). This orogeny, which occurred from the late Devonian through the Carboniferous, was characterized by transpressional deformation related to the oblique collision between the two terranes. This oblique tectonism led to the development of a major volcanic belt, the IPB, hosted within a highly compartmentalized sedimentary basin situated along the outermost margin of the SPZ.

The SPZ likely represents an exotic terrain, possibly Avalonia, and its location correlates with zones of the Variscan Arc such as SW Ireland and England, as well as the Rheno-Hercynian and Moravo-Silesian Zones.

The regional stratigraphy of the IPB is marked by a relatively straightforward sequence consisting of three lithostratigraphic units (Schermerhorn, 1971). From footwall to hanging wall: the Phyllite-Quartzite Group (PQ Group), the Volcanic-Sedimentary Complex (VS Complex), and the Baixo Alentejo Flysch Group (also known as the Culm Group), in total exceeding more than 5000 m in thickness. These units reflect an evolutionary depositional setting, transitioning from a stable platform during the sedimentation of the PQ Group to deposition of the VS Complex within continental pull-apart basins, and later collision with the formation of a synorogenic flysch sequence (Silva et al., 1990; Quesada, 1997). However, it has been noted that this simplified stratigraphy is not uniformly applicable across the belt, as the sequence can exhibit significant changes over short distances.

The PQ Group is a monotonous sequence of shale, sandstone, and quartzite of Middle to Late Devonian age (Moreno et al., 1996). Near the top, it includes discontinuous layers of limestone, debris flows, and delta deposits, which mark the start of the basin instability and the configuration of the second-order basin(s) where the VS Complex was deposited (Tornos, 2006). The VS Complex exhibits significant variability in thickness, ranging from 0 to 1300 m, shows a very heterogeneous distribution of rocks, and is the main host unit for the massive sulfides. The VS Complex includes both felsic and mafic rocks ranging in composition from basalt to rhyolite, volcaniclastic sediments, and shale. Its age ranges from the Late Famennian to the early Late Visean. Within the VS Complex, felsic volcanic rocks are predominantly dacite with lesser amounts of rhyolite and andesite, forming (crypto-) dome and sill complexes, all with a calc-alkaline composition. Mafic volcanic rocks include basaltic sills or small submarine lava flows, with local occurrences of pillow lavas observed in some areas, such as at Cerro Colorado; their composition is alkaline to tholeiitic (Mitjavila et al., 1997; Thieblemont et al., 1998; Valenzuela et al., 2011a, 2011b; Conde, 2016; Conde and Tornos, 2020). There are different types of glass, pumice- and crystal-rich breccias and volcaniclastic sandstones, commonly forming large mass-flow deposits derived from monomictic or polymictic sources. These deposits show evidence of subaqueous deposition and soft-sediment deformation.

The uppermost unit is the Baixo Alentejo Flysch Group, a thick and monotonous sequence of turbiditic carbonaceous shale, greywacke, and conglomerate, and its age ranges from the Late Visean to Middle/Late Pennsylvanian. This unit represents extensive sediment accumulation in a foreland basin, with sedimentation progressing southwards synchronously with the advancement of the Variscan front (Oliveira, 1983; Moreno, 1993).

Several interpretations exist regarding the precise geotectonic environment of the formation of the IPB. Proposed settings include an intracontinental pull-apart basin or a forearc trench basin developed during the oblique collision with the Iberian Autochthonous Massif (Silva et al., 1990; Quesada, 1997; Simancas et al., 2003). Another view suggests the IPB formed in an intra-to back-arc basin associated with southward subduction, a geodynamic scenario considered more akin to typical global Volcanogenic Massive Sulfide (VMS) deposits (Tornos et al., 2023).

The currently cryptic Variscan suture is marked by remnants of a small marine basin (represented by the Beja-Acebuches Ophiolite) and an accretionary prism (the Pulo de Lobo Unit). To the south, the IPB is thrust over the Baixo-Alentejo Unit, another unit of the SPZ.

Polycyclic Variscan compressive deformation led to the development of thin-skinned tectonics, which propagated above a midcrustal basal detachment surface (Silva et al., 1990). This resulted in a south-verging fold-and-thrust belt. Key structural elements include south-verging recumbent folds that evolve into low-angle thrust faults. Some of these structures are interpreted as inverted synvolcanic extensional faults. Later, major Variscan structures are crosscut by east-west and northwest-southeast trending normal and strike-slip faults of both Late Variscan and Alpine age (Ribeiro et al., 1990; Silva et al., 1990; Leitão, 1997; Quesada, 1997).

Regional metamorphism throughout the IPB is consistently low-grade, ranging from prehnite-pumpellyite to greenschist facies (Schermerhorn, 1975; Munhá, 1983). The intensity of metamorphism generally decreases from north to south but increases in proximity to major thrust zones (Sanchez España and Velasco, 1999).

Local geological framework of the Riotinto deposit

The orebodies of Riotinto are hosted within a thick volcano-sedimentary sequence that forms part of the Volcano-Sedimentary Complex recognized throughout the whole Iberian Pyrite Belt (Fig. 1.2 a). Structurally, the Riotinto Antiform, an E–W trending, south-verging antiformal stack that exposes hydrothermally altered felsic and mafic volcanic rocks, with extensive stockwork mineralization, dominates the district; it includes several tectonically stacked units (Fig. 1.2). The southern limb includes the Corta Atalaya Syncline, a subvertical tight fold. The interpretation of Mellado et al. (2006) and Díez-Montes et al. (2015) suggest that this unit is located between two out-of-sequence thrust systems (Fig. 1.3). The Culm Formation occupies the uppermost position and overlies the VS Complex along a major extensional detachment surface. This detachment truncates portions of the underlying stratigraphy, forming a footwall ramp geometry.

Lithostratigraphy

The stratigraphic framework of the Riotinto mining district, central to the Iberian Pyrite Belt (IPB), is characterized by a complex volcano-sedimentary sequence structured into four principal lithostratigraphic units.

Although the PQ Group does not crop out directly in the Riotinto area, it underlies the VS Complex regionally and is observable in adjacent zones (Rodríguez et al., 2002). In the Zarandas sector, shales intruded by glassy rhyolite sills, and overlain by mafic volcanic rocks, mark the PQ Group-VS transition (Tornos, 2006). The VS Complex itself reaches up to 1000 m in thickness, though its internal stratigraphy is often obscured by deformation and alteration (García Palomero, 1980, 1992). The composite section described integrates data from the Atalaya open pit, Zarandas, and the Odiel River areas.

From bottom to top, the four main units are:

The *Mafic-Siliciclastic Unit*, up to 450 m thick, consists of interbedded basaltic flows, volcaniclastics (sandstone and siltstone), and shale (see Fig. 1.4a). Basalt dominates the lower part and crops out in a continuous E–W band at the core of the Riotinto antiform. The overlying shale is late Strunian in age and correlates chronologically with sulfide-hosting shales elsewhere in the IPB, though in this area it is barren and lacks hydrothermal alteration (Pereira et al., 1996; Rodríguez et al., 2002; Conde, 2016).

The coherent basalt occurs mainly as submarine flows, featuring pillow structures, peperitic margins, and hyaloclastic breccias, indicative of quenching during seafloor extrusion. The basalt shows ophitic to porphyritic textures with plagioclase, pyroxene, and a chlorite–quartz $\pm$ epidote matrix. Vesicles (0.2–1 cm) are commonly filled with chlorite, carbonate, zeolites, and occasionally quartz and sulfides (Halsall, 1989; Conde, 2016).

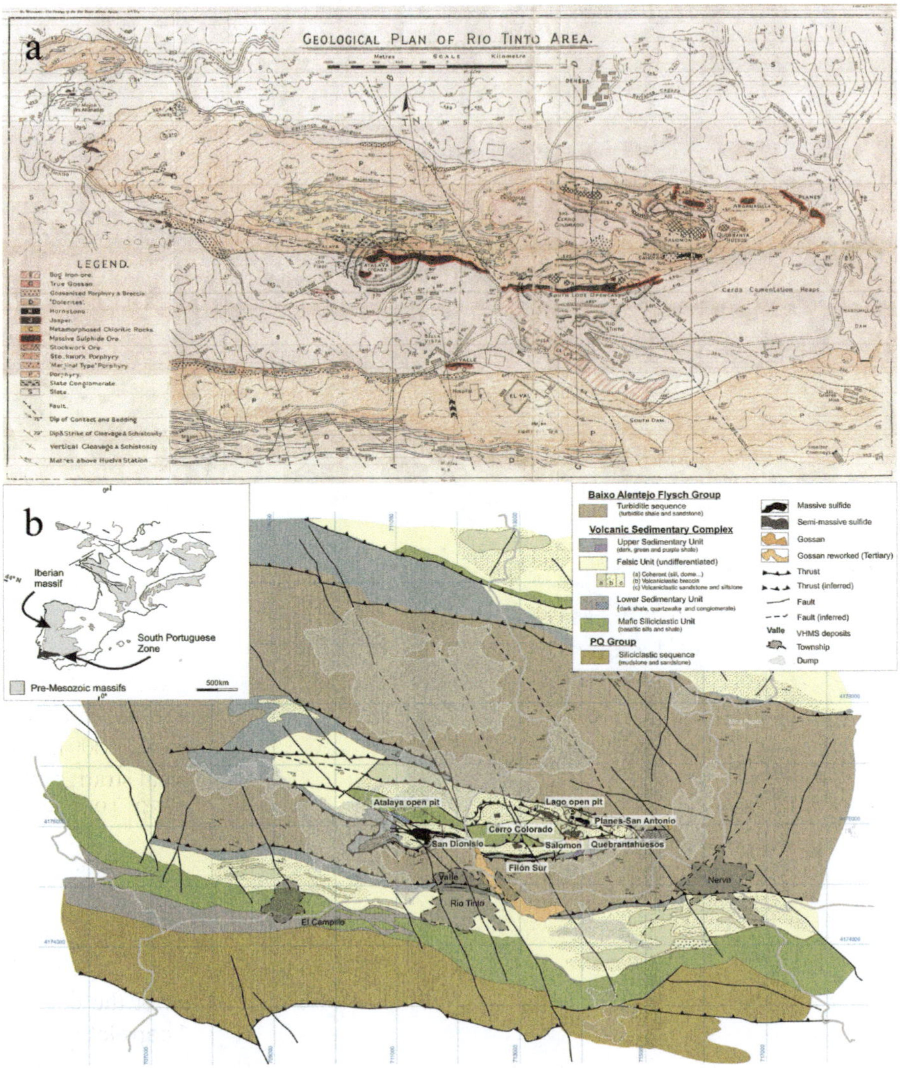

FIGURE 1.2 Geological Maps of the Río Tinto Area.

(a) The first complete geological map of the area was made by Williams (1934), and (b) the latest version illustrating the general geology and main orebodies. ***Map lines delineate study areas and do not necessarily depict accepted national boundaries.***

- Modified after Mellado, D., Clavijo, Tornos, F., Conde, C. 2006. Geología y estructura de la Mina de Río Tinto (Faja Pirítica Ibérica, España). Geogaceta. *40, 231–234; González Clavijo, E., Montes, A.D., 2010. Propuesta de estructura geológica para el depósito gigante de sulfuros masivos volcanogénicos de Río Tinto, Faja Pirítica Ibérica.* Geogaceta. *48, 204–206 and Conde, C. 2016.* Geology and Hydrothermal Evolution of Massive Sulfides of the Iberian Pyrite Belt, Spain. *PhD Thesis.*

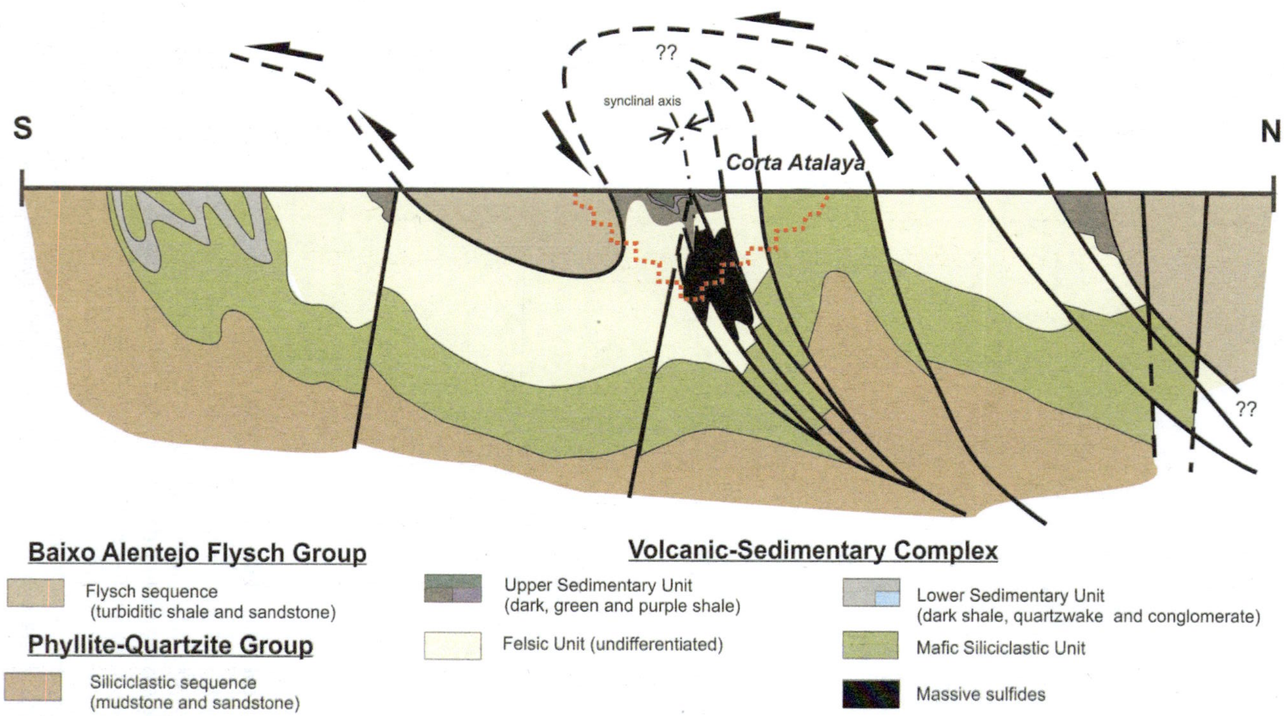

FIGURE 1.3 Simplified geological cross-section of the Río Tinto area.

Showing the main structural and geological features.

- *Modified after Mellado, D., Clavijo, Tornos, F., Conde, C. 2006. Geología y estructura de la Mina de Río Tinto (Faja Pirítica Ibérica, España).* Geogaceta. *40, 231–234; Díez-Montes, A., García-Crespo, J., Ayala, C., Luis García-Lobón, J., Sánchez-García, T., Rey-Moral, C., Bellido, F., Rubio, F., Mediato, J.F., Tornos, F. 2015.* Modelling of the Río Tinto Area. *Springer Science and Business Media LLC, 209–229. https://doi.org/10.1007/978-3-319-17428-0_10 and Conde, C. 2016.* Geology and Hydrothermal Evolution of Massive Sulfides of the Iberian Pyrite Belt, Spain. *PhD Thesis.*

FIGURE 1.4 Photographs of representative outcrops of the volcanic units.

(a) Interbedded shale and mafic volcaniclastic rocks from the Mafic Siliciclastic Unit (Cerro Colorado); (b) porphyritic felsic rocks (Felsic Unit II) with sericitic alteration and postdated pervasive chloritization, (Corta Atalaya); (c) peperitic texture formed during the interaction of a felsic melt with shale, in detail, it observed fragments of rhyodacite and individual phenocrysts of quartz within the shale; (d) rhyolite dome or cryptodome crosscut pervasively altered rhyodacite (Felsic Unit II) and the associated mineralization (Filón Norte).

Some basaltic bodies have intrusive relations, forming shallow-level sills and dykes with peperitic contacts, while a minor group of dykes of the Zarandas area display contact metamorphism, suggesting deeper emplacement. These microgabbro dykes may represent late-stage mafic intrusions into the volcanic-sedimentary complex.

In the upper parts of the unit, shale becomes more abundant and is interbedded with mafic epiclastics derived from erosion of the underlying flows (Conde, 2016). Despite early recognition of these mafic rocks (Williams, 1934), some maps misclassified them as chloritized felsic rocks (Junta de Andalucía, 2004).

Beneath this mafic unit, in the Odiel River area, a subvolcanic porphyritic dacite sill has been identified. Although not observed elsewhere in the Riotinto mine, it shows tectonic contacts with the overlying Mafic Siliciclastic Unit and underlying PQ Group (Valenzuela et al., 2011b).

The *Lower Sedimentary Unit* conformably overlies the Mafic-Siliciclastic Unit and reaches up to 90 m in thickness. It comprises interbedded dark shale, siltstone, and quartz wacke, with a distinctive polymictic conglomerate horizon near the top (García Palomero, 1980). This unit is mainly exposed in the Atalaya open pit and areas to the east. Its absence in Cerro Colorado is likely due to tectonic attenuation between the mafic and felsic units (Junta de Andalucía, 2004).

The conglomerate is shale-supported and poorly sorted, containing subrounded clasts of shale, greywacke, quartz-rich sandstone, felsic volcanic glass (dacite–rhyolite), and lesser mafic lithics. Toward the western part of the basin, the conglomerate becomes coarser and more polymictic, incorporating both mafic and felsic volcaniclastic debris, black shale, and quartz-clast clasts (Badham, 1982). Clast size can reach up to 1 m^3, as reported by Boulter (1993), implying deposition adjacent to a fault-controlled basin margin such as a half-graben.

The *Felsic Unit*, ranging from 10 to 500 m in thickness, is composed of strongly altered dacitic to rhyodacitic volcanic rocks. It is interpreted as a large dome complex with coherent facies, sills, hyaloclastites, and breccias formed in a submarine environment. Primary structures are the best preserved in the Odiel and Zarandas areas, whereas in the Riotinto mine, they are obliterated by the hydrothermal alteration. This unit hosts several replacement massive sulfide orebodies such as the Salomon, Lago and Quebrantahuesos, globally included in the Filón Norte sector.

The felsic volcanic sequence in the Riotinto region comprises at least three distinct magmatic units (de Mello et al., 2022), designated as Felsic Units I through III. Felsic Unit I, predominantly volcaniclastic, is primarily exposed in the northern Riotinto and Zarandas areas and includes transported hyaloclastite and glass- and crystal-rich mass flows. Felsic Unit II, which constitutes the main volume of the felsic sequence, is characterized by coherent volcanic facies, including (crypto)-domes and sills, and is most prevalent in the Atalaya and Cerro Colorado open-pits (Fig. 1.4b). Locally, the intrusive emplacement of felsic rocks into sedimentary sequences led to the formation of well-developed peperite textures along the margins of dykes and coherent intrusions (Fig. 1.4c). The peperites consist of cm-sized rhyodacite fragments embedded in the shale, occurring at distances of up to 1 meter from the contact. Isolated quartz phenocrysts are commonly observed floating within the shale groundmass.

Felsic Unit III, a late-stage rhyolitic intrusion, is observed in the northern Cerro Colorado open-pit where it intrudes hydrothermally altered rhyodacite of Unit II and related mineralized zones (Fig. 1.4d). This rhyolite is distinguished by its dark, glassy, nearly aphyric texture with sparse albite-plagioclase phenocrysts in a fine-grained, quartz-rich matrix, resembling other high-silica (>80 wt% SiO_2) rhyolite domes of the Iberian Pyrite Belt. This Felsic Unit is similar to that described by Valenzuela et al. (2011b) in the Odiel River, ca. 5 km east; however, facies in the Odiel River seem to represent the lateral volcaniclastic aprons of the Felsic Unit II.

The *Upper Sedimentary Unit*, which structurally overlies the Felsic Unit, consists of gray shale with volcaniclastic sandstone interbeds and hosts the Filón Sur orebody. Above the massive sulfides, there is a chert layer and Mn-rich purple shale—indicative of a condensed sedimentary sequence associated with exhalative activity and sea-level rise (Routhier et al., 1980; Badham, 1982; Ohmoto, 1996). This unit reflects a late-stage, localized third order basin, probably in a dome caldera (de Mello et al., 2022) and is assigned an early Tournaisian age (Rodríguez et al., 2002).

Finally, the Baixo Alentejo Flysch Group tectonically overlies the sequence and is composed of turbiditic shale and sandstone. Its >500 m thickness marks the onset of synorogenic sedimentation linked to Variscan basin inversion (Moreno, 1993).

This multi-unit stratigraphy, with clear volcanic, sedimentary, and tectonic controls, defines the Riotinto district as a type example of VHMS-hosting basins within the IPB. It integrates early mafic and felsic volcanism, sedimentation in dynamic subaqueous conditions, and later tectonic stacking—all key elements in the metal endowment of this globally significant mining area.

Lithogeochemistry of the host sequence

The host sequence in Riotinto is dominated by felsic rocks. Chemically, they range from dacite to rhyolite, with massive samples showing a high SiO_2 content (60–84 wt%), the highest values probably related to hydrothermal silicification, and volcaniclastic samples a more restricted range within the dacitic field. Chondrite-normalized REE patterns for felsic rocks display a steep slope with LREE enrichment over HREE and a pronounced negative Eu anomaly, consistent for both massive and volcaniclastic types. Volcaniclastic rocks derived from these domes are also widespread. Geochemically, the andesite to rhyolite compositions are characterized as calc-alkaline with low Al and high Nb. These magmas are interpreted as being of crustal origin, generated by crustal melting driven by underlying mafic intrusions and crustal thinning. Crucially, the felsic unit is geochemically differentiated into two main types based on Zr content: high Zr (>200 ppm) and low Zr (<200 ppm). This distinction is significant for mineralization, as the massive sulfide mineralization is related to the low-Zr felsic rocks, while the high-Zr type is consistently barren (Conde, 2016; de Mello et al., 2022). The high-Zr rhyodacite shows a minor Eu anomaly compared to the more accentuated anomaly in other felsic rocks.

Hydrothermal alteration

The hydrothermal alteration system at Riotinto is a defining feature of the hydrothermal system, and altered rocks are best exposed in the core of the antiform structure, covering approximately 8 km^2 (García Palomero, 1980; Costa, 1996). This alteration system extends to at least 600 m depth and is intimately associated with sulfide-rich stockwork zones. The alteration at Riotinto is irregular and structurally controlled, influenced by lithologic boundaries and the protolith composition (Solomon et al., 1980; Gumiel et al., 2001; Tornos, 2006).

Three primary alteration types are recognized: chloritization, phyllic alteration, and localized silicification. Phyllic alteration dominates in felsic rocks and appears as Ba-illite-quartz replacement with occasional ankerite veining (Costa, 1996). Chloritic zones are prevalent in basalt and shale and comprise ripidolite-picnochlorite ± quartz ± illite assemblages; however, chlorite-rich alteration overprints phyllic alteration of the felsic rocks in the most altered areas, i.e., near the feeder structures. Silicification is limited to major fluid pathways, with the host rocks replaced by almost massive quartz and pyrite.

The stockwork, often regarded as the feeder zone for VHMS systems, includes pervasive veining and disseminated sulfides in the altered volcanic and shale. The vein morphology varies from 1 to 5 cm, sometimes up to 1 m thick, and consists mainly of quartz and pyrite with minor ankerite. Larger veins show banded textures indicative of crack-seal processes (Nehlig et al., 1997). These veins, generally lacking displacement or crosscutting, likely formed during a long-lived extensional hydrofracturing regime under high fluid pressure (Solomon et al., 1980; Nehlig et al., 1997).

Geochemical zoning is evident within the alteration zones: chloritized shale tends to be enriched in copper, while sericitic rocks are more associated with lead and silver. Zn enrichment is variable. Key stockwork areas include Planes-San Antonio, Quebrantahuesos, and San Dionisio, where high-grade zones of pyrite, chalcopyrite, sphalerite, and galena occur within altered rhyodacite, basalt, and shale (Williams et al., 1975; García Palomero, 1980). Stockwork Cu grades at San Dionisio average 1.7% Cu (Solomon et al., 1980), with additional enrichment in Sn, Bi, and Au (Badham, 1982; Marcoux and Leistel, 1996).

A second generation of veins formed during Variscan deformation is also present. These veins are larger, quartz-rich, cut the regional cleavage, and often host coarse-grained pyrite and chalcopyrite. Unlike early veins, they are not associated with alteration halos but may be economically significant, with higher gold and copper contents (Gumiel et al., 2001). Fluid inclusions within late quartz suggest a distinct, possibly metamorphic or basinal fluid origin (Nehlig et al., 1997).

Geochemically, altered rocks show depletion in mobile elements such as Ca, Na, and K, and enrichment in elements like Al, Fe, and Mg. Phyllic and chloritic alteration show different trends. Phyllic alteration results in depletion of Na and Ca, while chloritization causes a specific enrichment in Fe and Mg. An external, less intense halo of albitization (±chlorite) in felsic rocks and widespread spilitization in mafic rocks are also noted, potentially related to seawater-rock interaction. Spilitization replaces primary minerals in mafic rocks with precipitation of albite, clinoamphibole, quartz, chlorite, epidote, and minor apatite, zircon, and ilmenite (Conde, 2016).

Geochronology

A recent U–Pb geochronological study of zircon from the felsic volcanic rocks provides insights into the timing of volcanism and mineralization (de Mello et al., 2022).

The felsic magmatism occurred in three distinct pulses (de Mello et al., 2022). The earliest felsic volcanism (Felsic Unit I), suggests that the high Zr volcaniclastic rhyodacite lateral apron to domes is of Early Tournaisian age (356 ± 1 Ma).

The main stage of massive sulfide mineralization is genetically related to a coherent rhyodacitic unit (Felsic Unit II). The representative samples from this unit, collected at Zarandas, Filón Norte, and the Atalaya open pits, yielded ages centered around 349.5 ± 0.4 Ma, corresponding to dome-related coherent rhyodacite.

The later, volumetrically minor pulse of unaltered and unmineralized intrusive rhyolite (Felsic Unit III) at Cerro Colorado has a younger concordia age of 340 ± 2 Ma (for autocrystic zircon); a population of older antecrystic zircon has yielded a concordia age of 350 ± 1 Ma (average 350 ± 3 Ma), suggesting some inheritance from the associated rocks. The main mineralizing event is therefore closely linked to the emplacement of the low-Zr felsic volcanic rocks between approximately 352 and 348 Ma.

Structural framework of the Riotinto deposit

The interpretation of Riotinto's structural setting has evolved over time, building upon early pioneering works by Williams (1934, 1966) and Williams et al. (1975), which proposed a simple synclinal-anticlinal structure with a vertical axis for the mine area (see Fig. 1.2a). These initial views were later refined by Rambaud (1969) and García Palomero (1980, 1992). More recent geological mapping (see Fig. 1.2b) and analysis have provided a more detailed understanding (Mellado et al., 2006; González Clavijo and Díez-Montes, 2010; Díez-Montes et al., 2015; de Mello et al., 2022). This current understanding suggests a geological evolution characterized by an earlier extensional event, which likely played a crucial role in conditioning the formation of the massive sulfides in third-order anoxic basins, followed by later, Variscan compressive events that resulted in thin-skinned tectonics. Most of the early syn-volcanic faults were inverted during Variscan times (Fig. 1.3).

A principal structural element identified within this framework is an extensional detachment located between the VC Complex and the Culm Group (González Clavijo and Díez-Montes, 2010). This detachment also extends into the upper parts of the VS Complex, affecting the purple shale and the top of the massive sulfide ores, where mylonite and ultramylonites formed. The detachment is particularly well exposed and clearly observable in the Corta Atalaya open pit (see Figs. 1.2 and 1.3). In addition, this detachment is interrupted by ductile thrusts, where stockwork mineralization parallels the main foliation. Both the detachment and the ductile thrusts are further interrupted by another set of E-W thrusts with a brittle-ductile character (Díez-Montes et al., 2015). During cortical compression, the superimposed units were affected by southward-vergent thrusts, which are more verticalized and largely responsible for the main structural features; a crenulation foliation (S2) dipping south is generated along these late thrusts, mainly in the hanging wall.

Locally, there are associated recumbent folds with laminated inverse limbs. This imbrication formed the antiformal stack that hosts the core of the Riotinto district. The shale band the most hydrothermally altered volcanic rocks, rich in phyllosilicates, have a systematic cleavage (S_1). These early structures are superimposed by later, also E-W trending folds, with vertical axial planes such as that in Corta Atalaya. The Variscan thrusts and folds are crosscut by various late fault systems predominantly trending NW-SE and E-W and of Late Variscan to Alpine age. They are very abundant and spatially coincide with large sulfide bodies, suggesting they might have been the principal extensional faults that controlled the volcanism and mineralization.

Mineralization

The Riotinto volcanic-hosted massive sulfide system exhibits a complex geological architecture, with its principal orebodies distributed along an east-west axis flanking the Riotinto structure. The Cerro Colorado stockwork, prominently exposed at the core of this structure, marks the central feeder system of the deposit (Williams et al., 1975). Initially interpreted as a singular, coherent, exhalative lens deposited on the seafloor and subsequently deformed (Williams, 1962; Badham, 1982, 2001; Solomon et al., 1980; Eastoe et al., 1986).

Modern studies (Tornos, 2006; de Mello et al., 2022) suggest that the mineralization comprises multiple genetically distinct subtypes rather than a single dismembered orebody.

The mineralization can be distinguished into four main groups: (a) (sub-) exhalative massive sulfides hosted by shale of the Upper Sedimentary Unit (Filón Sur, Atalaya, and Planes-San Antonio), see Fig. 1.5a; (b) lensoidal and replacive sulfides hosted by altered felsic volcanic rocks of Felsic Unit II in the northern limb (including Salomón, Lago, Dehesa, and Quebrantahuesos), see Fig. 1.5b; (c) extensive stockwork mineralization cropping out at Cerro Colorado (see, Fig. 1.5c); and, (d) secondary mineralization that developed through oxidative weathering and leaching of preexisting sulfides after the Alpine exhumation, likely during the Miocene. The largest gossan was located at Cerro Colorado, originally capping the stockwork therein; see Fig. 1.5d (Williams et al., 1975; Sobol et al., 1997; Velasco et al., 2013).

The (sub-) exhalative massive sulfide lenses in the Upper Sedimentary Unit are up to 40 m thick and traceable along strike for over 3 km. These are hosted in the shale of the Upper Sedimentary Unit, overlying the rhyodacitic of the Felsic Unit II. At the Atalaya open pit, the San Dionisio orebody occupies the core of a syncline, where sulfide thicknesses reach up to 130 m due to tectonic thickening. The basal contact is often sheared, obscuring the original depositional context. Structural-hydrothermal processes have led to localized enrichment in Cu $\pm$ Zn $\pm$ Pb above a sulfide-mylonite zone at the contact. The ores are typically capped by a siliceous layer composed of pyrite-bearing white and gray chert, reminiscent of VMS capping

FIGURE 1.5 Representative photographs of the mineralization in the Riotinto district.

(a) Massive sulfide hosted by shale characterized by pyrite-rich composition and showing layering and graded bedding (Corta Atalaya); (b) replacement of the pervasively sericitized rhyodacite led to the formation of massive sulfide that preserves the brecciated texture of the pre-existing stockwork (Quebrantahuesos); (c) post-Variscan veins formed by larger veins, enriched in quartz coarse-grained pyrite and sometimes with chalcopyrite and barite (Quebrantahuesos); (d) large red gossan capping overlying an irregular pyrite and chalcopyrite stockwork (Quebrantahuesos).

units described in regions such as the Hokuroko district in Japan (Ohmoto, 1996), and other Iberian Pyrite Belt deposits like Tharsis and Aljustrel (Barriga and Fyfe, 1997; Tornos et al., 1998).

These massive sulfides consist mainly of fine-grained pyrite with variable amounts of chalcopyrite, sphalerite, galena, and trace phases including tetrahedrite-tennantite, arsenopyrite, pyrrhotite, and cassiterite. Recrystallization is common, resulting in a general obliteration of primary features, although occasional sedimentary structures, intraclasts of shale, and sulfide-rich sedimentary breccias are observed (García Palomero, 1992). The massive sulfides are rooted in a stockwork system hosted by rhyodacite (Felsic Unit II) and the mafic units (Mafic-Siliciclastic Unit). This stockwork represents the feeder system that channelized the circulation of deep upflowing fluids.

In contrast, the Filón Norte orebodies represent subseafloor replacive mineralization. These subvertical lenses, formerly mined in Salomón and Lago pits, show sharp replacive contacts with hydrothermally altered rhyodacite both at the footwall and hanging wall. This mineralization is unrelated with shale and lacks of sedimentary structures (Conde, 2016). Massive sulfides are coarse-grained, have relatively abundant amounts of quartz, white mica and chlorite, have no sedimentary structures and show gradual replacive contacts with the host altered coherent or volcaniclastic rocks. This supports their formation through the replacement of glassy or porous facies of the felsic dome beneath the seafloor (Tornos, 2006). Such processes are consistent with those documented in other VHMS settings dominated by volcanic glass (Allen and Cas, 1990; Allen et al., 1996; Piercey, 2015; Tornos et al., 2015).

Postdepositional supergene processes significantly altered the original sulfide assemblages. The Cerro Colorado gossan, originally between 10 and 70 m thick, and now largely removed by mining, was highly enriched in iron oxides (up to 79%), gold (1.8–2.5 ppm), silver (35–45 ppm), lead (1.2%), and trace elements such as antimony and bismuth, while being notably depleted in copper and zinc. A secondary gossan at Alto de la Mesa, interpreted as a fluvial precipitate, contains negligible metal concentrations and is composed primarily of transported clasts and secondary iron oxides formed in a fault-controlled basin (Sobol et al., 1997; Velasco et al., 2013).

The mineralogical composition of the Riotinto gossan is characterized by the predominance of goethite, hematite, and complex solid solutions within the beudantite–plumbojarosite–potassium jarosite series. Silver occurs in a variety of mineralogical forms, including halides, sulfides, mercury–silver sulfohalides, and as a dilute solid solution of argentojarosite within the jarosite–beudantite mineral assemblage (Amorós et al., 1981; Viñals et al., 1995).

Discussion

The Riotinto deposit presents a unique case within the Iberian Pyrite Belt (IPB), not only for its size and geological complexity but also for the timing and duality of its mineralizing processes. Unlike the majority of the shale-hosted IPB deposits, which are of Late Devonian (Strunian) age, evidence from radiometric dating and palynological studies clearly indicates that the main phase of mineralization at Riotinto occurred during the Early Carboniferous (Tournaisian) (Pereira and Clavijo, 2007; Barrie et al., 2002), with the underlying Strunian shale at Riotinto being barren. This temporal distinction implies that the Riotinto hydrothermal system developed significantly later than other major VMS systems in the region, such as Neves Corvo or Tharsis (Pereira et al., 2008; Tornos, 2006) and Aznalcóllar-Los Frailes.

The occurrence of both exhalative and replacive styles of mineralization within the same deposit is an exceptional feature. The shale-hosted massive sulfides at Filón Sur and Planes-San Antonio exhibit characteristics consistent with seafloor exhalation, including sedimentary layering, graded bedding, and a capping chert horizon. These deposits formed in an anoxic, restricted basin environment that was likely governed by brine pooling in third-order basins, as evidenced by the stratiform geometry and the geochemical signatures of the sulfur isotopes (Solomon et al., 1980, 2002; Eastoe et al., 1986). The isotopic compositions ($\delta^{34}S$ between −6 and +8‰ in massive sulfides vs. +6 to +12% in stockworks) suggest a mixed sulfur source involving both leached basement rocks and biologically reduced seawater sulfate (Velasco et al., 1998; Tornos and Heinrich, 2008).

In contrast, the subvertical, coarse-grained pyrite bodies found in the rhyodacitic rocks of Filón Norte are clearly replacive, likely formed below the seafloor by interaction between hot hydrothermal fluids and modified seawater in the most permeable and reactive facies of the felsic domes. The presence of widespread silicification and sericitization, coupled with the absence of primary sedimentary features, supports a model of subseafloor replacement driven by focused fluid flow along preexisting faults and lithologic boundaries and later fluid mixing (Tornos, 2006; Conde, 2016).

The spatial and temporal distribution of felsic volcanic rocks further supports a genetic link with the ore formation processes. Geochemical discrimination of felsic units shows two distinct rhyolitic compositions based on Zr content: a barren high-Zr rhyolite (>225 ppm) and a mineralized low-Zr rhyolite (<200 ppm), indicating that only specific volcanic pulses were conducive to hydrothermal activity (Valenzuela et al., 2002; Conde and Tornos, 2014). These volcanic domes likely acted as both structural and thermal flow for hydrothermal circulation, providing pathways for fluid migration and sites for ore precipitation.

Hydrothermal alteration assemblages in the district show a clear zonation. Chloritic alteration dominates in mafic units and is spatially associated with copper-rich zones, while sericitization prevails in felsic lithologies and is correlated with lead-silver enrichment. The stockwork systems, particularly those at Cerro Colorado and Quebrantahuesos, reveal a progression from disseminated to veined sulfide mineralization, consistent with the structural focusing of hydrothermal fluids along a deep-seated, long-lived feeder zone (Solomon et al., 1980; Nehlig et al., 1997).

Conclusions

The Riotinto district stands out as the largest and most emblematic of the giant VMS deposits within the Iberian Pyrite Belt. It is recognized as the foremost massive sulfide province globally, hosting one of Earth's largest concentrations of sulfides, with original estimated resources at Riotinto exceeding 2500 Mt of combined stockwork and massive sulfides.

Historically, Riotinto is also significant as one of the world's oldest mining districts, with operations spanning over 4500-5000 years. Mining operations, currently managed by Atalaya Mining, were restarted in 2015, and Riotinto remains one of the four currently active mines in the Spanish IPB.

Stratigraphic and lithological controls have played a crucial role in determining the geometry and distribution of ore bodies. In Riotinto, the main volume of the felsic sequence, which comprises coherent volcanic facies, crypto-domes, and sills, is notably associated with the principal VMS mineralization. These felsic volcanic domes likely acted as both structural and thermal conduits for hydrothermal circulation, providing essential pathways for fluid migration and optimal sites for ore precipitation. A critical geochemical differentiation identifies two main types based on Zr content: a barren high-Zr rhyolite (>200 ppm) and a mineralized low-Zr rhyolite (<200 ppm).

Geochronologically, the main mineralizing event at Riotinto is distinct and significantly younger than most other major IPB VMS systems. This event occurred during the Early Carboniferous (Tournaisian), specifically between approximately 352 and 348 Ma, and is directly linked to the emplacement of the low-Zr felsic volcanic rocks.

The hydrothermal alteration system, which is extensive, structurally controlled, and intimately associated with sulfide-rich stockwork zones, further characterizes the deposit. Chloritization predominates in mafic units and is spatially associated with copper-rich zones, while sericitization prevails in felsic lithologies and correlates with lead-silver enrichment. The stockwork systems, particularly at Cerro Colorado, represent deep-seated, long-lived feeder zones that focused hydrothermal fluids.

The mineralization at Riotinto is exceptional due to the coexistence of two genetically distinct styles. Exhalative mineralization is represented by (sub-)exhalative stratiform massive sulfides hosted within shale of the Upper Sedimentary Unit, such as Filón Sur, Atalaya, and Planes-San Antonio. These sulfides exhibit characteristics consistent with seafloor exhalation. In contrast, replacive mineralization is found in subvertical, lensoidal massive sulfide bodies hosted within altered felsic volcanic rocks, including Salomón, Lago, Dehesa, and Quebrantahuesos. These coarse-grained sulfides show sharp, replacive contacts with the host rhyodacite and lack primary sedimentary structures, indicating their formation below the seafloor through the interaction of hot hydrothermal fluids with permeable and reactive felsic dome facies.

The intense Variscan deformation has led to the folding and dismemberment of the VMS mineralization and the underlying copper stockworks from their original conduits within the volcanic series, with observed banding and mylonitic foliation in massive sulfides being of tectonic origin. This tectonic stacking produced by the detachment and thrusts can either increase or reduce the tonnage of VMS deposits.

Cropping out, weathering, and physical and (bio)chemical erosion of these mineralized rocks over more than 50 million years have led to the development of the unique biological system described in the following chapters.

Acknowledgments

This work is a synthesis of the research carried by our group in the Riotinto area. We would like to express our gratitude to all those who have consistently supported us through their questions and insightful discussions during long fieldwork campaigns, especially: Nick Badham, Mike Solomon, Emilio González Clavijo, Alejandro Díez Montes and David Mellado. We also thank the laboratory directors and staff who, over the years, have contributed to our efforts to understand this unique and anomalous mining area.

We would further like to extend our sincere thanks to Atalaya Mining for consistently providing us with access and logistical support to carry out our work at their facilities. We are also grateful to the entire team of exploration geologists who have worked at the company over the years, thank you.

References

Allen, R.L., Cas, R.A.F., 1990. The Rosebery controversy: distinguishing prospective submarine ignimbrite-like units from true subaerial ignimbrites in the Rosebery–Hercules Zn–Cu–Pb massive sulfide district, Tasmania. In: Contentious Issues in Tasmanian Geology. Geological Society of Australia, Tasmania Division, pp. 107–108.

Allen, R.L., Weihed, P., Svenson, S.A., 1996. Setting of Zn-Cu-Au-Ag massive sulfide deposits in the evolution and facies architecture of a 1.9 Ga marine volcanic arc, Skellefte district, Sweden. Econ. Geol. 91 (6), 1022–1053. https://doi.org/10.2113/gsecongeo.91.6.1022Norway.

Amorós, J.L., Lunar, R., Tavira, P., 1981. Jarosite: a silver bearing mineral of the gossan of Rio Tinto (Huelva) and La Union (Cartagena, Spain). Miner. Deposita 16 (2), 205–213. https://doi.org/10.1007/BF00202735.

Avery, D., 1974. Not on Queen Victoria's Birthday: The Story of the Río Tinto Mines. Collins.

Badham, J.P.N., 2001. Physical and chemical paleogeography of the Río Tinto area during ore deposition. In: GEODE Workshop: Massive Sulfide Deposits in the Iberian Pyrite Belt: New Advances and Comparison with Equivalent Systems, pp. 411–412.

Badham, J.P.N., 1982. Further data on the formation of ores of Rio Tinto, Spain. Trans. Inst. Min. Metall. 91 (February), 26–32.

Barrie, T.C., Amelin, Y., Pascual, E., 2002. U-Pb geochronology of VMS mineralization in the Iberian Pyrite belt. Miner. Deposita 37 (8), 684–703. https://doi.org/10.1007/s00126-002-0302-7.

Barriga, F.J.A.S., Fyfe, W.S., 1997. Multi-phase water-rhyolite interaction and ore fluid generation at Aljustrel, Portugal. Miner. Deposita 33 (1–2), 188–207. https://doi.org/10.1007/s001260050140.

Boulter, c.A., 1993. High level peperitic sills at Río Tinto, Spain: implications for stratigraphy and mineralization. Trans. Inst. Min. Metall. 102, 3825–3830.

Conde, C., 2016. In: Geology and Hydrothermal Evolution of Massive Sulfides of the Iberian Pyrite Belt, Spain. Universidad de Salamanca, Salamanca, España, p. 379.

Conde, C., Tornos, F., 2014. Volcanic stratigraphy and geochemistry of the VS complex in the northern Iberian Pyrite Belt. Macla – Revista de la Sociedad Española de Mineralogía 19, 39–40.

Conde, C., Tornos, F., 2020. Geochemistry and architecture of the host sequence of the massive sulfides in the northern Iberian Pyrite Belt. Ore Geol. Rev. 127, 103042. https://doi.org/10.1016/j.oregeorev.2019.103042.

Costa, I.M.S.R., 1996. In: Efectos mineralógicos e geoquímicos da alteraçao mineralizante en rochas volcánicas félsicas de Río Tinto (Faixa Piritosa Ibérica, Espanha). Universidade de Lisboa.

de Mello, C.R., Tornos, F., Conde, C., Tassinari, C.C.G., Farci, A., Vega, R., 2022. Geology, geochemistry, and geochronology of the giant Rio Tinto VMS deposit, Iberian Pyrite Belt, Spain. Econ. Geol. 117 (5), 1149–1171. https://doi.org/10.5382/econgeo.4907.

Díez-Montes, A., García-Crespo, J., Ayala, C., Luis García-Lobón, J., Sánchez-García, T., Rey-Moral, C., Bellido, F., Rubio, F., Mediato, J.F., Tornos, F., 2015. Modelling of the Río Tinto Area. Springer Science and Business Media LLC, pp. 209–229. https://doi.org/10.1007/978-3-319-17428-0_10.

Eastoe, C.J., Solomon, M., Garcia Palomero, F., 1986. Sulphur isotope study of massive and stockwork pyrite deposits at Rio Tinto, Spain. Trans. Inst. Min. Metall. 95 (November), 201–207.

Franklin, J.M., Lydon, J.W., Sangster, D.F., 1981. Volcanic-Associated Massive Sulfide Deposits. Society of Economic Geologists, Inc., pp. 485–627. https://doi.org/10.5382/av75.15

García Palomero, F., 1980. Caracteres geológicos y relaciones maorfológicas y genéticas de los yacimientos del anticlinal de Río Tinto. Instituto de Estudios Onubenses "Padre Marchena".

García Palomero, F., 1992. In: Mineralizaciones de Riotinto (Huelva): Geología, génesis, y modelos geológicos para su explotación y evaluación de reservas minerales, vol 15. Recursos Minerales de España (CSIC), Textos Universitarios, pp. 1325–1352.

González Clavijo, E., Díez-Montes, A., 2010. Propuesta de estructura geológica para el depósito gigante de sulfuros masivos volcanogénicos de Río Tinto, Faja Pirítica Ibérica. Geogaceta 48, 204–206.

González, F., Moreno, C., Sáez, R., Clayton, G., 2002. Ore genesis age of the Tharsis mining district (Iberian Pyrite Belt): a palynological approach. J. Geol. Soc. 159 (3), 229–232. https://doi.org/10.1144/0016-764901-142.

Gumiel, P., Sanderson, D.J., Campos, R., Roberts, S., 2001. Massive sulphide deposits in the Iberian Pyrite Belt: new advances and comparison with equivalent systems. In: Fractal Geometry of the Cerro Colorado Stockwork, Rio Tinto Mine (Iberian Pyrite Belt) GEODE Workshop, pp. 23–24.

Halsall, C.E., 1989. In: The Relationship between Intrusive Magmatism, Volcanism and Massive Sulphide Mineralisation at Río Tinto, Spain.

Junta de Andalucía, 2004. Mapa de Edad Geológica de Andalucía, E 1:400.000, Conserjería de Medio Ambiente. Junta de Andalucía, Sevilla, España.

Laznicka, P., 1999. Quantitative relationships among giant deposits of metals. Econ. Geol. 94 (4), 455–473. https://doi.org/10.2113/gsecongeo.94.4.455.

Leitão, J., 1997. Geology of the Aljustrel massive sulfide deposits. Soc. Econ. Geol. Guideb. Ser. 27, 82–97.

Marcoux, E., Leistel, J.M., 1996. Mineralogy and geochemistry of massive sulfide deposits, Iberian Pyrite Belt. Bol. Geol. Min. 107, 117–126.

Mellado, D., Clavijo, Tornos, F., Conde, C., 2006. Geología y estructura de la Mina de Río Tinto (Faja Pirítica Ibérica, España). Geogaceta 40, 231–234.

Mitjavila, J., Marti, J., Soriano, C., 1997. Magmatic evolution and tectonic setting of the Iberian Pyrite Belt volcanism. J. Petrol. 38 (6), 727–755. https://doi.org/10.1093/petroj/38.6.727.

Moreno, C., 1993. Postvolcanic Paleozoic of the Iberian Pyrite Belt: an example of basin morphologic control on sediment distribution in a turbidite basin. J. Sediment. Petrol. 63 (6), 1118–1128. https://doi.org/10.1306/D4267CBC-2B26-11D7-8648000102C1865D.

Moreno, C., Sierra, S., Saez, R., 1996. Evidence for catastrophism at the Famennian-Dinantian boundary in the Iberian Pyrite Belt. In: Recent Advances in Lower Carboniferous Geology, pp. 153–162.

Munhá, J., 1983. Hercynian magmatism in the Iberian Pyrite Belt. Memorias Servicio Geologico Portugal. 29, 39–81.

Nehlig, P., Cassard, D., Marcoux, E., 1997. Geometry and genesis of feeder zones of massive sulphide deposits: constraints from the Rio Tinto ore deposit (Spain). Miner. Deposita 33 (1–2), 137–149. https://doi.org/10.1007/s001260050137.

Ohmoto, H., 1996. Formation of volcanogenic massive sulfide deposits: the Kuroko perspective. Ore Geol. Rev. 10 (3–6), 135–177. https://doi.org/10.1016/0169-1368(95)00021-6.

Oliveira, J.T., 1983. The marine Carboniferous of South Portugal: a stratigraphic and sedimentological approach. In: The Carboniferous of Portugal, pp. 3–37.

Pereira, Z., Matos, J.X., Fernandes, P., Oliveira, J.T., 2008. Palynostratigraphy and systematic palynology of the devonian and carboniferous successions of the South Portuguese Zone. Portugal. Memórias INETI 34, 1–176.

Pereira, Z., Clavijo, d G., 2007. New palynological ages improving the geological background of the Río Tinto area (Spanish part of the Iberian Pyrite Belt). In: Spores/Pollen and Acritarch Subcomissions, Joint Meeting, pp. 105–107.

Pereira, Z., Sáez, R., Pons, J.M., Oliveira, J.T., Moreno, C., 1996. Edad devónica (Struniense) de las mineralizaciones de Aznalcóllar (Faja Pirítica Ibérica) en base a palinología. Geogaceta 20, 1609–1612.

Piercey, S.J., 2015. A semipermeable interface model for the genesis of subseafloor replacement-type volcanogenic massive sulfide (VMS) deposits. Econ. Geol. 110 (7), 1655–1660. https://doi.org/10.2113/econgeo.110.7.1655.

Pinedo, I., 1963. In: Piritas de Huelva. Su historia, minería y aprovechamiento.

Quesada, C., 1997. A reappraisal of the structure of the Spanish segment of the Iberian Pyrite Belt. Miner. Deposita 33 (1–2), 31–44. https://doi.org/10.1007/s001260050131.

Rambaud, F., 1969. In: El sinclinal carbonífero de Riotinto (Huelva) y sus mineralizaciones asociadas. Memorias del Instituto Geológico y Minero de España (IGME).

Ribeiro, A., Quesada, C., Dallmayer, R.D., 1990. Geodynamic evolution of the Iberian Pyrite Belt. In: Pre-Mesozoic Geology of Iberia. Springer-Verlag, pp. 339–409.

Rodríguez, R.M., Montes, Leyva, F., Matas, J., Almarza, J., Donaire, M., 2002. Datación palinoestratigráfica del volcanismo en la sección de la Ribera del Jarama (Faja Pirítica Ibérica. Geogaceta 32, 247–250.

Rothenberg, B., Blanco-Freijeiro, A., 1981. Ancient Mining and Metallurgy in South-West Spain. IAMS.

Rothenberg, B., Pérez, J.A., 1987. In: Excavaciones de urgencia en la necrópolis de Stock de Gossa (Riotinto, Huelva). Consejería de Cultura. Dirección General de Bienes Culturales, pp. 187–191.

Routhier, F.A., Boyer, C., Lecólle, M., Moliere, P., Picot, P., Roger, G., 1980. Le ceinture sud-ibérique à amas sulfures dans sa partie espagnole médiane. Mémoire du Bureau de Recherches Géologiques et Minières. 94.

Sanchez España, J., Velasco, F., 1999. Constraints on the Hercinian metamorphism in the Northern IPB: ore textures and phyllosilicate crystallinity. Boletín de la Sociedad Española de Mineralogía 22 (August), 103–104.

Sawkins, F.J., 1990. Integrated tectonic-genetic model for volcanic-hosted massive sulfide deposits. Geology 18 (11), 1061–1064. https://doi.org/10.1130/0091-7613(1990)018<1061:ITGMFV>2.3.CO;2UnitedStates.

Schermerhorn, L.J.G., 1975. Spilites, regional metamorphism and subduction in the Iberian pyrite belt: some comments. Geol. Mijnbouw 54 (1–2), 23–35.

Schermerhorn, L.J.G., 1971. An outline of the stratigraphy of the Iberian Pyrite Belt. Bol. Geol. Min. 82, 239–268.

Silva, J.B., Oliveira, J.T., Ribeiro, A., 1990. Structural Outline. Springer Science and Business Media LLC, pp. 348–362. https://doi.org/10.1007/978-3-642-83980-1_24.

Simancas, J.F., Carbonell, R., González Lodeiro, F., Pérez Estaun, A., Juhlin, C., Ayarza, P., Kashubin, A., Azor, A., Martínez Poyatos, D., Almodóvar, G.R., Pascual, E., Sáez, R., Expósito, I., 2003. Crustal structure of the transpressional Variscan orogen of SW Iberia: SW Iberia deep seismic reflection profile (IBERSEIS). Tectonics 22 (6), 1–19. https://doi.org/10.1029/2002TC001479.

Sobol, F., Toscano, M., Castro, J.A., Sáez, R., 1997. A field guide to the Riotinto mines. Geology and VMS deposits of the Iberian Pyrite Belt. Soc. Econ. Geol. Fieldb. Ser. 27, 158–164.

Solomon, M., Tornos, F., Gaspar, O.C., 2002. Explanation for many of the unusual features of the massive sulfide deposits of the Iberian pyrite belt. Geology 30 (1), 87–90. https://doi.org/10.1130/0091-7613(2002)030<0087:EFMOTU>2.0.CO;2.

Solomon, M., Walshe, J.L., García Palomares, F., 1980. Formation of massive sulfide deposits at Río Tinto, Spain. Trans. Inst. Min. Metall. 89, 16–24.

Thieblemont, D., Pascual, E., Stein, G., 1998. Magmatism in the Iberian Pyrite Belt: petrological constraints on a metallogenic model. Miner. Deposita 33, 98–110.

Tornos, F., 2006. Environment of formation and styles of volcanogenic massive sulfides: the Iberian Pyrite Belt. Ore Geol. Rev. 28 (3), 259–307. https://doi.org/10.1016/j.oregeorev.2004.12.005.

Tornos, F., Clavijo, E.G., Spiro, B., 1998. The Filón Norte orebody (Tharsis, Iberian Pyrite Belt): a proximal low-temperature shale-hosted massive sulphide in a thin-skinned tectonic belt. Miner. Deposita 33, 150–169.

Tornos, F., Conde, C., Chiaradia, M., Velasco, F., 2015. Evolving subduction-related basins control the formation of VMS deposits in the Iberian Pyrite Belt. In: Society of Economic Geologists and Centre for Ore Deposit and Earth Sciences, Joint Conference.

Tornos, F., Conde, C., Mello Tassinari, C.C.G., 2023. A new vision of the geodynamic evolution of the Iberian Pyrite Belt: VHMS in an intra-arc basin. In: Mineral Resources in a Changing World, 17th SGA Biennial Meeting 2023, pp. 20–23.

Tornos, F., Heinrich, C.A., 2008. Shale basins, sulfur-deficient ore brines and the formation of exhalative base metal deposits. Chem. Geol. 247 (1–2), 195–207. https://doi.org/10.1016/j.chemgeo.2007.10.011.

Valenzuela, A., Donaire, T., González-Roldán, M.J., Toscano, M., Pascual, E., 2011b. Volcanic architecture in the Odiel River area and the volcanic environment in the

Riotinto–Nerva Unit, Iberian Pyrite Belt, Spain. J. Volcanol. Geoth. Res. 202 (1–2), 29–46. https://doi.org/10.1016/j.jvolgeores.2010.12.018.

Valenzuela, A., Donaire, T., Pin, C., Toscano, M., Hamilton, M.A., Pascual, E., 2011a. Geochemistry and U-Pb dating of felsic volcanic rocks in the Riotinto-Nerva unit, Iberian Pyrite belt, Spain: crustal thinning, progressive crustal melting and massive sulphide genesis. J. Geol. Soc. 168 (3), 717–732. https://doi.org/10.1144/0016-76492010-081Spain.

Valenzuela, A., Donaire, T., Pascual, E., 2002. Secuencia de facies volcánicas en el área del río Odiel (Faja Pirítica Ibérica, España). Geogaceta 32, 131–134.

Velasco, F., Herrero, J.M., Suárez, S., Yusta, I., Alvaro, A., Tornos, F., 2013. Supergene features and evolution of gossans capping massive sulphide deposits in the Iberian Pyrite Belt. Ore Geol. Rev. 53, 181–203. https://doi.org/10.1016/j.oregeorev.2013.01.008.

Velasco, F., Sánchez-España, J., Boyce, A.J., Fallick, A.E., Sáez, R., Almodóvar, G.R., 1998. A new sulphur isotopic study of some Iberian Pyrite Belt deposits: evidence of a textural control on sulphur isotope composition. Miner. Deposita 34 (1), 4–18. https://doi.org/10.1007/s001260050182.

Viñals, J., Roca, A., Cruells, M., Núñez, C., 1995. Characterization and cyanidation of Rio Tinto gossan ores. Can. Metall. Q. 34 (2), 115–122. https://doi.org/10.1179/0008-4433(94)00022-c.

Williams, D., 1934. The geology of the Rio Tinto mines. Trans. Inst. Min. Metall. 43, 593–678.

Williams, D., 1962. Further reflections on the origin of the porphyries and ores of Rio Tinto. Trans. Inst. Min. Metall. 71, 265–266.

Williams, D., 1966. Volcanism and ore deposits. In: Freiberger Forschungshefte, Reihe C, vol 210, pp. 93–111.

Williams, D., Stanton, R.L., Rambaud, F., 1975. The planes-San Antonio pyritic deposit of Rio Tinto, Spain: its nature, environment and genesis. Econ. Geol. 70 (5), 943–961.

CHAPTER

2 A historical perspective on the Rio Tinto ecosystem: Sedimentology, paleontology, and geobiology

David C. Fernández-Remolar[1, 2], Lara Maldanis[3], Laurence Lemelle[4], Alexandre Simionovici[5], David Gómez-Ortiz[6, 7], Jorge Sanjurjo-Sánchez[8], Ting Huang[1, 2], Andrew H. Knoll[9]

[1]*Macau University of Science and Technology, SKL Lunar and Planetary Sciences, Taipa, Macau;* [2]*CNSA Macau Center for Space Exploration and Science, Macau, China;* [3]*Vrije Universiteit Amsterdam, Earth Science Department, Science Faculty, Amsterdam, the Netherlands;* [4]*LGL-TPE, ENS de Lyon, Univ. de Lyon, CNRS, Lyon, France;* [5]*ISTerre, Univ. Grenoble Alpes, Univ. Savoie Mont Blanc, CNRS, IRD, IFSTTAR, Grenoble, France;* [6]*Universidad Rey Juan Carlos, Departamento de Geologia, Fisica y Quimica inorganica, Mostoles, Spain;* [7]*Research Group 'Geofísica y Geoquímica Ambiental', Universidad Rey Juan Carlos, Mostoles, Spain;* [8]*Universidade da Coruña, Instituto Universitario de Xeología, Campus de Elviña, A Coruña, Spain;* [9]*Harvard University, Department of Organismic and Evolutionary Biology, Cambridge, MA, United States*

The Rio Tinto ecosystem is renowned for its strongly acidic and vibrantly red river water, for the distinctive nature of its biodiversity, for its economically important resources of sulfuric acid and iron, and for its importance as an environmental analog for sedimentary rocks exposed in Meridiani planum, Mars. All aspects of this system are enhanced by the fact that the Rio Tinto ecosystem preserves a record of its own history, inscribed in goassans and sedimentary rocks deposited regionally over the past 25 million years. Rio Tinto's acidic brines, which result from meteoric waters interacting with sulfide minerals in adjacent hills, have long generated a distinctive hydrochemistry, with seasonal changes in water availability and flow driving precipitation of distinct iron oxyhydroxide and sulfate minerals. Fungi, protists, microbes and both plant and insect remains have been preserved by mineralization, as have chemical biosignatures. Collectively, the physical, chemical and paleobiological features of Rio Tinto rocks reveal how this distinctive ecosystem has changed through time.

Geomicrobiology of Río Tinto. https://doi.org/10.1016/B978-0-443-32948-7.00016-3

General geological setting

Located in the Iberian Pyrite Belt (IPB), Rio Tinto is situated within a massive sulfide deposit that formed during the late Paleozoic Era, specifically during the Carboniferous Period (ca. 359–299 million years ago; Leistel et al., 1997; Tornos, 2006). The primary bedrock geology of Rio Tinto consists of massive polymetallic sulfide deposits embedded within volcano-sedimentary sequences (Fig. 2.1). These deposits are predominantly composed of pyrite (FeS_2), with significant amounts of chalcopyrite ($CuFeS_2$), sphalerite (ZnS), and galena (PbS) (Sáez et al., 1999). The formation of these sulfide-rich deposits occurred in a submarine volcanic environment associated with hydrothermal fluid flow along the margins of the now extinct Rheic Ocean (Tornos, 2006). These deposits represent one of the world's largest concentrations of massive sulfide mineralization, forming what is known as the Iberian Pyrite Belt, which extends across parts of southern Spain and Portugal (Leistel et al., 1997; Tornos, 2006).

The Carboniferous materials that serve as primary sources for the sulfur and iron utilized in microbial metabolism include massive sulfide bodies, black shales, and volcano-sedimentary complexes (Fig. 2.1). The massive sulfide bodies are stratiform volcanogenic deposits containing up to 90% pyrite, which provides abundant reduced sulfur and iron (Tornos, 2006). These bodies formed through submarine hydrothermal processes associated with tectonic activity that characterized the region during the Carboniferous Period (Leistel et al., 1997). The black shales are rich in organic matter and early diagenetic pyrite that formed in anoxic marine environments (Sáez et al., 1999). These fine-grained sedimentary rocks accumulated in deep, oxygen-depleted basins where organic matter preservation was enhanced. The volcano-sedimentary complexes consist of felsic and mafic volcanic rocks interbedded with sedimentary units that contain disseminated sulfides (Tornos, 2006). These complexes establish the broader geological context in which the massive sulfide deposits formed, reflecting the complex interplay between volcanic activity, hydrothermal activity and sedimentary processes during the Carboniferous.

These Carboniferous materials have undergone various degrees of weathering and oxidation processes since their formation, creating the distinctive acidic environment of Rio Tinto. When exposed to oxygen and water, the sulfide minerals undergo oxidation reactions that release sulfuric acid and ferric iron, at least in part under microbial influence, creating the extremely acidic conditions (pH 0.8–2.3) that characterize the Rio Tinto system today (Fernández-Remolar et al., 2005; Amils et al., 2007). The primary oxidation reaction involves the conversion of pyrite to ferric iron and sulfuric acid, which can be represented by the general equation: $FeS_2 + 3.5O_2 + H_2O \rightarrow Fe^{2+} + 2SO_4^{2-} + 2H^+$. This process is significantly enhanced by the metabolic activities of acidophilic microorganisms, particularly iron- and sulfur-oxidizing bacteria and archaea such as *Acidithiobacillus ferrooxidans*, *Leptospirillum ferrooxidans*, and various species of *Ferroplasma*, all of which catalyze weathering reactions (González-Toril et al., 2003; Sánchez-Andrea et al., 2011). These microorganisms derive energy from the oxidation of reduced iron

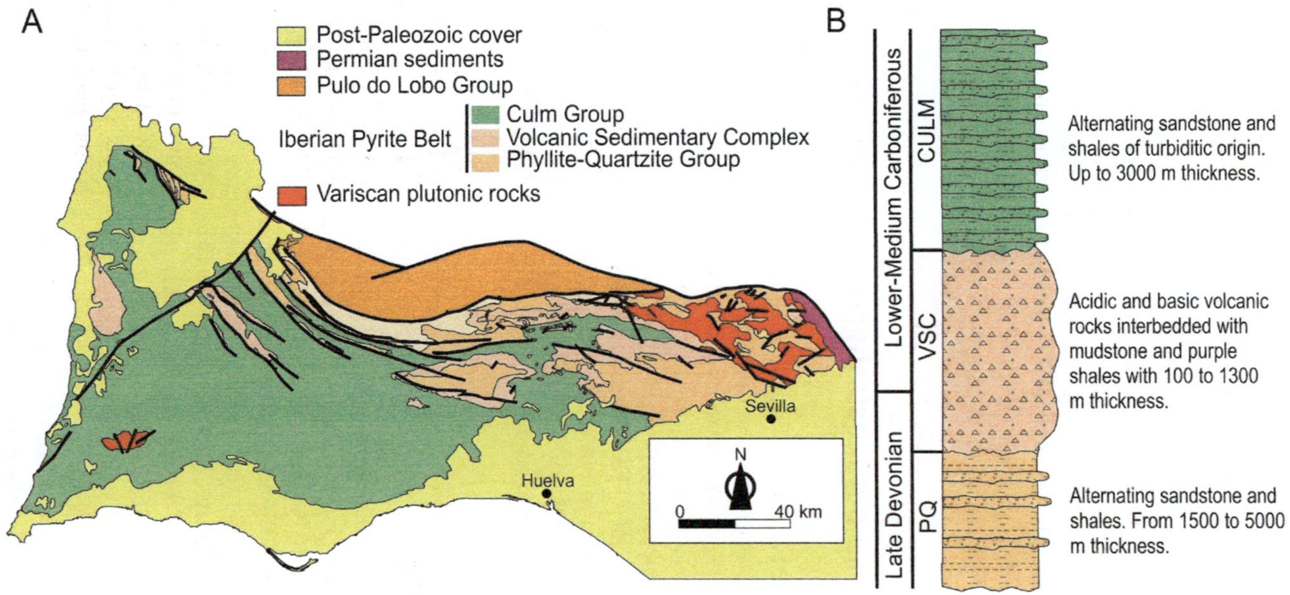

FIGURE 2.1

Geological framework and stratigraphy of the Iberian pyrite Belt in the context of the Rio Tinto fluvial basin. (A) Simplified geological map of the southwestern Iberian Peninsula showing the distribution of major lithotectonic units associated with the Iberian Pyrite Belt (IPB), including the Culm group, volcanic sedimentary complex (VSC), and Pulo do Lobo group, as well as surrounding units such as the Variscan plutonic rocks, Permian sediments, and post-Paleozoic cover. The distribution of these units reflects the complex tectono-sedimentary evolution of the region during the late Devonian to Carboniferous periods. (B) Generalized stratigraphic column of the IPB, highlighting the three principal lithostratigraphic units: The Phyllite–Quartzite group (PQ), composed of alternating sandstones and shales (1500–5000 m thick); the volcanic sedimentary complex (VSC), consisting of acidic and basic volcanic rocks interbedded with mudstones and purple shales (100–1300 m thick); and the Culm group, made up of turbiditic sandstones and shales up to 3000 m thick.

- Modified from Sáez, R., Pascual, E., Toscano, M., Almodóvar, G.R., 1999. The Iberian type of volcano-sedimentary massive sulphide deposits. Miner. Deposita. 34 (5–6), 549–570, https://doi.org/10.1007/s001260050220 and Tornos, F., 2006. Environment of formation and styles of volcanogenic massive sulfides: the Iberian Pyrite Belt. Ore Geol. Rev. 28 (3) (2006), 259–307, https://doi.org/10.1016/j.oregeorev.2004.12.005.

and sulfur compounds, accelerating the weathering process by several orders of magnitude compared to purely abiotic oxidation (Amils et al., 2007).

The distinctive geochemical, mineralogical, and sedimentary characteristics of the Río Tinto system are captured in a succession of gossan formations and iron-rich fluvial terraces, which are notably enriched in iron oxyhydroxides (Essalhi et al., 2011; Fernández-Remolar et al., 2004, 2021a). These geological features document a series of discontinuous weathering episodes, primarily driven by warm and humid climatic intervals that enhanced microbial alteration of exposed sulfide ore bodies. The resulting chronostratigraphic framework—comprising well-preserved and geochemically distinct terrace levels—offers an exceptional archive of the temporal evolution of geological and biological processes that have shaped this extreme acidic environment over millions of years.

Sedimentary record and chronostratigraphy of Rio Tinto Gossan and Terraces

The diagenetic and sedimentary record of the Rio Tinto area presents a complex chronostratigraphic sequence that spans from the Late Oligocene to the Holocene, as evidenced by the presence of various terraces and the gossan formation (Fernández-Remolar et al., 2005; Fernández-Remolar and Knoll, 2008). This record provides crucial insights into the paleoenvironmental evolution of the region and the timing of major alteration episodes. The chronostratigraphic framework has been established through a combination of radiometric dating, stratigraphy, and correlation with regional and global climatic events, allowing for a detailed reconstruction of the area's geological history (Fernández-Remolar et al., 2003, 2005).

Upper Gossan (>25 Ma)

The Upper Gossan represents the oldest preserved unit formed by acidic solutions in the headwaters of the Río Tinto, with an estimated age exceeding 25 million years (Late Oligocene; Essalhi et al., 2011; Fernández-Remolar et al., 2005; Velasco et al., 2013). At Peña de Hierro, this distinctive formation is exposed at approximately 500 m elevation, forming the iconic reddish outcrop that dominates the landscape. The unit is characterized by a massive, iron oxide-rich cap (Fig. 2.2A), which developed through intense oxidation of the underlying sulfide deposits. This oxidative weathering transformed primary sulfide minerals into a complex assemblage of secondary ferric phases, creating the distinctive weathering profile that extends downward toward the unaltered sulfide body.

The Peña de Hierro Upper Gossan displays a characteristic reddish-brown to yellowish coloration resulting from the predominance of iron oxides and oxyhydroxides, particularly hematite (Fe_2O_3) and goethite ($FeO(OH)$). These minerals

FIGURE 2.2

Field features of the upper Gossan unit at Río Tinto. (A) Panoramic view of the upper Gossan outcrop, showing a prominent ridge composed of intensely weathered, iron oxide-rich rock, with massive to brecciated textures and characteristic reddish tones. This expression represents the uppermost alteration zone developed above the primary massive sulfide bodies of the Iberian Pyrite Belt. (B) Detail of laminated iron oxide material displaying blister-like textures and finely preserved fungal filaments. These structures are hosted within iron oxide laminations formed during subaerial precipitation of oxysulfates transformed into iron oxides during early post-depositional processes. The presence of fungal filaments within these iron-rich layers suggests microbial activity played a role in mineralizing the gossan and mediating mineral precipitation during late-stage weathering.

formed through the diagenetic transformation of various ferric oxysulfates such as jarosite, copiapite, and schwertmannite, as well as amorphous to nanophase ferric oxyhydroxides with varying degrees of crystallinity. A remarkable feature of the Peña de Hierro gossan is the preservation of original rock textures through pseudomorphic replacement, where secondary ferric minerals have substituted for sulfate-rich precursors while maintaining the structural integrity of the host rock (Fig. 2.2B). This preservation enables geologists to reconstruct the nature and distribution of the original sulfide mineralization despite extensive chemical alteration.

The formation of the Upper Gossan at Peña de Hierro coincided with the Late Oligocene warming period (27–25 Ma), which likely accelerated weathering processes through increased precipitation and temperature. This significant climatic

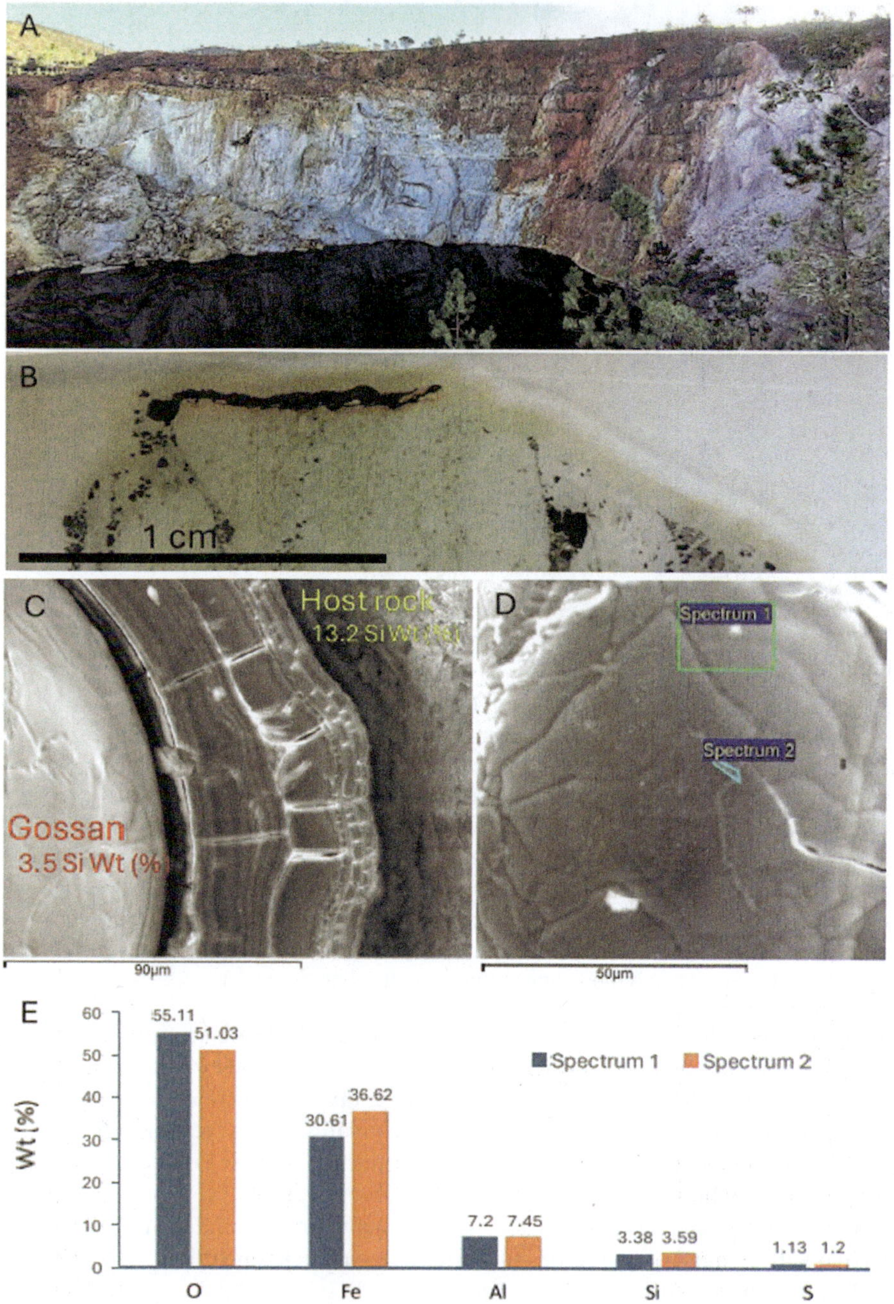

FIGURE 2.3

Structural and microscale characteristics of the lower Gossan in the Río Tinto system. (A) Panoramic view of the lower Gossan outcrop, illustrating its development along fault

shift enhanced chemical weathering rates, facilitating the oxidation and leaching of the sulfide body. The considerable vertical extent of the gossan, reaching depths corresponding to ages greater than 6 Ma, indicates that its development was not a single event but rather a protracted process spanning millions of years across varying climatic conditions. This prolonged formation history has resulted in the complex mineralogical and textural characteristics that make the Peña de Hierro gossan an exceptional natural laboratory for studying supergene processes and a valuable analog for understanding similar weathering environments on Mars.

Lower Gossan (6 Ma)

Beneath the Upper Gossan at Peña de Hierro lies the Lower Gossan, a transitional weathering zone that preserves much of the original structural and lithological fabric of the volcanic host rocks while exhibiting clear geochemical evidence of early-stage oxidation. This unit occurs below approximately 480 m in elevation and underlies the massive iron-rich cap, forming a discrete weathering horizon characterized by structurally controlled alteration fronts (Fig. 2.3A). The irregular morphology of this front reflects the influence of basement tectonic structures unique to the Peña de Hierro deposit, which channeled oxidizing fluid flow during supergene alteration.

The Lower Gossan is distinguished in drill core by the presence of ferric oxide infillings within fractures and microcracks that permeate the altered volcanic substratum (Fig. 2.3B). These vein-like features delineate the migration paths of oxidizing fluids, which exploited pre-existing discontinuities formed during post-uplift reorganization of the local phreatic regime. This uplift is linked to late Alpine tectonic reactivation, which influenced regional fluid dynamics by rejuvenating Paleozoic fault systems.

The Lower Gossan at Peña de Hierro exhibits a pronounced compositional contrast between the volcanic host rock and the fracture-infilled gossan material. While the host rock retains a relatively high silicon content, the gossan is

planes that acted as conduits for sulfide oxidation. The outcrop displays a stepped, terrace-like surface morphology at a lower elevation than the upper Gossan, suggesting an evolutionally younger stage of surficial weathering. (B) Photomicrograph of a thin section showing a fracture in the volcanic host rock, now infilled by lower Gossan material. The gossan infill marks an oxidized vein-like feature cutting across altered volcanic substrates. (C) Backscattered SEM image of the same crack fill, showing a sharp compositional contrast between the Si-rich host rock (13.2 wt% Si) and the Si-depleted gossan material (3.5 wt% Si). (D) SEM image of gossan material containing morphological structures consistent with mineralized microbial filaments, suggesting that microbial entombment occurred during gossan development. (E) EDS analysis from the region shown in panel D, with both spectra indicating sulfur concentrations exceeding 1 wt%, consistent with the typical sulfur concentration of the lower Gossan.

significantly silica-depleted, indicating intense chemical alteration driven by oxidizing fluid flow (Fig. 2.3C). This transformation reflects prolonged fluid–rock interaction along structurally controlled pathways. Within the ferric infill, filamentous structures consistent with mineralized microbial forms are preserved, suggesting microbial colonization and subsequent entombment during mineral precipitation (Fig. 2.3D). Their presence points to biologically active conditions during gossan formation. Supporting this, sulfur concentrations in the gossan exceed 1 wt%, consistent with secondary mineral assemblages involving iron sulfates. Together, these features indicate that the Lower Gossan records not only the geochemical footprint of fluid-driven alteration but also biosignatures of microbial activity. This highlights its value as both a geochemical boundary and a potential archive of past life (Fig. 2.3C–E).

Mineralogical analyses of Lower Gossan samples indicate a complex assemblage composed of goethite as the dominant iron oxyhydroxide, jarosite in fluid-concentrated zones, and relic pyrite grains with oxidized rims. These features typify an intermediate weathering stage—between the relatively unaltered, sulfide-rich protolith at depth and the intensely leached assemblage of the Upper Gossan.

The Lower Gossan at Peña de Hierro thus represents a pivotal stage in the vertical evolution of the weathering profile, preserving both structural conduits and early diagenetic biosignatures. Structural mapping has revealed a network of NW–SE fractures that served as primary pathways for oxidizing fluid ingress (Gómez-Ortiz et al., 2014), facilitating gossan development.Radiometric dating of supergene minerals from the Peña de Hierro gossan profile suggests that the upper portions began forming around 6 million years ago, during the Late Miocene Epoch (Velasco et al., 2013). This development coincides with a phase of late Alpine tectonic reactivation in southwestern Iberia, which caused the uplift of the Peña de Hierro area and exposed the sulfide bodies to meteoric weathering (Velasco et al., 2013). These tectonic movements reactivated local Paleozoic fault systems. allowing oxidizing fluids to penetrate deep into the ore system. Long-term chemical weathering was further modulated by Neogene–Quaternary climatic fluctuations, in which alternating humid and semi-arid intervals controlled the depth of oxidation and the formation of secondary mineral assemblages. Recent $^{40}Ar/^{39}Ar$ dating indicates multiple episodes of fluid infiltration between 5.8 and 2.3 Ma, with intense weathering during humid Pliocene intervals.

Upper Terrace (2.1 Ma)

The stepped terraces of the Río Tinto reflect successive phases of sedimentation and incision linked to climatic fluctuations and eustatic sea-level changes during the Late Neogene and Quaternary. The Upper Terrace, dated to the Late Pliocene (~2 Ma), likely formed during a relatively humid period associated with increased sediment supply and chemical weathering. Subsequent incision and terrace abandonment coincide with episodes of climatic cooling and sea-level fall during the Early

Pleistocene, which enhanced river downcutting. In contrast, the Intermediate and Young Terraces of the Río Tinto were formed during the last interglacial period (<12 Ka). Their formation reflects a combined influence of eustatic sea-level rise, which established new base levels for fluvial sedimentation, and localized tectonic activity associated with the Iberian Pyrite Belt, which contributed to fluvial incision and terrace stabilization. This interplay of eustatic forcing and tectonic uplift accounts for the stepped morphology observed in the upper Río Tinto valley today. The chronology and geodynamic context of these terraces will be discussed below.

Rio Tinto's upper terrace, prominently exposed at Alto de la Mesa (Fig. 2.4A), comprises a thick sequence of ferruginous conglomerates and quartz-rich sandstones interbedded with finely laminated to massive ironstones (Fernández-Remolar et al., 2005). These sediments, deposited more than 60 m above the modern river level, record a transition from high-energy fluvial conditions to quieter depositional settings. Within this succession, sedimentary structures such as cross-bedding, desiccation cracks, wrinkled surfaces, and lamination are well preserved, reflecting both episodic water flow and evaporation under seasonally arid conditions. Notably, some laminated horizons display morphological analogs to modern schwertmannite crusts and efflorescent sulfate structures, indicating that similar physicochemical processes were at work during terrace formation in the early Pliocene. The terrace is underlain by a kaolinized shale, likely representing a paleosol horizon that records a major episode of chemical alteration. The chronostratigraphic position of this kaolinitic layer (Fig. 2.4A) at the base of the upper terrace aligns with a lateritic and kaolinitic horizon exposed in the lower fluvial basin of the Río Tinto (Fernández-Caliani and Cantano, 2010). This correlation suggests the presence of relatively warm and humid conditions during the early Pleistocene, which would have favored the development of the upper terrace itself.

Mineralogically, the Upper Terrace is dominated by well-crystallized iron oxides, notably unrelaxed sextet goethite (urs-Gt) and hematite, representing the most diagenetically mature phases in the Río Tinto sequence. X-ray diffraction analyses reveal sharp, well-defined peaks for these minerals, in contrast to the broader, less distinct signatures observed in modern precipitates and younger terraces. Mössbauer spectroscopy confirms this high degree of crystallinity, with spectra characterized by fully developed, unrelaxed sextets. This mineralogical maturity reflects the extended diagenetic history of the Upper Terrace, where nanophase goethite has progressively transformed to well-crystallized goethite and, in more advanced stages, been partially replaced by hematite.

Biosedimentary features are equally striking in the upper terrace. Iron oxides replicate a range of biological textures, including casts of leaves and stems, fungal hyphae, and microbial streamers (Fernández-Remolar et al., 2005). Biological remains are preserved with remarkable fidelity, frequently appearing as iron-coated voids that likely formed following the decay of organic matter. In addition, small domal blisters—resembling modern gas-escape structures—are observed interrupting some surface laminae, suggesting in situ microbial gas production during

FIGURE 2.4

Morphological and microstructural features of the upper Terrace at Alto de la Mesa (Riotinto town). (A) General view of the upper Terrace outcrop resting atop a highly weathered horizon developed through the alteration of the Carboniferous volcanic-sedimentary host rock. This paleosurface, rich in kaolinite, likely formed by intense chemical weathering under warm climatic conditions during the early Pleistocene. (B) Surface expression of blister-like textures, interpreted as the result of gas release during microbial degradation of organic matter trapped within iron-rich sediments. (C) Popcorn-textured pseudomorphs, preserving the morphology of ancient sulfate efflorescences, now mineralogically stabilized as iron oxides. (D) Network of animal traces preserved through the mineralization of oxysulfates and Fe-oxyhydroxides, formed during the sediment infilling and precipitation of acidic Río Tinto waters. These filaments reflect arthropod activity and secondary mineralization within the oxidized terrace environment.

sediment deposition (Fig. 2.4B). Other common sedimentary features include salt pseudomorphs resembling popcorn textures and traces attributable to invertebrate activity, further highlighting the dynamic interplay of chemical, biological, and physical processes within the depositional environment (Fig. 2.4C and D). On the microscale, goethite-rich laminae contain casts and molds of coccoidal and filamentous microorganisms, which, in some horizons, have been partially overprinted by diagenetic hematite. These observations demonstrate that despite the oxidative and acidic conditions of the Río Tinto system, the upper terrace retains a robust archive of both environmental and biological processes, providing a valuable analog for the preservation of biosignatures in ancient Martian environments.

The Upper Terrace deposits date to approximately 2.1 Ma (Fernández-Remolar et al., 2005). Dating has been accomplished through multiple methods, with this age being slightly younger than the isochron ages of 2.58 ± 1.33 Ma from the Tharsis area that provide additional chronological control (Velasco et al., 2013). These radiometric dates are supported by stratigraphic relationships and correlation with regional geological events. This timing corresponds to the Gelasian Stage and the Piacenzian–Gelasian boundary at 2.59 Ma, marking the formal transition from the Pliocene to the Pleistocene. It aligns with evidence of climatic conditions inferred from alteration features and the development of paleosol horizons, which suggest a period of intensified chemical weathering under warm and humid conditions (Fernández-Caliani and Cantano, 2010; Fernández-Remolar et al., 2005).

Intermediate Terrace and the ochre level (early to mid-Holocene)

The Intermediate Terrace of the Río Tinto system is a well-preserved fluvial deposit located approximately 8 to 5 m above the current river level, with its most prominent exposure at the Nerva locality in the upper basin (Fig. 2.5A). Stratigraphically positioned between the Holocene lower terrace and older fluvial deposits reaching the early Holocene, the Intermediate Terrace records a distinct episode of geomorphic and geochemical activity linked to climatic and hydrological changes during the Holocene Epoch (Fernández-Remolar et al., 2005).

The terrace comprises a 3–4 m thick sequence of ferruginous conglomerates, pebbles, and sandstones embedded in an iron-rich cement (Fig. 2.5). Sedimentary structures include laminated and brecciated facies, with some levels displaying botryoidal crystalline fills, suggesting episodic deposition under fluctuating flow conditions (Fernández-Remolar et al., 2005). These features reflect variable energy regimes and possibly transient ponding within the fluvial system, likely modulated by seasonal variations in discharge.

FIGURE 2.5

Outcrop views and microbial textures of the intermediate Terrace and the associated ochre level at Río Tinto. (A) Overview of the intermediate Terrace, showing its distinct stratigraphic expression at Nerva. This terrace, formed under warm climatic conditions around or before 7 ka, developed after the stabilization of the upper Terrace and records continued sedimentation and iron oxidation. (B) Detail of surface textures on the intermediate Terrace, showing a variety of biosedimentary structures, including blisters (bl) resulting from gas escape during microbial degradation, partially degraded popcorn

Mineralogically, the Intermediate Terrace is dominated by well-crystallized iron oxides, specifically unrelaxed sextet goethite (urs-Gt) and hematite, with quartz occurring as a detrital component. X-ray diffraction (XRD) data show sharp goethite peaks, while Mössbauer spectroscopy confirms a high degree of magnetic ordering. This reflects a clear mineralogical progression from schwertmannite, nanophase oxides (np-Ox), nanophase goethite (np-Gt) in modern precipitates and young terraces to more diagenetically mature phases in older deposits (Fernández-Remolar et al., 2005). Although ferrihydrite and schwertmannite—often observed as nanophase iron oxides (np-Ox) and sulfate-bearing precipitates in modern Río Tinto sediments—likely formed during early diagenesis as precursors to goethite and hematite under acidic, iron- and sulfate-rich conditions (Borch and Fendorf, 2007; Cornell and Schwertmann, 2003; Fernández-Remolar et al., 2005), these metastable minerals are not preserved in the Intermediate Terrace, having been transformed during progressive diagenetic stabilization. Their absence suggests that transformation to more crystalline iron oxides is nearly complete, and localized geochemical heterogeneity—if once present—has been largely overprinted by diagenetic stabilization. The absence of sulfate salts, which are abundant in modern crusts and efflorescences, further indicates either dissolution during diagenesis or early mineral transformation during sediment stabilization. Despite a lower density of preserved biological structures compared to the Upper Terrace, the Intermediate Terrace retains key biosedimentary features, including laminated textures, blisters, pseudomorphs of sulfate efflorescences, and streamer-related morphologies (Fig. 2.5C and D) preserved in iron oxides (Fernández-Remolar et al., 2005).

In the spring system feeding the modern Rio Tinto, there exists a distinctive yellowish clay-rich horizon composed mainly of kaolinite (Fig. 2.5D). This unit, likely contemporaneous with the Intermediate Terrace, consists of clayey ochre deposits dated to >6.5 ka. It is interpreted as a product of intense chemical weathering, probably driven by the same climatic episode that influenced the formation of the Intermediate Terrace. Both the Intermediate Terrace and the ochre level record landscape responses to the Holocene Climatic Optimum—a warm and humid interval peaking between ~8000 and 5000 years ago, characterized by temperatures 1–2°C higher than present and significant shifts in precipitation patterns (Rohling et al., 2014). Although these features occupy different topographic positions, their similar ages and environmental signatures indicate that they formed under essentially the same climatic conditions.

textures (pc) representing efflorescence pseudomorphs, and a microbial filament preserved in situ between both structures. (C) Concentration of mineralized microbial filaments oriented in a similar direction, likely representing coated streamers that developed under flowing acidic waters. (D) Outcrop of the ochre level, occurring at a higher topographic position than the intermediate terrace, within the spring area of the Río Tinto source zone. This level, also formed during the warm post-glacial phase, records intense weathering in proximity to the discharge zone.

Lower Terrace (<4 ka)

The Lower Terrace of the Río Tinto system represents the youngest fluvial deposit in the sequence, situated just 10–50 cm above the current river level (Fig. 2.6A–C) (Fernández-Remolar et al., 2005). At Las Zarandas, this terrace is well-developed and originated as abandoned bars or minor terraces shouldering the current channel. Stratigraphically, the Lower Terrace comprises two distinct units: a lower section of strongly cemented, grain-supported polymictic conglomerates 1–2 m thick, with imbricated pebbles and internal erosional surfaces, overlain by well-sorted quartzose sandstones with trough cross-stratification and ripples, topped by ferric sands (Fig. 2.6D) rich in plant debris and microbial filaments. These quiet zones were seasonally oversaturated with ferric minerals, facilitating the precipitation of iron oxides that preserved delicate biological remains. Among the structures recorded are fungal filaments, plant stems, and complex biofilm textures coated by goethite, forming finely laminated ferric crusts. These features reflect a depositional setting closely analogous to the modern Río Tinto system, where seasonal drying and acidic conditions drive mineral precipitation and the early stages of fossilization.

Mineralogically, the Lower Terrace is dominated by poorly crystalline goethite (prs-Gt) and detrital quartz, with minor hydronium jarosite and/or natrojarosite detected in some samples. In contrast to the more mature mineralogy of the older terraces, the goethite in the Lower Terrace displays Mössbauer spectra that are intermediate between nanophase goethite doublets and unrelaxed goethite sextets, reflecting its relatively early stage of diagenetic maturation. This partially relaxed sextet goethite represents a transitional phase in the diagenetic sequence from the nanophase goethite (np-Gt) found in modern precipitates to the well-crystallized goethite typical of older deposits. The original mineral assemblage likely included ferrihydrite (np-Ox) and schwertmannite, which are common in actively precipitating, acidic, Fe- and sulfate-rich environments. While these metastable phases are no longer preserved, their early diagenetic transformation likely contributed to the formation of the current goethite and jarosite mineralogy. Minor hematite is present in some samples; it is considered primarily detrital in origin rather than a product of diagenetic processes.

Goethitic laminae that coat detrital grains are a distinctive feature of the Lower Terrace. These thin coatings effectively preserve fossils in remarkable detail, ranging from microscopic cocci and filaments of bacteria, algae, and fungi to insect cuticles with micron-scale seta canals intact, and plant tissues replicated in cellular detail (Fernández-Remolar et al., 2005; Fernández-Remolar and Knoll, 2008). The organic constituents of bioclasts typically decay over time, leaving void spaces that become partially filled by concentric laminated precipitates or inwardly directed crystals. These processes result in exceptional preservation of biological structures as goethite pseudomorphs, providing valuable insights into the microbial communities that inhabited the Río Tinto system during the Holocene.

Radiometric dating using ^{14}C has provided age constraints for the Lower Terrace, with dates ranging from approximately 4 ka to as young as 0.7 ka. These Holocene

FIGURE 2.6

Geomorphological and sedimentological features of the young Terrace in the Río Tinto system. (A) General view of the young Terrace in the Las Zarandas area (indicated by white arrows), showing its position entrenched by the modern Río Tinto channel, reflecting recent fluvial incision following terrace formation. (B) Stratigraphic exposure of the young Terrace at Barranco de los Locos, where a ferruginous deposit dated to ~2 ka caps a sequence of older sedimentary breccias and clay-rich laminated deposits dated up to ~4 ka. The deposits are cemented by iron oxides, forming resistant strata typical of this terrace level. (C) View of a thick ferruginous crust developed within the young Terrace, indicating intense iron precipitation and early lithification during episodic fluvial activity and terrace aggradation. (D) Outcrop of yellowish iron-rich deposits in the Berrocal area, overlying the upper portion of the young Terrace and dated to approximately 0.7 ka. These deposits are characterized by abundant fossil content, including mineralized plant remains and microbial fungi observed in the Las Zarandas area. They also preserve well-defined sedimentary structures such as ripple marks, reflecting fluvially influenced deposition under shallow flow conditions.

ages align well with stratigraphic considerations and estimates of channel cutting rates, confirming its recent formation relative to the other terraces in the Río Tinto sequence. The Lower Terrace preserves many of the sedimentary structures observed in modern non-clastic sediments, supporting the conclusion that it formed under environmental conditions similar to those observable in the current Río Tinto system. The presence of minor hydronium jarosite in the Lower Terrace, unlike in the older terraces, indicates incomplete hydrolysis of this phase before terrace formation and lithification, further confirming its relatively young age and limited diagenetic alteration.

The Lower Terrace thus represents an important transitional record between modern Río Tinto sedimentary processes and the more extensively diagenetically altered intermediate and upper terraces, providing critical insights into the early stages of mineral transformation and fossil preservation in this exceptional acid-sulfate system.

Gossan and Terrace diagenesis

The Río Tinto system provides an exceptional terrestrial analog for investigating the geochemical and mineralogical pathways that control biosignature preservation in acidic and oxidizing environments. Both the gossan at Peña de Hierro and the fluvial terrace deposits downstream originated from the oxidative weathering of pyritic ores, a process fundamentally driven by chemolithotrophic microbial communities that catalyze the oxidation of Fe^{2+} and sulfide to ferric iron and sulfate (Amils et al., 2007; Fernández-Remolar et al., 2003). These reactions generate hyperacidic, ferric sulfate-rich solutions that lead to the precipitation of metastable iron minerals such as schwertmannite and jarosite (Bigham et al., 1996; Fernández-Remolar et al., 2005).

Upon exposure to less acidic meteoric waters, a diagenetic transformation is initiated (Fig. 2.7A), marked by desulfatation (loss of structural sulfate) and subsequent dehydration, converting ferric oxysulfates into more stable iron oxyhydroxides such as goethite (FeOOH), and ultimately into hematite (Fe_2O_3) (Fernández-Remolar et al., 2021a, 2021b; Morris et al., 2000). These mineralogical changes are driven by shifts in pH and redox conditions and may be enhanced by elevated temperature or microbial activity. The transformation pathway has been clearly documented in the gossan at Peña de Hierro (Fernández-Remolar et al., 2021a, 2021b), but similar trends occur throughout the fluvial terraces, which span from the Late Miocene to the Holocene (Fernández-Remolar and Knoll, 2008; Fernández-Remolar, 2011).

In particular, the diagenetic evolution of goethite is well expressed across the terrace stratigraphy (Fig. 2.7A). In modern settings, goethite typically precipitates as nanophase goethite (np-goethite)—a poorly crystalline, superparamagnetic form resulting from the hydrolysis of ferric iron in acidic waters. As these sediments become buried, they transform into partially relaxed sextet goethite (prs-goethite),

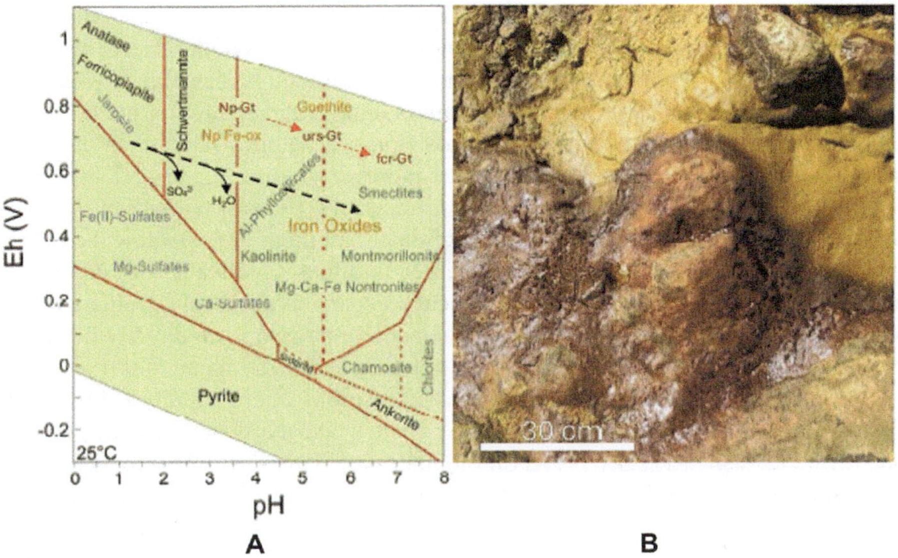

FIGURE 2.7

Diagenetic evolution of iron-bearing phases in the acidic environments of the Río Tinto system. (A) Eh–pH stability diagram at 25 °C showing the transformation of ferric phases under varying redox and acidity conditions. The schematic pathway (dashed red arrows) illustrates the progression from nanophase iron oxides (np-Fe-ox) and nanophase goethite (np-Gt) to increasingly ordered forms: Partially relaxed sextet goethite (prs-Gt), unrelaxed sextet goethite (urs-Gt), and ultimately fully crystalline goethite (fcr-Gt). Ferric oxysulfates such as jarosite, schwertmannite, and ferricopiapite transform into iron oxides through desulfatation and dehydration, driven by decreasing Eh, rising pH, and sulfate loss (Bigham et al., 1996; Regenspurg et al., 2004; Cornell and Schwertmann, 2003). This sequence reflects the mineralogical trends observed across the Río Tinto terrace system with increasing age and depth. (B) Field photo showing active desiccation and phase transformation in ferric precipitates. Yellowish zones rich in oxysulfates and ferrihydroxides transition into reddish crusts where dehydration and recrystallization toward goethite are occurring. These changes proceed over days to weeks, especially in summer, illustrating the rapid early diagenesis of Fe phases in acidic, sulfate-rich settings.

indicating increased crystallinity and grain size, as observed in the younger terrace levels. With continued diagenetic progression, goethite evolves into unrelaxed sextet goethite (urs-goethite) and may ultimately recrystallize or be replaced by hematite, particularly in the intermediate and older terraces. These mineralogical changes, tracked through X-ray diffraction and Mössbauer spectroscopy, reflect a shift from depositional to post-depositional control on iron mineralogy, governed by time, burial, and geochemical evolution.

The transformation of ferric oxysulfates such as jarosite and schwertmannite into stable iron oxides follows a similar path. The first step—desulfatation—is favored by rising pH (typically >3–3.5), fluctuating redox conditions, and sulfate dilution in porewaters (Bigham et al., 1996; Regenspurg et al., 2004). In minerals like schwertmannite, which incorporate sulfate into tunnel-like structures, desulfatation leads to rapid structural destabilization. Once sulfate is removed, dehydration promotes the reorganization of Fe(III) hydroxides into nanophase goethite, which then proceeds along the same maturation sequence into more crystalline phases (Cornell and Schwertmann, 2003). In Río Tinto's acidic terraces, this transformation is well documented: jarosite and schwertmannite dominate modern and shallow horizons, whereas crystalline goethite and hematite become more abundant with depth and age (Fernández-Remolar et al., 2005; Sánchez España et al., 2011). Under slightly less acidic or more oxidizing conditions, hematite may also form directly from ferrihydrite or by dehydration of goethite (Cornell and Schwertmann, 2003).

Crucially, although these mineralogical transitions occur rapidly (Fig. 2.7B), they are not wholly destructive to organic matter. On the contrary, both goethite and hematite can serve as preservation matrices. Time-of-flight secondary ion mass spectrometry (ToF–SIMS), Fourier transform ion cyclotron resonance mass spectrometry (FTICR-MS), and GC–MS analyses of gossan and terrace materials reveal a wide spectrum of preserved biomolecular fragments, including lipids, peptides, and sphingolipids (Fernández-Remolar et al., 2021a). Remarkably, peptidic chains have been identified in terrace deposits dated to 2.1 Ma, indicating that even complex molecular structures can survive long-term oxidative and acidic conditions when protected by early mineral entrapment (Colin-García et al., 2011).

The evolution of the Río Tinto acidic basin: From Oligocene to present

The Río Tinto basin represents one of Earth's most remarkable natural laboratories for studying long-term acid rock drainage (ARD) processes, offering critical insights into similar environments that may have existed on ancient Mars. Its chronostratigraphic sequence spans from the late Oligocene to the present, preserving a continuous record of acidic, oxidizing conditions that have persisted through significant climatic and tectonic changes (Fernández-Remolar et al., 2005; Amils et al., 2007).

The system originated when Alpine tectonic uplift exhumed the massive sulfide deposits of the Iberian Pyrite Belt during the late Oligocene. This exhumation coincided with a significant warming period, creating ideal conditions for accelerated oxidative weathering of massive sulfide deposits (Fernández-Remolar et al., 2021a; Zachos et al., 2001). As meteoric waters infiltrated these exposed sulfides, both microbial and abiotic oxidation mechanisms generated acidic, ferric sulfate-rich solutions that precipitated metastable minerals including schwertmannite, jarosite, and copiapite—establishing the foundational gossan cap that would later evolve into the distinctive landscape seen today (Fernández-Remolar et al., 2003; Essalhi et al., 2011).

The transformation of these initial precipitates followed well-defined mineralogical pathways, with hydrolysis and dehydration reactions progressively converting ferric sulfates to nanophase goethite, then to increasingly crystalline goethite, and ultimately to hematite in the oldest deposits (Schwertmann and Murad, 1983; Fernández-Remolar et al., 2005). This diagenetic sequence, driven by fluctuating water tables and shifting redox conditions, created the mineral evolution pattern recorded in the different Gossan and terrace levels (Fernández-Remolar et al., 2005).

The Río Tinto's fluvial terrace system provides a chronological framework for understanding how acidic conditions evolved from the Pleistocene through the Holocene. The Upper Terrace (2.1 Ma) formed during the Early Pleistocene climatic transition, coinciding with intensified tectonic activity in the Iberian Peninsula that triggered significant fluvial incision (Fernández-Remolar et al., 2005; Moreno et al., 2003). The Intermediate Terrace (>7 Ka) developed during the early-mid Holocene and correlates with the onset of more humid conditions following the Younger Dryas cold period, while the Lower Terrace (4–0.7 Ka) formed during the mid-late Holocene, capturing the transition from the Holocene Climatic Optimum toward cooler, more variable conditions of the Neoglacial period (Fernández-Remolar and Knoll, 2008; Rohling et al., 2014).

These terraces record distinct phases of fluvial incision and deposition modulated by regional tectonic pulses and global climate oscillations. Rather than developing continuously, acidic weathering and mineral formation proceeded in discrete episodes corresponding to specific climatic windows favorable for enhanced chemical weathering (Fernández-Remolar et al., 2005; Velasco et al., 2013).

Particularly notable is the system's response to the Holocene Climatic Optimum (~8–5 Ka), which generated contemporaneous but distinctly different facies in the Intermediate Terrace and the ochre level of the headwaters (Fernández-Remolar et al., 2005). The Intermediate Terrace formed during peak warmth and humidity around 7–6 Ka, coinciding with maximum Holocene sea levels and enhanced winter precipitation in Iberia driven by a northward shift of the North Atlantic storm tracks (Davis et al., 2003; Val-Peón et al., 2021). The yellowish clay-rich kaolinite horizon (ochre level) of the Lower Terrace, dated to >6.5 Ka, developed synchronously under these same climatic conditions but in a different topographic and hydrological setting, demonstrating how local factors can produce diverse depositional outcomes under uniform climatic forcing (Fernández-Remolar et al., 2005; Fernández-Caliani and Cantano, 2010).

The younger portions of the Lower Terrace (4–0.7 Ka) correspond to the establishment of Mediterranean climate cyclicity with pronounced seasonality in the Iberian Peninsula, punctuated by the Roman Warm Period and Medieval Climate Anomaly (Martín-Puertas et al., 2008). This period also coincides with the initiation of the present-day NAO (North Atlantic Oscillation) pattern, which strongly influences precipitation regimes in southwestern Spain. The latest phase of terrace development (<1 Ka) reflects both climatic factors and possibly early anthropogenic influences on the landscape through deforestation and mining activities (Davis et al., 2000).

The Paleontological record

The Río Tinto system challenges conventional assumptions about biosignature preservation in acidic environments. Contrary to the expectation that such harsh conditions would destroy organic matter, fossils are common (Preston et al., 2011; Maldanis et al., 2024; Fernández-Remolar and Knoll, 2008), and recent mass spectrometric studies have revealed exceptional preservation of molecular biosignatures throughout the system. Peptides, lipids, and sphingolipids persist within both the gossan and terrace units, suggesting that iron oxides may provide protective matrices for organic compounds over geological timescales (Fernández-Remolar et al., 2021a).

Morphological fossils

The oldest terrace deposits preserve the remains of leaves, fungal streamers (Fig. 2.8A and B), and microbial populations (Fig. 2.8C and D), all entombed in iron oxide minerals (Fig. 2.8D). As noted above, bedding surfaces also show evidence of gas blisters (Fig. 2.5B), known from the modern ecosystem to be biogenic (Fernández-Remolar and Knoll, 2008). Petrographic analysis, in turn, reveals abundant casts and molds of coccoidal and filamentous microorganisms, mostly bacterial (Fig. 2.8C and D). While transformation of earlier forming iron minerals can obliterate biological textures, local preservation can be surprisingly good, demonstrating the retention of morphological remains throughout the diagenetic transition from jarosite and poorly ordered goethite to hematite.

The modern Tinto system supports a high diversity of acid-tolerant bacteria and microscopic eukaryotes, including algae (diatoms, euglenids, chlorophytes and rhodophytes), heterotrophic protists and fungi (Amaral Zettler et al., 2002; Sabater et al., 2003; Aguilera et al., 2006); in particular, fungi constitute locally dominant populations in rivulets that feed the Rio Tinto, forming dense carpets of streamers (Fig. 2.8A). Fossils appear largely to reflect organisms similar to those observable today (Fig. 2.8B), with fossilization favoring organisms with decay resistant cell walls or cuticle. Some preserved microfossils are sufficiently simple to challenge fine-scale taxonomic interpretation.

Like their older counterparts, the younger terraces also preserve a range of fungal and other microbial morphologies, as well cellularly preserved plant debris (Fig. 2.8C) and insect cuticle (Fig. 2.8D) washed into the river from the surrounding woodlands. Again, the strong impression from fossil assemblages in the three terrace levels is that acid-tolerant organisms, much like those found today, have populated the Rio Tinto ecosystem for a long time.

The application of synchrotron X-ray nanoimaging has revealed an extraordinary level of structural preservation in microbial fossils from the Rio Tinto Upper Terrace. Maldanis et al. (2024) used ptychographic X-ray computed laminography

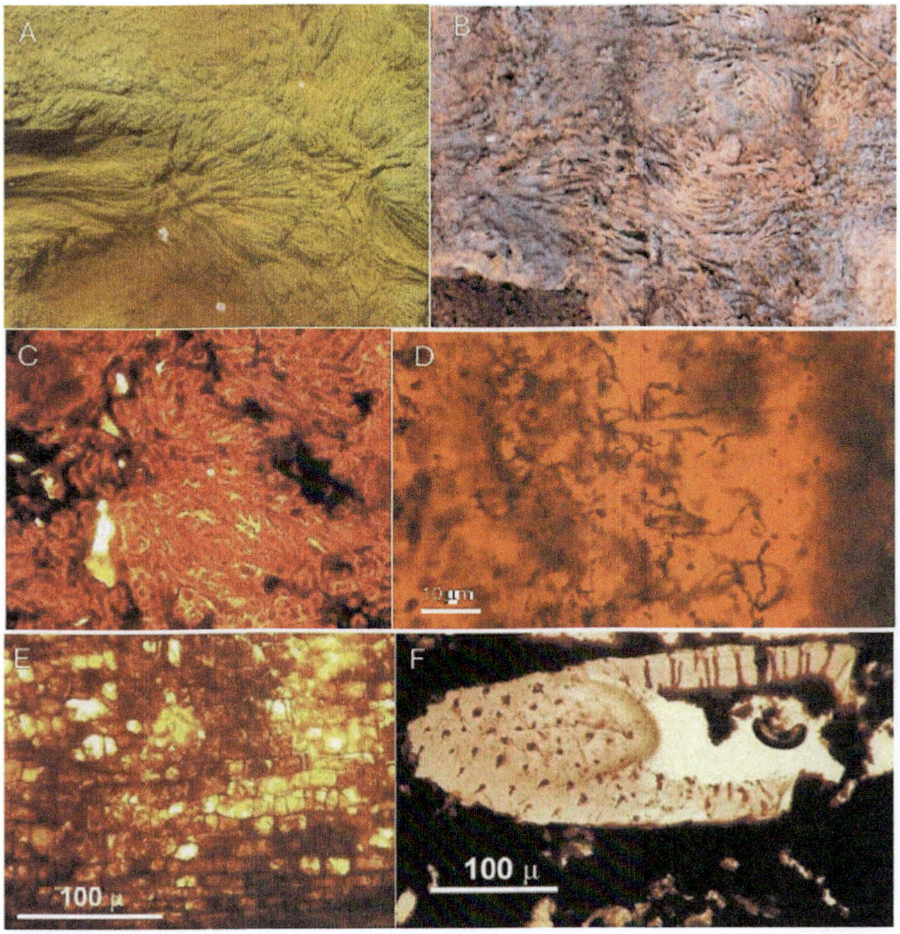

FIGURE 2.8

Morphological fossils in Rio Tinto terraces. (A) Fungal streamers in the modern river compared with (B) macroscopic streamers, with individuals preserved as castes in iron oxides. (C) Insect and (D) plant epidermal tissues preserved in terraces iron oxides. These remains are interpreted as organisms from the surrounding hillsides that washed into the depositional system. (E) Larger and (F) smaller filamentous bacteria preserved as iron oxide casts.

(PyXL) and nano-XRF to visualize micron-to submicron-scale microbial structures preserved within goethite- and jarosite-rich sediments. These techniques uncovered not only elongated filaments but also subcellular rod-like structures and spheroids, many of which were invisible to conventional optical microscopy due to their small size and embedding within opaque iron minerals (Figs. 2.8D and 2.9A–G).

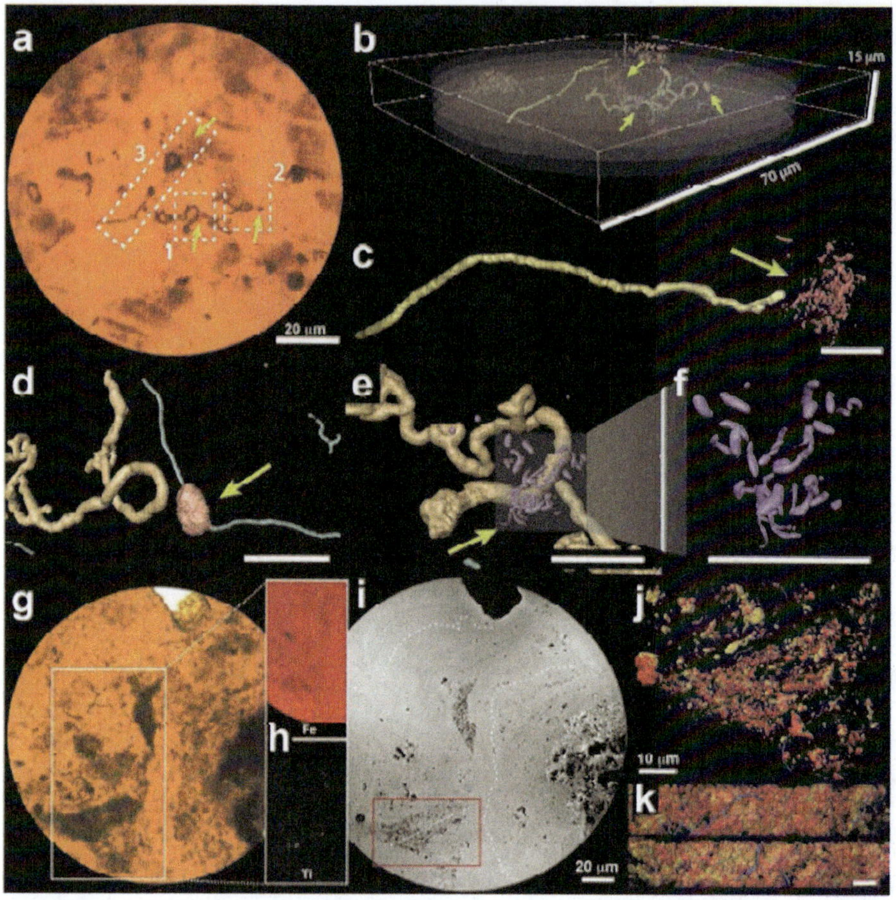

FIGURE 2.9

Morphological and compositional diversity of preserved microstructures from Rio Tinto 2 Mya terraces as revealed through high-resolution 3D imaging using ptychographic X-ray computed laminography (PyXL). Yellow arrows point to structures that are not visible in optical microscopy. (A) Transmitted optical micrograph showing dark brownish filaments and features within an orange goethite matrix. (B) 3D reconstruction of the same region observed in A, showing the distribution of filaments and other micro and nano structures within the rock slice. (C) Filament observed in the region outlined as 3, showing some branching and close association of rod-shaped structures. (D) Detail region 2 showing a thicker filament (~1 mm thick) with loop features and a thinner filament (~200 nm thick) displaying an associated globular structure. (E) Detail of region 3 with a filament displaying twists and loops and the closely associated rod-shaped structures. (F) Detail of the rod-shaped structures. (G) Transmitted optical micrograph of a larger area (180 mm) showing different mineral structures and filaments within the goethite matrix. (H) Elemental maps of Fe and Ti of the area outlined in G. Fe is widespread in the matrix and

Among the most remarkable findings was the exceptional three-dimensional preservation of microbial filaments, which displayed consistent cylindrical morphologies, hollow interiors, and no signs of mineral infill or metal enrichment. This supports the interpretation that they are fossilized microbial casts rather than mineral pseudomorphs. The morphological complexity—including septation, branching, looping, and nodulation—strongly suggests a biological origin (Fig. 2.9C–E). The authors identified three main filamentous morphotypes, the most intricate of which exhibited features comparable to branched Actinobacteria, a group known to inhabit Rio Tinto's subsurface (Amils et al., 2023). Morphotype 1, in particular, shows short, hook-like branches consistent with the vegetative hyphae of Actinobacteria (Fig. 2.9B–E), while nodular features resemble reproductive or stress-related structures known in this group (Fig. 2.9D).

Additionally, filaments were often found interwoven with nanometer-sized rod-shaped structures, possibly representing spore-like forms or different taxa (Fig. 2.9E and F). Their distribution within botryoidal goethite laminations and clusters of Fe-depleted, Ti-rich minerals suggests that these microbes colonized the sediments post-deposition and were fossilized during early diagenesis (Fig. 2.9G–K). These textural associations provide insights into both the ecological structure of the microbial community and the taphonomic conditions that facilitated preservation.

Preservation of organic molecules in the acidic deposits of Río Tinto

Contrary to traditional assumptions that acidic porewaters rapidly degrade biomolecules, research conducted over the past 2 decades demonstrates that Río Tinto's iron-rich mineral systems facilitate the stabilization and long-term preservation of diverse molecular and isotopic biosignatures. Initial insights into this phenomenon came from organic geochemical analyses of both modern and ancient Río Tinto deposits. Using solid phase microextraction coupled with GC–MS, Fernández-Remolar et al. (2007) reported a diverse suite of preserved organic

deprived on regions with cluster of minerals, which are richer in Ti. (I) Virtual PyXL slice of the same region displayed in G. Dotted lines mark the botryoidal laminations which seem to have compacted the crystals into clusters. (J) 3D reconstruction of the red-outlined area marked in I. Filaments (blue) are observed closely associated with the cluster of minerals, and are also not visible in visible light due to the opaque nature of these minerals. (K) Detail of the well preserved filaments cross-cutting the crystals, indicating their post-diagenetic scavenging behavior. Scale bars when not indicated are 5 mm.

- Modified from Maldanis, L., Fernandez-Remolar, D., Lemelle, L., Knoll, A.H., Guizar-Sicairos, M., Holler, M., Francisco Mateus Cirilo da Silva, Magnin, V., Mermoux, M., Simionovici, A., 2024. Unveiling challenging microbial fossil biosignatures from Rio Tinto with micro-to-Nanoscale chemical and Ultrastructural imaging. Astrobiology 24 (7), 721–733, https://doi.org/10.1089/ast.2023.0127.

molecules—including aldehydes, ketones, long-chain carboxylic acids, esters, fatty acids, and sulfur-bearing compounds such as dimethyl disulfide—within modern acidic waters and ferric sediments from the Río Tinto system. Although certain contaminants (e.g., phthalates and PVC-derived compounds) were detected and carefully annotated, the high spectral correlation (>85%) with NIST reference standards, the consistency of detected compounds with microbial membrane degradation products, and the spatial association with microbially active zones all strongly support the authigenic origin of the majority of the analytes. Additionally, variations in organic composition between upstream and downstream sites—as well as between waters and jarosite-bearing sediments—further reflect in situ microbial processes, plant input, and organic-sediment interactions, rather than generalized contamination. Together, these observations confirm that the molecular signatures captured in the analyses are representative of the modern biogeochemical environment of the Río Tinto system. These compounds were identified in jarosite-rich deposits precipitated under $pH < 1.5$, suggesting that rapid mineralization processes can entomb organic matter before complete degradation occurs. Remarkably, even modern jarosites contain compounds of plant origin, such as abietic acid and phytanal, indicating both external input and the preservation of structurally complex biomolecules.

Building on this geochemical evidence, Colín-García et al. (2011) provided direct confirmation of molecular preservation in the form of short-chain peptide sequences extracted from Río Tinto's fluvial terraces ranging in age from 2000 years to over 2.1 million years. Using Fourier Transform Ion Cyclotron Resonance Mass Spectrometry (FTICR-MS), they detected peptide masses exceeding 1000 Da in the oldest terrace samples. These peptides were not only present but showed systematic decline in abundance with age, supporting the view that they were not contaminants but remnants of ancient biological communities once thriving in the acidic sediments. Despite the oxidative and acidic conditions, the mineral matrix—rich in iron sulfates and oxyhydroxides—likely played a crucial role in stabilizing these biomolecules.

Further supporting this model, Fernández-Remolar et al. (2021a) used ToF–SIMS and other spectrometric methods to demonstrate the preservation of diverse lipid classes, including sphingolipids, acylglycerols, and terpenoids (Fig. 2.10A and B), alongside degraded macromolecular compounds such as polyphenols and organo-sulfur chains. Many of these compounds were spatially associated with filamentous structures encased within hematite and goethite suggesting that microbial remains served as nucleation sites for mineral precipitation, thereby promoting entombment and preservation. FTICR–MS analysis further confirmed the presence of CHO-, CHNO-, and CHOS-functionalized compounds, including lipidic and polyphenolic structures, distributed across the gossan matrix. The molecular diversity and complexity preserved in these ancient deposits indicate multiple biological sources, including fungi, bacteria, algae, and higher plants.

A key aspect of preservation in Río Tinto is the mineralogical context. Jarosite and schwertmannite, though commonly thought of as unstable under neutral pH conditions, act as early encapsulation agents. These iron-bearing sulfates eventually

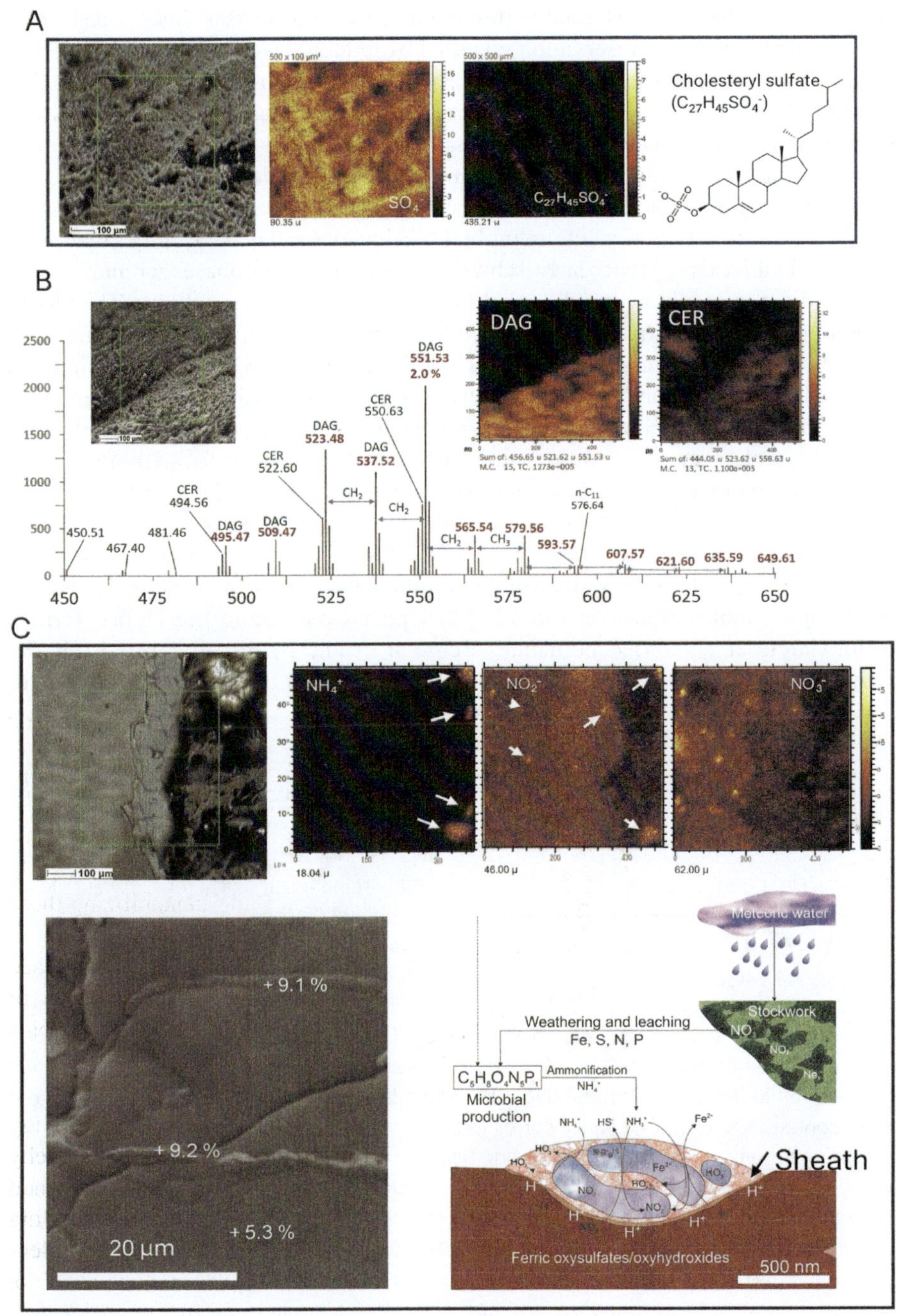

FIGURE 2.10

Evidence of molecular preservation and nitrogen cycling in the upper Gossan of Peña de Hierro, integrating ToF–SIMS molecular imaging, SEM-EDS elemental mapping, and

transform into goethite and hematite through diagenetic processes. Such transformations appear to enhance preservation potential by reducing permeability, limiting oxygen diffusion, and immobilizing organic molecules within nanoporous iron oxide matrices. As observed in several terrace samples, fossil filaments and plant tissues are initially coated with nanophase ferric minerals that act as both protective barriers and structural molds, sometimes preserving cellular-level detail.

Rapid mineralization driven by highly concentrated ferric and sulfate-rich waters plays a critical role in preserving microbial biosignatures. The swift precipitation of ferric oxyhydroxides, particularly schwertmannite and nanophase goethite, under hyperacidic and oxidizing conditions drastically reduces water activity. This dehydration effect, combined with the accumulation of toxic heavy metal cations (e.g., Cu^{2+}, Zn^{2+}, Pb^{2+}, and others), creates a hostile chemical environment that inhibits microbial metabolism and enzymatic activity, effectively halting the degradation of organic matter. These metals, in oversaturated concentrations, exert a biocidal effect that further enhances early fossilization. Additionally, the transformation of amorphous and nanophase iron phases into more crystalline goethite—and eventually hematite—can occur within weeks in modern deposits. These mineral transitions act as both chemical and physical barriers, shielding organic compounds from later diagenetic overprinting. As demonstrated by the detection of intact lipids, peptides, and complex biomolecules in the ~25 Ma Upper Gossan and the Upper Terrace (Colín-García et al., 2011; Fernández-Remolar et al., 2021a), this rapid mineral entombment under extreme geochemical stress provides a unique mechanism for

conceptual modeling. (A) ToF–SIMS ion maps showing the distribution of cholesteryl sulfate ($C_{27}H_{45}SO_4$-) and SO_4^{2-} within laminated ferric sediments. Cholesteryl sulfate aligns with thick filamentous structures resembling fungal hyphae, indicating the selective entrapment of eukaryotic membrane-derived steroids during early mineralization by ferric oxysulfates. (B) ToF–SIMS spectrum and molecular ion maps highlighting the spatial distribution of diacylglycerols (DAGs) and ceramides (CERs) in two distinct microbial microstructures. DAGs are localized in textures consistent with bacterial or algal biofilms, while CERs are concentrated in filamentous forms likely derived from fungal elements, reflecting differential lipid preservation according to microbial origin. (C) (Top) ToF–SIMS ion maps showing the distribution of ammonium (NH_4^+), interpreted as a relic of amino acids or pectic substances, alongside nitrite (NO_2^-) and nitrate (NO_3^-)—products of microbial redox cycling of nitrogen. These compounds appear spatially correlated with preserved microbial structures, suggesting activity of nitrogen-metabolizing bacteria in association with Fe^{3+} minerals, potentially using iron as a terminal electron acceptor. (Left bottom) SEM-EDS analysis reveals localized nitrogen enrichment (5.3–9.2 at%) within filamentous bacterial structures. (Right bottom) conceptual model illustrating a proposed Feammox-like microbial nitrogen cycle active within the gossan, where Fe(III) reduction may support anaerobic ammonium oxidation (Fernández-Remolar et al., 2021b). The Carboniferous hydrothermal host rock is proposed as the primary nitrogen reservoir, supplying reduced N species to microbial communities active during gossan formation and early diagenesis.

long-term biosignature preservation, even in oxidizing, low-pH environments that are typically considered destructive.

A record of nitrogen cycling at Rio Tinto

The Río Tinto system offers a remarkable example of how microbially mediated nitrogen cycling can be recorded and preserved in mineralized subsurface and surface deposits. Through the use of advanced molecular imaging techniques, such as ToF-SIMS (Fig. 2.10C), it has been possible to detect the mineralized by-products of nitrogen transformations—including ammonium (NH_4^+), nitrite (NO_2^-), and nitrate (NO_3^-)—within microstructures such as filaments and micronodules embedded in ferruginous laminations (Fernández-Remolar et al., 2021b, 2021c). These compounds are indicative of complex microbial processes, including ammonification, nitrification, and possibly Feammox-type metabolisms, where Fe^{3+} acts as an electron acceptor in the oxidation of reduced nitrogen species (Fig. 2.10C). Like other biosignatures, the preservation of these nitrogen-bearing species is facilitated by the rapid entombment of microbial communities in iron oxyhydroxides and oxysulfates, which act as both chemical stabilizers and physical barriers, inhibiting post-depositional alteration.

This molecular imprint of nitrogen cycling is not confined to the deep subsurface. Similar NOx-bearing structures and associated N-rich biomolecular fragments have been documented in the Upper Terrace and Upper Gossan units, indicating that surface and near-surface microbial communities were also actively engaged in nitrogen transformations. In the Upper Gossan, which dates back over 25 million years, NO_2^- and NO_3^- ions are found within preserved microfilaments and micronodules alongside lipids, amino acids, and sulfate-rich mineral matrices (Fernández-Remolar et al., 2021b; Fernández-Remolar et al., 2021c). Their co-occurrence with ferric oxides, silica, and secondary phosphate minerals points to localized microenvironments where microbial metabolism and mineralization converged under acidic, oxidizing conditions. The persistence of these nitrogen-related structures across multiple sedimentary generations highlights the resilience of biosignatures associated with the nitrogen cycle, and reinforces the potential of iron-rich terrestrial systems like Río Tinto to inform the search for ancient life on Mars and other planetary bodies (Fernández-Remolar et al., 2021b, 2021c).

Implications for Mars

In 2004, a discovery by NASA's MER rover Opportunity provided a new reason to study the geologic record of the Rio Tinto system. Mössbauer spectroscopy of sedimentary rocks deposited at Meridiani planum more than three billion years ago (Grotzinger et al., 2005; Squyres and Knoll, 2005) revealed the presence of jarosite and other sulfate minerals, as well as hematite, documenting acid-sulfate waters on the ancient Martian surface (Klingelhöfer et al., 2004; McLennan et al., 2005). Subsequent analysis of Rio Tinto terrace beds, using a Mössbauer instrument identical to the one deployed on Mars, demonstrated the importance of Rio Tinto as a terrestrial analog to the sedimentary record at Meridiani: Mössbauer spectra of Rio Tinto sedimentary rocks commonly proved to be nearly identical to those taken on Mars (Fernández-Remolar et al., 2005).

No terrestrial analog is precisely identical to its Martian counterpart, but Rio Tinto sheds considerable light on conditions in the Meridiani basin at a time life was just getting started on Earth. The analogy does not mean that life existed on Mars when Meridiani sediments were deposited, but it does tell us that if life was indeed present at that time and place, a record of its presence might well be preserved in iron oxide crusts and the small hematitic concretions (commonly called "blueberries" because of their appearance in false color images) that formed abundantly during early diagenesis (Calvin et al., 2008). The low water activity inferred for the Meridiani environment might also have facilitated the preservation of organic molecules, although at its most extreme it could have been challenging for life (Tosca et al., 2008).

Combined visible-near infrared reflectance spectra of the Rio Tinto system and X-ray diffraction measurements of selected samples have enabled planetary scientists to resolve ambiguities in orbital spectra of Meridiani planum (Kaplan et al., 2016). Moreover, the nondestructive multiscale imaging techniques developed for the study of Rio Tinto fossils hold promise for the study of other mars analog environments and, importantly, samples from Mars itself. The Río Tinto system thus provides not only a mineralogical and geochemical analog for Mars but also a biological preservation model that informs strategies for astrobiological exploration. Its complex record of climate-modulated weathering, mineral evolution, and biological preservation offers a sophisticated framework for interpreting similar deposits on the red planet and designing targeted searches for evidence of ancient Martian life (Knoll et al., 2005; Fernández-Remolar et al., 2021a; Maldanis et al., 2024).

AI disclosure

During the preparation of this work the author(s) used ChatGPT in order to Correcting writing and improving clarity in the text (DCFR). After using this tool/service, the author(s) reviewed and edited the content as needed and take(s) full responsibility for the content of the publication.

References

Amils, R., González-Toril, E., Fernández-Remolar, D., Gómez, F., Aguilera, A., Rodríguez, N., Malki, M., García-Moyano, A., Fairén, A.G., de la Fuente, V., Luis Sanz, J., 2007. Extreme environments as Mars terrestrial analogs: the Rio Tinto case. Planet. Space Sci. 55 (3), 370–381. https://doi.org/10.1016/j.pss.2006.02.006.

Aguilera, A., Manrubia, S.C., Gómez, F., Rodríguez, N., Amils, R., 2006. Eukaryotic community distribution and its relationship to water physicochemical parameters in an extreme acidic environment, Río Tinto (Southwestern Spain). Appl. Environ. Microbiol. 72 (8), 5325. https://doi.org/10.1128/AEM.00513-06.

Amaral Zettler, L.A., Gómez, F., Zettler, E., Keenan, B.G., Amils, R., Sogin, M.L., 2002. Eukaryotic diversity in Spain's River of Fire. Nature 417. https://doi.org/10.1038/417137a, 137–137.

Amils, R., Escudero, C., Oggerin, M., Sánchez, F.P., Rodríguez, A.A., Remolar, D.F., Rodríguez, N., Villadangos, M.G., Sanz, J.L., Briones, C., Román, M.S.-, Gómez, F., Leandro, T., Moreno-Paz, M., Ballesteros, O.P.-, Molina, A., Tornos, F., Andrea, I.S.-, Timmis, K., Pieper, D.H., Parro, V., 2023. Coupled C, H, N, S and Fe biogeochemical cycles operating in the continental deep subsurface of the Iberian Pyrite Belt. Environ. Microbiol. 25 (2), 428–453. https://doi.org/10.1111/1462-2920.16291.

Bigham, J.M., Schwertmann, U., Traina, S.J., Winland, R.L., Wolf, M., 1996. Schwertmannite and the chemical modeling of iron in acid sulfate waters. Geochim. Cosmochim. Acta 60 (12), 2111–2121. https://doi.org/10.1016/0016-7037(96)00091-9.

Borch, T., Fendorf, S., 2007. Phosphate interactions with iron (hydr)oxides: mineralization pathways and phosphorus retention upon bioreduction. In: Barnett, M.O., Kent, D.B. (Eds.), Developments in Earth and Environmental Sciences. Elsevier, Amsterdam, pp. 321–348.

Calvin, W.M., Shoffner, J.D., Johnson, J.R., Knoll, A.H., Pocock, J.M., Squyres, S.W., Weitz, C.M., Arvidson, R.E., Bell, J.F., Christensen, P.R., de Souza, P.A., Farrand, W. H., Glotch, T.D., Herkenhoff, K.E., Jolliff, B.L., Knudson, A.T., McLennan, S.M., Rogers, A.D., Thompson, S.D., 2008. Hematite spherules at Meridiani: results from MI, mini-TES, and Pancam. J. Geophys. Res. Planets 113 (E12). https://doi.org/10.1029/2007je003048.

Colín-García, M., Kanawati, B., Harir, M., Schmitt-Kopplin, P., Amils, R., Parro, V., García, M., Fernández-Remolar, D., 2011. Detection of peptidic sequences in the ancient acidic sediments of Río Tinto, Spain. Orig. Life Evol. Biosph. 41 (6), 523–527. https://doi.org/10.1007/s11084-011-9258-x.

Cornell, R.M., Schwertmann, U., 2003. The Iron Oxides: Structure, Properties, Reactions, Occurrences and Uses. Wiley-VCH Verlag GmbH & Co. KGaA, Weinheim. https://doi.org/10.1002/3527602097.

Davis, B.A.S., Brewer, S., Stevenson, A.C., Guiot, J., 2003. The temperature of Europe during the Holocene reconstructed from pollen data. Quat. Sci. Rev. 22 (15–17), 1701–1716. https://doi.org/10.1016/s0277-3791(03)00173-2.

Davis, R.A., Welty, A.T., Borrego, J., Morales, J.A., Pendon, J.G., Ryan, J.G., 2000. Rio Tinto estuary (Spain): 5000 Years of pollution. Environ. Geol. 39 (10), 1107–1116. https://doi.org/10.1007/s002549900096.

Essalhi, M., Sizaret, S., Barbanson, L., Chen, Y., Lagroix, F., Demory, F., Nieto, J.M., Sáez, R., Capitán, M.A., 2011. A case study of the internal structures of gossans and weathering processes in the Iberian Pyrite Belt using magnetic fabrics and paleomagnetic dating. Miner. Depos. 46 (8), 981–999. https://doi.org/10.1007/s00126-011-0361-8.

Fernández-Caliani, J.C., Cantano, M., 2010. Intensive kaolinization during a lateritic weathering event in South-West Spain. Mineralogical and geochemical inferences from a relict paleosol. Catena 80 (1), 23–33. https://doi.org/10.1016/j.catena.2009.08.005.

Fernández-Remolar, D., Gómez-Elvira, J., Gómez, F., Sebastian, E., Martin, J., Manfredi, J. A., Torres, J., González Kesler, C., Amils, R., 2004. The Tinto River, an extreme acidic environment under control of iron, as an analog of the Terra Meridiani hematite site of Mars. Planet. Space Sci. 52 (1–3), 239–248. https://doi.org/10.1016/j.pss.2003.08.027.

Fernández-Remolar, D.C., Gómez-Ortiz, D., Huang, T., Anglés, A., Shen, Y., Hu, Q., Amils, R., Rodríguez, N., Escudero, C., Banerjee, N.R., 2021b. The molecular record of metabolic activity in the subsurface of the Río Tinto mars analog. Astrobiology 21, 1325–1449.

Fernández-Remolar, D.C., Prieto-Ballesteros, O., Gómez-OrtIz, D., Fernández-Sampedro, M., Sarrazin, P., Gailhanou, M., Amils, R., 2011. Río Tinto sedimentary mineral assemblages: a terrestrial perspective that suggests some formation pathways of phyllosilicates on Mars. Icarus 211 (1), 114–138. https://doi.org/10.1016/j.icarus.2010.09.008.

Fernández-Remolar, D.C., Rodriguez, N., Gómez, F., Amils, R., 2003. Geological record of an acidic environment driven by iron hydrochemistry: the Tinto River system. J. Geophys. Res. Planets 108 (7), 1–16. https://doi.org/10.1029/2002je001918.

Fernández-Remolar, D.C., Morris, R.V., Gruener, J.E., Amils, R., Knoll, A.H., 2005. The Río Tinto Basin, Spain: mineralogy, sedimentary geobiology, and implications for interpretation of outcrop rocks at Meridiani Planum, Mars. Earth Planet Sci. Lett. 240 (1), 149–167. https://doi.org/10.1016/j.epsl.2005.09.043.

Fernández-Remolar, D.C., Carrizo, D., Harir, M., Huang, T., Amils, R., Schmitt-Kopplin, P., Sánchez-García, L., Gomez-Ortiz, D., Malmberg, P., 2021a. Unveiling microbial preservation under hyperacidic and oxidizing conditions in the Oligocene Rio Tinto deposit. Sci. Rep. 11 (1). https://doi.org/10.1038/s41598-021-00730-8.

Fernández-Remolar, D.C., Gómez-Ortiz, D., Malmberg, P., Huang, T., Shen, Y., Anglés, A., Amils, R., 2021c. Preservation of underground microbial diversity in ancient subsurface deposits (> 6 Ma) of the Rio Tinto basement. Microorganisms 9 (8). https://doi.org/10.3390/microorganisms9081592.

Fernández-Remolar, D.C., Knoll, A.H., 2008. Fossilization potential of iron-bearing minerals in acidic environments of Rio Tinto, Spain: implications for Mars exploration. Icarus 194 (1), 72–85. https://doi.org/10.1016/j.icarus.2007.10.009.

Fernández-Remolar, D.C., Salván, C.M., Bermejo, M.R., Knoll, A.H., 2007. The fate of biological materials in acidic environments of the Río Tinto, Southwestern Spain. In: Seckbach, J. (Ed.), Algae and Cyanobacteria in Extreme Environments. Springer Netherlands, Dordrecht, pp. 697–710.

Gómez-Ortiz, D., Fernández-Remolar, D.C., Granda, A., Quesada, C., Granda, T., Prieto-Ballesteros, O., Molina, A., Amils, R., 2014. Identification of the subsurface sulfide bodies responsible for acidity in Río Tinto source water, Spain. Earth Planet. Sci. Lett. 391, 36–41. https://doi.org/10.1016/j.epsl.2014.01.022.

González-Toril, E., Llobet-Brossa, E., Casamayor, E.O., Amann, R., Amils, R., 2003. Microbial ecology of an extreme acidic environment, the Tinto river. Appl. Environ. Microbiol. 69 (8), 4853–4865. https://doi.org/10.1128/aem.69.8.4853-4865.2003.

Grotzinger, J.P, Arvidson, R.E., Bell, J.F., Calvin, W., Clark, B.C., Fike, D.A., Golombek, M., Greeley, R., Haldemann, A., Herkenhoff, K.E., Jolliff, B.L., Knoll, A.H., Malin, M., McLennan, S.M., Parker, T., Soderblom, L., Sohl-Dickstein, J.N., Squyres, S.W., Tosca, N.J., Watters, W.A., 2005. Stratigraphy and sedimentology of a dry to wet eolian depositional system, Burns formation, Meridiani Planum, Mars. Earth Planet. Sci. Lett. 240 (1), 11–72. https://doi.org/10.1016/j.epsl.2005.09.039.

Kaplan, H.H., Milliken, R.E., Fernández-Remolar, D., Amils, R., Robertson, K., Knoll, A.H., 2016. Orbital Evidence for Clay and Acidic Sulfate Assemblages on Mars Based on Mineralogical Analogs from Rio Tinto, Spain. Academic Press Inc., United States Icarus, pp. 45–64. https://doi.org/10.1016/j.icarus.2016.03.019 vol. 275.

Klingelhöfer, G., Morris, R.V., Bernhardt, B., Schröder, C., Rodionov, D.S., de Souza, P.A., Yen, A., Gellert, R., Evlanov, E.N., Zubkov, B., Foh, J., Bonnes, U., Kankeleit, E., Gütlich, P., Ming, D.W., Renz, F., Wdowiak, T., Squyres, S.W., Arvidson, R.E., 2004. Jarosite and Hematite at Meridiani Planum from Opportunity's Mössbauer Spectrometer. Science 306, 1740–1745. https://doi.org/10.1126/science.1104653.

Knoll, A.H., Carr, M., Clark, B., Des Marais, D.J., Farmer, J.D., Fischer, W.W., Grotzinger, J. P., McLennan, S.M., Malin, M., Schröder, C., Squyres, S., Tosca, N.J., Wdowiak, T., 2005. An astrobiological perspective on Meridiani Planum. Earth Planet. Sci. Lett. 240 (1), 179–189. https://doi.org/10.1016/j.epsl.2005.09.045.

Leistel, J.M., Marcoux, E., Thiéblemont, D., Quesada, C., Sánchez, A., Almodóvar, G.R., Pascual, E., Sáez, R., 1997. The volcanic-hosted massive sulphide deposits of the Iberian pyrite Belt: review and preface to the thematic issue. Miner. Depos. 33 (1–2), 2–30. https://doi.org/10.1007/s001260050130.

Maldanis, L., Fernandez-Remolar, D., Lemelle, L., Knoll, A.H., Guizar-Sicairos, M., Holler, M., , Francisco Mateus Cirilo da Silva, Magnin, V., Mermoux, M., Simionovici, A., 2024. Unveiling challenging microbial fossil biosignatures from Rio Tinto with micro-to-nanoscale chemical and ultrastructural imaging. Astrobiology 24 (7), 721–733. https://doi.org/10.1089/ast.2023.0127.

Martín-Puertas, C., Valero-Garcés, B.L., Mata, M.P., González-Sampériz, P., Bao, R., Moreno, A., Stefanova, V., 2008. Arid and humid phases in southern Spain during the last 4000 years: the Zoñar Lake record, Córdoba. Holocene 18 (6), 907–921. https://doi.org/10.1177/0959683608093533.

McLennan, S.M, Bell, J.F., Calvin, W.M., Christensen, P.R., Clark, B.C., de Souza, P.A., Farmer, J., Farrand, W.H., Fike, D.A., Gellert, R., Ghosh, A., Glotch, T.D., Grotzinger, J.P., Hahn, B., Herkenhoff, K.E., Hurowitz, J.A., Johnson, J.R., Johnson, S. S., Jolliff, B., Klingelhöfer, G., knoll, A.H., Learner, Z., Malin, M.C., McSween, H.Y., Pocock, J., Ruff, S.W., Soderblom, L.A., Squyres, S.W., Tosca, N.J., Watters, W.A., Wyatt, M.B., Yen, A., 2005. Provenance and diagenesis of the evaporite-bearing Burns formation, Meridiani Planum, Mars. Earth Planet. Sci. Lett. 240 (1), 95–121. https://doi.org/10.1016/j.epsl.2005.09.041.

Moreno, C., Capitán, M.A., Doyle, M., Nieto, J.M., Ruiz, F., Sáez, R., 2003. Edad mínima del gossan de Las Cruces: implicaciones sobre la edad de inicio de los ecosistemas extremos en la Faja Pirítica Ibérica. Geogaceta 33, 75–78.

Morris, R.V., Golden, D.C., Bell, J.F., Shelfer, T.D., Scheinost, A.C., Hinman, N.W., Furniss, G., Mertzman, S.A., Bishop, J.L., Ming, D.W., Allen, C.C., Britt, D.T., 2000. Mineralogy, composition, and alteration of Mars Pathfinder rocks and soils: evidence from multispectral, elemental, and magnetic data on terrestrial analogue, SNC meteorite, and Pathfinder samples. J. Geophys. Res. Planets 105 (E1), 1757–1817.

Preston, L.J., Fernández-Remolar, D., Banerjee, N.R., Osinski, G.R., Southam, G., 2011. The preservation and degradation of filamentous bacteria and biomolecules within iron oxide deposits at Rio Tinto, Spain. Geobiology 3, 233–249. https://doi.org/10.1111/j.1472-4669.2011.00275.x.

Regenspurg, S., Brand, A., Peiffer, S., 2004. Formation and stability of schwertmannite in acidic mining lakes. Geochim. Cosmochim. Acta 68 (6), 1185–1197. https://doi.org/10.1016/j.gca.2003.07.015.

Rohling, E.J., Foster, G.L., Grant, K.M., Marino, G., Roberts, A.P., Tamisiea, M.E., Williams, F., 2014. Sea-level and deep-sea-temperature variability over the past 5.3 million years. Nature 508 (7497), 477–482. https://doi.org/10.1038/nature13230.

Sabater, S., Buchana, T., Guasch, H., Ivorra, N., Muñoz, I., Navarro, E., Real, M., Romaní, A., 2003. Structure and function of benthic algal communities in an extremely acid river. J. Phycol. 39, 481–489.

Sáez, R., Pascual, E., Toscano, M., Almodóvar, G.R., 1999. The Iberian type of volcano-sedimentary massive sulphide deposits. Miner. Deposita 34 (5–6), 549–570. https://doi.org/10.1007/s001260050220.

Sánchez-Andrea, I., Rodríguez, N., Amils, R., Sanz, J.L., 2011. Microbial diversity in anaerobic sediments at Río Tinto, a naturally acidic environment with a high heavy metal content. Appl. Environ. Microbiol. 77 (17), 6085–6093. https://doi.org/10.1128/AEM.00654-11Spain.

Schwertmann, U., Murad, E., 1983. Effect of pH on the formation of goethite and hematite from ferrihydrite. Clays Clay Miner. 31 (4), 277–284. https://doi.org/10.1346/ccmn.1983.0310405.

Squyres, S.W., Knoll, A.H., 2005. Sedimentary rocks at Meridiani Planum: origin, diagenesis, and implications for life on Mars. Earth Planet. Sci. Lett. 240 (1), 1–10. https://doi.org/10.1016/j.epsl.2005.09.038.

Tornos, F., 2006. Environment of formation and styles of volcanogenic massive sulfides: the Iberian Pyrite Belt. Ore Geol. Rev. 28 (3), 259–307. https://doi.org/10.1016/j.oregeorev.2004.12.005.

Tosca, N.J., Knoll, A.H., McLennan, S.M., 2008. Water activity and the challenge for life on Early Mars. Science 1204–1207. https://doi.org/10.1126/science.1155432.

Val-Peón, C., Santisteban, J.I., López-Sáez, J.A., Weniger, G.-C., Reicherter, K., 2021. Environmental changes and cultural transitions in SW Iberia during the Early-Mid Holocene. Appl. Sci. 11. https://doi.org/10.3390/app11083580.

Velasco, F., Herrero, J.M., Suárez, S., Yusta, I., Alvaro, A., Tornos, F., 2013. Supergene features and evolution of gossans capping massive sulphide deposits in the Iberian Pyrite Belt. Ore Geol. Rev. 53, 181–203. https://doi.org/10.1016/j.oregeorev.2013.01.008.

Zachos, J., Pagani, M., Sloan, L., Thomas, E., Billups, K., 2001. Trends, rhythms, and aberrations in global climate 65 Ma to present. Science 292, 686–693. https://doi.org/10.1126/science.1059412.

CHAPTER

Hydrological and hydrogeochemical characteristics of the source of the Río Tinto

3

David Gómez-Ortiz[1,2], David C. Fernández-Remolar[3,4]

[1]*Universidad Rey Juan Carlos, Departamento de Geologia, Fisica y Quimica inorganica, Mostoles, Spain;* [2]*Research Group 'Geofísica y Geoquímica Ambiental', Universidad Rey Juan Carlos, Mostoles, Spain;* [3]*Macau University of Science and Technology, SKL Lunar and Planetary Sciences, Taipa, Macau;* [4]*CNSA Macau Center for Space Exploration and Science, Macau, China*

Introduction

The Río Tinto river extends approximately 100 km through southwestern Spain, distinguished by its extreme acidity (pH < ~3) and elevated redox potential (Eh > ~400 mV), resulting in exceptionally high concentrations of dissolved heavy metals (Davis et al., 2000; Fernández-Remolar et al., 2003; Gónzalez-Toril et al., 2003). Originating from acidic springs in the Peña de Hierro area, the river flows southwestward for 92 km before discharging into the Huelva estuary in the Gulf of Cadiz, creating one of the few estuarine environments globally characterized by persistently low pH values and toxic metal contamination.

The Río Tinto mining district, situated within the Iberian Pyrite Belt (IPB), hosts some of the world's largest massive and stockwork sulfide deposits (Ruiz Cánovas, Olías and Nieto, 2014). This region's 5000-year history of intermittent mining activity has exposed substantial sulfide outcrops to meteoric water, accelerating oxidation processes (Hudson-Edwards et al., 1999; Sáez et al., 1996; Sáez et al., 1999). While mining operations have traditionally been considered a primary contributor to the river's acidic condition, comprehensive geological, geophysical, and microbiological investigations conducted over the past 15 years suggest that these extreme conditions have persisted for approximately 6 million years, predating anthropogenic mining activities (Essalhi et al., 2011; Velasco et al., 2013).

The acidic springs feeding the Río Tinto emerge in the Peña de Hierro area. Despite the cessation of mining operations in this locality in 1966, these springs have maintained stable hydrochemical characteristics, indicating a continuous supply from a sulfide-rich aquifer system. Previous geophysical investigations (Gómez-Ortiz et al., 2014) identified a deep aquifer with a recharge zone northwest of the Peña de Hierro pit lake. Tritium analysis indicates most spring waters have an approximate age of 60 years, with certain exceptions.

Geomicrobiology of Río Tinto. https://doi.org/10.1016/B978-0-443-32948-7.00011-4

This study integrates geological, hydrogeological, and geophysical data to characterize the groundwater circulation system and identify subsurface sulfide ore bodies in the Peña de Hierro area. Through comprehensive geochemical and isotopic analysis of the springs, we evaluate the interconnectivity of the groundwater system and determine the origins of heavy metals in solution. Our findings provide further evidence that the interaction between groundwater and sulfide ore bodies represents a natural process generating the acidic conditions that characterize the Río Tinto.

Geological setting

Peña de Hierro (Fig. 3.1) is situated at the contact between two distinct tectonic units within the Iberian Pyrite Belt (IPB). The IPB constitutes a geological formation approximately 250 km long and 60 km wide, representing the northernmost paleogeographic division of the South-Portuguese Zone within the Variscan Iberian Massif (Gómez-Ortiz et al., 2014; Sáez et al., 1996). Oceanic-affinity rocks separate the South-Portuguese Zone from the remainder of the Iberian Massif, suggesting an exotic origin relative to the Gondwanan-affinity Iberian Autochthon.

The stratigraphic sequence of the IPB comprises three primary units (Sáez et al., 1999; Davis et al., 2000; Tornos, 2006; Ruiz Cánovas, Olías and Nieto, 2014).

- **Phyllite-Quartzite (PQ) Group**: Frasnian to Late Famennian age (382.7–358.9 Ma), predominantly composed of sandstones and shales, with heterogeneous sedimentation in the upper sections including interbedded lenticular limestones.
- **Volcanosedimentary Complex (VSC)**: Late Famennian to Early Late Viséan (358.9–330.9 Ma), consisting of felsic and mafic volcanic rocks associated with detrital sedimentary sequences, including black shales and volcanically derived sandstones. This unit is economically significant due to its high metal content, formed during pre-collisional stages of the Variscan orogeny.
- **Culm Group**: Late Viséan to Upper Pennsylvanian (330.9–303.7 Ma), representing foreland basin infill during the collisional stage of the Variscan orogeny, characterized by turbiditic deposits.

The structural architecture of the IPB and the entire South-Portuguese Zone forms a large imbricate fan of thrust sheets rooted in the middle crust. The basal detachment within the Late Devonian PQ group explains the absence of older rocks throughout the region. This imbricate thrust structure incorporates significant oblique-slip components resulting from the oblique conditions of the Variscan collision.

The Río Tinto mining district covers approximately 20 km^2 and represents the most prominent mining operation in the IPB. Mining activities commenced around 2500 BCE, targeting the rich sulfide ore deposits. The British Rio Tinto mining

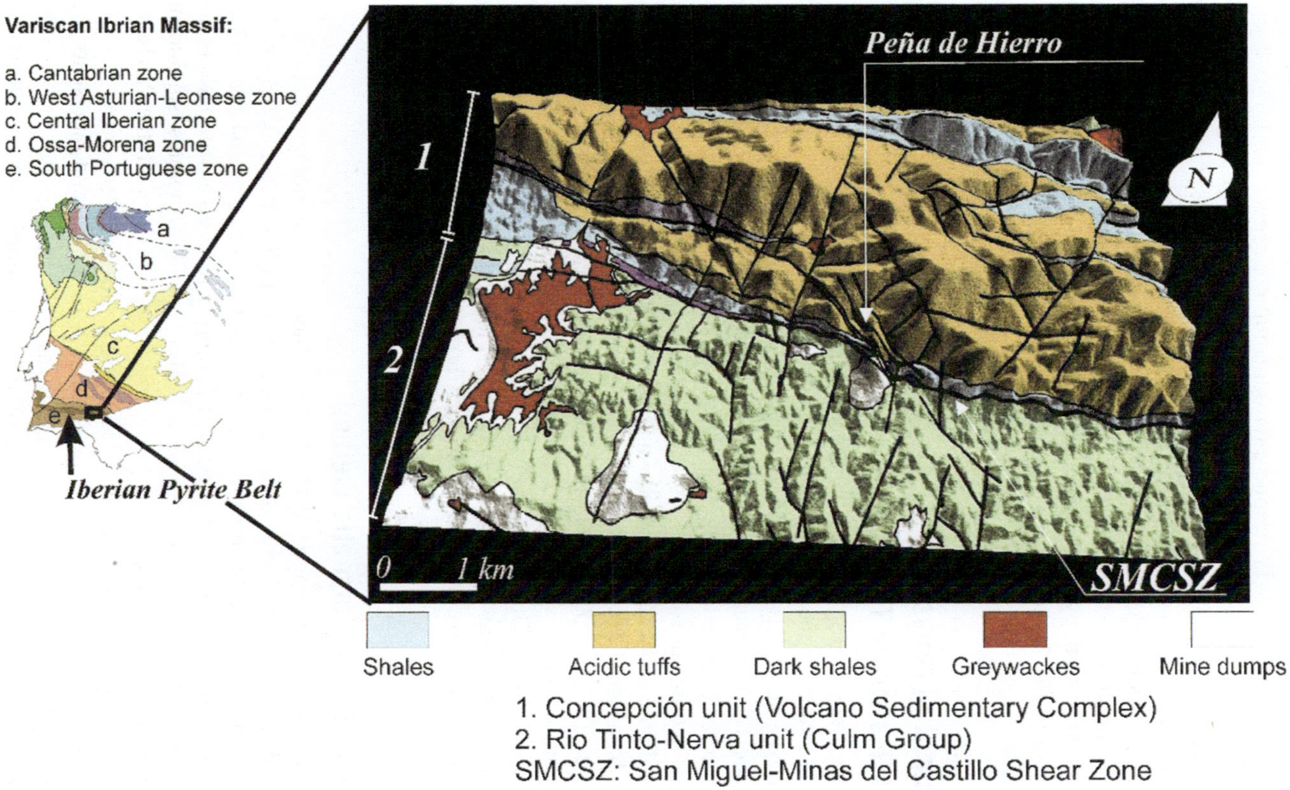

FIGURE 3.1

Simplified geological and tectonic setting of the study area. Peña de hierro is located in the Iberian pyrite belt at the boundary between two different units, concepcion unit and RioTinto-Nerva unit, mainly composed of shales and volcanosedimentary materials.

company acquired the mine in 1873, and extensive exploitation continued until 2001. Rising copper prices between 2001 and 2012 renewed interest in the mine, leading to its reopening in 2017 despite environmental concerns.

The Río Tinto ore body consists of thin, discontinuous massive deposits underlain by cross-cutting stockworks, hosted within acidic tuffs (Sáez et al., 1996). The ore body averages 5 km in length, 750 m in width, and 40 m in depth, with extensively developed stockwork mineralizations. The sulfide ore in Peña de Hierro is oriented within an east-west anticline (Fig. 3.1), characterized by fine-grained, brecciated, or fractured pyritic textures that are highly reactive and associated with a hydrothermally altered matrix of quartz and feldspar.

Geophysical surveys

Electrical Resistivity Tomography

Electrical resistivity tomography (ERT) was employed to investigate the complex geological structures in the study area (e.g Ducut et al., 2022). This technique involves acquiring resistivity profiles using a computer-controlled electrode array. By systematically increasing the separation between electrodes, information at greater depths is obtained, facilitating the creation of a true resistivity cross-section of the geological substrate.

The pole-dipole array configuration was selected for this investigation due to its optimal combination of horizontal and vertical resolution and significant investigation depth capabilities (~200 m). Data collection was performed using an ELREC 10 system (Iris Instruments), with two profiles extending 1425 m and utilizing 96 electrodes at 15 m intervals (Gómez-Ortiz et al., 2014) (Fig. 3.2). The maximum penetration depth achieved was approximately 200 m. Field data were processed using RESIP2DINV software to convert apparent resistivity measurements into true resistivity sections.

Time-Domain Electromagnetic method

The Time-Domain Electromagnetic (TDEM) method was implemented to map groundwater distribution and locate metallic ore bodies at greater depths. This technique utilizes a transmitter connected to a square wire loop and a receiver coil positioned centrally. When the transmitter current is rapidly reduced to zero, a voltage pulse is induced in the ground that decays over time, providing information about subsurface resistivity variations.

Measurements were recorded at 20 to 30 discrete time gates following the primary inducing pulse, with each gate corresponding to different frequencies and investigation depths. The Turam configuration was selected to facilitate measurements along traverse lines perpendicular to the transmitter loop. A PROTEM receiver (Geonics Ltd) was employed to measure the Hz component of the

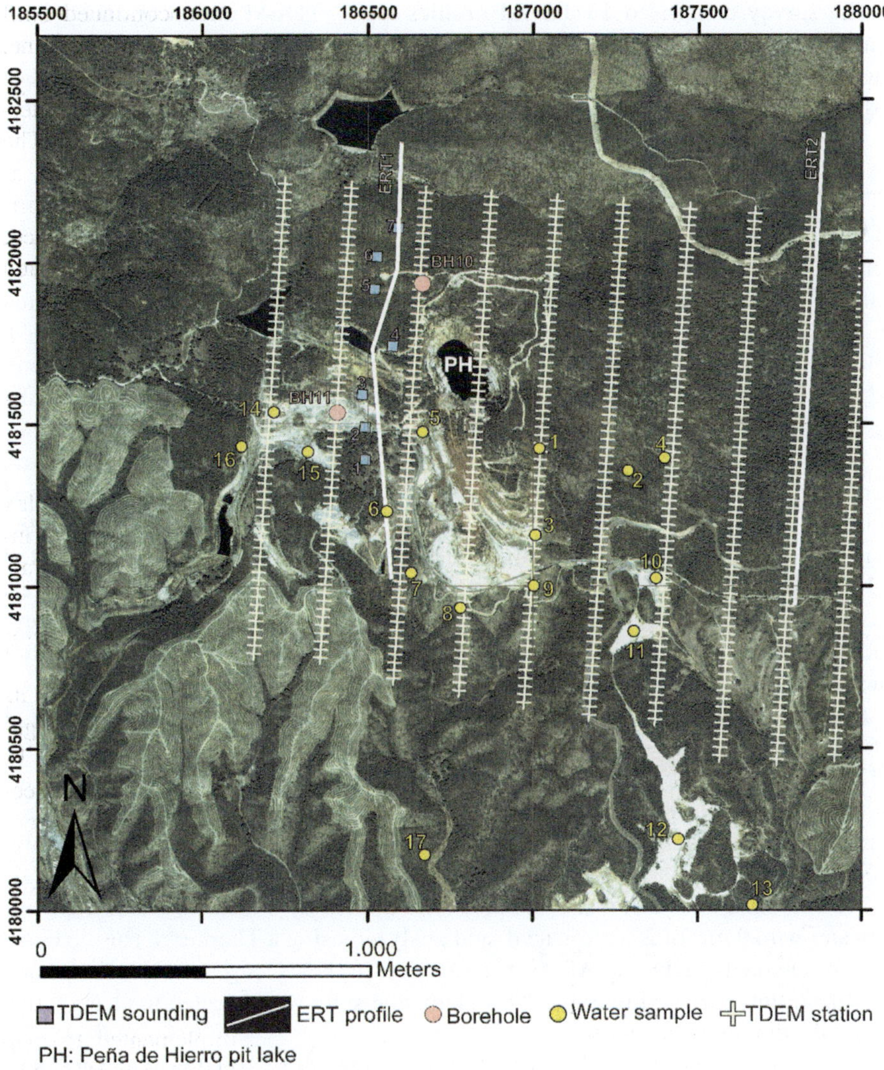

FIGURE 3.2

Location of water sampling sites, boreholes, electrical resistivity tomography (ERT) profiles and Time-Domain Electromagnetic (TDEM) measurements. UTM coordinates in meters, zone 29N.

secondary magnetic field at 20 discrete time gates. The transmitter loop dimensions were 1200 × 600 m, with profile lines extending 1500 m and recording points spaced at 25 and 50 m intervals.

The survey comprised 13 TDEM profiles and 7 TDEM soundings (Fig. 3.2), achieving an exploration depth of approximately 600 m. Anomalies in the electromagnetic response were interpreted to identify faults and conductive metallic bodies in the subsurface.

Hydrogeochemical study

Field campaigns conducted in 2004 and 2011 (Allman et al., 2021; Gómez-Ortiz et al., 2014) involved comprehensive sampling of acidic springs and streams around the Peña de Hierro area (Fig. 3.2). Detailed site observations were recorded, including flow rate, water clarity, precipitate formation, surrounding environment, and in-situ pH/Eh measurements (Table 3.1). Sample locations were determined using GPS and subsequently integrated into GIS and Google Earth for spatial analysis.

Water samples were collected using a 30 mL syringe fitted with a 20 μm filter. Sterile sample bottles (125, 250, and 500 mL) were triple-rinsed with filtered spring water before collection to eliminate potential contaminants. Selected springs were additionally sampled in 500 mL containers that were refrigerated and stored as reserves. Containers were filled to capacity to minimize air bubbles, thus reducing isotopic exchange between trapped gas and water. Samples were maintained at low temperatures during storage and transport to minimize further reactions.

In-situ pH and Eh measurements were performed using a multiparametric YSI 556MPS meter. Daily calibration was conducted using Hanna buffer solutions (HI70004C 4.01 and 7.01) to ensure measurement accuracy. Laboratory analyses included elemental determinations via Inductively Coupled Plasma Optical Emission Spectroscopy (ICP-OES) and anion quantification through Ion Chromatography (IC).

For ICP-OES analysis, water samples were diluted by a factor of 10 (1 mL sample water with 9 mL of 2% nitric acid) and analyzed using a Thermo Scientific iCAP 6300 for elements including Al, As, Ca, Cd, Co, Cu, Fe, K, Mg, Mn, Na, Ni, Si, and Zn, with results reported in mg/L. This elemental suite was selected to characterize potential lithological sources.

IC analysis required samples to be diluted by a factor of 100 due to elevated sulfate concentrations. A Metrohm 930 Compact Flex IC with a 150 mm Metrosep 5 column was employed, using 3.1 mM Na_2CO_3 and 1.1 mM $NaHCO_3$ eluent at a flow rate of 0.7 mL/min. Each sample was analyzed in triplicate to ensure analytical precision. Samples with extremely high sulfate concentrations were further diluted and reanalyzed.

Geochemical data were processed using PHREEQC software to model mineral saturation indices and to characterize groundwater-lithology interactions. GIS tools facilitated spatial representation and pattern identification in the hydrochemical characteristics of the springs.

Table 3.1 Compilation of physicochemical parameters (pH and Eh) as well as concentration of major cations and anions (mg·L^{-1}) of spring and stream solutions in the Peña de Hierro area.

Sample	Locality	pH	Eh (mV)	Al	As	Ca	Cu	Fe	K	Mg	Mn	Na	Ni	Si	Zn	Cl^-	SO_4^{2-}
1	Cave left spring	3.38	489	0.0	0.0	7.2	0.0	0.0	8.7	9.3	0.0	12.0	0.0	14.4	0.0	8.31	176.14
2	Right spring	6.38	171	0.0	0.0	369.9	0.0	499.4	9.0	407.0	99.4	30.3	0.7	5.7	5.7	6.75	50.46
3	Left Left spring	2.19	478	3053.1	13.0	18.7	15.7	15,428.8	7.2	1728.3	29.0	4.6	3.5	66.8	13.4	2.76	64,547.85
4	Top of the right	–	–	0.0	0.0	16.1	0.0	0.0	9.6	4.2	2.9	9.4	0.0	1.2	0.0	3.76	3088.09
5	Green spring	2.37	470	140.4	2.6	71.5	8.2	818.4	8.1	222.1	13.6	11.3	0.4	42.1	29.2	5.12	2913.09
6	Nuria spring	2.57	492	2062.0	24.9	135.9	74.2	14,952.9	8.7	501.7	26.9	29.7	3.2	73.2	112.7	1.77	98,645.26
7	UMA spring	1.35	484	2044.6	24.5	138.4	74.1	15,200.6	8.8	506.7	26.9	29.3	3.2	72.5	111.5	1.66	100,546.59
8	Ricci spring	1.85	452	2494.7	37.7	54.5	107.4	21,767.9	9.6	543.5	26.4	24.4	4.0	59.7	82.9	2.34	91,214.69
9	Angeles source	2.22	449	751.7	2.5	81.0	9.7	288.2	7.7	354.9	33.5	22.0	1.9	86.9	19.0	2.74	6584.3
10	New spring	1.55	533	235.5	32.0	26.7	31.2	5149.6	7.8	67.7	11.9	9.7	1.3	54.1	7.6	2.38	21,780.05
11	New New spring	1.02	489	180.4	40.8	7.6	81.4	5991.1	9.8	48.7	7.5	6.4	1.4	60.1	12.3	1.59	19,812.78
12	Fox spring	1.67	459	179.1	12.5	16.4	30.8	3390.5	11.5	53.8	6.9	10.5	1.0	60.6	7.7	3.69	12,935.9
13	Gypsum spring	2.81	411	0.0	0.0	562.7	0.0	426.6	8.6	297.0	47.8	52.7	0.5	6.5	2.0	6.11	3127.21
14	Left hand side anabels garden	2.72	362	126.1	2.6	436.6	19.0	1821.2	7.8	426.9	169.5	38.4	1.5	57.0	126.4	7.61	10,204.99
15	Right hand side anabels garden	1.99	420	279.8	4.7	354.9	24.9	2690.3	8.8	407.8	144.6	32.5	1.5	43.5	115.1	8.56	6981.06
16	Anabel confluence spring	2.38	449	116.2	2.4	150.5	11.0	328.9	7.8	149.4	45.8	21.5	0.5	48.4	32.3	10.5	2452.61
17	Isolated pool valley of the crazies	1.32	638	3438.9	21.2	517.6	125.8	17,579.2	7.2	1447.3	115.8	69.1	4.5	157.8	177.5	6.31	22,003.87
18	Valley of the crazies	1.76	662	994.0	3.9	145.9	29.1	4384.6	7.6	407.6	34.5	20.4	1.4	56.5	41.5	38.51	92,550.55

Isotopic analysis

Isotopic analyses of $\delta^{18}O$ and $\delta^{2}H$ were conducted using a Picarro L2130-I isotope analyzer to determine water origin and age. Eighteen samples (Table 3.2) were prepared for $\delta^{2}H$ and $\delta^{18}O$ determination (Gómez-Ortiz et al., 2014; Allman et al., 2021), with one sample analyzed in duplicate to assess analytical precision. Sample preparation involved transferring 2 mL of spring water into capped analysis tubes, minimizing atmospheric exposure. Each sample underwent nine measurement cycles over approximately 90 min, with the initial six results discarded to mitigate memory effects. The remaining three results were averaged and reported with uncertainties expressed as twice the standard deviation. Results are presented in delta notation relative to V-SMOW2, calibrated using certified reference materials IA-RO52, IA-RO53, and IA-RO54. Tritium (^{3}H) analysis provided information regarding water age. This radioactive isotope of hydrogen occurs naturally in the atmosphere at

Table 3.2 Compilation of measured oxygen 18 and deuterium isotopic values for both spring and stream solutions in the Peña de Hierro area. Results are reported as a mean ratio and an error expressed as two times the standard deviation of the analysis.

Sample	Locality	$\delta^{18}O$ (‰)	Error 2σ (‰)	$d^{2}H$ (‰)	Error 2σ (‰)
1	Cave left spring	−4.72	0.20	−24.44	0.22
2	Right spring	−4.10	0.04	−31.73	0.09
3	Left Left spring	−4.22	0.11	−20.16	0.11
4	Top of the right	−5.13	0.10	−27.11	0.20
5	Green spring	−1.56	0.02	−12.29	0.04
7	UMA spring	−5.05	0.30	−27.77	0.36
8	Ricci spring	−4.60	0.66	−24.35	0.73
9	Angeles source	−3.21	0.10	−18.89	0.10
10	New spring	−5.19	0.24	−28.38	0.22
11	New New spring	−4.78	0.15	−26.00	0.31
12	Fox spring	−5.04	0.17	−27.01	0.17
13	Gypsum spring	−4.54	0.08	−24.17	0.03
14	Left hand side Anabels garden	−4.59	0.10	−26.97	0.16
15	Right hand side Anabels garden	−3.51	0.18	−21.28	0.13
16	Anabel confluence Spring	−3.12	0.05	−19.11	0.18
17	Isolated pool valley of The crazies	−3.28	0.04	−18.94	0.09
18	Valley of the crazies	17.43	0.36	46.16	0.60

concentrations below 20 Tritium Units (TU), where 1 TU corresponds to one atom of ^{3}H per 10^{18} atoms of ^{1}H (FONTES, 1980). Thermonuclear testing between 1952 and 1963 increased atmospheric ^{3}H levels to approximately 1000 TU; despite cessation of testing, current levels remain elevated above natural background. Consequently, tritium measurements primarily distinguish pre-1952 water from more recent sources, with pre-1952 samples exhibiting ^{3}H levels near detection limits, while younger water contributions increase ^{3}H to detectable concentrations.

Geological structure of the Río Tinto source from geophysical data

Extensive geophysical surveys conducted in 2010 and 2011 employed multiple independent techniques to characterize groundwater systems and identify potential deep massive and/or stockwork sulfide bodies in the study area (Gómez-Ortiz et al., 2014). The integration of electrical resistivity tomography (ERT) and time-domain electromagnetic (TDEM) methods focused on Variscan sequences, exploiting the distinctive electrical property variations associated with water-bearing and metallic structures.

ERT profiles reaching depths of approximately 200 m (Fig. 3.3) revealed two distinct geoelectrical units (Gómez-Ortiz et al., 2014). The upper unit, characterized by medium-to-high resistivity values (~2000–2500 Ω·m), corresponds to the Tournaisian metarhyolites of the VSC, which outcrop in the northern (hanging wall) portion of the Peña de Hierro area. The lower unit exhibits low resistivity and correlates with the younger Culm group exposed in the southern (foot wall) region. The boundary between these units appears as a narrow north-dipping zone, coinciding with the previously mapped San Miguel-Minas del Castillo Shear Zone (Leistel et al., 1997; Quesada, 1997). This shear zone (Fig. 3.1) juxtaposes the Tournaisian VSC unit against the Culm geological unit, with the massive sulfide ore body at Peña de Hierro representing one of several structural imbricates (horses) along this major tectonic feature.

While the ERT profiles did not detect very low resistivity bodies indicative of massive sulfides, likely due to depth limitations, they provided valuable insights into the tectonic structure of the Paleozoic basement (Fig. 3.3). The major tectonic thrust is clearly delineated at depth and associated with several subvertical brittle strike-slip normal faults, representing secondary structures from a Late Variscan brittle deformation phase (Quesada, 1997). The reactivation of these faults during Alpine deformation is evidenced by deepening of the gossanization front in strike-slip normal faults exposed at the Peña de Hierro open-pit walls, indicating their function as preferential water flow paths (Velasco et al., 2013).

The TDEM method provided resistivity information to depths exceeding 600 m through individual soundings (1D profiles, see Fig. 3.4 for an example of the TDEM sounding 7) and interpolated resistivity maps (2D slices showing horizontal resistivity variations at specific depth ranges). Seven TDEM soundings distributed along a roughly N–S transect were acquired (Fig. 3.2), facilitating interpolation to create a

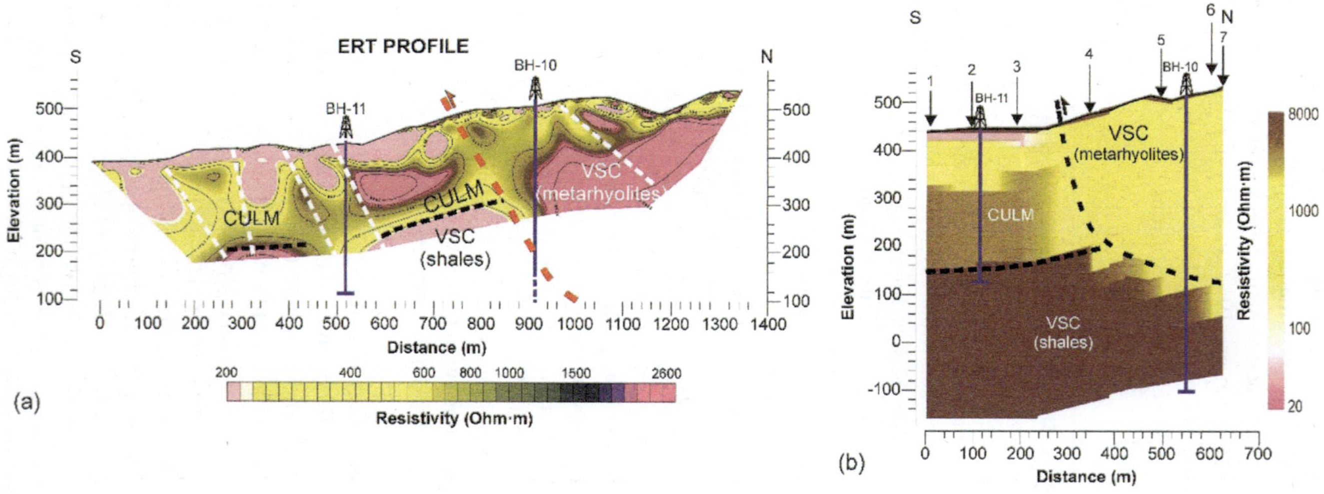

FIGURE 3.3

Electrical resistivity tomography (ERT) and Time-Domain Electromagnetic (TDEM) data: (a) Shows ERT profile 1, highlighting the different geological units and the presence of multiple brittle faults. The fault locations (dashed white lines) are inferred from lateral resistivity changes caused by weathering and groundwater flow; (b) continuous resistivity profile derived from the interpolation of TDEM soundings one to 7. Variations in resistivity allow the identification of boundaries between the culm group and the two distinct units of the volcanosedimentary complex (VSC) (shales and metarhyolites). The locations of two exploratory boreholes (BH10 and BH11), which were used to interpret the geophysical data, are also indicated.

- Modified from (D. Gómez-Ortiz, D.C. Fernández-Remolar, A. Granda, C. Quesada, T. Granda, O. Prieto-Ballesteros, A. Molina, R. Amils, Identification of the subsurface sulfide bodies responsible for acidity in Río Tinto source water, Spain. Earth Planet Sci. Lett. 391 (2014), 36–41, https://doi.org/10.1016/j.epsl.2014.01.022).

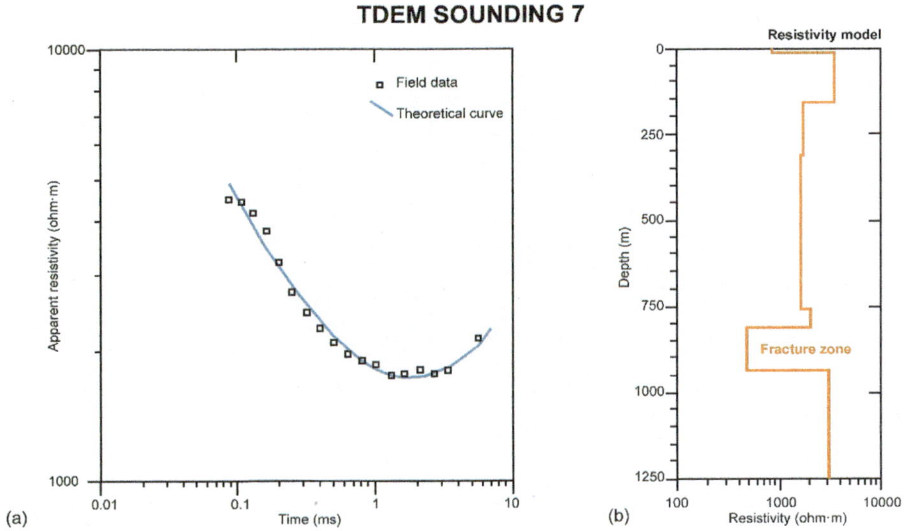

FIGURE 3.4

1D vertical resistivity variations obtained from the TDEM sounding number 7: (a) Field data showing the apparent resistivity values for every time window, together with the theoretical curve corresponding to the best 1D true resistivity model, and (b) 1D model of the vertical true resistivity variation at depth after inversion of the field data. A major fracture located at about 900 m depth is clearly imaged due to a large decrease in resistivity.

2D resistivity depth profile (Fig. 3.3). Two exploratory boreholes (BH10 and BH11) drilled subsequent to the geophysical survey provided ground-truth data for subsurface interpretation, enabling correlation between geoelectrical and tectonic units.

The listric geometry of the major thrust fault separating the VSC and Culm group tectonic units is clearly evident in the TDEM profiles (Fig. 3.3), consistent with the shallower ERT profile results. For subsurface lithological and hydrological analysis, resistivity data were examined at shallow (300–400 m), intermediate (400–600 m), and deep (>800 m) depth ranges (Fig. 3.5). The resistivity maps for shallow and intermediate depths reveal a high resistivity geoelectrical unit (2000–3000 Ω·m) extending to approximately 600 m depth. At greater depths (>800 m), a different geoelectrical unit with medium-to-high resistivity (500–2000 Ω·m) becomes apparent.

By integrating surface geological observations, geophysical soundings, and borehole data, the upper high resistivity unit is correlated with the outcropping metarhyolites of the VSC unit, which typically display resistivity values exceeding ~1500 Ω·m, characteristic of unweathered igneous rocks (e.g. (Telford et al., 1990; Reynolds, 1997)). The medium-to-high resistivity lower unit corresponds to the low-grade metamorphic rocks of the PQ group, which generally exhibit lower resistivity values than igneous rocks (e.g. Telford et al., 1990; Reynolds, 1997).

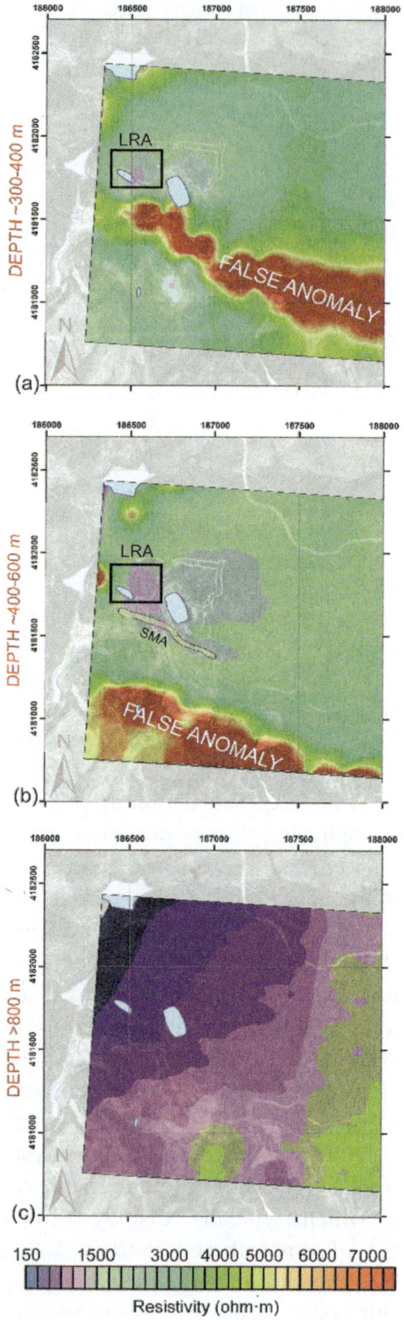

FIGURE 3.5

Three resistivity slices at varying depths around the peña de hierro pit lake, interpreted from TDEM data. It is important to highlight the occurrence of a false anomaly that is an

The TDEM resistivity maps (Fig. 3.5) highlighted two distinctive subsurface features in the Peña de Hierro basement: one of high resistivity and one of low resistivity. The high-resistivity feature, recorded in the southern portion of the study area with a WNW-ESE orientation, represents a methodological artifact unrelated to geological structures. The low-resistivity area (LRA), characterized by values ranging from 150–1000 Ω·m, is located at shallow to intermediate depths (within the VSC group) in a narrow zone northwest of the Peña de Hierro pit lake. This LRA extends to at least 600 m depth but disappears at greater depths (>800 m), where resistivity progressively increases southeastward. This trend has been interpreted (Gómez-Ortiz et al., 2014) as reflecting lithological variations within the underlying PQ group. The low resistivity values indicate elevated water content, suggesting this area functions as the recharge zone for the Peña de Hierro aquifer, from which subsurface water flows southeastward to supply the acidic springs feeding the Río Tinto headwaters. This interpretation aligns with previous hydrogeological findings (Fernández-Remolar et al., 2008) indicating an acidic aquifer discharging through springs south-southeast of the Peña de Hierro pit lake.

In addition to aquifer characterization, the TDEM data identified anomalies potentially indicating concealed sulfide deposits. A notable anomaly (Sulphide Massive Anomaly, SMA) extends in a WNW-ESE orientation near the southern margin of the Peña de Hierro pit lake (Fig. 3.5), suggesting that remnants of the historically mined sulfide ore body persist at intermediate depths (~500 m). This anomaly's location coincides with the major listric thrust fault identified in both TDEM and ERT profiles, indicating that the sulfide body was emplaced and fragmented into separate tectonic horses along this significant fault zone, with deeper portions remaining unexploited due to technical limitations of historical mining operations.

artifact inherent to the TDEM method and is unrelated to any geological feature. (a) Shallow (~300–400 m) resistivity slice, highlighting a low resistivity zone (LRA) located northwest of the pit lake, with an apparent false anomaly to the southeast. The LRA is interpreted as the recharge zone of the peña de hierro aquifer; (b) intermediate depth (~400–600 m) map, where the LRA remains visible but is shifted slightly northwest, and the false anomaly is located at the southern edge of the study area. SMA: Sulphide massive anomaly; (c) deep (>800 m) resistivity slice showing the disappearance of the LRA, with resistivity values decreasing toward the southeast, following a regional planar trend.

- Modified from (D. Gómez-Ortiz, D.C. Fernández-Remolar, A. Granda, C. Quesada, T. Granda, O. Prieto-Ballesteros, A. Molina, R. Amils, Identification of the subsurface sulfide bodies responsible for acidity in Río Tinto source water, Spain. Earth Planet Sci. Lett. 391 (2014), 36–41, https://doi.org/10.1016/j.epsl.2014.01.022).

Hydrogeological characteristics of the springs and streams

The chromatic characteristics of waters in the Río Tinto fluvial system serve as effective indicators of pH and potential mineral composition. Elevated temperatures and variable hydrochemistry promote the formation of precipitates including copiapite, nanophase oxyhydroxides, and gypsum (Fernández-Remolar et al., 2005; Allman et al., 2021). Left Richi Spring exhibited significant evaporite deposits, likely resulting from capillary action enhanced by high summer temperatures. Right Spring displayed extensive precipitates of hydrated ferric iron oxides coating all submerged surfaces. Ferric iron functions as a buffer, stabilizing water pH at approximately 2.3 through proton release during precipitation, counteracting any pH increases.

Certain springs, including RHS Anabel's Garden and Gypsum Spring, demonstrated substantial gypsum precipitation, potentially indicating distinctive hydrochemical conditions (Allman et al., 2021). These gypsum deposits frequently exhibited black staining, attributed to organic films on settled water surfaces. The presence of evaporites in dry channels confirmed the ephemeral nature of specific watercourses. Nanophase ferric oxyhydroxides were observed at the confluence of streams with extreme pH differentials. In streams such as VOTC, reduced turbulence resulted in clearer waters through settling of ferric complexes and mineral aggregates. Conversely, the lower ferric ion concentrations in Left Richi Spring contributed to its turbidity, facilitating clay suspension and transport.

Eh and pH measurements

The Río Tinto headwaters are characterized by consistently oxic and acidic conditions, with pH values typically below three and Eh values exceeding 400 mV (Aguilera et al., 2006; Amils et al., 2014; Fernández-Remolar et al., 2003; Gómez-Ortiz et al., 2014; Allman et al., 2021). Spring waters exhibited average pH and Eh values of 2.33 and 456.7 mV, respectively (Table 3.1). Most springs displayed pH values between one and three and Eh measurements ranging from 300 to 600 mV, with recorded extremes of 1.02–6.38 for pH and 110–662 mV for Eh.

Notable outliers include New Spring, which registered the minimum pH of 1.02, followed by VOTC, Uma, and Left Left Springs. Conversely, Top of Right Spring and Right Spring exhibited elevated pH values of 6.38 and 5.0, respectively. These variations reflect localized subsurface conditions and contribute to an overall standard deviation of 1.31 in pH measurements (Amils et al., 2014; Fernández-Remolar et al., 2003; Gómez-Ortiz et al., 2014; Allman et al., 2021).

Chemical composition

Chemical analysis included quantification of cations (Fe, Al, Mg, Ca, Na, K, Zn, Ni, Mn, Cu, Cd, Co, Si, and As) as summarized in Table 3.1 (Allman et al., 2021). Among the springs investigated, Uma, Ricci, and Left Left demonstrated the highest total cation concentrations, with Ricci and Left Left exhibiting particularly elevated values. Iron dominated the aqueous chemistry, constituting an average of 77% of the total elemental concentration. The sequence of elemental abundance, in descending order, is Fe > Al > Mg > Ca > Zn > Si > Cu > Mn > Na > Co > Ni > Cd, differing slightly from earlier investigations that reported Fe > Mg > Al > Ca > Na > Zn > Cu > Mn > K (Cánovas et al., 2014). Notably, calcium, zinc, and manganese concentrations were elevated in certain springs outside the Uma, Ricci, Left Left, and Top VOTC group.

Anion analysis (Table 3.1) included F^-, Cl^-, Br^-, NO_2^-, NO_3^-, SO_4^{2-}, and PO_4^{3-}. Sulfate dominated the anion composition in all samples, constituting 99.88% of the total anion content. Average sulfate concentration was 23.861 mg L^{-1}, with a maximum value of 100.542 mg L^{-1} recorded in UMA Spring. Fluorine and phosphate concentrations were generally minimal throughout the study area, though elevated levels were detected in Left Left Spring and Green Spring, respectively. Bromine and nitrite were undetectable in all samples, while nitrate was present in significant concentrations only in VOTC Spring and the Old Río Tinto River.

Isotopic composition

Isotopic values for δ^2H and $\delta^{18}O$ were measured and reported in delta notation relative to V-SMOW2 (Table 3.2). Comparative analysis was conducted between Río Tinto springs and regional precipitation based on Global Network on Isotopes in Precipitation (GNIP) data from the Morón de la Frontera Military Base (2000–06) (Allman et al., 2021). The global meteoric water line (GMWL) provided context, alongside local isotopic data from Sevilla (Mediterranean climate) and Río Tinto.

Precipitation collected at Río Tinto on June 24, 2015, exhibited δ^2H and $\delta^{18}O$ values of −23.6‰ and −3.7‰, respectively. GNIP data for Sevilla demonstrated broader variability: $\delta^2H = -62.4$–5.6‰ and $\delta^{18}O = -9.14$–4.8‰, with annual averages calculated using both arithmetic and weighted methods. The Río Tinto springs displayed a more constrained isotopic distribution. The spring water line was characterized by the equation $\delta^2H = 4.072 \cdot \delta^{18}O - 7.066$ ($R = 0.76$), with a gradient lower than both the GMWL and the local meteoric water line ($\delta^2H = 5.937 \cdot \delta^{18}O - 0.582$).

Seven springs exhibited deuterium enrichment, with Green Spring showing the maximum δ^2H value (−12.29‰) compared to the mean (−23.05‰). Right Spring displayed the minimum δ^2H value (−31.73‰). Similarly, $\delta^{18}O$ enrichment was observed in seven springs, with Green Spring exhibiting the highest value (−1.56‰) versus an average of −3.95‰. The lowest $\delta^{18}O$ value (−5.19‰) was recorded at Angeles Source. Evaporative effects were particularly evident in isolated pools such as Top VOTC.

Controls on pH and Eh

The spatial heterogeneity of pH values in the Río Tinto study area is particularly noteworthy given the relatively limited geographic extent (Allman et al., 2021).

All springs exhibit acidity resulting from sulfide oxidation, but the intensity varies considerably, primarily influenced by geological factors and microbial activity (Langmuir, 1997; Plumlee et al., 1997). Subsurface microbial interactions significantly enhance acid rock drainage (ARD) processes by accelerating sulfide oxidation through Fe^{3+} regeneration. This mechanism increases oxidation rates by 18–170 times compared to oxygen alone, reducing reaction times from years to minutes (Plumlee et al., 1997; Sánchez-Andrea et al., 2011). However, ferric iron regeneration is limited by oxygen diffusion into the aquifer system, which constitutes the rate-limiting step (Sánchez España, 2007). Oxygen saturation measurements from March 2013 indicate predominantly oxic conditions in most streams, with widespread bacterial distribution in ferrous iron-rich environments, though biological activity alone cannot fully account for pH variability (Fernández-Remolar et al., 2005).

The hydrochemical characteristics of spring waters are largely determined by geological heterogeneity resulting from intense deformation during Variscan and Cenozoic tectonic phases (Amils et al., 2014; Gómez-Ortiz et al., 2014). Chemical interactions with lithologies containing variable sulfide concentrations and neutralizing agents directly influence pH (Ruiz Cánovas, Olías and Nieto, 2014). Pyrite, the predominant sulfide mineral in the region, represents the principal contributor to acidity, with oxidation reactions accounting for 75% of acid production (Dold, 2010). Chalcopyrite generates comparatively less acidity due to its greater oxidation resistance, as evidenced in springs with elevated Cu/Fe ratios (Dold, 2010). The relatively low arsenic concentrations may reflect co-precipitation with ferric oxides rather than limited arsenopyrite availability (Cánovas et al., 2007; Dold, 2010).

Calcite functions as a critical neutralizing agent, dissolving under acidic conditions to buffer pH. Calcite lenses identified during the MARTE drilling project at depths ranging from 20 to over 600 m potentially contribute to pH elevation in specific springs (Fernández-Remolar et al., 2018). The reaction $CaCO_3 + 2H^+ \rightarrow Ca^{2+} + H_2CO_3$ demonstrates its effectiveness in neutralizing acidity at pH values below 6.3 (Dold, 2010). However, limited carbonate volumes and flow path complexity suggest only partial neutralization, with most waters maintaining acidic signatures.

Ferric oxyhydroxides, including ferrihydrite and schwertmannite, further modulate pH through precipitation and dissolution processes, releasing or consuming protons depending on prevailing conditions (Sánchez España, 2007). These minerals precipitate more readily at elevated pH, as observed in springs such as RHS Anabel's Garden, where higher saturation indexes (SI) reflect enhanced buffering capacity. Conversely, in highly acidic streams, acidity generation is dominated by sulfide oxidation rather than mineral hydrolysis (Dold, 2010). Aluminosilicate dissolution also consumes acidity but proceeds at significantly slower rates and exerts minimal impact compared to carbonate dissolution (Plumlee et al., 1997).

While biological activity facilitates ferric iron regeneration and influences oxidation rates, Río Tinto waters remain predominantly oxic, with Eh values typically ranging from 420 to 608 mV (Fernández-Remolar et al., 2003). Exceptions include springs such as Top of Right, with Eh values as low as 110–171 mV, potentially resulting from reducing agents encountered along flow paths (Langmuir, 1997). The variability in pH and Eh measurements, combined with geochemical and SI data, indicates that local geological factors and hydrological dynamics exert primary control over acidity. For instance, LHS and RHS Anabel's Garden springs, separated by only 10 m, exhibit pH values of 2.72 and 1.99, respectively, emphasizing the significance of localized lithological variations and limited subsurface connectivity (Bartlett et al., 2010). These findings highlight the complexity of ARD processes in the Río Tinto system, shaped by the interplay of sulfide oxidation mechanisms, microbial activity, and geological heterogeneity.

Lithological controls on aqueous chemistry

The geological framework within the study area exhibits substantial heterogeneity resulting from extensive faulting during Variscan and Tertiary deformational episodes, juxtaposing contrasting lithological units. Geophysical investigations employing ERT and TDEM methods identified a major thrust structure at depth associated with minor subvertical brittle strike-slip normal faults corresponding to Late Variscan deformation. The basement lithology is extensively fractured, facilitating subsurface water circulation through various stratigraphic units at different depths, with emergence occurring primarily along strike-slip and normal fault systems (Fernández-Remolar et al., 2008). The geochemical signatures of groundwater typically reflect water-rock interactions with specific lithologies. The study area's geological complexity, characterized by rapid lateral and vertical facies transitions, results in diverse hydrochemical compositions.

Lithological composition and hydrogeological properties significantly influence water chemistry by controlling flow pathways and element availability. Secondary permeability and porosity features, including fractures and solution joints, play crucial roles in water-rock interaction. Even volumetrically minor but highly reactive mineral phases can substantially impact hydrochemistry.

Río Tinto waters are characterized by extreme acidity and elevated metal concentrations (Allman et al., 2021). Springs including Richi Left, UMA Spring, and Left-Left exhibit exceptionally high sulfate concentrations, indicating extensive interaction with sulfide ore bodies. Mass balance calculations indicate that over 93% of dissolved sulfate originates from pyrite oxidation, with consistently low $Ca:SO_4^{2-}$ ratios across springs eliminating gypsum or anhydrite as significant contributors. Iron, copper, and zinc constitute 99.6% of the sulfide-derived metals (on a molar basis) and correlate strongly with sulfate ($R = 0.93$) (Allman et al., 2021). These metals and sulfate reach maximum concentrations in Richi Left, UMA Spring, and Left-Left, reflecting intensive sulfide interaction. Additionally, cobalt and nickel demonstrate strong positive correlation ($R = 0.93$), indicating a common source, while sulfate correlates positively with cobalt and zinc ($R = 0.90$), further linking these trace metals to sulfide mineralization.

Aluminum and fluorine concentrations show strong correlation ($R = 0.94$), with maximum values recorded in Left-Left, Richi Left, UMA, and Top VOTC springs. These elements likely derive from rhyolites and black shales abundant in the VSC, which typically contain elevated Al and F concentrations. The high sulfate levels in these springs further suggest interactions with sulfide deposits hosted within the VSC (Allman et al., 2021).

Sodium concentrations correlate with sandstone and shale interactions or hydrothermal influence. Release through hydrolysis of Na-plagioclase-rich rhyolites often coincides with elevated fluorine levels. Springs including LHS and RHS Anabel's Garden, Green Spring, and Anabel's Interception exhibit high Na^+/SO_4^{2-} ratios and low aluminum concentrations, indicating interactions with Culm-hosted sandstones and shales. RHS Anabel's Spring, however, displays low pH and high aluminum levels, suggesting contact with both sulfides and silicates. Its relatively low sulfate and metal concentrations likely result from gypsum precipitation, supported by high saturation index (SI) values for gypsum and anhydrite (−0.23 and 0.53) (Allman et al., 2021).

New New, New, and Fox Springs demonstrate low pH values, Na^+/SO_4^{2-} ratios, and total Fe, Cu, and Zn concentrations, indicating limited sulfide interaction and absence of neutralizing agents, evidenced by minimal calcium concentrations. Other springs with low pH and elevated aluminum but low Na^+/SO_4^{2-} ratios suggest interactions with igneous and sulfide minerals characteristic of the VSC.

Top of Right Spring and Cave Left contain negligible aluminum and low metal concentrations, with pH values of 5, 6.38, and 3.38, respectively. Gypsum Spring exhibits distinctive chemistry with elevated calcium and sodium concentrations and a pH of 2.81. These waters likely traversed sulfide-bearing zones before interacting with carbonate lenses within VSC deposits, such as those identified in BH10 cores. Gypsum Spring's supersaturation with respect to gypsum and near-equilibrium with anhydrite explain its diminished metal concentrations through precipitation mechanisms (Allman et al., 2021).

Isotopic fractionation patterns

Isotopic analyses of $\delta^{18}O$ and $\delta^{2}H$ were conducted on various Río Tinto springs and a precipitation sample collected on June 24, 2014 (Allman et al., 2021). Supplementary data from the Morón Military Base in Seville (2000–06) demonstrated that winter precipitation, characteristically depleted in heavy isotopes, contributes significantly to spring recharge. The field-collected rainwater sample exhibited $\delta^{18}O$ and $\delta^{2}H$ values of −3.70‰ and −23.61‰, respectively, compared to Seville's weighted averages of −4.78‰ and −28.57‰. This disparity suggests elevated seasonal temperatures during sampling, reducing the rainout effect and enabling heavier isotopes to penetrate further inland.

The distinctive isotopic fractionation patterns observed in Río Tinto springs result from subsurface processes. Groundwater recharge predominantly occurs during intense precipitation events, with water residing in the unsaturated zone for approximately 3 months before displacement according to the piston flow model. Evaporation in this zone enriches isotopic signatures, as reflected in the spring waters' meteoric water line gradient of 4.07, indicating substantial evaporation and derivation from arid sources. Most springs exhibit isotopic enrichment relative to Seville's weighted average, suggesting significant evaporation prior to infiltration. Heterogeneities in substrate permeability and soil composition further influence evaporation rates and isotopic retention.

Fractionation patterns in $\delta^{18}O$ and $\delta^{2}H$ vary considerably across springs. Some locations, such as Green Stream, demonstrate substantial enrichment attributed to intense evaporation in abandoned mine shafts, while others, including Right Spring, exhibit $\delta^{18}O$ enrichment likely resulting from calcite interactions during recrystallization. Springs including Ángeles, UMA, and Top of Right display $\delta^{18}O$ depletion, while others, such as Left-Left and Richi Left, show mixed isotopic behaviors. These variations indicate different recharge pathways and flow patterns with varying evaporation intensities and subsurface residence times.

Interactions between groundwater and sulfide deposits minimally influence isotopic signatures, as sulfate formation involves limited isotopic exchange (requiring approximately 350 years at ambient temperatures). Sulfite (SO_3^{2-}), a critical intermediate in sulfide oxidation, exchanges oxygen isotopes rapidly at low pH (0.88–1.4) in the presence of Fe^{3+}, with water serving as the primary oxygen source for sulfate under acidic conditions. Springs such as New New and Fox exhibit $\delta^{18}O$ depletion relative to $\delta^{2}H$, while other springs suggest different oxidation mechanisms and intensities. The isotopic data indicate limited hydraulic interconnectivity among springs, with fractionation patterns influenced by pH, temperature, oxidant availability, and flow path characteristics. Tritium measurements confirm that most spring waters originated from pre-1950 precipitation, though analytical coverage remains incomplete.

Concluding remarks

The integrated geophysical and hydrogeochemical investigation of the Río Tinto headwaters provides comprehensive insights into the natural processes generating its characteristic acidic conditions. The following key conclusions emerge from this study.

1. Geophysical surveys have delineated the structural architecture of the Peña de Hierro area, revealing a major listric thrust fault juxtaposing the Volcanosedimentary Complex against the Culm Group. This fault system serves as a primary control on groundwater flow and sulfide distribution.
2. Time-domain electromagnetic data identified a low-resistivity anomaly northwest of the Peña de Hierro pit lake, interpreted as the principal recharge zone for the aquifer system supplying the acidic springs. Additionally, the detection of

conductivity anomalies along the major thrust zone indicates the presence of unexploited sulfide ore bodies at intermediate depths (~500 m).

3. Hydrochemical analyses demonstrate significant spatial variability in acidity (pH 1.02–6.38) and redox conditions (Eh 110–662 mV) over relatively short distances, reflecting the strong influence of localized geological factors on water chemistry. The heterogeneous distribution of sulfide minerals, particularly pyrite, and neutralizing agents such as carbonates and aluminosilicates, primarily controls this variability.
4. Elemental associations and correlations, particularly between Fe–Cu–Zn–SO_4^{2-} and Al–F, indicate specific lithological interactions. Springs with elevated sulfate and metal concentrations reflect extensive contact with sulfide deposits within the VSC, while those exhibiting high Na^+/SO_4^{2-} ratios suggest interactions with sandstones and shales of the Culm Group.
5. Isotopic analyses reveal limited hydraulic interconnectivity among springs, with distinctive $\delta^{18}O$ and δ^2H signatures reflecting variable recharge mechanisms, evaporation intensities, and flow paths. The spring water line gradient (4.07) indicates substantial evaporative influence, while $\delta^{18}O$ depletion in specific springs is consistent with the proposed sulfatic-water model for the Río Tinto hydrogeological system.
6. Tritium data confirm that most spring waters originated from pre-1950 precipitation, indicating long residence times and minimal recent recharge contribution, consistent with the system's relative stability despite the cessation of mining activities in 1966.

These findings collectively support the hypothesis that the extreme acidity of the Río Tinto headwaters results from natural biogeochemical processes involving groundwater interaction with sulfide ore bodies, a phenomenon that has persisted for millions of years independent of anthropogenic mining activities. The integrated methodology employed in this study provides a valuable framework for investigating similar acid rock drainage systems worldwide, with implications for understanding both natural geochemical processes and mining-related environmental impacts.

References

Aguilera, A., Manrubia, S.C., Gómez, F., Rodríguez, N., Amils, R., 2006. Eukaryotic community distribution and its relationship to water physicochemical parameters in an extreme acidic environment, Río Tinto (Southwestern Spain). Appl. Environ. Microbiol. 72 (8), 5325–5330. https://doi.org/10.1128/AEM.00513-06.

Allman, C.J., Gómez-Ortiz, D., Burke, A., Amils, R., Rodriguez, N., Fernández-Remolar, D., 2021. Hydrogeochemical variability of the acidic springs in the rio tinto headwaters. Water 13 (20), 2861. https://doi.org/10.3390/w13202861.

Amils, R., Fernández-Remolar, D., Parro, V., Rodríguez-Manfredi, J.A., Oggerin, M., Sánchez-Román, M., López, F.J., Fernández-Rodríguez, J.P., Puente-Sánchez, F.,

Briones, C., Prieto-Ballesteros, O., Tornos, F., Gómez, F., García-Villadangos, M., Rodríguez, N., Omoregie, E., Timmis, K., Arce, A., Sanz, J.L., Gómez-Ortiz, D., 2014. Río Tinto: a geochemical and mineralogical terrestrial analogue of Mars. Life 4 (3), 511–534. https://doi.org/10.3390/life4030511.

Bartlett, R., Bottrell, S.H., Sinclair, K., Thornton, S., Fielding, I.D., Hatfield, D., 2010. Lithological controls on biological activity and groundwater chemistry in Quaternary sediments. Hydrol. Process. 24 (6), 726–735. https://doi.org/10.1002/hyp.7514.

Cánovas, C.R., Olías, M., Nieto, J.M., Sarmiento, A.M., Cerón, J.C., 2007. Hydrogeochemical characteristics of the Tinto and Odiel Rivers (SW Spain). Factors controlling metal contents. Sci. Total Environ. 373 (1), 363–382. https://doi.org/10.1016/j.scitotenv.2006.11.022.

Cánovas, C.R., Olías, M., Nieto, J.M., 2014. Metal(loid) attenuation processes in an extremely acidic river: the rio tinto (sw Spain). Water Air Soil Pollut. 225 (1). https://doi.org/10.1007/s11270-013-1795-7.

Davis, R.A., Welty, A.T., Borrego, J., Morales, J.A., Pendon, J.G., Ryan, J.G., 2000. Rio Tinto estuary (Spain): 5000 Years of pollution. Environmental Geology 39 (10), 1107–1116. https://doi.org/10.1007/s002549900096.

Dold, B., 2010. Basic Concepts in Environmental Geochemistry of Sulfidic Mine-Waste Management. InTech. https://doi.org/10.5772/8458.

Ducut, J.D., Alipio, M., Joanne Go, P., Concepcion II, R., Vicerra, R.R., Bandala, A., Dadios, E., 2022. A review of electrical resistivity tomography applications in underground imaging and object detection. Displays 73, 102208. https://doi.org/10.1016/j.displa.2022.102208.

España, S., 2007. The behavior of iron and aluminum in acid mine drainage. In: Speciation, Mineralogy, and Environmental Significance Thermodynamics, Solubility and Environmental Issues. Elsevier, Spain, pp. 137–150. https://doi.org/10.1016/B978-044452707-3/50009-4.

Essalhi, M., Sizaret, S., Barbanson, L., Chen, Y., Lagroix, F., Demory, F., Nieto, J.M., Sáez, R., Capitán, M.A., 2011. A case study of the internal structures of gossans and weathering processes in the Iberian Pyrite Belt using magnetic fabrics and paleomagnetic dating. Miner. Deposita 46 (8), 981–999. https://doi.org/10.1007/s00126-011-0361-8.

Fernández-Remolar, D.C., Rodriguez, N., Gómez, F., Amils, R., 2003. Geological record of an acidic environment driven by iron hydrochemistry: the Tinto River system. J. Geophys. Res.: Planets 108 (7), 1–16. https://doi.org/10.1029/2002je001918.

Fernández-Remolar, D.C., Morris, R.V., Gruener, J.E., Amils, R., Knoll, A.H., 2005. The Río Tinto Basin, Spain: mineralogy, sedimentary geobiology, and implications for interpretation of outcrop rocks at Meridiani Planum, Mars. Earth Planet Sci. Lett. 240 (1), 149–167. https://doi.org/10.1016/j.epsl.2005.09.043.

Fernández-Remolar, D.C., Gómez, F., Prieto-Ballesteros, O., Schelble, R.T., Rodríguez, N., Amils, R., 2008. Some ecological mechanisms to generate habitability in planetary subsurface areas by chemolithotrophic communities: the Ro Tinto subsurface ecosystem as a model system. Astrobiology 8 (1), 157–173. https://doi.org/10.1089/ast.2006.0022.

Fernández-Remolar, D.C., Banerjee, N., Gómez-Ortiz, D., Izawa, M., Amils, R., 2018. A mineralogical archive of the biogeochemical sulfur cycle preserved in the subsurface of the Río Tinto system. Am. Mineral. 103 (3), 394–411. https://doi.org/10.2138/am-2018-6288.

Fontes, J.C., 1980. Environmental Isotopes in Groundwater Hydrology. Elsevier BV, pp. 75–140. https://doi.org/10.1016/b978-0-444-41780-0.50009-2.

Gómez-Ortiz, D., Fernández-Remolar, D.C., Granda, A., Quesada, C., Granda, T., Prieto-Ballesteros, O., Molina, A., Amils, R., 2014. Identification of the subsurface sulfide bodies responsible for acidity in Río Tinto source water, Spain. Earth Planet Sci. Lett. 391, 36–41. https://doi.org/10.1016/j.epsl.2014.01.022.

Gónzalez-Toril, E., Gómez, F., Rodríguez, N., Fernández-Remolar, D., Zuluaga, J., Marín, I., Amils, R., 2003/10. Geomicrobiology of the Tinto River, a model of interest for biohydrometallurgy. Hydrometallurgy 71 (1–2), 301–309. https://doi.org/10.1016/S0304-386X(03)00169-5.

Hudson-Edwards, K.A., Schell, C., Macklin, M.G., 1999. Mineralogy and geochemistry of alluvium contaminated by metal mining in the Rio Tinto area, southwest Spain. Appl. Geochem. 14 (8), 1015–1030. https://doi.org/10.1016/S0883-2927(99)00008-6.

Langmuir, D., 1997. Aqueous Environmental Geochemistry. Prentice Hall.

Leistel, J.M., Marcoux, E., Thiéblemont, D., Quesada, C., Sánchez, A., Almodóvar, G.R., Pascual, E., Sáez, R., 1997. The volcanic-hosted massive sulphide deposits of the Iberian Pyrite Belt. Miner. Deposita 33 (1–2), 2–30. https://doi.org/10.1007/s001260050130.

Plumlee, G.S., Smith, K.S., Montour, M.R., Ficklin, W.H., Mosier, E.L., 1997. Geologic Controls on the Composition of Natural Waters and Mine Waters Draining Diverse Mineral-Deposit Types. Society of Economic Geologists, Inc., pp. 373–432. https://doi.org/10.5382/rev.06.19

Quesada, C., 1997. A reappraisal of the structure of the Spanish segment of the Iberian Pyrite Belt. Miner. Deposita 33 (1–2), 31–44. https://doi.org/10.1007/s001260050131.

Reynolds, J.M., 1997. An introduction to applied and environmental geophysics An introduction to applied and environmental geophysics.

Ruiz Cánovas, C., Olías, M., Nieto, J.M., 2014. Metal(loid) attenuation processes in an extremely acidic river: the rio tinto (SW Spain). Water, Air, Soil Pollut. 225 (1). https://doi.org/10.1007/s11270-013-1795-7.

Sáez, R., Almodóvar, G.R., Pascual, E., 1996. Geological constraints on massive sulphide genesis in the Iberian Pyrite Belt. Ore Geology Reviews 11 (6), 429–451. https://doi.org/10.1016/s0169-1368(96)00012-1.

Sáez, R., Pascual, E., Toscano, M., Almodóvar, G.R., 1999. The Iberian type of volcano-sedimentary massive sulphide deposits. Miner. Deposita 34 (5–6), 549–570. https://doi.org/10.1007/s001260050220.

Sánchez-Andrea, I., Rodríguez, N., Amils, R., Sanz, J.L., 2011. Microbial diversity in anaerobic sediments at Río Tinto, a naturally acidic environment with a high heavy metal content. Appl. Environ. Microbiol. 77 (17), 6085–6093. https://doi.org/10.1128/AEM.00654-11Spain.

Telford, W.M., Geldart, L.P., Sheriff, R.E., 1990. Applied Geophysics. Cambridge University Press.

Tornos, F., 2006. Environment of formation and styles of volcanogenic massive sulfides: the Iberian Pyrite Belt. Ore Geology Reviews 28 (3), 259–307. https://doi.org/10.1016/j.oregeorev.2004.12.005.

Velasco, F., Herrero, J.M., Suárez, S., Yusta, I., Alvaro, A., Tornos, F., 2013. Supergene features and evolution of gossans capping massive sulphide deposits in the Iberian Pyrite Belt. Ore Geology Reviews 53, 181–203. https://doi.org/10.1016/j.oregeorev.2013.01.008.

SECTION

Microbial and eukaryotic diversity

2

CHAPTER

Bacterial and archeal diversity along the Río Tinto

4

Elena González-Toril[1], Linda Amaral-Zettler[2,3], Graciela de Diego-Castilla[1], Antonio García-Moyano[4]

[1]*Centro de Astrobiología (CAB), INTA-CSIC, Torrejón de Ardoz, Madrid, Spain;* [2]*Department of Marine Microbiology and Biogeochemistry, NIOZ Royal Netherlands Institute for Sea Research, Texel, Netherlands;* [3]*Department of Freshwater and Marine Ecology, Institute for Biodiversity and Ecosystem Dynamics, University of Amsterdam, Amsterdam, Netherlands;* [4]*NORCE Research, Bergen, Norway*

The combined use of classical microbiology methodologies (enrichment and isolation of strains, and phenotypic characterization of isolates) and culture-independent molecular ecology has led to the identification of the most representative bacterial and archeal species in Río Tinto. In the water column, more than 80% of the cells belong to the bacterial domain, with only a small fraction corresponding to the archeal domain. The most abundant bacterial genera are: *Acidithiobacillus*, *Leptospirillum*, and *Acidiphilium*. All of them are involved in the iron and/or sulfur cycles. The constant and abundant presence of *Acidithiobacillus* and *Leptospirillum* indicates the important role of iron in this extreme river ecosystem.

Introduction

Extremophilic environments and extremophiles

The exploration of the planet has led to continuous discoveries of life in environments that were initially deemed uninhabitable. Indeed, it has been demonstrated that life can emerge and thrive under extremely harsh environmental conditions. This has given rise to a new discipline within biology: extremophily, which examines the diversity, ecology, and physiology of extreme environments and the organisms inhabiting them. These ecosystems are defined as locations characterized by one or more environmental factors inhospitable to most forms of life, particularly human life—thus rendering the definition inherently anthropocentric.

Despite the apparent hostility of such habitats, extreme ecosystems have emerged as capable of supporting higher levels of biodiversity than previously assumed. The environment we thrive in contains oxygen, is neither excessively cold nor hot (most organisms survive within a relatively narrow temperature range of 5 °C–40 °C), and is subject to moderate levels of radiation, filtered by the

Geomicrobiology of Río Tinto. https://doi.org/10.1016/B978-0-443-32948-7.00018-7

atmosphere. This familiar world shapes our perception of environmental normality, and organisms that exist at or beyond these boundaries are categorized as extremophiles. It is worth noting that extremophiles not only endure such adverse conditions to carry out their metabolic activity, but actually require them for proper development. Microorganisms capable of tolerating these conditions, yet also able to grow in moderate habitats, are not considered true extremophiles; rather, they are referred to as extreme-tolerant organisms.

The number of identified extremophiles has increased significantly in recent years. Examples include microbial communities that inhabit high-temperature environments, or hyperthermophiles, such as hyperthermophilic archaea, which may thrive at temperatures exceeding 100 °C (Kashefi and Lovley, 2003). Other organisms inhabit cold polar regions, such as Antarctic psychrophiles, which are capable of living in brackish waters with salinities several times higher than that of seawater, at temperatures below −10°C, and beneath ice layers up to 20 m thick (Price, 2000). Similarly, some organisms thrive in highly alkaline environments (up to pH 12), such as the larva of the rose moth (Madigan et al., 2021). At the opposite end of the pH spectrum are acidophiles, such as the archeon *Picrophilus oshimae*, which exhibits an optimal growth pH of 0.7, or *Cyanidium caldarium*, a microalgae inhabiting areas with acidic pH values ranging from 0 to 4.

Many microbial community members have been isolated from hypersaline environments, including lakes or salterns containing saturated saline solutions, such as those found in the Dead Sea. Others are capable of growth under intense ionizing radiation, such as that found within nuclear reactors. Some microorganisms thrive under extremely arid conditions, while others colonize the deepest parts of the oceans, where they are subjected to hydrostatic pressures exceeding 1000 bars. In order to survive, these organisms develop mechanisms enabling them to withstand freezing, complete desiccation, nutrient deprivation, elevated levels of radiation, and other physical or chemical challenges. These adaptations may be structural—conferring resistance—or molecular, involving metabolic adjustments that permit survival in conditions lethal to most life forms (Rothschild and Mancinelli, 2001).

Interest in the investigation of extreme environments and their microbial inhabitants has grown considerably in recent years, engaging various scientific disciplines. On the one hand, certain extreme environments reflect conditions characteristic of early Earth, thereby facilitating the study of life's origins. On the other hand, they help define the limits of life and the essential variables required for its persistence. Moreover, some of these environments hold significant value for space exploration, as they represent useful analogs for planetary and extraterrestrial settings that remain beyond our current reach. Lastly, though no less importantly, the considerable biotechnological potential of these microorganisms attracts considerable attention, as they constitute a promising source of novel microorganisms, enzymes and other cellular products with important applications in all sectors of the bioeconomy.

Microorganisms in extreme environments

When referring to extremophiles, bacteria and archaea are typically the foremost organisms that come to mind. The pioneering work of Thomas Brock on extremophiles, conducted in the hydrothermal environments of Yellowstone National Park in the USA, focused attention on the role of bacteria and archaea and their metabolism in extreme conditions. Additionally, the structural simplicity of bacteria and archaea has facilitated their superior adaptation to a wide range of environmental physicochemical parameters. For this reason, bacteria and archaea hold the majority of records for survival in extreme environments. Some of the most notable examples are summarized in Table 4.1.

Among the hyperthermophiles, the archeon *Geogemma barossii* is currently recognized as the microorganism capable of growth at the highest recorded temperatures, with an optimum of 121 °C and a maximum of 130 °C. Also noteworthy among hyperthermophiles is *Pyrolobus fumarii*, which, despite its slightly lower optimum growth temperature of 105 °C, remains the best-characterized species within this group to date. Both archaea have been isolated from submarine hydrothermal vents, where such elevated temperatures are possible due to the high pressure exerted by the overlying water column.

In the category of psychrophiles, the bacterium *Polaromonas vacuolata* is of particular interest. It exhibits an optimum growth temperature of 4 °C and retains metabolic activity even at 0 °C. This organism was found in cold marine waters, where increased salinity prevents complete freezing at sub-zero temperatures (Madigan et al., 2021).

In certain cases, microorganisms are capable of adapting simultaneously to multiple extreme environmental factors. The thermophilic archeon *Picrophilus oshimae*, which grows optimally at 60 °C, has the lowest recorded optimum pH for growth (0.7), and it can maintain metabolic activity at even lower, negative pH values. At the opposite end of the pH spectrum, the alkaliphilic archeon *Natronobacterium gregoryi* thrives in carbonate-rich lakes and soils with an optimum pH of 10. These environments are typically also saline, rendering this microorganism halophilic; it is capable of growing in the presence of up to 20% sodium chloride.

Archaea of the genus *Halobacterium*, frequently found in salt flats, grow optimally at a sodium chloride concentration of 20% (six times the salinity of sea water) and are able to survive even under saturated conditions exceeding 30%. Another notable example is the bacterium *Moritella yayanosii*, identified as the microorganism capable of growth under the highest known pressure, tolerating over 1000 atm, with an optimum at 500 atm. This species is found in abyssal sediments and oceanic depths, environments that combine extreme pressure with low temperatures. Consequently, *M. yayanosii* is also psychrophilic, with an optimum growth temperature of 4 °C.

It is important to highlight that such extreme conditions are not merely tolerated by extremophiles but are essential for their growth. The absence of these conditions frequently results in cell death. For instance, *Geogemma barossii* cannot grow below

Table 4.1 Examples of prokaryotic extremophiles, including several record-holding microorganisms for extreme environmental conditions. The table presents the genus and species, domain, habitat, and temperature or pressure limits at which these microorganisms thrive, highlighting their ability to survive in environments with extreme heat, cold, acidity, alkalinity, salinity, and pressure.

Extremophile	Genus/Species	Domain	Habitat	Extreme condition	Minimum	Optimal	Maximum
Hyperthermophile	*Pyrolobus fumarii*	Archaea	Submarine hydrothermal vents	Temperature	90 °C	106 °C	113 °C
Hyperthermophile	*Geogemma barossi*	Archaea	Submarine hydrothermal vents	Temperature	90 °C	121 °C	130 °C
Psychrophile	*Polaromonas vacuolata*	Bacteria	Sea ice	Temperature	0 °C	4 °C	12 °C
Acidophile	*Picrophilus oshimae*	Archaea	Acidic hydrothermal vents	Low pH	−0.06	0.7	4
Alkaliphile	*Natronobacterium gregoryi*	Archaea	Carbonate lakes	High pH	8.5	10	12
Halophile (NaCl)	*Halobacterium*	Archaea	Salt flats	Salinity (% NaCl)	15%	25%	32% (Saturation)
Barophile	*Moritella yayanosii*	Bacteria	Ocean sediments	Pressure	500 atm	700 atm	>1000 atm

90 °C, *Picrophilus oshimae* fails to develop at pH values above 4, and *Halobacterium* cannot survive if salinity drops below 15%. The evolutionary adaptation mechanisms of these microorganisms have led to a high degree of specialization, whereby extreme conditions represent their optimal ecological niche (Coker, 2023; Madigan et al., 2021; Schmid et al., 2020).

As demonstrated in the examples above, extreme environments are often characterized by more than one extreme physicochemical factor. For example, submarine hydrothermal vents not only exhibit high temperatures but also steep pH gradients and, due to their depth, elevated hydrostatic pressure. Highly acidic environments often contain elevated concentrations of heavy metals, as exemplified by the Tinto system, which also exhibits highly oxidizing conditions. Similarly, hypersaline ecosystems are frequently strongly alkaline. As a result, the majority of extremophilic microorganisms are either multi-extremophilic or at least multi-resistant (Madigan et al., 2021).

Although extremophiles require such harsh conditions for survival, these environments inevitably, induce cellular stress. It is evident that microorganisms able to proliferate under such conditions must possess specialized mechanisms to overcome this stress. Indeed, the number of adaptive strategies corresponds to the variety of stressors imposed by extreme physicochemical conditions—potentially even exceeding it, considering that a single stressor may be mitigated by multiple distinct mechanisms.

Nature and origin of extreme acidic ecosystems

Extreme ecosystems often exhibit multiple environmental factors typically deemed incompatible with most life forms. Despite this apparent hostility, such ecosystems frequently harbor higher-than-expected levels of biodiversity. Initially considered mere scientific curiosities, extremophiles have since become focal points in biotechnology research due to their potential applications across various sectors, including industry, agriculture, environmental management, and pharmacology (Schiraldi and De Rosa, 2002; Schmid et al., 2020). In recent years, a diverse array of bacteria (including cyanobacteria), archaea, and eukaryotes (including algae and fungi), have been isolated from these environments, highlighting their significance not only in biotechnology but also as models for studying biomolecular stability and the origins of life.

Extreme environments are particularly conducive to the generation of biodiversity and novel adaptive mechanisms. These settings, characterized by intense evolutionary pressures and spatial confinement, foster unique biological diversity and specific adaptive strategies. Consequently, extreme ecosystems are invaluable for studying and preserving biological heritage. However, their uniqueness and specificity also render them exceptionally fragile; disruptions to their delicate balance or the reintroduction of 'normal' conditions can lead to the loss of specialized biodiversity.

Acidic extreme environments and their resident microorganisms, known as acidophiles, have garnered significant attention due to their biotechnological and environmental implications. Acidophiles thrive in conditions with pH levels below 3, necessitating adaptations to maintain a neutral cytoplasmic pH and protect extracellular proteins and membranes from denaturation and disintegration.

Various ecosystems, including soils and freshwater bodies, can exhibit pH values below neutrality (pH 7). Environments with pH values below five are considered acidic, further classified into extremely acidic ($pH < 3$) and moderately acidic (pH 3–5). While moderately acidic conditions are common in peat lands, soils, and freshwater systems, extremely acidic environments are rarer and can result from natural phenomena (e.g., geothermal areas and exposed metal sulfides) or anthropogenic activities, particularly metal and coal mining (Johnson, 2009; Johnson and Schippers, 2017). Notably, in many extremely acidic environments, the low pH is a consequence of microbial metabolism rather than an inherent environmental condition, distinguishing them from other extreme habitats where factors like temperature or radiation are externally imposed (González-Toril et al., 2003). These environments also present challenges such as temperature variations ($<0°C$ to $>80°C$), elevated concentrations of solutes and heavy metals and oxidants, posing significant stressors to resident biota beyond the high concentrations of hydronium ions (H_3O^+) (Johnson and Schippers, 2017).

Acidic environments primarily originate from two distinct processes, each imparting unique physicochemical characteristics and influencing microbial ecology (Johnson and Aguilera, 2015). The first type is associated with geothermal and volcanic activity. Here, acidity arises from the microbial oxidation of reduced sulfur compounds, leading to sulfuric acid production. Volcanic gases like hydrogen sulfide and sulfur dioxide condense to form elemental sulfur, which can be observed as crystalline deposits around volcanic vents. Chemolithoautotrophic archaea and bacteria oxidize sulfur compounds, generating sulfuric acid and contributing to the formation of acidic mud pools, especially in high-temperature regions. Such environments, termed solfataras or fumaroles, are found globally, including in New Zealand (Rotorua), USA (Yellowstone), Iceland (Krýsuvík), Russia (Kamchatka), the Caribbean (Saint Lucia and Montserrat), the Azores, and Djibouti.

Fumaroles result from subsurface fractures emitting gases heated by nearby magma chambers. These fractures may fill with high-pressure, boiling water, forming geysers or, if below the surface, hydrothermal springs. The central region of a fumarole ("hot spot") exhibits the highest temperatures (up to 120°C). Sulfur-rich minerals dominate these areas, favoring sulfur-oxidizing microorganisms. The oxidation process lowers the pH, rendering many fumaroles acidic or highly acidic.

Surrounding the fumarole, temperature and pH gradients allow colonization by various microorganisms adapted to specific conditions. Typically, hyperthermophilic, sulfur-oxidizing, acidophilic archaea such as *Sulfolobus* or *Acidianus* inhabit the hottest zones. Moving away from the emission point, decreasing temperatures and acidity permit the succession of thermophilic acidophilic bacteria (e.g., *Acidithiobacillus caldus*), followed by non-thermophilic species (e.g., *Thiobacillus*). Farther from the "hot spot," as sulfur decreases, mesophilic and neutrophilic microbial communities dominate. In areas where temperatures remain below 50°C, the acidophilic eukaryotic microalga *Cyanidium caldarium* is commonly found.

Geothermal acidic ecosystems also exist in deep-sea hydrothermal vent systems, such as the East Pacific Rise, the Mid-Atlantic Ridge, and the Guaymas Basin. These areas drainages substantial amounts of reduced sulfur from magmatic sources. The resultant acidity from sulfur oxidation is rapidly neutralized by seawater, confining acidic conditions to vent conduits, which serve as ecological niches for acidophilic microorganisms (Johnson and Aguilera, 2015). Microorganisms in these habitats are not only acidophilic but also moderate to hyperthermophilic, such as archaea of the genus *Picrophilus* which exhibit optimal growth at pH 0.7 and can maintain metabolic activity at pH levels as low as −0.06, unable to survive above pH 4, representing the most acidophilic organisms described to date.

The second main origin of extremely acidic ecosystems involves the oxidation of metal sulfide deposits in the Earth's crust. When these deposits, primarily composed of pyrite (FeS_2), are exposed to the atmosphere and interact with water, oxygen, and chemolithoautotrophic bacteria, ferric iron and sulfuric acid are released, drastically lowering the pH in the surrounding water and soil. These ecosystems also exhibit high concentrations of heavy metals (e.g., cadmium, cobalt, copper, nickel, zinc) in millimolar ranges, significantly exceeding levels found in neutral pH waters (Barrie Johnson and Aguilera, 2019; Johnson and Aguilera, 2015).

In this context, the Río Tinto stands out as one of the most remarkable examples of extreme acidic ecosystems ever documented, due to its vast extent, consistently low pH, elevated concentrations of heavy metals, and significant microbial diversity, especially among eukaryotic organisms (Aguilera et al., 2006; Aguilera et al., 2007; Aguilera et al., 2007; Amaral-Zettler et al., 2011; Amaral Zettler et al., 2002). Due to its large-scale presence and ease access, Río Tinto is regarded as an exemplary model for investigating the microbial ecology of extreme acidic environments, and it holds significant relevance for astrobiological studies, as will be discussed further.

Río Tinto and its environment

Geological context: Iberian Pyrite Belt

Río Tinto originates in the province of Huelva (Southern Spain) and flows along almost 100 km through the Iberian Pyrite Belt (IPB), before discharging into the Atlantic Ocean. The IPB is one of the most significant polymetallic sulfide deposits globally, extending approximately 200 km in length and 40 km in width. It is part of the Southern Portuguese Zone of the Iberian Hercynian Massif and spans from the southwest of Spain to the Portuguese Atlantic coast. The belt consists of three lithological groups from the Upper Paleozoic (419–252 My): The Phyllite-Quartzite Group, which comprises a thick sequence of shales and sandstones; the Volcanic-

Sedimentary Complex, characterized by a mafic-felsic volcanic sequence intercalated with shales; and the Culm Group, predominantly composed of shales, sandstones, and conglomerates (Cánovas et al., 2007; Nieto et al., 2013).

Associated with the materials of the Volcanic-Sedimentary Complex are numerous underground massive sulfide deposits, with original reserves estimated to be 1700 Mt. Among these is the Río Tinto mining district, which constitutes the largest individual polymetallic sulfide deposit documented to date. The mineralization is dominated by pyrite (FeS_2), which makes up more than 90% by volume, and serves as the principal mineral of these deposits. This is accompanied by smaller amounts of sphalerite (ZnS), galena (PbS), chalcopyrite ($CuFeS_2$), arsenopyrite (FeAsS), and other sulfides, alongside trace quantities of cadmium, selenium, silver, gold, copper, mercury, and others. Pyrite is found in various textures, ranging from fine-grained crystals (colloquially referred to as framboidal) to coarse-grained crystals (Cánovas et al., 2007; Nieto et al., 2013).

The metallic sulfide deposits of the IPB have been subject to industrial exploitation since 3000 BCE, primarily for the extraction of gold, silver, and copper. During the Roman period, mining activity experienced resurgence, and traces of this can still be found throughout the IPB, where numerous archeological remains, such as hydraulic wheels, have been uncovered and can be viewed in the British Museum and the Archeological Museum of Huelva, Spain. Mining activity subsequently declined but continued sporadically during the Visigothic (fifth-eighth centuries), Arab, and Medieval periods (8th to 15th centuries). In the mid-19th century, coinciding with the Industrial Revolution, mining activity intensified. Throughout the 19th and 20th centuries, nearly 100 mines were operational, producing a total of 300 million tons of polymetallic ores primarily for the extraction of sulfur, copper, lead, and zinc (Cánovas et al., 2007; Nieto et al., 2013). Today, several mining operations remain active, with the most notable being Cobre Las Cruces mine in Seville province and the Atalaya Mining Project in Río Tinto, both extracting copper.

Microbial biodiversity of Río Tinto

The biodiversity present in the waters of Río Tinto is primarily microbial. While some plants can be observed in certain parts of the river, as is common in other extreme acidic ecosystems, their roots typically grow in sediments where the interstitial water has a pH significantly higher than 2. Microscopic metazoans such as rotifers are also found in Río Tinto where they are among the "top predators" in this ecosystem (Amaral-Zettler, 2013). In general, microbial biodiversity decreases with more acidic pH levels, although other factors (e.g., concentrations of soluble metals and other solutes) can also exert a significant influence. Studies conducted over the past 30 years on acidophilic extremophiles have demonstrated the presence of these microorganisms across the entire "tree of life", with representatives found in the three domains: Bacteria, Archaea, and Eukarya. This suggests that acidophily is a trait that has evolved multiple times during the history of life on Earth, rather than being a single evolutionary event (González-Toril et al., 2010; Johnson and Aguilera, 2015; Johnson and Schippers, 2017).

Like other extremophiles, acidophiles are highly specialized life forms adapted to very specific conditions, unable to grow in neutral environments. This is particularly true for bacterial and archeal species, whereas most eukaryotic acidophiles can also thrive at pH levels close to neutrality. Organisms that tolerate other extreme physical-chemical conditions (e.g., very high temperatures or hypersalinity) in addition to extreme acidity tend to be less tolerant of acidic pH than those species that can only withstand extreme acidity. This is particularly noticeable among eukaryotic acidophiles (Johnson and Aguilera, 2015; Johnson and Schippers, 2017).

The microorganisms in the river can also be classified according to the ecological niche in which they develop. We distinguish between those in the water column, those found in the riverbed, and those found in the mining lakes. Additionally, based on their location, we can further categories them as planktonic or benthic. Planktonic organisms remain suspended or floating in the water column and, although some can move or swim, they cannot resist currents or turbulence. Benthic organisms inhabit surfaces in aquatic ecosystems and can be found among sediments or, more commonly, form complex associations known as biofilms or microbial mats, which are easily distinguished at the riverbed. Filaments can also be observed floating above the surfaces or in the water column of stratified lakes (Souza-Egipsy et al., 2008, 2011; García-Moyano et al., 2007, 2012).

An important factor influencing the distribution of organisms is the presence of oxygen and light. In flowing waters, anaerobic zones are scarce, as the flow allows for air exchange, keeping the water oxygenated, especially in shallow areas. However, in areas of the riverbed below photosynthetic biofilms, oxygen levels can drop dramatically, as occurs in the sediments beneath the surface. In stagnant waters (as in acidic pits and lakes), anaerobic zones develop when the depth increases, sometimes affected by changes in temperature and the presence of photosynthetic filaments at certain times of the year (García-Moyano et al., 2007, 2012; Souza-Egipsy et al., 2008, 2011).

Bacterial and archeal biodiversity

The combined use of conventional microbial ecology methodologies (cultivation, species isolation, and phenotypic characterization of isolates) and molecular ecology techniques has led to the identification of the most representative bacterial and archeal microorganisms in Río Tinto (García-Moyano et al., 2007, 2012; González-Toril et al., 2010; López-Archilla et al., 2001). In the water column, over 80% of the cells belong to the domain Bacteria, with only a small fraction belonging to the domain Archaea. The most abundant bacterial genera are: *Acidithiobacillus*, *Leptospirillum*, and *Acidiphilium*. All of these are involved in the iron and/or sulfur cycles (Fig. 4.1) (González-Toril et al., 2003). The constant and abundant presence of *Acidithiobacillus* and *Leptospirillum* indicates the central role of iron in the river's ecosystem. Within the *Acidithiobacillus* genus,

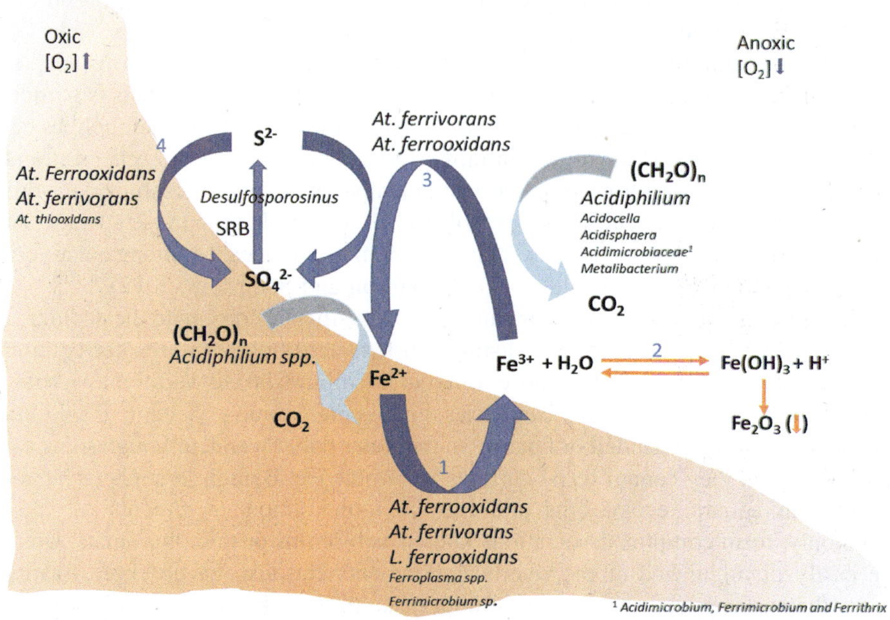

FIGURE 4.1 Geomicrobiological model of Río Tinto.

The central element that defines the characteristics of Río Tinto is iron. (1) ferrous iron is oxidized by chemolithotrophic bacteria, primarily by autotrophic microorganisms such as *At. ferrooxidans*, *At. ferrivorans*, and *Leptospirillum*, and to a lesser extent, by heterotrophic microorganisms. As a result of this metabolism, the environment becomes acidic and the iron is oxidized, leading to the red and acidic waters that characterize the river. (2) the high concentration of ferric iron in solution is responsible for buffering the water. (3) In anaerobic zones, the iron cycle will close, being reduced autotrophically by bacteria of the genus *Acidithiobacillus* and, heterotrophically, by other microorganisms such as *Acidiphilium*, *Acidocella*, *Acidisphaera*, etc. Additionally, *Acidiphilium* can perform this function under microaerobic conditions, with up to 60% dissolved oxygen. In most cases, this iron reduction is coupled with sulfur oxidation. (4) sulfur can also be aerobically oxidized by all species of the genus *Acidithiobacillus*. (5) despite the large amounts of sulfate, sulfate reduction is minor in these waters, being confined to a few microenvironments where *Desulfosporosinus* might be carrying out this reaction.

representatives of iron-oxidizing bacteria are found: *At. ferrooxidans*, *At. ferridurans*, and *At. ferrivorans*. Among these, *At. ferrooxidans* stands out for its ability to oxidize ferrous iron aerobically and reduce ferric iron under anaerobic conditions (Table 4.2) (Malki et al., 2006; González-Toril et al., 2003). Also of particular interest is *Acidithiobacillus ferrivorans*, which is characterized by its high rate of iron oxidation, as well as its ability to maintain an active metabolism at very low temperatures. This microorganism can remain active down to 4°C and is responsible for iron oxidation processes in colder times and places (Malki et al., 2006). *At.*

Table 4.2 Microbial diversity in Río Tinto. Main microorganisms classified by their phylogeny, carbon and energy sources, facultative iron reduction ability, optimal temperature, and presence in ***Río Tinto (RT)***. The table includes bacterial and archeal species, detailing their respective sources of carbon and energy, as well as their thermophilic, mesophilic, or psychrotolerant characteristics. The column **Facultative Iron Reducer** indicates whether each microorganism has the ability to reduce iron, while the **presence in *Río Tinto*** column notes how commonly the microorganism is found in ***Río Tinto*** environments.

	Phylogeny	Microorganism	Carbon source	Energy source	Facultative iron reducer	Optimal temperature	Presence in RT
Bacteria	Phylum: *Pseudomonadota* Clase: *Acidithiobacillia*	*Acidithiobacillus ferrooxidans*	O.A.	Fe^{2+}, S, H_2	YES	Mesophilic	Very abundant
		At. ferridurans	O.A.	Fe^{2+}, S, H_2	YES	Mesophilic	
		At. ferrivorans	O.A.	Fe^{2+}, S	YES	Psychrotolerant	
		At. ferriphilus	O.A.	Fe^{2+}, S	YES	Mesophilic	
		At. thiooxidans	O.A.	S	NO	Mesophilic	Sporadically
		At. caldus	O.A.	S, H_2	NO	Thermotolerant	Sporadically in some mine drainage
	Phylum: *Pseudomonadota* Clase: *Alphaproteobacteria*	*Acidiphilium* spp.	O.H.	O.C.	YES	Mesophilic	Very abundant
		Acd. organivorans	O.H.	O.C.	NO	Thermotolerant	
		Acidocella spp.	O.H.	O.C.	YES	Mesophilic	
	Phylum: *Pseudomonadota* Class: *Betaproteobacteria*	*Ferrovum*	O.A.	Fe^{2+}	NO	Psychrotolerant	Mainly found in streamers and macrofilaments
	Phylum: *Nitrospirota* Class: *Nitrospiria*	*Leptospirillum ferrooxidans*	O.A.	Fe^{2+}	NO	Mesophilic	Very abundant
		L. ferriphilum	O.A.	Fe^{2+}	NO	Thermotolerant	
		L. rubarum	O.A.	Fe^{2+}	NO	Mesophilic	
		L. ferrodiazotrophum	O.A.	Fe^{2+}	NO	Mesophilic	

Continued

Table 4.2 Microbial diversity in Río Tinto. Main microorganisms classified by their phylogeny, carbon and energy sources, facultative iron reduction ability, optimal temperature, and presence in ***Río Tinto (RT)***. The table includes bacterial and archeal species, detailing their respective sources of carbon and energy, as well as their thermophilic, mesophilic, or psychrotolerant characteristics. The column **Facultative Iron Reducer** indicates whether each microorganism has the ability to reduce iron, while the **presence in *Río Tinto*** column notes how commonly the microorganism is found in ***Río Tinto*** environments.—*cont'd*

	Phylogeny	Microorganism	Carbon source	Energy source	Facultative iron reducer	Optimal temperature	Presence in RT
	Phylum: *Actinomycetota* Class: *Acidimicrobiia*	*Ferrimicrobium*	O.H.	Fe^{2+}	YES	Mesophilic	Mainly found in streamers and macrofilaments. The presence of one genus or another is mostly determined by their optimal temperature.
		Acidimicrobium	F.A.	Fe^{2+}	YES	Thermotolerant	
		Ferrithrix	O.H.	Fe^{2+}	NO	Thermotolerant	
		Acidithiomicrobium	O.H.	Fe^{2+}	NO	Thermotolerant	
		Acidithrix	O.H.	Fe^{2+}	NO	Thermotolerant	
	Phylum: *Bacillota* Class: *Clostridia*	*Sulfobcillus* spp.	F.A.	O.C., Fe^{2+}, S, H_2	YES	Thermotolerant	In sediments and sometimes in macrofilaments
		Alyciclobacillus spp.	O.H o F. A.	O.C., some species Fe^{2+} ó S	Some species	Thermotolerant	
		Desulfosporosinus spp.	O.H.	O.C.	YES	Thermotolerant	
Archaea	Phylum: *Thermoplasmatota* Class: *Thermoplasmata*	*Ferroplasma* spp.	O.H.	Fe^{2+}	YES	Thermotolerant	Sporadically in some mine drainage
		Acidiplasma spp.	O.H.	Fe^{2+}	Not tested	Thermotolerant	
		Thermoplasma spp.	O.H.	O.C.	No	Thermophilic	

ferrivorans has even been identified in acid mine drainage in both the Arctic and Antarctica, where it is the primary agent responsible for the alteration of iron sulfides exposed due to the ongoing rise in global temperatures and the consequent melting (Dold et al., 2013; García-Moyano et al., 2015). Furthermore, this species is highly efficient as an iron reducer under anaerobic conditions (it is a facultative anaerobe, like *At. ferrooxidans*). These two physiological traits mean that in the specific case of Río Tinto, *At. ferrivorans* plays a significant role in deeper areas, such as mining ponds, where temperature and oxygen concentration are lower. It is also prominent in winter, when river water temperatures can drop to 6°C.

Regarding the *Leptospirillum* species, as the name suggests, these are comma-shaped or spiral or cork-screw shaped bacteria. Their cells often do not separate during division, leading to long spirals up to 5 microns in size. It is the only bacterium in Río Tinto with a spiral-shaped morphology, making it easily identifiable under an optical microscope. All *Leptospirillum* isolated or detected in the river are iron oxidizers and strictly aerobic. Although they can withstand low oxygen concentrations, they can form resistant structures when environmental conditions are adverse. These peculiar resistance forms are individual *Leptospirillum* cells that become coated with exotracellular polymeric substances (EPS) to the point where, under the microscope, they resemble small cocci. Representatives of all the *Leptospirillum* groups, such as *L. ferrooxidans*, *L. ferriphilum*, *L. ferrodiazotrophum*, and *L. rubarum*, have been identified, with *L. ferriphilum* predominating, as it can withstand higher temperatures, maintaining metabolic activity up to 40 °C (Table 4.2). As previously mentioned, *Leptospirillum* species are efficient producers of extracellular polymeric substances (EPSs), enabling them to remain together, forming streamers. These filaments are typically small (1–2 cm), with one end attached to the riverbed and the other floating with the current, forming an ideal ecological niche where microorganisms benefit from oxygenation, attachment, and easy access to their energy source. Although in the Río Tinto, these filamentous structures are predominantly made up of *Leptospirillum*, other associated genera, such as *Acidiphilium*, *Acidocella*, *Sulfobacillus*, *Acidiferrobacter*, and *Ferrimicrobium*, also appear, forming complex communities (García-Moyano et al., 2008, 2012).

Another noteworthy feature of the *Leptospirillum* genus is that its optimal pH is lower than that of *Acidithiobacillus*, making it more common in the more acidic zones of the river. Additionally, both *Acidithiobacillus* and *Leptospirillum* produce ferric ion as the primary product of their metabolism, and at high concentrations, this ion inhibits their growth. In the river and other acid mine drainages, ferric ions complex with sulfates and hydroxides form precipitates such as jarosite, schwertmannite, and limonite. These precipitates form around the iron-oxidizing microorganisms and are deposited in the sediments. The *Leptospirillum* genus appears to better tolerate growth inhibition by ferric ions or iron hydroxides and sulfates. This may be due to its propensity to excrete EPSs to protect itself from an adverse environment or its lower optimal pH, which means it inhabits areas where the formation of precipitates is less pronounced (García-Moyano et al., 2008, 2012).

As seen, *At. ferrooxidans*, *At. ferrivorans*, and the *Leptospirillum* species are iron oxidizers and therefore compete for the same ecological niche, appearing alternately in the river. The predominance of one genus over another depends on their specific physiological traits, as mentioned earlier. Generally, *Leptospirillum* bacteria dominate in the upper reaches of the river (near the source and around the former Peña de Hierro mine). This is likely due to the more extreme conditions of these zones, where the pH is consistently lower, and the ferric ion concentration is higher, conditions to which *Leptospirillum* is better adapted than other iron- and sulfur-oxidizing chemolithoautotrophs. Moreover, in general, the water temperature is higher in the upper reaches of the river, as the water column is generally smaller than in the rest of the river, making this area more suitable for *Leptospirillum* growth than for *Acidithiobacillus*, which is less abundant. This area is frequently covered by *Leptospirillum*-rich filaments, especially in summer, when water temperatures rise (García-Moyano et al., 2008, 2012).

In the middle section of the river, at the height of the locality of Berrocal, *At. ferrooxidans* predominates. In this area, the pH, redox potential, and not excessively high concentrations of ferric iron approach the optimal conditions for the growth of this bacterium (Table 4.2). On the other hand, the concentrations of zinc and copper are higher, which favors the presence of *At. ferrooxidans* over other iron-oxidizing bacteria. This species is found across a broader range of locations due to its metabolic versatility, occupying ecological niches that are suboptimal for *Leptospirillum* and other iron- or sulfur-oxidizers, which thrive in more extreme and restricted pH conditions. In environments such as dams, mine cuts, sediments, or wells, temperatures are typically low due to limited solar radiation, with *At. ferrivorans* being the most abundant microorganism. This is especially evident in winter, when this species often displaces *At. ferrooxidans* along much of the river's course. In contrast, mineral heaps, with their abiotic exothermic reactions and lower water content, tend to have higher core temperatures, fostering the growth of other iron-reducing bacteria, such as the actinobacteria *Acidimicrobium*, *Ferrithrix*, and *Acidithiomicrobium*, the clostridia *Sulfobacillus* and *Alicyclobacillus*, and certain archaea from the class Thermoplasmata. Consequently, *At. ferrooxidans* occupies the greatest number of sites along the river, exploiting ecological niches that are less suitable for other iron- or sulfur-oxidizers that require more extreme pH conditions (García-Moyano et al., 2008, 2012).

In locations where the water is colder due to greater depth, *At. ferrivorans* appears to predominate. This becomes particularly apparent in winter, when it completely displaces *At. ferrooxidans* in much of the river. There is no competition between both genera in the anaerobic zones of the river, such as the bottoms of dams, deep sediments, mining wells, mineral heaps, or the bottoms of mine cuts. In these areas, *Acidithiobacillus*, together with other iron-reducing bacteria that will be discussed below, takes center stage.

The third most abundant genus in the water column, *Acidiphilium*, is an acidophilic heterotroph with a diverse metabolism (Table 4.2). Under aerobic conditions, it obtains energy from organic compounds, whereas under anaerobic conditions, it is capable of oxidizing sulfur using iron as an electron acceptor and organic compounds as a carbon source. This genus frequently appears alongside *Acidithiobacillus* or *Leptospirillum*, as it removes organic compounds that are toxic to acidophilic iron oxidizers. This interspecies association is highly effective, allowing *Acidiphilium* to be distributed along the entire course of the river, both in the water column and in the sediments or in biofilms. In the upper part of the river, various *Acidiphilium* species are present, including *A. cryptum*, *A. organovorum*, and *A. multivorum*, while in the lower section, between the localities of Berrocal and Palma del Condado, *Acidiphilium* is less abundant. Its presence is replaced by other Alphaproteobacteria that are phylogenetically very close, such as *Acidocella* and *Acidisphaera*. This may be related to the relative sensitivity to copper exhibited by most *Acidiphilium* species (Johnson and Schippers, 2017; García-Moyano et al., 2008, 2012).

Although these three bacterial genera are the most abundant in the Río Tinto water column, the advent of high-throughput sequencing techniques has enabled the identification of numerous bacterial and archeal species that had previously gone undetected due to their very low abundance (González-Toril et al., 2003; Amaral-Zettler et al., 2011). Among these, both photosynthetic and bacterial biofilms have proven to be particularly diverse. Within them, the three dominant genera are present, along with other acidophiles such as the Betaproteobacterium *Ferrovum*, the Gammaproteobacterium *Acidiferrobacter*, the actinobacteria *Ferrimicrobium*, *Acidimicrobium*, *Ferrithrix*, *Aciditerrimonas*, *Acidithrix*, *Acidithiomonas*, the clostridia *Sulfobacillus* and *Alicyclobacillus*, or the archaea *Ferroplasma* and *Acidiplasma*. All of these acidophiles are associated with the iron or sulfur cycle, and their presence is determined by specific factors such as differences in optimal growth temperature or pH. Among them, actinobacteria are particularly interesting because, due to their morphology (long, very fine, interwoven filaments), they serve the function of binding cells together in microbial biofilms, with macrofilaments of up to several meters in length being observable. These macrofilaments tend to appear in calm water zones, such as dams located in the upper reaches of the river, and most likely originate from the bottoms of these dams, where oxygen concentration is very low. For this reason, the genera that form them are usually facultative anaerobes and iron reducers, such as *At. ferrooxidans*, *At. ferrivorans*, *Acidiphilium*, *Alicyclobacillus*, *Sulfobacillus*, or the actinobacteria themselves, *Acidisphaera*, *Ferrovum* or *Acidiferrobacter* (García-Moyano et al., 2007).

Archaea from the class *Thermoplasmata*, such as *Acidiplasma* and *Ferroplasma*, are also found in the upper reaches of the river, where the pH is lower and the influence of the mine is greater. Although the water temperature in these zones is not initially favorable for these moderately thermophilic archaea, it must be considered that during the summer, temperatures can be significantly higher. Moreover, in the mine outlets and in mineral heaps, the water flowing into the river is warmer due to the lithotrophic activity occurring within them.

The bacterial communities associated with the sediments have also been found to be more diverse than those in the water column. Chemolithoautotrophic iron oxidizers remain dominant, but the lesser homogeneity characteristic of sediments increases the likelihood of encountering microenvironments. In the upper riverbed, precipitates of iron hydroxides and sulfates are found that maintain a colloidal structure, in which the diversity does not differ significantly from that of the water column. In the middle course, however, sediments often exhibit very low oxygen concentrations. In these locations, we encounter the facultative anaerobes already mentioned: *Acidithiobacillus*, *Acidiphilium*, *Acidisphaera*, *Sulfobacillus*, *Alicyclobacillus*, *Metallibacterium*, *Ferrimicrobium*, etc. However, an additional bacterium becomes prominent: *Desulfosporosinus*. These organisms tolerate pH values below four and are capable of forming endospores, making them highly persistent even when environmental conditions are not always favorable. It is a fermentative genus, reduces iron, and can perform sulfate reduction. The high acidity of the river does not favor sulfate reduction, and it is probable that this microorganism is either fermenting or reducing iron in the river sediments. Nevertheless, the possibility that it contributes to the sulfur cycle cannot be ruled out. Indeed, it has been reported that this bacterium, under acidic conditions, is capable of participating in a process of sulfidogenesis in collaboration with acidophilic heterotrophs (*Acidocella*, *Alicyclobacillus*, *Acidisphaera*, and *Acidiphilium*) (García-Moyano et al., 2012).

In summary, the water column reflects the most abundant genera present in the biofilms and sediments, *Acidithiobacillus*, *Leptospirillum*, and *Acidiphilium*, though it should not be overlooked that the system's biodiversity extends well beyond these three genera.

Adaptation mechanisms to extremely acidic environments

Despite their morphological simplicity, the microorganisms inhabiting Río Tinto are highly diverse, both genetically and in the mechanisms, they employ to adapt to the surrounding environmental conditions. Acidophiles possess unique resistance mechanisms that are absent in organisms from non-extreme ecosystems. For this reason, research on extremophiles in general, and acidophiles in particular, provides essential insights into adaptation to acidic pH and resistance to metal toxicity. In this regard, recent advances in the genomic study of these species have yielded new knowledge about the versatile metabolism of these extremophiles. In addition to being of significant interest for evolutionary and ecological studies, acidophiles may represent a potential source of novel enzymes with possible applications in biotechnology and bioremediation.

Adaptation mechanisms in acidophilic bacteria and archaea

When referring to the optimal growth pH of a microorganism, it is always in reference to the extracellular pH, as the intracellular pH must remain near neutrality in order for cytoplasmic enzymes to maintain their structure and function. In microorganisms that grow at an optimal pH between six and 7, the internal pH remains close to neutrality, whereas in acidophiles and alkaliphiles, the internal pH can vary, this variation being limited by the sensitivity of biomolecules to pH. For instance, *Picrophilus oshimae*, the most acidophilic microorganism described to date, maintains an intracellular pH of 4.6, while internal pH values as high as 9.5 have been recorded in alkaliphiles. It is not known with certainty whether these values represent the absolute intracellular pH limits, but they are likely very close, since DNA undergoes hydrolysis at low pH and RNA at high pH (Madigan et al., 2021).

Although it is difficult to study intracellular pH in vivo in microorganisms, studies conducted thus far confirm the existence of a pH gradient across the cell membrane. The mechanisms that allow microorganisms to maintain their intracellular pH are related to modifications in their cell membranes that render them less permeable to proton influx, as well as the development of proton pumps that help sustain this gradient. An indirect indication that the intracellular pH of acidophiles is not as acidic as their environment is that, upon cell lysis, the pH of the medium increases. Furthermore, most cytoplasmic enzymes of acidophilic bacteria and archaea studied to date exhibit optimal activity at a pH close to 7. However, enzymes involved in iron oxidation processes, such as rusticyanin, which are located in the periplasmic region of Gram-negative acidophiles, have an optimal activity pH similar to that of the external environment. Likewise, enzymes that are actively secreted outside the cell must, by necessity, be active at acidic pH (Johnson, 2009; Johnson and Schippers, 2017).

The proton motive force (PMF) is employed by acidophiles in much the same way as by their neutrophilic counterparts. The conversion of ADP to ATP by membrane-bound ATPases facilitates substrate transport across the membrane and also powers flagellar rotation. To generate this PMF, most bacteria require the establishment of proton gradients, i.e., a difference in pH between the cytoplasm and the external medium (Madigan et al., 2021). In acidic environments, such a difference exists naturally, and in theory, acidophiles could generate ATP without energy expenditure; however, this would rapidly lead to severe cytoplasmic acidification. Two main mechanisms have been proposed for maintaining this gradient in acidophilic microorganisms.

The first involves preventing the entry of protons into the cell. Acidophiles are capable of distinguishing between water molecules (H_2O) and hydronium ions (H_3O^+). Entry of the latter is blocked, thereby preventing acidification of the intracellular pH. Like other extremophiles, acidophiles actively pump hydronium ions out of the cell. However, unlike their neutrophilic counterparts, they must do so against an enormous proton concentration gradient that can exceed six orders of magnitude compared to neutral environments. Nevertheless, due to their cell membranes being remarkably impermeable to passive H_3O^+ influx, pH gradients can be maintained both in resting (energy-deprived) cells and in metabolically active ones (Barrie Johnson and Aguilera, 2019; Johnson and Aguilera, 2015).

The second mechanism involves generating a positively charged cell membrane, which significantly reduces passive proton influx into the cytoplasm. Acidophilic microorganisms produce positive membrane potentials by the uptake of cations such as potassium. In metabolically active cells, ATP is synthesized via the controlled influx of hydronium ions driven by the PMF. In inactive cells, PMF tends to zero, and in order to maintain the same pH differential between the external medium and the cytoplasm, the membrane potential increases to compensate for the drop in PMF. Thus, in resting cells, the positive membrane potentials provide protection against the driving force for proton influx, which would otherwise lead to rapid cytoplasmic acidification and consequent cell death (Barrie Johnson and Aguilera, 2019; Johnson and Aguilera, 2015).

Proton influx must be tightly regulated to prevent cytoplasmic acidification. However, ferrous iron oxidation must necessarily occur outside the cytoplasm to avoid precipitation of ferric iron, which is highly insoluble above pH 2.5. In Gram-negative iron oxidizers, ferrous iron oxidation is mediated by redox-active proteins (such as rusticyanin in *Acidithiobacillus ferrooxidans*) retained in the acidic periplasm (Barrie Johnson and Aguilera, 2019; Johnson and Aguilera, 2015; Johnson and Schippers, 2017). Electrons are transferred via soluble and/or membrane-bound cytochromes and ultimately to cytochrome oxidase, where they are used along with protons to reduce oxygen to water.

The molecular adaptations that maintain the strong pH gradient have further implications that help explain some characteristics of microorganisms in extremely acidic environments. While these cells can withstand direct acidification of their cytoplasm by hydronium ions, they are sensitive to indirect acidification mediated by weak organic acids. Many organic acids have a pKa above pH three (e.g., acetic acid ~4.7), meaning they exist predominantly in undissociated forms in extremely acidic environments. As many of these organic acids are also lipophilic, they can passively diffuse across the acidophile's cell membranes. Once inside the cell, the near-neutral pH causes the acids to dissociate. This disequilibrium leads to continued influx of organic acids, resulting in severe acidification and often cell death. Consequently, acidophiles are paradoxically far more sensitive to weak acids than their neutrophilic counterparts. Additionally, these organic acids do not participate in classical sulfate reduction, as they are not in assimilable forms. Sulfide regeneration would thus be restricted to microniches with slightly higher pH, as may occur in sediments or in thick biofilms (Barrie Johnson and Aguilera, 2019; Johnson and Aguilera, 2015; Johnson and Schippers, 2017).

Due to the presence of positive charges on their membranes, acidophiles exhibit greater resistance to high concentrations of metallic cations, which are toxic to most organisms. Conversely, acidophiles tend to be sensitive to most anions, which can readily penetrate the cells. Therefore, they are generally sensitive to chloride (Cl^-), nitrate (NO_3^-), or anionic metals such as molybdate (MoO_4^{2-}), although they do not appear to be affected by sulfate ions, which are ubiquitous in their environment (Johnson and Schippers, 2017). This is the primary reason why, from an industrial perspective, bioleaching processes cannot be conducted using acidified seawater, requiring instead the use of freshwater.

Interestingly, despite these molecular adaptations, the cell walls of acidophilic bacteria and archaea are not markedly different from those of neutrophiles. An exception to this is found in the archaea of the order *Thermoplasmatales*, where the principal mechanism of adaptation to extremely acidic and thermophilic environments lies in modifications to the cell envelope. These archaea lack a cell wall yet thrive in environments where the external proton concentration greatly exceeds the internal concentration and, in many cases, at thermophilic or hyperthermophilic temperatures. Among the *Thermoplasmata*, the genus *Thermoplasma* has been extensively studied and is known to possess a uniquely structured membrane that allows it to withstand acidic and high-temperature conditions. This membrane is composed of a lipopolysaccharide known as lipoglycan, forming a tetraether monolayer with glucose and mannose and also containing glycoproteins, though lacking sterols. In contrast, *Picrophilus*, unlike other *Thermoplasmatales*, does possess a cell wall and DNA with a high GC content, rendering it more stable. The special composition and structure of its membrane lipids provide resistance to low pH, but the membrane loses integrity at pH levels above 4, which is why this archeon only survives in extremely acidic environments (Barrie Johnson and Aguilera, 2019; Johnson and Aguilera, 2015; Madigan et al., 2021).

Horizontal gene transfer as a mechanism of adaptive resistance

Phylogenetic analyses of metatranscriptomic libraries derived from eukaryotic microorganisms in the Río Tinto have indicated that many of the transcripts identified show high similarity to those of bacterial species. This suggests that horizontal gene transfer (HGT) may constitute one of the principal adaptive forces in these organisms, enabling them to colonize new environments which are, at first glance, highly adverse to their development. It has long been known that genomic rearrangements—including gene duplication and HGT—play critical roles in driving the adaptive evolution of microbial genomes in response to environmental change (Soucy et al., 2015). Accordingly, gene and/or genome duplication has attracted considerable attention as a dominant evolutionary force behind functional innovation. Furthermore, HGT is considered the predominant evolutionary mechanism contributing to genome diversification (Ullrich et al., 2016).

Simultaneously, the substantial increase in available genomic data in recent years has strengthened the hypothesis that genome reduction events are a widespread source of genetic modification, potentially responsible for much of the adaptive phenotypic diversity observed across species (Albalat and Cañestro, 2016). Gene loss has been reported across a wide range of organisms, from all domains of life, providing robust evidence in support of the ubiquity of gene loss (Albalat and Cañestro, 2016). Nonetheless, the precise contribution of each of the major evolutionary forces to microbial genome organization, structure, and diversification remains unclear.

The rapid development of high-throughput sequencing technologies and bioinformatics has enabled researchers to access an increasing number of complete genomes from acidophilic microorganisms. These genomes are yielding the first insights into the key genomic characteristics associated with life in acidic environments. To date, most of the sequenced genomes of acidophilic species pertain to bacteria and archaea, including *Sulfobacillus acidophilus* (Anderson et al., 2012; Li et al., 2011), *S. thermosulfidooxidans* (Guo et al., 2014; Travisany et al., 2012), *S. benefaciens*, and various strains of *Ferrovum*, *Acidithiobacillus caldus*, and *At. ferrooxidans* (Selkov et al., 2000; Valdes et al., 2009). Comparative genomic studies among various *Ferrovum* strains have identified genes originating from other species in all the strains analyzed. The HGT events were associated with genes involved in phagocytosis-related processes, transposases, integrases, and genes encoding secretion system proteins. Conversely, certain strains have lost gene clusters involved in motility, chemotaxis, and the secretion of EPSs, which are important for biofilm formation. Although gene loss may be detrimental during the early stages of colonizing a new habitat, once the population becomes established within the new ecosystem, a reduced genome size may confer a selective advantage (Ullrich et al., 2016).

In the case of acidophilic eukaryotes, even fewer species have been sequenced. To date, genomic data are available for three closely related thermoacidophilic microalgae from the order Cyanidiales: *Cyanidioschyzon merolae*, *Galdieria sulphuraria*, and *Galdieria phlegrea* (Matsuzaki et al., 2004; Qiu et al., 2013; Schönknecht et al., 2013). As in bacteria, comparative genomic studies have revealed evidence of HGT; however, in this case, the events involved organisms from different phyla, with gene transfer occurring between environmental domains. These HGT events, through the acquisition of new gene families and the loss of others, have played a decisive role in the adaptation of these red algae to acidic environments (Qiu et al., 2013). The Cyanidiales algae have acquired bacterial membrane transporters that confer arsenic resistance through biotransformation, as well as archeal ATPases, which probably contribute to the thermal tolerance of these microalgae (Schönknecht et al., 2013). It has been estimated that at least 5% of the genes in *G. sulphuraria* were acquired via HGT-related processes (Qiu et al., 2013). Furthermore, the reduction in genes encoding voltage-gated ion channels, along with the acquisition of new gene families associated with chloride transporters, has significantly contributed to acid pH resistance in these microorganisms (Schönknecht et al., 2013).

The first draft genome of the acidophilic green alga *Chlamydomonas eustigma* has also been published (Hirooka et al., 2017). When this genome was compared to that of its neutrophilic counterpart, *C. reinhardtii*, it was found—as expected—that genes encoding heat shock proteins and plasma membrane H^+-ATPase transporters were much more highly expressed in the acidophilic species. Additionally, *C. eustigma* has lost genes involved in major fermentation pathways that would otherwise acidify the cytoplasm, while acquiring genes encoding intracellular pH-buffering systems and arsenic detoxification mechanisms. The observation that these features are also present in other red and green acidophilic algae suggests the existence of common mechanisms underlying acid environment adaptation in eukaryotes.

In the case of Río Tinto, metatranscriptomic studies on cultures of *Chlamydomonas acidophila* and *Dunaliella acidophila* have demonstrated the presence of phytochelatin synthase genes of bacterial origin, which have played a significant role in their metal tolerance (Olsson et al., 2017; Puente-Sánchez et al., 2016, 2018). Heavy metals are generally toxic due to their displacement of endogenous enzymatic metal cofactors from their binding sites, and their abnormal reactivity with thiol groups in proteins, leading to the generation of free radicals that cause cellular damage (Diaz and Plummer, 2018). In response, cells typically activate various detoxification pathways, including the increased synthesis of sulfur-containing defense compounds such as cysteine, glutathione, and phytochelatins (PCs). This defense mechanism is essential for metal tolerance and survival in plants. Moreover, the synthesis of these compounds also increases in algae and fungi exposed to metals.

In plants, PCs are enzymatically synthesized from γ-glutamylcysteine by the enzyme dipeptidyl transpeptidase, commonly known as phytochelatin synthase, which is constitutively expressed but can also be regulated at the transcriptional and post-transcriptional levels. Glutathione serves as the substrate for PC synthesis, and its biosynthesis can be affected by oxidative stress. Genes encoding phytochelatin synthase have been identified in all vascular plants, as well as in some algae, fungi, and invertebrates. In addition to eukaryotes, similar but shorter genes have been found in Bacteria (Hasan et al., 2017).

In cultures of *Chlamydomonas acidophila* isolated from Río Tinto and treated with excess copper, numerous transcripts related to oxidative stress response and heavy metal detoxification were identified that were highly similar to those of bacteria (Olsson et al., 2015). Specifically, two phytochelatin synthases that were completely distinct from those previously described in green algae, designated *CaPCS1* and *CaPCS2*. Similarly, in *Dunaliella acidophila* cultures from the river subjected to metal excess, another variant was identified and named *DaPCS* (Olsson et al., 2017). To determine the evolutionary relationships between these novel phytochelatin synthase genes and those described in plants, bacteria, and green algae, a phylogenetic analysis was conducted. The three new phytochelatin synthases showed the highest similarity values to sequences from bacterial sources. Furthermore, when these genes were introduced into *Escherichia coli* cultures, it became evident that the expression of the *CaPCS2* gene enhanced cadmium resistance.

Based on Olsson et al. results (Olsson et al., 2015, 2017), the phytochelatin synthase genes identified in these acidophilic microalgae are likely of bacterial origin and were later acquired through horizontal gene transfer, in some cases through multiple events. The hypothesis of a bacterial origin is further supported by the fact that these genes are smaller than their plant counterparts, a characteristic also observed in bacteria. HGT has also been suggested to account for the presence of phytochelatins in metazoans and fungi, and even among vascular plants. Results confirm that the evolutionary history of these genes is, indeed, complex and highlights the necessity of including bacterial sequences in such analyses in order to resolve their phylogeny.

Geomicrobiological model of Río Tinto

Geomicrobiological studies conducted in Río Tinto and its surroundings have focused on three principal environments, each characterized by distinct hydrogeochemical properties. Firstly, there is the main course of the river itself, where the majority of biodiversity studies have been carried out. Numerous acidic mine drainages discharge into the main course, as well as streams originating from waste rock piles and outflows from open-pit mines. These latter, not being properly drained, overflow into previously used tunnels. These mine outflows are of particular interest as, within just a few meters, the waters transition from completely anoxic to oxygenated conditions, leading to a rapid microbial succession. Finally, attention must also be given to the lakes formed in open-pit mines. Within each of these environments, further subdivisions can be made, including the water column, sediments, and either floating or sessile biofilms. The presence of aerobic and anaerobic zones is also a key factor in describing the geomicrobiology of each environment.

Geomicrobiological model of the main stream

The model explaining the microbial ecology of Río Tinto is fundamentally based on the physicochemical and geomicrobiological properties of iron and sulfur. These elements may serve both as electron donors and acceptors in the metabolism of most bacterial and archeal organisms present in the river water. In its reduced form, iron acts as the primary energy source for the most quantitatively abundant microbes in the water, such as *Leptospirillum* spp., *At. ferrooxidans*, and *At. ferrivorans*, as well as other less abundant microbes including *Ferromicrobium*, *Acidimicrobium*, *Ferrovum*, and the archeon *Ferroplasma acidiphilum*. Oxidized iron can be utilized as an electron acceptor under anaerobic conditions by bacteria belonging to the genera *Ferromicrobium*, *Acidimicrobium*, *Acidithiobacillus*, *Ferrovum*, and *Acidiferrobacter*. Bacteria of the genus *Acidiphilium* can perform iron reduction under anaerobic or microaerobic conditions (with oxygen concentrations up to 60%) (Fig. 4.1) (González-Toril et al., 2003; Barrie Johnson and Hallberg, 2008).

These anaerobic or microaerobic processes are typically confined to deeper zones of stagnant waters, sediments, and biofilms, and play a key role in the regeneration of ferrous ions, which can then be reused as an energy source. The regeneration of iron through anaerobic reduction was first observed in species of the genus *Acidiphilium* but has since been found to be common among many acidophiles, including *Acidithiobacillus*, particularly abundant in the river's anoxic zones (also including *Ferromicrobium*, *Acidimicrobium*, *Ferrovum*, or *Acidiferrobacter* (Table 4.2). Beyond its metabolic role, iron is also responsible for maintaining a constant pH throughout the river, due to its buffering effect. Lastly, ferric ion appears to be an effective protector against ultraviolet radiation, offering an undeniable ecological advantage to microorganisms adapted to these environments (Madigan et al., 2021).

Although iron plays a predominant role in Río Tinto, sulfur must not be overlooked as another crucial metabolite, present in very high concentrations in the river water. It serves as an energy source for chemolithoautotrophic bacteria of the genus *Acidithiobacillus*, which oxidize sulfur under aerobic conditions. Within this genus, *At. ferrooxidans* and *At. ferrivorans* play a central role, being the most quantitatively important species. As for *At. thiooxidans* and *At. caldus*, these have not been detected in the river using molecular techniques, although they have been identified using conventional microbial ecology techniques based on enrichment cultures (López-Archilla et al., 2001).

The species *At. caldus* may be restricted to zones of higher temperature, such as areas close to mine outflows or abandoned waste heaps and thus may be difficult to detect within the water column. The presence of microorganisms such as *At. thiooxidans* and *At. caldus* also depends on the availability of ferric ion, which renders sulfur in sulfide compounds accessible to strict sulfur oxidizers (Barrie Johnson and Aguilera, 2019; Barrie Johnson and Hallberg, 2008). As a consequence of iron oxidation by chemolithoautotrophic bacteria from polymetallic sulfides, numerous intermediate sulfur compounds are formed, which can be used by sulfur-oxidizing organisms (Barrie Johnson and Aguilera, 2019; Barrie Johnson and Hallberg, 2008). In this way, the action of iron-oxidizing microorganisms, such as *L. ferrooxidans* and *At. ferrooxidans*, facilitates the activity of sulfur oxidizers like *At. thiooxidans* or *At. caldus*. Furthermore, as observed in the iron cycle, in anoxic environments the oxidation of sulfur coupled with iron reduction is carried out by bacteria such as *At. ferrooxidans* or *Acidiphilium* (Table 4.2).

Regarding Sulfur Reduction, despite the abundance of sulfate in acidic environments, whether of geothermal origin or derived from sulfide oxidation, such settings are not typically dominated by sulfate-reducing bacteria (SRB). The most plausible explanation for this is that, under acidic pH conditions, classical electron donors are in a non-assimilable state for these bacteria. Only limited evidence exists for classical sulfate reduction in acidic anaerobic environments, possibly occurring within microniches that permit the stability of electron donors (Johnson and Aguilera, 2015).

There is only one bacterial genus found in acidic environments—particularly within sediments—that may be capable of sulfate reduction: *Desulfosporosinus*. This genus comprises Gram-positive, spore-forming Firmicutes, which are mesophilic (25–40 °C, with an optimum at 30 °C) and moderately acidophilic (optimal pH 5.2, minimum pH 3.6). These are obligate anaerobes capable of heterotrophic respiration using compounds such as H_2, glucose, pyruvate, lactate, glycerol, and fructose, with sulfate as the electron acceptor. They exhibit considerable metabolic versatility, being capable of fermentation or of using ferric iron as an electron acceptor. The redox potentials under extremely acidic conditions suggest that ferric iron is a much more favorable electron acceptor than sulfate, which implies that bacteria capable of utilizing both may preferentially perform iron reduction rather than sulfate reduction. This suggests that sulfate reduction in extremely acidic waters is highly unlikely, despite the presence of *Desulfosporosinus*.

Nevertheless, sulfate reduction and sulfidogenesis under acidic conditions have been extensively studied for their potential application in the bioremediation of acid mine drainage contaminated with heavy metals. To date, the results have not been promising, as the metabolism of SRBs tends to raise the pH of the medium, which could complicate remediation efforts. Laboratory reactor studies have shown that co-cultures of *Desulfosporosinus* and *Acidocella* are effective in sulfidogenesis coupled with iron reduction. Both genera have been identified in Río Tinto, although it remains uncertain whether sulfidogenesis is actively taking place. If it is, it may represent a form of sulfur regeneration more compatible with extreme acidic environments than classical sulfate reduction (Barrie Johnson and Aguilera, 2019; Barrie Johnson and Hallberg, 2008; Johnson, 2012; Ňancucheo and Johnson, 2012).

The three most quantitatively dominant genera in the river hold particular significance within the proposed model. *Leptospirillum* is responsible for iron oxidation; *Acidithiobacillus* carries out both iron and sulfur oxidation under aerobic conditions and is involved in sulfur oxidation and ferrous ion regeneration under anaerobic conditions; while *Acidiphilium*, together with *Acidithiobacillus*, performs iron reduction coupled with sulfur oxidation in anaerobic zones. The activity of just these three dominant genera is sufficient to complete the iron cycle, resulting in an ecosystem governed by iron metabolism, characterized by low phylogenetic diversity but high metabolic versatility (González-Toril et al., 2003).

Although, as previously stated, many of the river's characteristics are due to the metabolism of iron-oxidizing chemolithoautotrophs, the role of acidophilic heterotrophic bacteria should not be overlooked. The most abundant are various species of the genus *Acidiphilium*, but other genera commonly found in acid mine drainage are also well represented, including *Acidocella*, *Acidimicrobium*, *Sulfobacillus*, and *Alicyclobacillus* (Table 4.2). In addition to their acidophily, these microorganisms share the trait of requiring only minimal amounts of organic matter for growth. In Río Tinto, due to its low pH, organic compounds are rapidly hydrolyzed and not readily available to heterotrophs. However, certain areas of the river form niches rich in organic matter derived from the EPSs of biofilms. As a result of the accumulation and metabolism of the constituent microorganisms, biofilms are intrinsically rich in carbon compounds, with concentrations high enough to support the development of acidophilic heterotrophs.

For most acidophilic iron oxidizers, organic compounds are toxic—even those derived from their own metabolism (i.e., secondary metabolites). Thus, the activity of heterotrophs contributes to improved growth of chemolithoautotrophic communities by consuming available organic matter. This explains why it is difficult to obtain and maintain pure, single-species cultures of chemolithoautotrophic iron oxidizers, and similarly why it is nearly impossible to find them alone in nature—they always develop more successfully in the presence of acidophilic heterotrophs. Accordingly, both in the river water, sediments, and especially in biofilms, the activity of oxidizers is enhanced by the presence of heterotrophic microorganisms that "cleanse" the environment of organic compounds and secondary metabolites (Barrie Johnson and Aguilera, 2019; Barrie Johnson and Hallberg, 2008; García-Moyano et al., 2012; González-Toril et al., 2003; Souza-Egipsy et al., 2011).

Most acidophilic heterotrophs are facultative anaerobes, capable of using organic compounds as an energy source and ferric iron as an electron acceptor. They are thus largely responsible for the regeneration of ferrous iron in the river's anaerobic zones (sediments, dam bottoms, or subsurface environments) (Barrie Johnson and Aguilera, 2019; Barrie Johnson and Hallberg, 2008).

Geomicrobiological model of mine drainages

Mine discharges and drainage from mineral waste dumps deserve special consideration, as they exhibit the complete geomicrobiological cycle. In the river, this cycle is represented across various ecological niches, which are often disconnected or spatially distant. In contrast, within drainage systems, all relevant reactions can be observed over just a few meters. Inside the waste dumps, oxygen concentration and pH are very low, while iron concentration and redox potential are high, and the temperature is moderately elevated. These conditions create a favorable environment for chemolithotrophic, facultatively anaerobic, acidophilic, and iron-reducing microorganisms. As previously described in the river model, one example is *Acidithiobacillus*, which under such conditions can utilize sulfur as an energy source and iron as an electron acceptor. In addition to this microorganism, other species capable of thriving in such an environment by using iron as an electron acceptor include *At. ferrivorans*, *Acidiphilium*, *Acidocella*, *Ferrimicrobium*, and the archeon *Ferroplasma*, among others (González-Toril et al., 2011).

It is worth highlighting that mine drainages represent the environment in which the greatest number of archaea, predominantly *Ferroplasma*, have been identified. The prevailing redox potential conditions facilitate the occurrence of this microorganism, which, although not abundant in the river, is frequently found in other extreme acidic waters. It is likely that *Ferroplasma* plays a significant role in iron reduction within mines, waste dumps, and ore piles. As a result of the metabolic activity of *Ferroplasma* and other iron-reducing organisms, the resulting waters are highly transparent and exhibit a greenish hue due to the elevated concentration of reduced iron. This water provides a substantial input of substrate for iron-oxidizing microorganisms, which are particularly abundant in the central stream of the Tinto River. Upon emergence, these small water streams are rapidly colonized by biofilms in which photosynthetic eukaryotes play a prominent role. Just a few meters (or even centimeters) downstream from the discharge point, the water becomes quickly oxidized. This is due to the presence of oxygen and the high concentration of ferrous iron, creating an ideal environment for the activity of iron-oxidizing chemolithoautotrophs such as *At. ferrooxidans*, *At. ferrivorans*, *Leptospirillum* spp., *Acidimicrobium*, and *Ferrimicrobium*. As such, it is common for water originating from mining residues that reaches the Tinto River to already be oxidized and to exhibit characteristics similar to those of the river's main course. An example of how such drainage systems operate is illustrated in Fig. 4.2, which depicts the geomicrobiological cycle at one of the most representative mine drainage sites within the Iberian Pyrite Belt, La Zarza-Perrunal (González-Toril et al., 2011).

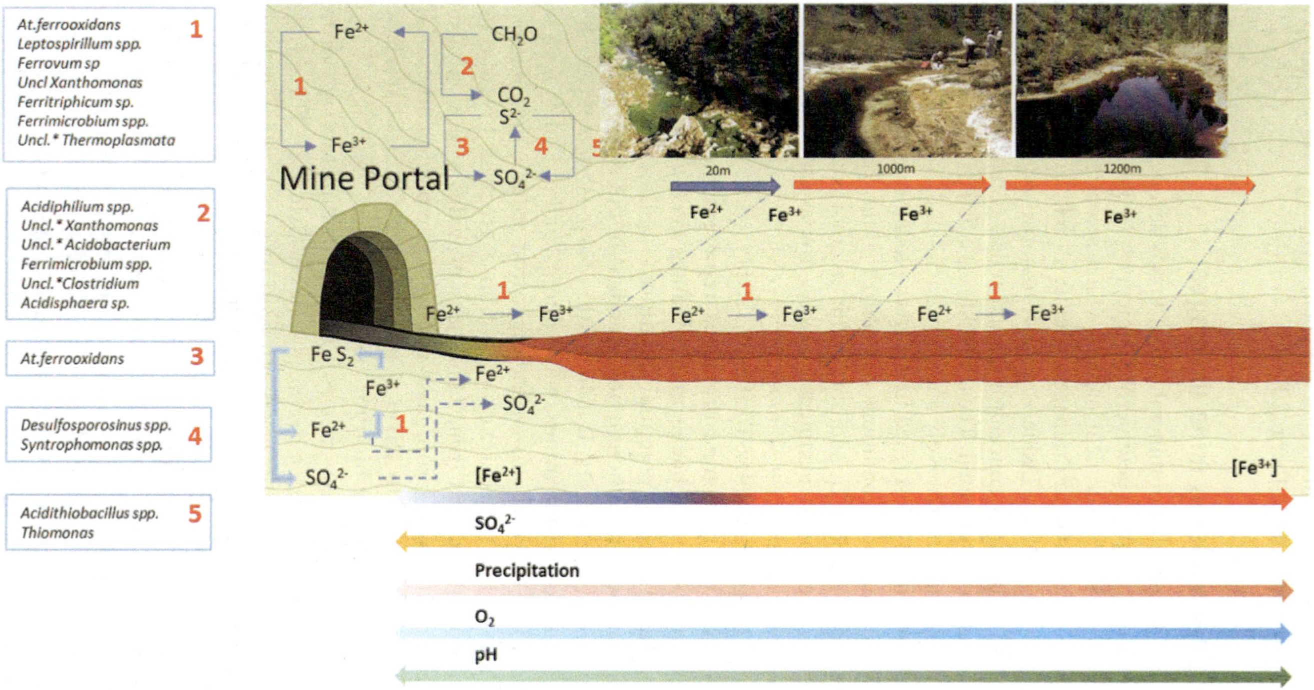

FIGURE 4.2 Geomicrobiological model of a typical acid mine drainage.

Acid mine drainages and effluents can often be explained by geomicrobiological models similar to that of Río Tinto. A remarkable feature of these environments is that, within just a few meters, all the reactions involved in the model can be observed. Inside the mine, oxygen concentration is low, and microorganisms are reducing iron, often coupled with sulfur oxidation. As a result, the water emerging from the mine portals is rich in ferrous iron, appears green in color, and is very transparent—making these places ideal habitats for acidophilic photosynthetic eukaryotes, which typically form dense mats of algae (*LZ1*). Conversely, the water, loaded with ferrous iron, when exposed to atmospheric oxygen, is rapidly oxidized by iron-oxidizing chemolithotrophic bacteria. Within just a few meters (*LZ2* and *LZ3*), the waters turn red. In the zones farther from the mine outlet, iron oxidation by chemolithotrophic bacteria continues, leading to an increase in sulfate concentration and the precipitation of iron hydroxides and sulfates, while photoautotrophs are nearly absent.

Geomicrobiological model of pit lakes

Similar processes to those occurring in mine discharges and drainage systems are also observed in the acidic lakes and reservoirs distributed throughout the Iberian Pyrite Belt (IPB). These environments likewise exhibit the complete geomicrobiological cycle previously described; however, rather than occurring horizontally, it unfolds vertically, since the water surface in these systems is aerobic and oxygen concentration decreases with depth. In general, a defining feature of acidic pit lakes is the frequent stratification of their iron- and sulfur-rich water columns in terms of pH, oxygen concentrations, and redox potentials. Surface zones tend to be oxic, highly acidic, and possess strongly positive redox potentials, whereas deeper zones are typically anoxic, with higher pH and lower redox potentials. Thus, in the upper layers of these acidic lakes, iron oxidation reactions are carried out by microorganisms such as *Acidithiobacillus* and *Leptospirillum*, while in the deeper layers, iron reduction coupled to sulfur oxidation is performed by microorganisms including *Acidithiobacillus*, *Ferrimicrobium*, *Acidiphilium*, *Acidocella*, and others, thereby completing the iron cycle (Fig. 4.3) (Santofimia et al., 2013).

These lakes may also exhibit a narrow chemocline as a transitional zone between these two layers. Most of the primary production in acidic lakes or pits is mediated by acidophilic microalgae, although chemolithotrophic bacteria also contribute to net CO_2 fixation. The availability of phosphate, which is closely linked to the reduction of iron oxides due to the low solubility of ferric phosphate, often limits primary production in such environments. This may result in certain apparent anomalies. For instance, in Cueva de la Mora pit lake (Jabugo, Huelva), algal presence and net primary production were observed to be concentrated just above the chemocline, at a depth of 7–8 m, where solar radiation was less than 2% of that available at the surface (López (Pamo et al., 2009). This phenomenon is attributed to increased phosphate bioavailability at that depth, due to its diffusion from the deeper anoxic zones and its release within the chemocline.

Acidic lakes in the IPB are less homogeneous than drainage waters or the river itself. Their physicochemical and geomicrobiological characteristics depend on the substrate in which they form, which determines the heavy metal load in their waters. In particular, iron concentration is the main factor influencing the geomicrobiology of a given lake. Some flooded pits, such as San Telmo (Cortegana, Huelva), have low iron concentrations, which prevents buffering of the water and precludes the occurrence of the same geomicrobiological cycle observed in the Rio Tinto. In contrast, other pits such as Nuestra Señora del Carmen (Paymogo, Huelva), Corta Atalaya (Minas de Río Tinto, Huelva) or Peña de Hierro (Nerva, Huelva), as well as the reservoirs located in the upper part of the river, exhibit high iron concentrations. In these waters, iron oxidation occurs rapidly at the surface and decreases with depth (Santofimia et al., 2013). The geomicrobiology of these types of lakes closely resembles that of the Río Tinto. Large macrofilaments, several meters in length, frequently appear at the surface. Molecular analysis of these filaments indicates that they are composed of facultatively anaerobic microorganisms (*At. ferrivorans*, *At.*

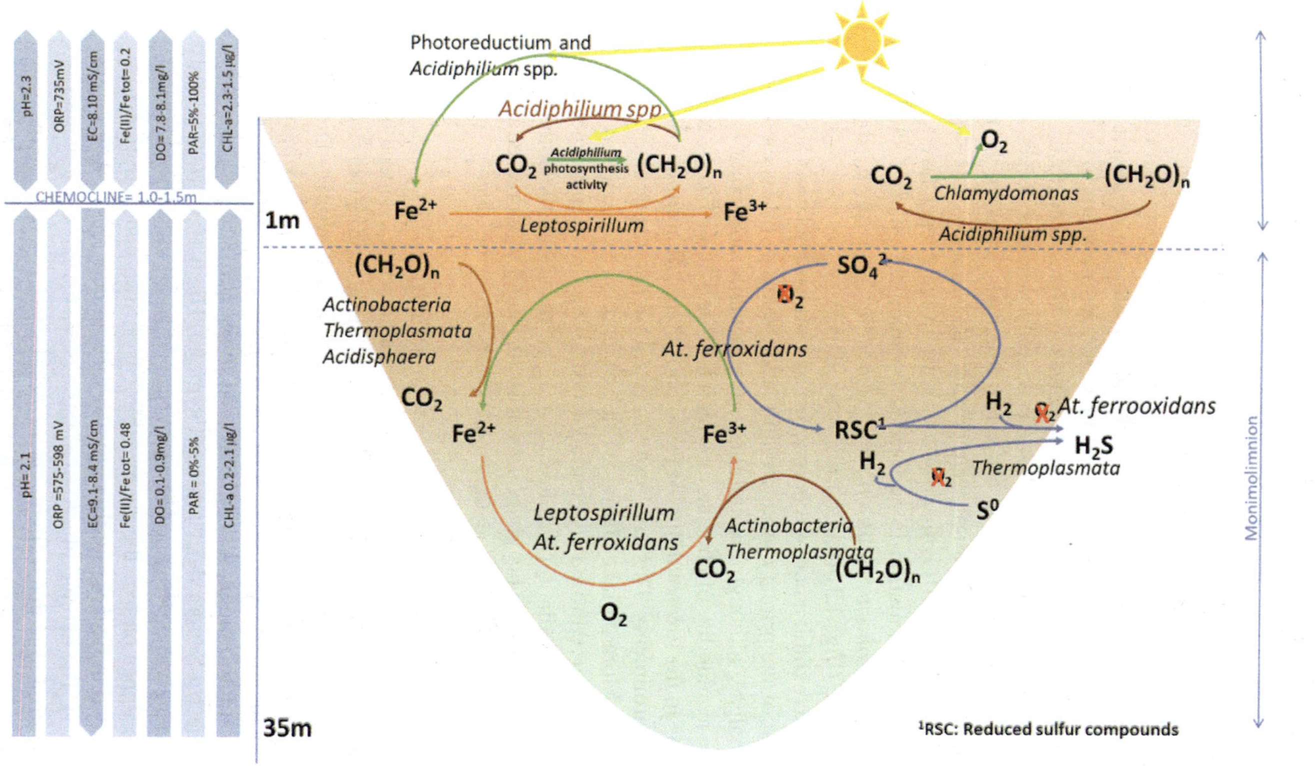

FIGURE 4.3 Geomicrobiological model of a typical open pit in the Iberian pyrite belt.

Among the more than 20 open pits or mining lakes found in the Iberian pyrite belt (IPB), those with high concentrations of ferric iron exhibit a model very similar to that of *Río Tinto*. The most remarkable aspect of these environments is that, within a single ecosystem, all reactions associated with the iron cycle in extreme acidic waters derived from the oxidation of polymetallic sulfides can be identified. At the surface of the lake, iron oxidation occurs through the activity of iron-oxidizing chemolithotrophic microorganisms. This is also the zone where photosynthetic microorganisms are found, primarily *Chlamydomonas*, which act as the primary producers of the system. Due to the dark red color of the waters, photosynthesis is limited to a shallow layer, as light penetration becomes nearly negligible beyond 20–30 cm. In the anaerobic zone, iron reduction takes place, regenerating ferrous iron, which becomes available again for oxidation. This reduction is often coupled to sulfur oxidation. The iron oxidation/reduction cycle creates a series of gradients, including a redox potential gradient (higher toward the surface), an electrical conductivity gradient (increasing with depth), and a color gradient: At the surface, the waters appear red due to ferric iron, while at depth they appear green due to ferrous iron. A marked stratification is often present, delineated by a chemocline typically found at depths of one–1.5 m

ferrooxidans, *Ferrimicrobium*), iron-reducing bacteria, and commonly also obligate anaerobes such as *Desulfosporosinus*. This suggests that the filaments form in the deeper regions of the lake and detach from the bottom once they reach considerable size, subsequently rising to the surface. With respect to the sulfur cycle, it is likely that sulfate reduction occurs in certain niches within lake sediments; however, this process is probably limited due to the inhibitory effect of low pH on this metabolism. As discussed in the river geomicrobiological model, it is most likely that microorganisms such as *Desulfosporosinus*, which are capable of both iron and sulfate reduction, are primarily engaged in iron reduction in these environments. Therefore, sulfate reduction appears to be neither prominent nor a major contributor to sulfur regeneration in these ecosystems (Fig. 4.3) (Santofimia et al., 2013).

Despite the inhibition of sulfate reduction in high-iron lakes, the diversity of lake types within the IPB is such that some cases have clearly demonstrated this phenomenon. One example is the Cantareras copper mine, where the upper zone of the lake is frequently filled with filaments down to a depth of approximately 15 cm (Pamo et al., 2009). These filaments consist of acidophilic algae growing on light brown bacterial filaments. Associated with these structures, a grey layer of copper sulfide can be observed. Most of the filament-associated bacteria are heterotrophs, many of which have been identified as iron-reducing. Phototrophic CO_2 fixation by microalgae within the filaments provides organic carbon for heterotrophic bacteria, fuelling dissimilatory ferric iron and sulfate reduction. This microbial ecosystem serves as a compelling example of natural attenuation of mine water pollution. Copper released from minerals is immobilized by sulfate-reducing bacteria present in the lower layer of the filaments. The contribution of acidophilic archaea to carbon and metal cycling in anoxic environments remains understudied, although the potential significance of these microbes in anoxic biofilms has been highlighted in studies of other acidic mine environments.

Role of eukaryotes in the geomicrobiological model of acidic waters

Although eukaryotic microorganisms do not directly participate in the geochemical transformations of iron and sulfur that lead to the dissolution of sulfide minerals and the formation of acid mine drainage, they nevertheless have a significant indirect impact on these geochemical processes (Barrie Johnson and Hallberg, 2008). For example, it has been observed that the presence of large biofilms of *Euglena mutabilis* can cause acid waters to become supersaturated with dissolved oxygen, reaching up to 200% (Brake et al., 2001), whereas in the absence of photosynthetic organisms, the waters draining underground mines and waste deposits are often devoid of dissolved oxygen, and soluble iron is predominantly present as reduced ferrous iron (Johnson, 2003; Johnson et al., 2012). Oxygenation of these waters by acidophilic algae facilitates the oxidation of ferrous iron and reduced sulfur compounds by bacteria such as *Acidithiobacillus* and *Leptospirillum* spp., in reactions that are net proton producing and may further acidify the environment downstream from the mine drainage outlet (Rowe et al., 2007).

A second way in which microalgae can influence the geochemical transformations in acid waters is by stimulating the growth of heterotrophic acidophiles that reduce iron and sulfur, through the release of organic exudates and lysates, since these waters generally contain very low concentrations of dissolved organic carbon (Das et al., 2009). Again, the presence of *Euglena* and *Chlorella* in these acidic environments significantly increases the concentration of dissolved organic carbon. Moreover, heterotrophic acidophilic bacteria such as *Acidiphilium* utilize sugars secreted by photosynthetic organisms—namely fructose, glucose, and mannitol of eukaryotic origin—helping to alleviate the nutrient scarcity typical of acidic waters. However, not all exudates produced by eukaryotes promote the growth of acidophilic bacteria; some may inhibit or even completely prevent their growth. A clear example of this is glycolate, produced and excreted as part of the metabolism of photosynthetic microorganisms in acidic waters. At acidic pH, glycolate exists as glycolic acid, which—along with acetic acid and many other low molecular weight aliphatic acids—is highly toxic to most acidophiles, with the exception of Firmicutes such as *Sulfobacillus* (Barrie Johnson and Hallberg, 2008). One of the traits of acidophilic microalgae is their ability to absorb, and presumably metabolize, glycolate that they had previously excreted. This capacity for heterotrophic growth among some microalgae has been well documented in other, non-extreme environments. For instance, *Chlorella pyrenoidosa* excretes glycolate under normal light conditions, which is then rapidly reabsorbed in the absence of light (Fogg et al., 1965; Miller et al., 1963)

The importance of acidophilic microalgae in sustaining heterotrophic bacterial populations in acidic environments has been demonstrated by comparing biofilms that grow within mines with those located outside in acidic drainage systems exposed to sunlight (Johnson, 2012). In subterranean sites, where primary production is exclusively mediated by chemolithotrophic acidophilic bacteria, biofilms are often dominated by autotrophic microbes such as *Ferrovum myxofaciens*. Conversely, in outdoor biofilms, it is the microalgae that foster the development of the underlying microbial communities. These latter communities are generally dominated by heterotrophic acidophiles, including many that catalyze the dissimilatory reduction of ferric iron, such as *Acidiphilium*, *Acidocella*, and *Acidobacterium* spp. Additionally, sulfidogenesis has been shown to be much more pronounced in the presence of acidophilic algae than when sulfate-reducing bacteria (SRB) are present alone. This supports the idea that acidophilic microalgae stimulate the growth of heterotrophic bacteria by providing organic substrates, and suggests that such interactions may underlie the strategies used by microorganisms to colonize new extreme environments.

In this regard, studies conducted in the Río Tinto on microbial colonization processes indicate that colonization begins with the accumulation of amorphous particles composed mainly of bacteria and inorganic mineral grains. After approximately 2 weeks, the organic matrix is also colonized by fungi and some heterotrophic eukaryotes, such as amebae and small flagellates. Diatoms only begin significant colonization in zones where fungal mycelia have already been established, whereas green flagellated microalgae such as *Dunaliella* or *Chlamydomonas*, as well as *Euglena*, appear during the earliest stages of biofilm formation, albeit in very low concentrations. Following the establishment of flagellated species, sessile microalgae such as *Chlorella* or *Cyanidium* emerge, and filamentous algae are generally the last to appear in biofilms. Once formed, these biofilms tend to remain relatively stable throughout the year, covering the same areas (Das et al., 2009; Aguilera et al., 2007)).

As with other eukaryotic species, the ecological role of fungi in acidic environments has not yet been adequately studied. It is believed that they may play a prominent role as primary decomposers of complex organic matter. It is important to note that acidic waters are typically poor in easily degradable organic carbon, such as that derived from leaf litter, as these environments lack invertebrates that actively shred plant material, thus facilitating organic carbon degradation. Under these conditions, fungi may become important primary decomposers of complex organic matter. At the same time, fungi contribute to oxygen consumption in the water, thereby reducing oxidative stress for sulfate-reducing bacteria, and they may also be directly involved in ferric iron or sulfur reduction, which are key electron transport processes in acidic waters, contributing to biological alkalinity (Aguilera et al., 2007).

AI disclosure

During the preparation of this work the author(s) used Chatgpt in order to correct the language. After using this tool/service, the author(s) reviewed and edited the content as needed and take(s) full responsibility for the content of the publication.

Acknowledgments

AGM received funding from the European Union's Horizon Europe Programme under Gran Agreement No. 101182278 (XTREAM). EGT was supported by the Spanish State Research Agency (AEI) of the Ministry of Science and Innovation under project REACTION (PID2023-146641NB-C21/AEI/10.13039/501100011033)

References

Aguilera, A., Manrubia, S.C., Gómez, F., Rodríguez, N., Amils, R., 2006. Eukaryotic community distribution and its relationship to water physicochemical parameters in an extreme acidic environment, Río Tinto (Southwestern Spain). Appl. Environ. Microbiol. 72 (8), 5325–5330. https://doi.org/10.1128/AEM.00513-06.

Aguilera, A., Souza-Egipsy, V., Gómez, F., Amils, R., 2007a. Development and structure of eukaryotic biofilms in an extreme acidic environment, río tinto (SW, Spain). Microb. Ecol. 53 (2), 294–305. https://doi.org/10.1007/s00248-006-9092-2.

Aguilera, A., Zettler, E., Gómez, F., Amaral-Zettler, L., Rodríguez, N., Amils, R., 2007b. Distribution and seasonal variability in the benthic eukaryotic community of Río Tinto (SW, Spain), an acidic, high metal extreme environment. Syst. Appl. Microbiol. 30 (7), 531–546. https://doi.org/10.1016/j.syapm.2007.05.003.

Albalat, R., Cañestro, C., 2016. Evolution by gene loss. Nat. Rev. Genet. 17 (7), 379–391. https://doi.org/10.1038/nrg.2016.39.

Amaral Zettler, L.A., Gómez, F., Zettler, E., Keenan, B.G., Amils, R., Sogin, M.L., 2002. Eukaryotic diversity in Spain's river of fire. Nature 417 (6885), 137. https://doi.org/10.1038/417137a.

Amaral-Zettler, L.A., Zettler, E.R., Theroux, S.M., Palacios, C., Aguilera, A., Amils, R., 2011. Microbial community structure across the tree of life in the extreme Río Tinto. ISME J. 5 (1), 42–50. https://doi.org/10.1038/ismej.2010.101.

Amaral-Zettler, L.A., 2013. Eukaryotic diversity at pH extremes. Front. Microbiol. 3. https://doi.org/10.3389/fmicb.2012.00441.

Anderson, I., Chertkov, O., Chen, A., Saunders, E., Lapidus, A., Nolan, M., Lucas, S., Hammon, N., Deshpande, S., Cheng, J.-F., Han, C., Tapia, R., Goodwin, L.A., Pitluck, S., Liolios, K., Pagani, I., Ivanova, N., Mikhailova, N., Pati, A., Palaniappan, K., Land, M., Pan, C., Rohde, M., Pukall, R., Göker, M., Detter, J.C., Woyke, T., Bristow, J., Eisen, J.A., Markowitz, V., Hugenholtz, P., Kyrpides, N.C., Klenk, H.-P., Mavromatis, K., 2012. Complete genome sequence of the moderately thermophilic mineral-sulfide-oxidizing firmicute Sulfobacillus acidophilus type strain (NALT). Standards in Genomic Sciences 6 (3), 293–303. https://doi.org/10.4056/sigs.2736042.

Barrie Johnson, D., Aguilera, A., 2019. Extremophiles and Acidic Environments Encyclopedia of Microbiology. Elsevier, United Kingdom, pp. 206–227. https://doi.org/10.1016/B978-0-12-809633-8.90687-3.

Barrie Johnson, D., Hallberg, K.B., 2008. Carbon, iron and sulfur metabolism in acidophilic micro-organisms. Adv. Microb. Physiol. 54, 201–255. https://doi.org/10.1016/S0065-2911(08)00003-9.

Brake, S.S., Dannelly, H.K., Connors, K.A., 2001. Controls on the nature and distribution of an alga in coal mine-waste environments and its potential impact on water quality. Environ. Geol. 40 (4–5), 458–469. https://doi.org/10.1007/s002540000181.

Cánovas, C.R., Olías, M., Nieto, J.M., Sarmiento, A.M., Cerón, J.C., 2007. Hydrogeochemical characteristics of the Tinto and Odiel Rivers (SW Spain). Factors controlling metal contents. Sci. Total Environ. 373 (1), 363–382. https://doi.org/10.1016/j.scitotenv.2006.11.022.

Coker, J.A., 2023. All about' extremophiles. Faculty Reviews 12. https://doi.org/10.12703/r/12-27.

Das, B.K., Roy, A., Koschorreck, M., Mandal, S.M., Wendt-Potthoff, K., Bhattacharya, J., 2009. Occurrence and role of algae and fungi in acid mine drainage environment with special reference to metals and sulfate immobilization. Water Res. 43 (4), 883–894. https://doi.org/10.1016/j.watres.2008.11.046.

Diaz, J.M., Plummer, S., 2018. Production of extracellular reactive oxygen species by phytoplankton: past and future directions. J. Plankton Res. 40 (6). https://doi.org/10.1093/plankt/fby039.

Dold, B., Gonzalez-Toril, E., Aguilera, A., Lopez-Pamo, E., Cisternas, M.E., Bucchi, F., Amils, R., 2013. Acid rock drainage and rock weathering in Antarctica: important sources for iron cycling in the southern ocean. Environ. Sci. Technol. 47 (12), 6129–6136. https://doi.org/10.1021/es305141b.

Fogg, G.E., Nalewajko, C., Watt, W.D., 1965. Extracellular products of phytoplankton photosynthesis. Proc. Roy. Soc. Lond. B Biol. Sci. 162 (989), 517–534. https://doi.org/10.1098/rspb.1965.0054.

García-Moyano, A., González-Toril, E., Aguilera, A., Amils, R., 2007. Prokaryotic community composition and ecology of floating macroscopic filaments from an extreme acidic environment, Río Tinto (SW, Spain). Syst. Appl. Microbiol. 30 (8), 601–614. https://doi.org/10.1016/j.syapm.2007.08.002.

García-Moyano, A., González-Toril, E., Aguilera, Á., Amils, R., 2012. Comparative microbial ecology study of the sediments and the water column of the Río Tinto, an extreme acidic environment. FEMS (Fed. Eur. Microbiol. Soc.) Microbiol. Ecol. 81 (2), 303–314. https://doi.org/10.1111/j.1574-6941.2012.01346.x.

García-Moyano, A., González-Toril, E., Moreno-Paz, M., Parro, V., Amils, R., 2008. Evaluation of Leptospirillum spp. in the Río Tinto, a model of interest to biohydrometallurgy. Hydrometallurgy 94 (1–4), 155–161. https://doi.org/10.1016/j.hydromet.2008.05.046.

García-Moyano, A., Austnes, A., Lanzén, A., González-Toril, E., Aguilera, Á., Øvreås, L., 2015. Novel and unexpected microbial diversity in acid mine drainage in svalbard (78° N), revealed by culture-independent approaches. Microorganisms 3 (4), 667–694. https://doi.org/10.3390/microorganisms3040667.

González-Toril, E., Águilera, A., Souza-Egipsy, V., Pamo, E.L., España, J.S., Amils, R., 2011. Geomicrobiology of La zarza-perrunal acid mine effluent (iberian pyritic belt, Spain). Appl. Environ. Microbiol. 77 (8), 2685–2694. https://doi.org/10.1128/AEM.02459-10Spain.

González-Toril, E., Llobet-Brossa, E., Casamayor, E.O., Amann, R., Amils, R., 2003. Microbial ecology of an extreme acidic environment, the Tinto River. Appl. Environ. Microbiol. 69 (8), 4853–4865. https://doi.org/10.1128/AEM.69.8.4853-4865.2003.

González-Toril, E., Aguilera, A., Rodriguez, N., Fernández-Remolar, D., Gómez, F., Diaz, E., García-Moyano, A., Sanz, J.L., Amils, R., 2010. Microbial ecology of Río Tinto, a natural extreme acidic environment of biohydrometallurgical interest. Hydrometallurgy 104 (3–4), 329–333. https://doi.org/10.1016/j.hydromet.2010.01.011.

Guo, X., Yin, H., Liang, Y., Qi, H., Zhou, X., Xiao, Y., Ma, L., Zhang, X., Qiu, G., Liu, X., Mormile, M.R., 2014. Comparative genome analysis reveals metabolic versatility and environmental adaptations of sulfobacillus thermosulfidooxidans strain ST. PLoS One 9 (6), e99417. https://doi.org/10.1371/journal.pone.0099417.

Hasan, M.K., Cheng, Y., Kanwar, M.K., Chu, X.Y., Ahammed, G.J., Qi, Z.Y., 2017. Responses of plant proteins to heavy metal stress—a review. Front. Plant Sci. 8. https://doi.org/10.3389/fpls.2017.01492.

Hirooka, S., Hirose, Y., Kanesaki, Y., Higuchi, S., Fujiwara, T., Onuma, R., Era, A., Ohbayashi, R., Uzuka, A., Nozaki, H., Yoshikawa, H., Miyagishima, S.-ya, 2017. Acidophilic green algal genome provides insights into adaptation to an acidic environment. Proc. Natl. Acad. Sci. 114 (39). https://doi.org/10.1073/pnas.1707072114.

Johnson, D.B., Aguilera, A., 2015. The Microbiology of Extremely Acidic Environments. American Society for Microbiology. https://doi.org/10.1128/9781555818821.ch4.3.1.

Johnson, D.B., 2003. Chemical and microbiological characteristics of mineral spoils and drainage waters at abandoned coal and metal mines. Water Air Soil Pollut. Focus 3 (1), 47–66. https://doi.org/10.1023/A:1022107520836.

Johnson, D.B., 2009. Extremophiles: Acidic Environments Encyclopedia of Microbiology, third ed. Elsevier, United Kingdom, pp. 107–126. https://doi.org/10.1016/B978-012373944-5.00278-9.

Johnson, D.B., 2012. Geomicrobiology of extremely acidic subsurface environments. FEMS (Fed. Eur. Microbiol. Soc.) Microbiol. Ecol. 81 (1), 2–12. https://doi.org/10.1111/j.1574-6941.2011.01293.x.

Johnson, D.B., Kanao, T., Hedrich, S., 2012. Redox transformations of iron at extremely low pH: fundamental and applied aspects. Front. Microbiol. 3. https://doi.org/10.3389/fmicb.2012.00096.

Johnson, D.B., Schippers, A., 2017. Editorial: recent advances in acidophile microbiology: fundamentals and applications. Front. Microbiol. 8, 428. https://doi.org/10.3389/fmicb.2017.00428.

Kashefi, K., Lovley, D.R., 2003. Extending the upper temperature limit for life. Science 301 (5635), 934. https://doi.org/10.1126/science.1086823.

Li, B., Chen, Y., Liu, Q., Hu, S., Chen, X., 2011. Complete genome analysis of sulfobacillus acidophilus strain TPY, isolated from a hydrothermal vent in the pacific ocean. J. Bacteriol. 193 (19), 5555–5556. https://doi.org/10.1128/jb.05684-11.

López-Archilla, A.I., Marin, I., Amils, R., 2001. Microbial community composition and ecology of an acidic aquatic environment: the Tinto River, Spain. Microb. Ecol. 41 (1), 20–35. https://doi.org/10.1007/s002480000044.

Madigan, M., Buckley, Bender, K., Sattley, W., 2021. Stahl, Brock Biology of Microorganisms, Global Edition. Pearson Global Editions, p. 16.

Malki, M., González-Toril, E., Sanz, J.L., Gómez, F., Rodríguez, N., Amils, R., 2006. Importance of the iron cycle in biohydrometallurgy. Hydrometallurgy 83 (1–4), 223–228. https://doi.org/10.1016/j.hydromet.2006.03.053.

Matsuzaki, M., Misumi, O., Shin-i, T., Maruyama, S., Takahara, M., Miyagishima, S.-ya, Mori, T., Nishida, K., Yagisawa, F., Nishida, K., Yoshida, Y., Nishimura, Y., Nakao, S., Kobayashi, T., Momoyama, Y., Higashiyama, T., Minoda, A., Sano, M., Nomoto, H., Oishi, K., Hayashi, H., Ohta, F., Nishizaka, S., Haga, S., Miura, S., Morishita, T., Kabeya, Y., Terasawa, K., Suzuki, Y., Ishii, Y., Asakawa, S., Takano, H., Ohta, N., Kuroiwa, H., Tanaka, Kan, Shimizu, N., Sugano, S., Sato, N., Nozaki, H., Ogasawara, N., Kohara, Y., Kuroiwa, T., 2004. Genome sequence of the ultrasmall unicellular red alga Cyanidioschyzon merolae 10D. Nature 428 (6983), 653–657. https://doi.org/10.1038/nature02398.

Miller, R.M., Meyer, C.M., Tanner, H.A., 1963. Glycolate excretion & uptake by Chlorella. Plant Physiol. 38 (2), 184–188. https://doi.org/10.1104/pp.38.2.184.

Ňancucheo, I., Johnson, D.B., 2012. Selective removal of transition metals from acidic mine waters by novel consortia of acidophilic sulfidogenic bacteria. Microb. Biotechnol. 5 (1), 34–44. https://doi.org/10.1111/j.1751-7915.2011.00285.x.

Nieto, J.M., Sarmiento, A.M., Canovas, C.R., Olias, M., Ayora, C., 2013. Acid mine drainage in the Iberian Pyrite Belt: 1. Hydrochemical characteristics and pollutant load of the Tinto and Odiel rivers. Environ. Sci. Pollut. Control Ser. 20 (11), 7509–7519. https://doi.org/10.1007/s11356-013-1634-9.

Olsson, S., Penacho, V., Puente-Sánchez, F., Díaz, S., Gonzalez-Pastor, J.E., Aguilera, A., 2017. Horizontal gene transfer of phytochelatin synthases from bacteria to extremophilic green algae. Microb. Ecol. 73 (1), 50–60. https://doi.org/10.1007/s00248-016-0848-z.

Olsson, S., Puente-Sánchez, F., Gómez, M.J., Aguilera, A., 2015. Transcriptional response to copper excess and identification of genes involved in heavy metal tolerance in the extremophilic microalga Chlamydomonas acidophila. Extremophiles 19 (3), 657–672. https://doi.org/10.1007/s00792-015-0746-1.

Pamo, Sánchez-España, J., Díez-Ercilla, M., Santofimia, E., Reyes, J., 2009. Cortas mineras inundadas de la Faja Pirítica: inventario e hidroquímica, 13. IGME.

Price, P.B., 2000. A habitat for psychrophiles in deep Antarctic ice. Proc. Natl. Acad. Sci. 97 (3), 1247–1251. https://doi.org/10.1073/pnas.97.3.1247.

Puente-Sánchez, F., Díaz, S., Penacho, V., Aguilera, A., Olsson, S., 2018. Basis of genetic adaptation to heavy metal stress in the acidophilic green alga Chlamydomonas acidophila. Aquat. Toxicol. 200, 62–72. https://doi.org/10.1016/j.aquatox.2018.04.020.

Puente-Sánchez, F., Olsson, S., Gómez-Rodriguez, M., Souza-Egipsy, V., Altamirano-Jeschke, M., Amils, R., Parro, V., Angeles, A., 2016. Solar radiation stress in natural acidophilic biofilms of Euglena mutabilis revealed by metatranscriptomics and PAM Fluorometry. Protist 167 (1), 67–81. https://doi.org/10.1016/j.protis.2015.12.003.

Qiu, H., Price, D.C., Weber, A.P.M., Reeb, V., Chan Yang, E., Lee, J.M., Kim, S.Y., Yoon, H. S., Bhattacharya, D., 2013. Adaptation through horizontal gene transfer in the cryptoendolithic red alga Galdieria phlegrea. Curr. Biol. 23 (19), R865. https://doi.org/10.1016/j.cub.2013.08.046.

Rothschild, L.J., Mancinelli, R.L., 2001. Life in extreme environments. Nature 409 (6823), 1092–1101. https://doi.org/10.1038/35059215.

Rowe, O.F., Sánchez-España, J., Hallberg, K.B., Johnson, D.B., 2007. Microbial communities and geochemical dynamics in an extremely acidic, metal-rich stream at an abandoned sulfide mine (Huelva, Spain) underpinned by two functional primary production systems. Environ. Microbiol. 9 (7), 1761–1771. https://doi.org/10.1111/j.1462-2920.2007.01294.x.

Santofimia, E., González-Toril, E., López-Pamo, E., Gomariz, M., Amils, R., Aguilera, Á., Balcazar, J.L., 2013. Microbial diversity and its relationship to physicochemical characteristics of the water in two extreme acidic pit lakes from the iberian pyrite belt (SW Spain). PLoS One 8 (6), e66746. https://doi.org/10.1371/journal.pone.0066746.

Schiraldi, C., De Rosa, M., 2002. The production of biocatalysts and biomolecules from extremophiles. Trends Biotechnol. 20 (12), 515–521. https://doi.org/10.1016/s0167-7799(02)02073-5.

Schmid, A.K., Allers, T., DiRuggiero, J., 2020. SnapShot: microbial extremophiles. Cell 180 (4), 818. https://doi.org/10.1016/j.cell.2020.01.018.

Schönknecht, G., Chen, W.H., Ternes, C.M., Barbier, G.G., Shrestha, R.P., Stanke, M., Bräutigam, A., Baker, B.J., Banfield, J.F., Garavito, R.M., Carr, K., Wilkerson, C., Rensing, S.A., Gagneul, D., Dickenson, N.E., Oesterhelt, C., Lercher, M.J., Weber, A.P. M., 2013. Gene transfer from bacteria and archaea facilitated evolution of an extremophilic eukaryote. Science 339 (6124), 1207–1210. https://doi.org/10.1126/science.1231707.

Selkov, E., Overbeek, R., Kogan, Y., Chu, L., Vonstein, V., Holmes, D., Silver, S., Haselkorn, R., Fonstein, M., 2000. Functional analysis of gapped microbial genomes: amino acid metabolism of Thiobacillus ferrooxidans. Proc. Natl. Acad. Sci. 97 (7), 3509–3514. https://doi.org/10.1073/pnas.97.7.3509.

Soucy, S.M., Huang, J., Gogarten, J.P., 2015. Horizontal gene transfer: building the web of life. Nat. Rev. Genet. 16 (8), 472–482. https://doi.org/10.1038/nrg3962.

Souza-Egipsy, V., González-Toril, E., Zettler, E., Amaral-Zettler, L., Aguilera, A., Amils, R., 2008. Prokaryotic community structure in algal photosynthetic biofilms from extreme acidic streams in Río Tinto (Huelva, Spain). Int. Microbiol. 11 (4), 251–260. https://doi.org/10.2436/20.1501.01.69.

Souza-Egipsy, V., Altamirano, M., Amils, R., Aguilera, A., 2011. Photosynthetic performance of phototrophic biofilms in extreme acidic environments. Environ. Microbiol. 13 (8), 2351–2358. https://doi.org/10.1111/j.1462-2920.2011.02506.x.

Travisany, D., Di Genova, A., Sepúlveda, A., Bobadilla-Fazzini, R.A., Parada, P., Maass, A., 2012. Draft genome sequence of the sulfobacillus thermosulfidooxidans cutipay strain, an indigenous bacterium isolated from a naturally extreme mining environment in northern Chile. J. Bacteriol. 194 (22), 6327–6328. https://doi.org/10.1128/JB.01622-12Chile.

Ullrich, S.R., González, C., Poehlein, A., Tischler, J.S., Daniel, R., Schlömann, M., Holmes, D.S., Mühling, M., 2016. Gene loss and horizontal gene transfer contributed to the genome evolution of the extreme acidophile "ferrovum. Front. Microbiol. https://doi.org/10.3389/fmicb.2016.00797.

Valdes, J., Quatrini, R., Hallberg, K., Dopson, M., Valenzuela, P.D.T., Holmes, D.S., 2009. Draft genome sequence of the extremely acidophilic bacterium Acidithiobacillus caldus ATCC 51756 reveals metabolic versatility in the genus Acidithiobacillus. J. Bacteriol. 191 (18), 5877–5878. https://doi.org/10.1128/JB.00843-09Chile.

CHAPTER

Microbial diversity of the anoxic sediments of Río Tinto

5

Irene Sánchez-Andrea
Science & Technology School, IE University, Segovia, Spain

Anoxic zones in Rio Tinto

Location and geochemistry of anoxic zones of the Tinto river

Based on geo-edaphological and climatic criteria, the Tinto River channel, from its origin in Peña de Hierro to its mouth in the city of Huelva, is divided into three main zones: upper course or northern zone, middle course or transition zone, and lower course or estuary (Rufo et al., 2007).

A big proportion of the work performed at Tinto River -as in other acid mine drainage (AMD) environments - focused on the description of the **oxic part** on the northern zone (from Peña de Hierro to La Palma del Condado) where the river waters exhibit the typical characteristics of the Tinto: high concentrations of ferric ion in solution and an acidic pH close to 2.3. Performed research includes either the microbial ecology of the water column (García-Moyano et al., 2007, 2008, 2012; Garrido et al., 2008; Gónzalez-Toril et al., 2003; López-Archilla et al., 2001; Souza-Egipsy et al., 2008), the diversity associated with macroscopic growths (García-Moyano et al., 2007) or the eukaryotic diversity (Aguilera et al., 2006; 2007a; 2007b).

The bed of the Tinto River is rocky. However, all around the basin we can find anoxic areas, both in sediments and in the stratified water column of acidic pit lakes (APTs) dispersed around the Iberian Pyrite Belt (IPB). These APTs, a legacy of mining activities, form in open-pit mines when mining ceases and dewatering stops, allowing groundwater to refill the pits at a rate controlled by local hydrogeology. These lakes exhibit a wide range of sizes, shapes, water chemistry, and ecological states due to variations in mine types (e.g., coal, metal, and hard-rock quarries) and geological settings. APLs can be considered unconventional AMD systems where depth becomes a critical factor, defining different ecological niches or biogeochemical zones. Some lakes display anoxic and reductive conditions at certain depths (van der Graaf et al., 2020). There has been an exhaustive review of research conducted in APLs in the IPB, which provides detailed information (Sánchez-España, 2024).

Besides APLs, several in-depth studies targeting anoxic zones identified anoxic conditions in sediments. Some sediment accumulations along the river iron gradient

Geomicrobiology of Río Tinto. https://doi.org/10.1016/B978-0-443-32948-7.00035-7

were detected (García-Moyano et al., 2012), and the upper 2 cm layer was sampled. An extensive search located other anoxic sediment locations around the basin (Sánchez-Andrea, 2012). In addition, a couple of dams reduce the flow rate to an extent that a proper sediment accumulation and zonation is possible, presenting physicochemical contrasting parameters in their waters and sediments (Sánchez-Andrea et al., 2011, 2012b).

This chapter will focus on **anoxic sediment zones** for which the coordinates were published. Those run through the northern zone of Tinto river (Fig. 5.1a): *TUN* (37.725685N,6.55634W), *SN* (37.72173N,6.557465W), *UMA* (37.716961, 6.554031W), and *AG* (37.724722N,6.56W) in the origins of the river; their union close to Nerva: *CEM* (37.699238,6.557379W), *JL* (37.691207N, 6.560587W) and at the middle course: *EST* (37.68832N,6.563773W), *SAL* (37.669622N, 6.550212W), and *BER* (37.601346 N, 6.549933W).

Near the origin, at the exit of the tunnel of the Peña de Hierro mine, there is the point called Tunnel (TUN), with ferrous waters with low oxygen content. Along its course, the water becomes oxygenated, and the mineral components oxidize until reaching Anabel's garden (AG), where the water already presents the typical characteristics of the Tinto, with low pH and high iron content in its oxidized form. This stream converges with another where Campo de Galdierias (GAL) is located, and downstream there is a dam named 2.9 (2.9). On the other hand, to the east of Peña de Hierro, three streams converge where the SN, UMA, and IZQ dams are located. Both groups of streams converge downstream, joining at a point near the cemetery (CEM). A few meters further, near the municipality of Nerva, there is a dam whose sampling point is known as JL, where the effluent from the municipal wastewater treatment plant of Nerva is discharged, and a few meters later is the sampling point station (EST). From this area, two stations were sampled: Salinas (SAL), named after the construction of evaporation ponds intended for copper recovery on the river terraces, and Berrocal (BER), near the town of the same name.

The Tinto River is considerably homogeneous in terms of parameters such as pH and redox potential, showing acidic and oxidizing conditions along its course, thanks to the buffering effect of ferric iron. However, the addition of neutral tributaries generates a decreasing gradient in metal concentration (especially iron) and, consequently, also a gradient in conductivity, which is evident—in the sampled areas—in a gradual decrease from the origins to the BER sampling point (Sánchez-Andrea, 2012). Certain sampling points stand out, such as TUN, near the origin, where anoxic conditions have not yet generated the extreme characteristics that will appear downstream, showing low redox potential, low sulfate content, and low conductivity; UMA, where the lowest pH values and the highest redox potential, conductivity, and arsenic values appear; or CEM, with the highest values of Ni, Cu, and Zn, possibly due to the influence of a large nearby gossan deposit.

The fact that the bed of the Tinto River is rocky means that most sampling points lack a real sediment structure: they consist of mineral particles of a certain size, such as sands and gravel, with high porosity. Their biogeochemical characteristics are strongly influenced by those of the water column. For instance, most of the

FIGURE 5.1

(a) Map of Tinto river sample locations with detected anoxic zones; (b) pictures of SN dam and a sediment core extracted; (c) pictures of JL dam and a sediment core extracted.

sediments collected in (Sánchez-Andrea, 2012) presented a pH range between 1.6 and 2.8, and the redox potential (E) ranged between 351.9 and 482.8 mV, positive and high, due to the presence of the strong oxidant ferric iron (Fe^{3+}).

Conductivity values vary widely and are directly related to metal concentration, particularly iron. Iron concentration decreases along the river course. Neutral tributaries added to the river dilute metal concentrations and indirectly influence the precipitation of ferric oxides, releasing protons and maintaining a constant pH. The highest values are recorded at *UMA* (8.9 g L^{-1} of iron and conductivity of 24.7 mS cm^{-2}), while the lowest are recorded at *BER* (0.3 g L^{-1} of iron and conductivity of 2.8 mS cm^{-2}). Sulfate concentrations measured along the river positively correlate with iron concentration, which is consistent with both being products of metallic sulfide oxidation.

Volatile fatty acids and nitrogen species such as nitrite and nitrate were detected in the Tinto water column (Table 5.1). These findings have ecological implications, indicating that substrates and intermediates for denitrification and carbon cycle-related metabolism are present.

In contrast with these sediments lacking structure, in the aforementioned small dams, the water rarely reaches a meter in depth, and this allows a stratified accumulation of sediments with smaller particle sizes, mainly silts and clays. These areas include the JL and SN points (Fig. 5.1b and c), where the lower porosity and stratification of the sediments allow differentiation of their physicochemical characteristics compared to those of the water column, resulting in true anoxic conditions. While in the SN dam the sediments appear more homogeneous, both spatially and temporally, as seen in Fig. 5.1b, in the JL dam the stratification has been repeatedly observed, although the location, thickness, and parameters of the strata vary over time (Fig. 5.1c).

In the SN sediments, the lowest pH recorded is 2.3 at 25 cm depth in Sánchez-Andrea et al. (2012a), while the highest reaches 4.0 at 40–45 cm in

Table 5.1 Physicochemical parameters in (mg/L) of water measured across samples. Unpublished data obtained in Chapter 6 of (Sánchez-Andrea, 2012).

Sample	Acetate	Propionate	Formate	Nitrite	Nitrate	Sulfate
BER	0.044	0	0.025	0.256	0.252	558.02
EST	0.177	0	0.025	0.284	0.397	2719.17
CEM	0.796	0.164	0	0.27	0.621	3473.5
SAL	0.044	0	0.027	0.438	0.349	2182.51
UMA	0	3.288	0	0.467	0	7160.53
TUN	0.068	0.1	0	0.12	0	562.9
AG	0.043	0	0.028	0.098	0	733.236
JL	0.469	0	0.102	0.103	0.182	1341.1

Sánchez-Andrea et al. (2011). These values reflect the more acidic and chemically homogeneous nature of the SN site, which is closer to the acidic water column and shows less vertical variability than JL. The redox potential at SN varies from a minimum of −13 mV, deep layer in Sánchez-Andrea et al. (2011), to a maximum of +312 mV, surface layer in Sánchez-Andrea et al. (2012a), underscoring its more oxidizing and iron-rich profile compared to JL.

When integrating all available data from the referenced studies, including those in Sánchez-Andrea et al. (2011, 2012b, 2012c); Sánchez-Andrea (2012), in the JL sediments, pH values range from a minimum of 3.6 at 5 cm depth (Sánchez-Andrea, 2012), to a maximum of 5.7 at 33 cm depth in Sánchez-Andrea et al. (2012a), indicating a relatively wide vertical gradient of acidity in this location. This variation reflects changes in microbial activity and chemical conditions with depth. The redox potential (E) in JL also spans a broad range, from a minimum of −141 mV at 45 cm depth (Sánchez-Andrea, 2012), to a maximum of +200 mV at 7 cm depth in Sánchez-Andrea et al. (2012a), though more typical values observed fall between −132 mV and +168 mV across multiple campaigns. These values reflect a clear stratification from oxidizing to reducing conditions, driven by processes like iron and sulfate reduction deeper in the sediment. In JL, two main zones generally appear: one brownish with pH values around 4 and oxidizing potential (up to 168 mV) and another blackish zone with pH values around 5.5 and reducing redox potentials (up to −246 mV) as can be observed in Fig. 5.1c.

Microbial diversity of the anoxic zones

The diversity found in the sediments of the Tinto River has proven to be much greater than that found in the water column in previous studies (Gónzalez-Toril et al., 2003). The 16SrRNA genes sequences analysed in this study correspond to the GenBank database accession numbers JQ815571-JQ815874 (Sánchez-Andrea, 2012). Overall, sequences belonging to 12 different bacterial phyla have been directly identified from sediments: *Pseudomonadota (classes Alpha-, Beta-, Gamma-, and Delta-), Bacillota, Bacteroidota, Acidobacteriota, Actinomycetota, Synergistota, Cyanobacteriota* (chloroplast-derived reads*), Saccharibacteria, Chloroflexota, Nitrospirota, Verrucomicrobiota,* and *Fibrobacterota*. The sequences have been phylogenetically affiliated with 72 different genera.

Unstructured sediments

Previous studies of the Tinto Sediments (García-Moyano et al., 2012; Sánchez-Andrea et al., 2011) have consistently demonstrated low archaeal diversity, with all sequences identified belonging to the *Thermoplasma* class. In an exhaustive analysis of bacterial diversity in the superficial sediments (Sánchez-Andrea, 2012), over 714 16S rRNA gene sequences clustered in 176 OTUs based on 97% sequence similarity. Between the 176 OTUs, 61 genera were identified, falling in 12 different phyla: *Cloacibacillus* (phylum Synergistota, formerly *Synergistetes*); chloroplast-derived reads from *Bacillariophyta*; *Saccharibacteria* (formerly "TM7 incertae sedis"); the family *Anaerolineaceae* within phylum *Chloroflexota* (formerly *Chloroflexi*); *Leptospirillum* (phylum *Nitrospirota*, formerly *Nitrospira*); *Prosthecobacter* (phylum *Verrucomicrobiota*, formerly *Verrucomicrobia*); *Fibrobacter* (phylum *Fibrobacterota*); a broad *Bacilli* assemblage—*Alkalibacterium*, *Bacillus* sensu stricto (many former species now placed in *Alkalihalobacillus*, *Peribacillus*, and related genera), *Paenibacillus*, *Oceanobacillus*, *Staphylococcus*, *Tumebacillus*, *Alicyclobacillus*—plus *Turicibacter* (class *Erysipelotrichia*), all now assigned to the phylum *Bacillota* (formerly *Firmicutes*); low-pH Clostridia including *Thermodesulfobium* and the former *Desulfotomaculum* group, now split into *Desulfallas*, *Desulfofundulus*, *Desulfofarcimen* and *Desulfohalotomaculum*; actinobacterial genera *Olsenella*, *Ferrimicrobium*, *Ferrithrix*, *Aciditerrimonas*, *Rothia*, *Brevibacterium*, *Leucobacter* and *Cutibacterium* (formerly *Propionibacterium* in skin isolates); alphaproteobacteria *Acidiphilium* and *Sphingomonas*; betaproteobacteria *Methylotenera*, *Aquabacterium*, *Ideonella*, *Acidovorax*, *Zoogloea* (type species now *Shinella zoogloeoides*), *Ferribacterium* and *Dechloromonas*; gammaproteobacteria *Legionella*, *Steroidobacter*, *Pseudomonas*, *Cellvibrio*, *Enhydrobacter*, *Acinetobacter*, and *Acidithiobacillus*; deltaproteobacteria *Geobacter*, *Desulfurella*, *Desulfosalsimonas* and *Desulfobulbus*; bacteroidotal genera *Flavobacterium*, *Fluviicola*, *Haliscomenobacter*, *Flectobacillus*, *Leadbetterella*, *Pedobacter* and *Microbacter*; additional *Anaerolineaceae* representatives (*Chloroflexota*); and acidobacterial classes *Blastocatellia* and *Acidobacteriia* (formerly subdivisions "Gp1" and "Gp3"). The distribution of the various phyla differed greatly across the sample sites.

Structured sediments

In the proper sediment accumulations, phylogenetic analyses revealed distinct microbial communities inhabiting the JL and SN sediment sites of the Tinto River (Sánchez-Andrea et al., 2011) study. At the JL site, a high taxonomic and functional diversity was observed, reflecting its stratified geochemical profile. In surface layers, *Acidobacteriota* dominated, whereas deeper, more-reduced zones contained many anaerobic and sulfur-cycling bacteria. Genera detected included *Alicyclobacillus*, *Bacillus*, *Clostridium* sensu stricto, *Desulfurella*, *Desulfosporosinus*, *Acididesulfobacillus*, *Sporacetigenium*, *Syntrophobacter* and *Microbacter* lineages. Other *Bacillota*—many involved in sulfur or metal cycling—included *Staphylococcus*, *Lysinibacillus*, *Cohnella*, *Sulfobacillus*, *Thermodesulfobium*, and *Veillonella*. Additional members of the phylum *Pseudomonadota* were present: *Acidiphilium* and *Pseudochrobactrum* (class *Alphaproteobacteria*); *Alcaligenes* and *Comamonas* (class *Betaproteobacteria*); and *Acidithiobacillus*, *Psychrobacter* and uncultured *Sinobacteraceae* representatives (class *Gammaproteobacteria*).

In contrast, the SN site, which is more acidic and geochemically homogeneous, showed a narrower range of taxa. In the surface layer, *Acidithiobacillus* spp. dominated, reflecting iron oxidation activity in the oxidized acidic zone. In the deeper layer, the sulfate-reducing genus *Desulfosporosinus* was detected. Some genera, such as *Acidiphilium*, were present at both sites, indicating their ecological flexibility across redox and pH gradients. Archaea from both JL and SN sites belonged exclusively to the order *Thermoplasmatales* (phylum *Thermoplasmatota*), regardless of depth, suggesting a conserved archaeal community adapted to the extreme environmental conditions of the Tinto River sediments.

Microbial niche differentiation across sediment depths

The Tinto River exhibits strong microbial niche differentiation across its sediment profile, particularly when comparing surface, larger-grained sediment accumulations to deeper deposits retained by artificial dams. In superficial accumulations, where physicochemical characteristics closely resemble those of the overlying water column, the microbial communities are similarly dominated by oxic habitat taxa. Common iron oxidizers such as the heterotrophic actinobacterium *Ferrimicrobium*, the strict aerobic lithotroph *Leptospirillum*, and the facultative anaerobe *Acidithiobacillus ferrooxidans* are frequently encountered. These coexist with strict aerobic heterotrophs like *Pseudomonas*, *Pedobacter*, and *Flectobacillus*. Despite the primarily aerobic environment, early colonization by facultative and obligate anaerobes is evident, with fermenters such as *Clostridium* spp. and sulfate-reducing bacteria (SRB) like *Desulfosporosinus* and *Desulfobulbus* also detected.

In contrast, sediment accumulations retained by dams (e.g., SN and JL) display well-developed vertical stratification, supporting increasingly anoxic and chemically reduced conditions with depth. The upper sediment layers in these zones serve as transitional interfaces, where aerobes such as *A. ferrooxidans* and *L. ferrooxidans*—well-adapted to high ionic strength and oxidative stress (Parro et al., 2007)—remain active. However, beneath these, a complex community unfolds, engaging in tightly coupled biogeochemical cycling of iron, sulfur, nitrogen, and carbon.

As redox potential declines with depth, a predictable sequence of anaerobic metabolisms emerges. The first to dominate are ferric iron reducers such as *A. ferrooxidans*, *Sulfobacillus*, *Geobacter*, *Acidiphilium*, and members of *Acidobacteria*, which contribute to redox buffering by reducing Fe(III) to Fe(II). As the environment becomes less oxidizing, bacterial taxa capable of nitrate reduction, such as *Dechloromonas*, *Propionibacterium*, *Pseudomonas*, and *Bacillus* become active, initiating denitrification processes that further increase pH and support milder conditions (Sánchez-Andrea et al., 2012c). Subsequently, sulfate-reducing bacteria—including *Desulfosporosinus*, *Desulfotomaculum*, *Desulfobulbus*, and *Thermodesulfobium* and the sulfur reducer *Desulfurella* become dominant, driving significant pH increases and redox decline. In the deepest, most reduced layers, low-energy-yield methanogenesis occurs, with methanogens such as *Methanosaeta* spp. identified via DGGE in Sánchez-Andrea et al. (2011) and in Sanz et al. (2011a).

This stratification fosters the development of distinct ecological niches. At the SN dam, located at the river's acidic headwaters (pH ~2.5, E ~+400 mV), iron remains highly soluble, making it the dominant electron acceptor. Accordingly, iron-reducing microorganisms such as *A. ferrooxidans* and *Acidiphilium* spp. predominate, regenerating ferrous iron and maintaining an oxidized, low-pH environment even within the sediment. Conversely, the JL dam, located ~5 km downstream, presents a more heterogeneous profile due to seasonal hydrological fluctuations, input from neutral tributaries, and increased organic matter—partly influenced by effluent from the Nerva wastewater treatment plant. This leads to lower iron concentrations and elevated organic carbon availability, favoring anaerobic processes like fermentation and sulfate reduction. As a result, JL exhibits a broader range of microbial taxa, from *Acidobacteria* in its oxidized layers to *Microbacter* and *Clostridium* in more reduced horizons, culminating in sulfate reducers such as *Desulfosporosinus* and sulfur reducers such as *Desulfurella* in the deepest, most reduced layers.

Evidence from cloning, DGGE, and CARD-FISH analyses highlights the stark contrast between the microbial communities at SN and JL, both in taxonomic diversity and abundance. These differences closely reflect their respective geochemical settings. At SN, the continuous acidic flow and high ferric concentrations buffer both water and sediment, supporting a stable, iron-centered microbial ecology. In contrast, JL sediments exhibit clear banding and heterogeneity, consistent with episodic changes in flow and redox conditions. Here, microbially mediated processes—particularly sulfate reduction—may originate in microniches and expand over time, creating broader zones of reduced pH and redox potential and contributing to the complex spatial structure of the microbial community.

Isolates from the Tinto River sediments

A systematic effort was done to uncover the sulfate reduction communities in acidic conditions using sediment samples from the Tinto River basin in southwestern Spain (Sánchez-Andrea et al., 2013). The initial enrichments, carried out at pH values ranging from 4.0 to 6.5 and with various electron donors including glycerol, lactate, methanol, succinate, and hydrogen, demonstrated active sulfate reduction even at the lowest tested pH 4 and yielded diverse consortia of fermentative and sulfate-reducing bacteria (SRB). Taxonomic analyses revealed the dominance of *Bacillota*, especially the genera *Desulfosporosinus*, a new taxon that was isolated and described as *Acididesulfobacillus acetoxydans* (Sánchez-Andrea et al., 2022), and *Desulfotomaculum* old cluster, as well as fermenters such as the latter isolated and described *Microbacter margulisiae* and *Oscillibacter*, highlighting the complexity and syntrophic nature of microbial communities in these sediments.

From these enrichments, a number of novel isolates were obtained and described in subsequent studies. *Desulfosporosinus acididurans*, described by Sánchez-Andrea et al. (2015), was the first acidophilic SRB isolated from Tinto River sediments and demonstrated optimal growth at pH 5.5, with the capacity to utilize a wide range of organic electron donors and acceptors, including sulfate, thiosulfate, ferric iron, and elemental sulfur. This strain clustered phylogenetically with *Desulfosporosinus* spp. but displayed sufficient physiological and genetic divergence to warrant species-level distinction. In parallel, a propionate-producing fermenter present in low abundance in the enrichments was isolated and described as *Microbacter margulisiae* (Sánchez-Andrea et al., 2014). Although not capable of sulfate reduction, this bacterium coexisted synergistically with SRB and likely contributed to carbon flux within the community.

Further diversification was observed in enrichments using succinate, which led to the discovery of *Lucifera butyrica*, a spore-forming fermenter described in Sánchez-Andrea et al. (2018). *L. butyrica* exhibited the ability to ferment glycerol and a wide variety of sugars, producing acetate and 1,3-propanediol, and was shown to stimulate sulfidogenesis when cocultured with the sulfur-reducing bacterium *Desulfurella amilsii*. The latter, isolated in a dedicated effort to enrich sulfur-reducing microbes, was described as a novel acidotolerant species within the *Desulfurella* genus by Florentino et al. (2016a). *D. amilsii* could respire elemental sulfur and thiosulfate and exhibited growth across a broad pH range (3.0–7.0). A follow-up proteomics study (Florentino et al., 2019) uncovered a noncanonical sulfur reduction pathway involving rhodanese like sulfurtransferases and suggested the presence of a modified Dsr system, opening new questions about sulfur metabolism under acid stress.

Perhaps the most striking discovery was the characterization of "strain I", a highly acidophilic SRB isolated from the same enrichment set. This strain, later named *Acididesulfobacillus acetoxydans* (Sánchez-Andrea et al., 2022), represents a novel genus and species within the *Bacillota*. *A. acetoxydans* grows optimally at pH 5, but it can grow at pH 3.8. In a latter study (Egas et al., 2026) this strain tolerates and remains active at pH as low as 2.5, making it the most acidophilic SRB described to date. Unlike many SRB, which only incompletely oxidize substrates, *A. acetoxydans* can completely oxidize acetate to CO_2, offering significant advantages in both environmental resilience and proton consumption. A series of follow-up studies explored the cellular adaptations that enable this functionality. A stress-response study (Egas et al., 2024b) demonstrated that upon exposure to acetic acid, *A. acetoxydans* downregulates glutamate and aspartate synthesis and alters membrane lipid composition, including an increase in saturated ether-bound lipids and a shift in fatty acid biosynthesis, adaptations likely critical to surviving membrane-permeant organic acid stress.

A. acetoxydans is also capable of dissimilatory nitrate reduction to ammonium (DNRA) via a previously undescribed mechanism (Egas et al., 2024a). Despite lacking canonical nitrite reductases such as NrfAH, NirB, or NasB, *A. acetoxydans* reduces nitrate to nitrite via a Nar-type nitrate reductase and further reduces nitrite to ammonium using either a novel function of the sulfite reductase complex AsrABC or a NirA-like ferredoxin-dependent enzyme. Transcriptomic and proteomic profiling confirmed the upregulation of these gene clusters during nitrate respiration, and resting cell experiments demonstrated complete reduction of both nitrite and hydroxylamine to ammonium. This novel pathway not only expands the known enzymatic strategies for DNRA but also has profound implications for nitrogen retention in the Tinto River basin, where denitrification was previously assumed to dominate under low pH, oxygen-limited conditions. The ability of *A. acetoxydans* to perform DNRA suggests that ammonium regeneration—and therefore nitrogen conservation—may be more prevalent than previously thought in these environments.

Remarkably, a nitrosative stress route was discovered in the same strain. Nitrite, the intermediate of DNRA, is a highly oxidant molecule that can pose stress to cells. It can diffuse out of the cell, and at low pH, extracellular nitrite disproportionate into nitric oxide, which can freely permeate the membrane and enter the cell. Cytoplasmic nitrite can also be converted (as a side activity) by NarG into nitric oxide. *A. acetoxydans* can reduce cytoplasmic nitric oxide to nitrous oxide by Hcp, which is subsequently reduced to dinitrogen gas by NosZ (and NosFYD) with PetABC. This mechanism would aid in reducing stress at low pH.

Altogether, this suite of studies represents a significant expansion in our understanding of microbial life in extreme acidic ecosystems. By uncovering novel species, describing their ecological interactions, and elucidating new biochemical pathways, especially those central to the sulfur and nitrogen cycles, this body of work not only challenges previous paradigms but also provides essential models for biotechnological applications in acid mine drainage bioremediation, metal recovery, and synthetic ecosystem design. The discovery of *A. acetoxydans* and its unconventional DNRA pathway particularly marks a turning point in our understanding of microbial nitrogen cycling under extreme conditions.

Geochemical cycles present in the sediments

Given the ecosystem's diverse metabolisms, it is worth reviewing the geochemical cycles in the sediments. The northern Tinto River's geological substrate contains high concentrations of iron and sulfur in various forms. Unlike the water, where the iron cycle dominates, the sediments also show significant activity related to nitrogen and carbon cycles.

Iron cycle

Iron, typically a limiting nutrient due to its low solubility at neutral pH, becomes highly bioavailable in the Tinto River ecosystem (~2.3 g L^{-1}) due to the acidic conditions generated by the IPB. Under these conditions, ferrous iron remains stable and resists chemical oxidation, creating an ecological niche for diverse acidophilic iron-oxidizing microorganisms that dominate both water and sediments.

Iron oxidation is primarily driven by chemolithotrophs such as *Acidithiobacillus ferrooxidans* and *Leptospirillum* spp., with the latter prevailing in the most acidic and oxidized zones due to its higher affinity for Fe^{2+} (Km = 0.025 vs. 1.34 mM for *At. ferrooxidans*) and tolerance to Fe^{3+} (Rawlings et al., 1999). In contrast, *At. ferrooxidans*, which thrive at lower redox potentials and can use hydrogen as an electron donor, compete effectively in less oxidant conditions. Cloning-based molecular surveys have expanded this picture, revealing a broader community of iron oxidizers in sediments, including autotrophs (e.g., *L. ferriphilum*), facultative autotrophs (*Sulfobacillus* spp., *Alicyclobacillus* spp.), and heterotrophs (*Ferrimicrobium*, *Ferrithrix*) (Zhuravleva et al., 2007).

Lithoautotrophic nitrate-reducing ferrous iron-oxidation (NRFeOx) has been reported, although the functional enzymes have not yet been identified (Grimm et al., 2025). Iron oxidation can also be chemically driven by nitrate derived from heterotrophic nitrate reducers. This activity has been detected in enrichment cultures of the Tinto and is supported by the presence of *Pseudomonas*, purple bacteria, and *Dechloromonas* spp. in Tinto sediments.

Dissimilatory iron reduction is widespread. Numerous iron-reducing acidophiles have been identified via DGGE and cloning techniques, and their activity has been confirmed in enrichment cultures, particularly at lower pH values. Among them, *At. ferrooxidans* can reduce Fe^{3+} using elemental sulfur or hydrogen (Ohmura et al., 2002) and constitute up to 17% of the microbial community at sites like SN (Sánchez-Andrea et al., 2011, 2012a). Other iron oxidizers—*Ferrimicrobium*, *Ferrithrix*, *Sulfobacillus*, and *Ferroplasma*—can also reduce heterotrophically Fe^{3+} using organic compounds (Dopson et al., 2004). *Alicyclobacillus* spp. may also reduce iron, though not always linked to growth.

Specialized iron-reducing bacteria are also present. *Acidiphilium cryptum* can reduce Fe^{3+} using various low molecular weight compounds—even in the presence of oxygen— and is abundant at sites like SN. Acidobacteria such as *Acidobacterium capsulatum* reduce Fe^{3+} while respiring glucose at pH 3–5 (Coupland and Johnson, 2008). Anaerobes like *Geobacter*, *Ferribacterium*, and *Aciditerrimonas ferrireducens* use organic acids or alcohols for Fe^{3+} reduction (Itoh et al., 2011). Moreover, sulfate-reducing bacteria (SRB), including *Desulfosporosinus orientis* or strain M1 of *Desulfosporosinus acididurans,* can also act as iron reducers (Ramamoorthy et al., 2006; Sánchez-Andrea et al., 2015).

Iron metabolism in the Tinto River is tightly linked to mineral transformations. For example, SRB like *Desulfosporosinus* sp. GBSRB4.2 promotes the transformation of schwertmannite into goethite as pH and Fe(II) concentrations increase (Bertel et al., 2012). *A. cryptum* JF-5 facilitates the dissolution of iron minerals such as jarosite and goethite via reductive mechanisms (Bridge and Barrie Johnson, 2000).

Sulfur cycle

In the anaerobic zones of ecosystems rich in pyrite and other metal sulfides—such as those found in the IPB—different oxidation state sulfur species accumulate, including thiosulfate, tetrathionate, elemental sulfur (ZVS), and hydrogen sulfide. These compounds can be used as electron donors by sulfur-oxidizing microorganisms. Several acidophilic iron oxidizers, such as *A. ferrooxidans* and *Sulfobacillus* spp., are also capable of oxidizing reduced sulfur species. In addition, *Acidiphilium* spp. (e.g., *A. acidophilum* and *A. cryptum*) can oxidize elemental sulfur using trace oxygen that diffuses into sediment layers (Rohwerder et al., 2007). Anaerobic phototrophic bacteria such as members of *Chromatiales* (purple sulfur bacteria) have also been detected and may oxidize sulfide under low-light, microoxic conditions.

While sulfate is the dominant stable sulfur species in acidic environments, the number of acidophilic microorganisms known to carry out dissimilatory sulfate reduction was historically limited. Traditionally, SRBs were thought to prefer circumneutral pH ranges (6–8) (Widdel and Pfennig, 1982). However, several studies—particularly those focused on the Tinto river—have demonstrated active sulfate reduction at pH values below 5. Acidophilic or acid-tolerant SRB taxa identified include *Desulfosporosinus*, *Desulfotomaculum*-like, *Desulfobulbus*, *Desulfosalsimonas*, *Acididesulfobacillus*, *Thermodesulfobium*, and others. Moreover, elemental sulfur reducers such as *Desulfurella*, *Acidithiobacillus*, and *Thermoplasma* have been found to contribute to sulfur cycling in these systems (Ayala-Muñoz et al., 2020; Johnson and Sánchez-Andrea, 2019).

Environmental and molecular data show that *Desulfosporosinus*, *Desulfurella*, and *Syntrophobacter* may account for up to 11%, 6%, and 5% of microbial communities in specific Rio Tinto sediment strata (Sánchez-Andrea et al., 2012a), especially at moderately acidic pH (4.2–6.2) and low redox potential. These areas show clear signs of biosulfidogenesis, including decreased sulfate concentrations and pH buffering due to sulfide production and precipitation of metal sulfides (Sánchez-Andrea et al., 2012b).

Enrichment cultures have confirmed sulfate reduction at pH as low as 4 using nonionic substrates such as glycerol, methanol, and hydrogen, and strains have been isolated under these conditions. Some of these strains are truly acidophilic, unlike previously known acidotolerant examples. Their isolation required careful selection of substrates and pH-stable media to avoid organic acid toxicity, a barrier that hindered earlier efforts at acidophilic SRB cultivation (Johnson and Sánchez-Andrea, 2019).

Recent work has significantly expanded our understanding of acidophilic sulfidogens. *Desulfosporosinus acididurans*, isolated from the Tinto (strain D and E) and White Rivers (strain M1), grows optimally at pH 5.5 and represents one of the first validated acidophilic SRB (Sánchez-Andrea et al., 2015). The most acidophilic SRB described to date, *Acididesulfobacillus acetoxydans*, remains active at pH 2.5, fully oxidizes acetate, and exhibits genomic and proteomic adaptations to low pH and organic acid stress (Sánchez-Andrea et al., 2022). These microorganisms also perform dissimilatory nitrate reduction to ammonium (DNRA), revealing a novel acidophilic pathway in nitrosative stress release (Egas et al., 2024a).

From the same sediments, the sulfur-reducing bacterium *D. amilsii* was isolated and shown to respire both elemental sulfur and thiosulfate. *D. amilsii* also seems to disproportionate elemental sulfur and expresses unique sulfur metabolism proteins under acidic conditions (Florentino et al., 2016a).

Given the acidophilic nature of these microbial consortia and their role in the natural attenuation of acidic mine waters, their use in biotechnological applications has been demonstrated. Coupled systems treating acid mine drainage and organic-rich wastewaters have achieved high efficiency in sulfate, metal, and COD removal in a single stage without the need for pH neutralization prior to inoculation (Johnson and Sánchez-Andrea, 2019; Sánchez-Andrea et al., 2012a). These findings underscore the ecological significance and applied potential of acidophilic sulfidogens in both natural and engineered systems.

Nitrogen cycle

Using molecular techniques, a range of nitrogen-transforming microorganisms has been identified in the Tinto River basin. Nitrogen-fixing species include the aerobic chemoorganotroph *Ideonella azotifigens* and *Anaerobacter polyendosporus*, an obligate anaerobic fermenter capable of reducing nitrite to ammonium and fixing atmospheric nitrogen (Madigan, 2021). Moreover, nitrogen fixation capacity has been described in various strains of *Leptospirillum* spp. isolated from acidic environments such as the Tinto River, suggesting this process plays a role even in extremely low pH niches (García-Moyano et al., 2008).

Denitrification, the dissimilatory reduction of nitrate to gaseous nitrogen species, is widespread among prokaryotes and includes both heterotrophic and autotrophic bacteria. Molecular analyses have revealed the presence of genera harboring denitrifiers such as *Alcaligenes*, *Pseudochrobactrum*, *Pseudomonas*, *Bacillus*, *Dechloromonas*, *Propionibacterium*, *Acidovorax*, *Delftia*, *Comamonas*, *Aquabacterium*, *Clostridium*, *Geobacter*, and *Anaerobacter*. These taxa span a broad spectrum of metabolic traits and oxygen tolerances, from strict anaerobes to facultative aerobes.

Denitrifying activity in Rio Tinto sediments has been demonstrated through enrichments across a pH range of 4.5–7, under both heterotrophic and autotrophic conditions. This activity is strongly pH-dependent: although nitrate reduction occurs across all tested conditions, nitrous oxide accumulates more readily at lower pH. This aligns with previous findings that denitrification is most complete between pH 6 and 8, with intermediate products like nitrite and N_2O accumulating under more acidic conditions (Šimek et al., 2002). DGGE analysis of these enrichments has revealed a wide diversity of denitrifiers, including *Bacillus*, *Paenibacillus*, *Lysinibacillus*, *Ammoniphilus*, *Rummelibacillus* (phylum *Bacillota*, *class* Bacilli); *Clostridium*, *Sedimentibacter* (class *Clostridia*, Bacillota); and *Alcaligenes* (class *Betaproteobacteria*, phylum *Pseudomonadota*), expanding our understanding of denitrifier diversity in the system.

In addition to classical denitrification, recent findings from the Tinto basin have revealed a novel route of dissimilatory nitrate reduction to ammonium (DNRA) in the acidophilic sulfate-reducing bacterium *Acididesulfobacillus acetoxydans*. As a reminder, *A. acetoxydans* lacks canonical nitrite reductases (e.g., NrfA, NirB), and instead reduces nitrite through a previously uncharacterized combination of AsrABC (anaerobic sulfite reductase), a ferredoxin-linked putative NirA-like protein, and Hcp, a hybrid cluster protein. Hcp also appears to act as a nitric oxide detoxifier under acidic, anoxic conditions—where NO accumulation is especially toxic. This unique pathway enables *A. acetoxydans* to reduce nitrate to ammonia via nitrite and intermediate nitric oxide, with downstream detoxification and potential conversion of NO to N_2O via NosZ. Importantly, this system appears adapted to function efficiently under the low pH, low redox conditions of the Tinto River, representing a significant adaptation for nitrogen cycling in extreme environments. Homologs of these DNRA-associated genes (particularly AsrABC and DEACI_1836) are now being identified in other acidophilic SRB, such as *Desulfosporosinus acididurans* and *D. acidiphilus*, suggesting this pathway may be more widespread than previously recognized (Egas et al., 2024a,b).

In parallel, new evidence supports the coupling of nitrate reduction to ferrous iron oxidation under anoxic conditions in the Tinto River. Isolates and enrichment cultures from the site, including those described by Kappler's group (Melton et al., 2012), have demonstrated that nitrate can drive the oxidation of iron, likely via both biotic and abiotic pathways. This process is especially relevant in sediments where oxygen is limited, but nitrate is available either from upstream inputs or microbial nitrogen fixation. These findings broaden our understanding of nitrate as a key terminal electron acceptor in the Rio Tinto system and strengthen the link between nitrogen and iron cycles in acidic, metal-rich environments.

Together, these insights reveal a more nuanced and diverse nitrogen cycle in the Tinto River than previously assumed, where traditional denitrification coexists with DNRA, nitrate-dependent iron oxidation, and robust mechanisms for NO detoxification. These microbial strategies highlight the evolutionary adaptations enabling nitrogen transformations under extreme geochemical stress and underscore the importance of acidic ecosystems in shaping novel biogeochemical pathways.

Carbon cycle

In acidic ecosystems such as acid mine drainage environments, the carbon cycle was historically thought to be tightly coupled to other biogeochemical processes, with heterotrophs playing a secondary role as organic matter detoxifiers for chemolithotrophs like *A. ferrooxidans* or *Leptospirillum* spp. (Baker and Banfield, 2003). However, studies in the Tinto River sediments challenge this view. Sequences and activity measurements confirm the presence of diverse carbon-processing metabolisms—heterotrophic oxidation, fermentation, acetogenesis, and even methanogenesis—indicating an active and self-sustaining carbon cycle in anaerobic strata.

Autotrophs, such as *Leptospirillum* spp., fix CO_2 via the Calvin or reductive tricarboxylic acid cycles. In contrast, heterotrophs assimilate organic carbon. Many identified aerobic or facultative heterotrophs—such as *Acidiphilium* (5.6% of sediment microbiota) (Sánchez-Andrea et al., 2011), *Aquabacterium*, *Pedobacter*, and *Sphingomonas*—are known to oxidize sugars and organic acids under microaerobic conditions. Facultative organisms like *Variovorax paradoxus*, capable of both heterotrophic and autotrophic growth, add flexibility to this carbon web.

Strict anaerobic fermenters play a particularly important role. These include members of *Clostridium*, *Veillonella*, *Turicibacter*, and *Dorea*, all capable of converting sugars into organic acids (acetate, butyrate, lactate, etc.), alcohols, CO_2, and H_2. Among them, *Clostridium* spp. are dominant in certain strata and were isolated as hydrogen producers under dark fermentation conditions (Rajhi et al., 2013; Sánchez-Andrea et al., 2011). One key isolate, *Microbacter margulisiae*, was obtained during enrichments for acidophilic SRBs (Sánchez-Andrea et al., 2014). It is a strictly anaerobic, propionigenic fermenter, capable of growing between pH 3 and 7 and producing propionate, lactate, and acetate from sugars and yeast extract.

It does not reduce external electron acceptors and forms a new genus within *Porphyromonadaceae*. Its frequent detection in acidic environments and bioreactors suggests a wider ecological role.

Similarly, *Lucifera butyrica*—another strict anaerobe isolated from Tinto sediments— is a spore-forming, acidotolerant fermenter capable of converting sugars, organic acids, and glycerol into acetate, butyrate, ethanol, hydrogen, and 1,3-propanediol (1,3-PDO) (Sánchez-Andrea et al., 2018). Its metabolic versatility and growth at pH as low as 3.5 underscore its ecological relevance in carbon and sulfur cycling. In coculture with the acetate-oxidizing sulfur reducer *D. amilsii*, *L. butyrica* boosts sulfidogenesis by providing fermentation-derived acetate, showcasing a model for syntrophic sulfur respiration in acidic systems.

Fermentation intermediates such as acetate, propionate, and butyrate can be further oxidized by acetogens like *Syntrophobacter* spp. and *Oxobacter pfennigii*. *Syntrophobacter*, capable of oxidizing propionate via the methylmalonyl-CoA pathway, account for up to 37% of microbial communities in certain sediment layers (Sánchez-Andrea et al., 2012b). Their syntrophic interaction with hydrogenotrophic partners is key to maintaining low H_2 partial pressures and favoring otherwise energetically unfavorable reactions.

Finally, methanogenesis—although thought to be inhibited at very low pH—is present and active in the Tinto system, especially around pH 5.0. Enrichment cultures confirmed methane production from acetate and hydrogen (Sanz et al., 2011a), and molecular analyses identified *Methanosaeta concilii* as the dominant methanogen in situ (JQ815591 in JL sediments), consistent with acetate being the most abundant volatile fatty acid in porewater. Methanogenic activity in such low-pH, metal-rich sediments has implications not only for carbon turnover but also for understanding the limits of life in analogous astrobiological settings such as Mars (Sanz et al., 2011a).

The high concentrations of sulfate further position the afore-mentioned SRBs as central players in mineralization, effectively bridging the sulfur and carbon cycles.

In conclusion, the carbon cycle in the Tinto River sediments is sustained by a diverse and robust microbial consortium, where fermenters like *Microbacter margulisiae* and *Lucifera butyrica* contribute significantly to organic matter degradation and secondary syntrophic processes, and sulfate-reducing bacteria would mineralize the shorter organic acids and glycerol. If at any point, organic substrates would be sufficient to deplete sulfate, methanogenesis would enter into action.

Geomicrobiological model of the Tinto River sediments

The integration of all this biogeochemical information results in a geomicrobiological model that incorporates the different elemental cycles operating in the Tinto River sediments (Fe, N, C, and S), as well as their interactions Fig. 5.2.

Unlike earlier studies of the Tinto River's water columnwhere microbial processes are primarily driven by iron oxidation—the sedimentary environment reveals

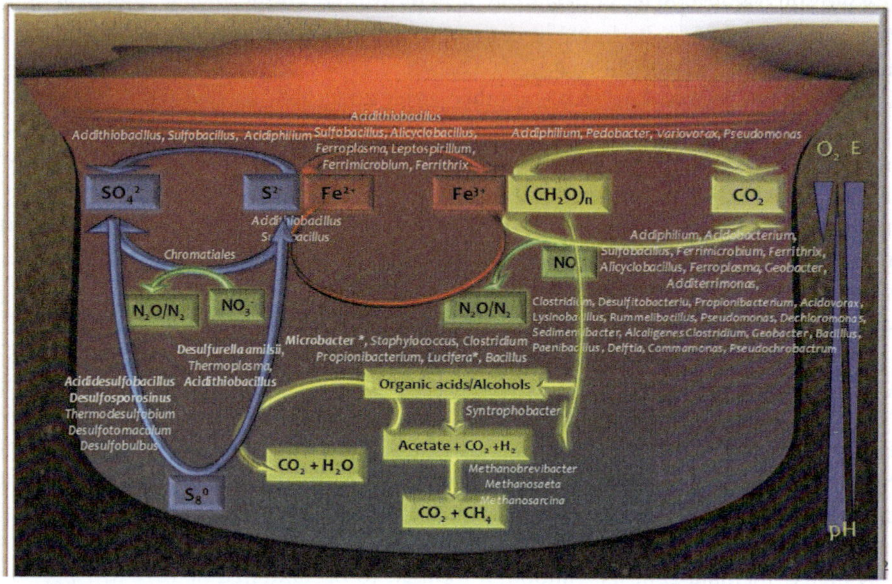

FIGURE 5.2 Geomicrobiological Model of the Tinto River Sediments.

The processes involved are indicated with blue arrows (sulfur cycle), red arrows (iron cycle), yellow arrows (carbon cycle), and green arrows (nitrogen cycle).

a fundamentally different ecosystem. In these anaerobic and low-pH layers, the sulfur cycle and organic carbon transformations are the central axes of microbial activity. These are intimately linked to active iron and nitrogen cycles, forming a tightly integrated geomicrobiological network, fueled by the oxidation of pyrite (FeS_2) and other sulfide minerals characteristic of the IPB.

Iron cycling in these sediments is highly dynamic. Ferrous iron, stable at low pH, supports chemolithotrophic growth under oxic and microoxic conditions by microorganisms such as *Acidithiobacillus ferrooxidans*, *Sulfobacillus* spp., *Ferrimicrobium* spp., and *Ferroplasma* spp. Under anoxic conditions, ferrous iron can be oxidized (indirectly) by nitrate reducers like *Pseudomonas* and *Alicyclobacillus* spp., revealing links between the iron and nitrogen cycles. In turn, ferric iron, which accumulates under acidic conditions, serves as a terminal electron acceptor for various heterotrophic acidophiles, including *Acidiphilium* spp., *A. capsulatum*, and *Ferrimicrobium* spp., or is reduced during sulfur oxidation by *At. ferrooxidans* and *Sulfobacillus* spp. These bidirectional redox processes regenerate ferrous iron, completing the cycle.

The sulfur cycle plays a dominant role in energy flow. Sulfide produced via dissimilatory sulfate or sulfur reduction is rapidly precipitated with present metals; or oxidized—either aerobically or anaerobically, including through nitrate-dependent pathways. *Acidithiobacillus ferrooxidans* and *Sulfobacillus* spp. are active oxidizers of reduced sulfur compounds. Sulfate is reduced under anoxic conditions by a diverse community of SRBs, including the acidophilic *D. acididurans* and *A. acetoxydans*. These bacteria are capable of growth below pH 4 and contribute significantly to metal sulfide precipitation and pH neutralization.

A key sulfur reducer in this environment is *D. amilsii*, an acidotolerant obligate anaerobe that uses elemental sulfur or thiosulfate as terminal electron acceptors, but not sulfate. It oxidizes acetate completely to CO_2 and plays a critical role in elemental sulfur respiration (Florentino et al., 2016a). Its metabolism complements that of fermentative bacteria and helps close the sulfur cycle in acidic, anoxic zones.

Among the fermenters, two novel isolates are particularly noteworthy for their roles in the carbon cycle. *Microbacter margulisiae* grows at pH values as low as 3.0, and ferments sugars into propionate, lactate, and acetate. Its consistent detection in natural acid mine environments and metal-rich bioreactors suggests it plays a crucial role in the anaerobic degradation of organic matter under extreme conditions (Sánchez-Andrea et al., 2014).

Lucifera butyrica ferment a wide range of substrates, including glycerol, sugars, and organic acids, producing acetate, butyrate, ethanol, hydrogen, and 1,3-propanediol. Isolated from the same Tinto sediments, it grows across a broad pH range (3.5–7.0) and displays resistance to high concentrations of zinc and iron. In coculture with *Desulfurella amilsii*, it boosts sulfidogenesis by supplying fermentation-derived acetate, demonstrating syntrophic cooperation between fermenters and elemental sulfur reducers (Sánchez-Andrea et al., 2018). This strategy has implications for biosulfidogenesis-based biotechnological processes.

The nitrogen cycle in the sediments involves multiple functional groups. Nitrogen fixation is attributed to *Leptospirillum* spp., *Ideonella azotifigens*, and *Anaerobacter polyendosporus*. Classical denitrification is supported by *Dechloromonas*, *Pseudochrobactrum*, and *Acidovorax*, and has been experimentally observed under both heterotrophic and autotrophic conditions. Denitrification rates are strongly pH-dependent, with accumulation of intermediates such as NO_2^- and N_2O under acidic conditions.

A- acetoxydans reduces nitrate to ammonium via a unique DNRA route involving AsrABC, a NirA-like ferredoxin, and Hcp, which also plays a role in nitric oxide detoxification—a particularly valuable trait under low pH and redox-stressed conditions.

The carbon cycle is driven by the interplay between heterotrophic respiration and fermentation. Aerobic or microaerophilic degradation of organic compounds is carried out by *Acidiphilium*, *Aquabacterium*, *Pedobacter*, and others. Anaerobic degradation is achieved through fermentation by a diverse group, including *Clostridium*, *Veillonella*, *Propionibacterium*, and *M. margulisiae*. These processes produce volatile fatty acids, ethanol, and H_2, which serve as substrates for SRB and syntrophic acetogens like *Syntrophobacter*. Acetate and hydrogen, in turn, feed methanogens such as *M. concilii*, *M. barkeri*, and *M. arboriphilus*. Methanogenesis in this system, though inhibited at very low pH, has been demonstrated around pH 5.0 and appears to be substrate-limited rather than pH-limited (Sanz et al., 2011b).

Previous geochemical studies on subsurface Tinto drillings also revealed the presence of CH_4, NO_2^-, and H_2 gases, potentially originating from both abiotic and microbial processes (Amils et al., 2004). The presence of fermenters, SRBs, methanogens, and nitrate reducers capable of producing or consuming these gases supports a biologically active deep biosphere.

In conclusion, the microbial communities in the Tinto River sediments do not follow the typical iron-dominated pattern of acidic surface waters. Instead, they constitute a sulfur- and carbon-centered ecosystem, modulated by iron and nitrogen dynamics. Recent advances in microbial isolation and genomic analyses confirm the activity of sulfate reduction, syntrophic sulfur respiration, low-pH DNRA, and acid-tolerant fermentation. These processes are performed by a metabolically versatile and acid-adapted community capable of sustaining complex interactions under conditions of extreme pH, redox, and metal load.

Mars analogues and the Tinto River: A window into anoxic surface habitability

Redefining Mars analogues in light of new discoveries

Mars has long captivated scientific interest as a potentially habitable world, primarily due to its history of water activity. Recent seismic and geochemical analyses suggest that the upper crust of Mars may still harbor high-porosity, water-saturated layers equivalent to hundreds of meters of global liquid water (Wright et al., 2024). While subsurface aquifers remain compelling targets for astrobiology, new findings are rapidly reshaping how we think about life-supporting environments on the Martian surface.

In a 2024 breakthrough, NASA's Curiosity rover accidentally cracked open a rock in the Gediz Vallis Channel, revealing bright yellow crystals of elemental sulfur - the first such discovery of native sulfur on Mars (NASA Science News). This contrasts sharply with previously detected sulfates and suggests a narrow geochemical window conducive to elemental sulfur precipitation, hinting at either localized reducing conditions or biological influences. What makes this discovery even more intriguing is that Gediz Vallis contains numerous rocks with similar textures, implying the potential abundance of native sulfur in surface layers exposed to light but largely deprived of oxygen.

The discovery of light-exposed but oxygen-poor niches opens new possibilities in analogue research. On Earth, the Tinto River basin in southwestern Spain provides an ideal comparison point: its sediments exhibit stable anoxic conditions just centimeters below the surface, while remaining fully exposed to light. This configuration closely mirrors what might be expected in Martian paleochannels like Gediz Vallis, where phototrophic or chemotrophic microbial communities could have evolved in surficial, oxygen-deprived zones. Moreover, Tinto sediments are rich in native sulfur forms (e.g., elemental sulfur and thiosulfate), iron oxides, and minerals such as jarosite—all key components thought to be on Mars. The recent identification of microbial sulfur respiration carried out by acidotolerant elemental sulfur reducers such as *D. amilsii*, alongside fermenters like *L. butyrica*, adds a functional layer to the analogy (Sánchez-Andrea et al., 2018; Florentino et al., 2016a).

Bridging biogeochemical cycles across planets

The geomicrobiological model established for Río Tinto sediments provides a coherent framework that could be applied to Mars. Sulfate- and sulfur-reducing bacteria, acidophilic fermenters, nitrate-reducing iron oxidizers, and syntrophic assemblages—all operating under low-pH, low-O_2, metal-rich conditions—have been documented and, in many cases, isolated and characterized. This includes acidophilic DNRA organisms like *A. acetoxydans*, which detoxify nitric oxide and retain nitrogen in ammonium form (Egas et al., 2024a).

These capabilities are particularly relevant in light of the presence of perchlorates and nitrates on Mars, and the possibility that oxidants like bromate or chlorate substituted for molecular oxygen in early planetary redox reactions. The detection of such oxidants—and the reinterpretation of manganese oxides as halogen-induced rather than oxygen-induced—reinforces the case for life strategies independent of O_2.

Tinto river's microbial communities also demonstrate metabolic flexibility under acidic, anoxic, and photic conditions—a crucial trifecta for Martian habitability analogues.

In conclusion, the discovery of elemental sulfur in Mars' Gediz Vallis Channel forces a rethinking of where life might have emerged—or persisted—on the Red Planet. Light-exposed but oxygen-deprived sediments, long overlooked in favor of deep aquifers or shielded caves, may in fact offer the best opportunity for surface habitability. Earth environments like the Tinto River sediments not only replicate these conditions but also demonstrate, through living microbial ecosystems, that life can thrive in them. By combining acidic pH, anoxia, heavy metals, sulfur cycling, and light exposure, they form what may be the most complete and accessible terrestrial analogue for studying potential life-supporting systems on ancient, and perhaps even present-day, Mars.

Bioremediation and metal recovery

Microbial contributions to AMD mitigation

Acid mine drainage (AMD) combines extremely low pH with high dissolved-metal burdens, posing acute toxicity to aquatic life and chronic risks to sediments, soils and, ultimately, human health (Foulds et al., 2014; Kossoff et al., 2014; Lumbroso et al., 2021). Catastrophic containment failures underscore its hazard, yet the long-term ecological footprint remains poorly quantified because many historic and abandoned sites are undocumented, especially in regions like China, Russia and India (Boldy et al., 2021; Kossoff et al., 2014; Macklin et al., 2023; Skousen et al., 2019). A recent global inventory of active and inactive metal mines suggests current estimates are conservative (Macklin et al., 2023).

AMD can be treated by abiotic, biotic or hybrid methods selected according to water chemistry and site logistics (Johnson and Hallberg, 2005; Simate and Ndlovu, 2014). Abiotic options such as adsorption, ion exchange, membrane filtration, evaporation and, most commonly, lime or other alkaline addition after forced aeration, rapidly raise pH and precipitate metals as hydroxides (Johnson and Hallberg, 2005; Skousen et al., 2019). Passive variants such as anoxic limestone drains and steel-slag beds avoid continuous chemical dosing but still generate alkaline metalliferous sludges requiring special disposal (Skousen et al., 2017).

Bioremediation aims to harness microbial metabolisms that generate net alkalinity and immobilise metals, often in passive systems such as permeable reactive barriers or anoxic wetlands (Skousen et al., 2017, 2019). These installations demand lower operating expenditure and consolidate metals in situ, yet they require substantial land, are harder to control, and their long-term stability is uncertain (Johnson and Hallberg, 2005; Skousen et al., 2017, 2019).

Several microbial pathways consume protons, but biosulfidogenesis, especially dissimilatory sulfate reduction to sulfide, most directly addresses AMD by simultaneously: (i) elevating pH, (ii) lowering sulfate concentrations and (iii) precipitating metals as sparingly soluble sulfides (Fig. 5.3). Depending on pH, the overall reaction can be proton-consuming, proton-producing or pH-neutral. Coupling sulfate ($pK_{a1} = 1.99$) reduction to complete oxidation of glycerol illustrates the extremes:

$$\text{Low pH: } 4\ C_3H_8O_3 + 7\ SO_4^{2-} + 14\ H\text{-}^{+} \rightarrow 12\ CO_2 + 7\ H_2S + 16\ H_2O$$

$$\text{Neutral pH: } 4\ C_3H_8O_3 + 7\ SO_4^{2-} \rightarrow 12\ HCO_3^{-} + 7\ HS^{-} + 4\ H_{-2}O + 5\ H^{+}$$

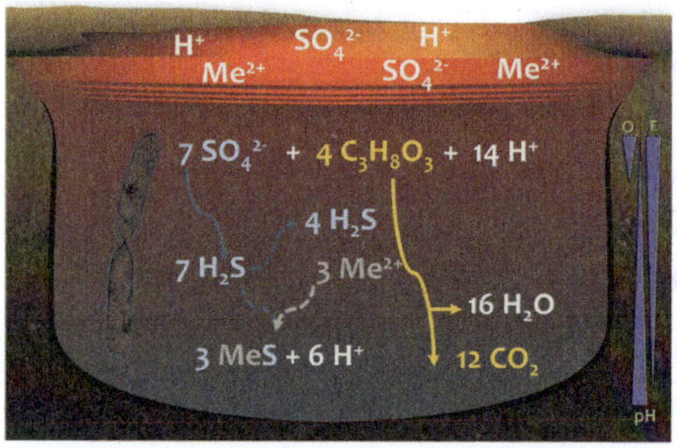

FIGURE 5.3

Bioremediation action of sulfate reducing bacteria inhabiting Tinto river sediments.

While sulfate itself is not toxic at AMD levels, its discharge is regulated to curb downstream sulfide formation. Provided sufficient electron donor, biosulfidogenesis can meet these limits; however, excess sulfide may accumulate if metals are depleted. Controlled donor dosing or integration with sulfur-oxidizing microbes that convert sulfide to elemental sulfur under low redox conditions—technology already proven in gas-desulfurisation systems—mitigates this risk and even enables sulfur recovery.

The acidophilic SRB of Tinto River sediments are capable of treating AMD without preliminary neutralization. In laboratory up-flow reactors receiving AMD and high-organic wastewater, these communities produced effluents with neutral pH and greatly reduced organics, sulfates, iron and heavy metals, demonstrating single-stage bioremediation under acid conditions (Sánchez-Andrea et al., 2012c).

From remediation to resource recovery

Industrial-scale biosulfidogenic processes such as SULFATEQ and BioSulphide (sulfate-reducing) and THIOTEQ (sulfur-reducing) employ neutrophilic organisms to generate sulfide off-line, shielding them from acidic, metal-rich feeds (Adams et al., 2008; Huisman et al., 2006; Muyzer and Stams, 2008; Santos and Johnson, 2018). This arrangement adds complexity and cost (Johnson and Santos, 2020; Nancucheo et al., 2017). Operating directly at low pH would allow a single reactor, suppress methanogenesis (conserving donor), produce gaseous H_2S readily sparged into contactors, eliminate lime addition and facilitate on-site metal selectivity (Nancucheo et al., 2017; Sánchez-Andrea et al., 2014; Van Houten et al., 2006). Early attempts to acid-adapt neutrophiles were unsuccessful (Hidalgo-Ulloa et al., 2020; Johnson and Santos, 2020), but recent work has established stable sulfidogenic reactors at pH 1.3–3.9 using naturally acidophilic or acid-tolerant consortia, achieving direct AMD treatment and selective Cu precipitation (Hedrich and Johnson, 2014; Holanda and Johnson, 2020; Nancucheo et al., 2017; Santos and Johnson, 2018; Ňancucheo and Barrie Johnson, 2014; Ňancucheo and Johnson, 2012).

Despite progress, only a handful of acidotolerant or acidophilic SRB are formally described—nine species across the bacterial and archaeal domains—including *D. acidiphilus*, *Thermodesulfobium acidiphilum* and *A. acetoxydans* (Alazard et al., 2010; Burggraf et al., 1997; Mori et al., 2003; Sánchez-Andrea et al., 2015, 2022). All are moderate acidophiles with growth ranges of pH 3–5; a true extreme acidophile would display an optimum $\leq$ pH 3 (Baker and Banfield, 2003).

A. acetoxydans, isolated from Rio Tinto sediments (Sánchez-Andrea et al., 2022), oxidizes acetate completely to CO_2—rare among acidophilic SRB—and dominated several low-pH reactors where it out-competed *Desulfosporosinus acididurans*, isolated from the Tinto (Sánchez-Andrea et al., 2015) at pH $\leq$ 4 (Santos and Johnson, 2018; Ňancucheo and Barrie Johnson, 2014). Proteomic analyses revealed few active proton-stress responses even at pH 3.9, suggesting alternative acid-tolerance strategies, possibly via increased saturated ether lipids (Sánchez-Andrea et al., 2022).

Sulfur cycling expands remediation options. Many AMD sites contain abundant xero-valent elemental sulfur - ZVS, a two-electron acceptor whose market price (~61 US $ t^{-1}) is far below that of organic donors (Florentino et al., 2016b). Replacing sulfate with S^0 can cut operational costs substantially (Florentino et al., 2015, 2016b). Acidic enrichments reduce S^0 between pH 3 and 5; the isolate *D. amilsii* grows optimally at pH 6–6.5 yet remains active down to pH 3 (Florentino et al., 2017, 2019). Sequences from *Desulfocapsa*-like disproportionators in the Tinto River imply that sulfide generation from sulfur disproportionation—a process needing no external electron donor—may also be widespread.

Most divalent "chalcophile" metals (Cu^{2+}, Ni^{2+}, Cd^{2+}, etc.) form highly insoluble sulfides:

$$Me^{2+} + HS^{-} \rightarrow MeS + H^{+}$$

Because solubility products (K_{sp}) differ by orders of magnitude and vary with pH, maintaining low but controlled sulfide concentrations allows sequential or selective precipitation, enhancing metal-recovery value (Hedrich and Johnson, 2014; Lewis, 2010).

Implications for a renewable energy-based economy

Throughout history, advances in metallurgy have defined societal milestones—Neolithic stone gave way to Eneolithic copper, and subsequent ages were shaped by ever-broader metal use. In modern terms, electricity underpins virtually every economic and technological process, and conductive metals—chiefly copper—constitute the essential infrastructure that distributes it. Fossil fuels currently supply the energy driving this system, yet their finiteness and environmental cost compel a transition to renewables—an epochal shift comparable to earlier age changes (Calvin et al., 2023). Governments and industry have therefore pledged steep carbon-emission cuts for the coming decades (United Nations and The World Bank Group).

Electrification and renewable technologies, however, are metal-intensive (Herrington, 2021). While copper demand is intuitively linked to grid expansion, other requirements are less obvious: replacing an internal-combustion car with an electric vehicle, for example, raises nickel consumption from <0.1 kg to ≈ 40 kg per unit. Recent models project three-to sixfold increases for base metals such as copper and nickel by 2050 (Herrington, 2021; Watari et al., 2020), and up to tenfold rises for renewables-critical elements like lithium and platinum (Herrington, 2021; Watari et al., 2018, 2020). Although forecasts cannot fully account for future recycling or metal substitution, a steep overall surge is clear.

Geological data show that economically recoverable ore bodies can indeed be exhausted (Henckens et al., 2016; Kerr, 2014; Sverdrup et al., 2019). Total crustal inventories remain vast (Lusty and Gunn, 2015), yet near-term supply is constrained by technological, environmental and geopolitical factors. Consequently, primary mining will stay indispensable until secondary sources, alternative materials or novel extraction methods mature (Henckens et al., 2016).

Current practice recovers only targeted metals, leaving most excavated material as waste (Parbhakar-Fox and Lottermoser, 2015; Tahmasebi and Hezarkhani, 2010). Waste rock—too low in grade for processing—and tailings, the slurry left after milling and chemical treatment, are stored on-site; their composition reflects ore mineralogy and process chemistry (Park et al., 2019). Producing 1 Mt of copper involves roughly 742 Mt of excavation, of which 538 Mt become waste rock and 204 Mt tailings. In 2016 alone, global operations generated ~72,000 Mt waste rock and ~8850 Mt tailings (Baker et al., 2020). Declining ore grades and the depletion of easily accessible deposits are driving these volumes upward (Calvo et al., 2016; Prior et al., 2012), expanding both active mines and waste repositories and intensifying pressures on long-term storage, environmental stewardship and regulatory compliance (Park et al., 2019; Werner et al., 2020).

Given the looming scarcity of high-grade ores, selective metal recovery from secondary streams—particularly the rapidly expanding flow of electronic waste—has become indispensable (Moncur et al., 2015; Tabelin et al., 2018). Simultaneously, Tinto River watershed and its pit lakes represent vast aqueous "urban mines," containing tonnes of Cu, Zn, Ni, and other strategic elements awaiting exploitation by next-generation biotechnologies (Kossoff et al., 2014). Extremophile consortia endemic to these sites already close the nitrogen, carbon, sulfur and iron cycles, naturally concentrating metals in forms amenable to capture. By deciphering and engineering these interconnected metabolisms, we can couple remediation with resource recovery, transforming environmental liabilities into sustainable supply chains. Nature thus supplies both catalyst and blueprint; our task is to harness these microbial processes responsibly to rectify the legacies of intensive mining. Advancing this frontier—where geomicrobiology, process engineering and materials science converge—promises some of the most exciting research avenues of the coming decade.

Conclusions

- Extremophile diversity: anoxic zones of the Río Tinto harbor a phylogenetically broad assemblage of *Bacteria* and *Archaea* that collectively catalyze the Fe, S, N and C cycles at pH 1.5–5 or more.
- Stratified biogeochemistry: vertical redox and pH gradients favor sequential iron reduction, denitrification, sulfate/sulfur reduction and, finally, methanogenesis, yielding natural attenuation of acidity and metal loads.
- Novel acidophiles: newly characterized taxa (e.g., *Acididesulfobacillus acetoxydans*, *Desulfurella amilsii*, *Lucifera butyrica*, *Microbacter margulisiae*) exhibit unprecedented acid tolerance and metabolic versatility, including

low-pH DNRA, complete acetate oxidation or industrially relevant metabolite production.

- Bioremediation promise: laboratory reactors inoculated with Tinto river consortia achieve single-stage AMD neutralization, metal sulfide precipitation and coupling to organic-waste treatment. SRB isolated from Rio Tinto show promising resistance to extreme low pH.
- Resource perspective: with declining ore grades and soaring demand for Cu, Ni, Li and others, AMD sites and e-waste streams represent untapped "urban mines" whose metals can be recovered biologically while closing elemental cycles. Acidophilic SRB inhabiting the Tinto River can serve as the main actors.

References

Adams, M., Lawrence, R., Bratty, M., 2008. Biogenic sulphide for cyanide recycle and copper recovery in gold–copper ore processing. Miner. Eng. 21 (6), 509–517. https://doi.org/10.1016/j.mineng.2008.02.001.

Aguilera, A., Manrubia, S.C., Gómez, F., Rodríguez, N., Amils, R., 2006. Eukaryotic community distribution and its relationship to water physicochemical parameters in an extreme acidic environment, Río Tinto (Southwestern Spain). Appl. Environ. Microbiol. 72 (8), 5325–5330. https://doi.org/10.1128/aem.00513-06.

Aguilera, A., Souza-Egipsy, V., Gómez, F., Amils, R., 2007a. Development and structure of eukaryotic biofilms in an extreme acidic environment, Río Tinto (SW, Spain). Microb. Ecol. 53 (2), 294–305. https://doi.org/10.1007/s00248-006-9092-2.

Aguilera, A., Zettler, E., Gómez, F., Amaral-Zettler, L., Rodríguez, N., Amils, R., 2007b. Distribution and seasonal variability in the benthic eukaryotic community of Río Tinto (SW, Spain), an acidic, high metal extreme environment. Syst. Appl. Microbiol. 30 (7), 531–546. https://doi.org/10.1016/j.syapm.2007.05.003.

Alazard, D., Joseph, M., Battaglia-Brunet, F., Cayol, J.L., Ollivier, B., 2010. *Desulfosporosinus acidiphilus* sp. nov.: a moderately acidophilic sulfate-reducing bacterium isolated from acid mining drainage sediments. Extremophiles 14 (3), 305–312. https://doi.org/10.1007/s00792-010-0309-4.

Amils, R., González-Toril, E., Gómez, F., Fernández-Remolar, D., Rodríguez, N., Malki, M., Zuluaga, J., Aguilera, A., Amaral-Zettler, L.A., 2004. Importance of Chemolithotrophy for Early Life on Earth: The Tinto River (Iberian Pyritic Belt) Case. Springer Science and Business Media LLC, pp. 463–480. https://doi.org/10.1007/1-4020-2522-x_28.

Ayala-Muñoz, D., Burgos, W.D., Sánchez-España, J., Couradeau, E., Falagán, C., Macalady, J.L., 2020. Metagenomic and metatranscriptomic study of microbial metal resistance in an acidic pit lake. Microorganisms 8 (9), 1350. https://doi.org/10.3390/microorganisms8091350.

Baker, B.J., Banfield, J.F., 2003. Microbial communities in acid mine drainage. FEMS Microbiol. Ecol. 44 (2), 139–152. https://doi.org/10.1016/S0168-6496(03)00028-X.

Baker, E., Davies, M., Fourie, Mudd, G., Thygesen, K., 2020. In: Towards Zero Harm: A Compendium of Papers Prepared for the Global Tailings Review, pp. 14–25.

Bertel, D., Peck, J., Quick, T.J., Senko, J.M., 2012. Iron transformations induced by an acid-tolerant *Desulfosporosinus* species. Appl. Environ. Microbiol. 78 (1), 81–88. https://doi.org/10.1128/AEM.06337-11.

Boldy, R., Santini, T., Annandale, M., Erskine, P.D., Sonter, L.J., 2021. Understanding the impacts of mining on ecosystem services through a systematic review. Extr. Ind. Soc. 8 (1), 457–466. https://doi.org/10.1016/j.exis.2020.12.005.

Bridge, T.A.M., Barrie Johnson, D., 2000. Reductive dissolution of ferric iron minerals by *Acidiphilium* SJH. Geomicrobiol. J. 17 (3), 193–206. https://doi.org/10.1080/01490450050121161.

Burggraf, S., Huber, H., Stetter, K.O., 1997. Reclassification of the crenarchaeal orders and families in accordance with 16S rRNA sequence data. Int. J. Syst. Bacteriol. 47 (3), 657–660. https://doi.org/10.1099/00207713-47-3-657.

Calvin, K., Dasgupta, D., Krinner, G., Mukherji, A., Thorne, P., Trisos, C., Romero, J., Aldunce, P., Barrett, K., Blanco, G., IPCC, 2023. In: Climate Change 2023: Synthesis Report. Contribution of Working Groups I.

Calvo, G., Mudd, G., Valero, A., Valero, A., 2016. Decreasing ore grades in global metallic mining: a theoretical issue or a global reality? Resources 5 (4), 36. https://doi.org/10.3390/resources5040036.

Coupland, K., Johnson, D.B., 2008. Evidence that the potential for dissimilatory ferric iron reduction is widespread among acidophilic heterotrophic bacteria. FEMS Microbiol. Lett. 279 (1), 30–35. https://doi.org/10.1111/j.1574-6968.2007.00998.x.

Dopson, M., Baker-Austin, C., Hind, A., Bowman, J.P., Bond, P.L., 2004. Characterization of *Ferroplasma* isolates and *Ferroplasma acidarmanus* sp. nov., extreme acidophiles from acid mine drainage and industrial bioleaching environments. Appl. Environ. Microbiol. 70 (4), 2079–2088. https://doi.org/10.1128/AEM.70.4.2079-2088.2004.

Egas, R.A., Kurth, J.M., Boeren, S., Sousa, D.Z., Welte, C.U., Sánchez-Andrea, I., 2024a. A novel mechanism for dissimilatory nitrate reduction to ammonium in *Acididesulfobacillus acetoxydans*. mSystems 9 (3). https://doi.org/10.1128/msystems.00967-23.

Egas, R.A., Bale, N.J., Koenen, M., Villanueva, L., Sousa, D.Z., Welte, C.U., Sánchez-Andrea, I., 2026. Proton stress adaptation in acidophilic sulfate-reducing bacteria: insights from *Acididesulfobacillus acetoxydans* for acid mine drainage bioremediation. Environ. Sci. Technol. In press.

Egas, R.A., Sahonero-Canavesi, D.X., Bale, N.J., Koenen, M., Yildiz, Ç., Villanueva, L., Sousa, D.Z., Sánchez-Andrea, I., 2024b. Acetic acid stress response of the acidophilic sulfate reducer *Acididesulfobacillus acetoxydans*. Environ. Microbiol. 26 (2). https://doi.org/10.1111/1462-2920.16565.

Florentino, A.P., Weijma, J., Stams, A.J.M., Sánchez-Andrea, I., 2015. Sulfur reduction in acid rock drainage environments. Environ. Sci. Technol. 49 (19), 11746–11755. https://doi.org/10.1021/acs.est.5b03346.

Florentino, A.P., Brienza, C., Stams, A.J.M., Sánchez-Andrea, I., 2016a. *Desulfurella amilsii* sp. nov., a novel acidotolerant sulfur-respiring bacterium isolated from acidic river sediments. Int. J. Syst. Evol. Microbiol. 66 (3), 1249–1253. https://doi.org/10.1099/ijsem.0.000866.

Florentino, A.P., Weijma, J., Stams, A.J.M., Sánchez-Andrea, I., 2016b. Ecophysiology and application of acidophilic sulfur-reducing microorganisms. In: Grand Challenges in Biology and Biotechnology. Volume 1, pp. 141–175. https://doi.org/10.1007/978-3-319-13521-2_5.

Florentino, A.P., Stams, A.J.M., Sánchez-Andrea, I., 2017. Genome sequence of *Desulfurella amilsii* strain TR1 and comparative genomics of *Desulfurellaceae* family. Front. Microbiol. 8. https://doi.org/10.3389/fmicb.2017.00222.

Florentino, A.P., Pereira, I.A.C., Boeren, S., Born, Stams, A.J.M., Sanchez-Andrea, I., 2019. Insight into the sulfur metabolism of *Desulfurella amilsii* by differential proteomics. Environ. Microbiol. 21 (1), 209–225.

Foulds, S.A., Brewer, P.A., Macklin, M.G., Haresign, W., Betson, R.E., Rassner, S.M.E., 2014. Flood-related contamination in catchments affected by historical metal mining: an unexpected and emerging hazard of climate change. Sci. Total Environ. 476–477, 165–180. https://doi.org/10.1016/j.scitotenv.2013.12.079.

García-Moyano, A., González-Toril, E., Aguilera, A., Amils, R., 2007. Prokaryotic community composition and ecology of floating macroscopic filaments from an extreme acidic environment, Río Tinto (SW, Spain). Syst. Appl. Microbiol. 30 (8), 601–614. https://doi.org/10.1016/j.syapm.2007.08.002.

García-Moyano, A., González-Toril, E., Moreno-Paz, M., Parro, V., Amils, R., 2008. Evaluation of *Leptospirillum* spp. in the Río Tinto, a model of interest to biohydrometallurgy. Hydrometallurgy 94 (1–4), 155–161. https://doi.org/10.1016/j.hydromet.2008.05.046.

García-Moyano, A., González-Toril, E., Aguilera, Á., Amils, R., 2012. Comparative microbial ecology study of the sediments and the water column of the Río Tinto, an extreme acidic environment. FEMS Microbiol. Ecol. 81 (2), 303–314. https://doi.org/10.1111/j.1574-6941.2012.01346.x.

Garrido, P., Toril, E.G.-, Moyano, A.G.-, Moreno-Paz, M., Amils, R., Parro, V., 2008. An oligonucleotide prokaryotic acidophile microarray: its validation and its use to monitor seasonal variations in extreme acidic environments with total environmental RNA. Environ. Microbiol. 10 (4), 836–850. https://doi.org/10.1111/j.1462-2920.2008.01477.x.

Gónzalez-Toril, E., Gómez, F., Rodríguez, N., Fernández-Remolar, D., Zuluaga, J., Marín, I., Amils, R., 2003. Geomicrobiology of the Tinto River, a model of interest for biohydrometallurgy. Hydrometallurgy 71 (1–2), 301–309. https://doi.org/10.1016/S0304-386X(03)00169-5.

Grimm, H., Lorenz, J., Straub, D., Joshi, P., Shuster, J., Zarfl, C., Muehe, E.M., Kappler, A., 2025. Nitrous oxide is the main product during nitrate reduction by a novel lithoautotrophic iron(II)-oxidizing culture from an organic-rich paddy soil. Appl. Environ. Microbiol. 91 (1). https://doi.org/10.1128/aem.01262-24.

Hedrich, S., Johnson, D.B., 2014. Remediation and selective recovery of metals from acidic mine waters using novel modular bioreactors. Environ. Sci. Technol. 48 (20), 12206–12212. https://doi.org/10.1021/es5030367.

Henckens, M.L.C.M., van Ierland, E.C., Driessen, P.P.J., Worrell, E., 2016. Mineral resources: geological scarcity, market price trends, and future generations. Resour. Policy 49, 102–111. https://doi.org/10.1016/j.resourpol.2016.04.012.

Herrington, R., 2021. Mining our green future. Nat. Rev. Mater. 6 (6), 456–458. https://doi.org/10.1038/s41578-021-00325-9.

Hidalgo-Ulloa, A., Sánchez-Andrea, I., Buisman, C., Weijma, J., 2020. Sulfur reduction at hyperthermoacidophilic conditions with mesophilic anaerobic sludge as the inoculum. Environ. Sci. Technol. 54 (22), 14656–14663. https://doi.org/10.1021/acs.est.0c02557.

Holanda, R., Johnson, D.B., 2020. Removal of zinc from circum-neutral pH mine-impacted waters using a novel "Hybrid" low pH sulfidogenic bioreactor. Front. Environ. Sci. 8. https://doi.org/10.3389/fenvs.2020.00022.

Huisman, J.L., Schouten, G., Schultz, C., 2006. Biologically produced sulphide for purification of process streams, effluent treatment and recovery of metals in the metal and mining industry. Hydrometallurgy 83 (1–4), 106–113. https://doi.org/10.1016/j.hydromet.2006.03.017.

Itoh, T., Yamanoi, K., Kudo, T., Ohkuma, M., Takashina, T., 2011. *Aciditerrimonas ferrireducens* gen. nov., sp. nov., an iron-reducing thermoacidophilic actinobacterium isolated from a solfataric field. Int. J. Syst. Evol. Microbiol. 61 (6), 1281–1285. https://doi.org/10.1099/ijs.0.023044-0.

Johnson, D.B., Hallberg, K.B., 2005. Acid mine drainage remediation options: a review. Sci. Total Environ. 338 (1–2), 3–14. https://doi.org/10.1016/j.scitotenv.2004.09.002.

Johnson, D.B., Sánchez-Andrea, I., 2019. Dissimilatory reduction of sulfate and zero-valent sulfur at low pH and its significance for bioremediation and metal recovery. Adv. Microb. Physiol. 75, 205–231. https://doi.org/10.1016/bs.ampbs.2019.07.002.

Johnson, D.B., Santos, A.L., 2020. Biological removal of sulfurous compounds and metals from inorganic wastewaters. In: Environmental Technologies to Treat Sulfur Pollution: Principles and Engineering, second ed. IWA Publishing, United Kingdom, pp. 215–246. https://doi.org/10.2166/9781789060966_0215.

Kerr, R.A., 2014. The coming copper peak. Science 343 (6172), 722–724. https://doi.org/10.1126/science.343.6172.722.

Kossoff, D., Dubbin, W.E., Alfredsson, M., Edwards, S.J., Macklin, M.G., Hudson-Edwards, K.A., 2014. Mine tailings dams: characteristics, failure, environmental impacts, and remediation. Appl. Geochem. 51, 229–245. https://doi.org/10.1016/j.apgeochem.2014.09.010.

Lewis, A.E., 2010. Review of metal sulphide precipitation. Hydrometallurgy 104 (2), 222–234. https://doi.org/10.1016/j.hydromet.2010.06.010.

López-Archilla, A.I., Marin, I., Amils, R., 2001. Microbial community composition and ecology of an acidic aquatic environment: the Tinto River, Spain. Microb. Ecol. 41 (1), 20–35. https://doi.org/10.1007/s002480000044.

Lumbroso, D., Davison, M., Body, R., Petkovšek, G., 2021. Modelling the Brumadinho tailings dam failure, the subsequent loss of life and how it could have been reduced. Nat. Hazards Earth Syst. Sci. 21 (1), 21–37. https://doi.org/10.5194/nhess-21-21-2021.

Lusty, P.A.J., Gunn, A.G., 2015. Challenges to global mineral resource security and options for future supply. Spec. Publ. 393 (1), 265–276. https://doi.org/10.1144/sp393.13.

Macklin, M.G., Thomas, C.J., Mudbhatkal, A., Brewer, P.A., Hudson-Edwards, K.A., Lewin, J., Scussolini, P., Eilander, D., Lechner, A., Owen, J., Bird, G., Kemp, D., Mangalaa, K.R., 2023. Impacts of metal mining on river systems: a global assessment. Science 381 (6664), 1345–1350. https://doi.org/10.1126/science.adg6704.

Madigan, M.T., 2021. Brock Biology of Microorganisms, sixteenth ed. International Microbiology, pp. 149–152.

Melton, E.D., Schmidt, C., Kappler, A., 2012. Microbial iron (II) oxidation in littoral freshwater lake sediment: the potential for competition between phototrophic vs. nitrate-reducing iron (II)-oxidizers. Front. Microbiol. 3, 197. https://doi.org/10.3389/fmicb.2012.00197.

Moncur, M.C., Ptacek, C.J., Lindsay, M.B.J., Blowes, D.W., Jambor, J.L., 2015. Long-term mineralogical and geochemical evolution of sulfide mine tailings under a shallow water cover. Appl. Geochem. 57, 178. https://doi.org/10.1016/j.apgeochem.2015.01.012.

Mori, K., Kim, H., Kakegawa, T., Hanada, S., 2003. A novel lineage of sulfate-reducing microorganisms: thermodesulfobiaceae fam. nov., *Thermodesulfobium narugense*, gen. nov., sp. nov., a new thermophilic isolate from a hot spring. Extremophiles 7 (4), 283–290. https://doi.org/10.1007/s00792-003-0320-0.

Muyzer, G., Stams, A.J.M., 2008. The ecology and biotechnology of sulphate-reducing bacteria. Nat. Rev. Microbiol. 6 (6), 441–454. https://doi.org/10.1038/nrmicro1892.

Ňancucheo, I., Barrie Johnson, D., 2014. Removal of sulfate from extremely acidic mine waters using low pH sulfidogenic bioreactors. Hydrometallurgy 150, 222–226. https://doi.org/10.1016/j.hydromet.2014.04.025.

Ňancucheo, I., Johnson, D.B., 2012. Selective removal of transition metals from acidic mine waters by novel consortia of acidophilic sulfidogenic bacteria. Microb. Biotechnol. 5 (1), 34–44. https://doi.org/10.1111/j.1751-7915.2011.00285.x.

Nancucheo, I., Bitencourt, J.A.P., Sahoo, P.K., Alves, J.O., Siqueira, J.O., Oliveira, G., 2017. Recent developments for remediating acidic mine waters using sulfidogenic bacteria. BioMed Res. Int. 2017. https://doi.org/10.1155/2017/7256582.

NASA Jet Propulsion Laboratory, 2024. NASA's Curiosity Rover discovers a surprise in a Martian rock'. https://www.jpl.nasa.gov/news/nasas-curiosity-rover-discovers-a-surprise-in-a-martian-rock/. (Accessed 18 July 2024).

Ohmura, N., Sasaki, K., Matsumoto, N., Saiki, H., 2002. Anaerobic respiration using Fe^{3+}, S^0, and H_2 in the chemolithoautotrophic bacterium *Acidithiobacillus ferrooxidans*. J. Bacteriol. 184 (8), 2081–2087.

Parbhakar-Fox, A., Lottermoser, B.G., 2015. A critical review of acid rock drainage prediction methods and practices. Miner. Eng. 82, 107–124. https://doi.org/10.1016/j.mineng.2015.03.015.

Park, I., Tabelin, C.B., Jeon, S., Li, X., Seno, K., Ito, M., Hiroyoshi, N., 2019. A review of recent strategies for acid mine drainage prevention and mine tailings recycling. Chemosphere 219, 588–606. https://doi.org/10.1016/j.chemosphere.2018.11.053.

Parro, V., Moreno-Paz, M., González-Toril, E., 2007. Analysis of environmental transcriptomes by DNA microarrays. Environ. Microbiol. 9 (2), 453–464. https://doi.org/10.1111/j.1462-2920.2006.01162.x.

Prior, T., Giurco, D., Mudd, G., Mason, L., Behrisch, J., 2012. Resource depletion, peak minerals and the implications for sustainable resource management. Glob. Environ. Change 22 (3), 577–587. https://doi.org/10.1016/j.gloenvcha.2011.08.009.

Rajhi, H., Conthe, M., Puyol, D., Díaz, E., Sanz, J.L., 2013. Dark fermentation: isolation and characterization of hydrogen-producing strains from sludges. Int. Microbiol. 16 (1), 53–62. https://doi.org/10.2436/20.1501.01.180.

Ramamoorthy, S., Sass, H., Langner, H., Schumann, P., Kroppenstedt, R.M., Spring, S., Overmann, J., Rosenzweig, R.F., 2006. *Desulfosporosinus lacus* sp. nov., a sulfate-reducing bacterium isolated from pristine freshwater lake sediments. Int. J. Syst. Evol. Microbiol. 56 (12), 2729–2736. https://doi.org/10.1099/ijs.0.63610-0.

Rawlings, D.E., Tributsch, H., Hansford, G.S., 1999. Reasons why '*Leptospirillum*'-like species rather than *Thiobacillus ferrooxidans* are the dominant iron-oxidizing bacteria in many commercial processes for the biooxidation of pyrite and related ores. Microbiology 145 (1), 5–13. https://doi.org/10.1099/13500872-145-1-5.

Rohwerder, T., Janosch, C., Sand, W., 2007. Elemental sulfur oxidation in *Acidiphilium* spp. Adv. Mater. Res. 20–21, 583. https://doi.org/10.4028/0-87849-452-9.583.

Rufo, L., Rodríguez, N., Amils, R., de la Fuente, V., Jiménez-Ballesta, R., 2007. Surface geochemistry of soils associated to the Tinto River (Huelva, Spain). Sci. Total Environ. 378 (1–2), 223–227. https://doi.org/10.1016/j.scitotenv.2007.01.051.

Sánchez-Andrea, I., Rodríguez, N., Amils, R., Sanz, J.L., 2011. Microbial diversity in anaerobic sediments at Río Tinto, a naturally acidic environment with a high heavy metal content. Appl. Environ. Microbiol. 77 (17), 6085–6093. https://doi.org/10.1128/AEM.00654-11Spain.

Sánchez-Andrea, I., 2012. *Diversidad microbiana de los sedimentos anaerobios de río Tinto* (Doctoral dissertation). Universidad Autónoma de Madrid.

Sánchez-Andrea, I., Knittel, K., Amann, R., Amils, R., Sanz, J.L., 2012a. Quantification of Tinto river sediment microbial communities: importance of sulfate-reducing bacteria and their role in attenuating acid mine drainage. Appl. Environ. Microbiol. 78 (13), 4638–4645. https://doi.org/10.1128/AEM.00848-12Spain.

Sánchez-Andrea, I., Rojas-Ojeda, P., Amils, R., Sanz, J.L., 2012b. Screening of anaerobic activities in sediments of an acidic environment: Tinto River. Extremophiles 16 (6), 829–839. https://doi.org/10.1007/s00792-012-0478-4.

Sánchez-Andrea, I., Triana, D., Sanz, J.L., 2012c. Bioremediation of acid mine drainage coupled with domestic wastewater treatment. Water Sci. Technol. 66 (11), 2425–2431. https://doi.org/10.2166/wst.2012.477Spain.

Sánchez-Andrea, I., Stams, A.J.M., Amils, R., Sanz, J.L., 2013. Enrichment and isolation of acidophilic sulfate-reducing bacteria from Tinto River sediments. Environ. Microbiol. Rep. 5 (5), 672–678. https://doi.org/10.1111/1758-2229.12066.

Sánchez-Andrea, I., Sanz, J.L., Stams, A.J.M., 2014. *Microbacter margulisiae* gen. nov., sp. nov., a propionigenic bacterium isolated from sediments of an acid rock drainage pond. Int. J. Syst. Evol. Microbiol. 64, 3936–3942. https://doi.org/10.1099/ijs.0.066241-0.

Sánchez-Andrea, I., Stams, A.J.M., Hedrich, S., Ňancucheo, I., Johnson, D.B., 2015. *Desulfosporosinus acididurans* sp. nov.: an acidophilic sulfate-reducing bacterium isolated from acidic sediments. Extremophiles 19 (1), 39–47. https://doi.org/10.1007/s00792-014-0701-6.

Sánchez-Andrea, I., Florentino, A.P., Semerel, J., Strepis, N., Sousa, D.Z., Stams, A.J.M., 2018. Co-culture of a novel fermentative bacterium, *Lucifera butyrica* gen. nov. Sp. nov., with the sulfur reducer desulfurella amilsii for enhanced sulfidogenesis. Front. Microbiol. 9. https://doi.org/10.3389/fmicb.2018.03108.

Sánchez-Andrea, I., van der Graaf, C.M., Hornung, B., Bale, N.J., Jarzembowska, M., Sousa, D.Z., Rijpstra, W.I.C., Sinninghe Damsté, J.S., Stams, A.J.M., 2022. Acetate degradation at low pH by the moderately acidophilic sulfate reducer *Acididesulfobacillus acetoxydans* gen. nov. sp. nov. Front. Microbiol. 13. https://doi.org/10.3389/fmicb.2022.816605.

Sánchez-España, J., 2024. The Geomicrobiology of Acidic Pit Lakes Geomicrobiology: Natural and Anthropogenic Settings. Springer Nature, Spain, pp. 95–121. https://doi.org/10.1007/978-3-031-54306-7_6.

Santos, A.L., Johnson, D.B., 2018. Design and application of a low pH upflow biofilm sulfidogenic bioreactor for recovering transition metals from synthetic waste water at a Brazilian copper mine. Front. Microbiol. 9. https://doi.org/10.3389/fmicb.2018.02051.

Sanz, J.L., Rodríguez, N., Díaz, E.E., Amils, R., 2011. Methanogenesis in the sediments of Rio Tinto, an extreme acidic river. Environ. Microbiol. 13 (8), 2336–2341. https://doi.org/10.1111/j.1462-2920.2011.02504.x.

Simate, G.S., Ndlovu, S., 2014. Acid mine drainage: challenges and opportunities. J. Environ. Chem. Eng. 2 (3), 1785–1803. https://doi.org/10.1016/j.jece.2014.07.021.

Šimek, M., Jíšová, L., Hopkins, D.W., 2002. What is the so-called optimum pH for denitrification in soil? Soil Biol. Biochem. 34 (9), 1227–1234. https://doi.org/10.1016/S0038-0717(02)00059-7.

Skousen, J., Zipper, C.E., Rose, A., Ziemkiewicz, P.F., Nairn, R., McDonald, L.M., Kleinmann, R.L., 2017. Review of passive systems for acid mine drainage treatment. Mine Water Environ. 36 (1), 133–153. https://doi.org/10.1007/s10230-016-0417-1.

Skousen, J.G., Ziemkiewicz, P.F., McDonald, L.M., 2019. Acid mine drainage formation, control and treatment: approaches and strategies. Extr. Ind. Soc. 6 (1), 241–249. https://doi.org/10.1016/j.exis.2018.09.008.

Souza-Egipsy, V., González-Toril, E., Zettler, E., Amaral-Zettler, L., Aguilera, A., Amils, R., 2008. Prokaryotic community structure in algal photosynthetic biofilms from extreme acidic streams in Río Tinto (Huelva, Spain). Int. Microbiol. 11 (4), 251–260. https://doi.org/10.2436/20.1501.01.69.

Sverdrup, H.U., Olafsdottir, A.H., Ragnarsdottir, K.V., 2019. On the long-term sustainability of copper, zinc and lead supply, using a system dynamics model. Resour. Conserv. Recycl. X 4. https://doi.org/10.1016/j.rcrx.2019.100007.

Tabelin, C.B., Igarashi, T., Villacorte-Tabelin, M., Park, I., Opiso, E.M., Ito, M., Hiroyoshi, N., 2018. Arsenic, selenium, boron, lead, cadmium, copper, and zinc in naturally contaminated rocks: a review of their sources, modes of enrichment, mechanisms of release, and mitigation strategies. Sci. Total Environ. 645, 1522–1553. https://doi.org/10.1016/j.scitotenv.2018.07.103.

Tahmasebi, P., Hezarkhani, A., 2010. Application of adaptive neuro-fuzzy inference system for grade estimation; case study, sarcheshmeh porphyry copper deposit, Kerman, Iran. Aust. J. Basic Appl. Sci. 4 (3), 408–420.

van der Graaf, C.M., Sánchez-España, J., Yusta, I., Ilin, A., Shetty, S.A., Bale, N.J., Villanueva, L., Stams, A.J.M., Sánchez-Andrea, I., 2020. Biosulfidogenesis mediates natural attenuation in acidic mine pit lakes. Microorganisms 8 (9), 1–26. https://doi.org/10.3390/microorganisms8091275.

Van Houten, B.H.G.W., Roest, K., Tzeneva, V.A., Dijkman, H., Smidt, H., Stams, A.J.M., 2006. Occurrence of methanogenesis during start-up of a full-scale synthesis gas-fed reactor treating sulfate and metal-rich wastewater. Water Res. 40 (3), 553–560. https://doi.org/10.1016/j.watres.2005.12.004.

Watari, T., McLellan, B., Ogata, S., Tezuka, T., 2018. Analysis of potential for critical metal resource constraints in the international energy agency's long-term low-carbon energy scenarios. Minerals 8 (4), 156. https://doi.org/10.3390/min8040156.

Watari, T., Nansai, K., Nakajima, K., 2020. Review of critical metal dynamics to 2050 for 48 elements. Resour. Conserv. Recycl. 155, 104669. https://doi.org/10.1016/j.resconrec.2019.104669.

Werner, T.T., Mudd, G.M., Schipper, A.M., Huijbregts, M.A.J., Taneja, L., Northey, S.A., 2020. Global-scale remote sensing of mine areas and analysis of factors explaining their extent. Glob. Environ. Change 60. https://doi.org/10.1016/j.gloenvcha.2019.102007.

Widdel, F., Pfennig, N., 1982. Studies on dissimilatory sulfate-reducing bacteria that decompose fatty acids II. Incomplete oxidation of propionate by *Desulfobulbus propionicus* gen. nov., sp. nov. Arch. Microbiol. 131 (4), 360–365. https://doi.org/10.1007/bf00411187.

Wright, V., Morzfeld, M., Manga, M., 2024. Liquid water in the Martian mid-crust. Proc. Natl. Acad. Sci. 121 (35). https://doi.org/10.1073/pnas.2409983121.

Zhuravleva, A.E., Tsaplina, I.A., Lsmailov, A.D., Zakharchuk, L.M., Karavaiko, G.I., 2007. Federation Metabolism peculiarities of bacteria of the genus *Sulfobacillus*. Adv. Mater. Res. 20–21 (469), 469–472. https://doi.org/10.4028/0-87849-452-9.

CHAPTER

Cyanobacteria at extreme acidic part of the Tinto system

6

Ignacio Carrasco-Ropero[1,2], Miguel Arribas-Tiemblo[1,3], Felipe Gómez[1]

[1]*Departmento de Planetología y Habitabilidad, Centro de Astrobiología (INTA-CSIC), Torrejón de Ardoz, Madrid, Spain;* [2]*Centro de Biología Molecular Severo Ochoa, Universidad Autónoma de Madrid, Cantoblanco, Madrid, Spain;* [3]*Universidad de Alcalá de Henares, Alcalá de Henares, Madrid, Spain*

Introduction

Río Tinto, located in the province of Huelva in southwestern Spain, is one of the most unusual and extreme natural environments on the Iberian Peninsula. With an extension of almost 100 km, the river is characterized by its extremely low pH and high levels of dissolved heavy metals. It raises in Peña de Hierro, in the core of the Iberian Pyrite Belt (IPB), one of the most massive deposit of metallic sulfides on Earth (Amils et al., 2002). The main characteristic of the river is its constant acidic pH, a direct consequence of the strong buffering capacity of the ferric ion (Fe^{3+}), which also is responsible for the river's characteristic red color (Amils et al., 2002).

Despite the extreme characteristics of the river, its diversity is really wide. Most of the bacterial species described in the extreme Rio Tinto ecosystem are represented by three genera of bacteria, *Acidithiobacillus*, *Leptospirillum* and *Acidiphilium*, involved in the maintenance of biogeochemical cycles in the Rio Tinto basin (González-Toril et al., 2003). Although the river water column presents a low prokaryotic diversity, which is consistent with what is expected from an extreme environment, it contains an unexpectedly high eukaryotic diversity, known as the eukaryotic paradox (Amaral Zettler et al., 2002).

Most of the eukaryotic organisms identified in the river are unicellular algae. These microalgae have traditionally been responsible for carrying out the photosynthetic activity of the ecosystem, sustaining primary production under such conditions (Amaral Zettler et al., 2002). Previous studies, based on both microscopic observations and molecular analysis, have consistently failed to detect cyanobacteria in acidic river waters, reinforcing the idea that they are unable to tolerate these extreme acidic environments (Introduction to the Cyanobacteria, 2002). Given the extreme conditions of Rio Tinto, could cyanobacteria persist in the microbial communities of the system?

Geomicrobiology of Río Tinto. https://doi.org/10.1016/B978-0-443-32948-7.00022-9

Río Tinto has long been established as an analogue of shallow, acidic freshwater environments that might have been abundant early in Mars' history, during the mid and late Noachian period (Grotzinger et al., 2022; Wordsworth, 2016; Kamada et al., 2021). It is during this time that minerals such as jarosite formed. This mineral is known to form under acidic conditions, and is present in the Martian surface and in Río Tinto (Amils et al., 2014; Oggerin et al., 2014; Klingelhöfer et al., 2004). Although acidic pockets might have existed in the past in the Martian surface, the Martian regolith currently exhibits a mostly neutral pH.

Acidotolerant cyanobacteria through history

Cyanobacteria are a group of photosynthetic prokaryotes that, due to their ability to carry out oxygenic photosynthesis, have played a fundamental role in the evolution of the terrestrial atmosphere and in the biogenesis of aquatic and terrestrial ecosystems (Introduction to the Cyanobacteria, 2002). They produce chlorophyll *a* and, in most cases, accessory pigments such as phycocyanin and phycoerythrin, responsible for their characteristic blue-green color (Saini et al., 2018). They have different lifestyles, being able to fix nitrogen and live freely (unicellular or multicellular filamentous) or in symbiosis with some fungi to form lichens and plants (Stanier et al., 1971). They are distributed in a wide variety of habitats, from oceanic waters to freshwater environments, soils, deserts, hypersaline environments, geothermal systems, even in the deep subsurface (Puente-Sánchez et al., 2018). This ecological versatility is supported by physiological adaptations, such as heterocyst formation for nitrogen fixation, tolerance to desiccation, and the ability to float through gas vacuoles, among others (Beversdorf et al., 2013). Most ecological reports suggest that cyanobacteria tend to be more diverse and abundant in pH-neutral or alkaline environments (Introduction to the Cyanobacteria, 2002). This is due to intracellular biochemistry that regulates photosynthesis and other essential metabolic processes. Because of these physiological limitations, cyanobacteria have historically been considered nonacidophilic and their presence in environments with pH < 5 is minimal or nonexistent (Introduction to the Cyanobacteria, 2002).

It has only been in the past few decades, with the advance of molecular and metagenomic techniques, that exceptional cases of cyanobacteria in acidic environments have begun to be identified (Gómez et al., 2024). Thus, certain species have been identified that manage to persist in acidic conditions. Under slightly acidic environments, with pH around 5.5–5.0, active *Synechococcus* strains have been recorded in hot springs such as Yellowstone (USA). Although their growth optimum resides in neutral conditions, these strains manage to maintain photosynthesis and ionic balance (Ward and Castenholz, 2025). Another example is found in *Chlorogloeopsis* strains, also isolated from hot springs with pH close to 5 (Keshari et al., 2022).

Between pH 4.5 and 4.1, cyanobacteria begin to show more specialized adaptations. In Scottish peat bogs, heterocyst genera such as *Hapalosiphon* and *Tolypothrix* have been observed as part of acidotolerant microbial communities (Introduction to the Cyanobacteria, 2002). Their ability to fix atmospheric nitrogen through heterocysts is attributed to their presence in these environments. These species also secrete exopolysaccharides (EPS) that form a protective matrix against acid and metal stress, acting as a physical and chemical barrier against environmental acidity (Zhang et al., 2007).

As the pH drops below 4, cyanobacteria are not that abundant, but certain filamentous genera such as *Gloeocapsa* and *Leptolyngbya* are still detected. These organisms form dense biofilms, whose EPS matrix acts as a buffering microenvironment (Brock, 1973). At pH around 3–2, some studies document the presence of cyanobacteria by optical microscopy in saline soils of peatlands and flooded paddy fields in Kerala, India, associated with soils originating from acid rock drainage (Introduction to the Cyanobacteria, 2002). Similarly, in the Los Rueldos mine (Asturias), cyanobacterial DNA sequences have been detected in water above pH 2 (Mesa et al., 2017).

Little is known yet about the relationship between the presence of cyanobacteria in extreme acidic environments and the molecular adaptations, which allow them to survive in such conditions. While traditional literature has held that low pH represents an impenetrable barrier for most cyanobacteria, recent studies have begun to change this view, revealing their presence in acidic systems, although at low densities or in association with complex microbial mats.

The role of microorganisms in humanity's path to Mars

Over the past several decades, Mars has been extensively studied through numerous unmanned missions launched by agencies such as NASA and ESA. Early orbiters such as Mariner 4 and 9, landers such as Viking 1 and 2 in the 1970s, and more recent rovers including Curiosity and Perseverance, have conducted in-depth analysis of Martian terrain and climate. These missions have significantly enhanced our understanding of Mars, offering evidence of an active geological past, a tenuous but dynamic atmosphere, and, perhaps most intriguingly, indications of both present and ancient water reservoirs (Clark et al., 2021; Kite and Conway, 2024).

Mars contains many of the essential building blocks for life. Elements that are critical for life on Earth—carbon, hydrogen, nitrogen, oxygen, phosphorus, and sulfur—have all been detected in Martian soils or atmosphere (Clark et al., 2021; Cockell, 2014). Additionally, trace nutrients such as iron, magnesium, nickel, and zinc have been found in Martian rocks (Bosak et al., 2024). These discoveries, along with the sedimentological evidence of ancient water flow, reinforce the idea that Mars may have once offered habitable conditions, and they strengthen the scientific rationale for future human missions.

One of the key breakthroughs in Martian science was the detection of organic molecules in samples analyzed by Curiosity's Sample Analysis at Mars (SAM) suite (Guzman et al., 2018). While the presence of such molecules does not definitively confirm biological activity, it highlights the potential for Mars to preserve biosignatures over long geological timescales. Other compounds of astrobiological interest—including amino acids (ten Kate et al., 2005; Megevand et al., 2021; Liu and Kounaves, 2021; Pavlov et al., 2022), lipids (Finkel et al., 2023; Moreno-Paz et al., 2023), and carotenoids (Baqué et al., 2016; Rzymski et al., 2022, 2024)—have not yet been identified on Mars, but studies suggest they could survive near the Martian surface for millions of years, particularly under certain mineral or regolith conditions.

Resupplying future Martian missions from the Earth will not be possible over short timescales, unlike current operations aboard the ISS, which rely on regular resupply missions for food, equipment, and life support systems (Koehle et al., 2023). Mars' distance from Earth, communication delays of up to 20 min, and the enormous cost and complexity of interplanetary transport will make consistent resupplies unfeasible. As a result, developing technologies that allow for autonomy and independence from Earth—referred to as In-Situ Resource Utilization (ISRU)—has become a critical area of research for space agencies around the world (Mapstone et al., 2022; Ramalho et al., 2022).

The potential of cyanobacteria on in situ resource utilization (ISRU)

To address the challenges of sustaining human presence on Mars, ISRU systems must be designed to produce vital resources such as oxygen, food, and fuel using only local materials and minimal external inputs. Among the most promising solutions for biological life support systems is the use of microorganisms—particularly cyanobacteria. These photosynthetic prokaryotes are uniquely suited to the Martian environment due to their metabolic versatility, simplicity, and resilience in extreme conditions.

The ability of cyanobacteria to produce significant amounts of oxygen through photosynthesis makes them a crucial component of many ecosystems, particularly in aquatic environments. In these, genera such as *Prochlorococcus* and *Synechococcus* contribute significantly to global oxygen production and primary productivity (Johnson et al., 2006; Bezalel-Hazony et al., 2023; Kim et al., 2018). *Prochlorococcus* alone accounts for an estimated 13% of photosynthetic activity in nutrient-poor ocean waters. In contrast to terrestrial plants, cyanobacteria require fewer nutrients and far less space, making them more efficient for closed-loop systems in space missions (Macário et al., 2022; Sánchez-Bayo et al., 2020).

In addition to oxygen, cyanobacteria produce a variety of useful biomolecules, which have applications in both nutrition and biotechnology, and in some cases, cyanobacteria can be used directly as food. Their high protein content, ease of cultivation, and broad spectrum of essential nutrients make them suitable for supplementing astronaut diets on long-duration missions. Species from the genera *Nostoc* and *Arthrospira* (commonly known as Spirulina) have long been consumed as nutrient-rich food sources in regions such as China and Peru (Johnson et al., 2008; Mutale-Joan et al., 2023).

Beyond nutrition, cyanobacteria are valuable for other ISRU applications. Some species can fix atmospheric nitrogen into bioavailable forms, reducing the need for chemical fertilizers. Others can produce biofuels, including third- and fourth-generation alternatives derived from lipids and carbohydrates (Farrokh et al., 2019; Grzesik et al., 2017). These traits open the door to integrated systems where cyanobacteria produce oxygen for breathing, organic matter for food or feedstock, and biofertilizers or energy carriers that support plant growth or auxiliary systems within a Martian habitat.

Besides producing organic matter and assisting in biofertilization, cyanobacteria can reliably provide another key limiting factor that any crewed mission outside of Earth needs: oxygen (Mapstone et al., 2022). Cyanobacteria are the first phototrophs to actively and abundantly produce oxygen in the history of Earth, and were the main oxygen producers for most of its history (Keller et al., 2023; Olsson-Francis and Cockell, 2010). Cyanobacteria can not only potentially grow well in the Martian regolith, but they can sustain their photosynthetic activity with little sunlight, and some can even perform far-red photosynthesis, which allows them to use parts of the spectrum typically inaccessible to photosynthetic organisms (Gan and Bryant, 2015). The resilience of cyanobacteria under sources of stress such as desiccation, radiation, and nutrient deprivation, along with their ability to efficiently produce oxygen and organic matter makes them highly compatible with the design of Bioregenerative Life Support Systems (BLSS), which will be necessary for space missions that aim to be sustainable for long periods of time away from Earth.

Cyanobacteria also demonstrate impressive resistance to environmental stressors. Exposure to ultraviolet (UV) radiation, which is intense on Mars due to the thin atmosphere and lack of a protective ozone layer, poses a serious challenge to any biological system. However, many cyanobacteria have evolved mechanisms to cope with UV stress. They produce specialized compounds that can absorb and dissipate harmful UV radiation (Rastogi et al., 2020; Dwivedi and Ahmad, 2023). Moreover, certain strains have robust DNA repair systems and antioxidant defenses that mitigate oxidative damage caused by ionizing radiation or desiccation.

Among the most resilient cyanobacteria studied is *Chroococcidiopsis*, a genus known for its tolerance to extreme desiccation, radiation, and even microgravity (Fagliarone et al., 2020; Billi et al., 2019). Its ability to survive under Mars-analog conditions has made it a model organism for astrobiology research. Specific desiccation-resistant cyanobacteria such as *Chroococcidiopsis* sp. CCMEE 029 have been used as models to assess the ability of specific cyanobacterial species to resist and recover from sources of damage common in the Martian surface, such as desiccation and UV radiation (Fagliarone et al., 2023).

Under experimental conditions, *Chroococcidiopsis* sp. CCMEE 029 can restart cellular division only 28 h after being rehydrated. That is the approximate time it takes to fully recover all DNA damage originated from radiation and desiccation. *Chroococcidiopsis* sp. CCMEE 029 has a lower growth rate than freshwater cyanobacteria such as *Desmonostoc* and *Anabaena*. However, these cyanobacteria are also considerably more sensitive to radiation and desiccation than the former, which we can consider to be the floor for cyanobacterial productivity under optimal conditions and the ceiling for survival and growth recovery under Martian surface conditions.

For these biological systems to be viable on Mars, they must be able to grow using only local resources. Martian regolith—a loose mixture of rock, dust, and salts—contains many of the elements required for cyanobacterial growth. However, due to limited access to actual Martian material, researchers rely on terrestrial analogs and commercial simulants such as MGS-1 and MMS-2 to study microbial colonization and viability (Cannon et al., 2019; Caporale et al., 2020). These simulants replicate the mineralogical and geochemical composition of Martian soil and can be supplemented with perchlorates, oxidizing compounds known to exist on the Martian surface (Hecht et al., 2009; Kounaves et al., 2014; McAdam et al., 2022; Clark and Kounaves, 2016).

While perchlorates are toxic to most life forms, some cyanobacteria demonstrate moderate tolerance. Experiments have shown that even in the presence of these compounds, species such as *Anabaena cylindrica* and *Desmonostoc muscorum* can grow in regolith simulants, albeit with reduced rates depending on the concentration and substrate composition (Moreno-Paz et al., 2023). These findings are promising, as they suggest that cyanobacteria can serve as pioneering organisms in the early stages of Martian bioengineering.

Cyanobacterial mats in Río Tinto: From bibliography to microbial confirmation

Cyanobacterial mats

Cyanobacterial mats represent one of the oldest and most structured forms of microbial life on Earth (Prieto-Barajas et al., 2018). These systems consist of multilayered communities that include photoautotrophic cyanobacteria, heterotrophic bacteria, chemolithotrophs, Archaea and often microalgae. Their fossil origin dates back to the Archean, with records of stromatolites older than 3.5 billion years, demonstrating their longevity as self-regulating functional consortia (Prieto-Barajas et al., 2018). The evolutionary advantage of this lifestyle resides in the close metabolic cooperation between the different members, the vertical compartmentalization of ecological functions and the resistance to extreme environmental perturbations such as These cyanobacterial mats are characterized by a marked vertical stratification of microbial communities, determined by physicochemical gradients of light, oxygen, sulfur, carbon dioxide and pH, which are largely generated by the metabolic activity of the organisms present. In this context, cyanobacteria play a key role as primary producers located in the surface layer, where irradiance allows oxygenic photosynthesis. Sometimes, cyanobacteria are covered by a layer of diatoms, fine sediments or a mucilaginous matrix rich in organic matter and protective pigments such as scytonemin, which protects the lower layers from UV radiation (Gómez et al., 2024).

Cyanobacteria's primary production introduces organic matter into the system, which is rapidly consumed by chemotrophic bacteria, causing anoxic conditions under the upper oxygen layer. This transition favors the establishment of anaerobic bacteria, particularly sulfate-reducing bacteria, which play an essential role in the degradation of organic matter, generating sulfide as a metabolic product. On the other hand, this sulfide is used as a substrate for anoxygenic phototrophic bacteria, such as purple sulfur bacteria. The deeper layer of the microbial mat is often colored black or gray, resulting from the formation of compounds such as ferrous sulfide (FeS) or pyrite (FeS_2) (Prieto-Barajas et al., 2018).

Daily environmental changes, such as sunlight, cause significant variations in the vertical profiles of light, oxygen, pH and redox compounds, demanding high metabolic plasticity from all microorganisms in the microbial mat. In this way, the idea that the microbial mat is neither static nor perfectly stratified is reinforced, acquiring the view that the structure is highly dynamic (Prieto-Barajas et al., 2018).

Can these cyanobacterial mats be found in Río Tinto?

The biodiversity of Río Tinto has been described in different studies in which the presence of both prokaryotes and eukaryotes has been reported (Amaral Zettler et al., 2002; Gómez et al., 2024; González-Toril et al., 2003). The only photosynthetic microorganisms described in the river have been eukaryotic algae (González-Toril et al., 2003), but the presence of cyanobacteria has not been reported so far.

The sampling area (37.721451° N, 6.551082° W) was located in an acidic water runoff, close to the origin of Río Tinto, over a mining tunnel wall. This runoff deposits mineral precipitates on the wall forming laminated crusts in which three different layers are visible, a yellow surface layer, a bluish greenish middle layer and an ochre brown lower layer (Fig. 6.1) (Gómez et al., 2024).

FIGURE 6.1

Cyanobacteria is located at the inner green layer. Several yellow and brown layers are covering the cyanobacteria ecosystem.

This stratigraphic crust is similar to the pattern of classic microbial mats described in neutral or alkaline environments, such as those observed in hypersaline systems or in alkaline thermal lakes. In the case of Río Tinto, this same structural principle is maintained, although adapted to an extremophile context. The microorganisms in the mat develop protected under a mineral crust, which reduces direct irradiance and acts as a physicochemical barrier against extreme pH. This three-dimensional structure not only buffers acidity and temperature gradients, but also creates microhabitats where metabolites accumulate and favorable redox conditions are stabilized. The extracellular matrix excreted by the higher mat bacteria and other members of the consortium also contributes to this compartmentalization, retaining moisture and nutrients, and allowing a biofilm-like organization in which metabolic processes are spatially coupled (Gómez et al., 2024).

To identify the presence of acidotolerant cyanobacteria in the Río Tinto microbial mats, a combination of complementary techniques was applied. Scanning electron microscopy (SEM), transmission electron microscopy (TEM) and Confocal laser scanning microcopy methods were used to analyze cellular morphology and structural microorganisms. X-ray diffraction (XRD) were served to analyze the mineral composition. Spectroscopy analyses, such as visible and near-infrared (VIS-NIR UV) and Raman spectroscopy helped characterize the pigment composition and optical properties of the microbial mat. Additionally, metagenomic sequencing provided comprehensive insights into the microbial diversity and functional potential of the community. Together, these multidisciplinary approaches enabled a robust assessment of cyanobacterial presence and adaptation in one of the most acidic natural environments on Earth (Gómez et al., 2024).

Acidotolerant cyanobacterial mat: Evidence of cyanobacteria in the microbial mats from Río Tinto

Cyanobacterial mat composition

As mentioned above, the microbial mat described in the source of Río Tinto presents a well-defined stratification in three visually and mineralogically distinct layers, whose composition has been studied by XRD, SEM, TEM and Raman spectroscopy techniques. The outermost layer, yellow-brownish color, is composed mostly of gypsum crystals ($CaSO_4$–$2H_2O$), as revealed by XRD analysis. This outer layer acts as a protective crust against direct solar irradiation and chemical leaching, which favors the structural preservation of the mat. In neutral microbial mats, similar structures have also been described as surface mineralized layers that stabilize the microbial matrix (Gómez et al., 2024).

The second layer, located immediately below the gypsum crust, has a bluish-green color and belongs to the metabolically active zone where the acidotolerant cyanobacteria were located. This layer was studied by SEM, observing spherical cell morphologies adhered to the mineral matrix, and detecting an abundance of iron (Fe) and sulfur (S), suggesting a matrix rich in natrojarosite ($NaFe_3(SO_4)_2(OH)_6$). This mineral was also confirmed by XRD and Raman. The fluorescence observed at 650 nm, together with Raman peaks at 1522, 1158 and 1106 cm^{-1}, indicated the presence of carotenoid pigments typical of cyanobacteria, integrated in a complex microbial mat. The association of microorganisms with minerals such as jarosite and goethite in acidic contexts has already been reported in similar environments, such as those of the IPB, where these precipitates act as protective matrices or even as mediators in biomineralization processes (Gómez et al., 2024).

Finally, the brownish ochre-colored layer closely overlaps with the blue-green layer. Mineralogical analysis revealed that this also consists mainly of natrojarosite, although with detectable traces of quartz (SiO_2). In contrast to mats in neutral or hypersaline environments, in the Río Tinto system the dominant gradients are of acidity, heavy metals and pH, which has conditioned a specialized and mineralogically consolidated organization (Gómez et al., 2024). The internal mineral layer acts as the structural base of the mat, promoting the physical and chemical stability of the microbial consortium, a characteristic also observed in mats of markedly geothermal environments described in the literature (Ward and Castenholz, 2025).

Physicochemical parameters of the cyanobacterial mat

The extreme conditions prevailing in the system do not prevent the microbial mat from being under stable conditions, which allow the development of a photosynthetic ecosystem in an environment with an average pH of 0.8 (Gómez et al., 2024). In situ measurements made with multiparameter probes (PD650, Eutech Instruments) and microelectrodes confirmed this constant acidity, both at the surface and inside the mineral deposit, which is consistent with the presence of secondary minerals such as natrojarosite, derived from the intense acid mine drainage characteristic of the Rio Tinto basin (Amann et al., 1990). The temperature also remained stable, around 12°C during the sampling campaign, with no relevant variations between mat layers. This thermal stability supports a low rate of cellular stress and allows the survival of microbial consortia.

Light intensity in the habitat, another key parameter for photosynthetic activity, was measured using VIS-NIR UV spectroscopy (200–850 nm). A progressive attenuation of irradiance was observed, with a 27% loss of light at 650 nm as it passed through the surface layers of the deposit (Gómez et al., 2024). However, the spectral fraction corresponding to photosynthetically active radiation (PAR) remained sufficient to allow photosynthetic activity, especially in the blue-green layer where the cyanobacteria were located. Specifically, this layer showed characteristic absorptions at 440, 620, and 675 nm, with a fluorescence peak at 650 nm, typical indicators of chlorophyll and carotenoids (Gómez et al., 2024). The mat structure, acting as a physical barrier, selectively filters harmful radiation and provides a photostable environment, a phenomenon also described in other microbial systems protected by mineral crusts in deserts. This also reduces pH and irradiance fluctuations within the system, generating favorable microenvironments for the coupled metabolic activity of cyanobacteria and heterotrophic bacteria (Gómez et al., 2024). Previous studies on microbial mats in neutral environments have shown that this type of environmental buffering is critical for maintaining photosynthetic efficiency and consortium integrity (Gómez et al., 2024). In this way, the mineral structure of the deposit acts not only as a physical shield, but also as an environmental regulator, allowing a biophysical equilibrium conducive to photosynthetic microbial life to be maintained even under conditions of extreme acidity (Gómez et al., 2024).

Cyanobacterial mat under microscopy

Structural characterization was approached by a combination of SEM, TEM and CARD-FISH. SEM images revealed spherical cells 0.5–1.4 μm in diameter, unicellular or forming small aggregates. The cell envelope exhibited a polysaccharide capsule visible as a thick band, followed by a layer of fibrous proteins and an externally mucilaginous matrix, a frequent feature in cyanobacteria adapted to osmotic and chemical stress (Gómez et al., 2024). SEM couplings confirmed the presence of Fe and S in these zones, aligning with the mineral composition of the inner layers of the mat (Gómez et al., 2024). TEM images examined the internal morphology in more detail, showing concentric peripheral thylakoids, carboxysomes, polyphosphate granules, glycogen inclusions and lipid droplets. These results indicate a subcellular organization typical of photosynthetic cyanobacteria, with capacity for energy accumulation and structural adaptation to extreme conditions (Gómez et al., 2024).

Through CARD-FISH it has been possible to taxonomically confirm the presence of cyanobacteria (Fig. 6.2), by positive hybridization of these microorganisms

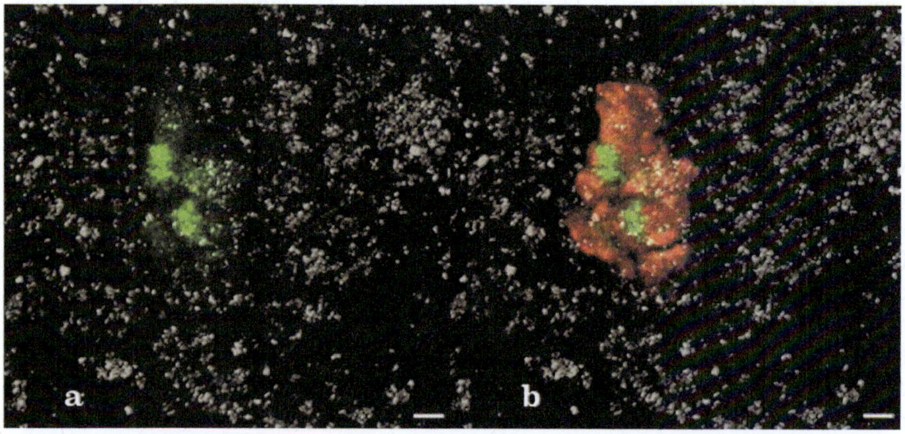

FIGURE 6.2 Cyanobacteria was identified and located using CARD-FISH techniques.

(A) Cyanobacteria detected by CYA 361 (green); (B): Merge of images: Bacteria detected by EUB338 I–III probe (orange) and cyanobacteria from (A) (green).

with the use of cyanobacterial DNA probe (CYA361) and the general bacterial DNA probe (EUB338 I-III) (Gómez et al., 2024), within the blue-green layer. These microorganisms are located closely related to other microorganisms. This connected cellular clustering allows to attenuate the extreme conditions of the system, allowing an optimal development within the microbial mat (Gómez et al., 2024).

Cyanobacteria are part of the living matrix in the mat. Their presence in protected microhabitats suggests adaptation to extreme stress by pH, metals and low light. Compared with mats in neutral conditions, such as those described by (Stal, 2012), the presence of concentric carboxysomes and thylakoids reinforces functional convergence at the subcellular level, even though the mineral and chemical context is radically different.

This approach has allowed us to determine the relationship of these cyanobacteria in situ, in addition to other microorganisms, being able to survive within protective mineral matrices, showing adaptations comparable with those of microbial mats in optimal environments, but under much more severe environmental conditions.

Genomic analysis of the cyanobacterial mat

The application of bioinformatics analysis was able to determine that 1.47% of the sequenced reads were assigned to the phylum Cyanobacteria, while the Bacteria domain predominated (~32%), with notable presence of Proteobacteria (current Pseudomonadota), Acidobacteria, and Firmicutes. These three phyla have been reported in the literature to be involved in cyanobacterial mats. Furthermore, sequences belonging to the genera *Synechococcus* and *Chlorogloeopsis* could be detected, showing their relationship with previously described thermoacidophilic strains (Gómez et al., 2024).

In order to understand the metabolic role of these microorganisms in the sample, a comprehensive genetic identification was performed. In this way, key genes of photosynthesis, such as *psaC, psbA* (D1 protein) and allophycocyanin, could be identified with more than 99% identity coverage, confirming the potential photoautotrophic capacity of cyanobacteria under extreme conditions (Gómez et al., 2024). Antioxidant genes such as *trxC* (thioredoxin) and *prx* (peroxiredoxin), associated with redox protection, were also found (Gómez et al., 2024).

An important ecological role reported for cyanobacteria is nitrogen fixation (Beversdorf et al., 2013; Gómez et al., 2024), especially in oligotrophic environments such as Río Tinto. By detecting the enzyme nitrogenase together with the gene *ntcA* (Gómez et al., 2024), which encodes for a regulatory protein required for the expression of the genes repressed by ammonium in cyanobacteria, the nitrogen-fixing capacity of cyanobacteria in this microbial mat is supported (Gómez et al., 2024). Membrane protein analysis revealed the presence of V-type ATPase (V-ATPase) variants, involved in proton pumping and extreme pH tolerance (Gómez et al., 2024). These pumps were identified with >95% identity to known acidophil sequences, supporting the hypothesis of active intracellular pH maintenance.

Understanding how these cyanobacteria may perform under extreme ecosystem conditions is also an important question. Little is known about the mechanisms of adaptation of cyanobacteria to such adverse environments as hyperacid pH. An ATP-dependent DNA helicase has been described to play a central role in the reaction that drives DNA helix unfolding and DNA repairing after DNA damage (Gómez et al., 2024). Cyanobacterial genomes are subject to damage by chemical agents in this hyperacidic environment with a high concentration of heavy metals. Thus, the identification of the ATP-dependent DNA helicase, RecG, in the metagenome indicates its possible role in different DNA repair systems (Gómez et al., 2024). Also, it has been possible to detect the enzyme YchF, an unconventional member of the universally conserved GTPase family, which preferentially hydrolyzes ATP rather than GTP. This GTPase has been associated with several cellular processes, including DNA repair. In particular, a possible role in the regulation of oxidative stress and the antioxidant response in bacteria has been suggested (Gómez et al., 2024). These metagenomic results clearly indicate that the identified genes are able to overcome part of the problems produced by the extreme conditions in which cyanobacteria develop.

Overall, the bioinformatics results demonstrate that, although the acidotolerant cyanobacteria of the Rio Tinto do not display a very complete metabolic repertoire, they can perform different metabolic pathways for energy production such as photosynthesis and nitrogen fixation.

Survival and growth of cyanobacteria under simulated Martian surface conditions

Growth in Martian regolith simulants

Several freshwater cyanobacteria—*Desmonostoc muscorum*, *Desmonostoc* sp., and *Anabaena cylindrica*— have been grown in Martian regolith analogs to assess the viability of their use through ISRU (Arribas Tiemblo et al., 2025). This was carried out under controlled conditions that simulate aspects of the Martian surface environment. Two main regolith simulants, MGS-1 and MMS-2, were tested, along with PRT, a highly acidic substrate based on rock samples from Río Tinto. These substrates differ in pH, mineral composition, and oxidative potential, simulating a range of geochemically diverse Martian terrains.

The first set of experiments focused on the ability of cyanobacteria to grow using only the soluble fraction of each regolith mixed with either distilled water (dH_2O) or a growth medium (MBL). All three species showed promising growth in extracts from both MGS-1 and MMS-2, although growth rates were higher in MGS-1. Growth in MMS-2 extracts was superior in those supplemented with MBL than in those purely in dH_2O. In contrast, MGS-1 supported similar growth levels regardless of supplementation, indicating that its intrinsic composition may provide sufficient soluble nutrients to support phototrophic life.

The acidic substrate PRT posed a significantly greater challenge. Only *Anabaena cylindrica* exhibited growth in this material, and at much lower rates compared with the other simulants. Unlike the cyanobacteria discussed previously, the three species used were nonacidotolerant freshwater cyanobacteria, which explains the absence of growth in an acidic substrate. Although some presumed ecosystems of Early Mars are presumed to have been acidic in nature, much of the Martian surface is expected to display a more neutral to slightly alkaline pH on much of the Martian surface (Gómez et al., 2010) at this time.

Colonization experiments on the solid phase of each regolith simulant yielded similar results. Using only dH_2O and sterilized regolith, *Desmonostoc muscorum* and *Desmonostoc* sp. effectively colonized both MGS-1 and MMS-2 within 70 days. The inflection point of their growth curves—an indicator of growth speed—was consistently lower in MGS-1, reinforcing the idea that its mineral composition facilitates the growth of cyanobacteria. When MBL was added to the substrate, differences between MGS-1 and MMS-2 nearly disappeared, suggesting that nutrient availability is the primary constraint in MMS-2. Interestingly, *A. cylindrica* formed patchy, inhomogeneous aggregates on all substrates and was the only species to exhibit limited growth in PRT.

Recovery of activity after desiccation and UV-B irradiation

The study also investigated cyanobacterial resistance to desiccation and UV-B radiation, simulating two of the main sources of stress on the Martian surface. Alongside MGS-1, MMS-2 and PRT, two clays (Montmorillonite and nontronite) were also used, as their hygroscopic properties are known to facilitate bacterial survival under desiccation (Fig. 6.3). Desiccated samples took approximately 2–4 times longer to reach full growth compared with wet controls. Still, most desiccated samples recovered their full viability and fully colonized the regolith within 40 days. Montmorillonite led to the fastest and most robust recovery, followed by nontronite. Both of these clay minerals display water retention capacities and their structure can offer protective pockets, which facilitate survival (Prieto-Barajas et al., 2018).

UV-B radiation experiments revealed more dramatic effects. Low doses ($<1000\ kJ/m^2$, roughly equivalent to 4 Martian sols of surface exposure) had little impact on cyanobacterial survival. However, at higher doses ($>24,000\ kJ/m^2$, or ~72 sols), growth was significantly delayed or completely absent, especially in MMS-2. This suggests that UV-B exposure is a cumulative and ultimately sterilizing force, while desiccation tends to consistently reduce—but not eliminate—viability. Importantly, short-term UV exposure alone is insufficient to fully sterilize cyanobacteria, while prolonged exposure may effectively do so. Therefore, specialized protective strategies—such as shielding with regolith, embedding in microniches, or cultivating under thin water layers—are essential.

Samples irradiated while submerged in a thin layer of water (1 mm) showed notably improved survival. Water is a UV shielder and helps moderate temperature fluctuations, both of which are relevant on the Martian surface. When irradiated

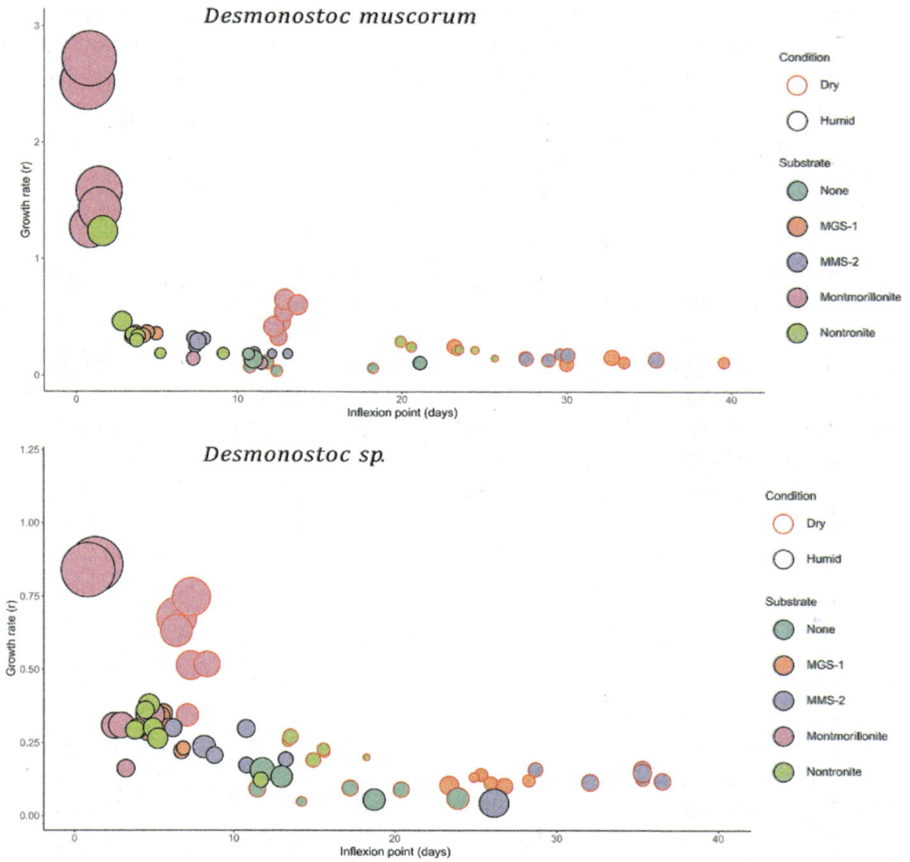

FIGURE 6.3

Representation of the growth rate (r) and inflection day (t_mid) of both cyanobacteria in all substrates but PRT (where no growth took place). Each point corresponds to the model derived from five replicates of each cyanobacterium, substrate and dose representation. The radius of each point is the standard error of the growth rate (y-axis).

under these conditions, growth rates were consistently higher, and inflection points were shorter. In these experiments, differences between regolith types also shifted. For example, nontronite—less favorable under dry conditions—outperformed MGS-1 in the presence of water, likely due to its hygroscopic properties and ability to buffer harsh environmental changes.

Throughout these trials, *Desmonostoc* sp. consistently showed faster growth and recovery than *D. muscorum*, despite being genetically close. This suggests that strain-level differences can play a significant role in selecting organisms for ISRU applications, especially under hostile environmental conditions.

The findings from this study highlight the considerable potential of filamentous cyanobacteria in ISRU strategies. All three tested strains grew in MGS-1 and MMS-2 using only their mineral composition, without supplemental nutrients. They colonized regolith surfaces efficiently, and although desiccation and UV-B stress reduced their growth rates and extended recovery periods, survival was still achieved under conditions relevant to Mars.

Particularly promising was the performance of *Desmonostoc* strains in MGS-1, which chemically resembles certain Martian terrains. Their ability to recover from both dehydration and radiation suggests they could be used as pioneer biofertilizers, preparing Martian soils for future plant cultivation by enriching them with organic matter, fixing nitrogen, and stabilizing surface materials.

Although the species used were freshwater and mesophilic—not true extremophiles—they exhibited robust tolerance to harsh conditions. This implies that more stress-tolerant genera, such as *Chroococcidiopsis* or *Thermosynechococcus*, may perform even better. Nevertheless, mesophilic strains are advantageous due to their faster growth, easier cultivation, and little need for compartmentalization.

Importantly, the deleterious effects of UV radiation appear to be cumulative rather than immediate, while desiccation, though impactful, does not preclude recovery if samples are rehydrated. Protecting cyanobacteria from UV exposure, whether through physical shielding, biofilms, or habitat engineering, is therefore a key step in preserving viability. In contrast, desiccation appears to be a manageable stressor, especially when cyanobacteria are housed in porous, moisture-retaining substrates such as clays.

Finally, the study underscores the low resource requirements and broad adaptability of cyanobacteria, which align perfectly with the goals of ISRU. Their use could help reduce payload mass, support sustainable life support systems, and even contribute to planetary terraforming efforts.

Freshwater cyanobacteria can not only sustain their own growth through the use of in situ resources in Mars (Arribas Tiemblo et al., 2025), but they can potentially drive the growth of complex plants through their use as biofertilizers. *Lemna minor*, a macrophyte, has been used as a vegetable model to assess the benefits of cyanobacteria and microalgae-based biofertilizers in its growth. Plants growing in Martian regolith simulants saw their growth and biomass production enhanced to levels comparable with plants grown in standard growth media, suggesting that the biofertilization of the Martian regolith by cyanobacteria may allow for consistent and efficient cultivation of commercial plant species. Substrates fertilized with *Arthrospira* cyanobacteria worked betters than those with *Desmonostoc* species however, while the latter colonized the regolith significantly faster than the former. *Desmonostoc* species can also successfully protect common commercial species such as the common bean (*Phaseolus vulgaris*), corn (*Zea mays*) and wheat (*Triticum aestivum*) from drought-derived sources of stress (Do Nascimento et al., 2019). Fertilized plants reached biomasses 150% higher than unfertilized ones under drought conditions. In a dry environment such as the Martian surface, sources of drought resistance will likely be invaluable.

Final remarks

The identification of microbial mats active in such an extremely acidic environment as the Tinto River, with pH values below 1, has represented a significant milestone in understanding the adaptation of cyanobacteria to extreme conditions. It was long assumed that cyanobacteria were unable to thrive in such extreme environments due to their apparent inability to maintain ionic and functional balance under severely acidifying conditions. However, this study has provided conclusive evidence for the presence of acidotolerant cyanobacteria organized within a highly stratified microbial structure protected by mineral crusts, composed mainly of natrojarosite. This physical arrangement provides a stable microenvironment that favors the persistence of these photosynthetic microorganisms, despite extreme external conditions.

The multidisciplinary methodological approach employed has allowed not only to confirm the presence of cyanobacteria by immunological, fluorescence in situ hybridization and metagenomic techniques, but also to explore the structural, functional and ecological properties of the organisms involved. The combination of SEM, TEM, XRD, ARD-FISH, VIS-NIR and Raman spectroscopy, as well as genomic analysis of the mat layers, have revealed a surprisingly complex ecosystem. The mat presents a stratified organization in layers, with a rich microbial biomass in the middle layer, where cyanobacteria are located.

From a morphological and structural point of view, the identified cyanobacteria show remarkable adaptations, well-developed polysaccharide capsules, peripherally organized thylakoid membranes and the presence of inclusions such as carboxysomes, lipid globules, polyphosphates and glycogen granules. These features indicate a protective strategy against external chemical stress and an optimization of photosynthetic metabolism under limited irradiation. In fact, spectral measurements demonstrated a remarkable attenuation of irradiance within the mat, selectively filtering harmful wavelengths and favoring the penetration of light useful for photosynthesis. In addition, Raman spectra analysis confirmed the presence of carotenoid pigments involved in both photosynthesis and photoprotection, supporting the adaptive function of these compounds under UV radiation and oxidative conditions.

At molecular level, metagenomic analyses identified a great diversity of typical cyanobacterial orders such as *Synechococcales, Nostocales, Oscillatoriales*, and *Chroococcales*, revealing the coexistence of multiple phylogenetic lineages in the same environment. Additionally, functional genes related to photosynthesis, carbon fixation, metal detoxification and antioxidant mechanisms were detected, which reinforces the hypothesis of an active and complex molecular adaptation to stress conditions.

Ecologically, the results suggest that these microbial communities could constitute a resilient and self-sufficient life form in extreme environments, acting as closed microecosystems. Thermal stability, mineral protection, and moderate access to light and CO_2 allow these cyanobacteria to perform oxygenic photosynthesis even at ultralow pH. This finding contradicts the widely accepted notion that cyanobacteria cannot develop below pH 3–4 and opens new perspectives on their ecological distribution and evolutionary potential.

The Río Tinto microbial mat study provides, therefore, robust evidence that microbial mats can not only exist in extreme environmental conditions, but that they do so in a highly organized, efficient and functional manner. These structures do not represent just microbial accumulations, but highly adapted ecological communities with functional and clear vertical specialization. The cyanobacterial layer acts as a primary producer, while lower layers, presumably dominated by chemotrophic, fermentative, and possibly archaeal bacteria, play key roles in nutrient recycling, sulfate reduction and mineral substrate transformation.

Finally, the results obtained in this work have relevant implications not only for the microbiology of extreme environments, but also for astrobiology. The Rio Tinto model consolidates as an excellent terrestrial analog of Martian environments, where similar mineral crust structures could have harbored life in the past. The survival of cyanobacteria at pH < 1, in conditions of reduced irradiance and protected in iron- and sulfate-rich crusts, is consistent with hypothetical scenarios on Mars, where analogous conditions could have existed in the subsurface. This finding not only extends the known range of habitability on Earth but also redefines the limits of microbial life and its resilience to chemically extreme conditions. It is this resilience that may have allowed cyanobacteria to colonize low-pH niches in Early Mars, and that enables their use as sources of organic matter and oxygen in the current oxidizing and slightly alkaline surface of Mars.

References

Amann, R.I., Binder, B.J., Olson, R.J., Chisholm, S.W., Devereux, R., Stahl, D.A., 1990. Combination of 16S rRNA-targeted oligonucleotide probes with flow cytometry for analyzing mixed microbial populations. Appl. Environ. Microbiol. 56 (6), 1919–1925. https://doi.org/10.1128/aem.56.6.1919-1925.1990.

Amaral Zettler, L.A., Gómez, F., Zettler, E., Keenan, B.G., Amils, R., Sogin, M.L., 2002. Eukaryotic diversity in Spain's river of fire. Nature 417 (6885), 137. https://doi.org/10.1038/417137a.

Amils, R., González-Toril, E., Fernández-Remolar, D., Gómez, F., Rodríguez, N., Durán, C., 2002. Interaction of the sulfur and iron cycles in the Tinto River ecosystem. Rev. Environ. Sci. Biotechnol. 1 (4), 299–309. https://doi.org/10.1023/a:1023232002312.

Amils, R., Fernández-Remolar, D., Parro, V., Rodríguez-Manfredi, J.A., Oggerin, M., Sánchez-Román, M., López, F.J., Fernández-Rodríguez, J.P., Puente-Sánchez, F., Briones, C., Prieto-Ballesteros, O., Tornos, F., Gómez, F., García-Villadangos, M., Rodríguez, N., Omoregie, E., Timmis, K., Arce, A., Sanz, J.L., Gómez-Ortiz, D., 2014. Río Tinto: a geochemical and mineralogical terrestrial analogue of Mars. Life 4 (3), 511–534. https://doi.org/10.3390/life4030511.

Arribas Tiemblo, M., Macário, I.P.E., Tornero, A., Yáñez, A., Andrejkovičová, S., Gómez, F., 2025. Survival of filamentous cyanobacteria through Martian ISRU: combined effects of desiccation and UV-B radiation. Microorganisms 13 (5). https://doi.org/10.3390/microorganisms13051083. Available from: www.mdpi.com/journal/microorganisms.

Baqué, M., Verseux, C., Böttger, U., Rabbow, E., de Vera, J.P.P., Billi, D., 2016. Preservation of biomarkers from cyanobacteria mixed with mars-like regolith under simulated Martian atmosphere and UV flux. Orig. Life Evol. Biosph. 46 (2–3), 289–310. https://doi.org/10.1007/s11084-015-9467-9.

Beversdorf, L.J., Miller, T.R., McMahon, K.D., Gilbert, J.A., 2013. The role of nitrogen fixation in cyanobacterial bloom toxicity in a temperate, eutrophic lake. PLoS One 8 (2), e56103. https://doi.org/10.1371/journal.pone.0056103.

Bezalel-Hazony, N., Zer, H., Nathanson, S., Tal, S.S., Ostersetzer-Biran, O., Keren, N., 2023. Functional flexibility of cyanobacterial light harvesting phycobilisomes enable acclimation to the complex light regime of mixing marine water columns. FEBS J. 290 (2), 400–411. https://doi.org/10.1111/febs.16597.

Billi, D., Verseux, C., Fagliarone, C., Napoli, A., Baqué, M., De Vera, J.P., 2019. A desert cyanobacterium under simulated mars-like conditions in low Earth orbit: implications for the habitability of mars. Astrobiology 19 (2), 158–169. https://doi.org/10.1089/ast.2017.1807.

Bosak, T., Shuster, D.L., Scheller, E.L., Siljeström, S., Zawaski, M.J., Mandon, L., Simon, J.I., Weiss, B.P., Stack, K.M., Mansbach, E.N., Treiman, A.H., Benison, K.C., Brown, A.J., Czaja, A.D., Farley, K.A., Hausrath, E.M., Hickman-Lewis, K., Herd, C.D.K., Johnson, J.R., Mayhew, L.E., Minitti, M.E., Williford, K.H., Wogsland, B.V., Zorzano, M.P., Allwood, A.C., Amundsen, H.E.F., Bell, J.F., Benzerara, K., Bernard, S., Beyssac, O., Buckner, D.K., Cable, M., Calef, F., Caravaca, G., Catling, D. C., Clavé, E., Cloutis, E., Cohen, B.A., Cousin, A., Dehouck, E., Fairén, A.G., Flannery, D.T., Fornaro, T., Forni, O., Fouchet, T., Gibbons, E., Gomez, F.G., Gupta, S., Hand, K.P., Hurowitz, J.A., Kalucha, H., Pedersen, D.A.K., Lopes-Reyes, G., Maki, J.N., Maurice, S., Nuñez, J.I., Randazzo, N., Rice, J.W., Royer, C., Sephton, M. A., Sharma, S., Steele, A., Tate, C.D., Uckert, K., Udry, A., Wiens, R.C., Williams, A., 2024. Astrobiological potential of rocks acquired by the perseverance rover at a sedimentary fan front in Jezero Crater, Mars. AGU Adv. 5 (4). https://doi.org/10.1029/2024AV001241. Available from: https://agupubs.onlinelibrary.wiley.com/journal/2576604x.

Brock, T.D., 1973. Lower pH limit for the existence of blue-green algae: evolutionary and ecological implications. Science 179 (4072), 480–483. https://doi.org/10.1126/science.179.4072.480.

Cannon, K.M., Britt, D.T., Smith, T.M., Fritsche, R.F., Batcheldor, D., 2019. Mars global simulant MGS-1: a Rocknest-based open standard for basaltic martian regolith simulants. Icarus 317, 470–478. https://doi.org/10.1016/j.icarus.2018.08.019.

Caporale, A.G., Vingiani, S., Palladino, M., El-Nakhel, C., Duri, L.G., Pannico, A., Rouphael, Y., De Pascale, S., Adamo, P., 2020. Geo-mineralogical characterisation of Mars simulant MMS-1 and appraisal of substrate physico-chemical properties and crop performance obtained with variable green compost amendment rates. Sci. Total Environ. 720. https://doi.org/10.1016/j.scitotenv.2020.137543. Available from: www.elsevier.com/locate/scitotenv.

Clark, B.C., Kounaves, S.P., 2016. Evidence for the distribution of perchlorates on Mars. Int. J. Astrobiol. 15 (4), 311–318. https://doi.org/10.1017/S1473550415000385.

Clark, B.C., Kolb, V.M., Steele, A., House, C.H., Lanza, N.L., Gasda, P.J., Vanbommel, S.J., Newsom, H.E., Martínez-Frías, J., 2021. Origin of life on mars: suitability and opportunities. Life 11 (6). https://doi.org/10.3390/life11060539. Available from: https://www.mdpi.com/2075-1729/11/6/539/pdf.

Cockell, C.S., 2014. Trajectories of martian habitability. Astrobiology 14 (2), 182–203. https://doi.org/10.1089/ast.2013.1106.

Do Nascimento, M., Battaglia, M.E., Rizza, L.S., Ambrosio, R., Di Palma, A.A., Curatti, L., 2019. Prospects of using biomass of N2-fixing cyanobacteria as an organic fertilizer and soil conditioner. Algal Res. 43, 101652. https://doi.org/10.1016/j.algal.2019.101652.

Dwivedi, S., Ahmad, I.Z., 2023. Evaluation of the effect of UV-B radiation on growth, photosynthetic pigment, and antioxidant enzymes of some cyanobacteria. Environ. Res. 218. https://doi.org/10.1016/j.envres.2022.114943. Available from: https://www.sciencedirect.com/journal/environmental-research.

Fagliarone, C., Napoli, A., Chiavarini, S., Baqué, M., de Vera, J.P., Billi, D., 2020. Biomarker preservation and survivability under extreme dryness and mars-like UV flux of a desert cyanobacterium capable of trehalose and sucrose accumulation. Front. Astron. Space Sci. 7. https://doi.org/10.3389/fspas.2020.00031. Available from: www.frontiersin.org/journals/astronomy-and-space-sciences.

Fagliarone, C., Mosca, C., Di Stefano, G., Leuko, S., Moeller, R., Rabbow, E., Rettberg, P., Billi, D., 2023. Enabling deep-space experimentations on cyanobacteria by monitoring cell division resumption in dried Chroococcidiopsis sp. 029 with accumulated DNA damage. Front. Microbiol. 14. https://doi.org/10.3389/fmicb.2023.1150224.

Farrokh, P., Sheikhpour, M., Kasaeian, A., Asadi, H., Bavandi, R., 2019. Cyanobacteria as an eco-friendly resource for biofuel production: a critical review. Biotechnol. Prog. 35 (5). https://doi.org/10.1002/btpr.2835.

Finkel, P.L., Carrizo, D., Parro, V., Sánchez-García, L., 2023. An overview of lipid biomarkers in terrestrial extreme environments with relevance for mars exploration. Astrobiology 23 (5), 563–604. https://doi.org/10.1089/ast.2022.0083.

Gan, F., Bryant, D.A., 2015. Adaptive and acclimative responses of cyanobacteria to far-red light. Environ. Microbiol. 17 (10), 3450–3465. https://doi.org/10.1111/1462-2920.12992. Available from: http://onlinelibrary.wiley.com/journal/10.1111/(ISSN)1462-2920.

Gómez, F., Mateo-Martí, E., Prieto-Ballesteros, O., Martín-Gago, J., Amils, R., 2010. Protection of chemolithoautotrophic bacteria exposed to simulated Mars environmental conditions. Icarus 209 (2), 482–487. https://doi.org/10.1016/j.icarus.2010.05.027.

Gómez, F., Rodríguez, N., Antonio Rodríguez-Manfredi, J., Escudero, C., Carrasco-Ropero, I., Martínez, J.M., Ferrari, M., De Angelis, S., Frigeri, A., Fernández-Sampedro, M., Amils, R., 2024. Association of acidotolerant cyanobacteria to microbial mats below pH 1 in acidic mineral precipitates in Río Tinto River in Spain. Microorganisms 12 (4), 829. https://doi.org/10.3390/microorganisms12040829.

González-Toril, E., Llobet-Brossa, E., Casamayor, E.O., Amann, R., Amils, R., 2003. Microbial Ecology of an extreme acidic environment, the Tinto River. Appl. Environ. Microbiol. 69 (8), 4853–4865. https://doi.org/10.1128/aem.69.8.4853-4865.2003.

Grotzinger, J.P., Sumner, K., Stack, K., Gupta, E., et al., 2022. A Habitable Fluvio-Lacustrine Environment at Yellowknife Bay.

Grzesik, M., Romanowska-Duda, Z., Kalaji, H.M., 2017. Effectiveness of cyanobacteria and green algae in enhancing the photosynthetic performance and growth of willow (Salix viminalis L.) plants under limited synthetic fertilizers application. Photosynthetica 55 (3), 510–521. https://doi.org/10.1007/s11099-017-0716-1.

Guzman, M., McKay, C.P., Quinn, R.C., Szopa, C., Davila, A.F., Navarro-González, R., Freissinet, C., 2018. Identification of chlorobenzene in the viking gas chromatograph-mass spectrometer data sets: reanalysis of viking mission data consistent with aromatic organic compounds on mars. J. Geophys. Res. Planets 123 (7), 1674–1683. https://doi.org/10.1029/2018JE005544. Available from: http://agupubs.onlinelibrary.wiley.com/agu/jgr/journal/10.1002/(ISSN)2169-9100/.

Hecht, M.H., Kounaves, S.P., Quinn, R.C., West, S.J., Young, S.M.M., Ming, D.W., Catling, D.C., Clark, B.C., Boynton, W.V., Hoffman, J., DeFlores, L.P., Gospodinova, K., Kapit, J., Smith, P.H., 2009. Detection of perchlorate and the soluble chemistry of martian soil at the phoenix lander site. Science 325 (5936), 64–67. https://doi.org/10.1126/science.1172466.

Introduction to the Cyanobacteria. The Ecology of Cyanobacteria: Their Diversity in Time and Space, 2002. Springer, pp. 1–11. https://doi.org/10.1007/0-306-46855-7_1.

Johnson, Z.I., Zinser, E.R., Coe, A., McNulty, N.P., Woodward, E.M.S., Chisholm, S.W., 2006. Niche partitioning among Prochlorococcus ecotypes along ocean-scale environmental gradients. Science 311 (5768), 1737–1740. https://doi.org/10.1126/science.1118052.

Johnson, H.E., King, S.R., Banack, Webster, C., Callanaupa, W.J., Cox, P.A., 2008. Cyanobacteria, Nostoc commune used as a dietary item in the Peruvian highlands produce the neurotoxic amino acid BMAA. J. Ethnopharmacol. 118, 159.

Kamada, A., Kuroda, T., Kasaba, Y., Terada, N., Nakagawa, H., 2021. Global climate and river transport simulations of early Mars around the Noachian and Hesperian boundary. Icarus 368. https://doi.org/10.1016/j.icarus.2021.114618. Available from: http://www.elsevier.com/inca/publications/store/6/2/2/8/4/3/index.htt.

Keller, R., Goli, K., Porter, W., Alrabaa, A., Jones, J.A., 2023. Cyanobacteria and algal-based biological life support system (BLSS) and planetary surface atmospheric revitalizing bioreactor brief concept review. Life 13 (3). https://doi.org/10.3390/life13030816. Available from. www.mdpi.com/journal/life/.

Keshari, N., Yang, Z., Kumar Das, S., Zhu, T., Lu, X., 2022. Cyanobacterial community structure and isolates from representative hot springs of Yunnan province, China using an integrative approach. Front. Microbiol. 13. https://doi.org/10.3389/fmicb.2022.872598.

Kim, Y., Jeon, J., Kwak, M.S., Kim, G.H., Koh, I.S., Rho, M., Dong, Q., 2018. Photosynthetic functions of Synechococcus in the ocean microbiomes of diverse salinity and seasons. PLoS One 13 (1), e0190266. https://doi.org/10.1371/journal.pone.0190266.

Kite, E.S., Conway, S., 2024. Geological evidence for multiple climate transitions on Early Mars. Nat. Geosci. 17 (1), 10–19. https://doi.org/10.1038/s41561-023-01349-2.

Klingelhöfer, G., Morris, R.V., Bernhardt, B., Schröder, C., Rodionov, D.S., De Souza, P.A., Yen, A., Gellert, R., Evlanov, E.N., Zubkov, B., Foh, J., Bonnes, U., Kankeleit, E., Gütlich, P., Ming, D.W., Renz, F., Wdowiak, T., Squyres, S.W., Arvidson, R.E., 2004. Jarosite and hematite at Meridiani Planum from opportunity's Mössbauer spectrometer. Science 306 (5702), 1740–1745. https://doi.org/10.1126/science.1104653.

Koehle, A.P., Brumwell, S.L., Seto, E.P., Lynch, A.M., Urbaniak, C., 2023. Microbial applications for sustainable space exploration beyond low Earth orbit. Npj Microgravity 9 (1). https://doi.org/10.1038/s41526-023-00285-0. Available from: https://www.nature.com/npjmgrav/.

Kounaves, S.P., Carrier, B.L., O'Neil, G.D., Stroble, S.T., Claire, M.W., 2014. Evidence of martian perchlorate, chlorate, and nitrate in Mars meteorite EETA79001: implications for oxidants and organics. Icarus 229, 206–213. https://doi.org/10.1016/j.icarus.2013.11.012.

Liu, D., Kounaves, S.P., 2021. Degradation of amino acids on mars by UV irradiation in the presence of chloride and oxychlorine salts. Astrobiology 21 (7), 793–801. https://doi.org/10.1089/ast.2020.2328.

Macário, I.P.E., Veloso, T., Frankenbach, S., Serôdio, J., Passos, H., Sousa, C., Gonçalves, F.J. M., Ventura, S.P.M., Pereira, J.L., 2022. Cyanobacteria as candidates to support mars colonization: growth and biofertilization potential using mars regolith as a resource. Front. Microbiol. 13. https://doi.org/10.3389/fmicb.2022.840098. Available from: https://www.frontiersin.org/journals/microbiology#.

Mapstone, L.J., Leite, M.N., Purton, S., Crawford, I.A., Dartnell, L., 2022. Cyanobacteria and microalgae in supporting human habitation on Mars. Biotechnol. Adv. 59. https://doi.org/10.1016/j.biotechadv.2022.107946. Available from: www.elsevier.com/inca/publications/store/5/2/5/4/5/5/index.htt.

McAdam, e C., Sutter, B., Archer, P.D., Franz, H.B., Wong, J.M.T.L., et al., 2022. Evolved gas analyses of sedimentary rocks from the Glen Torridon clay-bearing unit, Gale crater, Mars: results from the Mars Science Laboratory Sample Analysis at Mars instrument suite. J. Geophys. Res. Planet 127 (9). https://doi.org/10.1029/2022JE007179.

Megevand, V., Viennet, J.C., Balan, E., Gauthier, M., Rosier, P., Morand, M., Garino, Y., Guillaumet, M., Pont, S., Beyssac, O., Bernard, S., 2021. Impact of UV radiation on the Raman signal of cystine: implications for the detection of S-rich organics on mars. Astrobiology 21 (5), 566–574. https://doi.org/10.1089/ast.2020.2340.

Mesa, V., Gallego, J.L.R., González-Gil, R., Lauga, B., Sánchez, J., Méndez-García, C., Peláez, A.I., 2017. Bacterial, archaeal, and eukaryotic diversity across distinct microhabitats in an acid mine drainage. Front. Microbiol. 8. https://doi.org/10.3389/fmicb.2017.01756.

Moreno-Paz, M., dos Santos Severino, R.S., Sánchez-García, L., Manchado, J.M., García-Villadangos, M., Aguirre, J., Fernández-Martínez, M.A., Carrizo, D., Kobayashi, L., Dave, A., Warren-Rhodes, K., Davila, A., Stoker, C.R., Glass, B., Parro, V., 2023. Life detection and microbial biomarker profiling with signs of life detector-life detector chip during a mars drilling simulation campaign in the hyperarid core of the Atacama desert. Astrobiology 23 (12), 1259–1283. https://doi.org/10.1089/ast.2021.0174.

Mutale-Joan, C., Sbabou, L., Hicham, E.A., 2023. Microalgae and cyanobacteria: how exploiting these microbial resources can address the underlying challenges related to food sources and sustainable agriculture: a review. J. Plant Growth Regul. 42 (1), 1–20. https://doi.org/10.1007/s00344-021-10534-9.

Oggerin, M., Rodríguez, N., Moral, C.d., Amils, R., 2014. Fungal jarosite biomineralization in Río Tinto. Res. Microbiol. 165 (9), 719–725. https://doi.org/10.1016/j.resmic.2014.10.001.

Olsson-Francis, K., Cockell, C.S., 2010. Use of cyanobacteria for in-situ resource use in space applications. Planet. Space Sci. 58 (10), 1279–1285. https://doi.org/10.1016/j.pss.2010.05.005.

Pavlov, A.A., McLain, H.L., Glavin, D.P., Roussel, A., Dwork2in, J.P., Elsila, J.E., Yocum, K. M., 2022. Rapid radiolytic degradation of amino acids in the Martian shallow subsurface: implications for the search for extinct life. Astrobiology 22 (9), 1099–1115. https://doi.org/10.1089/ast.2021.0166.

Prieto-Barajas, C.M., Valencia-Cantero, E., Santoyo, G., 2018. Microbial mat ecosystems: structure types, functional diversity, and biotechnological application. Electron. J. Biotechnol. 31, 48–56. https://doi.org/10.1016/j.ejbt.2017.11.001.

Puente-Sánchez, F., Arce-Rodríguez, A., Oggerin, M., García-Villadangos, M., Moreno-Paz, M., Blanco, Y., Rodríguez, N., Bird, L., Lincoln, S.A., Tornos, F., Prieto-Ballesteros, O., Freeman, K.H., Pieper, D.H., Timmis, K.N., Amils, R., Parro, V., 2018. Viable cyanobacteria in the deep continental subsurface. Proc. Natl. Acad. Sci. 115 (42), 10702–10707. https://doi.org/10.1073/pnas.1808176115.

Ramalho, T.P., Chopin, G., Salman, L., Vincent, B., Heinicke, C., Verseux, C., 2022. On the growth dynamics of the cyanobacterium Anabaena sp. PCC 7938 in Martian regolith. Npj Microgravity 8 (1). https://doi.org/10.1038/s41526-022-00240-5.

Rastogi, R.P., Madamwar, D., Nakamoto, H., Incharoensakdi, A., 2020. Resilience and self-regulation processes of microalgae under UV radiation stress. J. Photochem. Photobiol. C Photochem. Rev. 43. https://doi.org/10.1016/j.jphotochemrev.2019.100322. Available from: http://www.elsevier.com/locate/jphotochemrev.

Rzymski, P., Poniedziałek, B., Hippmann, N., Kaczmarek, Ł., 2022. Screening the survival of cyanobacteria under perchlorate stress. Potential implications for mars in situ resource utilization. Astrobiology 22 (6), 672–684. https://doi.org/10.1089/ast.2021.0100.

Rzymski, P., Losiak, A., Heinz, J., Szukalska, M., Florek, E., Poniedziałek, B., Kaczmarek, Ł., Schulze-Makuch, D., 2024. Perchlorates on mars: occurrence and implications for putative life on the red planet. Icarus 421. https://doi.org/10.1016/j.icarus.2024.116246. Available from: https://www.sciencedirect.com/science/journal/00191035.

Saini, D.K., Pabbi, S., Shukla, P., pigments, C., 2018. Perspectives and biotechnological approaches. Food Chem. Toxicol. 120, 616–624. https://doi.org/10.1016/j.fct.2018.08.002.

Sánchez-Bayo, A., Morales, V., Rodríguez, R., Vicente, G., Bautista, L.F., 2020. Cultivation of microalgae and cyanobacteria: effect of operating conditions on growth and biomass composition. Molecules 25 (12). https://doi.org/10.3390/molecules25122834. Available from: https://www.mdpi.com/1420-3049/25/12/2834/pdf.

Stal, L.J., 2012. Cyanobacterial Mats and Stromatolites. Ecology of Cyanobacteria II: Their Diversity in Space and Time, 9789400738553, pp. 65–125. https://doi.org/10.1007/978-94-007-3855-3_4.

Stanier, R.Y., Kunisawa, R., Mandel, M., Cohen-Bazire, G., 1971. Purification and properties of unicellular blue-green algae (order Chroococcales). Bacteriol. Rev. 35 (2), 171–205. https://doi.org/10.1128/br.35.2.171-205.1971.

ten Kate, I.L., Garry, J.R.C., Peeters, Z., Quinn, R., Foing, B., Ehrenfreund, P., 2005. Amino acid photostability on the Martian surface. Meteoritics Planet Sci. 40 (8), 1185–1193. https://doi.org/10.1111/j.1945-5100.2005.tb00183.x.

Ward, D.M., Castenholz, R.W., 2025. Cyanobacteria in Geothermal Habitats. Springer Science and Business Media LLC, pp. 37–59. https://doi.org/10.1007/0-306-46855-7_3.

Wordsworth, R.D., 2016. The climate of early mars. Annu. Rev. Earth Planet Sci. 44 (1), 381–408. https://doi.org/10.1146/annurev-earth-060115-012355.

Zhang, Y., Chi, Z., Lu, W., 2007. Exopolysaccharide production by four cyanobacterial isolates and preliminary identification of these isolates. J. Ocean Univ. China 6 (2), 147–152. https://doi.org/10.1007/s11802-007-0147-x.

CHAPTER

7

Two distinct microbial communities at close proximity in the deep subsurface of the Iberian Pyrite Belt Mars analogue

Fernando Puente-Sánchez

Department of Aquatic Sciences and Assessment, Swedish University of Agricultural Sciences, Uppsala, Sweden

Río Tinto and the Peña de Hierro area

Peña de Hierro is an abandoned mine site located in the catchment area of the Tinto River (Fig. 7.1) that bears deposits comprised of a pyrite ore with variable concentrations of copper (between 1 and 4 wt%), and lower concentrations of Zn and Pb. The mine was working between the middle of the nineteenth century and 1966, and it produced about three million tons of waste rock and other residues that were dumped around the pit, covering 25 ha. As a result of the mining activity there is a 120 m-deep pit, of which the last 30 m are covered in acidic waters (Stoker et al., 2008).

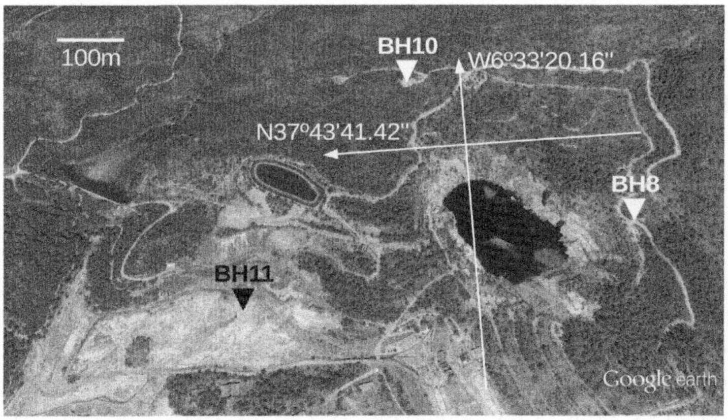

FIGURE 7.1

Map of the Peña de Hierro area, showing the locations of BH8 and BH10.

Geomicrobiology of Río Tinto. https://doi.org/10.1016/B978-0-443-32948-7.00008-4

The Tinto River origin is located just south of Peña de Hierro, where waters with high ionic loads and low pH emerge from the underground as seeps. Unlike in typical acid drainage systems, where acid is generated by the exposure of sulphide minerals to the surface during mining activities (Jacobs and Testa, 2014), the acidity in the Tinto River is thought to be originated in an underground biological reactor located in the subsurface of Peña de Hierro (Gómez-Ortiz et al., 2014).

The Iberian Pyrite Belt (IPB) is an excellent environment to study deep subsurface ecosystems and their influence in surface processes. Additionally, due to its unique geomineralogical characteristics, it can be used as a model system to test hypotheses about potential past and present Martian subsurface habitats and the iron-sulphur-driven early Earth biosphere. The overall heterogeneity of the continental crust subsurface translates into a higher possibility of finding distinct habitats when sampling both horizontally (different boreholes within the same area) and vertically (different depths within the same borehole), increasing the potential scientific return of a drilling campaign.

The scientific interest of the Tinto River area has been widely recognized by the scientific community, resulting in a large body of work (Amils, Fernández-Remolar and Team, 2014). This includes two large deep subsurface drilling campaigns: the Mars Analog Rio Tinto Experiment (MARTE) project campaign (2003–05), a joint project between the Centro de Astrobiología (CAB; INTA-CSIC) and NASA; and the Iberian Pyrite Belt Subsurface Life (IPBSL) project campaign (2011–12). The two campaigns had different scientific goals (see below) and in some cases used different analysis techniques (due to advances in the state-of-the-art from 2003 to 2012). However, both of them yielded an extensive biological, chemical and geological characterisation of the selected sampling sites.

Microbially mediated iron, sulfur and methane cycles associated to fractures and oxygen input in a stockwork pyrite complex

The Mars Analog Río Tinto Experiment (MARTE) project

The MARTE project was a joint collaboration between CAB and the NASA Ames Research Center that aimed to simulate a robotic drilling mission to search for subsurface life on Mars (Stoker et al., 2008). As such, the emphasis was put in technology development and protocol testing, culminating in a fully robotic characterisation of a shallow subsurface Mars analogue (Bonaccorsi and Stoker, 2008; Parro et al., 2008). Several boreholes were performed near the Tinto River headwaters at Peña de Hierro (Spain), aiming for the sites in which groundwater intesected the massive sulphide minerals. Four of the boreholes (BH1, BH4, BH7 and BH8, reaching 40, 160, 10 and 160 m below surface floor [mbsf] respectively) were further analysed in order to search for evidences of subsurface life.

Fernández-Remolar et al. (2008) found evidences of biological alteration in the subsurface pyrite of BH4 and BH8, which led to the modelling of a hypothetical Martian subsurface biosphere and its mechanisms to increase the habitability of a seemingly hostile environment (Fernández-Remolar et al., 2008). Their conclusions were however based only on geochemical and mineralogical data, since little biological information was available at the time of publication. Here we will present the results from a later study that explicitly sought to characterize the microbiology of BH8 using the molecular tools available at the time (Puente-Sánchez et al., 2014).

Borehole 8

The BH8 borehole is 160 m deep, and is located near the Peña de Hierro pit (Fig. 7.1), at the GPS coordinates of N37° 43′ 35.55″, W6° 33′ 11.60″ and at an elevation of 510 m above the sea level (masl). It cuts through a 40 m thick unit of gossan, below which sits a stockwork complex with massive veins of pyrite. At the time of drilling, the water table was situated at approximately 90 mbsf, and the aquifer was composed of oxidised waters in equilibrium with the meteoric waters recharging the Peña de Hierro pit (Fernando Tornos, personal communication). Iron oxides were detected in the first 10 metres below the water table, followed by 50 m of non-weathered stockwork. In the lowest part of the borehole (>150 mbsf), the stockwork complex was fractured, and the pyrite showed important signs of oxidation.

The microbial communities at various depths of BH8 below the water table (>80 mbsf) were characterized by a multi-technique approach that included DNA extraction followed by 16S rRNA gene amplification (and its subsequent analysis by either cloning and sequencing or hybridization into an oligonucleotide microarray described in (Garrido et al., 2008), CARD-FISH, and immunoprofiling with a microarray bearing 300 antibodies against ecologically relevant microbial strains, functional proteins and environmental extracts (Parro et al., 2011; Puente-Sánchez et al., 2014; Rivas et al., 2008).

Microbial processes operating in the BH8 subsurface

Based on the geochemical, immunological, DNA hybridization, and sequencing results, we could distinguish between three main compartments in BH8 (Puente-Sánchez et al., 2014). The three were dominated by microorganisms related to the iron, sulfur and methane cycles, while oxygen availability (itself dependant on depth and rock permeability) determined whether the main metabolisms were aerobic or anaerobic (Fig. 7.2). Compartments with oxygen influx still hosted putatively anaerobic microorganisms (e.g., methanogens, iron and sulfate reducers ...), we attribute this to the heterogeneity inherent to rock samples, which facilitates the existence of anaerobic microniches. The main electron donors found in the borehole were ferrous iron and sulfur (from pyrite, FeS_2), small organic acids (acetate and formate), hydrogen (putatively produced from rock-water interactions (Stevens and McKinley, 1995; Freund et al., 2002) and methane, while oxygen (where available), ferric iron,

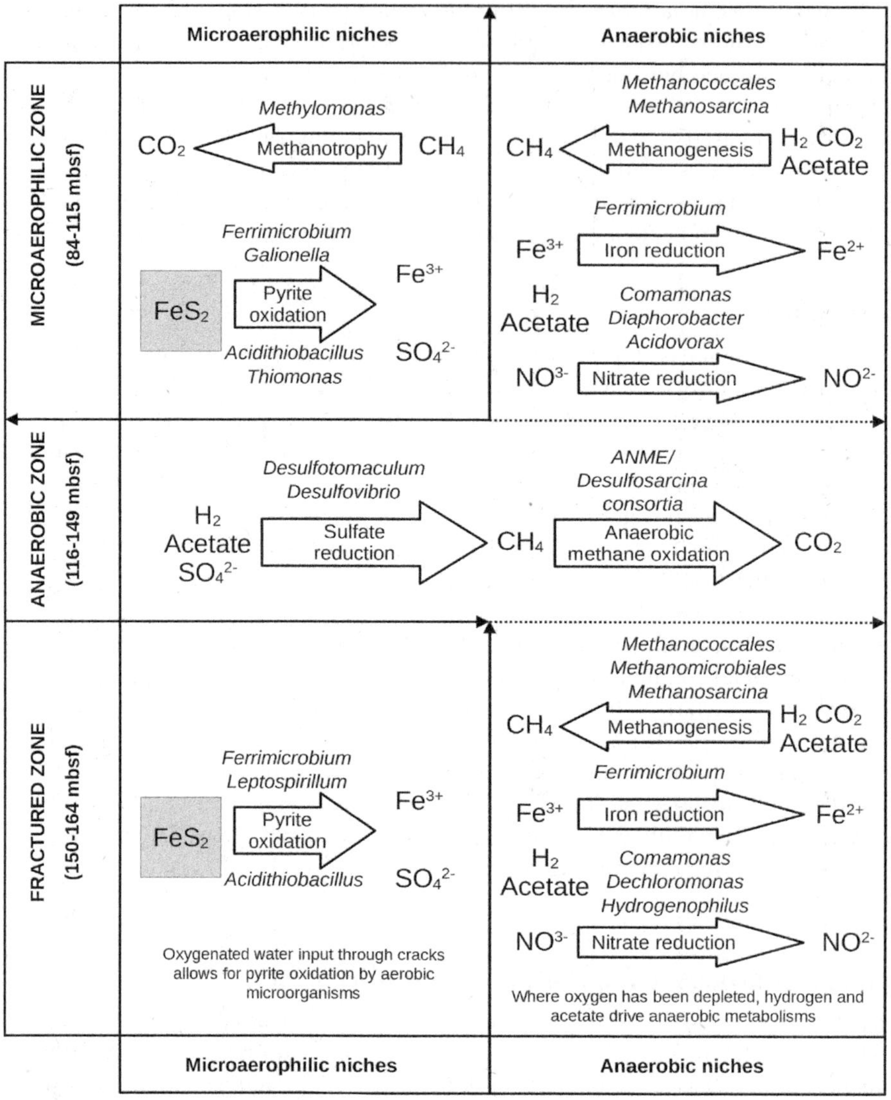

FIGURE 7.2

Proposed biogeochemical model for the BH8 subsurface.

sulfate and nitrate were the main electron acceptors. The three compartments are as follows.

(a) A microaerophilic zone (84–115 mbsf, Fig. 7.2, top) in which pyrite-rich aerobic microniches enable the oxidation of iron (by microaerophiles such as *Ferrimicrobium* and *Gallionella*), sulfur (*Acidithiobacillus, Thiomonas*) and methane (*Methylomonas*), while denitrification and methanogenesis linked to ferric iron reduction (*Methanococcales, Methanosarcina, Ferrimicrobium*) occur in anaerobic microniches, as previously described in (Sanz et al., 2021).

(b) An intermediate anaerobic zone (116–149 mbsf, Fig. 7.2, middle), dominated by sulfate reducers (*Desulfatomaculum, Desulfovibrio*) which may potentially be associated to anaerobic methane oxidizers. These consortia may be the main methane sink in the borehole (Knittel and Boetius, 2009; Orphan et al., 2001; Parkes et al., 2007; Pedersen, 2013).

(c) A fractured zone (150–164 mbsf, Fig. 7.2, bottom) with increased water and oxygen flow. A diverse prokaryotic community was detected in that zone, along with alterations in the geochemical profiles, such as an increase in sulfate and methane concentrations, and a decrease of ferrous iron, which could be products of an increased microbial activity. We hypothesize that pyrite oxidizers (*Leptospirillum, Ferrimicrobium, Acidithiobacillus*), consume all the available oxygen, creating anaerobic microniches that can be used by iron reducers (*Ferrimicrobium*), methanogens (*Methanococcales, Methanomicrobiales, Methanosarcina*) and nitrate reducers. This was the only zone in which methane could be detected, in spite of methanogens being present throughout the whole borehole, we attribute this to it being consumed by aerobic and anaerobic methane oxidizers in the shallower compartments. The absence of sulfate reducers in this compartment is puzzling, as sulfate was present at high concentrations and sulfate reducers usually compete favorably with methanogens for electron donors (Abram and Nedwell, 1978; Winfrey and Zeikus, 1977), however it has been described that sulfate reducers can be inhibited by high heavy metal concentrations and the precipitation of metal sulfides (Utgikar et al., 2002).

Overall, the microbiology of BH8 is resemblant to that of an Acid Mine Drainage, dominated by iron and sulfur oxidizers. However, the presence of anaerobic microniches and the spatial heterogeneity of the borehole results in a more complex community in which the methane cycle is also operative. BH8 would therefore be a good candidate for the underground biological reactor that has been proposed as the putative origin for the exceptional Tinto River (Amils et al., 2013; Amils, Fernández-Remolar and Team, 2014; Gómez-Ortiz et al., 2014).

A putative holobiont sustained by hydrogen and buried organic matter in an anoxic, low permeability area

The Iberian Pyrite Belt Subsurface Life (IPBSL) project

The MARTE campaigns successfully gathered evidences of an active microbial ecosystem in the IPB subsurface, but its focus was strongly biased towards technological development. The IPBSL project (CAB—Technische Universität

Braunschweig) was designed to answer the questions left open by the MARTE project by using state-of-the-art techniques. Two boreholes (BH10 and BH11, reaching 613 and 340 mbsf respectively) were drilled in the Peña de Hierro area. As of today, the geochemistry of the BH10 has been fully characterised by mineralogical (XRD), elemental (ICP-MS), pore water (spectrophotometry, ion chromatography, gas chromatography), petrographic and stable isotope analyses. The microbiota was in turn studied by using enrichment cultures, antibody microarrays, massive sequencing (of the 16S rRNA gene and of two deep subsurface shotgun metagenomes) and CARD-FISH (Amils et al., 2023; Escudero et al., 2018; Puente-Sánchez et al., 2014; Sanz et al., 2021).

Borehole 10

The BH10 borehole is 613 m deep, and is located farther from the Peña de Hierro pit than the BH8 (Fig. 7.1). Its GPS coordinates are N37° 43′ 45.74″, W6° 33′ 22.37″, and its elevation is of 525 masl. It cuts the upper part of the Volcanosedimentary Complex, including four superimposed dacite to rhyodacite dome complexes that host a large zone of hydrothermal alteration with widespread replacement by quartz, sericite, chlorite and pyrite. The deepest part of the borehole cuts through the shales and greywackes of the Culm Group. At the time of drilling, the water table was situated at 90 mbsf (Ricardo Amils, personal communication).

Unlike the BH8, the BH10 aquifer hosts anoxic waters with a low redox potential. Stockwork pyrite does exist, but it is unoxidised and in lower concentrations. Being farther from the Peña de Hierro pit, the rocks in the BH10 are generally less fractured, resulting in a lower permeability. Localized fracture zones do however exist at the interfaces between the different stratigraphic units.

The BH10 is more recent than the BH8 borehole, and thus it could be studied using high-throughput sequencing. Only samples below 100 mbsf were analyzed, as the first 100 m were drilled using destructive techniques which preclude the retrieval of intact cores. The information regarding the composition and potential metabolisms of the microbial communities found in the BH10 borehole was in turn integrated with the geochemical data obtained from the different samples in order to formulate hypotheses about the role, viability and survival mechanisms of the IPB deep subsurface microbiota.

Geological, physicochemical and organic compound profiling

Core samples were analysed by several techniques in order to identify relevant substrates or products of different microbial metabolisms, as well as organic compounds that may be related to microbial presence or activity. This information was integrated with a stratigraphic profile of the BH10, including details on the presence of faults, shears and pyrite content throughout the borehole (Puente-Sánchez et al., 2018).

Several faults were present between 100 and 150 mbsf. More definite fracture zones were found at 300, 420 and 500 mbsf. No faults were found below 500 mbsf, but rocks in that zone were more fractured than in the upper sections of the borehole. Bromide, which was added to the drilling water as a contamination tracer, remained at low concentrations except for a sample at 487 mbsf. Soluble iron was found at very low concentrations, except for a high peak between 250 and 350 mbsf, which was associated to an ammonium peak. Hydrogen was present in variable amounts, but significantly descended between 250 and 350 mbsf (being associated with the aforementioned increase in soluble iron and ammonium), between 400 and 450 mbsf (being associated with a fault) and below 500 mbsf (being associated with the more fractured Culm zone). An increase on methane concentrations was found between 250 and 350 mbsf, with another smaller increase being associated with the fault that marked the transition between the dark shale and the Culm zone at 500 mbsf. Sulphate, which is a known product of pyrite oxidation, was present at higher concentrations in the upper zone of the borehole, in which the presence of oxygen traces is more likely. The concentration of nitrate and nitrite generally descended with depth. Interestingly, nitrate and nitrite concentrations were negatively correlated between 400 and 500 mbsf.

Regarding organic compounds, formate and acetate, which are widespread electron donors for microbial metabolisms, were distributed in an irregular fashion, but followed a similar profile. Propionate was abundant at the shallower depths of the borehole, but was not detected below 300 mbsf. Oxalate was present at small concentrations, with peaks at 420 and 487 mbsf. Total organic carbon was generally low (1%–2%), with a small peak at 420 mbsf, while sugar and proteins were distributed irregularly. Finally, palmitic and stearic acids were detected at 353, 420 and 607 mbsf (Puente-Sánchez et al., 2018).

16S rRNA gene microbial community profiling

Post-sequencing assessment of potential contamination sources

Two different controls were performed in order to assess if our samples had been contaminated during the drilling, DNA extraction, amplification or sequencing processes. First, we amplified and sequenced 16S rRNA amplicons from the drilling fluids (DW control), to trace contamination events introduced during drilling (from the drilling fluids, but also other components or the machinery and also shallower parts of the borehole). Secondly, we performed a blank DNA extraction and 16S rRNA gene amplification in parallel to the real samples (IC control), to identify contamination events introduced during laboratory manipulation. Operational Taxonomic Units (OTUs) found in at least one of those controls were excluded from subsequent analysis (Puente-Sánchez et al., 2018).

Phylotype-level profile of borehole BH10

The phylotype-level profile of borehole BH10 is shown in (Puente-Sánchez et al., 2018). The most abundant phyla were *Actinomycetota*, *Pseudomonadota*

(*Alphaproteobacteria*) and *Cyanobacteriota*. *Actinomycetota* were more abundant in the shallower samples, and cyanobacteria were more abundant in the deeper samples; apart from this, no other clear trend could be visually identified. Regarding the two controls, both were distinguishable from the rest of the samples: the DW control was by far the most diverse, while the IC control was less diverse and dominated by gammaproteobacteria.

Individual OTU analysis

The most abundant OTUs in borehole BH10 belonged to the *Cyanobacteriota* phylum, and were related to known endolithic and hypolithic cyanobacterial genera such as *Calothrix* (Pentecost, 2005), the filamentous *Crinalium* (Stal and Moezelaar, 1997), *Chroococcidiopsis* (Baqué et al., 2013) or the cave-dwellers *Toxopsis* (Lamprinou et al., 2012) and *Leptolyngbia* (Bruno et al., 2009). Several non-cyanobacterial OTUs were also present in high abundances, including one related to an iron-oxidizing, nitrate-reducing *Acidovorax* (Straub et al., 1996), and the ubiquitous soil and subsurface bacteria *Arthrobacter* (Crocker et al., 2000; Kageyama et al., 2008; Van Waasbergen et al., 2000), *Bradyrhizobium* (Wang et al., 2013) and *Sediminibacterium* (Bollmann et al., 2010). Interestingly, an OTU related to the *Tessaracoccus* genus, which had been previously isolated from the nearby borehole BH11 (Puente-Sánchez et al., 2014) was also found among the most abundant OTUs.

Confirmation of cyanobacterial presence and association with decreased hydrogen concentrations

The presence of cyanobacteria in samples 420 and 607 was further confirmed by catalyzed reported deposition fluorescent in situ hybridization (CARD-FISH) with specific probes against the cyanobacterial 16S ribosomal RNA (Puente-Sánchez et al., 2018). This is the preferred technique to search for viable cells in deep subsurface settings (Lloyd et al., 2013; Schippers et al., 2005; Wilkins et al., 2014). This revealed the presence of clusters of cyanobacterial cells tightly attached to the mineral matrix and associated with other microorganisms.

Cyanobacterial abundance showed an apparent inverse correlation with hydrogen concentrations throughout BH10. For the whole *Cyanobacteriota* phylum, the correlation was weak but significant ($R^2 = 0.33$, $P = .03$). A more complex model including the cyanobacterial families *Rivulareaceae* and *Xenococcaceae*, as well as the alpha-proteobacterial families *Sphingomonadaceae* and *Bradyrhizobiaceae* showed a stronger negative correlation with hydrogen concentration ($R2 = 0.80$, $P = .003$). This supports the notion that *Cyanobacteria* (together with other taxa) may act as hydrogen sinks in the deep subsurface of the IPB.

Metagenomic analysis of samples from 420 to 607 mbsf

Due to the strong negative correlation found between hydrogen concentrations and the presence of certain cyanobacteria, the samples retrieved at 420 and 607 mbsf were chosen for further metagenomic analysis. The taxonomy of the metagenomic reads was consistent with the one found for those samples during the 16S rRNA amplicon analysis, with a predominance of *Cyanobacteriota*, *Alphaproteobacteria* and *Bacteroidota* (Puente-Sánchez et al., 2018). However, around 25% of the reads had an eukaryotic origin, most of which were further classified as ascomycotal. The presence of archaeal or viral reads was negligible. According to MG-RAST classification, the cyanobacterial population was mainly composed of reads from the *Nostocales* and *Chroococcales* orders; the ascomycotal population was mainly composed of reads from *Eurotiomycetes*, *Sordariomycetes* and *Dothideomycetes* classes; the alphaproteobacterial population was mainly composed of reads from the *Rhizobiales* order, and bacteroidetes were mainly composed of reads from the *Cytophagales* and *Sphingobacteriales* orders. The results regarding order-level bacterial taxonomy were again consistent to the ones found by 16S rRNA gene amplicon sequencing of those samples, except for the cyanobacterial population, for which the most abundant 16S OTU was a member of the *Leptolyngbia* (order *Oscillatoriales*) genus. A broad overview of the relative abundance of different metabolic subsystems in the metagenomes is shown in (Puente-Sánchez et al., 2018). After manually inspecting all the annotations, many specific functions were found to be of interest for discussing the potential origin and survival strategies of the microbial community (Puente-Sánchez et al., 2018).

In addition to this, we were especially interested in the potential of the microbial community from those samples for degrading complex organic matter, particularly lignin derivatives. A step-by-step reconstruction of all the aromatic degradation pathways found in the metagenomes can be found in (Puente-Sánchez et al., 2018). Central pathways for the degradation of aromatic compounds were reconstructed completely, while key enzymes for several peripheral pathways were supported by the metagenomic data. Almost all the reads classified as belonging to an enzyme involved in the degradation of aromatic compounds were assigned to the *Alphaproteobacteria* class. Interestingly, several alternative pathways for the degradation of benzoate were found in the metagenome, including an alternative pathway which does not require an oxygenolytic ring cleavage step, as well as a completely anaerobic pathway (Puente-Sánchez et al., 2018).

A versatile endolithic microbial consortium thriving in the deep subsurface of the IPB

The presence of cyanobacterial reads in most of our samples was unexpected, especially after the possibility of sample contamination was discarded (see previous section). Furthermore, the abundance of some cyanobacterial groups was found to be negatively correlated with hydrogen concentrations (Puente-Sánchez et al., 2018), suggesting that their presence might be associated with hydrogenotrophic communities. The existence of endolithic cyanobacteria such as the ones found in this study has been known for long (Friedmann, 1980; Friedmann and Meyer, 1987; Knoll et al., 1986), and their presence in deep subsurface environments has been reported several times (Kormas et al., 2003; Newberry et al., 2004; Rastogi et al., 2010); however, to the best of our knowledge their potential role in such ecosystems has never been discussed. Several dark metabolisms, such as different types of fermentation, coupled in some cases to non-respiratory sulfur reduction, have been described in *Cyanobacteriota* (see Stal, 2012 for a thorough review). In general they are supposed to only able to degrade reserve polysaccharides previously obtained by photosynthesis (Oren and Shilo, 1979; Stal and Moezelaar, 1997; Troshina et al., 2002), but at least one case exists in which anaerobic uptake and utilization of extracellular carbohydrates has been described (Stuart et al., 2016; Bagchi et al., 1990).

In order to elucidate the metabolic strategies of the microbial community present in BH10, as well as the potential roles of cyanobacteria in deep subsurface environments, we chose to perform shotgun metagenomic sequencing on samples 420 and 607, as our results indicated that it contained a relatively diverse microbial community in which the presence of cyanobacteria was correlated to a decrease in hydrogen concentration. Furthermore, organic carbon and intact fatty acids were found in those samples, indicating the presence of an active microbiota with access to buried organic matter, while the presence of fractures at that area suggested that water and nutrient input might be higher than in other depths of the borehole.

According to our metagenomic data, the bacterial communities were dominated by *Cyanobacteriota*, *Alphaproteobacteria* and *Bacteroidota*. Ascomycotal reads were also highly abundant (Puente-Sánchez et al., 2018). While the taxonomic classification of the metagenomic reads might suggest a broader diversity, the 16S rRNA amplicon data leads us to believe that each of the bacterial groups is mainly represented by a single genus. Other taxonomic groups such as *Actinomycetota*, *Deinococcota* and *Gammaproteobacteria* were also present in the metagenome, but their small abundance precluded a more detailed analysis of their potential metabolic roles. The community was found capable of resisting a wide array of environmental stresses such as desiccation, oxidative stress, salt stress or metal toxicity, while having the potential to perform several different metabolisms: hydrogenotrophy (*Alphaproteobacteria*, *Cyanobacteriota*), carboxidotrophy (*Alphaproteobacteria*), methylotrophy (*Alphaproteobacteria*), heterotrophy (*Ascomycota*, *Bacteroidota*), and degradation of refractory organic matter (*Ascomycota, Alphaproteobacteria*). Based on these results, we propose that these organisms are associated together to form a metabolically versatile endolithic holobiont (Fig. 7.3), in which each partner provides essential functions for the survival of the community.

In order to thrive in deep subsurface environments, organisms must gain access to water and nutrients. Given the non-porous nature of the silicates found in BH10, this likely requires the ability of burrowing into rocks in order to unveil buried resources and increase the available space for colonization. Note that, even when below the water table, endolithic organisms can be water-limited if they have no available pore space; conversely, biological rock weathering will lead to additional circulation of water and nutrients. Deep subsurface microorganisms also need to be able to exploit the different biotic and abiotic energy sources available to them. The heterogeneous nature of the continental subsurface makes metabolic versatility a

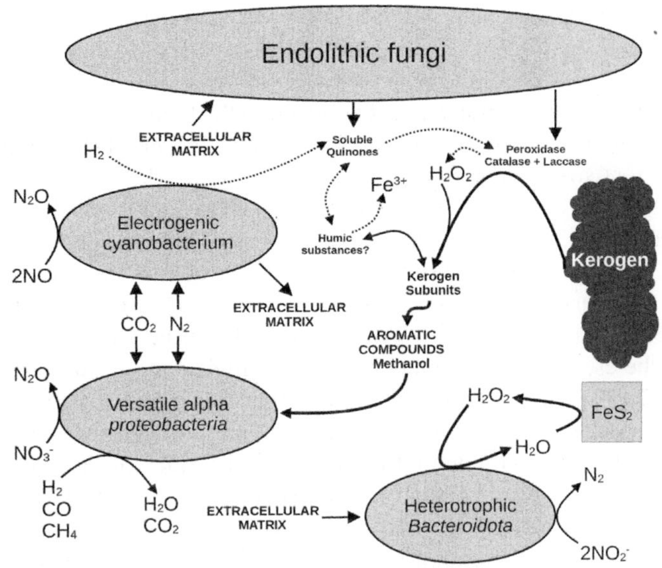

FIGURE 7.3 Proposed model of the microbial community found in BH10 at 420/607 mbsf.

Dotted arrows correspond to electron flow; solid arrows correspond to mass flow.

valuable treat in order to colonize different neighbouring microenvironments. Finally, microorganisms need to be able to survive with the very low energy fluxes characteristic of deep subsurface ecosystems, while withstanding a wide array of environmental stresses. In the following paragraphs we will discuss how the different partners in the holobiont could contribute towards the achievement of these goals.

Endolithic Ascomycota and the potential for kerogen degradation

Fungi are well-known for their ability to form complex symbiotic associations, in which they provide water, nutrients and colonizable niches to their bacterial partners in exchange for fixed nitrogen and carbon (Grube and Berg, 2009). The fungal metagenomic reads obtained from samples 420 and 607 were further classified as belonging to the *Eurotiomycetes*, *Sordariomycetes* and *Dothideomycetes* classes, which are known to harbour many extremotolerant rock-inhabiting species (Muggia et al., 2016; Ruibal et al., 2009). Those non-lichenized rock inhabiting ascomycetous black fungi are thought to have an ancient origin, being the ancestors of both pathogenic fungi and mutualistic lichens (Gueidan et al., 2008, 2011). They are able to penetrate igneous hard rocks, like granite and basalt as well as sedimentary soft rock such as limestone, sandstone or marble (Sterflinger and Krumbein, 1997; Sterflinger, 2000). Nutritionally, they can rely exclusively on sparse, airborne, low molecular weight nutrients (Gueidan et al., 2008; Gostinčar et al., 2012), although some of them can also form mutualistic interactions with photobionts (Muggia et al., 2013; Gorbushina et al., 2005) or degrade aromatic hydrocarbons (Gostinčar et al., 2012).

Unfortunately, the metabolic reconstruction of the fungal population present in our metagenomes was incomplete, probably due to the higher size of eukaryotic genomes and/or to the presence of a consortia of different lichenized and non-lichenized fungal species (Muggia et al., 2013). The available carbon and energy sources for the fungi present in sample 420 would be either small organic compounds released by their autotrophic partners (see below), or buried kerogen. Some black fungi have been hypothesised to be capable of lignin and refractory organic matter degradation (Kietäväinen and Purkamo, 2015; Nai et al., 2013). The refractory organic matter present in the IPB subsurface is mainly type III kerogen (Barberes, 2015) which is derived from lignic debris (Vandenbroucke and Largeau, 2007). Lignin depolymerization and the subsequent degradation of lignin monomers could be therefore good model processes for assessing the refractory organic matter degradation potential in the deep subsurface of the IPB.

Many fungal species (especially basidiomycetes, but also ascomycetes) are able to degrade lignin via extracellular oxidative enzymes such as laccases or peroxidases (Bugg et al., 2011) for a thorough review on lignin degradation pathways). These enzymes use either oxygen or hydrogen peroxide to oxidatively cleave lignin or other complex organic molecules. Fungal peroxidases and multicopper-oxidases (of which catalase is a representative) were indeed detected in the metagenome of samples 420 and 607 (Puente-Sánchez et al., 2018). Another possible path for lignin degradation involves the generation of reactive oxygen species via extracellular quinone-driven Fenton chemistry (Bugg et al., 2011; Guillén et al., 1997; Kerem et al., 1999). Interestingly, fungal quinate permeases were found in the metagenome (Puente-Sánchez et al., 2018). While they are usually involved in quinate degradation (with quinate being an intermediate in lignin synthesis and breakdown), they could also allow the excretion of other structurally related polar compounds.

These two paths for lignin oxidation ultimately require molecular oxygen, and oxygenated water from shallower regions of the borehole could indeed be entering the system through the fractures detected in the area surrounding sample 420. However, anaerobic lignin degradation linked to the presence multicopper oxidases and peroxidases has also been described (Van Waasbergen et al., 2000). We hypothesize that oxidative kerogen breakdown could be occurring in the deep subsurface of the IPB even in the absence of an oxygenated water influx. Water can react with defect surface sites on pyrite to ultimately produce hydrogen peroxide, even under dark, anoxic conditions (Borda et al., 2001, 2003). Its production by pyrite is dependent on the exposed crystal area, and is not limited to the initial contact with water (Borda et al., 2001). This hydrogen peroxide could be used by extracellular peroxidases—or a combination of catalases and oxidases—to break down refractory organic compounds. The ability of endolithic fungi to actively penetrate in hard rocks would increase the access of the community to buried carbon sources, while also exposing more pyrite surfaces to water, with the subsequent release of hydrogen peroxide (Fig. 7.3).

According to our results, it is unlikely that the fungi is able to fully degrade refractory organic matter by itself, as it appears to lack most of the pathways for aromatic compound degradation. Instead, we propose that it contributes to the initial unveiling and oxidative breakdown of kerogen as described above, while the resulting aromatic molecules are degraded by an alphaproteobacterium (see below). In this scenario, the ascomycetous fungi would feed on small organic compounds provided by its bacterial partners (Fig. 7.3). Assuming that not enough oxygen is available to sustain fully aerobic metabolisms, the fungi could obtain energy via fermentation (Puente-Sánchez et al., 2018). Ammonia fermentation couples oxidation of ethanol to non-respiratory nitrate reduction (Takasaki et al., 2004). Several fungi has been shown to perform respiratory denitrification (Kobayashi et al., 1996; Shoun et al., 1992), generally associated to formate oxidation (Kuwazaki et al., 2003; Ma et al., 2008). While a fungal formate dehydrogenase was found in the metagenome of sample 420, there were no evidences of fungal enzymes for denitrification. As discussed above, we can not rule out their presence completely due to the low coverage of the fungal part of the metagenomes.

Endolithic cyanobacteria, biofilm formation and potential hydrogenotrophy

As discussed above, the presence of cyanobacteria has been reported in other deep subsurface environments, but to the best of our knowledge their potential roles in such dark and ecosystems have not been addressed so far. Our physicochemical, taxonomic and metagenomic results showed that cyanobacteria are abundant in the subsurface of the BH10 area, and suggest that, at least in samples 420 and 607, they might be able to obtain energy by oxidizing molecular hydrogen (Fig. 7.3).

Hydrogenases are widespread in cyanobacteria, in which they are used either to minimize energy losses during nitrogen fixation (oxidizing molecular hydrogen and allocating the resulting electrons to the plastoquinone pool) or to dispose of excess reducing power (by reducing protons and releasing hydrogen) (Bothe et al., 2011; Stal, 2012). We found a cyanobacterial NADH evolving hydrogenase in the metagenomes of samples 420 and 607 (Puente-Sánchez et al., 2018). NADH could be converted to NADPH or ferredoxin via NAD(P) transhydrogenase and Ferredoxin-NADP(+) reductase in order to provide reducing power for biosynthetic pathways. NADH could also be oxidised by NADH dehydrogenase (NDH), generating proton-motive force (PMF) and transferring electrons to plastoquinone. NDH is part of the standard cyanobacterial electron transport chain, in which the reduced plastoquinone is usually reoxidised by cytochrome b6f (see Vermaas, 2001 for a simple review). However, cytochrome b6f derives electrons via plastocyanin to either photosystem I or an oxygen-consuming terminal oxidase. While the genes for both pathways were detected in the cyanobacterial subset of the metagenome, it is unlikely that they are operative in the dark, potentially anoxic deep subsurface of the IPB. Based on our metagenomic results and on the existing literature, we are able to propose two putative electron chains that would couple hydrogen oxidation to energy transduction in the conditions found in the BH10 subsurface (Puente-Sánchez et al., 2018).

A nitric oxide reductase (*nor*) was found in the cyanobacterial fraction of the metagenome of sample 420 (Puente-Sánchez et al., 2018). Cyanobacterial nitric oxide reductases connected to an electron transport chain have been described by previous studies (Büsch et al., 2002). This enzyme reduces nitric oxide (NO) by accepting electrons from reduced quinones, albeit it is unable to oxidise cytochrome *c* or generate PMF. Some *Cyanobacteria* are able to generate NO under anoxic conditions via nitrite reduction with cyanoglobin (Mallick et al., 1999; Sturms et al., 2011), which was also found in our metagenome. Nitrite could be obtained directly from the environment, or by an assimilatory nitrate reductase (Mallick et al., 1999). In order for this reaction to progress, the resulting oxidised cyanoglobin must be reduced again, and NO must be quickly scavenged so as not to inactivate the reduced cyanoglobin (Sturms et al., 2011). Since cyanoglobin is located in the membrane, it's co-localization with NOR would minimize cyanoglobin inhibition by NO. The in vivo mechanism for reducing oxidised cyanoglobin is yet unknown (Sturms et al., 2011). If cyanoglobin was to be reduced by NADH (as is the case for methemoglobin), only one in three hydrogen molecules consumed would result in PMF generation, and PMF would only be generated by NDH. Assuming that NDH translocates 2 protons per NADH oxidised, and that 3.33 protons are required for synthesizing 1 ATP molecule, this mechanism would yield 0.2 of mol ATP per mol of hydrogen consumed, consuming 0.33 mol of nitrate in the process. This would be still more efficient than, for example, hydrogenotrophic methanogenesis, which yields 0.12 mol ATP per mol of hydrogen consumed (as calculated by (Kaster et al., 2011), and the yield would drastically improve if nitrite or NO were available instead of nitrate. In any case, let the reader note that the calculations presented here are rough estimates. While the required enzymes were indeed detected in the metagenome (Puente-Sánchez et al., 2018), a detailed thermodynamic investigation would be needed in order to determine the actual feasibility of the proposed pathway.

Another more efficient energy transduction mechanism would involve the transfer of electrons to extracellular acceptors. This has been observed in several cyanobacterial genera (Hasan et al., 2014; Nishio et al., 2015; Pisciotta et al., 2010; Yongjin et al., 2009) including the *Leptolyngbya* genus (Hasan et al., 2014; Pisciotta et al., 2010), which was found in our samples. This process is thought to regulate the redox state of plastoquinone by shedding excess energy to the environment when it becomes over reduced in high light (Pisciotta et al., 2010). Electron transfer from hydrogen to plastoquinone via hydrogenase and NDH would also result in an excess of reduced plastoquinol, since cytochrome b6f, its usual electron acceptor (Trubitsin et al., 2005; Vermaas, 2001) would also be in its reduced state due to the absence of both light and oxygen (Pisciotta et al., 2011). showed that, in the absence of cytochrome b6f activity, plastoquinol is instead reduced by a cytochrome bd quinol oxidase, which is in turn responsible for electron transfer to the environment. We found reads similar to cytochrome D/bd ubiquinol oxidase (*cydAB*) in the cyanobacterial fraction of the metagenomes (Puente-Sánchez et al., 2018). It is not yet clear if bd quinol oxidase transfers electrons directly to extracellular electron acceptors or if additional downstream components are involved (Pisciotta et al., 2011), however the fact that natural *Leptolyngbya* biofilm are capable of electrogenic activity (Pisciotta et al., 2010) suggests that this activity is performed by some cyanobacteria under in vivo conditions.

Under the same assumptions as described for the putative nitrous oxide reducing electron transport chain, hydrogen oxidation coupled to extracellular acceptor reduction would yield 1.2 mol ATP per mol of hydrogen consumed. The ultimate fate of these extracellular electrons can not be elucidated solely from our metagenomic data, however several pathways would be possible according to our community model (Fig. 7.3). They would involve an initial electron transfer to extracellular quinones, which were also proposed as participants in kerogen degradation (see previous section). Extracellular quinones have been found to be mediators to other final acceptors (Newman and Kolter, 2000; Yamazaki et al., 2002); these quinones can be produced by microorganisms, but can also be present in the environment, such as the quinone moieties present in humic acids (Kappler et al., 2004). Humic acids are indeed expected to appear during the breakdown of the type III kerogen present in the deep subsurface of the IPB. These intermediate compounds would donate their electrons to several possible ultimate acceptors. Reduced quinones are substrates for fungal extracellular peroxidases (Valli et al., 1992; Valli and Gold, 1991). In this case, hydrogen peroxide would be the final electron acceptor. Similarly, the Fenton chemistry-based lignin degradation pathway discussed in previous section would involve transfer of electrons from reduced quinones to hydrogen peroxide, with iron acting as an intermediate, and the resulting hydroxyl radical attacking the lignin/kerogen molecules. This process would result in additional oxidised quinone moieties, which can be in turn used as additional electron acceptors (DeAngelis et al., 2013). This is especially interesting from the holobiont perspective, as it links the cyanobacterial energy transduction pathway to fungal kerogen degradation. Finally, humic acids also can shuttle electrons to poorly soluble oxidised iron or manganese minerals (Kappler et al., 2004).

By coupling hydrogenotrophy with the reduction of either nitrogen compounds or extracellular electron acceptors, the cyanobacterium found in the IPB subsurface would be able to act as a primary producer for the ecosystem, fixing carbon via the Calvin–Benson cycle and releasing part of as an extracellular polysaccharide from which the rest of the members of the community can benefit. This cyanobacterium would also potentially be able to fix nitrogen, since a cyanobacterial nitrogenase was found in both metagenomes (Puente-Sánchez et al., 2018). In practice, it is unlikely that this would be happening, as assimilatory reduction of the nitrate present in the sample would require less energy. The cyanobacterial sheath is probably playing a key role in the maintenance of the holobiont: cyanobacterial EPS contributes significantly to the structure of known endolithic biofilms and their interaction with the mineral substrate (Barker and Banfield, 1996). EPS matrices contribute to rock weathering by imbibing water, increasing their volume and causing rocks to crack (Ascaso et al., 2002), but also by trapping particles in the vicinity of the cells (Gorbushina and Krumbein, 2000; Wolfaardt et al., 1999). In addition to direct physical damage, they are also involved in chemical biomobilization and biomineralization processes (Barker et al., 1998; Barker and Banfield, 1996). Through these mechanisms, endolithic cyanobacteria increase the available pore space, and facilitate its colonization by other organisms, including heterotrophs (Blackhurst et al., 2005; De Los Ríos et al., 2007). Under our model (Fig. 7.3), the cyanobacterium would be the main contributor to biofilm formation (see Rossi and De Philippis, 2015 for a thorough review on the roles of cyanobacteria in mixed biofilms), providing heterotrophic organisms with low molecular weight compounds that can be used as either carbon sources or protectors against desiccation.

Aromatic compound-degrading alphaproteobacteria

The *Alphaproteobacteria* class is ecologically versatile, harbouring species adapted to a broad range of environments and lifestyles. Many of them are able to stablish trophic interactions with other organisms, ranging from the highly specific endosymbiosis found in legume nodules to a wide array of less specific and ecologically heterogeneous associations (Pini et al., 2011). This includes their presence in fungal-bacterial communities such as lichens or ectomycorrhizal symbionts (Barbieri et al., 2007; Grube and Berg, 2009).

According to the 16S rRNA amplicon results, the majority of the alphaproteobacterial population belonged to a single *Bradirhizobium* species closely related to *B.elkanii* (Puente-Sánchez et al., 2018), which has been also found to be the most representative species associated with some ascomycetous fungi such as *T. magnatum* (Barbieri et al., 2007). This was consistent with the metagenomic results, which showed a large proportion of reads classified as belonging the *Rhizobiales* order in the alphaproteobacterial fraction of the community. The alphaproteobacteria found in sample 420 were metabolically diverse, being potentially capable of growing via hydrogenotrophy, carboxydotrophy, methylotrophy, or by degrading aromatic compounds. Nitrate was found to be the most likely electron acceptor under anaerobic conditions.

Hydrogenotrophy and carboxydotrophy frequently appear together in alphaproteobacteria (Lorite et al., 2000; Meyer and Schlegel, 1983). They have been shown to occur aerobically, but also coupled to nitrate reduction (King, 2006; Neal et al., 1983), albeit nitrate-dependent anaerobic CO oxidation is thought to only provide energy for maintenance metabolism (King, 2006). Since both the uptake hydrogenase and the aerobic carbon monoxide dehydrogenase are membrane bound and transfer electrons to the quinone pool (Bernhard et al., 1997; Wilcoxen et al., 2011), under a purely hydrogenotrophic/carboxidotrophic metabolism NADH would have to be produced via reverse electron flow. The Calvin–Benson cycle was also detected in the alphaproteobacterial fraction of the metagenome, meaning that this organism would be potentially able to grow autotrophically.

Almost-complete pathways for the degradation of lignin derivatives and other aromatic compounds were found in the alphaproteobacterial fraction of our metagenomes (Puente-Sánchez et al., 2018). Benzoate and protocatechuate are central intermediates in the degradation of aromatic compounds, and their degradation usually requires oxygen for ring hydroxylation and cleavage, regardless of the actual electron acceptor used in respiration. Three different pathways for benzoate degradation were present: the canonical catechol pathway, requiring two molecules of oxygen per molecule of benzoate degraded; the aerobic *Azoarcus* pathway which includes a non-oxygenolytic ring cleavage step and thus consumes only one oxygen molecule per molecule of benzoate (Gescher et al., 2005); and the anaerobic *Azoarcus* pathway (Barragán et al., 2004). The two *Azoarcus* pathways were incomplete in our metagenome, albeit key enzymes exclusive of both of them, such as benzoyl-CoA-dihydrodiol lyase or cyclohex-1-ene-1-carboxyl-CoA hydratase, were detected. Regarding protocatechuate degradation, only the canonical pathway (consuming one molecule of oxygen per molecule of protocatechuate degraded) was found. It has been suggested that anaerobic protocatechuate degradation to glutaryl-CoA might proceed via 3-hydroxybenzoyl-CoA using enzymes from the anaerobic benzoate degradation pathway (Laempe et al., 2001; Philipp et al., 2002). To the best of our knowledge, the complete anaerobic protocatechuate degradation pathway has not yet been fully elucidated. Nonetheless, it has been shown that some of the participant enzymes are actually promiscuous enzymes from other pathways. For example, protocatechuate conversion to 3-hydroxy-1,5-cyclohexadien-1-carbonyl-CoA is carried out by the non-specific enzymes 3-hydroxybenzoate-CoA ligase, 4-hydroxybenzoyl-CoA reductase and benzoyl-CoA reductase (Ding et al., 2008), all of which were found in our metagenomes (Puente-Sánchez et al., 2018).

These results suggest that the alphaproteobacteria found in our metagenomes possess alternative pathways for aromatic hydrocarbon degradation that could be active under different oxygen concentrations. In addition to the potential for completely anaerobic pathways, the presence of trace amounts of oxygen (originating either by the infiltration of oxygenated waters from above, or by the action of peroxidases over pyrite-derived hydrogen peroxide) could support the more energetically favourable oxygenolitic breakdown of aromatic rings, with the resulting compounds being catabolysed anaerobically (Lucey and Leadbetter, 2014). Nitrate is a known electron acceptor in anaerobic aromatic compound degradation (Barragán et al., 2004).

The ability to grow on methanol is also relevant, as it can be a by-product of lignin degradation (Ander and Eriksson, 1985). The alphaproteobacteria present in sample 420 had all the enzymes required for methanol oxidation to CO_2, as well as malyl-CoA synthase, a key enzyme in the serine-glyoxylate cycle for the assimilation of C1 compounds. Genes for polyhydroxybutyrate (PHB) metabolism were also found (Puente-Sánchez et al., 2018). Most methylotrophs lack isocitrate lyase, and therefore use a modified version of the serine-glyoxylate cycle in which glyoxylate is regenerated via a pathway involving enzymes and intermediates from the polyhydroxybutyrate cycle (Korotkova et al., 2002). The enzymes for both pathways were present in our metagenome. The complete glyoxylate cycle is also used by isocitrate lyase-positive organisms to generate metabolic intermediates from acetyl-CoA, which is one of the main products of aromatic hydrocarbon degradation. It is therefore possible that the co-existence of both the glyoxylate shunt and the PHB pathway for glyoxylate regeneration indicates the presence of two different alphaproteobacterial species: a hydrocarbon degrader using the glyoxylate cycle for biosynthesis and a methylotroph using the serine-PHB cycle for C1 assimilation.

Finally, a soluble methane monooxigenase was found in the alphaproteobacterial metagenome for sample 420. If small amounts of oxygen were present in sample 420 (see above), the methylotrophic alphaproteobacterium would also be able of oxidizing methane to methanol. Methanol oxidation could then be coupled to nitrate reduction in order to obtain energy and fixed carbon.

Together, our results indicate that the alphaproteobacterial community found in sample 420 is linked with the breakdown of refractory organic matter, with a predominant species degrading kerogen monomers, and a less abundant species consuming the methanol generated during kerogen breakdown. The fact that alphaproteobacterial peroxidases and oxidases were also detected raises the question of whether the fungi is actually required for kerogen degradation. Alphaproteobacteria have been proposed to be one of the main contributors to lignin degradation in anoxic soils instead of fungi (DeAngelis et al., 2011), while poplar wood has been shown to be anaerobically degraded by a microbial consortium composed of members from *Alphaproteobacteria*, *Bacteroidetes* and *Cyanobacteria*, among others (van der Lelie et al., 2012). For example, *Phanaerochaete sordida*, one of the most efficient known lignin-degrading fungi, is deeply associated with a bacterial partner to the point of being unable to grow in axenic cultures (Lim et al., 2003). This suggests that, at least in some cases, the degradation of refractory organic matter proceeds through a physiologically obligate—at least for the fungi—syntrophic interaction, which strengthens our deep subsurface holobiont hypothesis. In any case, regardless of whether they actually participate in kerogen breakdown or not, the ability of the fungal lineages found in samples 420 and 607 to actively bore into rocks would result in their hydrocarbon-degrading alphaproteobacterial partners having increased access to buried kerogen deposits.

Biopolymer-degrading Bacteroidota

Members of the *Bacteroidota* phylumare thought to be specialized in degrading complex organic matter, especially in the form of biopolymers such as polysaccharides and proteins (Thomas et al., 2011). They have been found in a wide array of environments, including anaerobic lignin degradation consortia (van der Lelie et al., 2012), in which they are key biomass degraders.

Our results suggest that the *Bacteroidota* found in our samples are playing a similar role, as they possessed a high abundance of genes related to sugar and peptide uptake and catabolism (Puente-Sánchez et al., 2018). This figure is especially relevant given their relatively low abundance in the community. According to our metagenomes, these bacteroidetes would be adept at degrading polysaccharides containing mannose, rhamnose, fucose, xanthosine, galacturonic acid and N-acetylglucosamine, using nitrate as the electron acceptor (Puente-Sánchez et al.,

2018) under anaerobic conditions. Mass spectrometry analysis confirmed the presence of some of these sugars in the BH10 subsurface (Victor Parro, personal communication). A sodium-translocating NADH-quinone oxidoreducetase was found in the bacteroidetes fraction of the metagenomes. Using sodium instead of protons for building a transmembrane gradient minimizes energy loss due to ion leakage (Van De Vossenberg et al., 1995), which might be relevant under slow metabolic conditions. Finally, the presence of a cytochrome-c peroxidase would also allow for hydrogen peroxide to be used as the terminal acceptor.

While in the lignin-degrading consortium described by (van der Lelie et al., 2012) the bacteroidetes were thought to degrade the polysaccharidic fraction of lignocellulose, it is unlikely that this is happening for the kerogen in the deep subsurface of the IPB, as polysaccharides would have been destroyed during early diagenesis (Mukhopadhyay, 1992). Instead, we propose that they are are feeding on cyanobacterial and fungal extracellular components (see Thomas et al., 2011). Cyanobacterial extracellular polysaccharide (EPS) consumption by bacteroideteshas been reported in microbial mats (Cole et al., 2014). EPS in natural biofilms represent a more refractory, partially degraded, remnant of the original polymer molecules, with a variable equilibrium between EPS production and consumption in the different biofilm layers (Decho et al., 2005). While it is possible that the role of bacteroidetes is merely commensalistic, limited EPS degradation could also have a positive effect in the community by controlling biofilm permeability, maturation stage, and making organic matter available to other, less versatile heterotrophs in the community.

Microbial processes operating in the BH10 subsurface

Unlike in BH8 (see previous section), no clear trends in microbial composition appeared with depth, apart from a somewhat higher abundance of actinobacteria at the shallower samples, and a higher abundance of cyanobacteria at the deepest part of the borehole (Puente-Sánchez et al., 2018). The fact that the rocks in borehole BH10 were much less fractured than the ones in borehole BH8 suggests that the presence of local fissures, which imply access to water and nutrients, might be a more relevant factor than depth or bulk mineralogy in determining the composition and activity of the BH10 subsurface microbiota. Consistently with this hypothesis, higher fatty acid levels were detected in the fault zones around 350 and 420 mbsf, as well as in the more fractured Culm zone beyond 500 mbsf (Puente-Sánchez et al., 2018). Total organic carbon showed a single peak at 420 mbsf; however, it probably represents buried organic compounds rather than extant biological materials.

In spite of the presence of pyrite, microorganisms related to the iron and sulphur cycles were not especially abundant in BH10 (Puente-Sánchez et al., 2018), with the exception of a potentially iron-oxidizing *Acidovorax*. Instead, the most abundant community members either belonged to the *Cyanobacteria* phylum, or to ubiquitous soil groups such as *Arthrobacter* (*Actinobacteria*), *Bradyrhizobium* (*Alphaproteobacteria*) or *Sediminibacterium* (*Bacteroidetes*). As discussed in the above sections, these results are not due to contamination during sample retrieval or processing, and instead truly represent the microbial community composition at several depths of the BH10 borehole. The absence of pyrite oxidizing microorganisms such as *Leptospirillum* or *Acidithiobacillus* can be attributed to a low oxygen availability in the saturated zone of the aquifer. It has been described that pyrite oxidation does not proceed under a water table unless there is a high input of oxygen-saturated water (Hammack and Watzlaf, 1990), which is unlikely to happen in the unfractured rocks of BH10. The absence of pyrite oxidation would also explain the much lower sulphate concentrations (ranging from 100 to 0 ppm) present in BH10 when compared to BH8 (Puente-Sánchez et al., 2014) (two orders of magnitude larger, see (Puente-Sánchez et al., 2018). Nonetheless, pyrite might be playing a different role in the BH10 ecosystem, by generating hydroxyl radicals and hydrogen peroxide even under dark, anoxic conditions (Borda et al., 2001, 2003). This might be able to support oxidative metabolisms to a certain extent (Davila et al., 2008), albeit the experimental validation of this hypothesis was beyond the scope of our work. Nitrate, another potential electron acceptor for anaerobic respiration, was present in somewhat similar concentrations in both boreholes (ca. 1.5 ppm). In general, the Tinto-Odiel aquifer has relatively high nitrate concentrations, which are attributed to the use of fertilizers for intensive farming in the area (Cavaller et al., 2005; López-Guerrero and Fernández-Uria, 1985). Sulfate reduction can be inhibited by nitrate, and nitrate reducers outcompete sulphate reducers under electron donor limitations (Achtnich et al., 1995; Percheron et al., 1999). This is likely occurring in the BH10 area, as nitrate reducing bacteria vastly outnumbered sulphate reducers, and suggest that nitrate, entering the Peña de Hierro area via groundwater flow, is the primary electron acceptor in the anaerobic parts of the BH10 system.

Hydrogen, methane and soluble ferric ion concentrations were considered as potential drivers of microbial community composition in BH10. The fact that ferric iron and methane concentrations were correlated can be interpreted as either iron reduction inhibiting methanogenesis (Lovley and Phillips, 1987; Van Bodegom, Scholten and Stams, 2004; Zhang et al., 2009) or the presence of ferric iron-dependent anaerobic methane oxidation (Amos et al., 2012). Both processes would require the presence of an archaea (performing either standard or reverse methanogenesis) and an iron-reducing bacteria. However, even though methanogenic and methanotrophic archaea were indeed found in BH8, the PCR primers used in BH10 did not allow us to target the archaeal diversity. At any rate, their presence in BH10 would not be surprising, as both methanogenesis and anaerobic methane oxidation have been previously described in deep subsurface environments (Bomberg et al., 2015; Chapelle et al., 2002; Puente-Sánchez et al., 2014).

No obvious ferric iron reducing microbia were found to be amongst the most abundant taxa in BH10. Some members of the *Arthrobacter* genus (the most abundant *Actinomycetota* taxa in BH10) have however been found capable of metal respiration (Lee and Newman, 2003; Ottow and Glathe, 1971; Jones et al., 1984). Interestingly, actinobacterial abundance was negatively correlated with iron and methane concentrations. On the other hand, the significant relationship between soluble ferric iron concentration upon the microbial community composition might be due to the presence of iron oxidizing microorganisms producing ferric iron as a product. However, if this were to be the case the potential iron oxidizer *Acidovorax* (*Betaproteobacteria*) would show a negative correlation with ferric iron concentration. These results are consistent with the lack of significant pyrite oxidation in BH10. Both microbial iron oxidation and reduction can be performed by directly transferring electrons from or to undissolved mineral surfaces (Hernandez and Newman, 2001; Kostka et al., 2002). Such processes are therefore often dependent upon the available mineral surface area rather than the dissolved mineral concentrations (Kostka et al., 2002). Unfortunately, this variable was impossible to measure with the array of techniques available to us.

Another potential acceptor for anaerobic methane oxidation is sulphate (Knittel and Boetius, 2009; Marlow et al., 2014; Orphan et al., 2001; Parkes et al., 2007; Pedersen, 2013). *Deltaproteobacteria*, a class which hosts most of the known sulphate-reducing bacteria, was negatively correlated with methane concentration. However, as discussed above, the actual abundance of sulphate reducing bacteria in BH10 was practically negligible.

Finally, hydrogen concentration was negatively correlated with the abundance of the cyanobacterial families *Rivulareaceae* and *Xenococcaceae*. Metagenomic analysis of samples 420 and 607 revealed the presence of a versatile endolithic microbial community potentially supported by cyanobacterial hydrogenotrophy and the degradation of refractory organic matter, probably similar to type III kerogen (see previous section). It might be possible that the cyanobacteria found in other samples are taking part in similar associations: samples 496 and 249, for example, shared with sample 420 the presence of *Cyanobacteria*, as well as lower hydrogen concentrations than the rest of the samples (Puente-Sánchez et al., 2018). On the other hand, the cyanobacteria present in the rest of the samples belonged to other taxa and their presence did not show a clear correlation with hydrogen concentration. In those cases, cyanobacteriamight thus be playing a different role than the one described here.

The influence of heterotrophic metabolisms in the deep subsurface IPB ecosystem is hard to assess from our data. Sequences related to heterotrophs, including *T. lapidicaptus*, which was isolated from the nearby BH11, were also retrieved from BH10 samples. Metagenomic analysis revealed the potential for kerogen-degrading activities in the microbial community of samples 420 and 607; however, those results can not be easily extended to the rest of the samples, given the heterogeneity of deep subsurface habitats. Organic carbon was present in all the samples tested, suggesting that kerogen might be a potential carbon and energy source throughout the borehole.

Organic acids, especially acetate, were detected in several samples. Their ultimate origin remains unclear, as they can be produced and consumed by many different microbial metabolisms. For example, in sample 420 (see previous section) carbon reserves (either extracellular EPS or internal reserves such as alphaproteobacterial PHBs) could be fermented to organic acids in the absence of oxygen or nitrate. Conversely, in the presence of electron acceptors acetate could be used to obtain energy via acetyl-CoA synthetase and tricarboxylic acid cycle, and used as a carbon source via the glyoxylate cycle. Therefore, our samples could be acting either as an acetate source or an acetate sink depending on electron acceptor availability, or acetate could be being cycled between fermentative and respiratory members of the community. This reasoning could also be extended to other metabolisms (e.g., acetogenesis versus acetoclastic methanogenesis). While small organic acids could be good carbon and electron shuttles, we find it difficult to believe that they are being exchanged between the different microbial communities found through the borehole, especially after considering the long distances involved, and the hydrogeological nature of the BH10 area (see below). We thus believe that the organic acids detected in the BH10 are the result of local fermentative metabolisms.

Overall, the BH10 subsurface has an extremely low permeability. This makes it unlikely that it is operating as a single ecological unit, as there is probably little opportunity for nutrient exchange between the different depths sampled in this study. This is supported by the lack of clear trends in community composition with depth, and also by the apprarent influence of hydrogen and methane, two gases which can be generated abiotically (Etiope and Sherwood Lollar, 2013; Okland et al., 2014; Hellevang et al., 2011) and diffuse through rocks with relative ease. Nonetheless, even if most of the transect drilled in this study cuts through unfractured regions, the fact that fatty acid concentrations increased at the localized fracture zones suggests that water flow channels also exist in BH10, at least to a certain extent.

Access to pore-space and water might thus be a critical factor for surviving in compact continental deep subsurface ecosystems. Even although our samples were taken below the water table and that all non-isolated pores are expected to be flooded, the adsorption of water to silica surfaces might result in reduced water availability, especially in small pores with a high surface-to-volume ratio. Fracture areas would provide water and soluble nutrients, but microorganisms might still need to bore through rocks in order to reach non-soluble compounds such as pyrite or kerogen. Our 16S rRNA amplicon results showed the presence of bacteria similar to those found in other oligotrophic, endolithic and arid environments. Those organisms, even if usually found in surface habitats such as deserts, tundra soils or permafrost, possess the necessary traits for colonizing deep subsurface habitats: they are adapted to inhabit rocks while operating under extremely low energy flows and withstanding severe environmental stresses.

Metagenomic analysis revealed the presence of a metabolically versatile endolithic microbial community in which several organisms cooperated in order to exploit the scarce resources available in the IPB subsurface (Fig. 7.3). We believe that this included not only metabolic coupling, but also the accomplishment of other tasks that can be beneficial to the whole community, such as the production of an extracellular matrix that facilitates water and nutrient retention, or the active weathering of rocks in order to increase available space and gain access to novel sources of carbon and energy. Their association might therefore be an obligate one, with cooperation being the only way to fulfill all the requirements for surviving in the harsh continental deep subsurface.

Conclusions

Several boreholes were performed in the Peña de Hierro area within the context of the MARTE and IPBSL projects (Amils et al., 2023; Puente-Sánchez et al., 2014, 2018) (Fig. 7.1). Boreholes BH8 and BH10 were thoroughly studied using an array of complementary techniques in order to gain a deeper understanding of the biogeochemical processes occurring in the deep subsurface of the IPB. This led to the characterisation of two distinct underground habitats located in a unique geological setting which has several implications for the study of subsurface-surface interactions, early life environments or Mars habitability.

The rocks in the BH8 borehole were fractured, and its oxidised waters were in equilibrium with the nearby Peña de Hierro mine pit (Fernando Tornos, personal communication). The co-occurrence of pyrite, water and oxygen leads to a microbial community focused on pyrite oxidation, which results in the production of high levels of sulphuric acid. Some of the sulphate is in turn used an electron acceptor in the anaerobic part of the borehole, but a large part of it probably remains on the aquifer, leading to groundwater acidification and the subsequent dissolution of ferric iron. BH8 is therefore an example of the underground biological reactor proposed by several authors (Amils et al., 2013; Amils, Fernández-Remolar and Team, 2014; Fernández-Remolar et al., 2008; Fernández-Remolar et al., 2008; Gómez-Ortiz et al., 2014) in order to explain the unique conditions the Tinto River, which actually originates near the Peña de Hierro mine pit.

In spite of their proximity (c.a. 500 m), and of the fact that they share a common geological setting, the BH10 borehole was very different from BH8. The BH10 area was generally non-fractured, and its aquifer hosted anoxic waters. The pyrite concentration was lower, and it was unoxidised. No evidences of pyrite oxidation or sulphate reduction were found below the water table. Instead, nitrate seemed to be the primary electron acceptor, and hydrogen, methane or buried kerogen were the most likely electron donors. While the BH8 microbial community was to a large extent aerobic and similar to that of an acid mine drainage, the BH10 conditions appeared to select for microorganisms with strong adaptations towards endolithic lifestyles and the tolerance of severe nutritional and environmental stresses. Similarly, where the BH8 subsurface ecosystem could be divided into three main compartments showing a certain degree of nutrient exchange, the microbial communities found in the different samples of the BH10 borehole could not be assembled together into a comprehensive model of the whole column. A detailed metagenomic analysis of one of these samples revealed a microbial consortium in which the partners established complex physiological and ecological interactions. However, according to our results this consortium relied on elements that were already buried (kerogen, pyrite surfaces), likely of an abiotic origin (hydrogen) or present in the whole Tinto-Odiel aquifer (nitrate), being thus independent from any other process potentially occurring in the rest of the borehole. We therefore concluded that, unlike BH8, the BH10 borehole is not operating as a single ecological unit, and that instead each sample probably contained its own, independent ecosystem.

Our results highlight the key importance of the short-scale geological context in shaping deep subsurface microbial ecosystems, with an emphasis on the influence of water availability. In compact rocks with low-degrees of fracture, microorganisms might be living within enclosed microniches with a high degree of isolation. Physical and physicochemical characterisation of such microniches, and the way the interact between themselves and the main fault zones, is required in order to deepen the understanding on the actual short-scale environment that deep subsurface microorganisms live in, and to better put in context our metagenomic results.

References

Abram, J.W., Nedwell, D.B., 1978. Inhibition of methanogenesis by sulphate reducing bacteria competing for transferred hydrogen. Arch. Microbiol. 117 (1), 89–92. https://doi.org/10.1007/BF00689356.

Achtnich, C., Bak, F., Conrad, R., 1995. Competition for electron donors among nitrate reducers, ferric iron reducers, sulfate reducers, and methanogens in anoxic paddy soil. Biol. Fertil. Soils 19 (1), 65–72. https://doi.org/10.1007/bf00336349.

Amils, R., Fernández-Remolar, D., Parro, V., Rodríguez-Manfredi, J.A., Timmis, K., Oggerin, M., Sánchez-Román, M., López, J.F., Fernández, J.P., Gómez-Ortiz D, P.F., Briones, C., Gómez, F., Omoregie, E., García, M., Rodríguez, N., Sanz, J.L., 2013. Iberian Pyrite Belt Subsurface Life (IPBSL), A Drilling Project of Biohydrometallurgical Interest, pp. 15–18, 825. www.scientific.net/AMR.825.15.

Amils, R., Escudero, C., Oggerin, M., Sánchez, F.P., Rodríguez, A.A., Remolar, D.F., Rodríguez, N., Villadangos, M.G., Sanz, J.L., Briones, C., Román, M.S.-, Gómez, F., Leandro, T., Moreno-Paz, M., Ballesteros, O.P.-, Molina, A., Tornos, F., Andrea, I.S.-, Timmis, K., Pieper, D.H., Parro, V., Coupled, C., 2023. H, N, S and Fe biogeochemical cycles operating in the continental deep subsurface of the Iberian pyrite belt. Environ. Microbiol. 25 (2), 428–453. https://doi.org/10.1111/1462-2920.16291.

Amils, R., Fernández-Remolar, D., The IPBSL Team, 2014. Río Tinto: a geochemical and mineralogical terrestrial analogue of mars. Life 4 (3), 511–534. https://doi.org/10.3390/life4030511.

Amos, R.T., Bekins, B.A., Cozzarelli, I.M., Voytek, M.A., Kirshtein, J.D., Jones, E.J.P., Blowes, D.W., 2012. Evidence for iron-mediated anaerobic methane oxidation in a crude oil-contaminated aquifer. Geobiology 10 (6), 506–517. https://doi.org/10.1111/j.1472-4669.2012.00341.x.

Ander, P., Eriksson, K.E., 1985. Methanol formation during lignin degradation by Phanerochaete chrysosporium. Appl. Microbiol. Biotechnol. 21 (1–2), 96–102. https://doi.org/10.1007/BF00252369.

Ascaso, C., Wierzchos, J., Souza-Egipsy, V., De Los Ríos, A., Rodrigues, J.D., 2002. In situ evaluation of the biodeteriorating action of microorganisms and the effects of biocides on carbonate rock of the Jeronimos Monastery (Lisbon). Int. Biodeterior. Biodegrad. 49 (1), 1–12. https://doi.org/10.1016/S0964-8305(01)00097-X.

Bagchi, S.N., Chauhan, V.S., Palod, A., 1990. Heterotrophy and nitrate metabolism in a cyanobacterium Phormidium uncinatum. Curr. Microbiol. 21 (1), 53–57. https://doi.org/10.1007/BF02090100.

Baqué, M., Viaggiu, E., Scalzi, G., Billi, D., 2013. Endurance of the endolithic desert cyanobacterium Chroococcidiopsis under UVC radiation. Extremophiles 17 (1), 161–169. https://doi.org/10.1007/s00792-012-0505-5.

Barberes, G., 2015. Organic geochemical results from the shales of the South Portuguese Zone, southern Portugal. In: Presented at the European Regional Conference and Exhibition.

Barbieri, E., Guidi, C., Bertaux, J., Frey-Klett, P., Garbaye, J., Ceccaroli, P., Saltarelli, R., Zambonelli, A., Stocchi, V., 2007. Occurrence and diversity of bacterial communities in Tuber magnatum during truffle maturation. Environ. Microbiol. 9 (9), 2234–2246. https://doi.org/10.1111/j.1462-2920.2007.01338.x.

Barker, W.W., Banfield, J.F., 1996. Biologically versus inorganically mediated weathering reactions: relationships between minerals and extracellular microbial polymers in lithobiontic communities. Chem. Geol. 132 (1–4), 55–69. https://doi.org/10.1016/s0009-2541(96)00041-1.

Barker, W.W., Welch, S.A., Chu, S., Banfield, J.F., 1998. Experimental observations of the effects of bacteria on aluminosilicate weathering. Am. Mineral. 83 (11–12 Part 2), 1551–1563. https://doi.org/10.2138/am-1997-11-1243.

Barragán, M.J.L., Carmona, M., Zamarro, M.T., Thiele, B., Boll, M., Fuchs, G., García, J.L., Díaz, E., 2004. The bzd gene cluster, coding for anaerobic benzoate catabolism, in Azoarcus sp. strain CIB. J. Bacteriol. 186 (17), 5762–5774. https://doi.org/10.1128/jb.186.17.5762-5774.2004.

Bernhard, M., Benelli, B., Hochkoeppler, A., Zannoni, D., Friedrich, B., 1997. Functional and structural role of the cytochrome b subunit of the membrane-bound hydrogenase complex of Alcaligenes Eutrophus H16. Eur. J. Biochem. 248 (1), 179–186. https://doi.org/10.1111/j.1432-1033.1997.00179.x.

Blackhurst, R.L., Genge, M.J., Kearsley, A.T., Grady, M.M., 2005. Cryptoendolithic alteration of Antarctic sandstones: pioneers or opportunists? J. Geophys. Res.: Planets 110 (E12). https://doi.org/10.1029/2005je002463.

Bollmann, A., Palumbo, A.V., Lewis, K., Epstein, S.S., 2010. Isolation and physiology of bacteria from contaminated subsurface sediments. Appl. Environ. Microbiol. 76 (22), 7413–7419. https://doi.org/10.1128/AEM.00376-10UnitedStates.

Bomberg, M., Nyyssönen, M., Pitkänen, P., Lehtinen, A., Itävaara, M., 2015. Active microbial communities inhabit sulphate-methane interphase in deep bedrock fracture fluids in olkiluoto. Finland . BioMed Research International 2015, 1–17. https://doi.org/10.1155/2015/979530.

Bonaccorsi, R., Stoker, C.R., 2008. Science results from a mars drilling simulation (Río Tinto, Spain) and ground truth for remote science observations. Astrobiology 8 (5), 967–985. https://doi.org/10.1089/ast.2007.0152.

Borda, M.J., Elsetinow, A.R., Schoonen, M.A., Strongin, D.R., 2001. Pyrite-induced hydrogen peroxide formation as a driving force in the evolution of photosynthetic organisms on an early earth. Astrobiology 1 (3), 283–288. https://doi.org/10.1089/15311070152757474.

Borda, M.J., Elsetinow, A.R., Strongin, D.R., Schoonen, M.A., 2003. A mechanism for the production of hydroxyl radical at surface defect sites on pyrite. Geochem. Cosmochim. Acta 67 (5), 935–939. https://doi.org/10.1016/S0016-7037(02)01222-X.

Bothe, H., Schmitz, O., Geoffrey Yates, M., Newton, W.E., 2011. Nitrogenases and Hydrogenases in Cyanobacteria. Springer Science and Business Media LLC, pp. 137–157. https://doi.org/10.1007/978-94-007-0388-9_6.

Bruno, L., Billi, D., Bellezza, S., Albertano, P., 2009. Cytomorphological and genetic characterization of troglobitic Leptolyngbya strains isolated from roman hypogea. Appl. Environ. Microbiol. 75 (3), 608–617. https://doi.org/10.1128/aem.01183-08.

Bugg, T.D.H., Ahmad, M., Hardiman, E.M., Rahmanpour, R., 2011. Pathways for degradation of lignin in bacteria and fungi. Nat. Prod. Rep. 28 (12), 1883–1896. https://doi.org/10.1039/c1np00042j.

Büsch, A., Friedrich, B., Cramm, R., 2002. Characterization of the norB gene, encoding nitric oxide reductase, in the nondenitrifying cyanobacterium Synechocystis sp. strain PCC6803. Appl. Environ. Microbiol. 68 (2), 668–672. https://doi.org/10.1128/aem.68.2.668-672.2002.

Cavaller, A., Robles, J.R., Rebollo, Otero, R., 2005. Caracterización cuantitativa y cualitativa de las masas de agua subterránea en las unidades hidrológicas en el ámbito del Plan Hidrológico del Guadiana II (cuencas de los ríos Tinto, Odiell y Guadiana). VI Simposio del Agua en Andalucía: 1 a 3 de junio 1, 407–416. Sevilla.

Chapelle, F.H., O'Neill, K., Bradley, P.M., Methá, B.A., Ciufo, S.A., Knobel, L.R.L., Lovley, D.R., 2002. A hydrogen-based subsurface microbial community dominated by methanogens. Nature 415 (6869), 312–315. https://doi.org/10.1038/415312a.

Cole, J.K., Hutchison, J.R., Renslow, R.S., Kim, Y.M., Chrisler, W.B., Engelmann, H.E., Dohnalkova, A.C., Hu, D., Metz, T.O., Fredrickson, J.K., Lindemann, S.R., 2014. Phototrophic biofilm assembly in microbial-mat-derived unicyanobacterial consortia: model systems for the study of autotroph-heterotroph interactions. Front. Microbiol. 5. https://doi.org/10.3389/fmicb.2014.00109. Available from:

Crocker, F.H., Fredrickson, J.K., White, D.C., Ringelberg, D.B., Balkwill, D.L., 2000. Phylogenetic and physiological diversity of Arthrobacter strains isolated from unconsolidated subsurface sediments the GenBank accession numbers for the sequences determined in this study are given in Methods. Microbiology 146 (6), 1295–1310. https://doi.org/10.1099/00221287-146-6-1295.

Davila, A.F., Fairén, A.G., Gago-Duport, L., Stoker, C., Amils, R., Bonaccorsi, R., Zavaleta, J., Lim, D., Schulze-Makuch, D., McKay, C.P., 2008. Subsurface formation of oxidants on Mars and implications for the preservation of organic biosignatures. Earth Planet Sci. Lett. 272 (1–2), 456–463. https://doi.org/10.1016/j.epsl.2008.05.015.

De Los Ríos, A., Grube, M., Sancho, L.G., Ascaso, C., 2007. Ultrastructural and genetic characteristics of endolithic cyanobacterial biofilms colonizing Antarctic granite rocks. FEMS (Fed. Eur. Microbiol. Soc.) Microbiol. Ecol. 59 (2), 386–395. https://doi.org/10.1111/j.1574-6941.2006.00256.x.

DeAngelis, K.M., Allgaier, M., Chavarria, Y., Fortney, J.L., Hugenholtz, P., Simmons, B., Sublette, K., Silver, W.L., Hazen, T.C., 2011. Characterization of trapped lignin-degrading microbes in tropical forest soil. PLoS One 6 (4). https://doi.org/10.1371/journal.pone.0019306UnitedStates. Available from.

DeAngelis, K.M., Sharma, D., Varney, R., Simmons, B., Isern, N.G., Markilllie, L.M., Nicora, C., Norbeck, A.D., Taylor, R.C., Aldrich, J.T., Robinson, E.W., 2013. Evidence supporting dissimilatory and assimilatory lignin degradation in Enterobacter lignolyticus SCF1. Front. Microbiol. 4. https://doi.org/10.3389/fmicb.2013.00280. Available from:

Decho, A.W., Visscher, P.T., Pamela Reid, R., 2005. Production and Cycling of Natural Microbial Exopolymers (EPS) within a Marine Stromatolite, vol 219. Elsevier BV, pp. 71–86. https://doi.org/10.1016/b978-0-444-52019-7.50008-5.

Ding, B., Schmeling, S., Fuchs, G., 2008. Anaerobic metabolism of catechol by the denitrifying Bacterium Thauera aromatica—a result of promiscuous enzymes and regulators? J. Bacteriol. 190 (5), 1620–1630. https://doi.org/10.1128/jb.01221-07.

Escudero, C., Vera, M., Oggerin, M., Amils, R., 2018. Active microbial biofilms in deep poor porous continental subsurface rocks. Sci. Rep. 8 (1). https://doi.org/10.1038/s41598-018-19903-z.

Etiope, G., Sherwood Lollar, B., 2013. Abiotic methane on earth. Rev. Geophys. 51 (2), 276–299. https://doi.org/10.1002/rog.20011.

Fernández-Remolar, D.C., Gómez, F., Prieto-Ballesteros, O., Schelble, R.T., Rodríguez, N., Amils, R., 2008a. Some ecological mechanisms to generate habitability in planetary subsurface areas by chemolithotrophic communities: the Ro Tinto subsurface ecosystem as a model system. Astrobiology 8 (1), 157–173. https://doi.org/10.1089/ast.2006.0022.

Fernández-Remolar, D.C., Prieto-Ballesteros, O., Rodríguez, N., Gómez, F., Amils, R., Gómez-Elvira, J., Stoker, C.R., 2008b. Underground habitats in the Río tinto basin: a model for subsurface life habitats on mars. Astrobiology 8 (5), 1023–1047. https://doi.org/10.1089/ast.2006.0104.

Freund, F., Dickinson, J.T., Cash, M., 2002. Hydrogen in rocks: an energy source for deep microbial communities. Astrobiology 2 (1), 83–92. https://doi.org/10.1089/153110702753621367.

Friedmann, E.I., 1980. Endolithic microbial life in hot and cold deserts. Orig. Life 10 (3), 223–235. https://doi.org/10.1007/BF00928400.

Friedmann, E.I., Meyer, M.A., 1987. Antarctic cryptoendolithic microbial ecosystem research, 1986–1987. National Science Foundation 22 (5), 240–241.

Garrido, P., Toril, E.G.-, Moyano, A.G.-, Moreno-Paz, M., Amils, R., Parro, V., 2008. An oligonucleotide prokaryotic acidophile microarray: its validation and its use to monitor seasonal variations in extreme acidic environments with total environmental RNA. Environ. Microbiol. 10 (4), 836–850. https://doi.org/10.1111/j.1462-2920.2008.01477.x.

Gescher, J., Eisenreich, W., Wörth, J., Bacher, A., Fuchs, G., 2005. Aerobic benzoyl-CoA catabolic pathway in Azoarcus evansii : studies on the non-oxygenolytic ring cleavage enzyme. Mol. Microbiol. 56 (6), 1586–1600. https://doi.org/10.1111/j.1365-2958.2005.04637.x.

Gómez-Ortiz, D., Fernández-Remolar, D.C., Granda, A., Quesada, C., Granda, T., Prieto-Ballesteros, O., Molina, A., Amils, R., 2014. Identification of the subsurface sulfide bodies responsible for acidity in Río Tinto source water, Spain. Earth Planet Sci. Lett. 391, 36–41. https://doi.org/10.1016/j.epsl.2014.01.022.

Gorbushina, A.A., Krumbein, W.E., 2000. Subaerial Microbial Mats and Their Effects on Soil and Rock. Springer Science and Business Media LLC, pp. 161–170. https://doi.org/10.1007/978-3-662-04036-2_18.

Gorbushina, A.A., Beck, A., Schulte, A., 2005. Microcolonial rock inhabiting fungi and lichen photobionts: evidence for mutualistic interactions. Mycol. Res. 109 (11), 1288–1296. https://doi.org/10.1017/S0953756205003631.

Gostinčar, C., Muggia, L., Grube, M., 2012. Polyextremotolerant black fungi: oligotrophism, adaptive potential, and a link to lichen symbioses. Front. Microbiol. 3. https://doi.org/10.3389/fmicb.2012.00390.

Grube, M., Berg, G., 2009. Microbial consortia of bacteria and fungi with focus on the lichen symbiosis. Fungal Biol. Rev. 23 (3), 72–85. https://doi.org/10.1016/j.fbr.2009.10.001.

Gueidan, C., R Villaseñor, C., S de Hoog, G., A Gorbushina, A., A Untereiner, W., Lutzoni, F., 2008. A rock-inhabiting ancestor for mutualistic and pathogen-rich fungal lineages. Stud. Mycol. 61, 111–119. https://doi.org/10.3114/sim.2008.61.11.

Gueidan, C., Ruibal, C., de Hoog, G.S., Schneider, H., 2011. Rock-inhabiting fungi originated during periods of dry climate in the late Devonian and middle Triassic. Fungal Biol. 115 (10), 987–996. https://doi.org/10.1016/j.funbio.2011.04.002.

Guillén, F., Jesus Martinez, M.A., Muñoz, C., Martinez, A.T., 1997. Quinone redox cycling in the ligninolytic FungusPleurotus eryngiiLeading to extracellular production of superoxide anion radical. Arch. Biochem. Biophys. 339 (1), 190–199. https://doi.org/10.1006/abbi.1996.9834.

Hammack, R., Watzlaf, G., 1990. The effect of oxygen on pyrite oxidation. J. Am. Soc. Mining Reclam. 1990 (1), 257–264. https://doi.org/10.21000/JASMR90010257.

Hasan, K., Bekir Yildiz, H., Sperling, E., Ó Conghaile, P., Packer, M.A., Leech, D., Hägerhäll, C., Gorton, L., 2014. Photo-electrochemical communication between cyanobacteria (Leptolyngbia sp.) and osmium redox polymer modified electrodes. Phys. Chem. Chem. Phys. 16 (45), 24676–24680. https://doi.org/10.1039/c4cp04307c.

Hellevang, H., Huang, S., Thorseth, I.H., 2011. The potential for low-temperature abiotic hydrogen generation and a hydrogen-driven deep biosphere. Astrobiology 11 (7), 711–724. https://doi.org/10.1089/ast.2010.0559.

Hernandez, M.E., Newman, D.K., 2001. Extracellular electron transfer. Cell. Mol. Life Sci. 58 (11), 1562–1571. https://doi.org/10.1007/pl00000796.

Jacobs, J.A., Testa, S.M., 2014. Acid Drainage and Sulfide Oxidation: Introduction. Acid Mine Drainage, Rock Drainage, and Acid Sulfate Soils: Causes, Assessment, Prediction, Prevention, and Remediation, 9781118749197. 9780470487860, pp. 1–8. https://doi.org/10.1002/9781118749197.ch1. Available from:

Jones, J.G., Davison, W., Gardener, S., 1984. Iron reduction by bacteria: range of organisms involved and metals reduced. FEMS (Fed. Eur. Microbiol. Soc.) Microbiol. Lett. 21 (1), 133–136. https://doi.org/10.1111/j.1574-6968.1984.tb00198.x.

Kageyama, A., Morisaki, K., Omura, S., Takahashi, Y., 2008. Arthrobacter oryzae sp. nov. and Arthrobacter humicola sp. nov. Int. J. Syst. Evol. Microbiol. 58 (1), 53–56. https://doi.org/10.1099/ijs.0.64875-0.

Kappler, A., Benz, M., Schink, B., Brune, A., 2004. Electron shuttling via humic acids in microbial iron(III) reduction in a freshwater sediment. FEMS (Fed. Eur. Microbiol. Soc.) Microbiol. Ecol. 47 (1), 85–92. https://doi.org/10.1016/s0168-6496(03)00245-9.

Kaster, A.-K., Moll, J., Parey, K., Thauer, R.K., 2011. Coupling of ferredoxin and heterodisulfide reduction via electron bifurcation in hydrogenotrophic methanogenic archaea. Proc. Natl. Acad. Sci. 108 (7), 2981–2986. https://doi.org/10.1073/pnas.1016761108.

Kerem, Z., Jensen, K.A., Hammel, K.E., 1999. Biodegradative mechanism of the brown rot basidiomycete Gloeophyllum trabeum: evidence for an extracellular hydroquinone-driven fenton reaction. FEBS (Fed. Eur. Biochem. Soc.) Lett. 446 (1), 49–54. https://doi.org/10.1016/S0014-5793(99)00180-5.

Kietäväinen, R., Purkamo, L., 2015. The origin, source, and cycling of methane in deep crystalline rock biosphere. Front. Microbiol. 6. https://doi.org/10.3389/fmicb.2015.00725. Available from:

King, G.M., 2006. Nitrate-dependent anaerobic carbon monoxide oxidation by aerobic CO-oxidizing bacteria. FEMS (Fed. Eur. Microbiol. Soc.) Microbiol. Ecol. 56 (1), 1–7. https://doi.org/10.1111/j.1574-6941.2006.00065.x.

Knittel, K., Boetius, A., 2009. Anaerobic oxidation of methane: progress with an unknown process. Annu. Rev. Microbiol. 63 (1), 311–334. https://doi.org/10.1146/annurev.micro.61.080706.093130.

Knoll, A.H., Golubic, S., Green, J., Swett, K., 1986. Organically preserved microbial endoliths from the late Proterozoic of East Greenland. Nature 321 (6073), 856–857. https://doi.org/10.1038/321856a0.

Kobayashi, M., Matsuo, Y., Takimoto, A., Suzuki, S., Maruo, F., Shoun, H., 1996. Denitrification, a novel type of respiratory metabolism in fungal mitochondrion. J. Biol. Chem. 271 (27), 16263–16267. https://doi.org/10.1074/jbc.271.27.16263.

Kormas, K.A., Smith, D.C., Edgcomb, V., Teske, A., 2003. Molecular analysis of deep subsurface microbial communities in Nankai Trough sediments (ODP Leg 190, Site 1176). FEMS (Fed. Eur. Microbiol. Soc.) Microbiol. Ecol. 45 (2), 115–125. https://doi.org/10.1016/S0168-6496(03)00128-4.

Korotkova, N., Chistoserdova, L., Kuksa, V., Lidstrom, M.E., 2002. Glyoxylate regeneration pathway in the methylotroph Methylobacterium extorquens AM1. J. Bacteriol. 184 (6), 1750–1758. https://doi.org/10.1128/JB.184.6.1750-1758.2002.

Kostka, J.E., Dalton, D.D., Skelton, H., Dollhopf, S., Stucki, J.W., 2002. Growth of iron (III)-reducing bacteria on clay minerals as the sole electron acceptor and comparison of growth yields on a variety of oxidized iron forms. Appl. Environ. Microbiol. 68 (12), 6256–6262. https://doi.org/10.1128/AEM.68.12.6256-6262.2002.

Kuwazaki, S., Takaya, N., Nakamura, A., Shoun, H., 2003. Formate-forming fungal catabolic pathway to supply electrons to nitrate respiration. Biosci. Biotechnol. Biochem. 67 (4), 937–939. https://doi.org/10.1271/bbb.67.937.

Laempe, D., Jahn, M., Breese, K., Schägger, H., Fuchs, G., 2001. Anaerobic metabolism of 3-hydroxybenzoate by the denitrifying Bacterium Thauera aromatica. J. Bacteriol. 183 (3), 968–979. https://doi.org/10.1128/jb.183.3.968-979.2001.

Lamprinou, V., Skaraki, K., Kotoulas, G., Economou-Amilli, A., Pantazidou, A., 2012. Toxopsis calypsus gen. nov., sp. nov. (Cyanobacteria, Nostocales) from cave 'Francthi', Peloponnese, Greece: a morphological and molecular evaluation. Int. J. Syst. Evol. Microbiol. 62 (Pt_12), 2870–2877. https://doi.org/10.1099/ijs.0.038679-0.

Lee, A.K., Newman, D.K., 2003. Microbial iron respiration: impacts on corrosion processes. Appl. Microbiol. Biotechnol. 62 (2–3), 134–139. https://doi.org/10.1007/s00253-003-1314-7.

Lim, Y.W., Baik, K.S., Han, S.K., Kim, S.B., Bae, K.S., 2003. Burkholderia sordidicola sp. nov., isolated from the white-rot fungus Phanerochaete sordida. Int. J. Syst. Evol. Microbiol. 53 (5), 1631–1636. https://doi.org/10.1099/ijs.0.02456-0.

Lloyd, K.G., May, M.K., Kevorkian, R.T., Steen, A.D., 2013. Meta-analysis of quantification methods shows that archaea and bacteria have similar abundances in the subseafloor. Appl. Environ. Microbiol. 79 (24), 7790–7799. https://doi.org/10.1128/AEM.02090-13UnitedStates.

López-Guerrero, P.N., Fernández-Uria, A., 1985. Sistemas acuíferos de la Cuenca del Guadalquivir, pp. 105–166.

Lorite, M.J., Tachil, J., Sanjuán, J., Meyer, O., Bedmar, E.J., 2000. Carbon monoxide dehydrogenase activity in Bradyrhizobium japonicum. Appl. Environ. Microbiol. 66 (5), 1871–1876. https://doi.org/10.1128/AEM.66.5.1871-1876.2000.

Lovley, D.R., Phillips, E.J.P., 1987. Competitive mechanisms for inhibition of sulfate reduction and methane production in the zone of ferric iron reduction in sediments. Appl. Environ. Microbiol. 53 (11), 2636–2641. https://doi.org/10.1128/aem.53.11.2636-2641.1987.

Lucey, K.S., Leadbetter, J.R., 2014. Catechol 2,3-dioxygenase and other meta-cleavage catabolic pathway genes in the 'anaerobic' termite gut spirochete Treponema primitia. Mol. Ecol. 23 (6), 1531–1543. https://doi.org/10.1111/mec.12598.

Ma, W.K., Farrell, R.E., Siciliano, S.D., 2008. Soil formate regulates the fungal nitrous oxide emission pathway. Appl. Environ. Microbiol. 74 (21), 6690–6696. https://doi.org/10.1128/aem.00797-08.

Mallick, N., Rai, L.C., Mohn, F.H., Soeder, C.J., 1999. Studies on nitric oxide (NO) formation by the green alga Scenedesmus obliquus and the diazotrophic cyanobacterium Anabaena Doliolum. Chemosphere 39 (10), 1601–1610. https://doi.org/10.1016/s0045-6535(99)00058-2.

Marlow, J.J., Steele, J.A., Ziebis, W., Thurber, A.R., Levin, L.A., Orphan, V.J., 2014. Carbonate-hosted methanotrophy represents an unrecognized methane sink in the deep sea. Nat. Commun. 5. https://doi.org/10.1038/ncomms6094. Available from. http://www.nature.com/ncomms/index.html.

Meyer, O., Schlegel, H.G., 1983. Biology of aerobic carbon monoxide-oxidizing bacteria. Annu. Rev. Microbiol. 37, 277–310. https://doi.org/10.1146/annurev.mi.37.100183.001425.

Muggia, L., Gueidan, C., Knudsen, K., Perlmutter, G., Grube, M., 2013. The lichen connections of black fungi. Mycopathologia 175 (5–6), 523–535. https://doi.org/10.1007/s11046-012-9598-8.

Muggia, L., Fleischhacker, A., Kopun, T., Grube, M., 2016. Extremotolerant fungi from alpine rock lichens and their phylogenetic relationships. Fungal Divers. 76 (1), 119–142. https://doi.org/10.1007/s13225-015-0343-8.

Mukhopadhyay, P.K., 1992. Maturation of organic matter as revealed by microscopic methods: applications and limitations of vitrinite reflectance, and continuous spectral and pulsed laser fluorescence spectroscopy. Dev. Sedimentol. 47 (C), 435–510. https://doi.org/10.1016/S0070-4571(08)70571-5.

Nai, C., Wong, H.Y., Pannenbecker, A., Broughton, W.J., Benoit, I., de Vries, R.P., Gueidan, C., Gorbushina, A.A., 2013. Nutritional physiology of a rock-inhabiting, model microcolonial fungus from an ancestral lineage of the Chaetothyriales (Ascomycetes). Fungal Genet. Biol. 56, 54–66. https://doi.org/10.1016/j.fgb.2013.04.001.

Neal, J.L., Allen, G.C., Morse, R.D., Wolf, D.D., 1983. Nitrate, nitrite, nitrous oxide and oxygen-dependent hydrogen uptake by Rhizobium. FEMS (Fed. Eur. Microbiol. Soc.) Microbiol. Lett. 17 (1–3), 335–338. https://doi.org/10.1111/j.1574-6968.1983.tb00431.x.

Newberry, C.J., Webster, G., Cragg, B.A., Parkes, R.J., Weightman, A.J., Fry, J.C., 2004. Diversity of prokaryotes and methanogenesis in deep subsurface sediments from the Nankai trough, ocean drilling program Leg 190. Environ. Microbiol. 6 (3), 274–287. https://doi.org/10.1111/j.1462-2920.2004.00568.x.

Newman, D.K., Kolter, R., 2000. A role for excreted quinones in extracellular electron transfer. Nature 405 (6782), 94–97. https://doi.org/10.1038/35011098.

Nishio, K., Pornpitra, T., Izawa, S., Nishiwaki-ohkawa, T., Kato, S., Hashimoto, K., Nakanishi, S., 2015. Electrochemical detection of circadian redox rhythm in cyanobacterial cells via extracellular electron transfer. Plant Cell Physiol. 56 (6), 1053–1058. https://doi.org/10.1093/pcp/pcv066.

Okland, I., Huang, S., Thorseth, I.H., Pedersen, R.B., 2014. Formation of H_2, CH_4 and N-species during low-temperature experimental alteration of ultramafic rocks. Chem. Geol. 387 (1), 22–34. https://doi.org/10.1016/j.chemgeo.2014.08.003.

Oren, A., Shilo, M., 1979. Anaerobic heterotrophic dark metabolism in the cyanobacterium Oscillatoria limnetica: sulfur respiration and lactate fermentation. Arch. Microbiol. 122 (1), 77–84. https://doi.org/10.1007/bf00408049.

Orphan, V.J., Hinrichs, K.-U., Ussler, W., Paull, C.K., Taylor, L.T., Sylva, S.P., Hayes, J.M., Delong, E.F., 2001. Comparative analysis of methane-oxidizing archaea and sulfate-reducing bacteria in anoxic marine sediments. Appl. Environ. Microbiol. 67 (4), 1922–1934. https://doi.org/10.1128/aem.67.4.1922-1934.2001.

Ottow, J.C.G., Glathe, H., 1971. Isolation and identification of iron-reducing bacteria from gley soils. Soil Biol. Biochem. 3 (1), 43–55. https://doi.org/10.1016/0038-0717(71)90030-7.

Parkes, R.J., Cragg, B.A., Banning, N., Brock, F., Webster, G., Fry, J.C., Hornibrook, E., Pancost, R.D., Kelly, S., Knab, N., Jørgensen, B.B., Rinna, J., Weightman, A.J., 2007. Biogeochemistry and biodiversity of methane cycling in subsurface marine sediments (Skagerrak, Denmark). Environ. Microbiol. 9 (5), 1146–1161. https://doi.org/10.1111/j.1462-2920.2006.01237.x.

Parro, V., Fernández-Calvo, P., Rodríguez Manfredi, J.A., Moreno-Paz, M., Rivas, L.A., García-Villadangos, M., Bonaccorsi, R., González-Pastor, J.E., Prieto-Ballesteros, O., Schuerger, A.C., Davidson, M., Gómez-Elvira, J., Stoker, C.R., 2008. SOLID2: an antibody array-based life-detector instrument in a Mars Drilling Simulation Experiment (MARTE). Astrobiology 8 (5), 987–999. https://doi.org/10.1089/ast.2007.0126.

Parro, V., De Diego-Castilla, G., Moreno-Paz, M., Blanco, Y., Cruz-Gil, P., Rodríguez-Manfredi, J.A., Fernández-Remolar, D., Gómez, F., Gómez, M.J., Rivas, L.A., Demergasso, C., Echeverría, A., Urtuvia, V.N., Ruiz-Bermejo, M., García-Villadangos, M., Postigo, M., Sánchez-Román, M., Chong-Díaz, G., Gómez-Elvira, J., 2011. A microbial oasis in the hypersaline atacama subsurface discovered by a life detector chip: implications for the search for life on mars. Astrobiology 11 (10), 969–996. https://doi.org/10.1089/ast.2011.0654.

Pedersen, K., 2013. Metabolic activity of subterranean microbial communities in deep granitic groundwater supplemented with methane and H2. ISME J. 7 (4), 839–849. https://doi.org/10.1038/ismej.2012.144.

Pentecost, A., 2005. Organisms Associated with Travertine. Springer Science and Business Media.

Percheron, G., Bernet, N., Moletta, R.Ã.©, 1999. Interactions between methanogenic and nitrate reducing bacteria during the anaerobic digestion of an industrial sulfate rich wastewater. FEMS (Fed. Eur. Microbiol. Soc.) Microbiol. Ecol. 29 (4), 341–350. https://doi.org/10.1111/j.1574-6941.1999.tb00625.x.

Philipp, B., Kemmler, D., Hellstern, J., Gorny, N., Caballero, A., Schink, B., 2002. Anaerobic degradation of protocatechuate (3,4-dihydroxybenzoate) by Thauera aromatica strain AR-

1. FEMS (Fed. Eur. Microbiol. Soc.) Microbiol. Lett. 212 (1), 139–143. https://doi.org/10.1111/j.1574-6968.2002.tb11257.x.

Pini, F., Galardini, M., Bazzicalupo, M., Mengoni, A., 2011. Plant-bacteria association and symbiosis: are there common genomic traits in alphaproteobacteria? Genes 2 (4), 1017–1032. https://doi.org/10.3390/genes2041017.

Pisciotta, J.M., Zou, Y., Baskakov, I.V., 2010. Light-dependent electrogenic activity of cyanobacteria. PLoS One 5 (5). https://doi.org/10.1371/journal.pone.0010821. Available from:

Pisciotta, J.M., Zou, Y., Baskakov, I.V., 2011. Role of the photosynthetic electron transfer chain in electrogenic activity of cyanobacteria. Appl. Microbiol. Biotechnol. 91 (2), 377–385. https://doi.org/10.1007/s00253-011-3239-x.

Puente-Sánchez, F., Sánchez-Román, M., Amils, R., Parro, V., 2014. Tessaracoccus lapidicaptus sp. nov., an actinobacterium isolated from the deep subsurface of the Iberian pyrite belt. Int. J. Syst. Evol. Microbiol. 64 (Pt_10), 3546–3552. https://doi.org/10.1099/ijs.0.060038-0.

Puente-Sánchez, F., Arce-Rodríguez, A., Oggerin, M., García-Villadangos, M., Moreno-Paz, M., Blanco, Y., Rodríguez, N., Bird, L., Lincoln, S.A., Tornos, F., Prieto-Ballesteros, O., Freeman, K.H., Pieper, D.H., Timmis, K.N., Amils, R., Parro, V., 2018. Viable cyanobacteria in the deep continental subsurface. Proc. Natl. Acad. Sci. 115 (42), 10702–10707. https://doi.org/10.1073/pnas.1808176115.

Rastogi, G., Osman, S., Kukkadapu, R., Engelhard, M., Vaishampayan, P.A., Andersen, G.L., Sani, R.K., 2010. Microbial and mineralogical characterizations of soils collected from the deep biosphere of the former homestake Gold mine, south Dakota. Microb. Ecol. 60 (3), 539–550. https://doi.org/10.1007/s00248-010-9657-y.

Rivas, L.A., García-Villadangos, M., Moreno-Paz, M., Cruz-Gil, P., Gómez-Elvira, J., Parro, V., 2008. A 200-antibody microarray biochip for environmental monitoring: searching for universal microbial biomarkers through immunoprofiling. Anal. Chem. 80 (21), 7970–7979. https://doi.org/10.1021/ac8008093.

Rossi, F., De Philippis, R., 2015. Role of cyanobacterial exopolysaccharides in phototrophic biofilms and in complex microbial mats. Life 5 (2), 1218–1238. https://doi.org/10.3390/life5021218.

Ruibal, C., Gueidan, C., Selbmann, L., Gorbushina, A.A., Crous, P.W., Groenewald, J.Z., Muggia, L., Grube, M., Isola, D., Schoch, C.L., Staley, J.T., Lutzoni, F., de Hoog, G.S., 2009. Phylogeny of rock-inhabiting fungi related to Dothideomycetes. Stud. Mycol. 64, 123–133. https://doi.org/10.3114/sim.2009.64.06.

Sanz, J.L., Rodriguez, N., Escudero, C., Carrizo, D., Amils, R., 2021. Biological production of H2, CH4 and CO2 in the deep subsurface of the Iberian pyrite belt. Environ. Microbiol. 23 (7), 3913–3922. https://doi.org/10.1111/1462-2920.15561. Available from:

Schippers, A., Neretin, L.N., Kallmeyer, J., Ferdelman, T.G., Cragg, B.A., Parkes, R.J., Jørgensen, B.B., 2005. Prokaryotic cells of the deep sub-seafloor biosphere identified as living bacteria. Nature 433 (7028), 861–864. https://doi.org/10.1038/nature03302.

Shoun, H., Kim, D.-H., Uchiyama, H., Sugiyama, J., 1992. Denitrification by fungi. FEMS (Fed. Eur. Microbiol. Soc.) Microbiol. Lett. 94 (3), 277–281. https://doi.org/10.1111/j.1574-6968.1992.tb05331.x.

Stal, L.J., 2012. Cyanobacterial Mats and Stromatolites. Ecology of Cyanobacteria II: Their Diversity in Space and Time, 9789400738553, pp. 65–125. https://doi.org/10.1007/978-94-007-3855-3_4.

Stal, L.J., Moezelaar, R., 1997. Fermentation in cyanobacteria. FEMS (Fed. Eur. Microbiol. Soc.) Microbiol. Rev. 21 (2), 179–211. https://doi.org/10.1016/S0168-6445(97)00056-9.

Sterflinger, K., 2000. Fungi as geologic agents. Geomicrobiol. J. 17 (2), 97–124. https://doi.org/10.1080/01490450050023791.

Sterflinger, K., Krumbein, W.E., 1997. Dematiaceous fungi as a major agent for biopitting on mediterranean marbles and limestones. Geomicrobiol. J. 14 (3), 219–230. https://doi.org/10.1080/01490459709378045.

Stevens, T.O., McKinley, J.P., 1995. Lithoautotrophic microbial ecosystems in deep basalt aquifers. Science 270 (5235), 450–454. https://doi.org/10.1126/science.270.5235.450.

Stoker, C.R., Cannon, H.N., Dunagan, S.E., Lemke, L.G., Glass, B.J., Miller, D., Gomez-Elvira, J., Davis, K., Zavaleta, J., Winterholler, A., Roman, M., Rodriguez-Manfredi, J. A., Bonaccorsi, R., Bell, M.S., Brown, A., Battler, M., Chen, B., Cooper, G., Davidson, M., Fernández-Remolar, D., Gonzales-Pastor, E., Heldmann, J.L., Martínez-Frías, J., Parro, V., Prieto-Ballesteros, O., Sutter, B., Schuerger, A.C., Schutt, J., Rull, F., 2008. The 2005 MARTE robotic drilling experiment in Río Tinto, Spain: objectives, approach, and results of a simulated mission to search for life in the martian subsurface. Astrobiology 8 (5), 921–945. https://doi.org/10.1089/ast.2007.0217.

Straub, K.L., Benz, M., Schink, B., Widdel, F., 1996. Anaerobic, nitrate-dependent microbial oxidation of ferrous iron. Appl. Environ. Microbiol. 62 (4), 1458–1460. https://doi.org/10.1128/aem.62.4.1458-1460.1996.

Stuart, R.K., Mayali, X., Lee, J.Z., Craig Everroad, R., Hwang, M., Bebout, B.M., Weber, P. K., Pett-Ridge, J., Thelen, M.P., 2016. Cyanobacterial reuse of extracellular organic carbon in microbial mats. ISME J. 10 (5), 1240–1251. https://doi.org/10.1038/ismej.2015.180.

Sturms, R., Dispirito, A.A., Hargrove, M.S., 2011. Plant and cyanobacterial hemoglobins reduce nitrite to nitric oxide under anoxic conditions. Biochemistry 50 (19), 3873–3878. https://doi.org/10.1021/bi2004312.

Takasaki, K., Shoun, H., Yamaguchi, M., Takeo, K., Nakamura, A., Hoshino, T., Takaya, N., 2004. Fungal ammonia fermentation, a novel metabolic mechanism that couples the dissimilatory and assimilatory pathways of both nitrate and ethanol: role of acetyl CoA synthetase in anaerobic ATP synthesis. J. Biol. Chem. 279 (13), 12414–12420. https://doi.org/10.1074/jbc.M313761200.

Thomas, F., Hehemann, J.H., Rebuffet, E., Czjzek, M., Michel, G., 2011. Environmental and gut Bacteroidetes: the food connection. Front. Microbiol. 2 (MAY). https://doi.org/10.3389/fmicb.2011.00093. Available from:

Troshina, O., Serebryakova, L., Sheremetieva, M., Lindblad, P., 2002. Production of H2 by the unicellular cyanobacterium Gloeocapsa alpicola CALU 743 during fermentation. Int. J. Hydrogen Energy 27 (11–12), 1283–1289. https://doi.org/10.1016/S0360-3199(02)00103-9.

Trubitsin, B.V., Ptushenko, V.V., Koksharova, O.A., Mamedov, M.D., Vitukhnovskaya, L.A., Grigor'ev, I.A., Semenov, A.Y., Tikhonov, A.N., 2005. EPR study of electron transport in the cyanobacterium Synechocystis sp. PCC 6803: oxygen-dependent interrelations between photosynthetic and respiratory electron transport chains. Biochim. Biophys. Acta Bioenerg. 1708 (2), 238–249. https://doi.org/10.1016/j.bbabio.2005.03.004.

Utgikar, V.P., Harmon, S.M., Chaudhary, N., Tabak, H.H., Govind, R., Haines, J.R., 2002. Inhibition of sulfate-reducing bacteria by metal sulfide formation in bioremediation of acid mine drainage. Environ. Toxicol. 17 (1), 40–48. https://doi.org/10.1002/tox.10031.

Valli, K., Gold, M.H., 1991. Degradation of 2,4-dichlorophenol by the lignin-degrading fungus Phanerochaete chrysosporium. J. Bacteriol. 173 (1), 345–352. https://doi.org/10.1128/jb.173.1.345-352.1991.

Valli, K., Brock, B.J., Joshi, D.K., Gold, M.H., 1992. Degradation of 2,4-dinitrotoluene by the lignin-degrading fungus Phanerochaete chrysosporium. Appl. Environ. Microbiol. 58 (1), 221–228. https://doi.org/10.1128/aem.58.1.221-228.1992.

Van Bodegom, P.M., Scholten, J.C.M., Stams, A.J.M., 2004. Direct inhibition of methanogenesis by ferric iron. FEMS (Fed. Eur. Microbiol. Soc.) Microbiol. Ecol. 49 (2), 261–268. https://doi.org/10.1016/j.femsec.2004.03.017.

Van De Vossenberg, J.L.C.M., Ubbink-Kok, T., Elferink, M.G.L., Driessen, A.J.M., Konings, W.N., 1995. Ion permeability of the cytoplasmic membrane limits the maximum growth temperature of bacteria and archaea. Mol. Microbiol. 18 (5), 925–932. https://doi.org/10.1111/j.1365-2958.1995.18050925.x.

van der Lelie, D., Taghavi, S., McCorkle, S.M., Li, L.L., Malfatti, S.A., Monteleone, D., Donohoe, B.S., Ding, S.Y., Adney, W.S., Himmel, M.E., Tringe, S.G., 2012. The metagenome of an anaerobic microbial community decomposing poplar wood chips. PLoS One 7 (5). https://doi.org/10.1371/journal.pone.0036740UnitedStates. Available from:

Van Waasbergen, L.G., Balkwill, D.L., Crocker, F.H., Bjornstad, B.N., Miller, R.V., 2000. Genetic diversity among Arthrobacter species collected across a heterogeneous series of terrestrial deep-subsurface sediments as determined on the basis of 16s rRNA and recA gene sequences. Appl. Environ. Microbiol. 66 (8), 3454–3463. https://doi.org/10.1128/AEM.66.8.3454-3463.2000.

Vandenbroucke, M., Largeau, C., 2007. Kerogen origin, evolution and structure. Org. Geochem. 38 (5), 719–833. https://doi.org/10.1016/j.orggeochem.2007.01.001.

Vermaas, W.F.J., 2001. Photosynthesis and Respiration in Cyanobacteria Encyclopedia of Life Sciences. Wiley. https://doi.org/10.1038/npg.els.0001670.

Wang, Y., Wiatrowski, H.A., John, R., Lin, C.C., Young, L.Y., Kerkhof, L.J., Yee, N., Barkay, T., 2013. Impact of mercury on denitrification and denitrifying microbial communities in nitrate enrichments of subsurface sediments. Biodegradation 24 (1), 33–46. https://doi.org/10.1007/s10532-012-9555-8.

Wilcoxen, J., Zhang, B., Hille, R., 2011. Reaction of the molybdenum- and copper-containing carbon monoxide dehydrogenase from Oligotropha carboxydovorans with quinones. Biochemistry 50 (11), 1910–1916. https://doi.org/10.1021/bi1017182.

Wilkins, M.J., Daly, R.A., Mouser, P.J., Trexler, R., Sharma, S., Cole, D.R., Wrighton, K.C., Biddle, J.F., Denis, E.H., Fredrickson, J.K., Kieft, T.L., Onstott, T.C., Peterson, L., Pfiffner, S.M., Phelps, T.J., Schrenk, M.O., 2014. Trends and future challenges in sampling the deep terrestrial biosphere. Front. Microbiol. 5. https://doi.org/10.3389/fmicb.2014.00481. Available from.

Winfrey, M.R., Zeikus, J.G., 1977. Effect of sulfate on carbon and electron flow during microbial methanogenesis in freshwater sediments. Appl. Environ. Microbiol. 33 (2), 275–281. https://doi.org/10.1128/aem.33.2.275-281.1977.

Wolfaardt, G.M., Lawrence, J.R., Korber, D.R., 1999. Function of EPS. Springer Science and Business Media LLC, pp. 171–200. https://doi.org/10.1007/978-3-642-60147-7_10.

Yamazaki, S.I., Kaneko, T., Taketomo, N., Kano, K., Ikeda, T., 2002. Glucose metabolism of lactic acid bacteria changed by quinone-mediated extracellular electron transfer. Biosci. Biotechnol. Biochem. 66 (10), 2100–2106. https://doi.org/10.1271/bbb.66.2100.

Yongjin, Z., Pisciotta, J., Billmyre, R.B., Baskakov, I.V., 2009. Photosynthetic microbial fuel cells with positive light response. Biotechnol. Bioeng. 104 (5), 939–946. https://doi.org/10.1002/bit.22466UnitedStates.

Zhang, L., Keller, J., Yuan, Z., 2009. Inhibition of sulfate-reducing and methanogenic activities of anaerobic sewer biofilms by ferric iron dosing. Water Res. 43 (17), 4123–4132. https://doi.org/10.1016/j.watres.2009.06.013.

CHAPTER

Eukaryotic diversity along the Tinto basin

8

Ángeles Aguilera[1], Virginia Souza-Egipsy[2], Graciela de Diego-Castilla[1], Erik Zettler[3]

[1]*Departmento de Evolución Molecular, Centro de Astrobiología (CAB), INTA-CSIC, Torrejón de Ardoz, Madrid, Spain;* [2]*Instituto Estructura de la Materia, CSIC, Madrid, Spain;* [3]*Department of Marine Microbiology and Biogeochemistry, NIOZ Royal Netherlands Institute for Sea Research, Horntje, the Netherlands*

Introduction

Extreme environments and extremophiles

The investigation of our planet has continually led to the discovery of life forms thriving in places once deemed uninhabitable. Remarkably, organisms can survive in extremely harsh environments. This realization has spurred the emergence of a biological field known as Extremophiles, which focuses on understanding the diversity, ecological roles, and physiological adaptations of organisms in extreme habitats. Such environments are typically characterized by one or more conditions that are hostile to the majority of life forms—particularly humans—highlighting the inherently anthropocentric nature of this classification. Although these settings seem highly inhospitable, they often harbor a surprising amount of biodiversity. The environments familiar to us are rich in oxygen, maintain moderate temperatures (generally between 5°C and 40°C), and are shielded from excessive radiation by the atmosphere. This "normal" range forms the basis of our perception of habitable conditions. Organisms that exist on the edge of, or well beyond, these norms are known as extremophiles. What makes extremophiles particularly fascinating is that they don't just tolerate extreme conditions—they rely on them for their biological processes. In contrast, microorganisms that can survive in extreme conditions but also thrive in more moderate environments are classified as extremo-tolerant, rather than true extremophiles.

In recent years, interest in these organisms and their environments has increased across multiple scientific disciplines. Some of these extreme habitats mirror the conditions of early Earth, offering insight into the origins of life. Others help scientists explore the boundaries of life and the key variables that sustain it. Furthermore, they are of growing significance in astrobiology, as they provide analogs for the types of environments that may exist on other planets. Finally, extremophiles are of great promise in biotechnology, as they may yield novel enzymes and bioactive compounds an area still in its infancy but full of potential.

Geomicrobiology of Río Tinto. https://doi.org/10.1016/B978-0-443-32948-7.00001-1

Eukaryotic microorganisms in extreme environments

While extremophilic prokaryotes are the most recognized and thoroughly investigated, eukaryotic microbes are also capable of thriving in nearly all extreme environments, provided there is an available energy source. Notable exceptions to this adaptability include environments with very high temperatures (above 70°C) and the deep subsurface biosphere. Advancements in molecular tools and their application in microbial ecology have greatly expanded our understanding of eukaryotic diversity across a range of habitats. This is especially important for studying life in extreme conditions, which are often difficult to replicate accurately in laboratory settings (Coker, 2019; Madigan et al., 2019; Schmid et al., 2020). Through molecular ecological research, it has become evident that eukaryotic organisms can be just as resilient and adaptable as their prokaryotic counterparts, and such studies have also uncovered previously unknown genetic variations in organisms living in these challenging environments.

Temperature plays a crucial role in shaping the distribution and population density of species because it directly influences enzymatic function (Madigan et al., 2019). The extremophiles that can endure very high temperatures, typically between 95°C and 115°C, mostly belong to the domains Archaea and Bacteria. On the other hand, eukaryotic microorganisms are limited to a maximum known growth temperature of 62°C, and most multicellular eukaryotes (metazoans) cannot develop at temperatures exceeding 50°C (Niehaus et al., 1999). Interestingly, no photosynthetic prokaryotes, such as cyanobacteria, have been identified in acidic, high-temperature aquatic environments. Instead, these niches are often densely populated by members of the order Cyanidiales, a group of unicellular red algae (Madigan et al., 2019). Species within the genera *Galdieria* and *Cyanidium* have been isolated from hydrothermal systems, displaying optimal growth at around 45°C and tolerating temperatures up to 57°C. These hot springs are typically very acidic (pH ranging from 0.05 to 4) and are enriched with heavy metals such as cadmium, nickel, iron, and arsenic—elements that are extremely toxic to most known forms of life.

Eukaryotic microorganisms have also successfully adapted to thrive in environments where temperatures are at or below the freezing point. Some algal species flourish on snow surfaces during the spring season, and intricate microbial ecosystems have been documented on glaciers—environments that are arguably among the most thoroughly examined after those found in sea ice. One notable example is the green microalga *Chlamydomonas nivalis*, commonly located in high-altitude snowfields that remain covered year-round. This alga continues to perform photosynthesis under freezing conditions and intense ultraviolet radiation (Madigan et al., 2019). Diverse microbial communities, including psychrophilic bacteria, microalgae, and protozoa, inhabit meltwater pools that form on ice surfaces or within brine-filled channels beneath Arctic sea ice, enduring even in winter when temperatures can drop as low as −20°C (Price, 2000).

Anaerobic environments also host a wide variety of eukaryotic organisms. These oxygen-deprived habitats are found globally, ranging from waste processing facilities to deep-ocean hydrothermal vent systems. Research into these anoxic ecosystems has significantly contributed to our understanding of eukaryotic evolution, particularly through the study of protists that inhabit such extreme conditions. Of special interest are anoxic sediments associated with hydrothermal activity and iron-abundant environments, which may resemble conditions of early Earth. Among the most prevalent eukaryotes in these settings are ciliates. Typically, free-living ciliates rely on oxygen and use mitochondrial respiration for energy. However, in anaerobic species, mitochondria have been replaced by hydrogenosomes—organelles that generate energy through fermentative pathways under oxygen-free conditions. This process is functionally similar to mitochondrial respiration in aerobic organisms. Within hydrogenosomes, pyruvate, a central metabolite in energy production, is oxidized. Although the precise evolutionary origin of hydrogenosomes remains debated, prevailing evidence suggests they are derived from mitochondria that have gradually lost their DNA and the enzymes essential for aerobic respiration. Interestingly, these anaerobic ciliates are not closely related in evolutionary terms, indicating that their adaptations to oxygen-free environments occurred multiple times independently. Despite this, they maintain striking morphological similarities to their oxygen-dependent relatives, implying that the transition to anaerobiosis happened relatively recently in evolutionary history. Beyond ciliates, hydrogenosomes are also found in a wide array of other anaerobic eukaryotes, including various protists and several yeast species.

Eukaryotic organisms have adapted to live in harsh terrestrial environments where water is limited and desiccation occurs frequently. In such settings, the ground is often coated with what are termed "biological soil crusts," which are intricate microbial communities formed through tight interactions between soil particles and organisms such as cyanobacteria, algae, mosses and fungi. In addition to deserts, Antarctica's Dry Valleys are among the planet's most arid and inhospitable environments. These valleys lack ice coverage and contain minimal amounts of liquid water. Annual temperatures typically remain below −30°C with precipitation levels under 10 mm per year. Despite these severe conditions, a variety of life forms, including both single-celled and multicellular organisms, can be found here. Microscopic nematodes dominate the food web in this region. During the frigid and dry winter months, these nematodes enter a survival mode called anhydrobiosis, where they expel nearly all of their body water—up to 99%—in order to withstand the extreme environment.

In a similar way, many eukaryotic organisms are present in hypersaline environments, such as the Dead Sea—one of the world's largest and most saline lakes, with a dissolved salt content reaching 34% (Madigan et al., 2019). The unicellular green alga *Dunaliella salina* thrives in these highly saline conditions, despite being exposed to intense sunlight. This alga is also found in a variety of saline aquatic environments, including oceans, brine-rich lakes, and lagoons with salt concentrations exceeding 2M, along with elevated magnesium levels. *Dunaliella* adapts by producing and storing large quantities of glycerol inside its cells, serving as an osmoprotectant to balance the osmotic stress from the surrounding ions. It also accumulates high levels of β-carotene, which helps shield it from damaging solar radiation.

Another type of extreme habitat where eukaryotes are well-represented is deep-sea hydrothermal vents. These are hot, sulfur-emitting underwater volcanic structures typically located along tectonic plate boundaries. Despite the absence of sunlight, these ecosystems host unique deep-sea life forms and are the only known biological communities on Earth that do not rely on solar energy as a direct source. In the Pacific Ocean, prominent vent-dwelling metazoan species include tube worms such as *Alvinella pompejana* and certain clam species, whereas eyeless shrimp are unique to the hydrothermal vents of the Atlantic Ocean. Remarkably, these organisms can withstand sustained exposure to temperatures as high as 80°C (Madigan et al., 2019).

Most eukaryotic life forms in extreme settings are unicellular, and studying them presents a major challenge due to the lack of distinct morphological features and effective cultivation methods. However, the emergence of next-generation DNA sequencing technologies and advanced bioinformatics tools has transformed our ability to investigate the diversity and biology of these organisms (Rappaport and Oliverio, 2023). Molecular ecology, a dynamic and expanding discipline, now allows scientists to explore a broad spectrum of questions relating to species diversity, ecological interactions, physiological adaptations, and evolutionary history (Baker et al., 2009). These cutting-edge techniques have revealed an unexpectedly vast eukaryotic diversity in numerous environments, with particularly profound insights emerging from the study of extreme habitats, which are often difficult to simulate in the laboratory. Genomic adaptations, such as the presence of specialized enzymes and evidence of horizontal gene transfer, help these organisms survive and compete in extreme conditions. Today, researchers can examine crucial aspects of microbial diversity that were previously beyond the reach of scientific inquiry.

Nature and origin of extreme acidic ecosystems

Extreme environments often present a combination of harsh conditions typically regarded as inhospitable for most life forms. Nevertheless, despite these challenging factors, such ecosystems often harbor a surprisingly high level of biodiversity. Initially, extremophilic organisms were viewed primarily as scientific novelties, but today, their value in biotechnology has become a key reason for their ongoing study and cultivation. These organisms have driven progress in various biotechnological fields, including the development of novel bioactive compounds with applications in industry, agriculture, environmental management, and medicine (Schmid et al., 2020; Schiraldi and De Rosa, 2002). A diverse array of bacteria, archaea, cyanobacteria, protists, algae, and fungi has been isolated from these environments in recent years, owing to their remarkable potential—not only for biotechnological exploitation but also as models for investigating biomolecular stability and the origins of life.

Moreover, extreme environments serve as hotspots for the emergence of biological diversity and innovative adaptive strategies. They are settings where evolutionary pressures are especially intense, promoting unique survival mechanisms. Due to their selective nature and spatial confinement, these environments foster distinctive forms of life and specialized adaptations. As such, extreme ecosystems act as incubators of biological uniqueness, offering valuable opportunities for scientific research and the conservation of genetic heritage. However, their uniqueness also makes them particularly vulnerable. Any disruption of the fragile equilibrium in these already marginal systems, or the return to "normal" environmental conditions, could result in the loss of biodiversity uniquely suited to extreme habitats.

Acidic extreme environments and the specialized microorganisms that inhabit them—known as acidophiles—are increasingly becoming a major area of scientific interest due to their potential applications in biotechnology and environmental science. Acidophiles thrive in highly acidic conditions, often with pH values below 2. In such environments, proteins typically denature and lose their functionality, meaning these organisms must have evolved sophisticated srystems to keep their cytoplasm at a neutral pH. Additionally, they possess mechanisms to protect their extracellular proteins from the damaging effects of extreme acidity. The ability to tolerate a wide pH range is believed to be linked to the regulation of specific genes already present in the genome, enabling rapid adaptation to environmental changes (Dhakar and Pandey, 2016).

Various natural habitats, such as soils and freshwater ecosystems, can exhibit acidic conditions with pH values below the neutral level of 7. Acidic environments are classified as those with a pH under 5 and are further divided into extremely acidic (pH less than 3) and moderately acidic (pH between 3 and 5). While moderately acidic conditions are fairly common in peatlands, soils, and freshwater bodies, extremely acidic environments are relatively rare. They can arise from natural causes, such as geothermal activity or exposed metal sulfides, or from human activities, especially in areas impacted by metal and coal mining (Johnson, 2009; Johnson and Schippers, 2017).

What sets extreme acidic ecosystems apart is that the low pH is often generated by the microorganisms themselves through their metabolic processes, rather than being imposed by external environmental factors, as is often the case with other types of extreme environments (such as high temperatures, radiation, or pressure) (González-Toril et al., 2003). In these acidic conditions, cellular proteins are at high risk of denaturation and functional loss. Moreover, these environments can vary widely in temperature—from below freezing to above 80°C—and tend to contain significantly higher concentrations of dissolved solutes and heavy metals than ecosystems with neutral pH levels. Consequently, life forms in these settings face more than just the challenge of high hydronium ion (H_3O^+) concentrations, which can be up to 10 million times greater than those found in neutral waters (Johnson and Schippers, 2017).

Microbial diversity in Río Tinto

The life forms found in Río Tinto's waters are predominantly microbial, encompassing both prokaryotes and eukaryotes. While some areas along the river host plant life, as is typical in other highly acidic environments, these plants usually anchor their roots in sediments where the pH of the interstitial water is notably higher than 2. Generally, microbial diversity tends to decline in more acidic conditions; however, other environmental factors—such as the concentration of dissolved metals and solutes—can also significantly impact biodiversity. Over the past 3 - decades, research on acidophilic extremophiles has confirmed that these microorganisms are distributed across all three domains of life—Bacteria, Archaea, and Eukarya—highlighting that acid tolerance has independently evolved multiple times throughout the history of life on Earth, rather than arising from a singular evolutionary event (González-Toril et al., 2010; Johnson and Aguilera, 2015; Johnson and Schippers, 2017; Amaral-Zettler et al., 2002).

Similar to other extremophiles, acidophilic organisms are highly adapted to very specific and harsh environmental conditions, making them incapable of thriving in neutral pH habitats. This specialization is especially pronounced in prokaryotes, whereas many acid-tolerant eukaryotes are also capable of surviving in environments with near-neutral pH levels. Organisms that must endure multiple environmental extremes, such as high temperature or salinity in addition to low pH, often have a lower tolerance for acidity than those adapted solely to acidic conditions—an observation that is particularly true among acidophilic eukaryotes (Johnson and Aguilera, 2015; Johnson and Schippers, 2017).

Microorganisms in Río Tinto are further categorized by the ecological niches they occupy. They can be found within the water column, along the riverbed, or in the mining lakes (Grettenberger et al., 2020). Depending on their location, these organisms are classified as either planktonic or benthic. Planktonic microorganisms float or remain suspended in the water column and, although some possess mobility, they are unable to resist currents or turbulence. In contrast, benthic organisms live attached to surfaces within aquatic ecosystems. These may reside among sediments or more commonly form structured communities known as biofilms or microbial mats, which are readily visible on riverbed surfaces. Additionally, filamentous forms can be observed anchored to the riverbed, floating on surfaces, or suspended in the water columns of stratified lakes (García-Moyano et al., 2007, 2012).

One of the key factors that influences how organisms are distributed in an environment is the availability of oxygen and light. In flowing water systems, anaerobic zones are uncommon because the constant movement of water promotes gas exchange with the atmosphere, keeping oxygen levels relatively high, especially in shallow regions. However, in parts of the riverbed located beneath photosynthetic biofilms, oxygen levels can drop significantly, similar to what occurs below the surface in sediment layers. In still or stagnant waters, such as reservoirs, anaerobic conditions tend to develop with increasing depth and are sometimes influenced by seasonal temperature changes and the presence of photosynthetic filaments at certain times of the year (García-Moyano et al., 2007, 2012).

Eukaryotic diversity

It has long been believed that the elevated metal concentrations typical of acidic extreme environments inhibit the growth and limit the diversity of eukaryotic organisms due to their toxic effects. Furthermore, the accumulation of metal hydroxides—characteristic of such ecosystems—can alter the chemical and physical properties of rock and sediment surfaces, potentially impeding eukaryotic microbial colonization. Nevertheless, dense biofilms of various colors covering extensive areas of the river bottom, along with large floating masses of filamentous microalgae, are commonly observed along the Río Tinto (Fig. 8.1).

Remarkably, eukaryotic algae are estimated to account for at least 60% of the river's total biomass (López-Archilla et al., 2001). This unexpectedly high abundance and diversity of eukaryotes make Río Tinto one of the most extraordinary and distinct extreme acidic environments known to science.

Acidophilic microalgae

The eukaryotic diversity in Río Tinto encompasses a wide range of species from several major eukaryotic lineages (Aguilera et al., 2006, 2007a, 2007b; Amaral-Zettler et al., 2011; López-Archilla et al., 2001; Amaral-Zettler et al., 2002). The majority of these organisms are photosynthetic microalgae. Among the most prevalent are chlorophytes from the genera *Chlamydomonas*, *Dunaliella*, and *Chlorella*, along with cyanidiales, diatoms, chrysophytes, and the photosynthetic protist *Euglena mutabilis*. These species often form dense green and brown biofilms that cover large areas of the riverbed (Fig. 8.2).

Diatoms from the genera *Eunotia* and *Pinnularia* are particularly widespread along the river's length (Aguilera et al., 2007a, 2007b), known for their ability to withstand high metal concentrations and acidic conditions, making them common inhabitants of similar environments. Compared with nearby freshwater bodies, Río Tinto exhibits a lower diversity of diatoms, supporting the idea that many diatom species cannot survive when pH drops below a critical range of 4.5 to 3.5.

Filamentous algae, particularly from the genera *Zygnemopsis* and *Klebsormidium*, are frequently observed during warm, dry seasons. These conditions lead to greater evaporation, pushing the river's water chemistry to more extreme levels. These microalgae typically attach to sediments at the river's bottom and produce long surface filaments, sometimes over a meter in length. In winter, these formations nearly disappear, likely due to the increased water flow, which makes it difficult for them to maintain their structure. The dominance of these specific taxa in Río Tinto and other acidic environments likely stems from their exceptional adaptability and high degree of ecological specialization. This allows them to thrive in settings where competition and predation pressures are minimal. A key survival strategy in acidic habitats—where carbon availability for photosynthesis is limited—is mixotrophy,

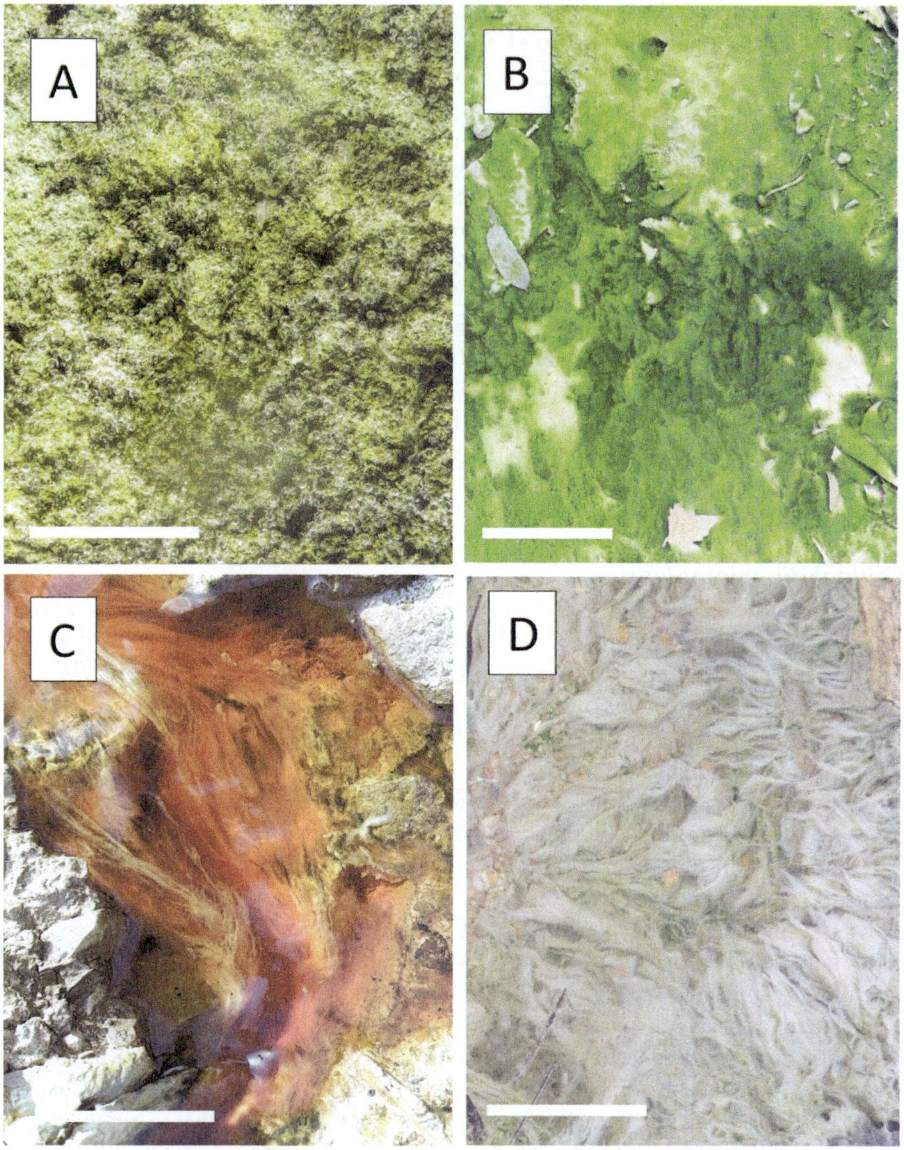

FIGURE 8.1 Microbial biofilms found at different Río Tinto locations

(A) Green biofilm formed mainly by *Chlorella acidophila*, (B) *Euglena mutabilis* biofilm, (C) *Leptospirillum* spp. Streamers, (D) *Acidiphilium* spp. Streamers. Scale bar 5 cm.

the capacity to switch between autotrophic and heterotrophic modes of nutrition. Mixotrophy, along with the ability to move vertically in the water column to seek out favorable light and nutrient conditions, are traits seen in prominent eukaryotic species in acid waters, such as *Euglena mutabilis* and *Chlamydomonas acidophila* (Roberts, 1999).

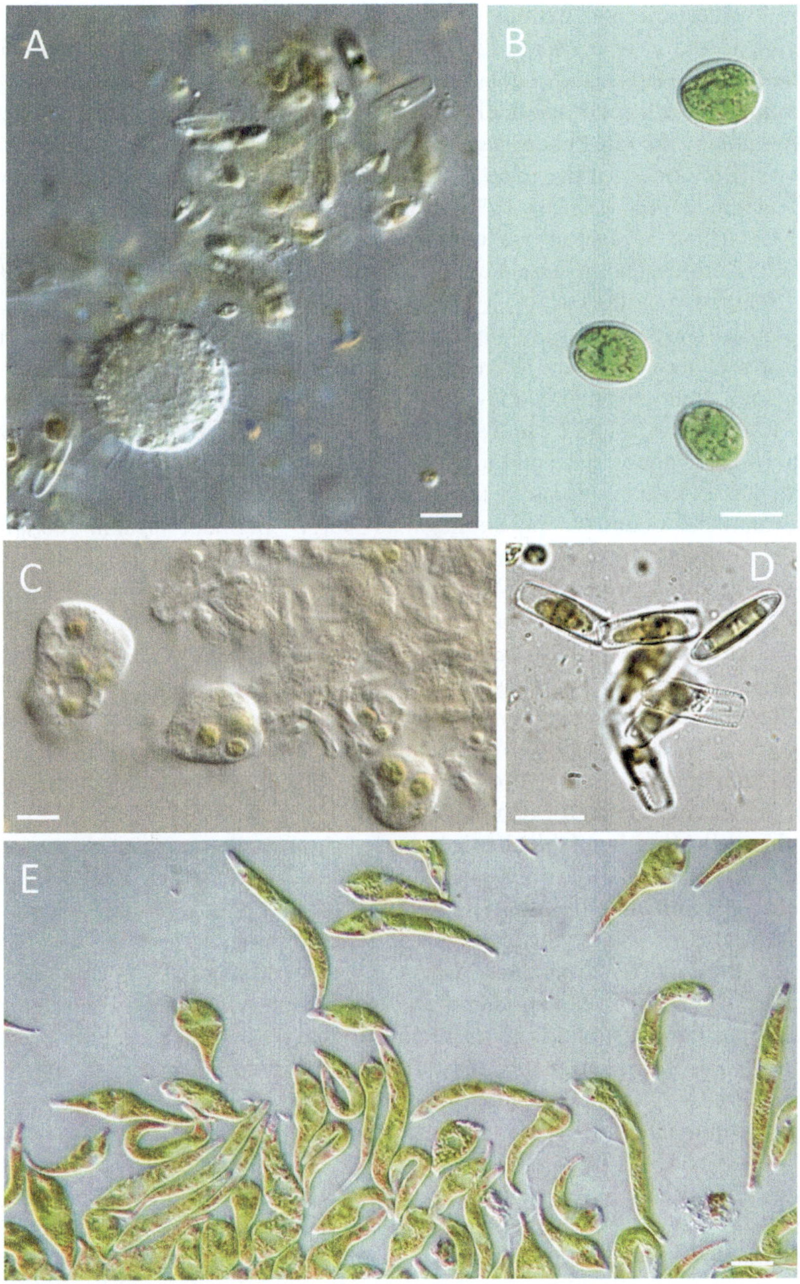

FIGURE 8.2 Light microscopy photographs of different eukaryotic species isolated from Río Tinto (Huelva, Spain):

(A) Heliozoa *Actinophrys* sp. (B) Green algae *Chlamydomonas acidophyla*, (C) Amoebas, (D) *Pinnularia* sp. diatoms, (E) photosynthetic protist *Euglena mutabilis*. Scale bar 20 μ.

The most acidic and extreme regions of the river are home to two species belonging to the genera *Dunaliella* (Chlorophyta) and *Cyanidium* (Rhodophyta). *Dunaliella acidophila* stands out as one of the most acid-tolerant organisms known, with optimal growth occurring at pH levels between 0 and 3. On the other hand, *Cyanidium* is among the rare eukaryotes—alongside some fungi—that can thrive at a pH as low as 1. Members of the order Cyanidiales are unicellular red algae that prefer highly acidic conditions (pH 0.5–3) and elevated temperatures (approximately 50–55°C). These organisms are commonly observed near acidic sulfur-rich hot springs, fumaroles, and in certain acid drainage sites, including Río Tinto (Aguilera et al., 2006, 2007a, 2007b).

Cyanidiales are structurally simple, characterized by spherical cells with thick walls, a chloroplast encircling the nucleus, up to three mitochondria, and a large central vacuole. Phylogenetically, they are considered one of the oldest lineages of red algae (Rhodophyta). Their photosynthetic machinery represents an intermediate form between cyanobacteria and modern eukaryotic chloroplasts, lacking the organized thylakoid structure found in higher plants and green algae. Both cyanobacteria and cyanidiales contain phycobilisomes—large pigment–protein complexes attached to thylakoid membranes—which function primarily as light-harvesting structures in these organisms. Additionally, *Cyanidium* demonstrates significant metabolic flexibility. It can grow heterotrophically in dark conditions, much like certain photosynthetic protists such as *Euglena*. It also exhibits high resistance to extreme levels of dissolved salts and heavy metals. Due to these adaptive traits, *Cyanidium* is a dominant eukaryotic component in most extremely acidic habitats (Seckbach and Joseph, 1999). Thus, extremophilic red algae has become models for understanding adaptation to hostile environments and the evolution of eukaryotic life on the early earth (Van Etten et al., 2023).

Light availability is a critical constraint for photosynthesis in these ecosystems, as high levels of dissolved metals—especially iron—give the water a deep red hue in many areas, significantly limiting the light that reaches photosynthetic microorganisms. Nevertheless, this challenge is not an issue for cyanidiales, which are highly adapted to low-light environments. These microalgae often grow endolithically—inside rocks—where light is naturally scarce. Interestingly, despite the limited light, their photosynthetic activity is comparable with that of other unicellular green algae such as *Chlorella*. Additionally, they display photoinhibition, suggesting a pronounced sensitivity to intense light exposure.

The exceptional photosynthetic performance of cyanidiales appears to be linked to their Photosystem I-light harvesting complex (PSI-LHCI), which differs notably in both size and structure when compared with that of other algae. Spectroscopic studies indicate that the LHCI proteins are more tightly integrated with Photosystem I in cyanidiales than in other known photosynthetic eukaryotes. This tighter configuration enhances their ability to perform photosynthesis efficiently in extremely low-light conditions, allowing them to thrive in ecological niches that are otherwise uninhabitable for most photosynthetic species (Seckbach and Joseph, 1999).

Euglenophytes and acidophilic protozoa

Euglena mutabilis is among the most widespread protists in acidic environments across the globe. It has been identified in microbial communities inhabiting acidic volcanic lakes, geothermal zones, and acid mine drainage sites, and is considered a reliable bioindicator of such ecosystems (Halter et al., 2012). Due to its photosynthetic metabolism and high abundance, this protist plays a pivotal role in primary production within these oligotrophic environments, potentially shaping microbial community diversity and metabolism through the release of metabolic byproducts.

E. mutabilis excretes a variety of metabolites into the surrounding environment, including numerous amino acids, polyamine compounds, urea, and certain sugars, though notably excluding fatty acids. This selective exudation contributes specific forms of organic matter to acidic ecosystems (Halter et al., 2012). These secretions can exert a significant influence on bacterial communities, as prokaryotes are capable of transporting and recycling the majority of metabolites released by *E. mutabilis*. As a result, this protist may exert an indirect yet important influence not only in acid mine drainage zones but also in other ecological niches often limited by nitrogen availability.*E. mutabilis* is motile, and through phototactic and aerotactic behavior, it is capable of forming dense, extensive biofilms. It can navigate through the layers of precipitates that intermittently cover microbial mats in acidic habitats, enabling it to maintain its position within the ecosystem structure.

Beyond euglenophytes, other protozoan groups are frequently found in acidic habitats and are recognized for their key roles within the pelagic food web of acidic mining lakes. These include several ciliate species (e.g., members of the genera *Urotricha*, *Oxytricha*, and Frontonia), flagellates (e.g., Ochromonas and Bodo), amoebae (e.g., Vahlkampfia and *Naegleria*), and heliozoans (e.g., *Actinophrys*) (Moser and Weisse, 2011; Weisse et al., 2013). Many of these organisms are not strictly acidophilic but rather acid-tolerant, capable of surviving and growing across a broad pH range. In acid mining pits, the phytoplankton community is often dominated by the mixotrophic flagellate *Ochromonas* spp. (Nixdorf et al., 1998). Ciliates appear to play a less prominent role in these environments compared with their ecological significance in neutral pH mesotrophic lakes (Deneke, 2000).

Acidophilic fungi and yeasts

Acidophilic microalgae and protozoa have received considerably more scientific attention than fungi and yeasts, despite the latter's acknowledged role in the acidification of sulfur-rich environments (Aguilera and González-Toril, 2019). Most studies involving these organisms have focused primarily on biodiversity, with limited knowledge available concerning their physiology or ecological interactions with other species. Among filamentous fungi, some of the most extreme acidophiles described to date can be found. Species such as *Acontium cylatium*, *Trichosporon cerebriae*, and *Cephalosporium* sp. are capable of growing at a pH close to 0 (Aguilera and González-Toril, 2019). Nonetheless, the majority of fungi inhabiting acidic environments are better characterized as acid-tolerant rather than strictly acidophilic, as they are typically able to grow at neutral or even alkaline pH levels. Most field studies on acidophilic fungi have been conducted with soil samples, leading to the identification of more than 80 distinct species.

Within the Ascomycetes, species from both Hemiascomycetes and Euascomycetes have been reported in the Iberian Pyrite Belt (IPB), including genera such as *Glomerella* and *Lecytophora*. Basidiomycetous yeasts belonging to the Hymenomycetes and Urediniomycetes classes have also been documented (Durán et al., 1999), with similar species observed in acidic geothermal areas (Aguilera and González-Toril, 2019). Fungi appear to play a significant ecological role in acidic environments by forming biofilms on rock surfaces, often in association with other microorganisms. These biofilms serve as sites for metal and mineral precipitation, while also providing structural support for other microbial populations. Fungi exhibit high metal resistance through sequestration mechanisms, facilitating the colonization of these habitats by less metal-tolerant microbial species (Durán et al., 1999).

A notable example is *Purpureocillium lilacinum*, a fungal species isolated from Río Tinto, which actively precipitates hydronium-jarosite. Mineral nucleation begins on the fungal cell wall of both live and dead cells, as well as within extracellular polymeric substances (EPS) secreted by the fungus. This organism functions as a crystallization nucleus, promoting passive jarosite precipitation and contributing to local ionic oversaturation, thereby influencing and regulating the geochemical properties of its environment (Oggerin et al., 2013).

Regarding yeasts, studies conducted in acidic environments are even scarcer and typically focus on biodiversity rather than on mechanisms of acid stress adaptation (Gadanho et al., 2006). Three novels asexual basidiomycetous yeast species have been identified in Río Tinto and Iberian Pyrite Belt ecosystems: *Cryptococcus aciditolerans*, *Cryptococcus ibericus*, and *Cryptococcus metallitolerans*. These species, classified within the order Filobasidiales, form a distinct phylogenetic clade (Aguilera and González-Toril, 2019). The *Cryptococcus* isolates appear to be specialists in acidic habitats, requiring low pH conditions for growth—an adaptation not previously observed in yeasts. Among the most frequently isolated fungi from acid mine drainage are members of the genus Rhodotorula, along with arthroconidial fungi such as *Trichosporon* sp. Other genera, including Candida, Cryptococcus, and Purpureocillium, have also been identified in highly acidic environments (Oggerin et al., 2013).

Although little is known about the ecological roles of yeasts in acidic ecosystems, mutualistic interactions have been described between *E. mutabilis* and *Cryptococcus*. When both species are cultured together, the tolerance of *E. mutabilis* to low pH and high metal concentrations increases by a factor of 10–100 compared with monocultures of the protist.

Plants

Another important consideration is that Río Tinto basin is not exclusively inhabited by unicellular eukaryotic organisms; various plant species are also found growing in the acidic soils along the riverbanks. These plants have developed a range of strategies to overcome the physiological challenges imposed by such extreme environmental conditions. Some exhibit resistance to the heavy metals concentrated in the surrounding soils (Amils et al., 2007; Rodríguez et al., 2005), while others actively accumulate specific metals in different plant tissues.

Imperata cylindrica is arguably the most representative plant associated with acidic environments in the Iberian Pyrite Belt. This perennial grass, characterized by its creeping rhizome, commonly establishes tall grass communities along springs and sandy banks, and is frequently observed along Río Tinto margins. Elemental distribution studies have confirmed its status as a hyperaccumulator of iron, which comprises more than 1% of its dry biomass. Iron is predominantly stored in the form of jarosite and ferritin, with significantly higher metal content found in the rhizomes compared with the leaves (Amils et al., 2007; Rodríguez et al., 2005). Jarosite is primarily deposited in the roots and rhizomes, whereas ferritin accumulates in the epidermal, parenchymatous, and vascular tissues of the shoots and leaves—mainly within cell walls and the cytoplasm. Iron deposits are visibly prominent on the rhizome surface and form extensive patches on both the abaxial and adaxial leaf epidermis. The presence of jarosite—a mineral that can only precipitate under acidic conditions—within *I. cylindrica* tissues suggests that iron metabolism in plants is more complex and adaptable than previously understood. It is believed that iron accumulation in the form of ferritin, at both structural and cellular levels, serves not only as a storage mechanism for this essential element but also plays a critical role in protecting cells against oxidative stress. Ferritin mitigates oxidative damage while simultaneously storing iron for cellular functions that depend on it (Rodríguez et al., 2005; de la Fuente et al., 2012).

Another widespread plant species found throughout the Iberian Pyrite Belt is the endemic heath *Erica andevalensis*. The soils supporting *Erica* populations are highly acidic and nutrient-poor, forcing the plants to tolerate a root environment that combines nutrient scarcity with heavy metal toxicity. This species can withstand the harsh conditions of mining areas without showing visible signs of metal toxicity, even when growing along the riverbanks with roots partially submerged in acidic water. Its survival in such soils relies on several strategies to cope with high metal concentrations, including the reduction of metal uptake or translocation, detoxification through binding to organic ligands, and intracellular compartmentalization (Oliva et al., 2012). *Erica* plants do not hyperaccumulate metals such as copper or lead; instead, the retention of metals in the roots, combined with their slow growth rate, appears to confer a degree of protection against heavy metal toxicity.

Taken together, these studies demonstrate that eukaryotes are capable of colonizing and even dominating biodiversity in extreme acidic environments. Given that some of these acidophilic eukaryotes are closely related to neutrophilic counterparts, it can be inferred that eukaryotes possess the capacity to adapt from neutral to acidic conditions over relatively short geological timescales. This suggests that they may be more broadly distributed and phylogenetically diverse than initially assumed.

Population dynamics of eukaryotic communities

Seasonal monitoring of eukaryotic microbial biomass in Río Tinto reveals a pronounced seasonal pattern. Biomass levels are substantially lower in winter compared with summer, a trend that closely aligns with the seasonal fluctuation in river discharge (Aguilera et al., 2006, 2007a, 2007b). The local climatic regime is characterized by alternating wet and dry seasons, with mild, rainy winters and hot, arid summers. While a high groundwater table sustains river flow throughout the dry season, elevated evaporation rates lead to the concentration of dissolved heavy metals. This dual-seasonality exerts a strong influence on the structure and productivity of the photosynthetic microbial community. In winter, the river's flow velocity can reach approximately 0.25 $m\ s^{-1}$ in certain segments, resulting in the scouring of biofilms that had accumulated during the preceding summer.

During the summer months, green algae are primarily responsible for the observed increase in eukaryotic biomass—an observation that is consistent with data from other acidophilic ecosystems. This proliferation coincides with seasonal rises in temperature and photoperiod, as well as reduced flow rates, all of which facilitate cell sedimentation and the establishment of microbial mats. Dissolved oxygen levels peak in winter and decline sharply during the arid summer and autumn months. In contrast, parameters such as conductivity, redox potential, and pH exhibit minimal seasonal variation.

Despite the seasonal variability in biomass, the overall taxonomic diversity of phototrophic biofilms remains relatively stable throughout the year. Notable exceptions include filamentous algal taxa such as *Zygnemopsis* and *Klebsormidium*, which display marked seasonality. These taxa form extensive mats on the river's surface during summer but are largely absent in winter, having been removed by increased runoff associated with seasonal rainfall. Heterotrophic flagellates also exhibit seasonal dynamics, although their abundance peaks during winter and spring, contrasting with most other eukaryotes. Ciliates, rotifers, and heliozoans, by contrast, tend to be more prevalent during late summer.

Spatial heterogeneity in the river's physicochemical conditions plays a key role in shaping the distribution of eukaryotic microorganisms along its course. Among these factors, the concentration of heavy metals appears to exert a stronger influence than pH (Aguilera et al., 2006, 2007a, 2007b). Sites with the harshest environmental conditions support significantly lower levels of biomass and taxonomic richness than more moderate sections of the river. Only *Dunaliella* and *Cyanidium*—two highly acidophilic eukaryotic taxa—are capable of colonizing the most extreme stretches of the river. In contrast, the highest levels of eukaryotic diversity are observed at downstream sampling stations, where water chemistry is comparatively less extreme with respect to pH and metal content.

These findings support a well-established ecological principle: environmental extremity is inversely correlated with biodiversity. Accordingly, gradients of increasing physicochemical stress along the river are accompanied by corresponding declines in community richness and abundance—a pattern consistently observed in Río Tinto system.

Geomicrobiological model of Río Tinto

Geomicrobiological investigations in Río Tinto system have focused on three primary environments, each defined by distinct hydrogeochemical characteristics. The main river channel, where most biodiversity studies have been concentrated, receives input from numerous sources of acid mine drainage. These include discharges from mine shafts, runoff from waste deposits, and overflows from abandoned mining pits that often reoccupy historical drainage tunnels. These discharge points are of particular interest, as the waters undergo rapid transitions from anoxic to oxygen-rich conditions over short distances, leading to microbial successions over short spatial scales. In addition to the main channel, researchers have also studied the artificial lakes formed within mining pits. Across all these settings, the aquatic column, sediment layers, and both floating and attached biofilms can be distinguished as microhabitats. The presence of oxic and anoxic zones further contributes to the complexity of geomicrobiological processes in each environment.

Role of eukaryotes in the geomicrobiological framework of acidic waters

Although eukaryotic microorganisms are not directly involved in the primary geochemical transformations of iron and sulfur that drive sulfide mineral dissolution and acid generation—a role typically played by chemolithotrophic bacteria—they exert significant indirect influence on these processes (Johnson, 2009). For example, large biofilms dominated by *Euglena mutabilis* can cause dissolved oxygen supersaturation levels of up to 200% (Brake et al., 2001). In contrast, waters draining from subsurface mines and reservoirs in the absence of photosynthetic organisms often lack dissolved oxygen and tend to contain iron predominantly in its reduced, ferrous form (Johnson, 2003; Johnson et al., 2012). Algal oxygen production thus enhances the oxidation of ferrous iron and reduced sulfur compounds by microorganisms such as *Acidithiobacillus* and *Leptospirillum* spp., catalyzing proton-releasing reactions that may further acidify downstream environments (Rowe et al., 2007).

A second mechanism through which algae shape acid water geochemistry is by fueling the activity of acidophilic heterotrophs—particularly iron- and sulfur-reducing bacteria—through the release of exudates and cellular debris. Given that these environments are typically depleted in dissolved organic carbon (Das et al., 2009), the presence of photosynthetic organisms such as *Euglena* and *Chlorella* significantly enriches the system with bioavailable carbon. *Acidophilic* heterotrophs such as *Acidiphilium* readily utilize eukaryote-derived sugars, including glucose, fructose, and mannitol, helping to mitigate nutrient limitations. However, not all algal exudates are growth-promoting. Some compounds, such as glycolate, can be inhibitory or even toxic to acidophilic bacteria. Under acidic conditions, glycolate exists as glycolic acid, which—alongside other small aliphatic acids such as acetic acid—is harmful to most acidophiles, with exceptions such as *Sulfobacillus* (Firmicutes) (Johnson, 2009). Interestingly, acidophilic algae can often reabsorb and possibly metabolize previously secreted glycolate. This metabolic flexibility has been observed in other algal taxa, including *Chlorella pyrenoidosa*, which releases glycolate under light conditions and reabsorbs it in the absence of light (Miller et al., 1963; Fogg et al., 1965).

The essential role of acidophilic algae in sustaining heterotrophic bacterial communities has been highlighted in comparative studies of biofilms growing in subterranean versus surface environments (Johnson, 2012; Nancucheo and Johnson, 2011). In underground systems, where primary production relies exclusively on chemolithoautotrophic bacteria, biofilms are typically dominated by autotrophic prokaryotes such as "*Ferrovum myxofaciens*". In contrast, sunlit biofilms near mine drainage outlets are shaped by microalgae, which promote the growth of diverse heterotrophic acidophiles, including ferric iron reducers such as *Acidiphilium*, *Acidocella*, and *Acidobacterium* spp. Sulfide production is also notably enhanced in these algal-associated systems, suggesting that acidophilic algae facilitate sulfate-reducing bacterial (SRB) activity by supplying essential organic substrates. These interactions may be critical for microbial colonization and persistence in extreme environments.

Colonization studies conducted in Río Tinto show that community development begins with the accumulation of amorphous aggregates composed of bacterial cells and mineral particles. Within approximately 2 weeks, fungi and small heterotrophic eukaryotes—such as amoebae and flagellates—colonize the organic matrix. Diatoms typically establish only in areas where fungal hyphae have already formed a supportive network. In contrast, flagellated green algae such as *Dunaliella*, *Chlamydomonas*, and *Euglena* are among the earliest colonizers, albeit in low abundance. These are followed by sessile algae such as *Chlorella* and *Cyanidium*, while filamentous algae generally appear in the final stages of biofilm development. Once established, the microbial communities remain relatively stable throughout the year, occupying consistent niches along the system (Aguilera et al., 2007a; Hujslová et al., 2019).

Fungal roles in acidic environments, such as those of other eukaryotes, have received limited attention. However, fungi are believed to serve as primary decomposers of complex organic matter—particularly in these carbon-poor systems where easily degradable organic inputs (e.g., leaf litter) are scarce, partly due to the absence of invertebrate shredders. In such settings, fungi may play a crucial role in breaking down recalcitrant carbon compounds. Additionally, fungal respiration contributes to oxygen consumption, potentially lowering oxidative stress for sulfate-reducing bacteria. Some fungi may also participate directly in ferric iron or sulfur reduction, key electron transfer processes that support biological alkalinity in acidic waters (Aguilera et al., 2007b).

Composition, structure, and rheological properties of biofilms

Research on microbial ecology in acidic aquatic systems has traditionally emphasized planktonic communities of chemolithotrophic bacteria, owing to their central role in mediating redox reactions that drive mineral dissolution and precipitation. However, eukaryotic organisms capable of withstanding low pH and high metal concentrations also play a substantial role in shaping these ecosystems. One of the most striking visual features of such environments is the reddish water and the yellow-brown coloration of sediments—caused respectively by dissolved ferric iron and the accumulation of iron hydroxides and sulfur-based precipitates. Also notable are the benthic or floating filaments and vividly colored biofilms that form on submerged sediments and along shorelines. These biofilms provide a structural niche for benthic organisms that rely on surface attachment for growth. Bacterial biofilms are often orange in color, while in shallow areas where light penetrates sufficiently to support photosynthesis, they may exhibit a variety of hues, from bright green *Euglena* to golden brown diatoms, reflecting the presence of microorganisms with diverse photopigments.

While many studies on extreme-environment biofilms focus on biodiversity or functional interactions, their microstructural architecture has often been overlooked. By preserving field samples as biological tissues, cross-sectional observations have revealed biofilm thickness, composition, and the spatial distribution of microbial cells. Scanning electron microscopy (SEM) has further uncovered the ultrastructure of photosynthetic biofilms in their natural state (Fig. 8.3).

To visualize biofilms with their water content, EPS networks, and embedded minerals, low-temperature scanning electron microscopy (CryoSEM or LTSEM) offers the most comprehensive resolution (Fig. 8.4).

This method revealed the structure of the matrix within the biofilms, and we could relate them with their associated rheological properties (Souza-Egipsy et al., 2011). The study found that biofilms with a denser and more interconnected network structure, as seen in LTSEM images, had higher values of cohesive energy and stiffness, indicating stronger mechanical properties. For example, bacterial and some algal biofilms (*Leptospirillum–Acidiphilium, Chlorella, Cyanidium*) showed compact, intertwined networks and correspondingly high elastic modulus and cohesive energy. In contrast, biofilms with a more open structure and less network development, such as those formed by *Euglena* and *Pinnularia*, exhibited lower values of both moduli and cohesive energy, making them softer and more mobile under mechanical stress. This demonstrates a clear relationship between the morphology

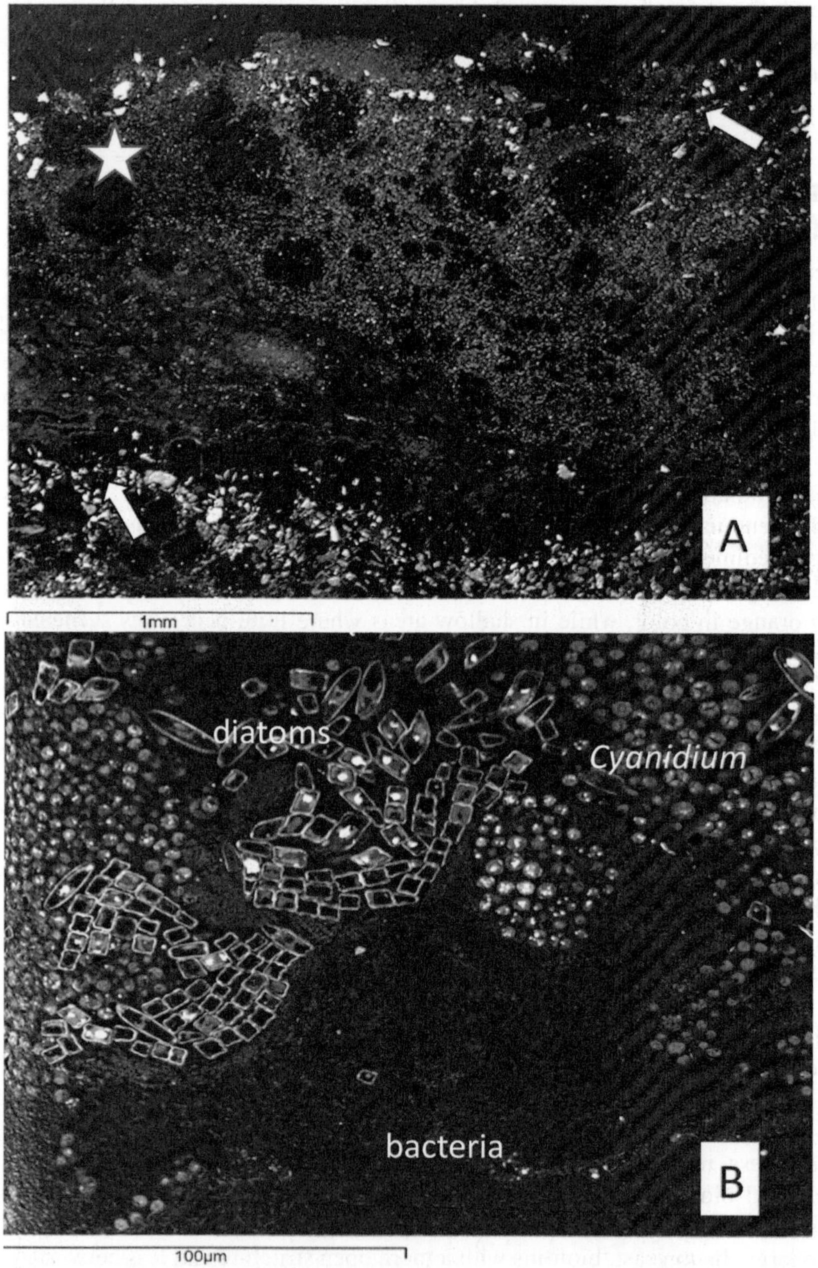

FIGURE 8.3 SEM-BSE photographs from biofilms collected in the summer months

(A) Biofilms are formed of several layers of microorganisms, separated by areas were humid acids accumulated, black lumps (Star), Bar 1 mm. (B) Detail view of the upper part of the biofilm, diatoms frustules and spherical shapes of *Cyanidium* cells are observed. The lower part is formed by a bacteria biofilm. Bar 100 μm.

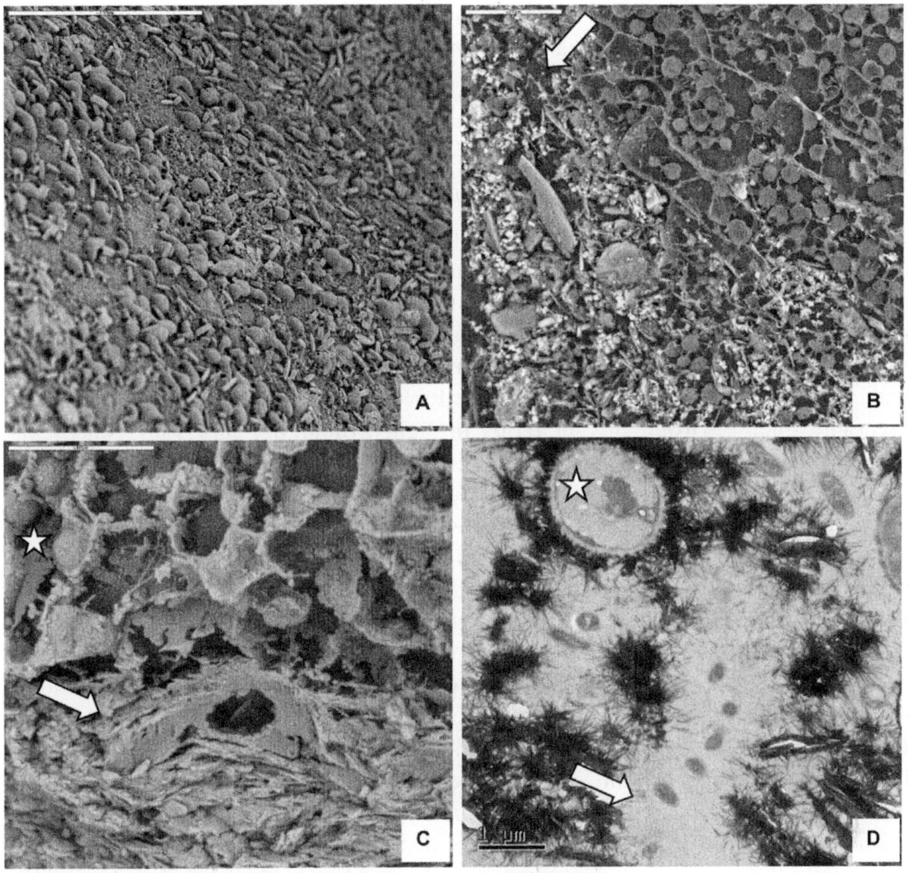

FIGURE 8.4 Low-Temperature scanning microscopy photograph

(A), Low-temperature scanning microscopy photograph of superficial view of a natural biofilm. Notice curved shapes from *Euglena* and diatom frustules embedded in the biofilm matrix. Bar 200 μm. (B). LTSEM, transversal section of biofilm close to the sediments (arrow). Note areas around cells without minerals in the upper right section. Bar 20 μm (C). Biofilm remains in the lower sediment interface (arrow). Cells with mineral deposits and minerals in the matrix around them (star). Bar 20 μm (D). Transmission electron microscopy photograph showing cells decay after mineral deposition (star). Note the clear areas around bacteria without mineral deposition (arrow). Bar 1 μm.

(network density and structure) and the rheological parameters (stiffness, cohesive energy, relaxation time) of the biofilms.

Transmission electron microscopy (TEM) provides complementary insight at the ultrastructural level, allowing precise localization of mineral particles within and on the surface of cells embedded in EPS matrices (Fig. 8.4D). Scanning electron microscopy (SEM) combined with energy-dispersive X-ray spectroscopy has proven useful for mapping the spatial distribution of elements surrounding cells embedded in biofilms. These analyses reveal that iron and sulfur are the most prevalent cations associated with the outermost layers of the biofilm matrix (Fig. 8.5).

Moreover, the silicification of cell walls—particularly in unicellular algae entrapped within sediments—has also been documented in these environments, indicating processes of permineralization involving silica deposition (Souza-Egipsy et al., 2010).

In Río Tinto, most eukaryotic biomass is concentrated within such biofilms along the entire river system. While certain taxa—such as *Chlamydomonas,* ciliates, and heliozoans—may be found in the plankton, the majority are associated with surface-attached communities. This cohesive growth form not only promotes ecological interactions but also provides a refuge from the most extreme conditions and enhances resistance to hydrodynamic forces, allowing organisms to remain in favorable microenvironments in terms of light availability and water chemistry (Souza-Egipsy et al., 2021). Some form elongated filaments that move with the current, adding further structure to the habitat. Photosynthetic biofilms are especially prominent in Río Tinto and can grow to considerable size during the summer months (Aguilera et al., 2007a). Fungi- and yeast-dominated biofilms are also common in streambed sediments, and these are typically accompanied by abundant bacterial populations—either as floating aggregates or sediment-associated colonies (Souza-Egipsy et al., 2008).

Río Tinto's photosynthetic biofilms are spatially complex, three-dimensional structures often dominated by a single microbial group (Fig. 8.3A). Initial colonization typically involves bacteria and fungi that settle along with mineral particles. Heterotrophic flagellates and ciliates appear early in the development process but remain in low abundance throughout. *Euglena* is the only protist capable of forming extensive monospecific structures, while diatom-dominated biofilms, especially those with *Pinnularia,* are second in prevalence after bacterial forms. Generally, distinct biofilms do not integrate but may exist side by side or overlap spatially (Fig. 8.3B). In the most chemically extreme habitats—characterized by low pH and high metal concentrations—biofilms are dominated by microalgae such as *Chlamydomonas*, *Cyanidium*, and *Dunaliella*, together with substantial fungal and bacterial populations. In contrast, biofilms in shallow waters are primarily composed of *Chlorella* and *Euglena*. Filamentous green algae such as *Zygnemopsis* and *Klebsormidium* are more seasonal in nature, peaking in abundance around June and declining sharply between October and November. Additionally, floating bacterial filaments are frequently observed on the surface of impounded waters (García-Moyano et al., 2007).

The ability of microorganisms to form biofilms is largely due to their production of extracellular polymeric substances (EPS), predominantly polysaccharides, which establish the cohesive matrix surrounding the cells (Sutherland, 2001). EPS matrices also incorporate proteins, nucleic acids, phospholipids, and humic acids, collectively forming the structured microhabitat in which the organisms live (Flemming and Wingender, 2010). This mode of life provides advantages over planktonic forms,

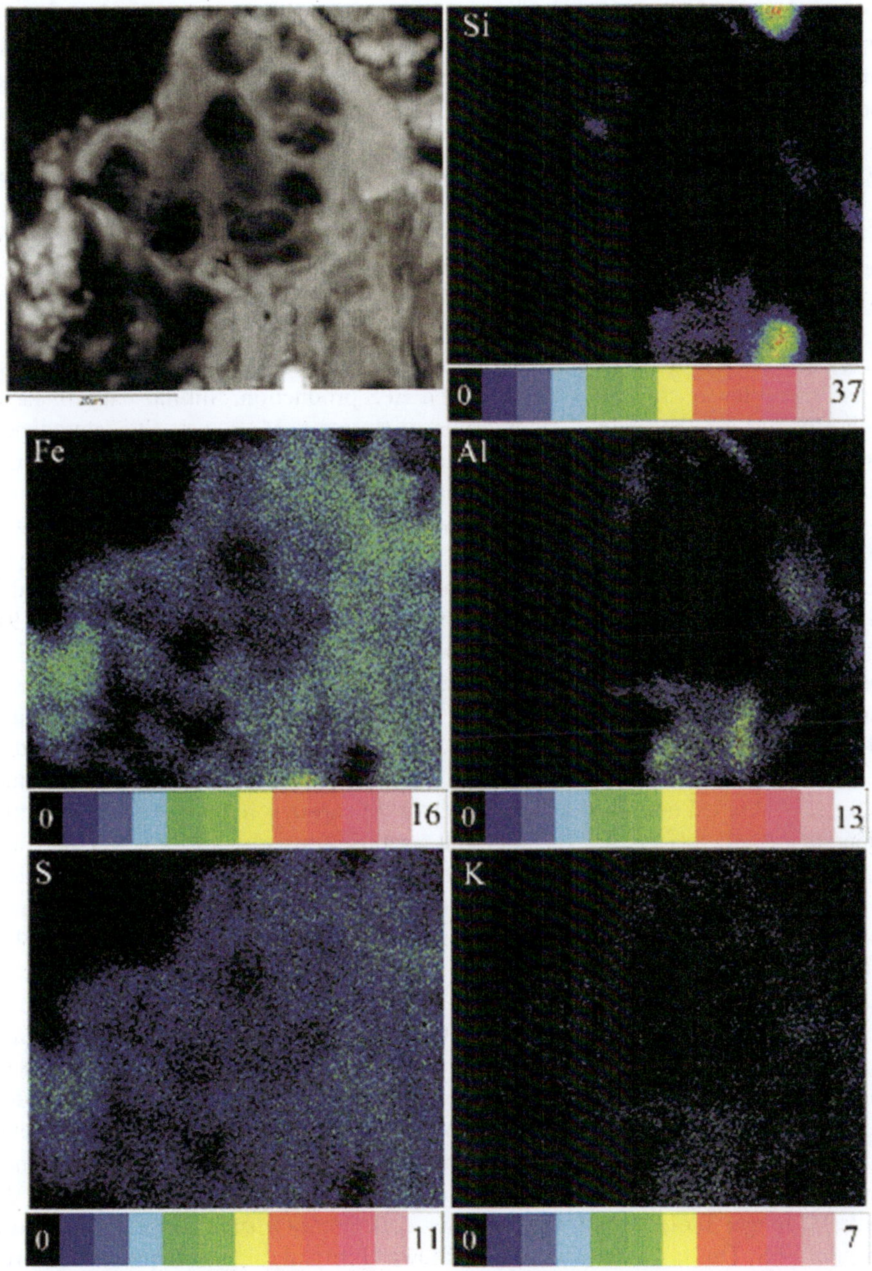

FIGURE 8.5 SEM-BSE photograph of an area of the biofilm where cells are surrounded by minerals

Bar 20 μm. EDX microanalytical maps of silica (Si), iron (Fe), aluminum (Al), Sulfur (S), and potassium (K). Scale in each graph show maximum relative intensities.

particularly in terms of resistance to environmental stressors (García-Meza et al., 2005). Although the presence of biofilms in extreme aquatic and terrestrial ecosystems is well documented, the underlying mechanisms governing their formation and persistence remain poorly understood.

Despite the ecological importance of eukaryotic biofilms in extreme systems such as Río Tinto, their characteristics remain underexplored. One particularly noteworthy finding is that the dry weight of EPS extracted from these biofilms reaches milligram levels, whereas in neutral pH environments, EPS yields are typically measured in micrograms—representing a difference of three orders of magnitude (Aguilera et al., 2008a, 2008b). High EPS production has also been documented in other extreme settings, from Arctic sea ice to Antarctic marine bacteria. In Río Tinto, total EPS quantity correlates positively with heavy metal concentrations: higher metal levels coincide with increased EPS production. Similar patterns have been reported in other acidic and extreme ecosystems. Moreover, EPS abundance also increases as pH decreases, reinforcing its link to environmental stress. Seasonal dynamics have been observed as well; during summer, filamentous algae such as *Zygnemopsis* and *Klebsormidium* rapidly increase their biomass, producing long floating strands. This growth pattern has been associated with elevated photosynthetic activity. Notably, these species produce greater EPS quantities in the acidic waters of Río Tinto than in colder or geothermal environments (Blanco et al., 2019).

Biochemical analysis of Río Tinto eukaryotic biofilms shows that their EPS is composed mainly of polysaccharides, heavy metals, and humic acids, with minor contributions from proteins and DNA (Aguilera et al., 2008b; Blanco et al., 2019). Overall, the extracellular matrix can comprise up to 44% of the total dry weight of a biofilm, though the cellular fraction generally remains predominant. EPS can be divided into two key components: the capsular fraction, which is tightly bound to the cell surface, and the colloidal fraction, consisting of loosely associated organic matter. The ratio between these fractions varies depending on the dominant organism. Most biofilms contain more colloidal than capsular EPS, except in fungal-dominated biofilms, where the capsular component prevails (Aguilera et al., 2008b). These differences are likely influenced by both the biology of the constituent species and the environmental context in which the biofilm develops.

As seen in other extreme environments, polysaccharides are the most abundant component of the extracellular matrix in Río Tinto biofilms, accounting for up to 85% of the total dry weight. In biofilms formed by microalgae, polysaccharides are particularly enriched in the colloidal fraction. Metal cations represent the second most abundant component, constituting roughly 10% of the dry weight and accumulating primarily in the colloidal EPS—i.e., the portion furthest from the cells. In contrast, fungal biofilms contain a higher concentration of metal cations in the capsular EPS. This is consistent with the known ability of fungi to sequester and precipitate metals, which likely explains the elevated metal content in this tightly bound matrix. Humic acids form the third most abundant component of the matrix, though they account for only around 4% of the total weight. These are especially enriched in the colloidal fraction of microalgal biofilms (Fig. 8.3A).

Proteins represent another component of the extracellular matrix, comprising around 2% of the dry weight, predominantly within the capsular layer. Total protein content remains below 3%, consistent with observations from bacterial biofilms and other extreme habitats where extracellular protein production is typically minimal. DNA is also detectable, though in trace amounts—below 0.5% of dry weight—and appears equally distributed between capsular and colloidal fractions. Its exact origin and ecological role remain uncertain.

The extracellular matrix is likely a key factor in the remarkable metal tolerance exhibited by these biofilms (Fig. 8.4C and D). EPSs can chelate toxic metal cations, providing cellular protection through complexation mechanisms (Kaplan et al., 1987). This capacity depends on functional groups present in proteins, lipids, and carbohydrates. At acidic pH, most metals remain as free cations, while in alkaline conditions they tend to form insoluble precipitates. Moreover, low pH alters the ionization of functional groups within EPS, particularly carboxyl groups—the most hydrophilic and acidic—by promoting their protonation and thus reducing their binding capacity. Phosphate and sulfate groups, however, tend to retain their negative charges even at low pH. The capsular fraction, being tightly associated with the cell wall, exhibits relatively low functional group diversity and is less efficient in metal sequestration. This is consistent with findings showing that the colloidal fraction—less structurally bound and spatially separated from cell membranes—accumulates the majority of metals.

Comparative analyses of dissolved metals in river water versus their concentrations in EPSs show a general trend of similar or slightly reduced accumulation, with notable exceptions. Lead is found in higher concentrations within *Euglena* and *Cyanidium* biofilms than in ambient water, and *Cyanidium* also exhibits elevated arsenic accumulation. These interspecific differences in metal-binding capacity are likely due to variations in the abundance of carboxyl, hydroxyl, sulfate, and phosphate groups in the EPS polysaccharides, as well as differences in protein composition of the cell wall. In unicellular algae such as *Chlorella* and *Cyanidium*, this effect is particularly pronounced (Gardea-Torresdey et al., 1990). Although these organisms bind copper efficiently at pH 5, this capacity diminishes sharply at pH 2 due to carboxyl group protonation. Consequently, the strength of the chelation effect varies not only by species but also by the acidity of the habitat and the availability of binding ligands.

Río Tinto waters contain a wide array of dissolved metals in significant concentrations. Electrostatic interactions—particularly short-range forces—are critical to metal binding and become more intense with increasing cationic charge and decreasing ionic radius. Thus, metals with higher valence states bind more readily than divalent ions. Geochemical conditions in these acidic waters also favor the formation of iron oxyhydroxides and hydroxy-sulfates, which can in turn facilitate the coprecipitation of other metals (Ferris et al., 2004). Interestingly, at sites with lower concentrations of dissolved iron, biofilms contained more iron but less copper and zinc than the surrounding water. While copper and zinc share similar ionic radii, ferric ions are smaller than their ferrous counterparts. These solubility differences are influenced more by geochemical dynamics than by microbial activity (González-Toril et al., 2011).

In summary, the structural organization of biofilms in Río Tinto offers compelling evidence of microbial adaptation to harsh environmental gradients. By aggregating into biofilms, diverse microbial taxa gain access to localized nutritional resources and form protective microenvironments that buffer them against the river's extreme physicochemical stressors. However, our understanding of the mechanisms that regulate biofilm development and interspecies cooperation remains incomplete. Further studies of the rheological properties of biofilms will help us explain how biofilms respond to mechanical forces in their environment, which affects their stability, resilience, and ability to survive. The mechanical properties, such as stiffness and cohesive energy, are linked to the biofilm's microstructure and determine how well the biofilm can withstand external stresses, detach, or maintain its structure. Understanding these properties provides insight into factors that control the ecological niche distribution of different biofilms and their interactions with the environment. Additionally, the energy required to disrupt the biofilm network (cohesive energy) is thought to influence the ecological mechanics for survival. Thus, rheological analysis provides valuable information about the survival strategies and ecological roles of biofilms in extreme environments.

Mechanisms of adaptation to extreme acidic environments

Although morphologically simple, the microorganisms inhabiting Río Tinto exhibit remarkable genetic diversity and a wide array of physiological strategies that enable them to thrive under harsh environmental conditions. Acidophilic species, in particular, possess specialized resistance mechanisms not found in organisms from more moderate ecosystems. As a result, research on extremophiles—especially acidophiles—provides valuable insights into biological adaptations to low pH and metal toxicity. Recent advances in genomics have significantly expanded our understanding of the flexible and resilient metabolic pathways these organisms employ. Beyond their scientific interest for evolutionary and ecological studies, acidophiles also hold promise as sources of novel enzymes with potential applications in biotechnology and environmental remediation.

Adaptation strategies in acidophilic eukaryotes

Eukaryotic microorganisms that inhabit highly acidic environments are expected to display molecular architectures and physiological controls that differ substantially from those of their neutrophilic counterparts. Historically, genomic sequencing required organisms to be cultured in the laboratory, limiting access to many acidophilic taxa. This barrier has been largely overcome with the advent of culture-independent approaches, such as metagenomics and single-cell genomics. Together with complementary "omics" techniques—including transcriptomics (mRNA expression profiling), proteomics (protein identification and quantification), and metabolomics (analysis of intracellular metabolites)—these tools have transformed our capacity to investigate the structure and function of microbial communities. The integration of these approaches within systems biology has made it possible to study not only the functional roles of individual species but also their interactions with one another and with the environment. In recent years, these methods have provided critical insight into the molecular strategies extremophiles use to survive in conditions that challenge conventional life.

Adaptation to acidic pH

Similar to acidophilic prokaryotes, eukaryotic acidophiles maintain a near-neutral cytoplasmic pH despite exposure to highly acidic surroundings. This requires managing steep proton gradients across the cell membrane. For instance, *Chlamydomonas acidophila* maintains an average cytosolic pH of 6.6, whether it is grown in an external pH of 2 or 7. By contrast, the closely related neutrophile *Chlamydomonas reinhardtii* maintains a cytosolic pH of 7.1 in neutral conditions (Messerli et al., 2005). Moreover, *C. acidophila* displays an unusually low transmembrane electrical potential—close to 0 mV—when measured with intracellular electrodes at both acidic and neutral pH, a feature rarely observed among plants, animals, or protists (Messerli et al., 2005). This ability to sustain a neutral cytosol in a pH 2 environment implies the maintenance of a proton gradient of approximately 10^5, meaning the concentration of protons outside the cell is 100,000 times higher than inside. Such extreme gradients are exceptional in nature and comparable only with those found in mammalian gastric fluids or in the acidic vacuoles of plant cells.

Sustaining this kind of gradient would be energetically prohibitive unless the cell membrane is adapted to limit passive proton influx. Evidence suggests that *C. acidophila* achieves this by actively exporting protons into intracellular vacuoles while maintaining a relatively neutral cytosol. This strategy likely reduces the burden on plasma membrane proton pumps and minimizes their exposure to highly acidic extracellular fluids. Interestingly, this suggests that the primary structure of *C. acidophila's* proton transporters may not differ markedly from those found in neutrophilic protists (Messerli et al., 2005).

In contrast, *Dunaliella acidophila*—one of the most extreme acidophiles known—employs a markedly different strategy. This microalga thrives at extraordinarily low pH values, with optimal growth occurring between pH 0 and 3, while still maintaining an internal pH of approximately 7. It achieves this through a combination of mechanisms designed to restrict proton entry and facilitate their active extrusion. These include generating a positive internal membrane potential, maintaining a positively charged cell surface, and synthesizing high levels of a highly efficient plasma membrane H^+-ATPase.

Under acidic conditions, *D. acidophila* dramatically increases production of this enzyme. Although structurally similar to the H^+-ATPases found in higher plants, the version expressed in *D. acidophila* is kinetically distinct, with greater molecular weight and two to three times higher enzymatic activity. Its optimal function occurs at pH 6.0, slightly lower than that of its plant counterparts, allowing it to operate more effectively in low-pH environments (Sekler et al., 1993, 1994).

Another important feature contributing to proton resistance in *D. acidophila* is its positively charged cell surface, which resembles that observed in acidophilic prokaryotes. Most cells typically have a net negative surface charge at neutral pH, due to the prevalence of acidic residues in membrane proteins and anionic lipids on the cell surface. However, at pH 1, these anionic groups become protonated, resulting in a net positive charge. This surface chemistry may not only reduce proton influx but also enhance resistance to multivalent cations and heavy metals—both common stressors in acidic ecosystems.

Adaptations to nutrient scarcity

Extreme acidic environments are typically oligotrophic, exhibiting low biological productivity due to limited concentrations of essential nutrients such as ammonium, phosphate, and nitrate (Spijkerman et al., 2007). One critical adaptive trait among acidophilic microalgae is their ability to synthesize phosphatases under nutrient-limited conditions. For example, *Chlamydomonas acidophila* not only increases overall phosphatase activity in response to phosphate deprivation but also produces a specialized phosphatase with an optimal activity at approximately pH 2 (Spijkerman, 2007). This adaptation likely contributes to the ecological success and high abundance of *C. acidophila* across a broad range of acidic habitats, including both mining-impacted areas and natural geothermal sites.

In Río Tinto, bicarbonate concentrations are negligible, making dissolved CO_2 the primary source of inorganic carbon for photosynthetic organisms. The efficiency of CO_2 uptake depends on both passive diffusion and the organism's capacity to actively concentrate CO_2 within the cell. Most aquatic phototrophs rely on carbon-concentrating mechanisms (CCMs) to elevate intracellular CO_2 levels in proximity to Rubisco, their primary carboxylating enzyme. Green algae and cyanobacteria are well known to employ such mechanisms to overcome the dual constraints of low dissolved CO_2 concentrations and its inherently slow diffusion in aquatic environments. *C. acidophila* possesses a functional CCM and exhibits high CO_2 affinity under limiting conditions, allowing it to perform well in acidic, low-flow habitats such as lakes and stagnant waters (Spijkerman et al., 2014).

Photosynthetic adaptations

Photosynthetic processes are highly susceptible to environmental stress, including fluctuations in salinity, acidity, and the presence of toxic compounds. Photosynthesis in acidophilic microorganisms remains understudied, with most investigations focused on overall primary productivity in acidic lakes. Current evidence suggests that reduced productivity in these ecosystems is largely due to elevated concentrations of dissolved metals and limited availability of reactive phosphate, rather than the acidic pH itself (Spijkerman et al., 2007).

Light availability is another major constraint on photosynthesis in acidified waters. Elevated iron concentrations contribute to a dark reddish water color, which alters both the intensity and spectral quality of light penetrating the water column (Aguilera et al., 2012). Nevertheless, benthic diatom biofilms inhabiting acid lakes have demonstrated a strong capacity for growth in low light conditions. Measurable photosynthetic activity under reduced irradiance suggests that these organisms possess pigment systems capable of efficiently harvesting red light—the dominant wavelength in these iron-rich, intensely colored environments (Koschorreck and Tittel, 2002).

To explore the photosynthetic strategies of microalgae and autotrophic protists from Río Tinto, two primary analytical techniques have been employed: oximetry and pulse-amplitude modulated (PAM) fluorometry. Oximetric analyses allow for the generation of photosynthesis–irradiance (P–I) curves, which help differentiate between light saturation and photoinhibition responses. These curves reflect the physiological and biochemical mechanisms underlying photosynthesis and enable the estimation of key parameters such as maximum photosynthetic rate (Pmax), compensation irradiance (Ic), photosynthetic efficiency (α), and photoinhibition rate (β).

PAM fluorometry, in contrast, measures chlorophyll fluorescence to assess the kinetics of electron transport within photosystem II. This technique yields indicators such as the maximum quantum yield (Fv/Fm) and electron transport rate (ETR), both of which are widely used to evaluate the photosynthetic performance of microbial and algal communities in situ.

In Río Tinto, photosynthetic biofilms composed of *Chlorella* spp., *Euglena mutabilis*, *Pinnularia* spp., and *Zygnemopsis* spp. exhibit distinct photophysiological responses. Most of these communities demonstrated strong photoinhibition in their photosynthesis–irradiance (P–I) curves, a typical feature of organisms adapted to low-light conditions and unable to manage excessive photon influx. The exception was *Zygnemopsis*, which displayed a photosaturation response, suggesting high-light tolerance (Aguilera et al., 2012; Souza-Egipsy et al., 2011). Notably, biofilms with photoinhibitory behavior were primarily located on the riverbed, beneath layers of highly pigmented red water. These organisms exhibited extremely low compensation (Ic) and saturation irradiance (Ik) values, highlighting their adaptation to dim environments. For instance, the diatom *Pinnularia* maintained detectable photosynthetic activity even under irradiance as low as 5 μmol m^{-2} s^{-1}. In contrast, floating filaments of *Zygnemopsis*—exposed to intense summer sunlight at the water surface—showed classic photosaturation behavior, indicating ecological specialization for high-irradiance environments. In the case of natural *E. mutabilis* biofilms we observed large-scale transcriptomic reprogramming during midday due to a dynamic photoinhibition and solar radiation stress. The transcription pattern was altered in genes involved in Photosystem II stability and repair, UV damaged DNA repair, nonphotochemical quenching and oxidative stress, supporting the photoinhibition detected by PAM fluorometry at midday (Puente-Sánchez et al., 2016). These divergent light niches help explain the species-specific differences in photophysiological traits.

Despite substantial contributions of phototrophs to total biomass—over 65% in Río Tinto (López-Archilla et al., 2001) —maximum photosynthetic rates (Pmax) remained relatively low compared with other aquatic systems (Aguilera et al., 2012). Unlike acid mine lakes, where low productivity is often attributed to limited phytoplankton biomass, the low Pmax observed here may stem from nutrient limitations, particularly reduced availability of ammonium, phosphate, and nitrate (Spijkerman et al., 2007).

Adaptations to metal toxicity

High concentrations of heavy metals are typical of acidic environments. Río Tinto environment affects the global budget of several heavy metals (Braungardt et al., 2003). As in many other acid rock drainage systems, small headwater streams of the RT have different groundwater sources that percolate through soils and rocks with distinct mineralogy leading to substantial variation in water chemistry (Amaral-Zettler et al., 2011; Palacios et al., 2008). Therefore, studies on acidophilic organisms under controlled conditions will improve our understanding on the evolution of heavy metal tolerance and detoxification in their natural habitats.

D. acidophila showed a high constitutive expression of genes related to oxidative stress and reactive oxygen species defense, even without metal exposure. Photosynthetic activity remained stable under both cadmium and natural metal-rich conditions, likely due to increased synthesis of structural photosynthetic components. Chronic exposure to natural metal-rich water leads to an overrepresentation of genes involved in metabolism, transcription, and protein degradation, suggesting a higher protein turnover and possible endoplasmic reticulum stress response. These findings highlight the molecular mechanisms underlying heavy metal tolerance in this extremophilic microalga (Puente-Sánchez et al., 2016).

A similar study with *C., acidophila* cultivated under control conditions and with high concentrations of copper (Olsson et al., 2015) showed that the photosynthetic activity was not adversely affected by addition of high amounts of copper. Up-regulated transcripts include several transcripts related to photosynthesis and carbohydrate metabolism, transcripts coding for general stress response, and a transcript annotated as homologous to the oil-body-associated protein HOGP coding gene. The first de novo assembly of *C. acidophila* significantly increases transcriptomic data available on extremophiles and green algae and thus provides an important reference for further molecular genetic studies. The differences between differentially expressed transcripts detected in our study suggest that the response to heavy metal exposure in *C. acidophila* is different from other studied green algae. Similarly, a study in the acidophilic green alga. C. *acidophila* showed that phytochelatin synthase genes of bacterial horizontal gen transfer origin played an important role in the tolerance to cadmium (Olsson et al., 2017).

To better understand heavy metal tolerance in *C. acidophila*, we assembled its transcriptome and measured transcriptomic expression before and after Cd exposure (Puente-Sánchez et al., 2016). Genes possibly related to heavy metal tolerance and detoxification were identified and analyzed as potential key innovations that enable this species to live in an extremely acid habitat with high levels of heavy metals. In addition, we were able to provide a data set of single orthologous genes from eight green algal species as a valuable resource for comparative studies including eukaryotic extremophiles. Our results based on differential gene expression, detection of unique genes and analyses of codon usage all indicate that there are important genetic differences in *C. acidophila* compared with *C. reinhardtii*. Exposure to cadmium strongly increases transposon expression in green algae, and that oil biosynthesis genes are induced in *Chlamydomonas* under heavy metal stress. This acidophilic alga reorganizes their metabolic pathways as an adaptive response to environmental stress from heavy metals and acidity (Cid et al., 2010). The comparison of the transcriptomes of several acidophilic and nonacidophilic algae showed that the *Chlamydomonas* genus is polyphyletic and that acidophilic algae have distinctive aminoacid usage patterns.

Cultures of *C. acidophila* RT46 have been used also to study Cd and As metal tolerance (Díaz et al., 2020). This strain was quite sensitive to Cd but very tolerant to As, although arsenite was almost four-fold more toxic than arsenate. Both As inorganic species induced the generation of remarkable superoxide anion levels in cellular populations. Intense vacuolization and a drastic increase of starch and lipid reserves were detected after Cd and As treatments. Moreover, As(III) and As(V) caused severe disorganization in the eyespot or stigma and thylakoids, but no ultrastructural modification of these organella was detected after Cd exposure. Metal (loid) bioaccumulation and biosorption might be implicated in the tolerance of *C. acidophila* to Cd and As. TEM–XEDS analysis and qRT–PCR data suggested that poly-phosphate and phytochelatins are involved in bioaccumulation. Additionally, a massive discharge of exopolymers on the cell surface occurred in the presence of the toxics, producing an increase in the thickness of the capsule, where metallic deposits were entrapped. Later genomic studies have provided new information on the function of phytochelatin synthase (CaPCS2) from *C. acidophila*. The heterologous expression of CaPCS2 in *E. coli* provided protection against oxidative stress produced by metals and exposure to different reactive oxygen species inducing agents. However, the function of this PCS was not related to heavy metal bioaccumulation (Díaz et al., 2020).

Outlook and applications

Over the past 25 years, our understanding of the phylogenetic and physiological diversity of acidophilic microorganisms has significantly increased. However, findings from biomolecular research conducted in highly acidic environments indicate that many acidophilic species are yet to be isolated and studied. There is considerable interest in these microorganisms, not only for insights into how they survive in environments that are inhospitable to most life forms, but also because of their relevance to environmental issues (such as mine waste and drainage) and their applications in biotechnology (particularly in biomining and the extraction of metals from polluted soils). Current research efforts are focused on identifying acidophiles that can withstand additional environmental challenges, including higher temperatures than those tolerated by known thermoacidophiles and varying salinity levels. It is likely that new opportunities will arise to utilize both existing and newly discovered acidophilic microorganisms in future biotechnological applications, leveraging their unique capabilities to thrive in moderately to extremely acidic conditions and facilitate the transformation of both inorganic and organic compounds.

Acknowledgments

This research has been funded by grants No. PID2022-138986OB-I00 by the Spanish Ministry of Science and Innovation/State Agency of Research MCIN/AEI/10.13039/501100011033 and by "ERDF A way of making Europe," as well as with funds from European Research Executive Agency (REA) XTREAM Project # 10118278.

AI disclosure

During the preparation of this work, the author(s) used ChatGPT in order to correct English. After using this tool/service, the author(s) reviewed and edited the content as needed and take (s) full responsibility for the content of the publication.

References

Aguilera, A., González-Toril, E., 2019. Eukaryotic Life in Extreme Environments: Acidophilic Fungi. Springer Science and Business Media LLC, pp. 21–38. https://doi.org/10.1007/978-3-030-19030-9_2.

Aguilera, A., Manrubia, S.C., Gómez, F., Rodríguez, N., Amils, R., 2006. Eukaryotic community distribution and its relationship to water physicochemical parameters in an extreme acidic environment, Río Tinto (Southwestern Spain). Appl. Environ. Microbiol. 72 (8), 5325–5330. https://doi.org/10.1128/AEM.00513-06.

Aguilera, A., Souza-Egipsy, V., Gómez, F., Amils, R., 2007a. Development and structure of eukaryotic biofilms in an extreme acidic environment, Río Tinto (SW, Spain). Microb. Ecol. 53 (2), 294–305. https://doi.org/10.1007/s00248-006-9092-2.

Aguilera, A., Zettler, E., Gómez, F., Amaral-Zettler, L., Rodríguez, N., Amils, R., 2007b. Distribution and seasonal variability in the benthic eukaryotic community of Río Tinto (SW, Spain), an acidic, high metal extreme environment. Syst. Appl. Microbiol. 30 (7), 531–546. https://doi.org/10.1016/j.syapm.2007.05.003.

Aguilera, A., Souza-Egipsy, V., Amils, R., 2008a. Biodiversity and Ecology of Eukaryotic Organisms in Extreme Acidic Environments, the Río Tinto Case Environmental Microbiology Research Trends. Nova Science Publishers, Inc., Spain, pp. 213–236.

Aguilera, A., Souza-Egipsy, V., San Martín-Úriz, P., Amils, R., 2008b. Extracellular matrix assembly in extreme acidic eukaryotic biofilms and their possible implications in heavy metal adsorption. Aquat. Toxicol. 88 (4), 257–266. https://doi.org/10.1016/j.aquatox.2008.04.014.

Aguilera, A., Souza-Egipsy, V., Amils, R., 2012. Photosynthesis in Extreme Environments. InTech. https://doi.org/10.5772/26940.

Amaral-Zettler, L.A., Gómez, F., Zettler, E., Keenan, B.G., Amils, R., Sogin, M.L., 2002. Eukaryotic diversity in Spain's River of Fire: this ancient and hostile ecosystem hosts a surprising variety of microbial organisms. Nature 417 (6885), 137. https://doi.org/10.1038/417137a.

Amaral-Zettler, L.A., Zettler, E.R., Theroux, S.M., Palacios, C., Aguilera, A., Amils, R., 2011. Microbial community structure across the tree of life in the extreme Río Tinto. ISME J. 5 (1), 42–50. https://doi.org/10.1038/ismej.2010.101.

Amils, R., de la Fuente, V., Rodríguez, N., Zuluaga, J., Menéndez, N., Tornero, J., 2007. Composition, speciation and distribution of iron minerals in Imperata cylindrica. Plant Physiol. Biochem. 45 (5), 335–340. https://doi.org/10.1016/j.plaphy.2007.03.020.

Baker, B.J., Tyson, G.W., Goosherst, L., Banfield, J.F., 2009. Insights into the diversity of eukaryotes in acid mine drainage biofilm communities. Appl. Environ. Microbiol. 75 (7), 2192–2199. https://doi.org/10.1128/AEM.02500-08UnitedStates.

Blanco, Y., Rivas, L.A., González-Toril, E., Ruiz-Bermejo, M., Moreno-Paz, M., Parro, V., Palacín, A., Aguilera, Á., Puente-Sánchez, F., 2019. Environmental parameters, and not phylogeny, determine the composition of extracellular polymeric substances in microbial mats from extreme environments. Sci. Total Environ. 650, 384–393. https://doi.org/10.1016/j.scitotenv.2018.08.440.

Brake, S.S., Dannelly, H.K., Connors, K.A., 2001. Controls on the nature and distribution of an alga in coal mine-waste environments and its potential impact on water quality. Environ. Geol. 40 (4–5), 458–469. https://doi.org/10.1007/s002540000181.

Braungardt, C.B., Achterberg, E.P., Elbaz-Poulichet, F., Morley, N.H., 2003. Metal geochemistry in a mine-polluted estuarine system in Spain. Appl. Geochem. 18 (11), 1757–1771. https://doi.org/10.1016/S0883-2927(03)00079-9.

Cid, C., Garcia-Descalzo, L., Casado-Lafuente, V., Amils, R., Aguilera, A., 2010. Proteomic analysis of the response of an acidophilic strain of Chlamydomonas sp. (Chlorophyta) to natural metal-rich water. Proteomics 10 (10), 2026–2036. https://doi.org/10.1002/pmic.200900592.

Coker, J.A., 2019. Recent advances in understanding extremophiles. F1000Res. 8, 1917. https://doi.org/10.12688/f1000research.20765.1.

Das, B.K., Roy, A., Koschorreck, M., Mandal, S.M., Wendt-Potthoff, K., Bhattacharya, J., 2009. Occurrence and role of algae and fungi in acid mine drainage environment with special reference to metals and sulfate immobilization. Water Res. 43 (4), 883–894. https://doi.org/10.1016/j.watres.2008.11.046.

de la Fuente, V., Rodríguez, N., Amils, R., 2012. Immunocytochemical analysis of the subcellular distribution of ferritin in Imperata cylindrica (L.) Raeuschel, an iron hyperaccumulator plant. Acta Histochem. 114 (3), 232–236. https://doi.org/10.1016/j.acthis.2011.06.007.

Deneke, R., 2000. Review of rotifers and crustaceans in highly acidic environments of pH values $\leq$3. Hydrobiologia 433 (1–3), 167–172. https://doi.org/10.1023/a:1004047510602.

Dhakar, K., Pandey, A., 2016. Wide pH range tolerance in extremophiles: towards understanding an important phenomenon for future biotechnology. Appl. Microbiol. Biotechnol. 100 (6), 2499–2510. https://doi.org/10.1007/s00253-016-7285-2.

Díaz, S., De Francisco, P., Olsson, S., Aguilera, Á., González-Toril, E., Martín-González, A., Toxicity, 2020. Physiological, and ultrastructural effects of arsenic and cadmium on the extremophilic microalga Chlamydomonas acidophila. Int. J. Environ. Res. Publ. Health 17 (5), 1650. https://doi.org/10.3390/ijerph17051650.

Durán, C., Marín, I., Amils, R., 1999. Specific metal sequestering acidophilic fungi. Process Metallurgy 9 (C), 521–530. https://doi.org/10.1016/S1572-4409(99)80141-2.

Ferris, F.G., Hallbeck, L., Kennedy, C.B., Pedersen, K., 2004. Geochemistry of acidic Rio Tinto headwaters and role of bacteria in solid phase metal partitioning. Chem. Geol. 212 (3–4), 291–300. https://doi.org/10.1016/j.chemgeo.2004.08.019.

Flemming, H.C., Wingender, J., 2010. The biofilm matrix. Nat. Rev. Microbiol. 8 (9), 623–633. https://doi.org/10.1038/nrmicro2415.

Fogg, G.E., Nalewajko, C., Watt, W.D., 1965. Extracellular products of phytoplankton photosynthesis. Proc. R. Soc. Lond. B Biol. Sci. 162 (989), 517–534. https://doi.org/10.1098/rspb.1965.0054.

Gadanho, M., Libkind, D., Sampaio, J.P., 2006. Yeast diversity in the extreme acidic environments of the Iberian pyrite belt. Microb. Ecol. 52 (3), 552–563. https://doi.org/10.1007/s00248-006-9027-y.

García-Meza, J.V., Barrangue, C., Admiraal, W., 2005. Biofilm formation by algae as a mechanism for surviving on mine tailings. Environ. Toxicol. Chem. 24 (3), 573–581. https://doi.org/10.1897/04-064R.1.

García-Moyano, A., González-Toril, E., Aguilera, A., Amils, R., 2007. Prokaryotic community composition and ecology of floating macroscopic filaments from an extreme acidic environment, Río Tinto (SW, Spain). Syst. Appl. Microbiol. 30 (8), 601–614. https://doi.org/10.1016/j.syapm.2007.08.002.

García-Moyano, A., González-Toril, E., Aguilera, Á., Amils, R., 2012. Comparative microbial ecology study of the sediments and the water column of the Río Tinto, an extreme acidic environment. FEMS Microbiol. Ecol. 81 (2), 303–314. https://doi.org/10.1111/j.1574-6941.2012.01346.x.

Gardea-Torresdey, J.L., Becker-Hapak, M.K., Hosea, J.M., Darnall, D.W., 1990. Effect of chemical modification of algal carboxyl groups on metal ion binding. Environ. Sci. Technol. 24 (9), 1372–1378. https://doi.org/10.1021/es00079a011.

González-Toril, E., Llobet-Brossa, E., Casamayor, E.O., Amann, R., Amils, R., 2003. Microbial ecology of an extreme acidic environment, the Tinto River. Appl. Environ. Microbiol. 69 (8), 4853–4865. https://doi.org/10.1128/aem.69.8.4853-4865.2003.

González-Toril, E., Aguilera, A., Rodriguez, N., Fernández-Remolar, D., Gómez, F., Diaz, E., García-Moyano, A., Sanz, J.L., Amils, R., 2010. Microbial ecology of Río Tinto, a natural extreme acidic environment of biohydrometallurgical interest. Hydrometallurgy 104 (3–4), 329–333. https://doi.org/10.1016/j.hydromet.2010.01.011.

González-Toril, E., Águilera, A., Souza-Egipsy, V., Pamo, E.L., España, J.S., Amils, R., 2011. Geomicrobiology of La Zarza-Perrunal acid mine effluent (Iberian Pyritic Belt, Spain). Appl. Environ. Microbiol. 77 (8), 2685–2694. https://doi.org/10.1128/AEM.02459-10Spain.

Grettenberger, C.L., Mccauley Rench, R.L., Gruen, D.S., Mills, D.B., Carney, C., Brainard, J., Hamasaki, H., Ramirez, R., Watanabe, Y., Amaral-Zettler, L.A., Ohmoto, H., Macalady, J.L., 2020. Microbial population structure in a stratified, acidic pit lake in the Iberian Pyrite Belt. Geomicrobiol. J. 37 (7), 623–634. https://doi.org/10.1080/01490451.2020.1751748.

Halter, D., Goulhen-Chollet, F., Gallien, S., Casiot, C., Hamelin, J., Gilard, F., Heintz, D., Schaeffer, C., Carapito, C., Van Dorsselaer, A., Tcherkez, G., Arsène-Ploetze, F., Bertin, P.N., 2012. In situ proteo-metabolomics reveals metabolite secretion by the acid mine drainage bio-indicator, Euglena mutabilis. ISME J. 6 (7), 1391–1402. https://doi.org/10.1038/ismej.2011.198.

Hujslová, M., Bystrianský, L., Benada, O., Gryndler, M., 2019. Fungi, a neglected component of acidophilic biofilms: do they have a potential for biotechnology? Extremophiles 23 (3), 267–275. https://doi.org/10.1007/s00792-019-01085-9.

Johnson, D.B., 2003. Chemical and microbiological characteristics of mineral spoils and drainage waters at abandoned coal and metal mines. Water Air Soil Pollut. Focus 3 (1), 47–66. https://doi.org/10.1023/A:1022107520836.

Johnson, D.B., 2009. Extremophiles: acidic environments. In: Encyclopedia of Microbiology, third ed. Elsevier, United Kingdom, pp. 107–126. https://doi.org/10.1016/B978-012373944-5.00278-9.

Johnson, D.B., 2012. Geomicrobiology of extremely acidic subsurface environments. FEMS Microbiol. Ecol. 81 (1), 2–12. https://doi.org/10.1111/j.1574-6941.2011.01293.x.

Johnson, D.B., Aguilera, A., 2015. The microbiology of extremely acidic environments. In: Manual of Environmental Microbiology, fourth ed. Bloomsbury Publishing Plc, United Kingdom, pp. 793–816. https://doi.org/10.1128/9781555818821.ch4.3.1.

Johnson, D.B., Schippers, A., 2017. Recent advances in acidophile microbiology: fundamentals and applications. Front. Microbiol. 8, 428. https://doi.org/10.3389/fmicb.2017.00428.

Johnson, D.B., Kanao, T., Hedrich, S., 2012. Redox transformations of iron at extremely low pH: fundamental and applied aspects. Front. Microbiol. 3. https://doi.org/10.3389/fmicb.2012.00096.

Kaplan, D., Christiaen, D., Arad, S.(M.), 1987. Chelating properties of extracellular polysaccharides from Chlorella spp. Appl. Environ. Microbiol. 53 (12), 2953–2956. https://doi.org/10.1128/aem.53.12.2953-2956.1987.

Koschorreck, M., Tittel, J., 2002. Benthic photosynthesis in an acidic mining lake (pH 2.6). Limnol. Oceanogr. 47 (4), 1197–1201. https://doi.org/10.4319/lo.2002.47.4.1197.

López-Archilla, A.I., Marin, I., Amils, R., 2001. Microbial community composition and ecology of an acidic aquatic environment: the Tinto River, Spain. Microb. Ecol. 41 (1), 20–35. https://doi.org/10.1007/s002480000044.

Madigan, M., Bender, K., Buckley, D., Sattley, W.M., Stahl, D., 2019. In: Brock Biology of Microorganisms. Volume 15, pp. 261–427.

Messerli, M.A., Amaral-Zettler, L.A., Zettler, E., Jung, S.K., Smith, P.J.S., Sogin, M.L., 2005. Life at acidic pH imposes an increased energetic cost for a eukaryotic acidophile. J. Exp. Biol. 208 (13), 2569–2579. https://doi.org/10.1242/jeb.01660.

Miller, R.M., Meyer, C.M., Tanner, H.A., 1963. Glycolate excretion & uptake by chlorella. Plant Physiol. 38 (2), 184–188. https://doi.org/10.1104/pp.38.2.184.

Moser, M., Weisse, T., 2011. Combined stress effect of pH and temperature narrows the niche width of flagellates in acid mining lakes. J. Plankton Res. 33 (7), 1023–1032. https://doi.org/10.1093/plankt/fbr014.

Nancucheo, I., Johnson, D.B., 2011. Significance of microbial communities and interactions in safeguarding reactive mine tailings by ecological engineering. Appl. Environ. Microbiol. 77 (23), 8201–8208. https://doi.org/10.1128/AEM.06155-11UnitedKingdom.

Niehaus, F., Bertoldo, C., Kähler, M., Antranikian, G., 1999. Extremophiles as a source of novel enzymes for industrial application. Appl. Microbiol. Biotechnol. 51 (6), 711–729. https://doi.org/10.1007/s002530051456.

Nixdorf, B., Mischke, U., Leßmann, D., 1998. Chrysophytes and chlamydomonads: pioneer colonists in extremely acidic mining lakes (pH <3) in Lusatia (Germany). Hydrobiologia 369–370, 315–327. https://doi.org/10.1023/A:1017010229136.

Oggerin, M., Tornos, F., Rodríguez, N., del Moral, C., Sánchez-Román, M., Amils, R., 2013. Specific jarosite biomineralization by Purpureocillium lilacinum, an acidophilic fungi isolated from Río Tinto. Environ. Microbiol. 15 (8), 2228–2237. https://doi.org/10.1111/1462-2920.12094.

Oliva, S.R., Mingorance, M.D., Leidi, E.O., 2012. Tolerance to high Zn in the metallophyte Erica andevalensis Cabezudo & Rivera. Ecotoxicology 21 (7), 2012–2021. https://doi.org/10.1007/s10646-012-0953-9.

Olsson, S., Puente-Sánchez, F., Gómez, M.J., Aguilera, A., 2015. Transcriptional response to copper excess and identification of genes involved in heavy metal tolerance in the extremophilic microalga Chlamydomonas acidophila. Extremophiles 19 (3), 657–672. https://doi.org/10.1007/s00792-015-0746-1.

Olsson, S., Penacho, V., Puente-Sánchez, F., Díaz, S., Gonzalez-Pastor, J.E., Aguilera, A., 2017. Horizontal gene transfer of phytochelatin synthases from bacteria to extremophilic green algae. Microb. Ecol. 73 (1), 50–60. https://doi.org/10.1007/s00248-016-0848-z.

Palacios, C., Zettler, E., Amils, R., Amaral-Zettler, L., Ahmed, N., 2008. Contrasting microbial community assembly hypotheses: a reconciling tale from the Río Tinto. PLoS One 3 (12), e3853. https://doi.org/10.1371/journal.pone.0003853.

Price, P.B., 2000. A habitat for psychrophiles in deep Antarctic ice. Proc. Natl. Acad. Sci. U.S. A. 97 (3), 1247–1251. https://doi.org/10.1073/pnas.97.3.1247.

Puente-Sánchez, F., Olsson, S., Aguilera, A., 2016. Comparative transcriptomic analysis of the response of Dunaliella acidophila (Chlorophyta) to short-term cadmium and chronic natural metal-rich water exposures. Microb. Ecol. 72 (3), 595–607. https://doi.org/10.1007/s00248-016-0824-7.

Rappaport, H.B., Oliverio, A.M., 2023. Extreme environments offer an unprecedented opportunity to understand microbial eukaryotic ecology, evolution, and genome biology. Nat. Commun. 14 (1). https://doi.org/10.1038/s41467-023-40657-4.

Roberts, 1999. The poikilotrophic micro-organism and its environment. In: Microbial Strategies of Establishment . Growth and Survival, pp. 165–173.

Rodríguez, N., Menéndez, N., Tornero, J., Amils, R., De La Fuente, V., 2005. Internal iron biomineralization in Imperata cylindrica , a perennial grass: chemical composition, speciation and plant localization. New Phytol. 165 (3), 781–789. https://doi.org/10.1111/j.1469-8137.2004.01264.x.

Rowe, O.F., Sánchez-España, J., Hallberg, K.B., Johnson, D.B., 2007. Microbial communities and geochemical dynamics in an extremely acidic, metal-rich stream at an abandoned sulfide mine (Huelva, Spain) underpinned by two functional primary production systems. Environ. Microbiol. 9 (7), 1761–1771. https://doi.org/10.1111/j.1462-2920.2007.01294.x.

Schiraldi, C., De Rosa, M., 2002. The production of biocatalysts and biomolecules from extremophiles. Trends Biotechnol. 20 (12), 515–521. https://doi.org/10.1016/s0167-7799(02)02073-5.

Schmid, A.K., Allers, T., DiRuggiero, J., 2020. SnapShot: microbial extremophiles. Cell 180 (4), 818. https://doi.org/10.1016/j.cell.2020.01.018.

Seckbach, J., Joseph, 1999. Enigmatic microorganims and life in extreme environments. In: Cellular Origin and Life in Extreme Habitats, vol. 1. Kluwer Academic Publishers.

Sekler, I., Remis, D., Gimmler, H., Pick, U., 1993. Inhibition of the plasma-membrane H+-ATPase from Dunaliella adidophila by omeprazole. Biochim. Biophys. Acta Bioenerg. 1142 (1–2), 88–92. https://doi.org/10.1016/0005-2728(93)90088-w.

Sekler, I., Weiss, M., Pick, U., 1994. Activation of the Dunaliella acidophila plasma membrane H+-ATPase by trypsin cleavage of a fragment that contains a phosphorylation site. Plant Physiol. 105 (4), 1125–1132. https://doi.org/10.1104/pp.105.4.1125.

Souza-Egipsy, V., González-Toril, E., Zettler, E., Amaral-Zettler, L., Aguilera, A., Amils, R., 2008. Prokaryotic community structure in algal photosynthetic biofilms from extreme acidic streams in Río Tinto (Huelva, Spain). Int. Microbiol. 11 (4), 251–260. https://doi.org/10.2436/20.1501.01.69.

Souza-Egipsy, V., Aguilera, A., Mateo-Martí, E., Martín-Gago, J.A., Amils, R., 2010. Fossilization of acidophilic microorganisms. Geomicrobiol. J. 27 (8), 692–706. https://doi.org/10.1080/01490450903564096.

Souza-Egipsy, V., Vega, J.F., González-Toril, E., Aguilera, Á., 2021. Biofilm mechanics in an extremely acidic environment: microbiological significance. Soft Matter 17 (13), 3672–3680. https://doi.org/10.1039/d0sm01975e.

Souza-Egipsy, V., Altamirano, M., Amils, R., Aguilera, A., 2011. Photosynthetic performance of phototrophic biofilms in extreme acidic environments. Environ. Microbiol. 13 (8), 2351–2358. https://doi.org/10.1111/j.1462-2920.2011.02506.x.

Spijkerman, E., 2007. Phosphorus acquisition by Chlamydomonas acidophila under autotrophic and osmo-mixotrophic growth conditions. J. Exp. Bot. 58 (15–16), 4195–4202. https://doi.org/10.1093/jxb/erm276.

Spijkerman, E., Barua, D., Gerloff-Elias, A., Kern, J., Gaedke, U., Heckathorn, S.A., 2007. Stress responses and metal tolerance of Chlamydomonas acidophila in metal-enriched lake water and artificial medium. Extremophiles 11 (4), 551–562. https://doi.org/10.1007/s00792-007-0067-0.

Spijkerman, E., Stojkovic, S., Beardall, J., 2014. CO2 acquisition in Chlamydomonas acidophila is influenced mainly by CO2, not phosphorus, availability. Photosynth. Res. 121 (2–3), 213–221. https://doi.org/10.1007/s11120-014-0016-6.

Sutherland, I.W., 2001. The biofilm matrix – an immobilized but dynamic microbial environment. Trends Microbiol. 9 (5), 222–227. https://doi.org/10.1016/S0966-842X(01)02012-1.

Van Etten, J., Cho, C.H., Yoon, H.S., Bhattacharya, D., 2023. Extremophilic red algae as models for understanding adaptation to hostile environments and the evolution of eukaryotic life on the early earth. Semin. Cell Dev. Biol. 134, 4–13. https://doi.org/10.1016/j.semcdb.2022.03.007.

Weisse, T., Moser, M., Scheffel, U., Stadler, P., Berendonk, T., Weithoff, G., Berger, H., 2013. Systematics and species-specific response to pH of Oxytricha acidotolerans sp. nov. and Urosomoida sp. (Ciliophora, Hypotricha) from acid mining lakes. Eur. J. Protistol. 49 (2), 255–271. https://doi.org/10.1016/j.ejop.2012.08.001.

CHAPTER

9 Unraveling the secrets of fungal life in the Río Tinto

Monike Oggerin de Orube, Catalina del Moral, Ricardo Amils
Centro de Biología Molecular Severo Ochoa, Madrid, Spain

Introduction

The Río Tinto, a natural enigma that flows through the province of Huelva in southwestern Spain (Fig. 9.1A), has long fascinated scientists, environmentalists, and lovers of natural extreme environments. Originating at Peña de Hierro in the Sierra de San Cristóbal, the river is anything but ordinary. For 100 km before reaching the Atlantic Ocean at the Huelva estuary, where it joins the Odiel River, the Río Tinto flows through a region rich in geological history and industrial heritage (Salkield, 1987). It crosses the Iberian Pyrite Belt (IPB), one of the most important mineral deposits in the world, known for its vast reserves of copper and other base metals. This geological formation, which formed during the Paleozoic era, contains extensive sulfide deposits composed primarily of pyrite, along with other sulfide minerals such as chalcopyrite, sphalerite, and galena (Tornos, 2006). These deposits are not only economically important but also chemically active, contributing to the distinctive characteristics of the river.

Perhaps the most striking feature of the Rio Tinto is the bright red color of its waters, from which it takes its name. This intense color is the result of high concentrations of dissolved iron, especially ferric iron (Fe^{3+}), produced by the chemolithotrophic subsurface activity of microorganisms, which oxidize the volcanogenic massive sulfide ores of the IPB (Amils et al., 2023; Gómez-Ortiz et al., 2014). These ores are also responsible for the river's other characteristics, such as its remarkably high and stable acidity, with a pH around 2.3. This low pH is regulated by the buffering effect of the ferric ion that despite the rainfall and droughts that are common in the area, plays a crucial role in regulating the chemistry of the river. Thus, when rain or tributaries dilute the river, ferric hydroxide is formed, releasing protons that keep the pH constant despite the contribution of external water. During dry periods, when water evaporation increases, ferric hydroxide precipitates dissolve, consuming protons. This dynamic equilibrium ensures that the river maintains an extremely constant acidic environment (Amils et al., 2002; Gónzalez-Toril et al., 2003).

Initially, the Río Tinto's harsh conditions, along with its extreme acidity, combined with high levels of dissolved heavy metals such as iron, copper, and zinc, led to the assumption that it was a hostile environment, inhospitable to most forms of life. The toxic nature of the water, with concentrations of iron ranging from 0.4 to

Geomicrobiology of Río Tinto. https://doi.org/10.1016/B978-0-443-32948-7.00012-6

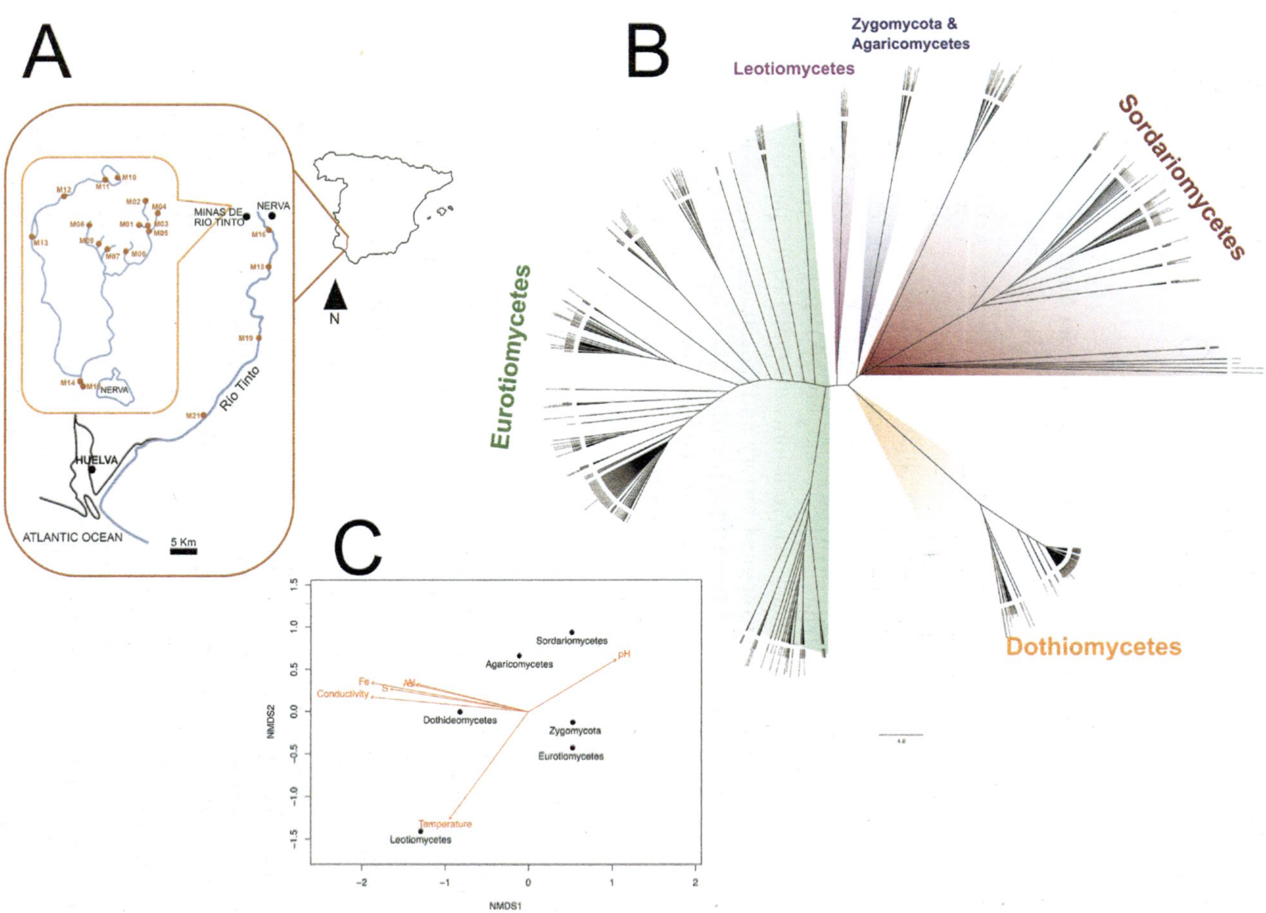

FIGURE 9.1

(A) Detailed map of Río Tinto indicating the various sampling stations along its course. (B) Phylogenetic tree constructed from the fungal isolates based on the ITS region, illustrating their clustering into distinct classes. (C) Nonmetric multidimensional scaling (NMDS) analysis, which groups the fungal isolates according to the distinct physicochemical parameters of the river, thereby highlighting the influence of environmental variables on the fungal community.

20.2 g/L and of copper and zinc reaching 0.7 g/L and 0.56 g/L, respectively (Gónzalez-Toril et al., 2003), seemed too severe for the survival of any organism. This belief was reinforced by the river's historical association with mining; for many years, it was thought that the river's extreme environment was a consequence of centuries of intense mining activity in the region. However, this perspective has changed significantly with advances in microbiology and microbial ecology. In particular, the use of molecular techniques has revealed that the Río Tinto is far from lifeless. In fact, eukaryotic biodiversity (algae and fungi) was found to be surprisingly high compared to the relatively low prokaryotic biodiversity detected (Aguilera et al., 2006; Amaral Zettler et al., 2002; Gónzalez-Toril et al., 2003; López-Archilla et al., 2001, 2004). Findings show that the fungal community belongs, in great measure, to the phylum Ascomycetes, with smaller numbers from groups such as Basidiomycetes and Zygomycetes. Within the Ascomycetes, the classes Eurotiomycetes, Dothideomycetes, and Sordariomycetes are especially well represented, exhibiting different degrees of metal tolerance, with Eurotiomycetes and Sordariomycetes revealing the highest levels of resistance, far exceeding the metal concentrations typically found in the river.

So, despite the extreme conditions of Río Tinto, its fungal community has developed strategies and mechanisms for metal tolerance and resistance as coping methods to deal with the harsh chemical environment of the river. These may include chelating agent synthesis, complexation and crystallization, decreased transport or impermeability by cell wall thickening, intracellular compartmentalization and precipitation, biosorption to cell walls, or biomineralization, among others (Brown and Hall, 1990; Gadd, 1990, 2008, 2010). The most well-documented mechanisms by which Río Tinto fungi adapt to high concentrations of toxic heavy metals include biosorption to cell walls, binding metals to the surface of their cells, which lowers the metal concentration in the surroundings (Oggerin et al., 2015), specific intracellular sequestration (Durán et al., 1999), active transport and compartmentalization, isolating toxic metals within specific cellular compartments, preventing them from interfering with essential cellular processes (López-Archilla et al., 2004; Oggerin et al., 2015), and biomineralization processes, facilitating the nucleation and growth of mineral deposits (Oggerin et al., 2013, 2016, 2017).

These processes and adaptations not only help the fungi to thrive in their environment but also contribute to the geochemical dynamics of the river, which is of enormous interest not only from an academic point of view but also from a biotechnological one, especially in areas such as bioremediation, where the ability to tolerate or even neutralize heavy metals is extremely valuable.

The diversity of fungal life in the Río Tinto is not only taxonomic but also functional, as different fungal groups contribute to the ecosystem in different ways, from decomposing organic matter to forming complex biofilms that cover large areas of the river. The diverse and resilient fungal community at Río Tinto challenges our understanding of the limits of life and demonstrates the adaptability of biological systems. Its discovery provides a unique opportunity to explore the interplay between biology and geochemistry in one of Earth's most extreme environments. It raises important questions about the evolutionary processes that have enabled these organisms to adapt to such extreme conditions and about the potential applications of these adaptations in biotechnology. For example, the ability of some Río Tinto fungi to sequester metals could be exploited to develop new methods for cleaning industrial effluents or recovering valuable metals from contaminated environments.

These findings have not only expanded our understanding of what types of life can survive in extreme environments but have also provided new insights into the potential of the adaptive mechanisms that allow these organisms to thrive where few others can.

This chapter aims to introduce the complex and fascinating world of Río Tinto fungi, exploring the unique adaptations that allow them to thrive in one of the most extreme environments on the planet, their role in the biogeochemical cycling of metals, and their potential applications in biotechnology and astrobiology. By unlocking the secrets of fungal life in Río Tinto, this chapter seeks to advance our understanding of the adaptability of life.

Fungal communities in the Río Tinto - diversity and ecology of acidophilic fungi

Despite the hostile conditions of extreme acidity, elevated concentrations of heavy metals, and oxidative stress, as mentioned above, the fungal community in the Río Tinto are not passive inhabitants; they show a remarkable level of biodiversity, displaying a series of adaptations to survive and live in these harsh conditions, contributing to the decomposition of organic matter, nutrient cycling, and biogeochemical processes. The distribution of these communities is not uniform, but varies along the river, influenced by factors such as metal concentration, pH, and sediment deposition, presenting each section of the river with unique environmental conditions that shape fungal communities, and so, the definition of the river's hydrological zones is the crucial first step.

The hydrological and geochemical context of the Río Tinto

The Río Tinto's course can be divided into three main hydrological zones, each defined by distinct geochemical profiles and environmental conditions. These influence fungal diversity in different ways: the upper course with the harshest physicochemical conditions, the middle course, representing a transitional environment, and the lower course with milder conditions. These zones create unique habitats that select for specific fungal taxa, influencing their abundance, diversity, and ecological roles.

The upper part of the river comprises three different areas, the two initial stretches and the confluence of both to originate the main body of the river (sites M01 to M09, sites M10 to M13, and sites M14 and M15, respectively) (A) with different physicochemical characteristics among them. The headwaters stretch (sites

M01 to M08) is defined by extreme concentrations of dissolved heavy metals, such as Fe exceeding 10,000 mg L^{-1}, while S, Al, Ni, and Cd reach levels of 12,000 mg L^{-1}, 30 mg L^{-1}, 1.1 mg L^{-1}, and 0.5 mg L^{-1}, respectively, with the lowest pH values (Table 9.1). The combination of acidity and oxidative stress limits microbial diversity but should support highly specialized fungi.

In the other initial stretch (sites M10 to M13) (A), although still extreme, the concentration of heavy metals is lower than in the headwaters stretch, peaking Fe up to 1400 mg $L^{-}1$ with slightly higher pH values around 2.6 (Table 9.1).

The confluence of both sections of the river (M14 and M15) (A), where small tributaries are incorporated as well, makes up this area, which presents physicochemical characteristics different from the previous ones, showing the highest Cu content of the entire river with values that can reach more than 300 mg L^{-1}, as well as Cd and Al with values of 1.7 mg L^{-1} and 1.3 mg L^{-1}, respectively, while the pH remains around 2.6 (Table 9.1).

As the river progresses into the middle stretch (sites M16 to M18) (A), the conditions begin to moderate. Heavy metal concentrations decrease slightly, while sediment accumulation increases. Organic matter carried from upstream starts to settle, creating microhabitats that can support a greater diversity of fungal species. The middle river serves as a transitional zone where species from the upper and lower sections overlap.

The lower river (sites M19 to M21) (A) experiences the least extreme conditions, with reduced metal concentrations, a wider riverbed, and a slightly higher pH. Organic matter deposition is more substantial here, supporting diverse fungal communities, including opportunistic and saprotrophic species less specialized for acidic environments. This stretch of the river highlights the interplay between ecological opportunity and adaptation.

Biodiversity and taxonomic composition

The fungal communities inhabiting the Río Tinto basin provide a fascinating example of microbial life thriving under extreme environmental conditions. Over the course of three years, 359 fungal isolates were obtained from 19 strategically chosen sampling sites (A), representing a gradient of physicochemical conditions along the river from sediments and water samples (Oggerin et al., 2024). The study revealed that 97.8% of these isolates belonged to the phylum Ascomycota, with Eurotiomycetes being the most abundant class, followed by Dothideomycetes and Sordariomycetes. Basidiomycota and Mucoromycota were present in smaller proportions, accounting for the remaining 2% of the diversity (B). Among the identified fungi, genera such as *Penicillium* (34%) and *Acidiella* (20.9%) were particularly dominant, showcasing their ability to adapt to the challenging conditions of the Río Tinto.

During these years, from 2009 to 2011, seasonality appeared to have minimal influence on fungal community composition, suggesting that the extreme conditions of the Río Tinto exert consistent selective pressures that override seasonal variation.

Table 9.1 Physicochemical parameters and heavy metal concentration across 21 Río Tinto sampling stations from 2009 to 2011.

							Heavy metal content								
		pH	Redox [mV]	Conductivity [mS cm^{-2}]	Temperature [°C]	d02 [mg L^{-1}]	S [mg L^{-1}]	Fe [mg L^{-1}]	Cu [mg L^{-1}]	Ni [mg L^{-1}]	As [mg L^{-1}]	Co [mg L^{-1}]	Cd [mg L^{-1}]	Pb [mg L^{-1}]	Al [mg L^{-1}]
Upper course	M01	2.2 ± 0.3	501.4 ± 22.2	22.7 ± 5.6	15.1 ± 6.6	6.9 ± 2.7	11,228.4 ± 7713.3	13,943.7 ± 8299.0	11.4 ± 8.6	1.2 ± 0.5	10.0 ± 6.1	17.2 ± 8.3	0.1 ± 0.0	0.0 ± 0.0	2372.6 ± 1870.2
	M02	2.6 ± 0.4	445.8 ± 44.9	6.2 ± 3.5	16.1 ± 6.8	7.5 ± 2.7	1535.5 ± 1229.6	618.7 ± 550.8	0.4 ± 0.1	0.6 ± 0.5	0.2 ± 0.3	2.3 ± 2.0	0.0 ± 0.0	0.0 ± 0.0	88.5 ± 78.8
	M03	2.4 ± 0.3	492.2 ± 19.2	13.1 ± 7.9	15.7 ± 7.4	7.4 ± 2.8	4692.0 ± 3822.9	5168.4 ± 3877.5	5.7 ± 5.4	1.0 ± 0.4	4.3 ± 3.4	7.5 ± 6.1	0.0 ± 0.1	0.0 ± 0.0	766.0 ± 641.7
	M04	4.7 ± 0.8	264.0 ± 112.7	3.9 ± 2.5	15.2 ± 5.1	8.4 ± 1.7	752.2 ± 587.2	296.4 ± 249.0	0.0 ± 0.0	0.4 ± 0.4	0.1 ± 0.2	1.8 ± 1.5	0.0 ± 0.0	0.0 ± 0.0	1.4 ± 1.0
	M05	2.6 ± 0.3	460.9 ± 25.3	9.47 ± 5.0	15.0 ± 6.1	8.6 ± 1.9	2450.7 ± 2022.5	2447.1 ± 2816.8	2.2 ± 1.4	0.6 ± 0.4	2.1 ± 2.4	4.5 ± 3.4	0.0 ± 0.0	0.0 ± 0.0	314.5 ± 226.9
	M06	2.1 ± 0.3	459.4 ± 12.2	27.6 ± 4.7	22.8 ± 1.5	3.9 ± 1.3	9606.4 ± 5947.4	12,847.8 ± 5692.9	93.5 ± 31.5	1.0 ± 0.4	18.5 ± 9.6	16.1 ± 4.6	0.4 ± 0.2	0.0 ± 0.1	2820.1 ± 1268.0
	M07	1.9 ± 0.5	441.2 ± 20.4	47.8 ± 16.1	19.0 ± 2.2	3.9 ± 1.5	9976.8 ± 5049.9	13,835.3 ± 7666.4	62.6 ± 27.3	0.4 ± 0.2	24.3 ± 14.1	13.7 ± 5.7	0.2 ± 0.1	0.1 ± 0.0	1459.5 ± 674.9
	M08	2.5 ± 0.4	492.8 ± 43.1	12.8 ± 3.2	17.8 ± 0.9	3.1 ± 1.7	3775.9 ± 1786.9	3546.3 ± 1898.6	33.4 ± 15.1	0.6 ± 0.3	5.8 ± 3.4	4.0 ± 1.5	0.2 ± 0.1	0.0 ± 0.0	448.3 ± 168.3
	M09	2.2 ± 0.5	438.6 ± 24.9	29.6 ± 9.8	19.1 ± 5.6	4.7 ± 2	8317.7 ± 6046.4	11,190.7 ± 7736.0	53.6 ± 34.7	0.7 ± 0.4	18.7 ± 12.4	11.6 ± 6.2	0.3 ± 0.2	0.1 ± 0.1	1132.8 ± 584.1
	M10	2.4 ± 0.3	472.9 ± 81.1	4.1 ± 0.8	15.2 ± 5.5	9.2 ± 1.1	881.4 ± 33.6	477.5 ± 98.2	6.3 ± 1.3	0.0 ± 0.0	0.2 ± 0.2	0.7 ± 0.1	0.1 ± 0.0	0.0 ± 0.1	68.1 ± 4.9
	M11	2.6 ± 0.4	386.0 ± 31.9	6.7 ± 1.0	15.3 ± 0.7	1.9 ± 1.8	1460.7 ± 122.3	1088.7 ± 200.7	9.3 ± 2.8	0.2 ± 0.3	0.6 ± 0.5	1.3 ± 0.2	0.3 ± 0.1	0.2 ± 0.0	168.0 ± 45.9
	M12	2.6 ± 0.3	410.0 ± 29.9	7.0 ± 1.5	189 ± 6.8	8.3 ± 2.6	1668.3 ± 390.2	1230.5 ± 327.8	22.2 ± 13.8	0.4 ± 0.4	1.0 ± 0.9	2.0 ± 0.9	0.3 ± 0.1	0.2 ± 0.1	185.2 ± 44.0
	M13	2.6 ± 0.2	465.0 ± 36.1	5.9 ± 2.3	17.1 ± 6.1	3.3 ± 2.9	1742.5 ± 758.0	1263.1 ± 1004.3	14.5 ± 6.5	0.2 ± 0.2	1.3 ± 1.2	1.3 ± 1.5	0.2 ± 0.3	0.3 ± 0.3	168.5 ± 99.0
	M14	2.6 ± 0.2	455.4 ± 47.0	9.7 ± 2.9	18.1 ± 7.8	6.9 ± 3.2	3041.9 ± 1109.0	876.4 ± 374.6	205.6 ± 113.3	1.1 ± 0.8	1.6 ± 0.6	4.1 ± 1.8	0.9 ± 0.6	0.0 ± 0.0	849.1 ± 416.9
	M15	2.5 ± 0.2	387.0 ± 22.4	16.6 ± 3.2	19.1 ± 1.9	3.7 ± 1.6	5019.6 ± 220.9	5620.5 ± 1034.8	369.6 ± 33.2	1.4 ± 0.9	17.3 ± 5.5	3.2 ± 0.6	1.5 ± 0.2	0.0 ± 0.1	760.1 ± 165.8
Middle course	M16	2.6 ± 0.4	443.0 ± 16.1	8.9 ± 3.2	19.4 ± 7.6	8.3 ± 2.9	2641.5 ± 1239.4	1221.4 ± 800.4	151.9 ± 91.8	1.0 ± 0.7	2.6 ± 1.5	3.1 ± 1.7	0.6 ± 0.3	0.1 ± 0.1	625.2 ± 322.9
	M18	2.6 ± 0.5	417.9 ± 22.1	9.0 ± 1.9	12.5 ± 4.7	9.4 ± 2.4	2778.5 ± 778.7	1548.4 ± 724.8	129.0 ± 40.4	0.9 ± 0.2	4.5 ± 2.3	3.3 ± 1.1	0.6 ± 0.2	0.3 ± 0.1	593.7 ± 315.8
Lower course	M19	2.8 ± 0.4	528.4 ± 73.4	4.1 ± 2.7	16.3 ± 6.1	9.5 ± 2.8	991.8 ± 866.6	525.9 ± 558.9	46.8 ± 42.7	0.3 ± 0.3	0.8 ± 0.7	1.2 ± 1.0	0.2 ± 0.2	0.2 ± 0.2	170.5 ± 135.8
	M21	2.7 ± 0.4	531.1 ± 69.9	2.9 ± 2.0	17.7 ± 6.5	10.0 ± 1.7	593.0 ± 529.0	253.9 ± 276.8	24.5 ± 22.9	0.2 ± 0.2	0.3 ± 0.4	0.7 ± 0.7	0.1 ± 0.1	0.2 ± 0.1	108.9 ± 88.5

Instead, factors such as pH, conductivity, and metal concentrations were the primary drivers of fungal distribution. These results are consistent with the hypothesis that environmental stressors in extreme habitats lead to a homogenization of microbial communities, favoring taxa with robust stress-response mechanisms. Nonmetric multidimensional scaling (NMDS) analysis further revealed strong correlations between fungal community structure and environmental parameters, with different fungal classes occupying specific ecological niches defined by pH and metal concentrations, as can be shown in the C.

In the headwaters stretch (sites M01 to M08) (A), the river is defined by extreme environmental stress, with Fe concentrations exceeding 10,000 mg/L, alongside significant levels of S (up to 12,000 mg/L), Al (30 mg/L), and Cd (0.5 mg/L) (Fig. 9.2A–C). The lowest pH values (<2) further exacerbate the oxidative stress and toxicity in this zone, severely limiting fungal diversity. Nevertheless, these conditions favor highly specialized taxa such as *Dothideomycetes*, including *Acidiella*, which dominate due to their melanized cell walls. Melanin is a pigment with important roles in fungal survival, as it provides protection by scavenging reactive oxygen species (ROS) and binding to toxic metal ions (Gadd and de Rome, 1988; Medina-Armijo et al., 2024). In this region, fungal communities in sediments are more prominent than in water, as sediments provide a relatively more stable environment with biofilm formation aiding in protection against extreme conditions (Oggerin et al., 2024). Thus, these sediment-associated fungi in biofilms may be active key players in biomineralization processes and sediment stabilization (Hujslová et al., 2019; Oggerin et al., 2013).

The second initial stretch (sites M10 to M13) (Fig. 9.1A), while still extreme, presents less harsh conditions. Fe concentrations drop significantly to peaks of 1400 mg/L, while pH values rise slightly to around 2.6 (Fig. 9.2A–C and Table 9.1). These conditions allow for a modest increase in fungal diversity, particularly from the *Eurotiomycetes* class, including *Penicillium* and *Talaromyces*. These genera exhibit enhanced resilience to acidic environments with moderated metal toxicity (Oggerin et al., 2024). Fig. 9.2C and Table 9.1 also reveal lower levels of Cd and Ni in this section, which aligns with the emergence of less specialized taxa. Fungal communities in this section show a balance between water and sediment habitats, with taxa in water being less specialized and more opportunistic compared to those in sediments, where biofilm formation continues to provide a critical ecological advantage (Hujslová et al., 2019; Oggerin et al., 2024).

At the confluence of the first sections of the river (sites M14 and M15) (Fig. 9.1A), the incorporation of tributaries introduces distinct physicochemical characteristics. This area has the highest concentrations of Cu (over 300 mg/L) and significant levels of As, Cd (1.7 mg/L), and Al (1.3 mg/L), while maintaining a pH of approximately 2.6 (Fig. 9.2A–C and Table 9.1). These conditions create a unique niche that supports taxa such as *Purpureocillium lilacinum*, which exhibit exceptional Cu tolerance and contribute to biofilm formation that stabilizes sediments. In this region, sediment fungal communities are particularly diverse due to

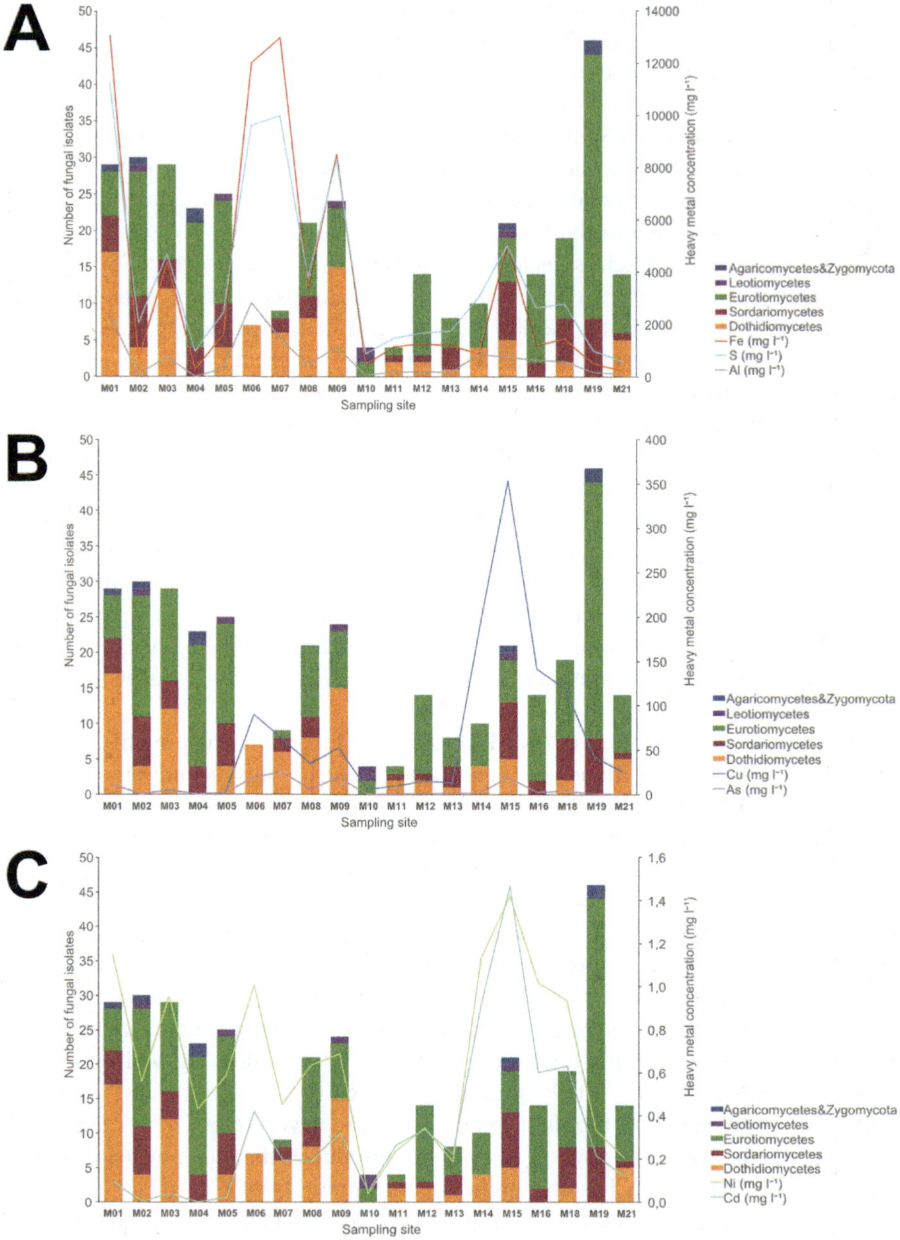

FIGURE 9.2

Relationship between heavy metal concentrations and the distribution of fungal isolates along Río Tinto. (A) Spatial variation in nickel (Ni) and cadmium (Cd) concentrations (mg L^{-1}) and their correspondence with the number of isolates from fungal classes, including the influence of metal-specific adaptations and the availability of microhabitats formed by biofilm matrices (Oggerin et al., 2024).

Regarding the yeasts, the study conducted by López-Archilla et al. (2004) identified species of *Cryptococcus* as notable contributors to the microbial diversity of the upper river, displaying remarkable metal tolerance. Its cell walls are rich in polysaccharides that bind metal ions, immobilizing them and minimizing their uptake. Together with *Cryptococcus*, *Candida fluviatilis* seemed to be part of the organic matter degradation in the upper river, playing a key role in the microbial food web.

Despite their low diversity, the fungi in the upper river fulfill essential ecological roles. They not only contribute to the biogeochemical cycling of metals by immobilizing them in biofilms but also initiate the breakdown of organic matter, providing resources for other microorganisms. These processes are crucial for maintaining the limited ecological balance in this extreme environment.

The middle stretch of the river (sites M16 to M18) (Fig. 9.1A) marks a transitional zone where environmental conditions begin to moderate. Heavy metal concentrations decrease slightly, with Fe levels reducing to below 5000 mg/L, and sediment accumulation increases, enriched with organic matter from upstream (Fig. 9.2A–C and Table 9.1). These changes support greater fungal diversity, mixing taxa from upstream with more opportunistic species from downstream. In this zone, sediment communities have a richer fungal composition than those in the water, reflecting the protective and nutrient-enriched characteristics of the sediments. *Penicillium* species dominate but coexist with saprotrophic fungi that exploit the richer organic environment. The overlap of specialized and generalist taxa makes this stretch a critical area for studying fungal community dynamics and ecological succession (Oggerin et al., 2024).

López-Archilla et al. (2004) showed that yeasts continue to play a significant role in the middle river, with *Cryptococcus humicola* and *Candida fluviatilis* being prominent. These yeasts are integral components of biofilms and sediments, where they contribute to organic matter decomposition and nutrient cycling. The presence of biofilms in the middle river enhances ecological stability by providing habitats that buffer environmental fluctuations and support interspecies interactions (Hujslová et al., 2019).

Finally, the lower stretch of the river (sites M19 to M21) (Fig. 9.1A) experiences the least extreme conditions. Metal concentrations, including Cu and Fe, are significantly reduced, and the pH rises slightly to levels approaching 3.0, creating a more stable and hospitable environment (Fig. 9.2A–C and Table 9.1). This part of the river supports a highly diverse fungal community dominated by taxa such as *Penicillium*

Agaricomycetes and Zygomycota, Leotiomycetes, Eurotiomycetes, Sordariomycetes, and Dothidiomycetes, across the sampling sites. (B) Copper (Cu) and arsenic (As) concentrations alongside the fungal isolate counts for the same classes. (C) Distribution of iron (Fe), sulfur (S), and aluminum (Al) concentrations (mg L^{-1}) and their relationship with the abundance of fungal isolates along the river.

simplicissimum that thrive in these relatively "benign" conditions. In this area, sediment communities continue to exhibit higher diversity than aquatic communities, highlighting that sediments are the primary habitat for the fungal community and serve as a reservoir for organic matter (Hujslová et al., 2019; Oggerin et al., 2024).

Yeasts such as *Rhodotorula mucilaginosa* were also prominent in this region (López-Archilla et al., 2004). This yeast is notable for its ability to oxidize Fe and tolerate oxidative stress, traits that enable it to colonize sedimented areas effectively. Other yeasts, including *Candida* and *Cryptococcus* species, persist in lower abundances, highlighting their ecological versatility and adaptability to a range of environmental conditions.

Ecological roles of acidophilic fungi in the Río Tinto

The acidophilic fungi of the Río Tinto play a central ecological role in maintaining the stability and functionality of this extreme ecosystem. Through complex interactions with the chemical environment of the river and with other microbial taxa, these fungi contribute significantly to metal cycling, organic matter decomposition, biofilm formation, and habitat stabilization. The fungal communities of the Rio Tinto serve not only as primary agents in biogeochemical processes, but also as essential stabilizers within the ecosystem, modulating extreme conditions in ways that support both fungal survival and the persistence of other microbial life forms.

Biofilm formation

The fungal biofilms in the Río Tinto represent complex, multilayered structures that support diverse microbial communities composed primarily of acidophilic fungi, bacteria, and algae. These biofilms are thought to play a critical role in fungal survival under the river's extreme conditions, providing structural integrity and functional advantages in environments characterized by low pH and high metal concentrations. Their formation is not only an adaptive response but also a key mechanism by which fungi interact with and modify their environment.

The biofilm matrix is primarily composed of extracellular polymeric substances (EPS), which create a stable structure that binds fungal cells and other microorganisms to substrates such as sediment particles and mineral surfaces or allows them to float on the water's surface. This EPS matrix, composed of polysaccharides, proteins, lipids, and nucleic acids, not only facilitates cell attachment but also provides protection against acidic conditions and heavy metal toxicity. Moreover, the EPS serves as a scaffold that supports a stratified or differentiated microbial community, where different microbial groups occupy specific microenvironments within the biofilm (Aguilera et al., 2007; Flemming and Wingender, 2010).

As described by Aguilera et al. (2007), biofilm colonization in the Río Tinto begins with the adhesion of primary fungal colonizers that establish the initial structure. These fungi secrete EPS to stabilize their attachment to mineral-rich substrates, and their filamentous morphology increases the available surface area for biofilm formation by providing a large surface area for adherence and forming a three-dimensional matrix that supports the subsequent colonization by other microorganisms.

In 2016, a floating brown biofilm (Fig. 9.3A and B) was sampled from station M11 (Fig. 9.1A) within the Río Tinto. Field emission scanning electron microscopy (FE-SEM) revealed fungal structures (Fig. 9.3C, white triangles) alongside bacterial components (Fig. 9.3C, red triangles). In addition, EPS-like structures and metal-rich mineralized formations, primarily composed of iron, were observed. A complementary experiment involved incubating a biofilm sample with fluorescently labeled wheat germ agglutinin (WGA), a lectin widely used for chitin detection in fungal cell walls (Goldstein and Hayes, 1978; Sharon and Lis, 2004; Sterflinger, 2000). The sample was counterstained with 4′,6-diamidino-2-phenylindole (DAPI) to visualize DNA, thereby allowing the visualization of microorganisms not stained by WGA, such as prokaryotes. Confocal microscopy analysis (Fig. 9.3D) confirmed the FE-SEM observations by revealing long, green-stained fungal structures with blue-stained nuclei, in addition to numerous prokaryotic cells detected exclusively by DAPI staining. Moreover, structures suggestive of biomineralization events were also evident.

NGS sequencing analysis of the 18S rRNA identified a range of fungal classes within the biofilm, including representatives of the Ascomycota, involving classes such as Eurotiomycetes, Dothideomycetes, and Sordariomycetes, among others, as fungal biofilm inhabitants, although the dominant species remained unassigned at a finer taxonomic level (Fig. 9.3E). Microscopy-based studies highlighted the critical role of EPS in biofilm organization, offering both physical cohesion and chemical buffering. EPS functions as a metal-binding medium, sequestering metal ions through chelation and adsorption, which reduces metal toxicity and may protect fungal and bacterial cells (Gadd, 2010; Oggerin et al., 2024). This metal-sequestering capability is especially critical in the Río Tinto, where metals such as iron and copper are present in exceptionally high concentrations.

In fact, the distinguishing feature of these biofilms was the big difference between their air-exposed and water-contacting sides. Spectroscopic analyses, including Fourier transform infrared spectroscopy (FTIR) and X-ray photoelectron spectroscopy (XPS), revealed geochemical differences between the two surfaces. Variations in the oxidation states of metal and nonmetal ions adsorbed to the EPS matrix as well as differences in the mineral phases present suggest that a differential distribution of microbial populations creates distinct micro-niches that may support very different bacterial metabolic processes, including oxidation–reduction reactions that may drive the formation of different mineral phases, a phenomenon facilitated by the active secretion of EPS by fungal communities (Flemming, 2011; Gadd, 2018). In this context, fungal communities facilitate redox transformations through interactions with iron-metabolizing bacteria, enabling the conversion between Fe^{3+} and Fe^{2+}, contributing to iron cycling in the Río Tinto, facilitating

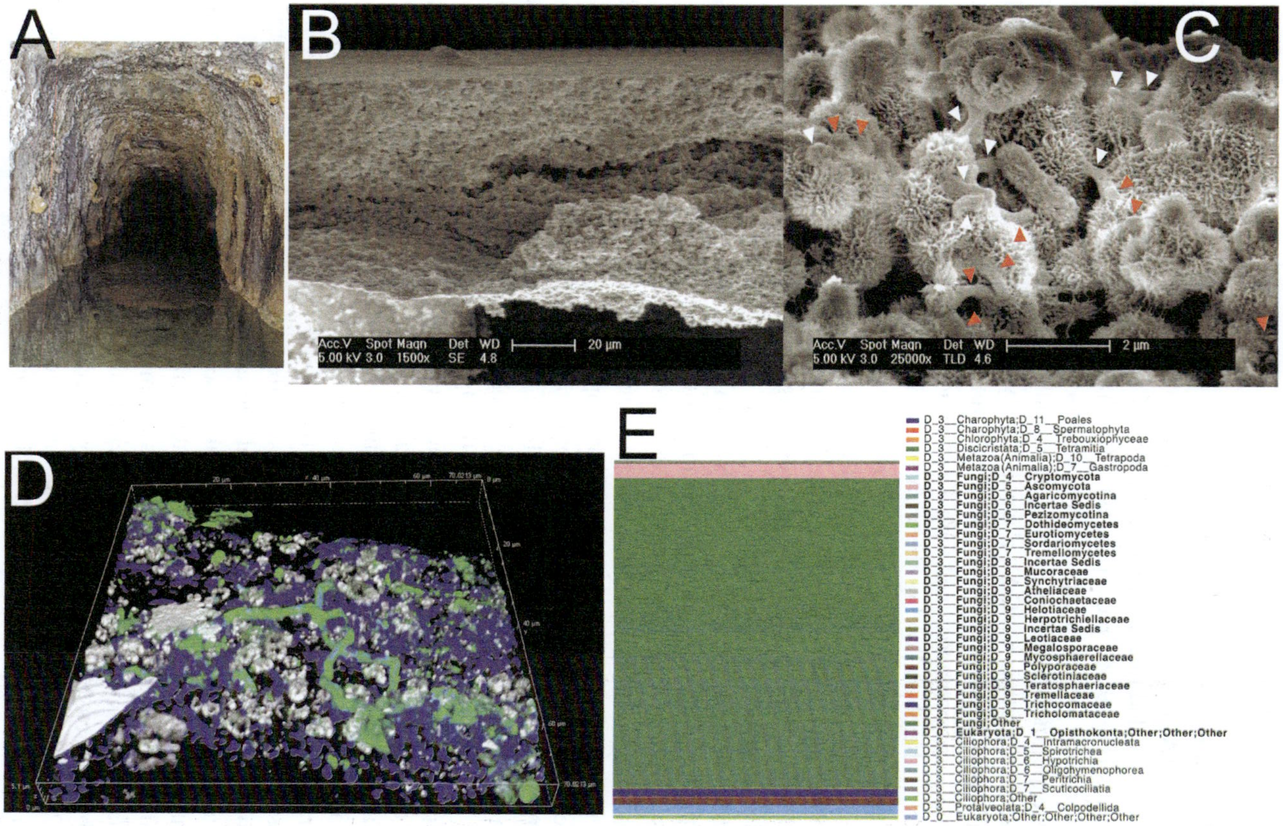

FIGURE 9.3

Characterization of fungal biofilms in Río Tinto. (A) The overall appearance of a floating brown biofilm collected at station M11. (B) FE-SEM image underlining a transversal section of the biofilm. (C) Image revealing distinct fungal structures (indicated by white triangles) combined with bacterial components (red triangles) and EPS. (D) Confocal microscopy image of the biofilm stained with wheat germ agglutinin (WGA; green) for chitin detection and DAPI (blue) for nucleic acids; gray color shows the presence of (bio)minerals. (E) Summary of the taxonomic composition of the fungal community as determined by 18S rRNA sequencing within the biofilm.

iron sequestration via their EPS matrices, and thereby allowing associated bacteria to mediate redox reactions and preventing their diffusion into deeper biofilm layers where more metal-sensitive microorganisms may reside (Branda et al., 2005; Breitenbach et al., 2022; Gadd, 2007).

In addition to their role in metal immobilization and detoxification, EPS contribute significantly to biofilm resilience by retaining water, maintaining hydration, and facilitating nutrient exchange between microbial layers. The water retention properties of EPS create microenvironments that buffer against pH fluctuations, thereby enabling fungal cells to maintain stable internal conditions despite the extreme acidity of the Río Tinto. This buffering capacity is essential for preserving the functional stability of the biofilm under variable flow conditions and seasonal changes (Flemming and Wingender, 2010).

But fungal communities in the Río Tinto ecosystem play a central role as primary decomposers, mediating the breakdown of organic matter within both sediment and biofilm layers. This decomposition process is critical for nutrient cycling, particularly in nutrient-poor, acidic environments, where the availability of carbon, nitrogen, and phosphorus is limited. The fungi inhabiting Río Tinto are adapted to thrive under extreme physicochemical conditions. Despite these harsh conditions, these organisms have evolved highly efficient enzymatic mechanisms to degrade complex biopolymers such as lignin, cellulose, and hemicellulose, thereby sustaining microbial community structure and ecosystem productivity (Gadd, 2017, 2018).

Recent studies have demonstrated that fungal isolates from Río Tinto exhibit remarkable lignocellulolytic activity under acidic conditions (Daddaoua et al., 2023). Of the 182 fungal strains analyzed, 84% belonged to the phylum Ascomycota. A significant proportion of these strains exhibited enzymatic activities relevant to plant biomass deconstruction: 44% showed endoglucanase (EG) activity above 4350 U/g protein, 56% showed cellobiohydrolase (CBH) activity above 90 U/g, and 19% were positive for β-glucosidase (BGL) activity exceeding 1940 U/g. Although β-xylosidase (BXL) and xyloglucanase (XYGL) activities were less prevalent, 6% and 16% of the strains, respectively, exhibited notable activity above defined thresholds (>1620 U/g for BXL; >10,770 U/g for XYGL). Notably, individual strains exhibited extraordinarily high enzymatic outputs. *Penicillium janthinellum* (strain M24005) produced 32,893 U/g of EG and 42,247 U/g of XYGL, while an unidentified strain (M02013) reached 33,469 U/g and 52,498 U/g for the same enzymes. These activities were measured under standard acidic assay conditions (pH 4.5, 50 °C), and several strains maintained over 80% of their enzymatic activity even at pH 3.0, confirming their acidophilic profile.

Meanwhile, shotgun proteomic analyses of selected high-performing strains revealed that between 50% and 75% of their secreted proteins corresponded to carbohydrate-active hydrolases (CAZymes), including enzymes from glycoside hydrolase families GH3, GH7, GH16, and GH11. These percentages are comparable to those observed in industrial reference strains such as *Myceliophthora thermophila* and *Trichoderma reesei*, widely used in second-generation (2G) bioethanol production (Daddaoua et al., 2023).

So, the decomposition process releases essential elements such as carbon, nitrogen, and phosphorus into the river, which are then assimilated by biofilm-dwelling bacteria and other fungi, reinforcing a self-sustaining nutrient cycle, increasing the bioavailability of substrates for bacterial communities, which further contribute to nutrient cycling via processes such as nitrification and denitrification (Falkowski et al., 2008; Gadd, 2017).

Heavy metal tolerance, sequestration, and detoxification

Microorganisms that inhabit extreme environments such as Río Tinto must develop their own specialized metabolic strategies to cope with both highly acidic conditions and elevated concentrations of dissolved heavy metals they encounter. Fungi are particularly skilled at surviving in environments where multiple stress factors occur simultaneously (Gadd, 2007; Sterflinger, 2000). One of the key characteristics of fungi is their ability to tolerate a broad pH range, being able to tolerate and thrive from very low pH to acidic and moderate to even alkaline conditions. This remarkable adaptability is largely due to their capacity to maintain intracellular pH homeostasis, which involves actively expelling protons to regulate cytoplasmic acidity (Gross and Robbins, 2000; Magan, 2007; Oggerin et al., 2024).

In addition to the extremely acidic pH, exposure to heavy metals constitutes another challenge for the fungi at Río Tinto. Certain metals can play a dual role, acting as cofactors in enzymatic processes at low concentrations, but becoming toxic when present in excess. Heavy metals such as copper, zinc, cobalt, and magnesium are essential for fungal metabolism, while others, including cadmium and lead, have no biological function and represent significant toxicity risks when accumulated. High concentrations of all these metals can severely affect fungal physiology by inhibiting growth, sporulation, and mycelial development. In addition, they can affect metabolic functions by reducing enzymatic activity, disrupting membrane stability, displacing essential metal ions, and interfering with antioxidant defense mechanisms, among other effects (Das et al., 2008; Gadd, 1993, 2010; Mehra and Winge, 1991).

Fungal survival in metal-rich environments is achieved through a sophisticated network of defense mechanisms operating at both the cell surface and within the cytoplasm at cellular and molecular levels, including reduced metal transport, intracellular chelation, vacuolar sequestration, extracellular immobilization, metal ion reduction, and metal adsorption onto the fungal cell wall, which function synergistically rather than independently to reduce metal-induced stress (Das et al., 2008; Gadd, 1993; Mehra and Winge, 1991; Sterflinger, 2000).

Primary defenses refer to the mechanisms that operate outside the fungal cell, such as extracellular immobilization and adsorption on the fungal cell wall. These processes are critical for fungal metal tolerance. The secretion of EPS and other organic metabolites promotes interactions with metal ions near the cell, facilitating their complexation or precipitation and thereby lowering local metal concentrations (Gadd, 1993, 2010; Gadd et al., 2012). In many instances, these processes lead to biomineralization on or near the cell surface (Gadd, 2018, 2021; Oggerin et al., 2013, 2016).

The fungal cell wall itself acts as a passive barrier against metal toxicity through adsorption. Composed mainly of polysaccharides such as chitin, β-glucans, and glycoproteins, the cell wall contains negatively charged functional groups, carboxyl, hydroxyl, and phosphate, that serve as binding sites for metal ions. This binding captures and immobilizes the metals at the cell wall, limiting their entry into the cytoplasm and reducing their toxic effects (Gadd, 2004; Sterflinger, 2000).

Another key defense mechanism is the regulation of metal uptake. Fungal cells reduce metal transport by downregulating specific transporters or modifying membrane permeability to limit the accumulation of metals inside the cell. Studies have shown that fungi adjust the expression of high-affinity metal transporters to reduce the influx of toxic metals such as Cu^{2+}, Cr^{3+}, and Cd^{2+} (Gadd, 1993, 2010). Furthermore, changes in membrane composition can decrease passive metal diffusion. Metal-resistant strains often exhibit reduced permeability to divalent metal cations, preventing the buildup of toxic concentrations in the cytoplasm (Baldrian, 2003; Gadd, 1993). Active efflux systems, including P-type ATPases and cation–proton antiporters, further assist by exporting excess metals from the cytoplasm (Fomina and Gadd, 2007; Gadd, 2010).

Once metal ions enter the cell, intracellular chelation becomes a crucial defense. Fungi synthesize various chelating molecules, metallothioneins, metal-binding proteins, and metal-binding peptides that form stable complexes with metal ions. This chelation neutralizes the metals' reactivity, preventing them from generating ROS or disrupting critical cellular processes, while also facilitating their safe storage (Gadd, 1990a,b, 1993, 2010; Gadd et al., 2012). In addition, vacuolar sequestration plays a complementary role. Metals are transported into the vacuole, where they are often chelated by polyphosphates, ensuring that they remain in a nontoxic form (Fomina and Gadd, 2007; Gadd, 1993, 2010).

Another important detoxification mechanism is the enzymatic reduction of metal ions. Fungi can catalyze redox reactions to convert metals from more toxic, soluble forms to less reactive, less soluble states. For instance, reducing hexavalent chromium (Cr^{6+}) to trivalent chromium (Cr^{3+}) significantly decreases its toxicity and enhances its subsequent chelation and sequestration. These redox transformations are mediated by specific fungal enzymes that use electron donors, such as glutathione, thereby reducing the oxidative stress associated with metal exposure (Gadd, 1993, 2010; Gadd et al., 2012; Lalueza et al., 2014).

Collectively, these mechanisms form an integrated defense network that enables fungi to survive and thrive in environments containing heavy metals at concentrations that would otherwise be lethal.

The integrative defense mechanisms that allow fungi to mitigate metal toxicity are exemplified by the fungal strains found at Rio Tinto. The detailed studies of chromium sequestration and removal, as well as copper tolerance and removal, demonstrate how these organisms have developed robust defense strategies. This adaptive capacity not only permits their survival but also enables them to thrive in environments with extremely high concentrations of heavy metals.

Adaptations for copper resistance in Purpureocillium lilacinum M15001 *isolate from Río Tinto*

The extreme environmental conditions of the Río Tinto ecosystem have driven the evolution of metal-tolerant microorganisms. Among the fungal species isolated from this environment, *Purpureocillium lilacinum* M15001 has emerged as an outstanding model for the study of copper tolerance. Isolated from the stations M15 with the highest Cu concentration, it exhibited an extraordinary ability to survive and proliferate in the presence of up to 1 M Cu. This strain not only withstands extreme metal exposure but also actively contributes to copper removal from its environment, with efficiency rates of 25% at 50 mM Cu, 32% at 100 mM Cu, and up to 51% at 500 mM Cu (Oggerin et al., 2015). The mechanisms underlying this detoxification strategy involve a complex interplay between extracellular immobilization, intracellular sequestration, and extensive proteomic adaptations that collectively mitigate copper toxicity.

A fundamental defense mechanism employed by *P. lilacinum* M15001 is the extracellular immobilization of copper ions, which limits their entry into the cytoplasm and reduces their immediate toxic effects. The fungal cell wall, as indicated above, is an intricate matrix primarily composed of chitin and β-glucans, serving as the first barrier against copper stress. These structural components are interwoven with proteins and display a high density of negatively charged functional groups, which exhibit a strong affinity for Cu^{2+} (Sterflinger, 2000). The interaction between copper and these charged groups results in biosorption, effectively preventing free copper ions from diffusing into the cytosol.

Quantitative analyses revealed that at lower copper concentrations (50 mM Cu), the majority (72%) of the removed metal remains adsorbed onto the cell wall. However, as the copper concentration increases, the fraction of copper retained extracellularly decreases, comprising only 36% at 100 mM Cu and further dropping to 9.2% at 500 mM Cu (Oggerin et al., 2015). This progressive decline suggests that at higher copper concentrations, the biosorption capacity of the cell wall becomes saturated, reducing its efficiency as a primary detoxification mechanism because pH decreases as Cu concentration increases. The saturation of binding sites likely results from competition between Cu^{2+} cations and protons for available functional groups, a phenomenon observed in various metal-biosorbing microorganisms (Gadd, 1993). Consequently, *P. lilacinum* M15001 must rely on additional intracellular strategies to maintain metal homeostasis under extreme conditions.

Once copper ions surpass the extracellular barrier and enter the cytoplasm, *P. lilacinum* M15001 employs a highly efficient intracellular sequestration system to neutralize their toxic effects through active transport into vacuoles, which serve as intracellular reservoirs for metal sequestration (Oggerin et al., 2015). Transmission electron microscopy (TEM) revealed the presence of electron-dense granules within vacuoles, and energy-dispersive X-ray (EDX) analysis confirmed that these deposits consist predominantly of copper (Fig. 9.4A).

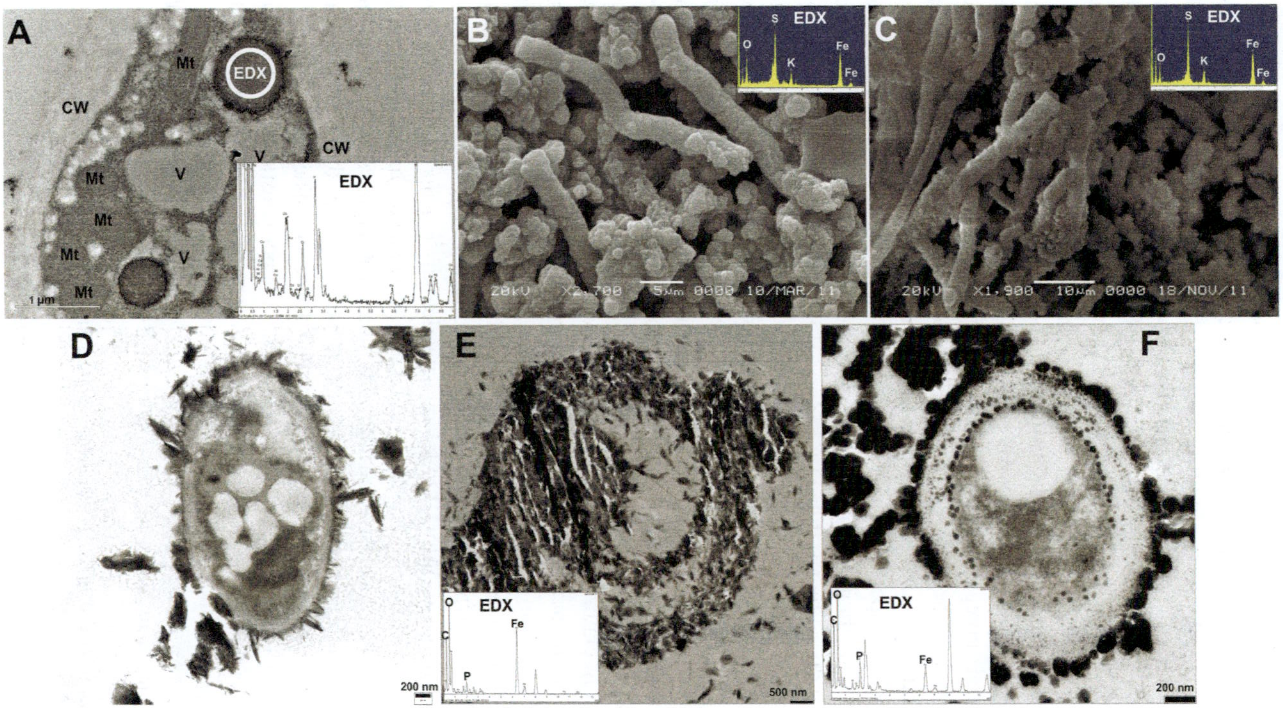

FIGURE 9.4

Characterization of metal sequestration and biomineralization processes in Río Tinto fungi. (A) Transmission electron microscopy (TEM) image of a fungal cell displaying internal Cu compartmentation as electron-dense deposits analyzed by EDX (white panel). Cell wall (CW), mitochondrion (Mt), vacuoles (V) (scale bar: 1 μm). (B) Jarosite biomineralized *P. lilacinum* alive cells with EDX spectrum of the jarosite analysis (scale bar: 5 μm). (C) Jarosite biomineralized *P. lilacinum* dead cells with EDX spectrum of the jarosite analysis (scale bar: 10 μm). (D) TEM image of jarosite biomineralization by *P. lilacinum* showing needle-like mineral structures growing on the fungal cell wall (scale bar: 200 nm). (E) TEM image of a biomineralized fungal cell inhabiting Rio Tinto M13 sediments; EDX analysis (white panel) suggested the formation of iron oxyhydroxides (scale: 500 nm). F) TEM image of a fungal cell inhabitant of M01 sediments; rounded electrodense structures represent biominerals located inside and outside the cell. EDX analysis (white panel) suggested the formation of iron phosphates (scale: 200 nm).

The efficiency of intracellular sequestration was further highlighted by the progressive increase in copper accumulation within fungal biomass as external copper concentrations rose. At 50 mM Cu, intracellular copper represented 8.2% of the total fungal dry weight, increasing to 13% at 100 mM Cu and reaching 22.4% at 500 mM Cu (Oggerin et al., 2015). These findings suggested that as external copper levels raised, the fungus shifted from extracellular biosorption to intracellular sequestration as the primary detoxification strategy. This shift underscored the adaptability of *P. lilacinum* M15001, enabling it to tolerate extreme metal concentrations without compromising cellular integrity.

In-depth proteomic analysis was performed to elucidate additional potential mechanisms used by *P. lilacinum* M15001 that were not visible as vacuole sequestration, identifying a substantial upregulation of metal-binding proteins under high copper conditions, emphasizing their role in intracellular detoxification. In parallel to an increased expression of antioxidant enzymes such as catalase, which counteract ROS generation induced by copper exposure. Additionally, the upregulation of several proteins related to the glutathione pathway suggested an active role in the prevention of oxidative damage induced by Cu (Oggerin et al., 2015; Pócsi et al., 2004).

Notably, proteomic data indicate a substantial remodeling of the fungal cell wall under copper stress. Most of the enzymes involved in chitin synthesis were overexpressed (Oggerin et al., 2015), resulting in an increased deposition of structural polysaccharides, possibly linked to the observed thickening of the fungal cell wall. This structural fortification enhances the biosorptive capacity of the cell wall, further limiting copper influx into the cytoplasm. Moreover, proteins involved in vesicular trafficking, including clathrin and SNARE proteins, were upregulated as well, facilitating the formation of vacuolar compartments enriched in copper deposits but also the cell wall synthesis (de León et al., 2013; Kirchhausen, 2000). This suggests a highly coordinated cellular response in which membrane dynamics are modulated to optimize metal sequestration and minimize cytosolic toxicity.

The ability of *P. lilacinum* M15001 to tolerate and detoxify copper through extracellular and proteomic adaptations illustrates a highly specialized fungal response to extreme metal stress. The progressive transition from biosorption to vacuolar sequestration as copper concentrations increase reflects an adaptive strategy that ensures metal detoxification without impairing essential metabolic functions. The upregulation of stress-response proteins further supports the notion that copper resistance in this strain is not a passive phenomenon but rather an active, regulated process involving multiple cellular pathways. These findings not only expand our understanding of fungal metal tolerance but also underscore the potential of *P. lilacinum* M15001 as a model system for bioremediation applications in metal-contaminated environments.

Chromium and arsenic metal tolerance and sequestration mechanisms in Río Tinto fungal strains

Chromium (Cr) and arsenic (As) are among the most persistent and toxic metal pollutants released through industrial activities. Trivalent chromium (Cr^{3+}), widely used in tanning, electroplating, and pigment production, is a common contaminant in wastewater from leather processing facilities. Due to its persistence in sediments and ability to bioaccumulate, Cr^{3+} poses significant risks to both ecosystems and human health. Similarly, pentavalent arsenic (As^{5+}), primarily derived from mining and metallurgical processes, is highly toxic and associated with carcinogenic, neurological, and developmental effects. The need for effective, low-cost remediation technologies is particularly urgent in acidic and metal-rich environments where conventional treatments often fail (Durán et al., 1999, 2001).

Acidophilic fungi, naturally adapted to extreme and high-metal-concentration environments, have emerged as promising candidates for bioremediation. These organisms exhibit intrinsic tolerance to metal stress and the ability to bioaccumulate or immobilize heavy metals through active mechanisms. A systematic screening of fungal isolates from acidic, metal-rich habitats identified several strains with notable resistance and sequestration capacities for both Cr^{3+} and As^{5+}.

Among these, *Penicillium* sp. V80 exhibited exceptional chromium tolerance, maintaining growth at Cr^{3+} concentrations up to 200 mM. In metal sequestration assays, V80 removed up to 75% of Cr^{3+} at 100 mM, indicating efficient metal uptake under extreme conditions. Transmission electron microscopy (TEM) of Cr-exposed V80 cells revealed electron-dense intracellular inclusions, which EDX spectroscopy confirmed to be rich in Cr. These results suggest that chromium sequestration involves energy-dependent uptake followed by intracellular compartmentalization or precipitation, rather than passive cell surface adsorption (Durán et al., 2001).

Other Cr^{3+}-tolerant fungi included *Aspergillus* sp. P51 (tolerant up to 100 mM) and *Cladosporium* sp. I18, which demonstrated growth at 400 mM Cr^{3+}. However, despite their resistance, these strains showed modest or unquantified Cr sequestration. Conversely, *Hormonema* sp. I12 achieved 35% Cr^{3+} removal at 200 mM despite lower tolerance, highlighting the importance of evaluating metal resistance and sequestration as distinct parameters (Durán et al., 1999).

In the case of arsenic (Durán et al., 1999), several fungal isolates displayed remarkable tolerance and removal capacity. Screening of 60 taxonomically diverse strains revealed growth at As^{5+} concentrations up to 400 mM. Noteworthy strains included *Bahusakala* sp. O66, *Cladosporium* sp. P72, *Hormonema* sp. I12, *Nodulisporium* spp. V56 and V58, *Penicillium* sp. P54, and *Scytalidium* spp. O64 and O74, all of which grew robustly under high arsenate stress. In sequestration assays, *Penicillium* sp. I25 and *Scytalidium* sp. P65 removed 72% and 90% of As^{5+}, respectively, at 10 mM, with removal closely linked to active cell growth. Comparative experiments using growing, nongrowing, and disrupted cells confirmed that arsenic sequestration is largely metabolism-dependent. Passive biosorption to the cell wall, a common mechanism in many fungi, was negligible. These findings imply the presence of active transport systems and intracellular sequestration mechanisms, possibly involving vesicle trafficking or chelation (Durán et al., 1999).

The observed correlation between high arsenic tolerance and effective sequestration suggests the presence of specialized, possibly inducible resistance pathways. Collectively, these results highlight the biotechnological potential of acidophilic fungi for the remediation of arsenic- and chromium-contaminated waters and effluents, especially in challenging environments where conventional methods are ineffective or economically unsustainable.

Biomineralization induced by acidophilic fungi

Overview of biomineralization processes

Fungal metal interaction extends beyond simple sequestration, playing a pivotal role in geochemical cycles through their ability to mediate both the formation and dissolution of minerals. These processes frequently involve mechanisms of metal mobilization (Gadd, 2010).

The biologically mediated formation of minerals is referred to as *biomineralization*. In both prokaryotic and eukaryotic systems, two distinct models of biomineralization have been characterized: biologically controlled mineralization (BCM) and biologically induced mineralization (BIM). The primary distinction between these models lies in the degree of biological control exerted over mineral nucleation, growth, composition, morphology, and localization. In the case of BCM, the organism tightly regulates these parameters, enabling the intracellular or extracellular precipitation of minerals in microenvironments that are isolated from ambient conditions. These biominerals often serve specific physiological or structural functions (Gadd and Raven, 2010; Konhauser, 1998; Lowenstam, 1981). Due to its dependence on tightly regulated metabolic pathways, BCM can only be carried out by viable, metabolically active cells.

Conversely, BIM results from modifications to the local microenvironment caused by microbial activity or by the mere presence of cell surface structures such as cell walls, EPS, or spores. These structures commonly exhibit charged functional groups capable of interacting with metal ions or other solutes in the surrounding milieu. Such interactions lead to the passive adsorption of ions and subsequently promote mineral nucleation and growth, without the organism exerting direct biochemical control over the process (Frankel and Bazylinski, 2003; Konhauser, 1998).

Within BIM, two mechanistic variants are recognized: active BIM and passive BIM (Frankel and Bazylinski, 2003). In active BIM, mineral precipitation occurs as a direct consequence of the organism's metabolic activity. In contrast, passive BIM involves nonspecific electrostatic interactions between negatively charged groups on the fungal cell wall or EPS and environmental cations, leading to mineral nucleation and growth in the absence of metabolic input (Frankel and Bazylinski, 2003; Sterflinger, 2000). The mineralogy of the resulting precipitate is thus influenced by physicochemical parameters such as growth conditions, the nature and surface charge of the microbial structures, the microenvironmental composition, and the availability of reactive anions. In passive BIM, due to its metabolic independence, cell viability is not a prerequisite. Nonviable cells may still act as nucleation sites through the retention of metal-binding functional groups (Beveridge and Murray, 1976; Tobin et al., 1984).

Beveridge and Murray (1976) proposed a two-step model to describe cell wall-associated biomineralization. In this model, the initial step involves the sequestration of metal ions from solution via electrostatic interactions with negatively charged moieties on the cell wall. This is followed by the formation of mineral precipitates, driven by the reduction of thermodynamic nucleation barriers. The initial ion adsorption facilitates the generation of additional binding sites, allowing for progressive mineral accretion.

Case studies: jarosite formation by P. lilacinum; phosphate and iron oxyhydroxide biomineralization in Río Tinto sediments

One of the most relevant biomineralization processes in Río Tinto is the formation of jarosite by *P. lilacinum*, a mineral that typically forms in acidic, iron-rich environments. Its significance extends beyond Earth, as it has been detected on Mars, making Río Tinto a valuable analog for studying potential extraterrestrial life and biomineralization processes (Fernández-Remolar et al., 2005; Knoll et al., 2005).

The occurrence of biomineralization was experimentally confirmed in fungal isolates obtained from the Río Tinto ecosystem. Notably, *P. lilacinum*, isolated from the river's source, exhibited the ability to specifically biomineralize jarosite (Oggerin et al., 2013). Jarosite is a hydrated potassium iron sulfate with the general formula $KFe_3(SO_4)_2(OH)_6$, in which the K^+ cation can be substituted by variable amounts of Na^+, Ag^+, Pb^{2+}, or H_3O^+. This mineral forms exclusively under acidic conditions, typically arising from the oxidative weathering of pyrite-bearing rocks or the interaction of acidic fluids with iron-rich lithologies (Bladh, 1982).

The study encompassed several fungal strains isolated from Río Tinto, representing the genera *Penicillium*, *Bispora*, and *Purpureocillium*. The growth media were prepared using Río Tinto water, yet the physicochemical parameters were adjusted to conditions that were thermodynamically unfavorable for jarosite precipitation.

Despite these constraints, a mineral precipitate was observed in *P. lilacinum* cultures, regardless of cellular viability, including the type strain $DSM846^T$ isolated from a distinct extreme environment. The comparison between X-ray diffraction (XRD) standard reference pattern for jarosite and the diffraction pattern of the biomineral formed by the Río Tinto isolate demonstrated that *P. lilacinum* was capable of inducing the selective formation of well-crystallized jarosite as the sole mineral phase, an ability not observed in the other tested fungal strains (Oggerin et al., 2013).

Scanning electron microscopy (SEM) analysis revealed that biomineralization occurred at the fungal cell wall in both live (Fig. 9.4B) and dead (Fig. 9.4C) cell cultures. The process involved the formation of globular mineral structures on the cell surface, eventually leading to complete encrustation. In live-cell cultures, lamellar crystals appeared to interact, forming rosette-like aggregates (Fig. 9.4B). EPS were shown to play a role in the biomineralization process by providing nucleation sites for metal ion complexation and subsequent mineral formation. The capacity of EPS to chelate metal ions has been well-documented (Comte et al., 2008), and their presence was also associated with the formation of larger mineral aggregates. EDX analysis (Fig. 9.4B) confirmed that the mineral phase coating the cell wall, as well as the globular and aggregate structures associated with EPS, were composed predominantly of O, S, Fe, and K, with stoichiometric ratios consistent with jarosite. Similar elemental compositions were detected in cultures of dead cells (Fig. 9.4C), where mineralized cell envelopes devoid of cytoplasmic content were observed. Notably, EPS were not detected in the dead-cell cultures, nor was the formation of mineral aggregates.

TEM observation of the samples (Fig. 9.4D) further corroborated these findings, showing that jarosite nucleation initiated at the cell wall through the formation of acicular (needle-like) crystals, likely corresponding to the lamellar structures observed under SEM, growing until the entire cell was encrusted. Importantly, EDX analysis of live-cell cross-sections revealed negligible intracellular concentrations of K, Fe, and S, indicating that jarosite formation was restricted to the extracellular space (Oggerin et al., 2013).

The occurrence of jarosite precipitation in both live and dead *P. lilacinum* cells clearly indicated that the biomineralization process was independent of metabolic activity. Given that the culture media were thermodynamically unfavorable for jarosite precipitation under bulk ionic conditions, a mechanistic model was proposed in which Fe^{3+}/Fe^{2+} microdomains form at the fungal cell wall. These microenvironments may facilitate local ionic supersaturation, thereby reducing the energy barrier for nucleation. This process likely involves electrostatic interactions mediated by negatively charged functional groups on the cell wall, which modulate the speciation and redox balance of dissolved iron. Such mechanisms were consistent with those proposed in other biomineralization systems (Konhauser, 1998). Collectively, these findings support the classification of jarosite formation by the acidophilic fungi *P. lilacinum* as a case of passive BIM.

Another particularly illustrative case was observed at sampling station M13, a small dam within the Río Tinto system, where the dam wall was covered by dense mineral deposits, mainly hydronium-jarosite and quartz (Oggerin et al., 2016). The geochemical modeling confirmed that the river water, at this site, was supersaturated with respect to iron oxyhydroxides such as goethite and lepidocrocite, among other mineral phases, although they were not detected by XRD, most likely due to kinetic constraints under acidic conditions, and it is known that supersaturation is not enough to overcome the kinetic barriers (Bethke, 2022).

A detailed study (Oggerin et al., 2016) performed with TEM and high-angle annular dark field scanning transmission electron microscopy (HAADF-STEM) from the dam wall sediments revealed biomineralization processes involving fungal structures (Fig. 9.4E), where the mineral nucleation initiated on the exterior surface of fungal cell walls, gradually growing and aggregating, and eventually forming fibrous structures that encrust the entire cell. The EDX analyses showed that these precipitates were composed primarily of Fe, O, P, and, to a lesser extent, Si. Interestingly, although the bulk sediment was dominated by hydronium jarosite, as indicated by XRD, EDX spectra from the fungal-associated minerals did not show significant sulfur signals, suggesting that these structures are not iron sulfates.

Instead, their composition and morphology point toward amorphous or poorly crystalline iron oxyhydroxides (Oggerin et al., 2016).

Further elemental mapping revealed localized enrichment in Fe, O, and P within the mineralized fungal matrices, suggesting the precipitation of an iron phosphate such as vivianite ($(Fe^{2+})_3(PO_4)_2{\cdot}8H_2O$). However, the Fe:P ratio obtained did not correspond to known stoichiometries of any iron phosphate mineral, indicating that phosphorus may instead be associated with the fungal cell wall or occur in trace amounts within the precipitated matrix. The fibrous texture of the mineral crusts was most likely consistent with lepidocrocite morphology, a mineral phase of the goethite group that, although supersaturated, is typically absent in the abiotic mineral record of Río Tinto due to its slow nucleation kinetics. Supersaturation does not indicate a mandatory precipitation because it depends on other factors such as slow precipitation kinetics and pH dependence (Schwertmann and Cornell, 2000a, 2000b). In fact, goethite minerals precipitate at pH close to neutrality (Bigham and Nordstrom, 2000), and they are not usually direct precipitates in acid mine drainages (AMD) (Fernández-Remolar et al., 2005) due to the mentioned factors above. Therefore, the presence of fungal structures in these sediments was proposed to play a key role in facilitating the nucleation and growth of these otherwise kinetically inhibited minerals (Oggerin et al., 2016).

These findings support a model of passive BIM, where fungal cell walls provide reactive surfaces capable of concentrating metal ions from solution through nonspecific electrostatic interactions. This surface-mediated mechanism reduces the thermodynamic barriers required for mineral nucleation and may operate independently of cell viability, as previously proposed for other microbial systems. In this context, fungal cells (alive or dead), along with their extracellular polymers or spores, act as catalysts of mineral precipitation by generating localized microenvironments conducive to supersaturation and crystal growth (Oggerin et al., 2016).

A similar pattern was observed with sediments from the M19 sampling station, where no crystalline minerals were detected by XRD, likely due to poor crystallinity or amorphous structures (Oggerin et al., 2017). Nevertheless, TEM images showed fungal cells coated in fibrous mineral aggregates. EDX confirmed Fe and O as dominant elements, with minor P and Si, and no sulfur, again suggesting the presence of iron oxyhydroxides. This further supports a BIM mechanism, whereby mineralization is driven by surface interactions rather than metabolic input.

But at station M01 (Oggerin et al., 2017), the most chemically extreme site (pH 2.0, Fe^{2+} ~2169 mg/L), TEM analysis of sediments revealed mineralized fungal cells containing rounded mineral particles both intracellularly and attached to the outer cell wall (Fig. 9.4F). EDX analyses showed Fe, O, and P in proportions compatible with iron phosphate phases. Importantly, phosphorus was absent from the surrounding water, indicating an internal biological origin. The observed Fe/P ratios (1.4–1.8) align with iron phosphate ratios (Cosmidis et al., 2014) and most likely with strengite ($Fe^{3+}(PO_4){\cdot}2(H_2O)$), an Fe^{3+} phosphate, already identified in biomineralization processes (Konhauser, 1998), supporting the classification of this process as BCM, dependent on living, metabolically active fungal cells.

Together, these findings demonstrate that fungi are distributed throughout the Río Tinto system and are actively involved in mineral formation. They are capable of mediating both highly regulated and passive mineralization pathways, depending on species, environmental conditions, and physiological state. Their ability to overcome kinetic and thermodynamic barriers to mineral precipitation, barriers that prevent abiotic mineral formation even under supersaturated conditions, highlights their ecological and geochemical significance. The fungal community in Río Tinto should be considered a key geological agent, not only for its contribution to biogenic mineral formation, but also for its potential involvement in the weathering of existing minerals and the broader geochemical cycling of iron, phosphorus, and other elements. The cooccurrence of both BCM and BIM processes in the same ecosystem underscores the functional versatility of fungi in shaping extreme environments and opens new perspectives in the fields of geomicrobiology, bioremediation, and astrobiology.

Environmental and biotechnological applications and future perspectives

Exploiting acidophilic Río Tinto fungi for the biotransformation of lignocellulose into high-value bioproducts

The conversion of lignocellulosic biomass into renewable biofuels and bioplastics is a central component of the circular bioeconomy (Mujtaba et al., 2023). However, this process remains technically challenging due to the intrinsic recalcitrance of lignin and the crystalline nature of cellulose, which hinder enzymatic access and saccharification efficiency (Daddaoua et al., 2023). Overcoming these structural barriers requires highly specialized enzymatic systems capable of deconstructing complex polysaccharides under industrial conditions.

Fungi, as primary decomposers in natural ecosystems, possess enzymatic toolkits uniquely adapted for biomass degradation. Their ability to secrete a diverse array of hydrolytic and oxidative enzymes enables them to convert structurally resistant plant material into fermentable substrates. In addition, their physiological adaptability to a broad range of environmental conditions makes them particularly well-suited for large-scale biomass processing (Burgaud et al., 2022; Daddaoua et al., 2023; Gadd et al., 2007; Oggerin et al., 2024). In industrial biotechnology, fungi are extensively exploited to produce enzymes, organic acids, and bioactive metabolites, with several strains optimized for high-yield fermentation under acidic or stress-prone conditions.

Recent studies have highlighted the remarkable capabilities of fungal isolates from the Río Tinto ecosystem, an extremely acidic environment enriched in heavy metals. Screening of more than 180 fungal strains from this site revealed that a large proportion secrete lignocellulolytic enzymes such as EGs, CBHs, and BGLs, as well as accessory enzymes including BXLs and XYGLs, exhibiting optimal catalytic activity at pH values between 3.0 and 5.0, well below the pH range of most commercial enzymatic cocktails. Additional proteomic analysis of Río Tinto fungal secretomes has further identified glycoside hydrolases from families GH3, GH7, GH16, and GH11, some of which are functionally equivalent, or even superior, to standard industrial enzymes, despite remaining structurally uncharacterized. Critically, these enzymes maintain high catalytic efficiency at pH ≤ 4.5, making them highly compatible with industrial acid pretreatment workflows typically used in second-generation (2G) bioethanol production, where lignocellulosic substrates are pretreated with dilute acids to enhance digestibility (Daddaoua et al., 2023). The experimental supplementation of enzymatic cocktails with Río Tinto fungal secretomes resulted in an 8%–15% increase in fermentable sugar release from pretreated agricultural residues such as corn stover and sugarcane straw, which is considered industrially significant, as it increases the efficiency of substrate conversion and boosts the yield of downstream products such as ethanol. These findings support the view that Río Tinto fungi are a promising biological resource for improving saccharification processes in 2G biorefineries (Daddaoua et al., 2023; Oggerin et al., 2024).

Beyond bioethanol, the sugars derived from fungal hydrolysis of lignocellulose can also serve as substrates for the microbial production of lactic acid, a key monomer in the synthesis of polylactic acid (PLA), a biodegradable thermoplastic increasingly used in packaging, agriculture, and biomedical applications (Melo et al., 2024; Ojo and de Smidt, 2023). Fungi such as *Rhizopus oryzae* and *Aspergillus* spp. have been shown to efficiently ferment hexoses and pentoses into lactic acid, even under low-pH conditions, which reduces the need for costly pH adjustment steps during industrial fermentation (Daly et al., 2021; Mujtaba et al., 2023; Reshmy et al., 2021). Given their acidophilic nature and enzymatic efficiency, Río Tinto fungi are uniquely suited for closed-loop bioconversion systems that valorize agricultural waste streams into high-value bioplastics. Their use could significantly lower the environmental and economic cost of PLA production, enabling scalable, sustainable alternatives to fossil-derived plastics.

The fungal biodiversity of Río Tinto represents a versatile and underexploited platform for the biotechnological valorization of lignocellulosic biomass. Their enzymatic systems not only improve the release of fermentable sugars under acidic conditions but also support the integrated production of bioethanol and biodegradable polymers. This dual capability aligns with global objectives to reduce fossil fuel dependency, minimize waste, and foster sustainable biomanufacturing within the circular bioeconomy framework.

Environmental applications of fungi from Río Tinto in metal pollution mitigation

A wide range of industrial processes, including leather tanning, asbestos and mirror manufacturing, pulp and paper production, cement synthesis, and fertilizer fabrication, generate significant amounts of chromium-containing waste, which is often discharged into the environment (Pawlisz et al., 1997). Although chromium can exist in multiple oxidation states (from Cr^{2-} to Cr^{6+}), in natural environments it is predominantly found as either trivalent chromium (Cr^{3+}) or hexavalent chromium (Cr^{6+}).

Regulatory frameworks across multiple countries, including the European Union, the United States, China, and India, have established maximum permissible discharge limits for chromium in water bodies. These regulations differentiate between hexavalent chromium (Cr^{6+}), which is highly toxic, mobile, and carcinogenic, and trivalent chromium (Cr^{3+}), which is more stable, less bioavailable, and significantly less toxic.

For Cr^{6+}, discharge limits into surface waters typically range from 0.05 to 0.1 mg/L, reflecting its classification as a priority pollutant in many jurisdictions, causing severe morbidity and mortality in humans (Garg et al., 2012). In contrast, Cr^{3+} discharge limits are generally less stringent, commonly ranging from 0.2 to 1.0 mg/L, depending on the receiving environment and national regulations. For instance, U.S. EPA effluent guidelines allow Cr^{3+} concentrations up to 1.0 mg/L for certain industries, while Chinese and Indian standards permit values in the range of 0.2–0.5 mg/L for treated industrial effluents.

These differentiated limits highlight the importance of chromium speciation in environmental monitoring and regulatory compliance. They also underscore the need for targeted remediation strategies that address both oxidation states, particularly in dynamic systems where redox transformations can occur.

As previously noted, the selective metal-sequestering capacity of fungal isolates from Río Tinto is being investigated as a biotechnological alternative for the remediation of chromium-contaminated wastewater from leather tanning processes. This research is conducted in collaboration with the School of Engineering of Igualada and a consortium of tanning companies. Pilot-scale experiments have demonstrated that this fungal-based technology is effective, achieving up to 89% Cr^{3+} removal from wastewater using three successive treatment cycles (Lalueza et al., 2010). Ongoing work aims to isolate new fungal strains that exhibit higher sequestration efficiencies and can tolerate the presence of biocides and preservatives commonly used during leather storage prior to processing.

Beyond chromium, efforts have also focused on the selective recovery of other metals. A *Penicillium* isolate from Río Tinto, able to recover between 47% and 86% of the copper, has shown promising potential in the biorecovery of Cu^{2+} from both concentrated industrial effluents and biohydrometallurgical leachates. And *P. lilacinum* M15001 is able to tolerate up to 1 M Cu with removal efficiency rates of 25% at 50 mM Cu, 32% at 100 mM Cu, and up to 51% at 500 mM Cu (Durán et al., 1999; Oggerin et al., 2015). Although the current recovery rates are still lower than those achieved by conventional solvent extraction methods used in the metallurgical industry, ongoing research seeks to discover fungal strains with improved copper-sequestering capabilities that could make this biological approach industrially competitive.

Similarly, acidophilic fungi capable of sequestering arsenate (As^{5+}) from contaminated waters have also been identified (Cánovas et al., 2003). The environmental impact of arsenic contamination has gained public attention since the 1998 tailings dam failure at the Aznalcóllar mine, which severely polluted the Guadiamar River and posed a serious threat to the Doñana National Park. The elevated arsenic concentrations in flotation residues from the IPB are largely due to the arsenopyrite-rich ores processed in the region. A targeted screening program identified *Penicillium* strains capable of growing in media containing up to 750 mM As^{5+}, highlighting their potential application in the bioremediation of arsenic-contaminated effluents.

Astrobiological perspectives

Knowing the fungal diversity identified in the Tinto basin and the different fungal strategies used to overcome the extreme conditions of acidity and toxic metals existing in the ecosystem, it is plausible to consider Río Tinto's fungi good candidates to thrive in the extreme conditions existing on Mars. Unfortunately, up to now, no Río Tinto fungal isolate has been exposed to Martian temperature, radiation, and oxidant conditions, so we do not know their level of tolerance to these extreme conditions, but we are planning to obtain this type of information in the near future.

As mentioned above, the most interesting astrobiological characteristic of the Tinto basin fungal diversity is their role in the formation of minerals identified on Mars. One of the most relevant biomineralization processes detected in Río Tinto is the formation of jarosite by *P. lilacinum*, a mineral detected on Mars by several space missions (Grotzinger et al., 2005; Rampe et al., 2017). The identification on Mars of jarosite as well as other iron minerals existing in the Tinto basin is the main reason for considering Río Tinto a good geochemical and mineralogical Mars analog essential to the study of potential life on the red planet (Fernández-Remolar et al., 2005; Knoll et al., 2005).

Río Tinto's fungi would not only be able to develop on Mars, but the record of their presence might be well preserved in the iron oxides and hematitic concentrations formed during diagenesis. Moreover, the nondestructive multiscale imaging technique developed for the study of Río Tinto fossils, described in Chapter 2, holds promise for the study of samples from Mars. The Rio Tinto system thus provides not only a mineralogical and geochemical analog for Mars but also a biological preservation model of interest for astrobiological exploration (Fernández-Remolar et al., 2021; Knoll et al., 2005; Maldanis et al., 2024).

Conclusion

The study of fungal communities inhabiting the Río Tinto presents convincing evidence that life can not only persist but thrive in poly-extreme environments characterized by low pH, high concentrations of heavy metals, and chronic oxidative stress. Far from being passive survivors, acidophilic fungi in this ecosystem demonstrate a range of adaptive strategies that allow them to actively contribute to biogeochemical cycles, environmental structuring, and mineral transformations.

Dominated by Ascomycota, particularly Eurotiomycetes, Dothideomycetes, and Sordariomycetes, the fungal communities demonstrate remarkable structural and functional flexibility. Adaptations such as biosorption at the cell wall, intracellular sequestration, vacuolar compartmentalization, and enzymatic redox detoxification enable them to tolerate and process extreme concentrations of metals like copper, chromium, and arsenic. *P. lilacinum* M15001 exemplifies this resilience, with tolerance to copper concentrations exceeding 1 M and active participation in copper removal through both extracellular and intracellular mechanisms. Similarly, other isolates display remarkable abilities to tolerate and immobilize chromium and arsenate, suggesting a high level of metal-specific specialization across taxa. Moreover, the capacity of these fungi to induce biomineralization processes, ranging from passive precipitation of jarosite and iron oxyhydroxides to controlled intracellular formation of iron phosphate minerals, positions them as key biogeochemical agents in the Río Tinto system. These processes not only influence local sediment mineralogy but also serve as valuable terrestrial analogs for Martian geochemistry, with implications for astrobiological research.

Beyond their ecological roles, Río Tinto fungi exhibit considerable biotechnological potential. Their acid-tolerant enzymatic systems are well-suited for lignocellulosic biomass degradation under industrial conditions, making them promising candidates for second-generation bioethanol and bioplastic production workflows. Furthermore, their metal-binding capacities can be harnessed for bioremediation, offering low-cost, sustainable alternatives to more conventional treatments of contaminated industrial effluents.

The Río Tinto fungal ecosystem thus challenges conventional assumptions about the limits of eukaryotic life and highlights the evolutionary ingenuity of fungi under persistent environmental stress. These organisms operate at the interface of biology and geochemistry, exploiting niche opportunities within constraints that would typically be considered incompatible with life. Their collective activities contribute to the formation of new mineral phases, the detoxification of their surroundings, and the cycling of essential nutrients in a setting traditionally viewed as inhospitable. Therefore, the extremophilic fungi of Río Tinto offer a model system for exploring microbial adaptation, ecological resilience, and biotechnological innovation. Their study underscores the value of looking to extreme environments not only to understand life's capacity for adaptation but also to uncover novel bioresources with transformative applications in environmental and industrial biotechnology.

AI disclosure

During the preparation of this work, the author(s) used DeepL and Gemini in order to improve the readability and language of the work. After using this tool/service, the author(s) reviewed and edited the content as needed and take(s) full responsibility for the content of the publication.

References

Aguilera, A., Manrubia, S.C., Gómez, F., Rodríguez, N., Amils, R., 2006. Eukaryotic community distribution and its relationship to water physicochemical parameters in an extreme acidic environment, Río Tinto (Southwestern Spain). Appl. Environ. Microbiol. 72 (8), 5325–5330. https://doi.org/10.1128/AEM.00513-06.

Aguilera, A., Souza-Egipsy, V., Gómez, F., Amils, R., 2007. Development and structure of eukaryotic biofilms in an extreme acidic environment, Río Tinto (SW, Spain). Microb. Ecol. 53 (2), 294–305. https://doi.org/10.1007/s00248-006-9092-2.

Amaral Zettler, L.A., Gómez, F., Zettler, E., Keenan, B.G., Amils, R., Sogin, M.L., 2002. Eukaryotic diversity in Spain's River of Fire. Nature 417 (6885), 137. https://doi.org/10.1038/417137a.

Amils, R., González-Toril, E., Fernández-Remolar, D., Gómez, F., Rodríguez, N., Durán, C., 2002. Interaction of the sulfur and iron cycles in the Tinto river ecosystem. Rev. Environ. Sci. Biotechnol. 1 (4), 299–309. https://doi.org/10.1023/a:1023232002312.

Amils, R., Escudero, C., Oggerin, M., Sánchez, F.P., Rodríguez, A.A., Remolar, D.F., Rodríguez, N., Villadangos, M.G., Sanz, J.L., Briones, C., Román, M.S.-, Gómez, F., Leandro, T., Moreno-Paz, M., Ballesteros, O.P.-, Molina, A., Tornos, F., Andrea, I.S.-, Timmis, K., Pieper, D.H., Parro, V., 2023. Coupled C, H, N, S and Fe biogeochemical cycles operating in the continental deep subsurface of the Iberian Pyrite Belt. Environ. Microbiol. 25 (2), 428–453. https://doi.org/10.1111/1462-2920.16291.

Baldrian, P., 2003. Interactions of heavy metals with white-rot fungi. Enzym. Microb. Technol. 32 (1), 78–91. https://doi.org/10.1016/s0141-0229(02)00245-4.

Bethke, C.M., 2022. Geochemical and Biogeochemical Reaction Modeling. Cambridge University Press.

Beveridge, T.J., Murray, R.G., 1976. Uptake and retention of metals by cell walls of Bacillus subtilis. J. Bacteriol. 127 (3), 1502–1518. https://doi.org/10.1128/jb.127.3.1502-1518.1976.

Bigham, J.M., Nordstrom, D.K., 2000. Iron and aluminum hydroxysulfates from acid sulfate waters. Rev. Mineral. Geochem. 40 (1), 351–403. https://doi.org/10.2138/rmg.2000.40.7.

Bladh, K.W., 1982. The formation of goethite, jarosite, and alunite during the weathering of sulfide-bearing felsic rocks. Econ. Geol. 77 (1), 176–184. https://doi.org/10.2113/gsecongeo.77.1.176.

Branda, S.S., Vik, A., Friedman, L., Kolter, R., 2005. Biofilms: the matrix revisited. Trends Microbiol. 13 (1), 20–26. https://doi.org/10.1016/j.tim.2004.11.006.

Breitenbach, R., Gerrits, R., Dementyeva, P., Knabe, N., Schumacher, J., Feldmann, I., Radnik, J., Ryo, M., Gorbushina, A.A., 2022. The role of extracellular polymeric substances of fungal biofilms in mineral attachment and weathering. NPJ Mater. Degrad. 6 (1). https://doi.org/10.1038/s41529-022-00253-1.

Brown, M.T., Hall, I.R., 1990. Metal tolerance in fungi. In: Shaw, J. (Ed.), Heavy Metal Tolerance in Plants: Evolutionary Aspects. CRC Press, Boca Raton.

Burgaud, G., Edgcomb, V., Hassett, B.T., Kumar, A., Li, W., Mara, P., Peng, X., Philippe, A., Phule, P., Prado, S., Quéméner, M., Roullier, C., 2022. Marine fungi. In: Stal, L.J., Cretoiu, M.S. (Eds.), The Marine Microbiome. Springer Science and Business Media LLC, pp. 243–295. https://doi.org/10.1007/978-3-030-90383-1_5.

Cánovas, D., Durán, C., Rodríguez, N., Amils, R., De Lorenzo, V., 2003. Testing the limits of biological tolerance to arsenic in a fungus isolated from the River Tinto. Environ. Microbiol. 5 (2), 133–138. https://doi.org/10.1046/j.1462-2920.2003.00386.x.

Comte, S., Guibaud, G., Baudu, M., 2008. Biosorption properties of extracellular polymeric substances (EPS) towards Cd, Cu and Pb for different pH values. J. Hazard Mater. 151 (1), 185–193. https://doi.org/10.1016/j.jhazmat.2007.05.070.

Cosmidis, J., Karim, B., Morin, G., Vincent, B., Lebeau, O., Jézéquel, D., Vincent, N., Dublet, G., Othmane, G., 2014. Biomineralization of iron-phosphates in the water column of Lake Pavin (Massif Central, France). Geochem. Cosmochim. Acta 126, 78–96. https://doi.org/10.1016/j.gca.2013.10.037.

Daddaoua, A., Álvarez, C., Oggerin, M., Rodriguez, N., Duque, E., Amils, R., Jean, A., Segura, A., Luis Ramos, J., 2023. Rio Tinto as a niche for acidophilus enzymes of industrial relevance. Microb. Biotechnol. 16 (5), 1069–1086. https://doi.org/10.1111/1751-7915.14192.

Daly, P., Cai, F., Kubicek, C.P., Jiang, S., Grujic, M., Rahimi, M.J., Sheteiwy, M.S., Giles, R., Riaz, A., de Vries, R.P., Akcapinar, G.B., Wei, L., Druzhinina, I.S., 2021. From lignocellulose to plastics: knowledge transfer on the degradation approaches by fungi. Biotechnol. Adv. 50. https://doi.org/10.1016/j.biotechadv.2021.107770.

Das, N., Vimala, R., Karthika, P., 2008. Biosorption of heavy metals - an overview. Indian J. Biotechnol. 7 (2), 159–169.

Durán, C., Sergeant, C., Rodríguez, N., Amils, R., 2001. Biohydrometallurgy: "fundamentals, technology and sustainable development". In: Proceedings of the International Biohydrometallurgy Symposium, IBS-2001 Elsevier Publishing Company Specific Cr 3+ Sequestering Using an Acidophilic Fungal Isolate.

Durán, C., Marín, I., Amils, R., 1999. Specific metal sequestering acidophilic fungi. In: Amils, R.,, Ballester, A. (Eds.), Process Metallurgy. Elsevier, Spain.

Falkowski, P.G., Fenchel, T., Delong, E.F., 2008. The microbial engines that drive earth's biogeochemical cycles. Science 320 (5879), 1034–1039. https://doi.org/10.1126/science.1153213.

Fernández-Remolar, D.C., Gómez-Ortiz, D., Malmberg, P., Huang, T., Shen, Y., Anglés, A., Amils, R., 2021. Preservation of underground microbial diversity in ancient subsurface deposits (>6 Ma) of the Rio Tinto basement. Microorganisms 9 (8), 1592. https://doi.org/10.3390/microorganisms9081592.

Fernández-Remolar, D.C., Morris, R.V., Gruener, J.E., Amils, R., Knoll, A.H., 2005. The Río Tinto Basin, Spain: mineralogy, sedimentary geobiology, and implications for interpretation of outcrop rocks at Meridiani Planum, Mars. Earth Planet Sci. Lett. 240 (1), 149–167. https://doi.org/10.1016/j.epsl.2005.09.043.

Flemming, H.C., 2011. The perfect slime. Colloids Surf. B Biointerfaces 86 (2), 251–259. https://doi.org/10.1016/j.colsurfb.2011.04.025.

Flemming, H.C., Wingender, J., 2010. The biofilm matrix. Nat. Rev. Microbiol. 8 (9), 623–633. https://doi.org/10.1038/nrmicro2415.

Fomina, M., Gadd, G.M., 2007. Metal and mineral transformations: a mycoremediation perspective. In: Robson, G.D., van West, P., Gadd, G.M. (Eds.), Exploitation of Fungi. Cambridge Cambridge University Press (CUP), Cambridge. https://doi.org/10.1017/cbo9780511902451.014.

Frankel, R.B., Bazylinski, D.A., 2003. Biologically induced mineralization by bacteria. Rev. Mineral. Geochem. 54 (1), 95–114. https://doi.org/10.2113/0540095.

Gadd, G.M., 2021. Fungal biomineralization. Curr. Biol. 31 (24), R1557. https://doi.org/10.1016/j.cub.2021.10.041.

Gadd, G.M., 2007. Geomycology: biogeochemical transformations of rocks, minerals, metals and radionuclides by fungi, bioweathering and bioremediation. Mycol. Res. 111 (1), 3–49. https://doi.org/10.1016/j.mycres.2006.12.001.

Gadd, G.M., Rhee, Y.J., Stephenson, K., Wei, Z., 2012. Geomycology: metals, actinides and biominerals. Environ. Microbiol. Rep. 4 (3), 270–296. https://doi.org/10.1111/j.1758-2229.2011.00283.x.

Gadd, G.M., 2010. Metals, minerals and microbes: geomicrobiology and bioremediation. Microbiology 156 (3), 609–643. https://doi.org/10.1099/mic.0.037143-0.

Gadd, G.M., 2017. The geomycology of elemental cycling and transformations in the environment. Microbiol. Spectr. 5 (1). https://doi.org/10.1128/microbiolspec.FUNK-0010-2016.

Gadd, G.M., Raven, J.A., 2010. Geomicrobiology of eukaryotic microorganisms. Geomicrobiol. J. 27 (6–7), 491–519. https://doi.org/10.1080/01490451003703006.

Gadd, G.M., de Rome, L., 1988. Biosorption of copper by fungal melanin. Appl. Microbiol. Biotechnol. 29 (6), 610–617. https://doi.org/10.1007/BF00260993.

Gadd, G.M., 2008. Bacterial and fungal geomicrobiology: a problem with communities? Geobiology 6 (3), 278–284.

Gadd, G.M., 2018. Geomycology: fungi as agents of biogeochemical change. Biol. Environ. 118B (4).

Gadd, G.M., 1990a. Heavy metal accumulation by bacteria and other microorganisms. Experientia 46 (8), 834–840.

Gadd, G.M., 1990b. Fungi and yeasts for metal binding. In: Ehrlich, H., Brierley, C.L. (Eds.), Microbial Mineral Recovery. McGraw-Hill, New York McGraw-Hill, New York.

Gadd, G.M., 1993. Interactions of fungi with toxic metals. New Phytol. 124 (1), 25–60.

Gadd, G.M., 2004. Microbial influence on metal mobility and application for bioremediation. Geoderma 122 (2–4), 109–119. https://doi.org/10.1016/j.geoderma.2004.01.002.

Gadd, G.M., Watkinson, S.C., Dyer, P.S., 2007. In: Fungi in the Environment. Cambridge Cambridge University Press (British Mycological Society Symposia), Cambridge.

Garg, S.K., Tripathi, M., Srinath, T., 2012. Strategies for chromium bioremediation of tannery effluent. Rev. Environ. Contam. Toxicol. 217, 75–140.

Goldstein, I.J., Hayes, C.E., 1978. The lectins: carbohydrate-binding proteins of plants and animals. Adv. Carbohydr. Chem. Biochem. 35, 127–340.

Gómez-Ortiz, D., Fernández-Remolar, D.C., Granda, A., Quesada, C., Granda, T., Prieto-Ballesteros, O., Molina, A., Amils, R., 2014. Identification of the subsurface sulfide bodies responsible for acidity in Río Tinto source water, Spain. Earth Planet Sci. Lett. 391, 36–41. https://doi.org/10.1016/j.epsl.2014.01.022.

Gónzalez-Toril, E., Gómez, F., Rodríguez, N., Fernández-Remolar, D., Zuluaga, J., Marín, I., Amils, R., 2003. Geomicrobiology of the Tinto River, a model of interest for biohydrometallurgy. Hydrometallurgy 71 (1–2), 301–309. https://doi.org/10.1016/S0304-386X(03)00169-5.

Gross, S., Robbins, E.I., 2000. Acidophilic and acid-tolerant fungi and yeasts. Hydrobiologia 433, 91–109. https://doi.org/10.1023/A:1004014603333.

Grotzinger, J.P., Arvidson, R.E., Bell, J.F., Calvin, W., Clark, B.C., Fike, D.A., Golombek, M., Greeley, R., Haldemann, A., Herkenhoff, K.E., Jolliff, B.L., Knoll, A.H., Malin, M., McLennan, S.M., Parker, T., Soderblom, L., Sohl-Dickstein, J.N., Squyres, S.W., Tosca, N.J., Watters, W.A., 2005. Stratigraphy and sedimentology of a dry to wet eolian depositional system, burns formation, Meridiani Planum, Mars. Earth Planet Sci. Lett. 240 (1), 11–72. https://doi.org/10.1016/j.epsl.2005.09.039.

Hujslová, M., Bystrianský, L., Benada, O., Gryndler, M., 2019. Fungi, a neglected component of acidophilic biofilms: do they have a potential for biotechnology? Extremophiles 23 (3), 267–275. https://doi.org/10.1007/s00792-019-01085-9.

Kirchhausen, T., 2000. Three ways to make a vesicle. Nat. Rev. Mol. Cell Biol. 1 (3), 187–198. https://doi.org/10.1038/35043117.

Knoll, A.H., Carr, M., Clark, B., Des Marais, D.J., Farmer, J.D., Fischer, W.W., Grotzinger, J. P., McLennan, S.M., Malin, M., Schröder, C., Squyres, S., Tosca, N.J., Wdowiak, T., 2005. An astrobiological perspective on Meridiani Planum. Earth Planet Sci. Lett. 240 (1), 179–189. https://doi.org/10.1016/j.epsl.2005.09.045.

Konhauser, K.O., 1998. Diversity of bacterial iron mineralization. Earth Sci. Rev. 43 (3–4), 91–121. https://doi.org/10.1016/S0012-8252(97)00036-6.

Lalueza, J., Puig, R., Rius, A., Martí, E., Martí, J.F., Rodríguez, N., Amils, R., 2014. Removal of Chromium III from tannery wastewaters with acidophilic fungi. J. Am. Leather Chem. Assoc. 109 (1), 14–24.

Lalueza, J., Rius, A., Puig, R., Martí, M.E., Rodriguez, N., Amils, R., Martí, J.F., 2010. Biotechnological sequestering of chromium (III) from post-tanning effluents: first results. J. Am. Leather Chem. Assoc. 105, 214–221.

de León, N., Sharifmoghadam, M.R., Hoya, M., Curto, M.A., Doncel, C., Valdivieso, M.H., 2013. Regulation of cell wall synthesis by the clathrin light chain is essential for viability in *Schizosaccharomyces pombe*. PLoS One 8 (8). https://doi.org/10.1371/journal.pone.0071510.

López-Archilla, A.I., Marin, I., Amils, R., 2001. Microbial community composition and ecology of an acidic aquatic environment: the Tinto River, Spain. Microb. Ecol. 41 (1), 20–35. https://doi.org/10.1007/s002480000044.

López-Archilla, A.I., González, A.E., Terrón, M.C., Amils, R., 2004. Ecological study of the fungal populations of the acidic Tinto River in southwestern Spain. Can. J. Microbiol. 50 (11), 923–934. https://doi.org/10.1139/w04-089.

Lowenstam, H.A., 1981. Minerals formed by organisms. Science 211 (4487), 1126–1131. https://doi.org/10.1126/science.7008198.

Magan, N., 2007. Fungi in extreme environments. In: The Mycota. Volume 4, pp. 85–103.

Maldanis, L., Fernandez-Remolar, D., Lemelle, L., Knoll, A.H., Guizar-Sicairos, M., Holler, M., da Silva, F.M.C., Magnin, V., Mermoux, M., Simionovici, A., 2024. Unveiling challenging microbial fossil biosignatures from Rio Tinto with micro-to-nanoscale chemical and ultrastructural imaging. Astrobiology 24 (7), 721–733. https://doi.org/10.1089/ast.2023.0127.

Medina-Armijo, C., Isola, D., Illa, J., Puerta, A., Viñas, M., Prenafeta-Boldú, F.X., 2024. The Metallotolerance and biosorption of As(V) and Cr(VI) by black fungi. J. Fungi 10 (1). https://doi.org/10.3390/jof10010047.

Mehra, R.K., Winge, D.R., 1991. Metal ion resistance in fungi: molecular mechanisms and their regulated expression. J. Cell. Biochem. 45 (1), 30–40. https://doi.org/10.1002/jcb.240450109.

Melo, N.T.M., de Oliveira Junqueira, A.C., Lima, L.F., de Oliveira, K.B.S., dos Reis, M.C.G., Franco, O.L., Paes, H.C., 2024. Just around the corner: advances in the optimization of yeasts and filamentous fungi for lactic acid production. J. Fungi 10 (3). https://doi.org/10.3390/jof10030207.

Mujtaba, M., Fernandes Fraceto, L., Fazeli, M., Mukherjee, S., Savassa, S.M., Araujo de Medeiros, G., do Espírito Santo Pereira, A., Mancini, S.D., Lipponen, J., Vilaplana, F.,

2023. Lignocellulosic biomass from agricultural waste to the circular economy: a review with focus on biofuels, biocomposites and bioplastics. J. Clean. Prod. 402, 136815. https://doi.org/10.1016/j.jclepro.2023.136815.

Oggerin, M., del Moral, C., Rodriguez, N., Fernandez-Gonzalez, N., Martínez, J.M., Lorca, I., Amils, R., 2024. Metal tolerance of Río Tinto fungi. Front. Fungal Biol. 5. https://doi.org/10.3389/ffunb.2024.1446674.

Oggerin, M., Tornos, F., Rodriguez, N., Pascual, L., Ricardo, A., 2016. Fungal iron biomineralization in Río Tinto. Minerals 6 (2), 37. https://doi.org/10.3390/min6020037.

Oggerin, M., Escudero, C., del Moral, C., Rodríguez, N., Amils, R., 2015. High copper tolerant *P. lilacinum* strain isolated from a rich environment in copper, Río Tinto (Sw, Spain). Adv. Mater. Res. 1130, 157–160. https://doi.org/10.4028/www.scientific.net/amr.1130.157.

Oggerin, M., Tornos, F., Rodríguez, N., del Moral, C., Sánchez-Román, M., Amils, R., 2013. Specific jarosite biomineralization by Purpureocillium lilacinum, an acidophilic fungi isolated from Río T into. Environ. Microbiol. 15 (8), 2228–2237. https://doi.org/10.1111/1462-2920.12094.

Oggerin, M., Rodríguez, N., Amils, R., 2017. Fungal mineralization processes in Rio Tinto. Solid State Phenom. 262, 354–357. https://doi.org/10.4028/www.scientific.net/ssp.262.354.

Ojo, A.O., de Smidt, O., 2023. Lactic acid: a comprehensive review of production to purification. Processes 11 (3). https://doi.org/10.3390/pr11030688.

Pawlisz, A.V., Kent, R.A., Schneider, U.A., Jefferson, C., 1997. Canadian water quality guidelines for chromium. Environ. Toxicol. Water Qual. 12 (2), 123–183. https://doi.org/10.1002/(SICI)1098-2256(1997)12:2<123::AID-TOX4>3.3.CO;2-O.

Pócsi, I., Prade, R.A., Penninckx, M.J., 2004. Glutathione, altruistic metabolite in fungi. Adv. Microb. Physiol. 49, 1–76. https://doi.org/10.1016/S0065-2911(04)49001-8.

Rampe, E.B., Ming, D.W., Blake, D.F., Bristow, T.F., Chipera, S.J., Grotzinger, J.P., Morris, R. V., Morrison, S.M., Vaniman, D.T., Yen, A.S., Achilles, C.N., Craig, P.I., Des Marais, D.J., Downs, R.T., Farmer, J.D., Fendrich, K.V., Gellert, R., Hazen, R.M., Kah, L.C., Morookian, J.M., Peretyazhko, T.S., Sarrazin, P., Treiman, A.H., Berger, J.A., Eigenbrode, J., Fairén, A.G., Forni, O., Gupta, S., Hurowitz, J.A., Lanza, N.L., Schmidt, M.E., Siebach, K., Sutter, B., Thompson, L.M., 2017. Mineralogy of an ancient lacustrine mudstone succession from the Murray formation, Gale crater, Mars. Earth Planet Sci. Lett. 471, 172–185. https://doi.org/10.1016/j.epsl.2017.04.021.

Reshmy, R., Thomas, D., Philip, E., Paul, S.A., Madhavan, A., Sindhu, R., Sirohi, R., Varjani, S., Pugazhendhi, A., Pandey, A., Binod, P., 2021. Bioplastic production from renewable lignocellulosic feedstocks: a review. Rev. Environ. Sci. Biotechnol. 20 (1), 167–187. https://doi.org/10.1007/s11157-021-09565-1.

Salkield, L.V., 1987. A Technical History of the Rio Tinto Mines: Some Notes on Exploitation from Pre-Phoenician Times to the 1950s.

Schwertmann, U., Cornell, R.M., 2000a. Goethite. In: Schwertmann, U., Cornell, R.M. (Eds.), Iron Oxides in the Laboratory: Preparation and Characterization. Wiley VCH Verlag, Weinheim, Germany.

Schwertmann, U., Cornell, R.M., 2000b. Lepidocrocite. In: Schwertmann, U., Cornell, R.M. (Eds.), Iron Oxides in the Laboratory: Preparation and Characterization. Wiley VCH Verlag, Weinheim, Germany.

Sharon, N., Lis, H., 2004. History of lectins: from hemagglutinins to biological recognition molecules. Glycobiology 14 (11), 53R. https://doi.org/10.1093/glycob/cwh122.

Sterflinger, K., 2000. Fungi as geologic agents. Geomicrobiol. J. 17 (2), 97–124. https://doi.org/10.1080/01490450050023791.

Tobin, J.M., Cooper, D.G., Neufeld, R.J., 1984. Uptake of metal ions by Rhizopus arrhizus biomass. Appl. Environ. Microbiol. 47 (4), 821–824. https://doi.org/10.1128/aem.47.4.821-824.1984.

Tornos, F., 2006. Environment of formation and styles of volcanogenic massive sulfides: the Iberian Pyrite Belt. Ore Geol. Rev. 28 (3), 259–307. https://doi.org/10.1016/j.oregeorev.2004.12.005.

CHAPTER

10

Biodiversity and biotechnological applications of plant species and communities adapted to an extreme acidic environment: Río Tinto basin

Vicenta de la Fuente[1], Lourdes Rufo[2], Daniel Sánchez Mata[3, 4, 5], Esteban Ramírez[4, 5]

[1]*Universidad Autónoma de Madrid, Campus Cantoblanco, Madrid, Spain;* [2]*Instituto de Investigaciones Biosanitarias, Facultad de Ciencias Experimentales, Universidad Francisco de Vitoria, Madrid, Spain;* [3]*Department of Plant Biology II, Faculty of Pharmacy, Universidad Complutense, Ciudad Universitaria, Madrid, Spain;* [4]*Department of Organismic and Evolutionary Biology, Harvard University Herbaria, Cambridge, MA, United States;* [5]*Real Colegio Complutense at Harvard University, Cambridge, MA, United States*

Introduction

Río Tinto is in Huelva province (Spain), in the southwest of the Iberian Peninsula. It is 92 km in length until its mouth into the Atlantic Ocean in the Huelva Estuary, next to the Odiel river. The river basin covers an area of 1730 km^2 and is bordered to the east by the Guadalquivir River basin (in the province of Seville) and to the west by the Odiel River basin.

The source of the Río Tinto is in the center of the IPB (Iberian Pyrite Belt), a territory considered one of the most important metallogenic and mining regions of the world. This river is considered an extreme habitat for life due to its acidity (average pH 2.3) and by the high concentration of heavy metals in its waters, constant along the length of its first 80 km. These characteristics extend to the sediments and soils. These features make Tinto ecosystem a unique habitat and an optimum territory to study the diverse biological adaptations to this type of extreme environment (Amils et al., 2002; López-Archilla et al., 2001).

Río Tinto is a typical Mediterranean river showing important variations in the flow during the year, therefore it is characterized by drought periods in summer (from June to September) and floods in the rainy periods that may occur in winter,

Geomicrobiology of Río Tinto. https://doi.org/10.1016/B978-0-443-32948-7.00002-3

autumn, and spring. The riparian natural vegetation series of the area is structured by woodlands that we can ascribe to the Ibero-Lusitan meso-thermomediterranean fluvial geoseries of riparian woods of *Salix pedicellata*, *Alnus glutinosa*, *Nerium oleander*, and *Fraxinus angustifolia*. The structure and maturity of these vegetation types are related to geomorphology, the continuity of the water flow, and the degree of the degradation of the area.

The Río Tinto runs through two distinct thermotypic zones, which can be classified as mesomediterranean and thermomediterranean. Rivas-Martínez (1988), in his comprehensive synthesis on bioclimatology, biogeography, and vegetation series of Western Andalucía, used several woody plant species as bioindicators of both thermotypes, based on thermoclimatic values. Notable woody species representative of the thermomediterranean thermotype include *Asparagus aphyllus*, *Calicotome villosa*, *Ceratonia siliqua*, *Chamaerops humilis*, *Juniperus macrocarpa*, *J. turbinata*, *Osyris quadripartita*, *Retama monosperma*, *Rhamnus oleoides*, and *Salix pedicellata*, among others.

Description of the study area/location of the Río Tinto

The river is divided into three artificial sections, reflecting its natural characteristics. The first one corresponds to the origin of the river and occupies the northern area from Peña de Hierro to La Palma del Condado. This section represents the core of mining activity in the region, although traces of mining are present across the entire study area. The second section or transitional area extends from La Palma del Condado to San Juan del Puerto. The last section is the estuary, which occupies the area between San Juan del Puerto and Huelva, where Tinto and Odiel rivers converge. The mouth is occupied by extensive salt marshes with a variable pH (from pH 3 to circumneutral) conditioned by the strong Atlantic Ocean tides.

Biogeography

This territory is framed within three biogeographic provinces: West Iberian Mediterranean, Betic and Coastal Lusitania, and West Andalusia (Rivas-Martínez et al., 2017a).

In terms of biogeography (Fig. 10.1), the Marianica Range sector, framed within the Lusitania and Extremadura Subprovince (West Iberian Mediterranean Province), is characterized by Paleozoic siliceous rocks and materials derived from their erosion. This sector, the most extensive and southernmost of the province, is the only one in the province that encompasses thermomediterranean zones and supports interesting endemics such as *Erica andevalensis*, *Genista polyanthos*, and *Ulex eriocladus*. The Río Tinto flows through this sector, defining the territorial boundaries of the municipalities of Nerva, Minas de Río Tinto, El Campillo, Zalamea la Real, Berrocal, Valverde del Camino, and Paterna del Campo.

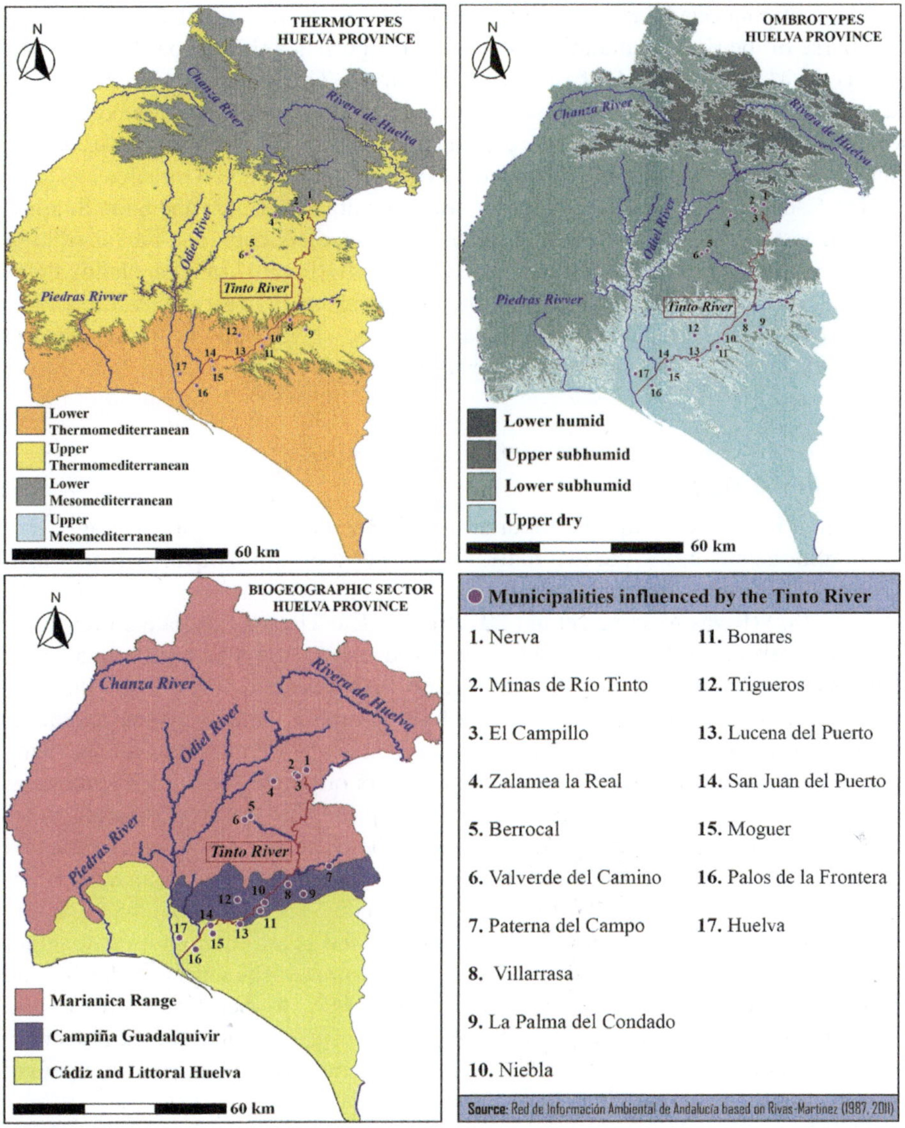

FIGURE 10.1

Biogeographic sectors, thermotypes, ombrotypes, and municipalities of the province of Huelva influenced by the Río Tinto. ***Map lines delineate study areas and do not necessarily depict accepted national boundaries.***

The Campiña Guadalquivir sector of the Betica Province is situated near the middle course of the river, maintaining a very sparse presence throughout the territory, and is consistently located far from the main channel. This sector is characterized by basophilic vegetation series interspersed with large sandy islets that are poor in bases. The principal municipalities exhibiting biogeographical traits of this sector include Villarasa, La Palma del Condado, Niebla, Bonares, and Trigueros.

The Cádiz and Littoral Huelva sector, framed within the Cádiz and Sado Subprovince (Coastal Lusitania and West Andalusia Province), encompasses areas of high oceanity, including Lucena del Puerto, San Juan del Puerto, Moguer, Palos de la Frontera, and Huelva, distinguished by its prominent sandy coastline and important coastal paleodune formations. Occasionally flooded clayey soils in some areas are also characteristic of this biogeographic sector territories.

Bioclimatology

The region is framed within the Mediterranean Macrobioclimate: the average rainfall in millimeters during the two hottest months of the summer quarter is less than twice the average temperature in Celsius degrees of those same months ($Ps2 < 2Ts2$) (Rivas-Martínez et al., 2017b). The Río Tinto bioclimate corresponds to the Mediterranean Pluviseasonal Oceanic type with thermomediterranean as main thermotype and subhumid and dry ombroclimatic types (Fig. 10.1).

The river lies within a transitional bioclimatic zone, situated between the mesomediterranean territories of the Sierra de Aracena—where the main sources of the Tinto, Odiel, Rivera de Huelva, and Chanza rivers originate—and the surrounding towns of Nerva, Minas de Riotinto, and El Campillo. The Río Tinto mainly flows through territories of the thermomediterranean thermotypes. Average temperatures are between 16 and 19°C in the thermomediterranean thermotype range and termicity index between 350 and 450. It is therefore an index related to the intensity of the coldest month, a limiting factor for many plants and plant communities, together with the average annual temperature. In the southwestern quadrant of the Iberian Peninsula, it is common for the months from May to September to be arid ($P < 2T$) and even for precipitation to be so scarce in July and August that these months become hyperarid ($P < 1/4$ T) or ultra-hyperarid ($P < 1/10$ T), which never happens in the Levante region (Rivas Martínez, 1988).

Ombrotermic index $Io > 2$, corresponds to ombrotype subhumid (Io 3.6–6, dominant in the first section of the river) and dry (2.2–3, in the transitional and last section). The rainfall regime in western Andalucía has a marked Mediterranean character, meaning it features a long, distinctly arid summer. Winter is the wettest season, followed by spring, autumn, and summer. It is also noteworthy that May can occasionally exhibit arid conditions, whereas September consistently does so (Rivas-Martínez, 1988).

Geology

The river's origin is nestled in one of the largest volcanic-sedimentary complexes of massive sulfides in the world, known as the IPB. The IPB is a geologically significant region both economically and environmentally, as it contains rich deposits of metallic sulfides that have been exploited for centuries. Its geological characteristics, primarily composed of pyrite (FeS_2) and other minerals such as chalcopyrite ($CuFeS_2$), sphalerite (ZnS), and galena (PbS), make it one of the most important areas for mining in Europe. Thus, the IPB includes one of the largest mining activity regions, starting with the deposits of Aznalcóllar and the Castillo de las Guardas in Seville and extending to 25 km from the Atlantic Ocean with the Caveria deposit in Portugal (Lescuyer et al., 1998; Fernández Remolar et al., 2005).

The geological composition of this section is of particular interest, as it contains the chemical bases by which the river acquires the characteristics of acidity and toxicity for which it is recognized as an extreme habitat. The IPB has a relatively simple geological record, which includes sequences of rocks dated from the Upper Paleozoic at depths ranging from 1000 to 5000 m. The oldest rocks are included within the PQ Group (Phyllitic-Quartzitic) and were formed during the period between Frasnian and Upper Famennian (Late Devonian, 375 Ma). This sequence, with a thickness exceeding 2000 m, consists of debris, mud, and sand typical of a continental shelf. This group is connected to the volcanic-sedimentary complex (VSC) through an irregular contact (Tornos, 2006).

The VSC is composed of a complex mafic-felsic volcanic sequence, with several intercalated layers of mud and other sediments. It is dated between the Upper Famenian and the Lower-Upper Visean (Upper Devonian-Lower Carboniferous, 369–333 Ma). The complex does not exhibit a continuous surface exposure, showing only 25% of the entire IBP. Its thickness varies from 0 to 1300 m, and the composition of the sequence changes from north to south. Specifically, in the Río Tinto district, the VSC sequence includes massive sulfide groups, intrusives in both shales and coherent volcanic rocks. The interrupted mass of sulfides corresponds to a layer covering an area of 4 km^2, containing a total of 500 Mt. In this area, the unit underlying the VSC consists of black shales with basaltic intrusions and lava flows at local points. Over this phase appears a heterogeneous sequence, which includes layers of mud with shale fragments and felsic volcanic rocks. This phase gives way to the so-called Felsic Unit, whose composition is highly variable. In the south and west of the Río Tinto, this unit is composed of felsic rocks, while in the heart of the syncline formed by the Río Tinto, the material is primarily subvolcanic. This unit is cut by a Transition Series, formed by shales with massive sulfide intrusions in the South Vein, which gradually transform into volcanoclastic sands containing fragments of volcanic rocks from the lower layer (Tornos, 2006). Above the VSC, rocks from the Culm Group appear primarily composed of shales, litharenites, and conglomerates with turbidite characteristics. This phase has an approximate thickness of 3000 m and is dated from the Upper Visean to the Middle-Upper Pennsylvanian (Carboniferous, 350–300 Ma) (Tornos, 2006). Once the upper course is surpassed, the river is entrenched in the sandy terraces of the Quaternary formation. At the height of the towns of La Palma del Condado and Niebla, small areas of marls, sands, sandstones, and conglomerates with limestones belonging to the Neogene, specifically the Miocene series (6.5 Ma), are included. These limestone-origin rocks are displaced at the height of Moguer by the mud and sandy terraces of the Quaternary (Pleistocene, 1.8 Ma) (IGME, 1928–1972).

Edaphology

Soils from the headwaters and middle course display extreme acidity, with average pH values of 3.0 and 2.6, respectively, due to elevated levels of iron (Fe) and sulfur (S). In the estuary , seawater input raises the pH to neutral levels, mirroring trends seen in river waters (Rufo et al., 2007).

Electrical conductivity (EC) is notably high at 1.8 dS/cm in both headwater and middle course, increasing to 5.8 dS/cm in the estuary. Fe and S dominate the soil composition, originating from pyrite weathering and intensified land exploitation, with toxic metal concentrations (Co, Cu, Zn, As, Pb, and Cr) exceeding reported averages (Bowen, 1979; Fernández-Remolar et al., 2005), while Mn stays constant. S concentrations exceed 1.5% in the headwaters, increase in the middle course, and decline in the estuary.

Nutrient deficiencies in Ca, Mg, K, Na, and P are common, although their levels rise downstream, with Na and Mg normalizing in the estuary. Mineral analysis reveals iron oxides (goethite, hematite) in headwater and middle course soils, while feldspar is more prevalent in the estuary. Jarosite presence indicates hydromorphy, acidic, oxidizing conditions, and gypsum reflects secondary mineralization from desiccation phases, prevalent at the river's headwaters.

Vegetation in Río Tinto

The topography of the land across Río Tinto is very rugged, featuring numerous low mountains with steep slopes where soil development cannot reach much depth. This entire area roughly corresponds to what is known as the upper course of the river. In the middle course, the terrain loses altitude, and the river becomes entrenched in the sandy terraces of the Quaternary formation after surpassing the slate cliffs. Starting from La Palma del Condado, areas of marls, sands, sandstones, and conglomerates with limestones from the Neogene are included. In the lower course or estuary (from San Juan del Puerto to Huelva), the riverbed and floodplain are composed of gravels, sands, silts, and clays from the Quaternary.

The Río Tinto soils exhibit an atypically low pH and high concentrations of sulfate ($SO4^{2-}$), iron (Fe), copper (Cu), zinc (Zn), lead (Pb), and arsenic (As) in the first two sections of the river (Rodríguez et al., 2007; Rufo et al., 2007). This phenomenon indicates that soils near the river are significantly influenced by the river waters and sediments, as well as the parent rock, which have been reported to maintain elevated levels of SO_4^{2-}, Fe, and other metals in these sections (López-Archilla et al., 2001; Cánovas et al., 2007). This influence only diminishes due to the presence of seawater in the estuary.

The scarcity of elements regarded as essential plant nutrients is a general characteristic across all groups. Elements such as calcium (Ca), potassium (K), and phosphorus (P) only increase in concentration in the presence of certain potential sources, such as lithological sources (limestone in the middle course) or industrial sources (effluents from the industrial hub of Huelva).

Concerning the vegetation of the area, we can recognize: climatophylous vegetation series, edaphohigrophylous vegetation series, and permanent vegetation.

The climax natural vegetation comprises sclerophyllous oak open forests (*Quercus rotundifolia*) and cork oak forests (*Quercus suber*) (Fig. 10.2) of the *Quercetea ilicis* phytosociological class (Rivas Goday, 1964). These open forests, along with their preforest stages, occur on the most evolved deep soils. When these ecosystems become disturbed, they are replaced by scrubland, pioneer therophytic plant communities, and perennial rupestrian stages.

Currently, these series are poorly preserved, with patches of proper forests found in areas where the topography makes access and exploitation difficult. Historically, the resources of this region have been intensively exploited in various ways. The large deposits of metallic sulfides have been mined since ancient times, extracting copper, gold, and silver from their various mines. This exploitation, both open-pit and underground, as well as the activity of the associated populations, which were itinerant depending on the area of active exploitation at any given time, have led to soil degradation, loss of vegetation cover, and an increase in acidity and metal concentration throughout the territory.

Mining, especially open pit mining, has removed a large part of the land for mineral extraction and the construction of necessary buildings for mining operations. Likewise, clearings and other landscape modifications have been necessary to provide for livestock and agricultural use of the area. To these activities, we can add the devastating effects of accidental fires and floods. Therefore, today, dehesas, scrub formations, and reforestations of *Eucalyptus* sp. pl., *Pinus pinea*, and *P. pinaster* dominate the landscape (Fig. 10.3A).

Vegetation series, geoseries, and geopermaseries

The concept of ecological series is used to classify plant communities according to their adaptation to climate, soils, and the specific conditions of their environment. These categories allow for a better study and understanding of how vegetation is distributed and adapts based on different environmental factors (Rivas Martínez, 2007). According to this author, we recognized the following series and geoseries.

FIGURE 10.2

Climatophilous vegetation of the Río Tinto. (A) Red waters bordered by holm oak woodlands (*Quercus rotundifolia*); (B) and (C) plant communities featuring *Arbutus unedo* and *Phillyrea angustifolia*; (D) Cork oak forests (*Quercus suber*) with *Chamaerops humilis*; (E) Heathlands dominated by *Erica australis*; (F) Secondary dry scrub communities characterized by *Ulex eriocladus* and *Cistus ladanifer*; (G) Pioneer annual vegetation dominated by *Agrostis subspicata*.

Climatophilous Series: These vegetation series adjust to the climatic conditions of a specific area. They are primarily determined by the climate, specifically by the amount and distribution of precipitation, as well as the characteristics of the soils based on the climate. These series can be affected by phenomena such as fog, shaded slopes, or the influence of certain relic vegetation formations.

FIGURE 10.3

Edaphic-hygrophilous vegetation of the Río Tinto. (A) and (B) Red waters at the river's source. (C) *Nerium oleander* (Apocynaceae) in bloom. (D) Endemic communities of *Erica andevalensis* with *N. oleander* and *Pteridium aquilinum*. (E) Details of *E. andevalensis* flowers.

Edafoxerophylous Series: These vegetation series are found in environments with extreme edaphic dryness. Their vegetation adapts to shallow or hard soils (such as lithosols and leptosols). These series are in rocky areas or exist under additional environmental factors such as wind, salinity, or heavy metals, which create dry conditions, making them xerophytic. They also include those where dryness is caused by more specific factors.

Geoseries and Edafohygrophylous Series: Geoseries are contiguous spatial sequences of different vegetation series within a defined area, following a toposequence. At the lower part of the toposequence, such as in valleys or foothills, we find the edafohygrophytic series, which are vegetations adapted to moist soils, such as in riparian zones (near rivers), lakes, or floodplain areas. ***Geopermaseries***: The vegetation of a geopermaserie, often termed geopermasigmeta, refers to a distinct catenal arrangement of adjacent permaseries or permasigmeta that exists in environments characterized by specific and variable topographic and soil conditions. Importantly, geopermaseries are defined by the absence of successional stages. These unique plant communities thrive in extreme climates—such as those found in high mountains and polar regions—and exceptional microtopographic settings, including cliffs, marine outcrops, salt flats, and rocky substrates. Ideal sites for geopermaseries encompass ledges, rock cracks, peat bogs, coastal saltmarshes and inland saline habitats, mobile sand dunes, and areas adjacent to lakes and streams, all of which contribute to the establishment of continuous vegetation permaseries that reflect consistent ecological persistence (Rivas Martínez et al., 2017).

Climatophilous vegetation series

The main climatophilous series of vegetation in the Río Tinto basin area are the open forests of *Quercus rotundifolia* (Fig. 10.2A) and *Q. suber* (Fig. 10.2D) and their preforest and secondary scrub stages, which are predominant in almost all the territory (Rufo et al., 2010).

- Evergreen mesomediterranean Luso-Extremaduran silicious dry to subhumid forest of *Quercus suber* (Sanguisorbo agrimonioides-Querco suberis S.)

In the study area, this series is confined to the deep soils of the Sierra de Aracena, Picos de Aroche, and their foothills, with elevations exceeding 400–500 m. Its presence in the Río Tinto basin is very sparse.

- Evergreen mesomediterranean Luso-Extremaduran silicious dry to subhumid forest of *Quercus rotundifolia* (Pyro bourgaeanae-Querco rotundifoliae S.).

This series is located on siliceous soils of the mesomediterranean thermotype in the Luso-Extremaduran Subprovince (Ladero, 1987). In the study area, which is essentially thermomediterranean, it is represented above 500 m on lower summits near the Sierra de Aracena.

The climax open forest of this series corresponds to the oak formations (*Quercus rotundifolia*) with *Pyrus bourgeana*. Thermophilic elements include species such as *Pistacia lentiscus*, *Smilax aspera*, *Olea europaea* subsp. *sylvestris*, *Teucrium fruticans*, *Cistus monspeliensis*, and *Arisarum simorrhinum*, among others. The initial stages of degradation are characterized by the communities of *Asparago albi-Rhamnetum oleoidis quercetosum cocciferae* and *Retamo sphaerocarpae-*

Cytisetum bourgaei in dry seasons, and *Phillyreo angustifoliae-Arbutetum unedonis pistacietosum lentisci* in subhumid seasons. The subsequent stage in the regressive dynamics are the scrublands (jarales), which in this region are characterized by the species *Ulex eriocladus*, and a greater presence of rockrose species such as *Cistus ladanifer* and *Cistus monspeliensis*, and in the humid seasons, the heath *Erica australis* prevails (Rufo and de la Fuente, 2010).

- Evergreen thermomediterranean Iberoatlantica silicious dry to subhumid forest of *Quercus rotundifolia.*

Myrto communis-Querco rotundifoliae S. Acidophilic and Thermomediterranean, this open forest is located on compact siliceous substrates (slates and quartzites, in our territory) from the dry to lower subhumid zone. It covers a large part of the Río Tinto basin, showing different facies depending on the moisture level.

The typical facies are led by the holm oak open forests with myrtle (*Myrto communis-Quercetum rotundifoliae*), which are usually found on hills and gentle slopes of the dry zone. At the entrances to valleys and areas with some moisture, the forest is enriched with cork oaks (*Quercus suber*), and on very steep slopes, where the river is embedded and where thermic compensation phenomena due to humidity occur, Portuguese oaks are frequent trees (*Quercus broteroi*). The characteristic species of this forest are *Myrtus communis*, *Pyrus bourgaeana*, *Hyacinthoides hispanica*, *Smilax aspera*, *Anemone palmata*, *Rubia peregrina*, and *Sanguisorba hybrida*, among others.

Rufo and de la Fuente (2010) described a subhumid variant of these open forests that grows in sharp gullies boxed in the first section of the river course. These situations cause fog and thermic inversions that allow certain thermic compensation by humidity, limiting the effect of the dry and hot summer. Thus, the forest formation is enriched with singular species such as *Quercus broteroi*, *Viburnum tinus*, *Phillyrea latifolia*, *Lonicera implexa*, *Pistacia terebinthus*, and *Asplenium onopteris*. Besides, it also includes some species typical of cork oak open forests, such as *Arbutus unedo* and *Erica arborea*, together with some thermophilous elements (*Quercus coccifera* and *Pistacia lentiscus*) and a decrease in the presence of *Chamaerops humilis*, *Myrtus communis*, and *Asparagus aphyllus*, among others. This group of taxa is more typical of cork oak open forests (Rivas-Martínez, 1988; Valle et al., 2003); however, the lack of deep soils and the steep topography limit the presence and establishment of cork oak trees, with *Quercus rotundifolia* being the dominant tree. Of course, an intermediate situation exists between these two commented situations: this circumstance occurs in valleys and smooth slopes, where a certain degree of humidity and deeper soils are present. Thus, a pool of indicators of humidity, such as *Quercus suber*, *Erica arborea*, *E. lusitanica*, *E. australis*, *Cistus populifolius*, *Rhamnus alaternus*, and *Lavandula viridis*, appears. This transitional community holds significant potential within the mining territory, contingent upon the fulfillment of its specific edaphic and climatic requirements.

It is uncommon to find a good representation of these open forests (less than 1% of the studied area), as their potential area was cleared up for livestock farming or has been altered by mining activity. The best examples are found on steep, hard-to-access slopes. Most of these open forests were destroyed by mining activities or partially cut down for raising pigs, cattle, plantations of *Eucalyptus* sp. pl., or reforestation of *Pinus pinea* and *P. pinaster* s.l.

The first stages of forest replacement are, in dry conditions, the scrublands of *Asparago albi-Rhamnetum oleoidis*, which in wetter situations can be replaced by strawberry tree thickets of *Phillyreo angustifoliae-Arbutetum unedonis*.

Preforest stages

Dry community of the European fan palm (Chamaerops humilis)

The vegetation stage before the dry type of open forest consists mainly of formations structured by the European fan palm (*Chamaerops humilis*) and the olive (*Olea europaea* subsp. *sylvestris*), which appear together with *Asparagus albus*, *A. aphyllus*, *Phlomis purpurea*, *Arisarum simorrhinum*, with *Phillyrea angustifolia*, *Pistacia lentiscus*, and *Osyris alba*. Rarely *Rhamnus oleoides* can be found. These woody formations normally appear in the driest locations, growing on typical acidic and compact substrates (Paleozoic slates). This kind of vegetation is included in the *Asparago albi-Rhamnetum oleoidis* phytosociological association (Rivas Goday et al., 1959) and is practically outside of the mining area, being limited to the rocky compact substrates.

Subhumid community of strawberry tree (*Arbutus unedo*)

Instead of the *Chamaerops humilis* community, the subhumid variety of the open forests presents a preforest stage typical of the humid cork oak forests of the Iberian Peninsula. It is structured mainly of the strawberry tree (*Arbutus unedo*), the tree heather (*Erica arborea*), and the Oleaceae shrub *Phillyrea angustifolia*. This plant association, which corresponds to the *Phillyreo angustifoliae-Arbutetum unedonis* phytosociological association (Rufo and de la Fuente, 2010), includes numerous shrubs as companion species such as *Cistus populifolius*, *C. ladanifer*, *C. monspeliensis*, *Erica scoparia*, *E. australis*, *Calluna vulgaris*, *Genista triacanthos*, and *Cytisus striatus* ssp. *eriocarpus*, among others (Fig. 10.2B and C).

The next stage in dry conditions corresponds to the gorse and broom scrublands of *Ulici eriocladi-Cistetum ladaniferi* subass. *Cistetosum monspeliensis*, which transforms into the subassociation *ericetosum australis* in wetter conditions (Rufo and de la Fuente, 2010). In this same situation, but in extremely acidic conditions, the community *Ulici eriocladi-Ericetum andevalensis* can be recognized in the mining area where a large amount of waste rock and tailings exists. These materials are deposited on the surface of the area and are the potential sources of some metals (Fe, Cu, Zn, As, and Pb) that might leach in the predominant extreme acid conditions of this territory. These environmental conditions have led to the establishment of plant communities with a high level of adaptation.

The scrub community of this habitat (fluvial terraces and gullies caused by hillside erosion of the tailings and mining areas of the Río Tinto basin) corresponds to the *Ulici eriocladi-Ericetum andevalensis* phytosociological association and is composed mainly of the endemic heather *Erica andevalensis*, accompanied by *E. australis*, and sometimes by *E. umbellata* (Rodríguez et al., 2007).

Heathland dwarf scrub serial communities are found on eroded acid soils. These formations are structured by the dwarf heather *Erica umbellata* together with *E. australis*, *Halimium ocymoides*, *Ulex eriocladus*, *Genista tridentata* susp. *lasiantha* and *Calluna vulgaris* (*Ulici eriocladi-Ericetum umbellatae*) (Rivas-Martínez, 1979); its distribution in the studied area is limited to small patches in the highest terrains of the northern section of the basin.

When the heath-gorse disappears, the most abundant community in the Tinto basin is *Loto hispidi-Chaetopogonetum fasciculati* (Rufo and de la Fuente, 2011). Occasionally, in some areas, perennial grasslands can be recognized: *Trifolium subterranei-Poetum bulbosae* in drier conditions and *Gaudinio fragilis-Agrostietum castellanae* in conditions of temporary hydromorphism. In strongly compacted soils, *Crassulo tillaeae-Saginetum apetalae* appears. In disturbed soils, xerophilic hemicryptophytic grasslands of *Andryalo laxiflorae-Hyparrhenietum hirtae* or *Inulo viscosae-Piptatheretum miliacei* can be recognized. On rocky slopes, *Phagnalo saxatilis-Rumicetum indurati* grows, while in areas with higher moisture concentration, *Selaginello denticulatae-Anogrammetum leptophyllae* develops (Rivas-Martínez, 2007; Valle et al., 2003).

Secondary dry scrub communities are characterized by the presence of *Ulex eriocladus* and *Cistus ladanifer* (Fig. 10.2F); these formations are rich in other *Cistus* species (*C. monspeliensis*, *C. crispus*, *C. salviifolius*) and other bushes (*Genista hirsuta*, *Erophaca baetica*, *Lavandula stoechas* subsp. *luisieri*). These formations correspond to the *Ulici eriocladi-Cistetum ladaniferi* phytosociological association, which is endemic to the southwestern Iberian Peninsula (Rivas-Martínez, 1979) and replaces the dry open forest.

Subhumid scrub of *Erica australis*

This plant community is the subhumid variety of the previous scrub. It differs from the dry type in the abundance of the heather *Erica australis*. This variety is known as subassociation *ericetosum australis*, and it represents one of the most common plant communities in the mining area of the Río Tinto basin (Fig. 10.2E).

Subhumid and acid scrub of *Erica andevalensis*

This endemic community, linked to the *Ulici eriocladi–Ericetum andevalensis* association, occupies acidic, metal-rich substrates such as mine tailings and eroded slopes of the Río Tinto basin. Dominated by *Erica andevalensis*, a metallophyte tolerant to extreme edaphic conditions, it represents a key pioneer stage in degraded landscapes. It develops under subhumid thermomediterranean conditions and shows a remarkable capacity to persist in oligotrophic, highly leached soils. Its ecological significance and restricted distribution make it a conservation priority and a potential asset for site-specific phytostabilization.

Dwarf scrub of *Erica umbellata*

Heathland dwarf scrub serial community on eroded acid soils. These formations are structured by the dwarf heather *Erica umbellata* together with *E. australis*, *Halimium ocymoides*, *Ulex eriocladus*, *Genista tridentata* subsp. *lasiantha* and *Calluna vulgaris* and framed within *Ulici eriocladi-Ericetum umbellatae* phytosociological association (Rivas-Martínez, 1979). In the studied area, its distribution is limited to small patches in the highest areas of the northern section of the basin.

Pioneer terophytic community of *Agrostis subspicata*

Pioneer annual and dwarf vegetation developed on bare soils, occasionally periodically flooded. This kind of vegetation is rather abundant in the mining basin, where it is found in bare soils between the scrub communities. The annual graminoid species *Agrostis subspicata*, *Micropyrum tenellum*, *Agrostis castellana*, and *Vulpia myuros* structured this vegetation type; other species such as *Jasione montana*, *Tolpis barbata*, *Corrigiola telephiifolia*, or *Trifolium campestre* are also frequent. This vegetation blooms in spring and lasts only until early summer (Fig. 10.2G). It corresponds to the phytosociological community *Loto hispidi-Chaetopogonetum fasciculati*.

Dwarf perennial chasmo-comophyte rupestrian community of *Rumex induratus*

Dwarf perennial chasmo-comophyte rupestrian community growing on walls and loose shifting stones. This vegetation corresponds to the *Phagnalo saxatilis-Rumicetum indurati* phytosociological association, which has a large distribution area in the Mediterranean West Iberian biogeographic province. Floristically, it is structured by *Rumex induratus* being *Phagnalon saxatile*, *Helichrysum stoechas*, *Dianthus lusitanus*, *Rumex acetosella* subsp. *angiocarpus* or *Arrhenatherum album* also frequent species.

This community has special relevance in the mining area, since it is the only one that can grow on the numerous slopes and cuttings of the terrain with loose stones common in the area.

Edaphohygrophilous vegetation series and geoseries

- Fluvial-alvear and fluvial Mediterranean Iberolusitan mesomediterranean dry-humid soft freshwater series of *Alnus glutinosa* and *Scrophularia scorodonia* forests. Faciation and geofaciation of thermophilic Iberolusitan of *Nerium oleander*. [Scrophulario scorodoniae-Alno glutinosae sigmetum et geosigmetum].

Riparian woodlands of *Alnus glutinosa* that grow on poor sandy or siliceous rock soil on riverbeds with running oligotrophic waters. No well-preserved examples of this community have been found in the red waters of the Río Tinto basin.

Normally, alder forests are found alongside the helophytic communities of sedges (*Caricetum tartessianae*) (Rufo and de la Fuente, 2008), which develop primarily in the stony beds of clear, flowing rivers. In small streams with clear waters, communities of *Glycerio declinatae-Oenanthetum crocatae* can be established, and on the banks, reed beds of *Juncetum rugoso-effusi* tend to develop (Rufo and de la Fuente, 2011). *Scrophulario scorodoniae-Alnetum glutinosae* may present a thorny border of *Lonicero hispanicae-Rubetum ulmifolii*. These alder forests typically connect with oleander willow groves of *Nerio oleandri Salicetum pedicellatae* toward the water, and toward drier areas, with oleander thickets of *Oenantho crocatae-Nerietum oleandri* (de la Fuente et al., 2007). Although well-developed alder forests have not been found in red waters, there are sections where this tree is present in the Agrio and Odiel rivers, as well as in clear tributaries of the Río Tinto, indicating the possible potential of this series in permanent red riverbeds. Despite the deforestation of the riverbanks in the mining area, it is an interesting fact that some alder trees have been found in the Tintillo River (UTM 29SQB 0578). This river is a tributary of the Odiel River, but its source comes from the sulgences and leachates of Corta Atalaya, the most striking mine in the entire area. It is likely absent due to poor conservation of these areas, being replaced by the oleander thicket in its acidic variant (*Oenantho-Nerietum oleandri ericetosum andevalensis*) or, in more degraded situations, by reed beds of *Junco rugosi-Ericetum andevalensis*. Oleander formations included scrubby communities dominated by *Nerium oleander*; this woody species is a micro-nano phanerophyte which forms woodland communities in the initial soils of riverbeds and creeks subdued to intense summer droughts in all the thermomediterranean area (Fig. 10.3). *Nerium oleander* (Fig. 10.3B and C) and *Erica andevalensis* (Fig. 10.3D and E), adapted to extremely acidic waters, constitute the main species of streams and rivers in the mining area of Río Tinto (Huelva, Spain) and the Iberian Pyritic Belt, often accompanied by the heliophilous fern *Pteridium aquilinum* (Fig. 10.3D).

- Fluvial and riverine marianic thermo-mesomediterranean slightly hard freshwater formations of *Salix pedicellata* with *Nerium oleander*. (Nerio oleandri-Salici pedicellatae S.)

This vegetation series can be recognized in the closest tareas to the waterways, which can be found occupying the banks and the promontories that emerge from the riverbed, where their roots remain in constant contact with the groundwater layer. Additionally, the willow groves exhibit certain adaptations that allow them to withstand the strong flows during the torrential rain periods typical of this climatic regime. This series is characteristic of the upper thermomediterranean and lower mesomediterranean levels of the Mariánico-Monchiquense sector. In the area under study, it is not frequent, as the upper course of the river, whose bioclimate would be optimal for this series, has many water courses that are deeply entrenched, on rocky and poorly developed soils that do not allow water retention during the summer. The few examples we have found appear in the branches of some reservoirs where water is guaranteed year-round. This series is structured by the willow groves of *Nerio oleandri-Salicetum pedicellatae*, which may be bordered by the thickets of *Lonicero hispanicae-Rubetum ulmifolii*. Toward the drier areas, they connect with alder groves, oleander thickets, or ash groves, depending on the hydromorphology. Toward the water, they give way to the reed beds (*Typho angustifoliae-Phragmitetum australis*), and in calm waters, the helophyte reeds of *Glycerio declinatae-Eleocharidetum palustris* and in the watercress communities of *Glycerio declinatae-Apietum nodiflori* may appear.

- Fluvial-alvear and riverine inwards Mediterranean Iberoatlantic soft freshwater thermo-mesomediterranean dry-humid series of *Nerium oleander* and *Oenanthe crocata* with *Flueggea tinctoria* and *Myrtus communis.* [Oenantho crocatae-Nerio oleandri sigmetum]

Vegetation series structured by oleander beds of *Oenantho crocatae-Nerietum oleandri*. They are usually located along waterways, in areas that experience drought during the summer months, and are enriched with tamujo (*Flueggea tinctoria*) in their drier phases. These formations connect with alder groves toward the water and with ash groves toward the dry land and may present a fringe of brambles characteristic of our territory (*Lonicero hispanicae-Rubetum ulmifolii*). The alteration of these communities gives way to wet rush meadows (*Juncetum rugoso-effusi*) toward the wet areas and to a *Trifolio resupinati-Holoschoenetum vulgaris* version toward the dry areas.

In streams with acidic waters, there is an acidophilic-metallic faciation characterized by *Erica andevalensis*. This variant has the oleander beds of *Oenantho crocatae-Nerietum oleandri ericetosum andevalensis* as its climax stage, which is replaced toward the wet areas by *Junco rugosi-Ericetum andevalensis* and, toward the dry areas, by an impoverished version of species from *Trifolio resupinati- Holoschoenetum vulgaris*.

In the upper section of the Tinto basin, the soil is poorly developed, and the waters flow through narrow ravines that dry up in the summer, preventing the development of willow and alder groves. Therefore, in these channels, the oleander bed is the potential edaphic vegetation for permanent communities.

- Fluvial Iberolusitan thermo-mesomediterranean dry subhumid freshwater soft or slightly hard series and geoserie of *Fraxinus angustifolia* and *Ranunculus ficariiformis* with *Lonicera hispanica* and *Erica scoparia.* [Ranunculo ficariae-Fraxino angustifoliae sigmetum et geosigmetum]

This vegetation series is structured by *Fraxinus angustifolia*, located in the drier bands of permanent water courses, where flooding is rare, on pseudogley sandy siliceous soils. In the Luso-Extremaduran faciation, the bramble community *Lonicero hispanicae-Rubetum ulmifolii* can act as a forest fringe and sometimes as its first stage of succession, although it is usually replaced by the oleander thickets of *Oenantho crocatae-Nerietum oleandri*, often enriched with *Flueggea tinctoria* (tamujo). In more degraded phases, the community is replaced in wetter areas by rushes of *Juncetum rugoso-effusi*, *Trifolio resupinati-Holoschoenetum*, or *Trifolio resupinati-Caricetum chaetophyllae* if the land is grazed; in drier areas by the perennial community of *Imperato cylindricae-Scirpetum holoschoeni*. When these ash groves grow in wetter soils, they often interface with willow thickets. At the bottom of the watercourse, with continuous water flow, we can find perennial communities of sedges (*Carex tartessiana*).

In the Tinto basin, there are no well-preserved ash groves except in some specific areas of the mid-course, although their stages of degradation are visible, occupying a large part of the streams in the mid-course.

Theoretically, the catenary arrangement of the different edaphohygrophilic vegetation series in decreasing order of humidity would consist of willow groves (Nerio oleandri-Salici pedicellatae S.), alder forests (Scrophulario scorodoniae-Alno glutinosae S.), oleander thickets (Oenantho crocatae-Nerio oleandri S.), and ash groves (Ficario ranunculoidis-Fraxino angustifoliae S.). In the Tinto basin, the potential of these geoseries covers the entire upper and middle course, and their development depends on the geomorphology and continuity of the waterways. Thus, in the upper course, where most of the waterways are deeply incised and experience prolonged low water levels, the conditions do not allow for the emergence of willow groves and alder forests, with oleander thickets developing preferentially (Rufo et al., 2011). In general, the banks of the Tinto and its tributaries show a high state of degradation, which is why good examples of willow groves, alder forests, and ash groves are rarely found, and when they do appear, they occupy small, poorly structured areas, typically dominated by their successional stages. The different characteristics of the waterways that comprise the Tinto basin (red waters and clear waters) result in variations in vegetation and flora. Specifically, according to our observations, the red water courses currently lack the following communities: *Caricetum tartessianae*, *Glycerio declinatae-Oenanthetum crocatae*, *Juncetum rugoso-effusi*, *Nerio oleandri-Salicetum pedicellatae*, and *Ficario ranunculoidis-Fraxinetum angustifoliae*. Rufo et al. (2010), after studying the vegetation of these geoseries, present a variant that, in addition to being thermophilic, is acidophilic-metallic and is therefore characterized by the presence of *Erica andevalensis*.

Erica andevalensis is the only species exclusive to red waters. It is important to emphasize that these observations are based on the current vegetation, which, as noted earlier, is seriously affected by industrial activities. Therefore, it is complicated to ascertain which of the species absent from red waters cannot inhabit them. In this sense, changes in the flora of both types of water that flow very close together are particularly relevant. For example, clear water courses that end in red water lose species such as *Carex reuteriana* subsp. *tartessiana* or *Oenanthe crocata*.

On the other hand, changes in the sizes, shapes, and other characteristics of many species common to both types of water are also of interest. For instance, oleander thickets in red waters are usually not as dense and tall as those in clear waters. Species such as *Alnus glutinosa*, and others such as *Scirpoides holoschoenus* or *Phragmites australis*, have their size significantly reduced when they grow in red waters.

Permanent vegetation of saline soils

This group encompasses the estuarine soils of the lower Río Tinto basin, where permanent halophytic vegetation develops under strong tidal influence and extreme edaphic constraints (Fig. 10.4). Within a narrow spatial gradient, distinct plant communities are arranged according to elevation, salinity, and flooding regime. These formations are primarily composed of perennial graminoids such as *Spartina maritima* and *S. densiflora*, along with succulent nanophanerophytes from the Chenopodiaceae family—including *Sarcocornia perennis*, *S. pruinosa*, *S. alpini*, and *Arthrocnemum macrostachyum*—which collectively define the physiognomy of the tidal marshes (Ramírez et al., 2025).

Vegetation cover is typically dense and continuous, forming monospecific or low-diversity canopies that occupy the soil surface almost entirely, even under recurrent submersion. Soil pH in this sector ranges from extremely acidic to slightly alkaline, reflecting the influence of tidal flushing and sedimentary heterogeneity. Salinity levels are moderate to high, with EC values that exceed those of other vegetated zones in the basin. The substrate is dominated by quartz and phyllosilicates (mainly kaolinite and illite), with accessory components such as Ca and Na feldspars, dolomite, calcite, and trace K feldspar (Rufo et al., 2010).

The daily tidal cycle not only moderates acidity but also drives the accumulation of sodium and magnesium, promotes saline enrichment, and triggers partial precipitation of dissolved metals in the water column. However, the geochemical behavior of metals in the Río Tinto estuary deviates markedly from that of other estuarine systems. Unlike typical metal-rich rivers, here the partitioning of dissolved metals to suspended particulates is minimal—except for lead—and metal retention through flocculation or adsorption remains limited even at elevated salinity (Elbaz-Poulichet et al., 2001). As a result, substantial loads of copper, zinc, and arsenic are transported downstream and ultimately discharged into the Atlantic Ocean (Achterberg et al., 2003; Braungardt et al., 2003; Morillo et al., 2004).

Halophytic intertidal Lusitanian-Andalusian thermomediterranean of the decumbent crassicaulous subshrubs geopermaserie of *Sarcocornia perennis* with *Puccinellia iberica*.

Puccinellio ibericae-Sarcocornio perennis geopermaserie.

FIGURE 10.4

Marshland plant community along the Río Tinto. (A) Tidal marshes dominated by *Spartina maritima*, strongly influenced by daily tidal fluctuations; (B) Zonation of halophilic vegetation along an ecological gradient, with *Sarcocornia perennis* occupying lower, more saline areas closer to the sea, and *Sarcocornia pruinosa* occurring slightly higher in microtopographic elevation; (C) Invasive plant communities of *Spartina densiflora* displacing native marsh vegetation; (D) Salt marsh communities in drier zones dominated by *Sarcocornia alpini* and *Arthrocnemum macrostachyum*; (E) Salt crusts (efforescences) with background patches of halophyte vegetation primarily composed of *A. macrostachyum*.

This geopermaserie is characteristic of estuaries and coastal marshes of the Lusitanian Andalusian Coast. It is commonly found in saltworks, saline soils, and salt marshes at the mouths of rivers.

The communities that make up this geopermaserie are dominated in biomass by species tolerant of excessive salinity and essentially belong to the classes *Spartinetea maritimae* and *Sarcocornietea fruticosae*.

The theoretical catenal arrangement of the different associations, varying in humidity, corresponds to the following scheme: *Spartinetum maritimae* (perennial meadows of halophytic grasses with complete daily flooding) (Fig. 10.4A) and *Puccinellio ibericae-Sarcocornietum perennis* (decumbent shrublands with partial daily flooding) (Fig. 10.4B). Slightly drier is characterized by communities of *Cistancho phelypaeae-Sarcocornietum pruinosae* (Fig. 10.4B) (Rufo et al., 2016), which are partially inundated by daily tides and experience complete inundation during spring tides. *Spartinetum densiflorae* (perennial meadows of halophytic grasses that withstand some degree of daily flooding, Fig. 10.4C), *Halimiono portulacoidis-Sarcocornietum alpini* (elevated but low shrublands that withstand partial flooding (Fig. 10.4D), *Inulo crithmoidis-Arthrocnemetum macrostachyi* (elevated nonflooded shrublands (Fig. 10.4E) (Ramírez et al., 2021), and *Polygono equisetiformis-Limoniastretum monopetali* (nanophanerophytic community of edges and elevated areas of nonflooded clay marsh).

Among the clearings formed by suffruticose and succulent Chenopodiaceae, therophytic communities characterized by high inundation levels and tidal influences thrive, specifically those classified under the Thero-Salicornietea class. In contrast, areas subjected to lower inundation levels are characterized by ephemeral spring communities classified within the Saginetea maritimae class (Rufo, 2009; Salazar-Mendías and Lendínez, 2021).

In the Tinto Basin, this geopermaserie occupies the entire lower course essentially, extending from the town of San Juan del Puerto to the mouth in Huelva. This area of Tinto is highly industrialized, which is why the geopermaserie is not fully represented. The communities primarily dominated by *Sarcocornia pruinosa* and, to a lesser extent, *Sarcocornia alpini* and *Arthrocnemum macrostachyum*, are being replaced by the invasive grass *Spartina densiflora* in several areas of the estuarine territory of the Río Tinto (Fig. 10.4C) (Ramírez et al., 2025).

Economic activities

Extractive activities, both historical and contemporary, have left a significant imprint on the landscape of the Andévalo region. The IPB, shared between Spain and Portugal, hosts one of the largest deposits of sulfide minerals in Western Europe (Martín-Méndez et al., 2023). Notable examples of these mining sites include São Domingos and Caveira in Portugal, as well as Río Tinto, Aznalcóllar, and Las Cruces in Spain; some of which are now abandoned with limited remediation efforts.

The region has undergone a notable industrial transformation, especially in the vicinity of the Río Tinto and the Río Odiel, converging at the port of Huelva. The Association of Chemical, Basic, and Energy Industries of Huelva (AIQBE) stands as the most important industrial hub in Andalusia. This industrial ecosystem, comprising 20 plants within the municipalities of Huelva and Palos de la Frontera, includes companies such as Air Liquide, Algry, Altersun, Atlantic Copper, Bioolis, CEPSA, Decal, Enagas, Endesa, Equo, Exolum, Fertiberia, Gunvor, Lipsa, Magnon,

Natrurgy, Repsol, and Venator. In 2022, the sector generated 20% of the provincial gross value added (GVA) and 6.1% of employment, representing 75.6% of the port traffic of Huelva, with a workforce exceeding 12,000 people. The agriculture, livestock, forestry, and fishing sectors are also very important in Andalusia, being strongly focused on foreign markets.

The phosphogypsum ponds, situated between the Estero del Rincón and the Río Tinto, occupy an extensive marshland characterized by sparse halophytic vegetation. These deposits in Huelva represent a major environmental concern due to their high contamination potential (González, 2022). In response, the Nuclear Safety Council and Fertiberia have initiated restoration plans targeting the most affected areas, particularly the ponds formed during phosphoric acid production from phosphate rock (European Parliament, 2023). The site is divided into five zones: Marismas del Pinar (Zone 1), Marismas del Rincón (Zones 2 and 5), and Marismas de Mendaja (Zones 3 and 4). Zones 1 and 4 have been partially restored, although specific degraded areas within Zone 4 require further intervention. Restoration strategies include capping the ponds with a soil layer. Zone 5, covering 154 ha, was not subject to waste disposal and presents no radiological risk, though selective restoration is planned in some areas.

The favorable climatic conditions of recent decades have favored the expansion of crops, especially strawberries, in municipalities such as Palos de la Frontera, Moguer, Lucena del Puerto, Rociana del Condado, and Bonares. These territories, sustained by a network of streams and rivers that flow into the Río Tinto, are critical water reserves. Notably, the Estero de la Rivera de Niebla and Rivera del Nicoba drain into the Río Tinto from the left bank, while Arroyo Roenero, Arroyo Hondo, Arroyo Mariana Gomez, Arroyo del Fresno (Bonares, Lucena del Puerto), Arroyo Horcajo, Arroyo de los Pilarejos, Arroyo de Valbuena, Arroyo de Galarin, Rio Pozo, and Rio Moguer contribute from the right bank.

Huelva represents 98% of strawberry production in Spain and 30% of the European Union (TNFD, 2024). Strawberry production has multiplied in the last 40 years, with Huelva producing 65% of all Spanish strawberries, which are characterized by their large size, conical shape, intense red color, and sweet, juicy flavor. These strawberries are exported to over 30 countries.

References

Achterberg, E.P., Herzl, V.C.M., Braungardt, C.B., Millward, G.E., 2003. Metal behaviour in an estuary polluted by acid mine drainage: the role of particulate matter. Environ. Pollut. 121, 283–292.

Amils, R., González-Toril, E., Fernández-Remolar, D., Gómez, F., Rodríguez, N., Durán, C., 2002. Interaction of the sulfur and iron cycles in the Rio Tinto ecosystem. Rev. Environ. Sci. Biotechnol. 1, 299–309.

Bowen, H.J.M., 1979. Environmental chemistry of the elements. Academic Press, London.

Braungardt, C.B., Achterberg, E.P., Elbaz-Poulichet, F., Morley, N.H., 2003. Metal biogeochemistry in a mine-polluted system in Spain. Appl. Geochem. 1.

Cánovas, C.R., Olías, M., Nieto, J.M., Sarmiento, A.M., Cerón, J.C., 2007. Hydrogeochemical characteristics of the Tinto and Odiel rivers (SW Spain): factors controlling metal contents. Sci. Total Environ. 373 (1), 363–382. https://doi.org/10.1016/j.scitotenv.2006.11.022.

de la Fuente, V., Rufo, L., Rodríguez, N., Amils, R., 2007. Los adelfares del suroeste de la Península Ibérica. Lazaroa 5–14.

European Parliament (2023) Parliamentary questions: Answer given by the Commission (Accessed: 30 March 2025).

Elbaz-Poulichet, F., Braungardt, C., Achterberg, E., Morley, N., Cossa, D., Beckers, J.M., Nomérange, P., Cruzado, A., Leblanc, M., 2001. Metal biogeochemistry in the Tinto-Odiel rivers (Southern Spain) and in the Gulf of Cadiz: a synthesis of the results of TOROS project. Cont. Shelf Res. 21, 1961–1973.

Fernández-Remolar, D., Morris, R.V., Gruener, J.E., Amils, R., Knoll, A.H., 2005. The Rio Tinto Basin, Spain: mineralogy, sedimentary geobiology, and implications for interpretation of outcrop rocks at meridiani planum, Mars. Earth Planet Sci. Lett. 240, 149–167.

González, F., 2022. InSAR-based mapping of ground deformation caused by industrial waste disposals: the case study of the Huelva phosphogypsum stack, SW Spain. Bull. Eng. Geol. Environ. 81, 304. https://doi.org/10.1007/s10064-022-02809-6.

I.G.M.E, 1928–1972. *Mapa geológico de España escala 1:50.000. Hojas: 938–960–982–999.* Servicio Geológico. Ministerio de Industria y Energía, Madrid.

Ladero, M., 1987. España lusoextremadurense. In: Peinado, M., Rivas-Martínez, S. (Eds.), La vegetación de España. Universidad de Alcalá, de Henares, pp. 453–486.

Lescuyer, J.L., Leistel, J.M., Marcoux, E., Moléis, J.P., Thiéblemont, D., 1998. Late devonian–early carboniferous peak sulphide mineralization in the western Hercynides. Miner. Deposita 33, 208–220.

López-Archilla, A.I., Marín, I., Amils, R., 2001. Microbial community composition and ecology of an acidic aquatic environment: the Rio Tinto, Spain. Microb. Ecol. 41, 20–35.

Martín-Méndez, I., Llamas Borrajo, J., Bel-lan, A., et al., 2023. Geochemical distribution in residual soils of Iberian Pyrite Belt (Spain). J. Iber. Geol. 49, 97–114. https://doi.org/10.1007/s41513-023-00210-0.

Morillo, J., Usero, J., Gracia, I., 2004. Heavy metal distribution in marine sediments from the southwest coast of Spain. Chemosphere 55, 431–442.

Ramírez, E., Rufo, L., Sánchez-Mata, D., Sánchez-Gavilán, I., de la Fuente, V., 2021. Arthrocnemum macrostachyum plant communities in the Iberian Peninsula, Balearic and Canary Islands (Spain and Portugal). In: Pedrotti, F., Box, E.O. (Eds.), Tools for Landscape-Scale Geobotany and Conservation. Springer International Publishing, pp. 231–245. https://doi.org/10.1007/978-3-030-74950-7_12.

Ramírez, E., Rufo, L., de la Fuente, V., 2025. Dynamics of perennial halophytic communities in the Salicornioideae subfamily: a case study from the Tinto river and Western Iberian Peninsula. Mediterranean Botany 46 (1), e98653. https://doi.org/10.5209/mbot.98653.

Rivas-Goday, F., Esteve, F., Fernández-Galiano, E., Rigual, A., Rivas-Martínez, S., 1959. Contribución al estudio de la Quercetea ilicis hispanica. Anales del Instituto Botánico A.J. Cavanilles 17 (2), 285–403.

Rivas-Goday, S., 1964. Vegetación y flórula de la cuenca extremeña del Guadiana. Diputación Provincial de Badajoz 777.

Rivas-Martínez, S., 1979. Brezales y jarales de Europa occidental (Revisión fitosociológica de las clases Calluno-Ulicetea y Cisto-Lavanduletea). Lazaroa 1, 5–127.

Rivas-Martínez, S., 1988. Bioclimatología, biogeografía y series de vegetación de Andalucía Occidental. Lagascalia 15 (Extra 1), 123–145.

Rivas-Martínez, S., 2007. Mapa de series, geoseries y geopermaseries de vegetación de España. Parte I. Itinera Geobotanica 17, 1–436.

Rivas-Martínez, S., Peñas, A., Díaz González, T.E., Canto, P., del Río, S., Costa, J.C., Herrero, L., Molero, J., 2017a. Biogeographic units of the Iberian Peninsula and Balearic Islands to district level. In: Loidi, J. (Ed.), The Vegetation of the Iberian Peninsula. Springer, pp. 131–188.

Rivas-Martínez, S., Peñas, A., del Río, S., Díaz González, T.E., Rivas-Sáenz, S., 2017b. Bioclimatology of the Iberian Peninsula and Balearic Islands. In: Loidi, J. (Ed.), The Vegetation of the Iberian Peninsula. Springer, pp. 29–80.

Rodríguez, N., Amils, R., Jiménez-Ballesta, R., Rufo, L., de la Fuente, V., 2007. Heavy metal content in *Erica andevalensis*: an endemic plant from the extreme acidic environment of Rio Tinto and its soils. Arid Land Res. Manag. 21, 51–65. https://doi.org/10.1080/15324980601074578.

Rufo, L., 2009. *Caracterización de las fitocenosis del Rio Tinto: Biodiversidad y aplicaciones biotecnológicas*. Tesis Doctoral. Universidad Autónoma de Madrid.

Rufo, L., de la Fuente, V., 2008. Carex reuteriana subsp. tartessiana (Luceño & Aedo) Rivas Martínez in the río Tinto basin (Huelva, España). Stud. Bot. 27, 193–196.

Rufo, L., de la Fuente, V., 2010. Vegetación arbórea y arbustiva de la cuenca del Rio Tinto (Huelva, España). Lazaroa 31, 39–58.

Rufo, L., Rodríguez, N., Amils, R., De la Fuente, V., Jimenez-Ballesta, R., 2007. Surface geochemistry of soils associated to the Tinto River (Huelva, Spain). Sci. Total Environ. 378, 223–227. https://doi.org/10.1016/j.scitotenv.2007.01.051.

Rufo, L., Rodríguez, N., de la Fuente, V., 2010. Chemical and mineralogical characterization of the soils of the main plant communities of the Río Tinto basin. Schironia 9, 5–12.

Rufo, L., de la Fuente, V., 2011. Vegetación herbácea vivaz de la cuenca del Rio Tinto (Huelva, España). Lazaroa 32, 73–89. https://doi.org/10.5209/rev_LAZA.2011.v32.37253.

Rufo, L., Rodríguez, N., Fuente, V., 2011. Plant communities of extreme acidic waters: the Rio Tinto case. Aquat. Bot. 95, 129–139.

Rufo, L., de la Fuente, V., Sánchez-Mata, D., 2016. *Sarcocornia* plant communities of the Iberian Peninsula and the Balearic Islands. Phytocoenologia 46 (4), 383–396. https://doi.org/10.1127/phyto/2016/0113.

Salazar-Mendías, C., Lendínez, M.L., 2021. Mediterranean halophytic flora and vegetation in the Iberian Peninsula (Spain and Portugal). In: Grigore, M.N. (Ed.), Handbook of Halophytes. Springer International Publishing, pp. 325–368. https://doi.org/10.1007/978-3-030-57635-6_12.

TNFD (2024) Case study: red Gold Spanish strawberries and German supermarkets (Accessed: 30 March 2025).

Tornos, F., 2006. Environment of formation and styles of volcanogenic massive sulfides: the Iberian Pyrite Belt. Ore Geology Reviews 28, 259–307.

Valle, F., et al., 2003. Mapa de Series de Vegetación de Andalucía. Editorial Rueda S.L., Madrid, p. 131 pp. + 1 map.

CHAPTER

Elemental concentration and biomineralization of Río Tinto flora

11

Lourdes Rufo[1], Esteban Ramírez[2], Irene Sánchez-Gavilán[2], Nuria Rodríguez[3], Vicenta de la Fuente[4]

[1]*Instituto de Investigaciones Biosanitarias, Facultad de Ciencias Experimentales, Universidad Francisco de Vitoria, Madrid, Spain;* [2]*Real Colegio Complutense at Harvard University, Cambridge, MA, United States;* [3]*Centro de Astrobiologïa (INTA-CSIC), Madrid, Spain;* [4]*Universidad Autónoma de Madrid, Campus Cantoblanco, Madrid, Spain*

Introduction

The Río Tinto represents a unique Mediterranean ecosystem characterized by its extreme acidity. The average pH of 2.3 over an area of 80 km and the high concentrations of heavy metals are two characteristics of its waters, which pose a challenge to the development of life (Amils et al., 2002, 2004). These characteristics are also evident in the sediments and soils along the river, from its source to its mouth, and condition the vegetation and flora that develop in this environment.

It is noteworthy that the Tinto ecosystem exhibits a number of distinctive characteristics that render it an ideal territory for the study of biological adaptations to acidity and metalliferous soils. The source of this river is found in the center of one of the most significant metallogenic regions of the world, the polymetallic complex sulfides known as the Iberian Pyritic Belt (IPB), which is situated in the south-west of the Iberian Peninsula (Tornos, 2006). Contrary to earlier assumptions attributing the origin of Río Tinto's extreme geochemical conditions primarily to anthropogenic mining activities, current evidence indicates that the river's acidity and high metal content are largely sustained by a natural subsurface bioreactor. This biogeochemical system, driven by the microbial oxidation of extensive sulfide deposits embedded within the Iberian Pyrite Belt, reflects a deep-seated and long-standing geological process. Thirty years of geomicrobiological research in the region (Amils, 2016) have revealed that, while mining may have amplified surface manifestations, the dominant mechanisms shaping the river's extreme environment remain fundamentally natural in origin.

It can thus be concluded that the flora that composes the Río Tinto vegetation must tolerate both acidity and high metal concentrations and must present adaptations to these features. The plant adaptations in question can be classified into one of three groups, as proposed by Baker (1981): exclusion, bioindication, and accumulation. The classification is determined by the concentrations of metals in the aerial

Geomicrobiology of Río Tinto. https://doi.org/10.1016/B978-0-443-32948-7.00021-7

tissues of the plants in relation to their substrates. Plants that function as excluders maintain metal concentrations in their aerial tissues at much lower concentrations than those present in their substrates. Of particular interest to phytostabilization techniques applicable to metal-contaminated areas are the plants included in this group. This trait is particularly advantageous for phytostabilization strategies in metal-rich environments, as it minimizes metal mobility and ecological risk while promoting soil stabilization and vegetation cover. Bioindicator plants, which accumulate metal in their aerial tissues depending on the metal concentration present in their substrates, can be used for mining prospection, among other utilities. Finally, accumulator and hyperaccumulator species have the potential to enhance the efficacy of phytoextraction techniques, which are employed to remediate metal-contaminated soils (Brooks, 1998).

Soils of the Río Tinto

Soil means "The layer(s) of generally loose mineral and/or organic material that are affected by physical, chemical, and/or biological processes at or near the planetary surface and usually hold liquids, gases, and biota and support plants", as stated in the latest definition provided by the Soil Science Society of America (van Es, 2017).

The presence of layers of inorganic substrate, i.e., soils, has been identified beneath the vegetation of the Río Tinto. These soils are of particular interest as, in conjunction with the local climate, they determine the plant species and communities that can thrive in the area.

The geomorphological characteristics of the river vary along its course, allowing for the identification of three distinct sections: source, middle course, and mouth. Corresponding changes in soil properties such as depth, texture, mineralogy, acidity, electrical conductivity, and inorganic and organic composition are observed throughout. These variations are influenced by factors including the type of substrate, water dynamics, topography, and, near the mouth, tidal effects. As a result, soil analyses are structured according to these three sections. In addition to the soils that develop under the plant communities located on the riverbanks, there are also soils in areas that are distant from the watercourse and flood zones, where climatophilous vegetation develops. This other group also merits separate consideration for study, as they are less influenced by the action of the river water.

In general, it can be stated that Río Tinto soils exhibit high concentrations of sulfur (S), iron (Fe), copper (Cu), zinc (Zn), arsenic (As), and lead (Pb); average concentrations of manganese (Mn); and low concentrations of potassium (K), magnesium (Mg), calcium (Ca), phosphorus (P), and sodium (Na). These findings are based on the analysis of soil samples obtained from the entire length of the river and the different plant communities observed (Rufo et al., 2007, 2010). The Río Tinto soils exhibit an atypical extremely low pH and high concentrations of S, Fe, Cu, Zn, Pb, and As in the two initial sections of the river (source and middle course). This observation indicates that the soils in the vicinity of the river are evidently influenced by river waters and sediments, in addition to the parent rock. However, the influence of these factors appears to diminish gradually toward the estuary (mouth of the river), as the influence of seawater becomes more dominant.

The soils of the forest and serial scrub, which are in areas distant from the river water, exhibit a marginal increase in pH and a decrease in sulfur concentrations. A comparison of riverside soils and those of the climatophilous vegetation series reveals common features, suggesting that the latter soils are also influenced by IPB substrates and mine waste lixiviates, albeit to a lesser extent.

Another common characteristic is the scarcity of elements considered essential plant nutrients. This is a general characteristic of all the groups of soils. Elements such as Ca, K, and P only increase their concentration in the presence of some potential sources such as lithologic sources (limestone in the middle course) or industrial ones (effluents from the industrial hub of Huelva).

Soils of the source of the river

This group includes soils associated with riparian plant communities dominated by *Nerium oleander* and *Erica andevalensis*, which colonize the acidic red streambanks of the region. Formed over slates, mine dumps, and other waste substrates, these soils display varying degrees of development depending on the extent of substrate alteration. Under *E. andevalensis* and *N. oleander*, soils can become relatively well developed and support dense vegetation dominated by these species. Vegetation cover ranges from 60% to 80% in mature communities to 20%–40% in early colonization stages.

The pH of this group of soils was found to be ultra-acid (median: 3.01), and their electrical conductivity varied between slightly saline and nonsaline (1.03 dS/m). The mean carbon concentrations were found to be $10.6 \pm 2.9\%$, with a significant disparity observed between the values obtained for horizons A and B in comparison to those of horizon C (0.66%–34%). *E. andevalensis* is a perennial bush that loses a large quantity of organic matter (mostly leaves) throughout the year. In these situations, an organic and mineral superficial horizon is observed. Levels from 9.8% to 25% of organic matter have been reported in horizons A and B from soils under well-established communities of this species (Rodríguez et al., 2007).

Regarding their elemental composition, Fe is the most significant element, with a median concentration of 6.04% and a maximum of 41% in the lower horizons, where it is generally found in higher concentrations than in the upper horizons. Other elements with high concentrations include S (1.32%), Cu (308.7 mg/kg), As (283.7 mg/kg), Pb (264.1 mg/kg), and Zn (106.3 mg/kg). Conversely, nutrients such as Mg (1492), K (1442 mg/kg), Ca (869.6 mg/kg), Na (654.1 mg/kg), and P (181.2 mg/kg) manifest themselves in negligible concentrations. The occurrence of elements characteristic of soils, including Al (0.664%), Cr (13.7 mg/kg), and Ti (130.5 mg/kg), is also observed. The mineralogical composition of the samples exhibits notable uniformity, primarily due to the dominance of silicate minerals, with quartz and phyllosilicates predominating across all analyzed soils. In addition, the presence of K and Ca and Na feldspars is evident. Among phyllosilicates, illite and kaolinite are distinctly present in most of the samples. Iron oxides and sepiolite were observed in a few cases (Rufo et al., 2010).

Soils of the middle course of the river

The riverside sediments and soils of the second section of the river are mostly bare. Where vegetation does occur, it typically consists of perennial helophytes such as *Phragmites australis*, *Typha domingensis*, and the grass *Imperata cylindrica*. These communities grow on alluvial soils, frequently flooded in rainy periods, which dry in summer. The recurrent flood and drought periods, typical of Mediterranean rivers, allow the continuous rejuvenation of the soils. Floods sweep away sediments and organic matter, leaving new sediments, and the soils usually present poor development, and horizons rarely differ. The predominant plant communities in this river section are characterized by rhizomatous growth, typically attaining vegetation cover levels exceeding 40%.

These soils are very acidic (median: 2.83) and slightly saline (1.64 dS/m). The elemental concentration of these soils reflects the transport dynamics of the river. The materials carried by the river waters contain large amounts of S and metals (Braungardt et al., 2003; Galán et al., 2003; Hudson-Edwards et al., 1999). The result is that the soils of this section have high concentrations of Fe (12.5%), S (3.45%), As (1456 mg/kg), Pb (1134 mg/kg), Cu (560 mg/kg), Zn (209.1 mg/kg) and Co (12.0 mg/kg), and their values are comparable to or exceed those observed in the previous group (river source). Conversely, low concentrations of K (2773 mg/kg), Ca (2314 mg/kg), Mg (1979 mg/kg), Na (942.1 mg/kg), Mn (259.3 mg/kg), and P (115.2 mg/kg) have been observed. Flood and drought periods produce a cycle of dissolution and precipitation of salts that controls to a great extent the transport of elements such as Fe, Cu, Zn, As, Pb, and the ion SO_4^{2-} along the course of the river (Buckby et al., 2003; Cánovas et al., 2007, 2008). The metals are transported along the river course through different phases: as an ionic form dissolved in the waters, adsorbed into solids, quelated in organic matter, or coprecipitated in small particles of Fe sulfides, oxides, hydroxides, oxyhydroxides, and oxyhydroxy-sulfates that compose the particulate suspended matter (Hudson-Edwards et al., 1999). The high concentrations of metals found in the soils of the middle course and mouth of the river are a consequence of the migration of the elements. The less mobile elements (As and Pb) are concentrated in the upper horizons of the second section of the river, while more mobile metals (Cu and Zn) reach their highest concentration in the estuary.

An evolution in the mineralogy of the soils is also observed along the river course. The proportion of clay increased in the second section of the river, where a trace of the influence of the neighboring limestone and dolomite also exists.

The dominant mineralogical groups are quartz, phyllosilicates, and feldspars, with illite being the most prevalent among the phyllosilicates. Other trace minerals present include kaolinite, smectite, K jarosite, gypsum, and dolomite (Rufo et al., 2010).

Soils of the mouth of the river

The group under consideration includes the soils of the Río Tinto estuary plant communities that are subject to the tide effect. The area is characterized by the presence of several plant communities within a few meters, placed along a flood gradient. The plant communities are composed mainly of perennial species of graminoids (*Spartina densiflora* and *S. maritima*) and succulent Chenopodiaceae dwarf shrubs (*Sarcocornia perennis*, *S. alpini*, *S. pruinosa*, and *Arthrocnemum macrostachyum*). Vegetation covers are nearly 100%.

The soils of the estuarine section are subjected to the daily tides. This fact causes an increase in the pH, the concentrations of Na and Mg, and the salinity and the precipitation of a part of the metals in the waters of the river. The pH of these soils ranges from extremely acidic to slightly alkaline (median: 6.87), with the mean electrical conductivity value classified as slightly saline (5.11 dS/m). The estuary soils share with the other groups the characteristic of high concentrations of Fe (5.73%) and S (0.615%), yet they differ from them in their notably high concentrations of Na (10,449 mg/kg), Mg (6748 mg/kg), and P (2728 mg/kg). Additionally, they exhibit the highest levels of Zn (1705 mg/kg) and Cu (1331 mg/kg), a consequence of the migration of the elements, as stated before. The predominant mineralogy is dominated by quartz and phyllosilicates (kaolinite and illite), with calcium-rich feldspars present in lower proportions. Trace mineral analysis reveals the presence of K feldspar, dolomite, and calcite (Rufo et al., 2010).

Soils of climatophilous vegetation series (forests and scrubs)

This group encompasses soils from the main climatophilous vegetation series within the area, including *Quercus rotundifolia* and *Q. suber*, along with their preforest and secondary scrub stages, which are dominant across most of the territory. Soils within this group are predominantly Entisols, with some Inceptisols present (Heredia et al., 1980). These soils are characterized by a thin superficial layer of organic matter, with the underlying rock, predominantly slates, remaining largely unaltered. The vegetation cover varies significantly (60%–80%) depending on the state of conservation of the vegetation and the topography of the terrain. Notably, certain plant communities within this vegetation series are able to thrive on steep slopes, where soils are susceptible to frequent erosion processes.

The pH of these samples is extremely acidic (median: 3.93), with values higher than those observed in the soil samples of the source and middle course of the river but lower than those of the soil samples of the estuary. The electrical conductivity of these samples is not saline, indicating the lowest salinity among the studied soil groups (0.076). Generally, the carbon concentration of these samples is less than 10%. These soils exhibit high concentrations of Fe (5.06%), S (0.16%), and As (233.9 mg/kg), which are lower than those found in the middle section of the river. Concentrations of Pb (230 mg/kg), Cu (168 mg/kg), and Zn (128 mg/kg) are also high. Concentrations of Mg (1519 mg/kg), K (1171 mg/kg), Ca (311.3 mg/kg), Mn (221.6 mg/kg), P (113.4 mg/kg), and Na (103.9 mg/kg) are low, as occurs in the soils of the source and middle course of the river.

Quartz and phyllosilicates are predominant in this group, with clays such as illite and smectite reaching high proportions. Clay proportions are much higher in the forest soils, indicating a higher degree of evolution. Among clays appear illite, a typical product of the degradation of the mica capable of retaining K^+, and smectite. Kaolinite and feldspars are present in smaller proportions. In soils developed over gossan (oxide massive sulfide) parent rock, a high concentration of hematite was observed. In soils taken from substrates subjected to prolonged acid washes, quartz was found to be the predominant mineral. Soils located over the limestone of the second section of the river had basic pH (8.1), high concentrations of Ca (10.3%) and Mg (1.21%), lower values of S (0.02%) and Fe (2.80%), and a dominance of calcite and phyllosilicates (Rufo et al., 2010).

General strategies to cope with metalliferous soils of the Río Tinto flora

Direct interaction between plants and soil occurs via the radicular system, with roots absorbing most of the mineral elements required by plants, in addition to water. As previously outlined, Río Tinto soils represent a distinctive environment due to their high pH, elevated concentrations of metals and metalloids, including Fe, Cu, Zn, As, and Pb, and low concentrations of nutrients (Ca, Mg, etc.). Therefore, any plant species growing naturally on these substrates must be considered tolerant to acidity and phytotoxic concentrations of the dominant metals in the area.

Typically, plants are divided into two categories according to their ability to avoid the entry of metals (excluders) or to take up and accumulate metals in different parts of their body at concentrations higher than those found in the soil (accumulators) (Baker, 1981). Regarding the level of accumulation, a distinction is made between so-called hyperaccumulators, i.e., plants that accumulate metal concentrations exceeding 0.1% of their dry weight in the wild (Brooks, 1998). Regardless of their accumulation capacity, the species that make up the flora of metalliferous soils can be divided into *strict or obligate metallophytes*, which grow only on this type of substrate, and *facultative or tolerant metallophytes*, which can grow on this and other types of substrates with different characteristics (without high concentrations of metal).

In studying the elemental concentration of 97 plant species growing on Río Tinto substrates, we found that all the species analyzed, taken as a whole, show a general pattern of absorption and accumulation in aerial tissues that is consistent with the nutritional requirements of any vascular plant (the general pattern of accumulation corresponds to the following: macronutrients (Ca, Mg) > micronutrients (Na > Fe, Mn > Cu, Zn) > toxic elements (As, Pb). An exception to this trend was observed in halophyte taxa from the estuary, such as *Sarcocornia pruinosa* and *Arthrocnemum macrostachyum*, which accumulated sodium in aerial tissues to levels surpassing those of key macronutrients, pointing to adaptive responses to hypersaline conditions. They are all tolerant to acidity and to high concentrations of heavy metals in the soil, given the characteristics of the substrates on which they grow. In general, despite the lack of macronutrients (Ca and Mg) in the soils, the concentrations of these elements in the plants correspond to the normal values reported for vascular plants (Reeves and Baker, 2000). When compared with soil nutrient concentrations, it can be concluded that Río Tinto plant species actively accumulate both macronutrients in aerial tissues. This accumulation strategy plays a key role in the internal recycling and conservation of essential nutrients. When leaves are shed, some of these elements return to the soil where they can be reabsorbed by the plants, thus preventing their loss in the acid washes that are frequent in this area (de la Fuente et al., 2010).

On the other hand, metal accumulation is the least common strategy in the Río Tinto flora. Accumulation is observed for some metals, such as Fe in the grass *Imperata cylindrica* and Mn in shrub and forest species, although most plants are excluders of Fe, Cu, Ni, Zn, As, and Pb, especially in the case of As and Pb. Furthermore, the concentrations measured for these elements slightly exceed the levels considered normal for vascular plants (Pendias and Pendias, 2001; Reeves and Baker, 2000).

Elemental concentration and biomineralization in Río Tinto plant species

Macronutrients

Calcium (Ca)

Calcium concentrations

Calcium is an alkaline earth element that is considered an essential macronutrient in plants. At the cellular level, calcium is found in the vacuoles of plant cells, as well as in the endoplasmic reticulum and mitochondria. Calcium is also an essential component of the plant cell wall, where it is associated with pectin. It plays a key role in cell signaling and in the transport and retention of other elements and has various functions, such as regulating the permeability of the cytoplasmic membrane or counteracting the effects of alkaline salts and organic acids in plants. In the vacuoles of certain plant species, calcium precipitates mainly in the form of insoluble calcium oxalate crystals, although calcium salts of phosphate, sulfate, or carbonate are also known. Calcium deficiency has been observed to affect tissue formation, especially that of root meristems, resulting in deformed tissues or stunted growth (Kathpalia and Bhatla, 2018).

On the other hand, calcium concentration in plants is partly determined by genetic factors, as families considered as monocotyledonous angiosperms typically have lower calcium concentrations than other groups, such as eudicots.

Several studies have shown that plant species growing on Río Tinto substrates accumulate calcium concentrations above the total values present in the soil (de la Fuente et al., 2010; Sánchez-Gavilán et al., 2021). In the study of 97 different species, a mean of 6492 mg/kg d.w. was found. Significant differences were found between the mean concentrations of monocotyledons (mean $\pm$ standard deviation: (2685 $\pm$ 164.3 mg/kg d.w.) and dicotyledons (8243 $\pm$ 631 mg/kg d.w.).

From the group of plants studied, the following stood out for their high calcium content (more than 10,000 mg/kg d.w.): *Mesembryanthemum nodiflorum* (Aizoaceae), *Nerium oleander* (Apocynaceae), *Ceratonia siliqua* (Leguminosae), and *Pulicaria paludosa* (Compositae). On the other hand, some species showed concentrations lower than 5000 mg/kg d.w.: *Arthrocnemum macrostachyum* (Chenopodiaceae), *Scirpoides holoschoenus* (Cyperaceae), *Cytisus striatus* subsp. *eriocarpus*, *Ulex eriocladus* (Leguminosae), all analyzed species of the genus *Erica* (Ericaceae) and of the families Gramineae and Juncaceae (de la Fuente et al., 2010). Analysis of *N. oleander* specimens growing on extremely acidic (red waters, with average pH 2.2) and less extreme acidic waters of the Río Tinto basin has shown no differences regarding concentration of calcium (12,700–16,490 mg/kg) (Franco, Rufo and de la Fuente, 2012).

Comparing the total calcium concentrations in the substrates with those present in the stems and leaves of the plants analyzed, it is observed that the species of the families Gramineae (*Imperata cylindrica*, *Micropyrum tenellum*, and *Spartina densiflora*) and Juncaceae (*Juncus rugosus*, *J. effusus*), together with some Chenopodiaceae (*Sarcocornia perennis*, *Halimione portulacoides*), maintain a ratio lower than or equal to 1, indicating that there is no active accumulation. On the other hand, there are many species with a high accumulation value, with plant concentrations more than 20 times higher than those found in the substrates. This is the case for *Chamaerops humilis* (Palmae), *Cistus ladanifer*, *C. populifolius*, *C. monspeliensis* (Cistaceae), *Myrtus communis* (Myrtaceae), *Olea europaea* subsp. *sylvestris* (Oleaceae), *Pulicaria paludosa* (Compositae), and *Pinus pinaster* (Pinaceae).

The calcium concentration in individuals of the same species showed variability, such as in the case of the succulent stems of several individuals of the Chenopodiaceae *Salicornia patula* (5394–20,600 mg/kg), *Sarcocornia pruinosa*

(4591–6329 mg/kg), or *Arthrocnemum macrostachyum* (1830–11,817 mg/kg), or in the leaves of *Erica andevalensis* (2012–5942 mg/kg). The seeds of *S. pruinosa* have also been found to contain nonnegligible amounts of calcium (7641 mg/kg) (de la Fuente et al., 2018; Ramírez, Rodríguez and de la Fuente, 2024; Rodríguez et al., 2007; Sánchez-Gavilán et al., 2021).

Distribution and biomineralization of calcium

Microscopic observations allow us to detect crystallizations, whose main inorganic component is calcium, inside cells of very different species, being more abundant in dicotyledons, although their size, shape, location, and abundance vary from one family to another. The most common crystals are druses (star-shaped aggregates of crystals), prisms, or spherical crystals located along xylem and phloem cells in the central vascular cylinder (Fig. 11.1F–I). When present, they are found in the mesophyll and leaf epidermis, and cortical, pith parenchyma in stems and the central cylinder. In some species of *Quercus* genus, polyhedral crystals of calcium oxalate are observed along the vascular bundles of the leaves, as shown in the images corresponding to leaf sections of *Q. suber* in Fig. 11.1B, D, and G. A similar pattern is found in *Erica andevalensis* leaves (Fig. 11.1E) and *Nerium oleander*, though in this last one, calcium oxalate crystallization changes its forms, with a predominance of rose-shaped druses instead of polyhedral prismatic crystals (Fig. 11.1C and I). These druses have also been found in the parenchyma cells of the pith of the stem of *E. australis* (Fig. 11.1F). Another type of calcium crystals are raphides, which are flat needle-type crystals with pointed ends. They are usually observed in clusters, bundled together, as in *Typha domigensis* parenchyma cells around the xylem vessels (Fig. 11.1H).

In some cases, a high concentration of crystals in the fruit, close to the seed, is found. Crystals can be found in the stamens of some species. The distribution of crystals in different plant tissues is maintained in the genus, suggesting genetic control. These crystals have also been found in grasses, although in smaller proportions (Franco, Rufo and de la Fuente, 2012; Rufo, 2009).

X-ray analysis of succulent stems of *S. pruinosa* and *A. macrostachyum* reveals the presence of the oxalate weddellite, $C_2CaO_4{\cdot}2H_2O$ (de la Fuente et al., 2018; Ramírez, Rodríguez and de la Fuente, 2024).

Magnesium (Mg)

Magnesium concentrations

Magnesium is another alkaline earth element, more mobile than calcium. It is known to have several functions related to photosynthesis (constituent of chlorophyll, enzyme cofactor to produce adenosine triphosphate (ATP), intervenes in the activation of ribulose bisphosphate carboxylase (Rubisco) and phosphoenolpyruvate carboxylase (PEP carboxylase)). Furthermore, magnesium is a requisite component for the activity of numerous enzymes implicated in both respiration and nucleic acid

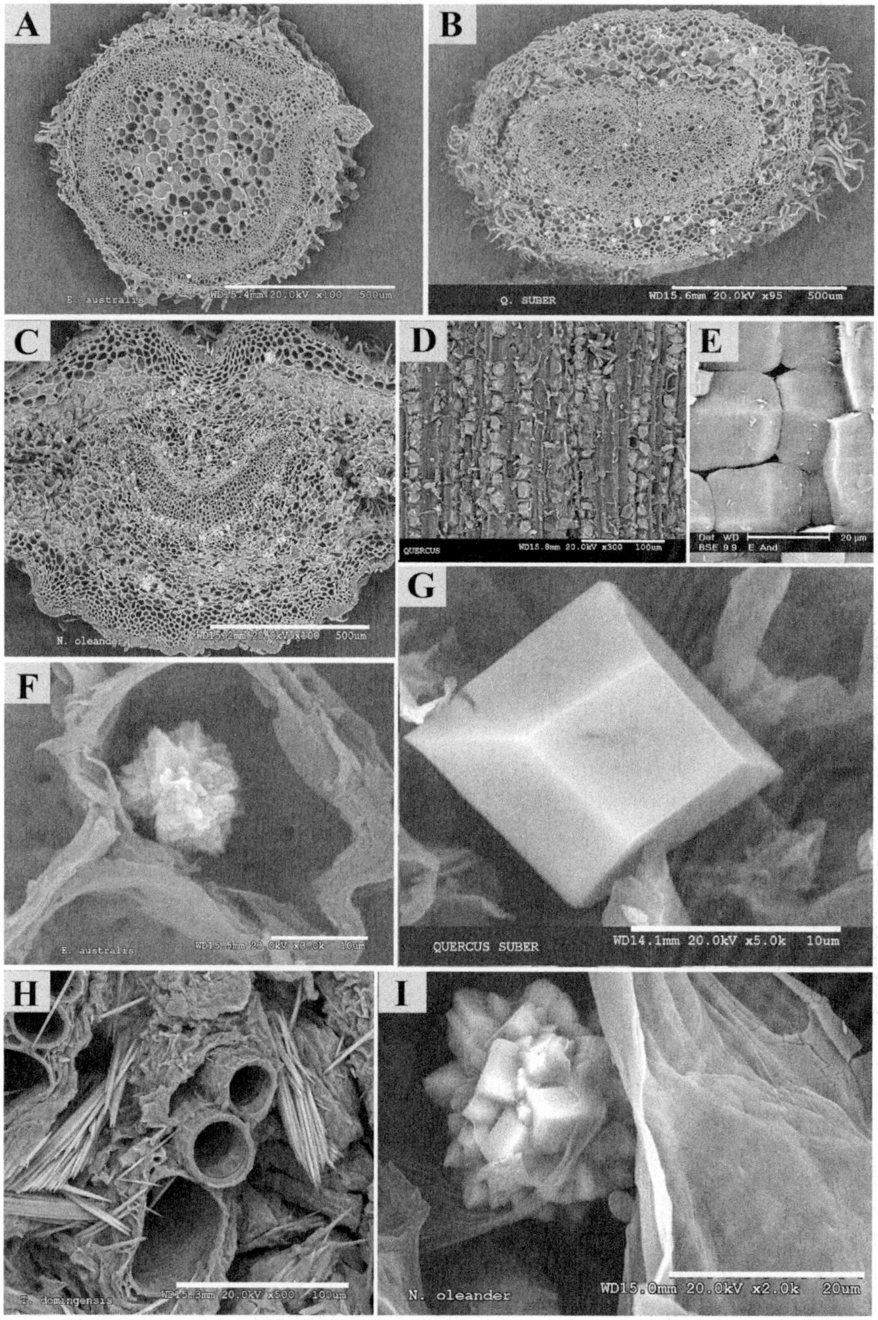

FIGURE 11.1 SEM images of Ca biomineralization.

Images showing biomineral accumulation in various species. *Erica australis* (A, F) A: Transverse section of *E. australis* leaf. F: Details of calcium crystals druse type. *Quercus suber* (B, D, G): B: Transverse section of the petiole showing prismatic calcium oxalate crystals. D and G: Crystal details. *Nerium oleander* (C and I): Another view of the transverse section of the petiole with calcium oxalate druses. *Erica andevalensis* (E): Prismatic calcium crystals. *Typha dominguensis* (H): calcium raphides around the xylem vessels.

biosynthesis. Its absence leads to interveinal chlorosis, anthocyanin pigment accumulation in older leaves, and premature leaf abscission (Kathpalia and Bhatla, 2018).

An overview of the magnesium content of 97 plant species belonging to Río Tinto reveals significant variations in concentration, which are predominantly shaped by phylogenetic lineage and the specific habitat of each species (Fig. 11.5). A notable disparity was observed between the mean concentrations of monocot (1760 mg/kg d.w.) and eudicot species (3194 mg/kg d.w.). Individuals from monocotyledonous families (Gramineae, Cyperaceae, and Juncaceae), in conjunction with certain *Erica* species (Fig. 11.1A and F), exhibited mean values below 2000 mg/kg d.w., concentrations that were lower than those observed in soils, except for *Erica scoparia*. Regarding the endemism of *E. andevalensis*, magnesium was detected in leaves and stems at concentrations ranging from 883 to 2566 mg/kg (Rodríguez et al., 2007), and in the case of *N. oleander* concentrations ranged from 1466 to 3717 mg/kg d.w., with no differences between roots and shoots (Franco, Rufo and de la Fuente, 2012).

The highest magnesium concentrations were observed in estuarine species, particularly within the succulent Chenopodiaceae, such as *Halimione portulacoides* and *Sarcocornia perennis*, with concentrations exceeding 6000 mg/kg d.w. Studies on other succulent species of this family have reported magnesium concentrations in seeds and succulent stems exceeding 6000 mg/kg for *S. pruinosa*, from 7200 to 15,000 mg/kg for succulent stems of *Salicornia patula*, and from 4700 to 9000 mg/kg for *Arthrocnemum macrostachyum* from the Río Tinto estuary (de la Fuente et al., 2010; Ramírez, Rodríguez and de la Fuente, 2024; Sánchez-Gavilán et al., 2021; de la Fuente et al., 2018).

Distribution and biomineralization of magnesium

Microscopic studies, in combination with energy-dispersive X ray diffration (EDXRD), have made it possible to visualize the presence of magnesium in the tissues of the Río Tinto plants. Given its higher concentration in estuarine plants, this element is particularly evident in the analysis of such plants, appearing in a generalized form in the cell walls and contents of almost all tissues, generally accompanied by other elements such as chlorine, potassium, and/or sodium. A higher concentration has been observed in epidermal cells, parenchyma cells (particularly those involved in water storage), and cortex cells (in cross sections of flowering and nonflowering succulent stems of *S. pruinosa* and *A. macrostachyum* (Ramírez, Rodríguez and de la Fuente, 2024; de la Fuente et al., 2018).

In some cases, crystallizations identified as magnesium oxalate (glushinkite) have been observed in these regions, particularly within the water storage tissue, as seen in the annual species *Salicornia* sp. (Fig. 11.4G). Although less frequently, magnesium accumulations were also detected coating the cells of the central cylinder in *Arthrocnemum macrostachyum* (Fig. 11.4H).

X ray diffraction analyses have been used to identify the presence of glushinkite ($C_2MgO_4{\cdot}2H_2O$).

Micronutrients

Iron (Fe)

Iron concentrations

Together with sulfur, iron is the most characteristic element of the Río Tinto ecosystem. It is found in high concentrations in substrates and waters throughout the ecosystem, mainly in the oxidized state (Fe^{3+}), which gives the medium its characteristic red color. Although it is often poorly bioavailable in substrates, high concentrations can be toxic and promote oxidative stress. In plants, iron is considered an essential micronutrient because it is required for the synthesis of many proteins (ferredoxin and cytochromes) involved in metabolic processes such as photosynthesis and respiration. It acts as a cofactor for catalases and peroxidases and is also required for the synthesis of chlorophyll. Iron deficiency results in symptoms such as chlorosis and necrosis, which first become visible in young leaves (Kathpalia and Bhatla, 2018).

However, the mean concentration found in the Río Tinto flora reported by de la Fuente et al. (2010) was 396 mg/kg d.w., which is slightly higher than that found in plants growing on nonmetalliferous soils. However, it is possible to find large differences between species.

In general, most of the species analyzed show undetectable leves of this element, although it is possible to highlight some species from different ecological environments because of their high content. This is the case of shrub species such as rockroses (*Cistus* sp.), species of the genus *Lavandula*, and some tree species such as *Q. suber*, which generally exceed 300 mg/kg d.w. Other shrub species, such as heathers (*Erica* sp.) and other Ericaceae, such as *A. unedo*, have lower Fe concentrations that do not reach 300 mg/kg d.w. This is particularly striking for *E. andevalensis*, which occupies large areas close to the riverbank and forms the scrub on rubble and bare substrates in the mine itself, where we find the most acidic pH and high concentrations of this metal (Rodríguez et al., 2007; de la Fuente et al., 2010).

Species growing in the estuary, adapted to the saline environment and to substrates with a slightly higher pH, have a higher concentration of this element in their tissues, especially succulent stems and leaves of some species of Chenopodiaceae, such as *Salicornia patula* and *A. macrostachyum* (from 77.8 mg/kg to 1005 mg/kg, with one sample from the Río Tinto estuary displaying a remarkably higher concentration (2135 mg/kg d.w.), and Aizoaceae, such as *Mesembryanthemum nodiflorum* (Fig. 11.5). Seed of *Sarcocornia pruinosa* showed an iron concentration of 418 mg/kg d.w. (de la Fuente et al., 2010; Sánchez-Gavilán et al., 2021; Ramírez, Rodríguez and de la Fuente, 2024; de la Fuente et al., 2018).

Of all the plants for which data are available, grasses stand out in terms of iron content, both in the upper and middle reaches and in the estuary (Fig. 11.5). The species *I. cylindrica* stands out. This grass, which is widespread throughout the world, grows on the banks of the middle reaches of the river and is considered to be a hyperaccumulator of this metal, since concentrations of up to 2.3% d.w. have been found in roots and rhizomes and 1.1% d.w. in leaves (Rodríguez et al., 2005).

Seeds are another organ where iron has been found. The concentration of iron in seeds of *S. pruinosa* and *A. macrostachyum* was found to vary between 93.4 and 418 mg/kg d.w. If the pericarp is included in the measurement, the concentration rises to more than 700 mg/kg d.w. in *A. macrostachyum* (de la Fuente et al., 2018; Ramírez, Rodríguez and de la Fuente, 2024).

Distribution and biomineralization of iron

The distribution of iron in different plant organs has been studied in some of the Río Tinto species. In general, all the plants studied showed a higher concentration of this element in their underground organs, especially in the root, although iron was detected in all organs. Microscopic observations and tissue staining (Prussian blue dye) show a heterogeneous distribution of Fe in the tissues of the plants analyzed, which also depends on the species.

In the roots of several species, a so-called iron plaque is formed. This is a mineral layer that envelops the root and sometimes the rhizome and is formed by the action of the plant. Its presence has been observed in both monocotyledons and dicotyledons, with Fe, S, and K as the dominant elements in its composition, and it has been attributed a protective effect against the toxicity of metals, acting as a physical barrier to the absorption of metals by retaining them precipitated on it (Franco, Rufo and de la Fuente, 2012; Franco et al., 2013). A representative selection of images of *E. andevalensis* root iron plaque is shown in Fig. 11.2A, B, and D, where the formation of polyhedric crystals composed of Fe, S, and lead was observed.

Once inside the roots, iron is transported to the rest of the organs. Studies on *I. cylindrica* have shown that while iron accumulates in various root tissues, dark precipitates are not observed within the endodermis cells themselves. This suggests that the endodermis acts primarily as a control point for transport rather than a major site of iron storage. However, iron accumulation has been observed in the parenchyma cell walls near the endodermis, indicating that the endodermis influences the localization of iron within the root tissues (de la Fuente et al., 2016).

Major differences have been found between monocotyledons and dicotyledons in iron management. In the case of grass accumulator *I. cylindrica*, iron is observed in the form of various types of minerals such as jarosite (Fig. 11.2E and F), iron oxide, and in the protein form ferritin. It is particularly visible in the xylem vessels and epidermal cells of roots and rhizomes and in the parenchyma around vascular bundles (iron particles) of leaves. In dicotyledons such as *Sarcocornia pruinosa*, *Salicornia patula*, and *Arthrocnemum macrostachyum*, iron is found in the stroma of the chloroplasts, mainly in the cells around the vascular bundle, often in the form of phytoferritin. In *A. macrostachyum*, ferritin was abundantly localized in the chloroplasts of cells and tissues in succulent stems and inflorescences. In monocotyledons such as *I. cylindrica*, *Cynodon dactylon*, and *Panicum repens*, phytoferritin was not observed in the chloroplasts. Instead, iron was found in special plastids located exclusively in the phloem cells of the central cylinder (sieve tube plastids) (Fig. 11.2G). These plastids contain crystalline units of iron (de la Fuente et al., 2016, 2018, 2020; Ramírez, Rodríguez and de la Fuente, 2024).

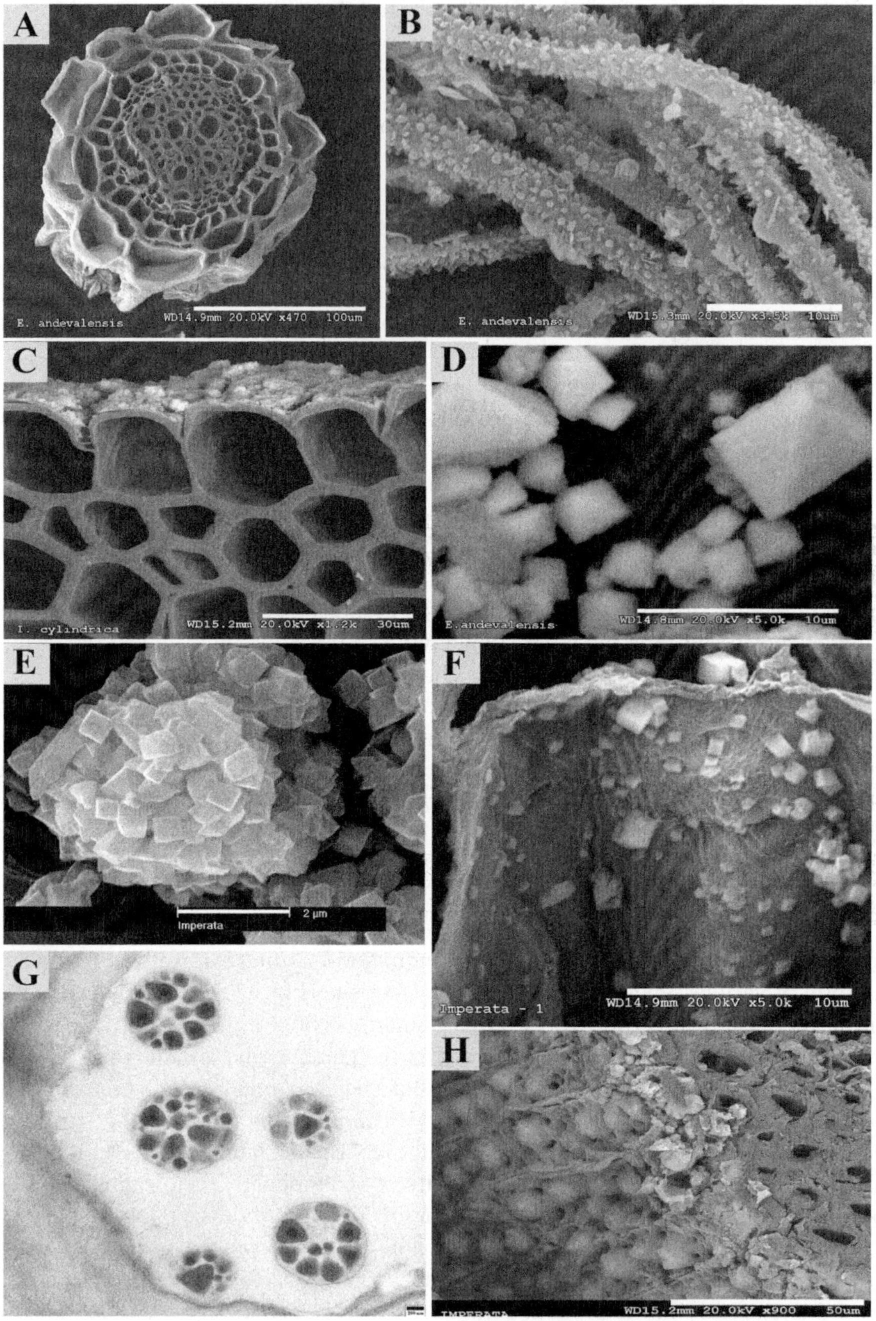

FIGURE 11.2 Representative SEM and TEM micrographs of iron biomineralization.

Representative SEM and TEM micrographs and biomineral accumulation in *Erica andevalensis* (A, B, D). A: Transversal section of a root with epidermis, endodermis,

Iron has also been found in the seeds of some marsh plants, such as *S. pruinosa*. Staining with Prussian blue shows that Fe is more concentrated in the seeds than in the surrounding tissues inside the plant. Within the plant, iron is preferentially concentrated in the cotyledons and radicle. Amorphous Fe oxides of nanometric size were found in the cell walls of the embryonic cells and in the cells that make up the root cap.

Iron concentration, distribution, and biomineralization in cultured Río Tinto species

Studies under controlled laboratory conditions have elucidated the iron accumulation capacity of some of the species that grow naturally in the Río Tinto, namely *Nerium oleander*, and the grasses *Cynodon dactylon* and *Imperata cylindrica*. In controlled experiments, these plants were grown in a nutrient solution enriched with a high concentration of ferrous iron (500 mg/kg).

Nerium oleander

Although adult *N. oleander* in the wild is considered to be an iron excluder, cultured seedlings have shown the capacity to accumulate large amounts of this element when grown under conditions of pH and elemental composition and concentration like those found in the waters of the Río Tinto. Indeed, the levels found in the cultivated plants exceed normal values for vascular plants and reach levels comparable to known iron accumulators such as *I. cylindrica* (Franco, Rufo and de la Fuente, 2012; Franco et al., 2013).

In cultivated seedlings, iron was detected in roots, stems, and leaves, with bioformations observed in stems and roots but not in leaves. Inside the roots and stems of cultured seedlings, iron appeared as amorphous deposits of Fe, K, S, and P covering the inner walls of xylem vascular cells (Franco et al., 2013).

Cynodon dactylon

Cultivated plants had high concentrations of iron in all their vegetative organs. These concentrations were higher in roots and leaves than in wild plants growing in the same environment, while rhizome concentrations were lower. During the first 30 days of the experiment, a rapid translocation of iron from roots to leaves was observed, with the iron concentration sequence being leaf > root > rhizome. This sequence later changed to root > leaf > rhizome between 45 and 60 days, but rhizome iron concentrations always remained the lowest (Franco et al., 2013).

cortical parenchyma, and vascular cylinder and iron plaque (B) with prismatic crystals (D) composed of S, Fe, and Pb (plumbojarosite). *Imperata cylindrica* (C, E, F, H) C: Iron plaque in the exodermis and epidermal cells. Jarosite crystal (E) in the intercellular space of an epidermal and a parenchyma cell (F), and Fe oxides, jarosite, and silicon collapsing the xylem vessels of the rhizome (H). *Cynodon dactylon* (G). Details of several cuneiform and electro-dense structures inside the P-type plastids of the sieve tubes with Fe.

Iron bioformations, mainly composed of Fe, S, and K, were detected in all organs of the cultivated plants, especially in roots and leaves. A thin, irregular layer of iron was observed inside and on the epidermis of both roots and rhizomes. Small Fe particles were also found on the cell walls of the epidermis and in the apoplast between these cells.

Imperata cylindrica

The iron accumulation and biomineralization capacities of *I. cylindrica* were confirmed under controlled laboratory conditions. Greenhouse-treated plants showed the same general pattern of iron distribution as wild plants, with iron accumulation found in the cell walls and intercellular spaces (Fig. 11.2G) of most tissues (except sclerenchyma) and the root endodermis. Iron transport from the root to the rest of the plant is also observed in cultivated plants, like wild populations.

Both jarosite and iron oxides were observed in both wild and cultivated plants. In the rhizomes of cultivated plants, jarosite was seen growing on the walls of xylem vessels, sometimes collapsing the connecting pores (Fig. 11.2H). Iron oxides were also found in the xylem of young rhizomes. X-ray absorption spectroscopy (XAS) analysis indicated that the forms of iron oxides may differ between fresh wild samples and cultivated plants. Fresh samples could be characterized by a combination of ferrihydrite (60%), jarosite (20%), and a mixture of magnetite, hematite, and maghemite (20%), whereas cultivated samples might have different percentages of these compounds. Heating the samples (calcination) leads to crystallization, mainly in the form of maghemite. Finally, Mössbauer measurements also suggest a decrease in jarosite content in cultivated samples compared to fresh samples (de la Fuente et al., 2016).

Manganese (Mn)

Manganese is considered an essential micronutrient because it is involved in chloroplast development, and its absence causes disorganization of the thylakoid membranes. It is also necessary for the activation of enzymes of photosynthesis, respiration, and nitrogen metabolism and is directly involved in photosynthetic processes (Kathpalia and Bhatla, 2018).

Screening of the Río Tinto Flora reveals that manganese concentrations are like iron concentrations, but variable and species dependent. Of note are three species of the genera *Pinus* and *Quercus*, as well as the introduced species *Eucalyptus globulus* and shrub species of the genera *Cistus*, *Cytisus*, *Erica*, *Genista*, *Lavandula*, or climbers such as *Lonicera hispanica*. The same applies to some species typical of scrubland growing on less developed soils, such as *Helichrysum stoechas*, or scrubland on dunes, such as *Halimium calycinum*. These have concentrations that sometimes exceed 1000 mg/kg d.w. and are therefore considered to be accumulators of this element. On the other hand, the lowest concentrations ($<$100 mg/kg d.w.) were found in some species in the estuary and in the middle part of the river (Fig. 11.5) (de la Fuente et al., 2010).

In the species with the highest concentrations of this element, microscopic studies indicate that in *Quercus* and *Lavandula* species, this element accumulates preferentially in the epidermis of the leaves (Fig. 11.3F and H). In the case of *L. stoechas* subsp. *luisieri*, traces of this element were found in the pollen grains (Rufo, 2009).

Zinc (Zn)

Zinc is another essential micronutrient with structural and physiological functions. It is involved in metalloproteins, tryptophan synthase, superoxide dismutase (SOD), and other enzymes. It participates in the formation of chlorophyll and prevents its destruction and also regulates the conversion of carbohydrates and sugar consumption. It plays an essential role in maintaining the structure and function of DNA transcription factors, including Zn finger, Zn cluster, and RING finger domains (Kathpalia and Bhatla, 2018).

The mean concentration of Zn detected in the screening of the Río Tinto flora was 78 mg/kg d.w. Unlike Mn, no plants were found to accumulate this element, although some species exceeded 200 mg/kg d.w. These were mainly scrub species such as some rock roses (*Cistus ladanifer*, *C. populifolius*, *C. monspeliensis*), the flax-leaved daphne (*D. gnidium*), species from poorly developed soils (*H. stoechas*, *P. saxatilis*, *P. paludosa*), and *L. viridis*. This element has been shown to have stable concentrations in plants such as *E. andevalensis*, despite the different concentrations in the soils on which it grows (de la Fuente et al., 2010).

Copper (Cu)

This metal is involved in photosynthetic processes and in the synthesis of structural molecules (lignin) and is also a component of some proteins, such as cytochrome oxidase, plastocyanin, and SOD. It is absorbed as a divalent cupric ion (Cu^{2+}) in aerated soils or as a monovalent ion (Cu^{+}) in wet soils (Kathpalia and Bhatla, 2018).

The mean concentration obtained for this metal in the Río Tinto flora test is 36 mg/kg d.w., which is slightly higher than the levels considered normal in vascular plants. Only 5% of the observations were above 100 mg/kg d.w. These were mainly estuarine species of the families Chenopodiaceae, Aizoaceae (*M. nodiflorum*), Poaceae (*S. densiflora*), and Plumbaginaceae (*L. algarvense*). Copper is also found in seeds of *S. pruinosa* (191 mg/kg d.w.) and *A. macrostachyum* (12.9–13.6 mg/kg d.w.) (de la Fuente et al., 2018; Ramírez, Rodríguez and de la Fuente, 2024; de la Fuente et al., 2010).

Other elements

Among the elements studied, which are not considered essential nutrients for the development of the plant, we have data on some toxic elements found in high concentrations in the soil, such as arsenic and lead.

Río Tinto's plants act as exclusion agents for both, maintaining strategies that allow the concentrations in the aerial parts of the plants to be kept at levels low enough not to be harmful. The averages found in the flora test are around 5 mg/kg d. w. in both cases. There is no pattern related to the taxonomic origin of the species or their location within the river basin (de la Fuente et al., 2010).

Silicon (Si)

Silicon is one of the most abundant elements in the Earth's crust. It is found in numerous living organisms and is an essential element in the cellular cycles of algae and plants. From a chemical perspective, silicon, being tetravalent, is the most similar analogue to carbon, allowing it to bond with different molecules (chemical species). Amorphous silica produced by plants is key in the global silicon cycle. Silicon is absorbed by plant roots in the form of silicic acid $Si(OH)_4$, and according to Kumar et al. (2025), the role of silicon (Si) in various abiotic and biotic mechanisms remains largely unclear. Several fundamental questions continue to persist, particularly regarding how Si functions at the physical, chemical, and genetic levels during its accumulation in shoot and root tissues under stress conditions. To improve our understanding, it is essential to take an integrative approach that combines hormonal, transcriptomic, lipidomic, proteomic, and other omics analyses in both silicon rich and silicon deficient tissues. Furthermore, more research is required to elucidate the interactive components of silicon (Si) involved in plant signaling pathways. This will help to guide future studies into plant growth, development, and stress response mechanisms.

Silica biomineralization has been a fundamental process in the evolutionary history of land plants for over 400 million years (Trembath-Reichert et al., 2015). Although grasses are currently the main mobilizers of silica in terrestrial environments, their appearance is relatively recent, raising questions about the origin and evolution of the biological silica cycle. To explore this process, Trembath-Reichert et al. (2015) studied the abundance of silica in land plants from ancient lineages. The results showed that silica biomineralization is widely distributed across all groups of land plants, with particularly high concentrations in lycophytes and primitive ferns. In contrast, gymnosperms, which evolved later, exhibit low silica levels, suggesting a complex evolutionary development of this process in seed plants.

Due to the beneficial functions of silicon in plants, the International Plant Nutrition Institute (IPNI) recently classified it as a beneficial element (Coskun et al., 2019; Oraee and Tehranifar, 2024).

Silicon concentration

Accurate quantification of Si in the Río Tinto flora required overcoming specific analytical challenges. We applied the digestion protocol described by Zuluaga et al. (2011) and performed by Inductively coupled plasma mass spectrometry (ICP-MS) analysis (ELAN-6000 PE-Sciex, Toronto, Canada) at the SIDI facility (Autonomous University of Madrid). Several of the Si concentrations reported here correspond to the following species: *Imperata cylindrica*, *Cynodon dactylon*, *Arundo donax*, *Typha* spp., *Chamaerops humilis*, *Nerium oleander*, *Erica andevalensis*, *Cistus populifolius*, *Cistus ladanifer*, *Halimione portulacoides*, *Sarcocornia pruinosa*, and *Salicornia* spp.

Guerriero et al. (2016) classified the plants with Si into three main categories: silicon excluders, silicon intermediate types, and silicon accumulators. Plants of the order Equisetales, Poales, and Cyperales belong to the first group, which accumulate more than 4% Si (of shoot dry weight) in their tissues; plants showing more than 1% Si (between 2% and 4%, e.g., Cucurbitales and Urticales) are of the intermediate type; the silicon excluders show less than 0.5% Si.

Like Guerriero et al. (2016), when we explored Si concentration in Río Tinto's flora, we found levels between 5% and 10% in fern species such as *Equisetum ramosissimum* and *Pteridium aquilinum*, with values of 51,646 mg/kg d.w. and 10,657 mg/kg d.w., respectively.

Silicon accumulation in Poaceae is generally high, though markedly heterogeneous among subfamilies and species. Studies on *Imperata cylindrica* growing in Río Tinto have reported silicon concentrations in organs of this plant ranging from 4282 in stems to 26,814 mg/kg d.w. in mature leaves. *Cynodon dactylon* ranges from 42,225 to 55,225 mg/kg d.w., while *Arundo donax* can reach up to 8110 mg/kg d.w. In the genus *Typha* (Typhaceae), Si content in roots, rhizomes, and leaves is approximately 15,200, 12,000, and 4000 mg/kg d.w., respectively. The widely distributed Mediterranean palm, *Chamaerops humilis*, contains between 2000 and 17,000 mg/kg d.w. in young or mature leaves.

In general, dicotyledonous species show lower silicon (Si) concentration levels compared to monocotyledonous species. For instance, *Nerium oleander* accumulates approximately 1125 mg/kg d.w. of Si, while *Erica andevalensis* reaches around 1202 mg/kg d.w.

Additionally, we have analyzed the Si content in seeds of some representative species from different ecosystems. We can confirm that, besides the high Si content in the seeds of lycophytes and grasses mentioned earlier, high levels are also found in seeds of the Cistaceae family, such as *Cistus populifolius* with 8198 mg/kg d.w. and *Cistus ladanifer* with 2930 mg/kg d.w.

In halophytic species from estuarine and saltmarsh environments, Si concentrations are also notable: *Halimione portulacoides* contains approximately 2651 mg/kg d.w., *Sarcocornia pruinosa* around 1000 mg/kg d.w., and *Salicornia* species reach 1052 mg/kg d.w.

Distribution and biomineralization of silicon

Electron microscopy and X ray spectroscopy reveal that tissues commonly mineralized include roots, stems, leaves, fruits, and seeds. Studies on the distribution of silicon (Si) in different plant organs have been conducted on some species from Río Tinto. The plants examined showed the presence of cells with varying Si content in pteridophytes, Poaceae, and a lower content in dicotyledonous angiosperms.

In the observed species, Si concentration has been detected in the plaques forming the roots, along with iron ions (Fe). Generally, the concentration of Fe is higher than that of Si. The epidermis of roots typically contains Si or may be silicified, indicating that silicon plays a role in the structure and mineralization of this tissue. In the specimens observed, no accumulations of silicon or iron were found within the layers of the endodermis, except in the outer layer in contact with the parenchyma cells, which do contain Fe and Si. These observations are consistent across both SEM (scanning electron microscopy) and TEM (transmission electron microscopy). In TEM images, more electron-dense precipitates can be seen in the intercellular spaces, surrounding the interior of cell membranes and some organelles. This indicates that mineral deposits of Fe and Si are present in specific areas outside the endodermis but not within its layers. Recent studies on rice (Carrasco-Gil et al., 2018) and cucumber (Hernández-Apaolaza et al., 2020) have shown that adding silicon to the growth medium can lead to increased iron plaque formation, which decreases iron absorption and activates iron deficiency responses in roots—even when iron is supplied adequately.

In the specimens studied, a Casparian strip forms next to the parenchyma cells, and its silicon content varies depending on the species. For example, in genera like *Imperata*, *Cynodon*, and *Thypha*, silicon and iron deposits occur in the outer layer, while the successive membranes lack these ions.

It is also important to highlight the role of silicon deposits in signaling, as discussed by Mandlik et al. (2020). Silicon transport involves aquaporins, such as proteins like Nod 26 or NIP (Deshmukh et al., 2013; Grégoire et al., 2012). In rice roots, the endodermis contains transport proteins Lsi1 and Lsi2, which control silicic acid movement. Lsi1 is located on the distal side of endodermis cells and facilitates passive transport of silicic acid into the cell, while Lsi2 is on the proximal side and helps move silicic acid across the Casparian strip (Ma et al., 2007). However, this polar distribution is not universal for all grasses, as seen in maize and barley (Mitani et al., 2009).

Additionally, ancient lycophytes and ferns like *Equisetum ramossisimum* and *Pteridium aquilinum* have silicified cell walls in almost all organs—roots, stems, and fronds—especially visible in the epidermis and stomata (Fig. 11.3E).

The autochthonous palm *Chamaerops humilis*, as mentioned in the previous section, has a high silicon content and presents silicon crystals in the parenchyma cells surrounding the vascular bundles of the leaves (Fig. 11.3B and C).

In Poaceae species, silicified cells in stems and leaves vary among genera and species. Long epidermal silicified cells are accompanied by short silica bodies, which serve as taxonomic characters to differentiate species (Fig. 11.3A). Silicon was found to be heterogeneously distributed. It is found as a component of cell walls in all tissues and is also concentrated in specific structures and cells, such as silica bodies in the epidermis of leaves and rhizomes, the endodermis of rhizomes and roots, where silicon deposits form thick, multilayered cell walls, bulliform cells, and bundle sheath cells (Figs. 11.2H and 11.3A–D). Cells with a silicon-collapsed lumen were found in the outer cortex of *I. cylindrica* and *Thypha domingensis* (Fig. 11.3D and G). Finally, stomata also showed a high concentration of silicon in their walls compared to other epidermal cells. TEM and electron diffraction analysis revealed electrically dense amorphous materials, broken into round fragments

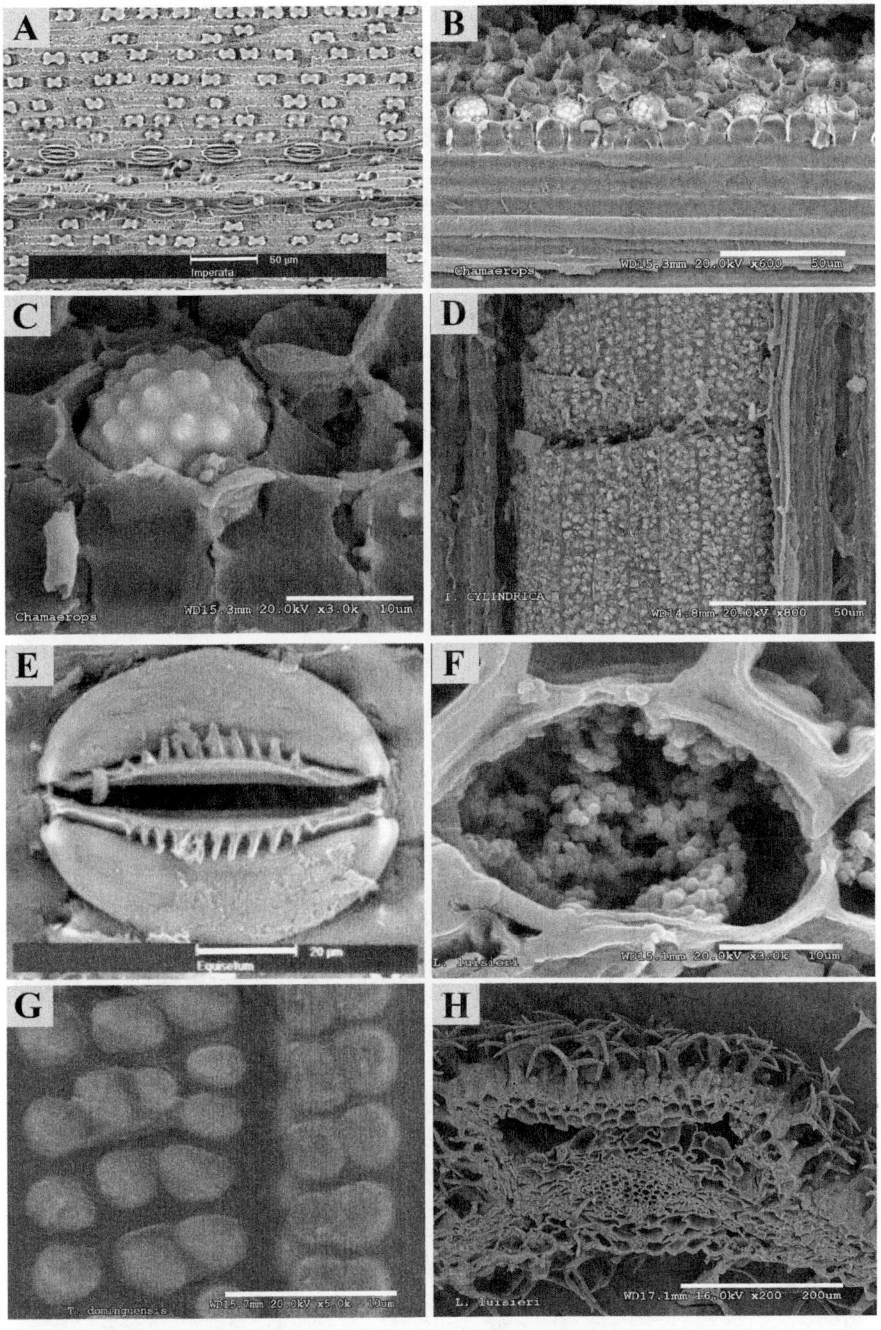

FIGURE 11.3 Representative SEM micrographs of silicon biomineralization.

Representative SEM micrographs and biomineral accumulation in *Imperata cylindrica* (A and D). A: Leaf epidermis with long epidermal cells, stomata, and short siliceous cells. D: Silicon-saturated conducting vessels of the rhizome. *Chamaerops humilis* (B and C): silicon crystals in the parenchyma cells of leaves. *Equisetum ramosissimum* (E): Details of siliceous stoma. *Typha domingensis* (G): Collapsed pores on the conducting vessels of the rhizome with Fe and Si. *Lavandula stoechas* subsp. *luisieri* (F and H). H: Leaf cross section. F: Epidermal cells with precipitates containing Mn and Zn.

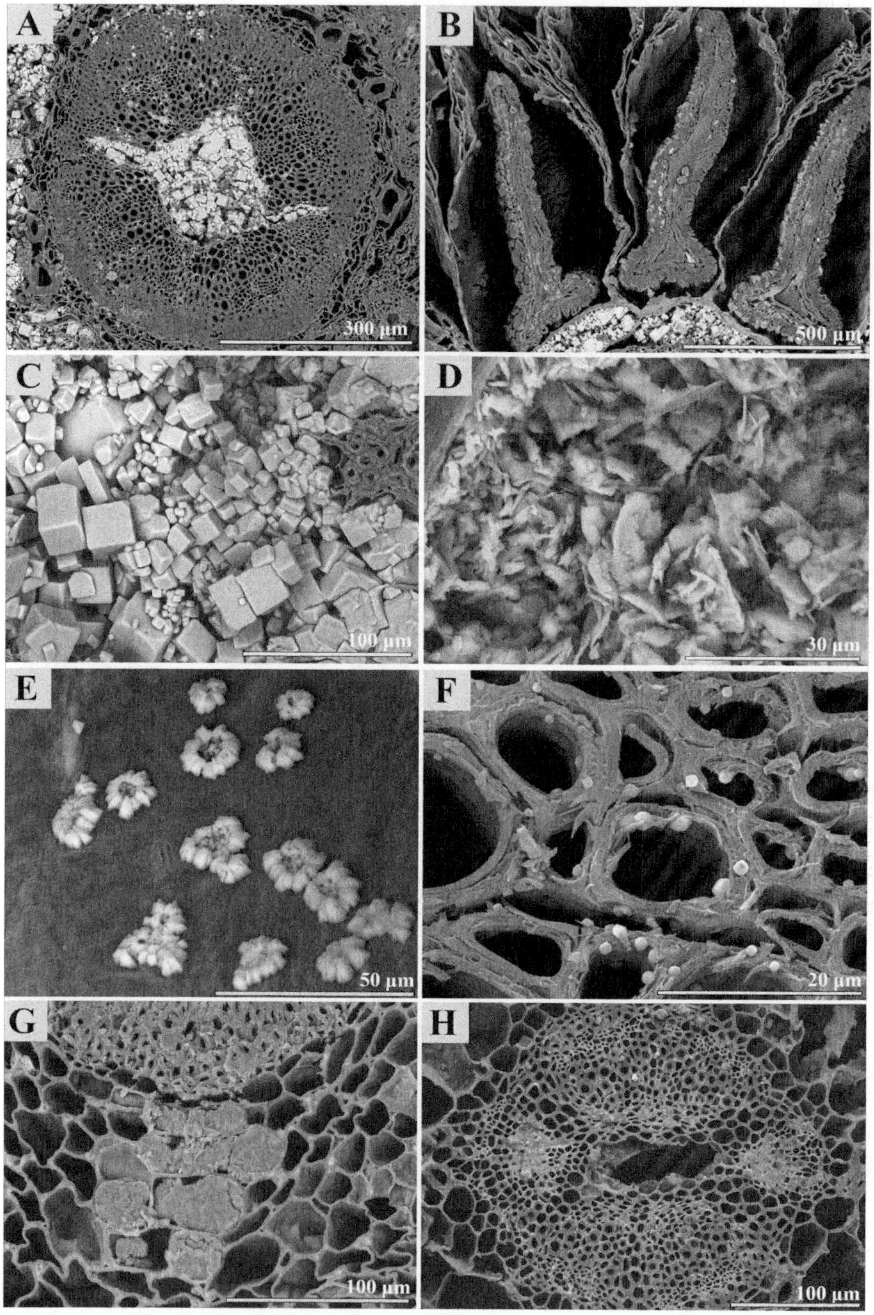

FIGURE 11.4 SEM images of sodium biomineralization.

SEM images showing biomineral accumulation in *Arthrocnemum macrostachyum* (A, B, H), *Salicornia* sp. (C, F, G), *Sarcocornia perennis* (D), and *Sarcocornia alpini* (E), collected

and sometimes crossed by fine strands, which may be cellulose fibrils (Rufo, Franco and de la Fuente, 2014).

Another element that is not essential for plant growth, but which plays a crucial role in the estuarine ecosystem of the estuary, sodium, deserves separate consideration.

Sodium (Na)

Sodium concentrations

A general trend of sodium accumulation above soil levels was observed in the plant species studied. The highest Na concentrations are typically found in salt marsh species. This is probably due to the need for osmotic adjustment to prevent water loss under saline conditions. These plants often accumulate salt in their tissues or in specialized salt glands as a strategy to survive on saline soils and to avoid dehydration.

Although the mean concentration of this element was 4489 mg/kg d.w. for all the plants analyzed, the mean value exceeds 30,000 mg/kg d.w. if only the plants growing in the estuary are considered. Thus, the sodium concentrations for most of the riverine plants in the upper and lower reaches of the estuary, as well as for the surrounding shrub and tree species, are usually below 2000 mg/kg d.w. In the specific case of the plants growing in the estuary, we find differences between monocotyledons and dicotyledons. The former act mainly as salt excluders, with concentrations below 5000 mg/kg d.w. However, most of the dicotyledons analyzed show much higher concentrations, reaching values above 90,000 mg/kg d.w., indicating active accumulation of this element in their tissues (de la Fuente et al., 2010).

The seeds of some of these species also contain high levels of sodium (*Arthrocnemum macrostachyum*: 3431–7542 mg/kg d.w. and *Sarcocornia pruinosa* 16,612 mg/kg d.w.). The values found vary when the pericarp is included in the measurement, reaching higher concentrations than those of the seeds alone (9952–25,498 mg/kg d.w. for *A. macrostachyum*) (de la Fuente et al., 2018; Ramírez, Rodríguez and de la Fuente, 2024).

from the Río Tinto salt marshes. A: Cubic and irregular salt crystals located in the water storage tissue and the central pith of the vascular cylinder. B: Three immature flowers with external salt accumulation around the central cylinder. C: Details of NaCl (halite) crystals; a few phloem-like cells may be distinguishable. D: Tenardite (Na_2SO_4) crystals forming dense aggregates. E: Spherical clusters composed of cubic and prismatic KCl crystals (sylvite), located along the cell walls of the epidermis. F: Spherical calcium crystals embedded within the cells of the central cylinder. G: Water storage parenchyma cells partially filled with amorphous magnesium biominerals identified as glushinskite. H: Magnesium biominerals (glushinskite) coating phloem- and xylem-like cells of the central cylinder.

Species	Mg	Ca	Na	Mn	Fe	Cu	Zn	As	Pb
Dominant Plants in Río Tinto's Riparian Ecosystems									
Nerium oleander	2093.4	12069.3	151.6	67.1	330.7	19.0	49.8	1.6	8.1
Scirpoides holoschoenus	2384.0	3104.0	2733.0	190.0	199.0	19.7	28.9	4.3	2.4
Equisetum ramosissimum	4697.7	17538.7	392.4	219.2	2694.4	16.9	66.5	31.6	84.3
Pteridium aquilinum	485.2	3282.8	1172.1	253.6	1304.0	89.2	32.5	4.3	11.3
Erica andevalensis	1566.3	1711.45	1326.9	284.33	811.08	4.87	16.54	2.59	5.79
Flueggea tinctoria	884.0	15324.0	-	62.9	284.0	23.2	32.4	5.0	2.8
Cynodon dactylon	2384.4	2750.1	4967.9	128.9	1611.5	27.3	51.1	4.1	17.2
Imperata cylindrica	746.0	1638.0	1026.0	77.7	1808.0	26.6	71.5	13.3	9.2
Phragmites australis	1528.2	1972.9	1140.0	138.6	898.7	21.4	17.2	5.7	12.9
Juncus effusus	1382.0	2677.0	1292.0	868.0	151.0	79.9	109.0	0.8	3.2
Juncus rugosus	2363.0	2320.0	4305.0	184.0	168.0	27.4	89.7	5.3	7.0
Tamarix gallica	5806.0	16571.0	2482.0	7.4	157.0	32.7	42.0	3.0	5.2
Typha domingensis	983.4	2728.0	3911.1	287.7	1335.8	23.8	25.2	16.1	39.3
Dominant Plants in Río Tinto's Saltmarsh Ecosystems									
Mesembryanthemum nodiflorum	5057.0	15068.0	54558.0	24.5	915.0	160.0	84.7	6.5	7.0
Halimione portulacoides	11166.0	5616.0	979.0	45.2	632.0	110.0	51.5	10.4	6.8
Arthrocnemum macrostachyum	6621.1	4710.6	91317.8	27.2	481.4	38.8	32.3	2.3	6.1
Salicornia patula	12092.3	12677.7	121425.0	95.2	1567.0	68.9	48.7	7.0	6.1
Sarcocornia perennis	7904.0	5158.0	1000.0	15.2	190.0	60.7	27.4	8.6	2.7
Sarcocornia pruinosa	6448.0	5460.0	61107.5	24.5	245.3	64.2	42.8	1.4	2.6
Spartina densiflora	2524.2	2157.7	3587.0	74.5	581.0	69.5	115.3	11.6	2.1
Spartina maritima	2370.9	1267.5	4879.5	31.8	514.2	33.0	49.0	7.9	1.6
Juncus acutus	1383.0	644.0	3016.0	32.7	71.0	16.3	18.9	6.2	0.6

A

Dominant Plants in Río Tinto's Forest Ecosystems and Seral Stages									
Cistus ladanifer	1865.0	5576.0	793.0	664.0	724.0	26.3	113.0	9.5	10.6
Cistus populifolius	4772.0	10960.0	1402.0	747.0	341.0	24.2	154.0	5.4	5.5
Cistus monspeliensis	2579.0	18902.0	1614.0	29.2	564.0	42.9	319.0	7.8	9.1
Cistus salviifolius	2364.0	10528.0	141.0	330.0	266.0	44.5	286.0	8.0	5.7
Cistus albidus	4323.0	8402.0	950.0	115.0	390.0	21.5	124.0	10.6	6.7
Halimium ocymoides	1833.0	5870.0	1744.0	432.0	377.0	29.2	86.5	7.4	5.0
Helichrysum stoechas	4397.0	7710.0	2393.0	1692.0	330.0	60.5	212.0	11.1	2.3
Arbutus unedo	2394.0	8522.0	684.0	125.0	204.0	13.1	57.0	4.1	4.1
Erica australis	1048.0	1447.0	984.0	573.0	274.0	17.5	29.2	1.0	5.4
Erica arborea	1614.0	2205.0	537.0	346.0	223.0	13.0	23.9	2.5	1.1
Erica lusitanica	3961.0	3634.0	2400.0	985.0	212.0	26.1	61.3	4.1	1.7
Erica scoparia	1975.0	2868.0	602.0	191.0	202.0	18.8	27.7	3.1	2.3
Quercus suber	2161.0	5223.0	399.0	1129.0	636.0	17.1	45.8	10.7	13.4
Quercus rotundifolia	2822.0	5047.0	378.0	1853.0	227.0	23.1	54.5	5.8	8.0
Piptatherum miliaceum	627.0	2310.0	769.0	14.3	550.0	156.0	57.0	29.7	4.7
Lavandula sampaioana	2578.0	9086.0	408.0	2391.0	474.0	45.2	188.0	3.6	3.1
Lavandula viridis	3442.0	4831.0	2256.0	2208.0	643.0	39.7	301.0	11.2	9.2
Ceratonia siliqua	2371.0	12817.0	2461.0	65.9	193.0	18.5	30.2	8.6	8.5
Cytisus eriocarpus	1169.0	2618.0	1004.0	403.0	260.0	25.4	66.9	6.2	4.6
Ulex eriocladus	1737.0	1222.0	796.0	192.0	251.0	11.6	44.5	5.2	2.8
Genista hirsuta	1024.0	5121.0	583.0	491.0	221.0	34.8	66.9	4.5	2.3
Genista polyanthos	2160.0	3375.0	719.0	884.0	592.0	35.5	105.0	5.1	4.8
Myrtus communis	4785.0	10475.0	871.0	772.0	127.0	23.3	74.4	2.5	3.8
Eucalyptus globulus	3967.0	17910.0	561.0	1671.0	231.0	22.1	71.0	8.0	2.4
Phillyrea angustifolia	7008.0	5753.0	448.0	184.0	153.0	9.3	73.1	3.3	3.9
Olea sylvestris	2870.0	19764.0	691.0	37.1	218.0	33.2	49.0	4.1	7.6
Chamaerops humilis	2484.0	3060.0	707.0	212.0	118.0	14.5	28.8b	9.2	3.1
Pinus pinea	7203.0	7589.0	197.0	1703.0	510.0	11.2	31.1	6.1	9.2
Pinus pinaster	5742.0	19999.0	1886.0	3531.0	197.0	18.9	111.0	4.3	4.1
Rumex induratus	6882.0	2273.0	2008.0	459.0	316.0	49.1	82.9	2.7	4.1
Sanguisorba hybrida	4257.0	8655.0	473.0	914.0	225.0	42.3	126.0	1.8	1.8
Daphne gnidium	3049.0	20248.0	356.0	113.0	228.0	12.1	441.0	10.0	-

B

FIGURE 11.5

Mean concentrations (mg/kg dry weight) of Mg, Ca, Na, Mn, Fe, Cu, Zn, As, and Pb in dominant plant species from Río Tinto's riparian, saltmarsh, and forest ecosystems. Most data are from de la Fuente et al. (2010), with updates from Ramírez et al. (2024) for *A. macrostachyum*, Sánchez-Gavilán et al. (2021) for *S. patula*, de la Fuente et al. (2018) for *S. pruinosa*, and unpublished new data for *N. oleander*, *E. ramosissimum*, *P. aquilinum*, *E. andevalensis*, *C. dactylon*, *T. domingensis*, *S. densiflora*, and *S. maritima*.

Distribution and biomineralization of sodium

In the dicotyledonous species, we found more data on sodium distribution and compartmentalization. The species studied (*Arthrocnemum macrostachyum*, *Sarcocornia pruinosa*, *S. alpini*, and *Salicornia* sp.) are all succulents with photosynthetic stems and water storing parenchyma.

Microscopic observations of succulent stems of *S. pruinosa* indicate that the greatest accumulation of salt crystals is in the parenchymal tissue, followed by the vascular bundle and finally the epidermis (de la Fuente et al., 2018). Salt accumulation occurs mainly in the vacuoles of all tissues, although with a greater presence in the tissues of the spongy parenchyma. A similar distribution was observed in *A. macrostachyum* (Fig. 11.4A and B), where sodium accumulation occurred mainly in biologically less active tissues and cells such as xylem, vascular bundles of flower stems, water-storing cells, and epidermis. Reduced sodium chloride crystals were also detected at the edges and inside the papillae of *A. macrostachyum* seeds (Ramírez, Rodríguez and de la Fuente, 2024). Massive cubic crystals are likewise observed in annual salt marsh species such as *Salicornia* sp., extending from the outer epidermis to the central vascular cylinder (Fig. 11.4C). X ray diffraction analyses confirmed the presence of biominerals containing primarily sodium, in the form of halite (NaCl), along with thenardite (Na_2SO_4) and potassium (as sylvite, KCl) in the analyzed species. Although thenardite and potassium-based crystals are generally less abundant, they occasionally form striking patterns within the tissues—such as the massive thenardite accumulations observed in *Sarcocornia perennis* (Fig. 11.4D) or the sylvite clusters embedded in the cell walls of *Sarcocornia alpini* (Fig. 11.4E).

Conclusion

Río Tinto's flora thrives under extreme acidity, metal toxicity, and nutrient scarcity. Most species act as excluders of toxic elements, yet actively accumulate macronutrients such as Ca and Mg, even in severely depleted soils. In the estuarine saltmarshes, halophytes like *Sarcocornia perennis*, *S. pruinosa*, *S. alpini*, and *Arthrocnemum macrostachyum* exhibit marked sodium accumulation, often exceeding 30,000 mg/kg d.w., reflecting strong osmotic adaptation.

Biomineralization is pervasive across environments: *Imperata cylindrica* hyperaccumulates Fe (>2% d.w.), while *Cynodon dactylon* and *Equisetum ramosissimum* reach Si concentrations above 50,000 mg/kg d.w. Crystalline deposits of Ca, Mg, Fe, and Si are differentially compartmentalized in roots, leaves, and reproductive tissues. Forest and scrub species such as *Quercus suber* and *Lavandula stoechas* show elevated Mn levels (>1000 mg/kg d.w.), highlighting selective micronutrient accumulation.

Together, these findings reveal a resilient plant assemblage capable of excluding, sequestering, or biologically immobilizing metals while conserving essential elements. Río Tinto offers a rare natural system to explore biomineralization, elemental homeostasis, and the ecological thresholds of plant survival under extreme geochemical stress.

References

Amils, R., 2016. Lessons learned from thirty years of geomicrobiological studies of Río Tinto. Res. Microbiol. 167 (7).

Amils, R., González-Toril, E., Gómez, F., Fernández-Remolar, D., Rodríguez, N., Malki, M., Zuluaga, J., Aguilera, A., Amaral-Zettler, L.A., 2004. Importance of Chemolithotrophy for Early Life on Earth: The Tinto River (Iberian Pyritic Belt) Case. Springer Science and Business Media LLC, pp. 463–480. https://doi.org/10.1007/1-4020-2522-x_28.

Amils, R., González-Toril, E., Fernández-Remolar, D., Gómez, F., Rodríguez, N., Durán, C., 2002. Interaction of the sulfur and iron cycles in the Tinto River ecosystem. Rev. Environ. Sci. Biotechnol. 1 (4), 299–309. https://doi.org/10.1023/a:1023232002312.

Baker, A.J.M., 1981. Accumulators and excluders—strategies in the response of plants to heavy metals. J. Plant Nutr. 3 (1–4), 643–654. https://doi.org/10.1080/01904168109362867.

Braungardt, C.B., Achterberg, E.P., Elbaz-Poulichet, F., Morley, N.H., 2003. Metal geochemistry in a mine-polluted estuarine system in Spain. Appl. Geochem. 18 (11), 1757–1771. https://doi.org/10.1016/S0883-2927(03)00079-9.

Brooks, R.R., 1998. Plants that Hyperaccumulate Heavy Metals: Their Role in Phytoremediation, Microbiology, Archaeology, Mineral Exploration and Phytomining. CAB International.

Buckby, T., Black, S., Coleman, M.L., Hodson, M.E., 2003. Fe-sulphate-rich evaporative mineral precipitates from the Río Tinto, southwest Spain. Mineral. Mag. 67 (2), 263–278. https://doi.org/10.1180/0026461036720104.

Cánovas, C.R., Hubbard, C.G., Olías, M., Nieto, J.M., Black, S., Coleman, M.L., 2008. Hydrochemical variations and contaminant load in the Río Tinto (Spain) during flood events. J. Hydrol. 350 (1–2), 25–40. https://doi.org/10.1016/j.jhydrol.2007.11.022.

Cánovas, C.R., Olías, M., Nieto, J.M., Sarmiento, A.M., Cerón, J.C., 2007. Hydrogeochemical characteristics of the Tinto and Odiel Rivers (SW Spain). Factors controlling metal contents. Sci. Total Environ. 373 (1), 363–382. https://doi.org/10.1016/j.scitotenv.2006.11.022.

Carrasco-Gil, S., Rodríguez-Menéndez, S., Fernández, B., Pereiro, R., de la Fuente, V., Hernandez-Apaolaza, L., 2018. Silicon induced Fe deficiency affects Fe, Mn, Cu and Zn distribution in rice (Oryza sativa L.) growth in calcareous conditions. Plant Physiol. Biochem. 125, 153–163. https://doi.org/10.1016/j.plaphy.2018.01.033.

Coskun, D., Deshmukh, R., Sonah, H., Menzies, J.G., Reynolds, O., Ma, J.F., Kronzucker, H. J., Bélanger, R.R., 2019. Defence of the selective transport and role of silicon in plants. New Phytol. 223 (2), 514–516. https://doi.org/10.1111/nph.15764. Available from: http://onlinelibrary.wiley.com/journal/10.1111/(ISSN)1469-8137.

Deshmukh, R.K., Vivancos, J., Guérin, V., Sonah, H., Labbé, C., Belzile, F., Bélanger, R.R., 2013. Identification and functional characterization of silicon transporters in soybean using comparative genomics of major intrinsic proteins in Arabidopsis and rice. Plant Mol. Biol. 83 (4–5), 303–315. https://doi.org/10.1007/s11103-013-0087-3.

de la Fuente, V., Rufo, L., Rodríguez, N., Ramírez, E., Sánchez-Gavilán, I., Amils, R., 2020. Differential iron management in monocotyledon and dicotyledon plants from the Río Tinto basin. Protoplasma 257 (3), 889–900. https://doi.org/10.1007/s00709-019-01476-x.

de la Fuente, V., Rufo, L., Rodríguez, N., Amils, R., Zuluaga, J., 2010. Metal accumulation screening of the Río Tinto flora (Huelva, Spain). Biol. Trace Elem. Res. 134 (3), 318–341. https://doi.org/10.1007/s12011-009-8471-1.

de la Fuente, V., Rufo, L., Juárez, B.H., Menéndez, N., García-Hernández, M., Salas-Colera, E., Espinosa, A., 2016. Formation of biomineral iron oxides compounds in a Fe hyperaccumulator plant: Imperata cylindrica (L.) P. Beauv. J. Struct. Biol. 193 (1), 23–32. https://doi.org/10.1016/j.jsb.2015.11.005.

de la Fuente, V., Rufo, L., Sánchez-Gavilán, I., Ramírez, E., Rodríguez, N., Amils, R., 2018. Plant tissues and embryos biominerals in Sarcocornia pruinosa, a halophyte from the Río Tinto salt marshes. Minerals 8 (11), 505. https://doi.org/10.3390/min8110505.

Franco, A., Rufo, L., Rodríguez, N., Amils, R., 2013a. Vicenta de la Fuente, Iron absorption, localization, and biomineralization of Cynodon dactylon, a perennial grass from the Río Tinto basin (SW Iberian Peninsula). J. Plant Nutr. Soil Sci. 176 (6), 836–842. https://doi.org/10.1002/jpln.201200640.

Franco, A., Rufo, L., Zuluaga, J., de la Fuente, V., 2013b. Metal Uptake and Distribution in Cultured Seedlings of Nerium oleander L. (Apocynaceae) from the Río Tinto (Huelva, Spain). Biol. Trace Elem. Res. 155 (1), 82–92. https://doi.org/10.1007/s12011-013-9761-1.

Franco, A., Rufo, L., de la Fuente, V., 2012. Metal concentration and distribution in plant tissues of Nerium oleander (Apocynaceae, Plantae) from extremely acidic and less extremely acidic water courses in the Río Tinto area (Huelva, Spain). Ecol. Eng. 47, 87–91. https://doi.org/10.1016/j.ecoleng.2012.06.024.

Galán, E., Gómez-Ariza, J.L., González, I., Fernández-Caliani, J.C., Morales, E., Giráldez, I., 2003. Heavy metal partitioning in river sediments severely polluted by acid mine drainage in the Iberian Pyrite Belt. Appl. Geochem. 18 (3), 409–421. https://doi.org/10.1016/s0883-2927(02)00092-6.

Grégoire, C., Rémus-Borel, W., Vivancos, J., Labbé, C., Belzile, F., Bélanger, R.R., 2012. Discovery of a multigene family of aquaporin silicon transporters in the primitive plant Equisetum arvense. Plant J. 72 (2), 320–330. https://doi.org/10.1111/j.1365-313X.2012.05082.x.

Guerriero, G., Hausman, J.F., Legay, S., 2016. Silicon and the plant extracellular matrix. Front. Plant Sci. 7 (2016). https://doi.org/10.3389/fpls.2016.00463. Available from: http://journal.frontiersin.org/article/10.3389/fpls.2016.00463/pdf.

Heredia, J., Bellinfante, N., Paneque, G., 1980. Alteración de rocas y formación de suelos en el área del Riotinto (Huelva) II. Estudio morfológico y propiedades generales de los suelos. An. Edaf. Agrobiol. 39 (11–12), 1845–1857.

Hernández-Apaolaza, L., Escribano, L., Zamarreño, Á.M., García-Mina, J.M., Cano, C., Carrasco-Gil, S., 2020. Root silicon addition induces Fe deficiency in cucumber plants, but facilitates their recovery after Fe resupply. A comparison with Si foliar sprays. Front. Plant Sci. 11. https://doi.org/10.3389/fpls.2020.580552.

Hudson-Edwards, K.A., Schell, C., Macklin, M.G., 1999. Mineralogy and geochemistry of alluvium contaminated by metal mining in the Rio Tinto area, southwest Spain. Appl. Geochem. 14 (8), 1015–1030. https://doi.org/10.1016/S0883-2927(99)00008-6.

Kathpalia, R., Bhatla, S.C., 2018. Plant Mineral Nutrition Plant Physiology, Development and Metabolism. Springer Singapore, India, pp. 37–81. https://doi.org/10.1007/978-981-13-2023-1_2. Available from: https://link.springer.com/book/10.1007/978-981-13-2023-1.

Kumar, A., Choudhary, A., Kaur, H., Singh, K., Guha, S., Choudhary, D.R., Sonkar, A., Mehta, S., Husen, A., 2025. Exploring the role of silicon in enhancing sustainable plant growth, defense system, environmental stress mitigation and management. Discov. Appl. Sci. 7 (5). https://doi.org/10.1007/s42452-025-06866-w.

Ma, J.F., Yamaji, N., Tamai, K., Mitani, N., 2007. Genotypic difference in silicon uptake and expression of silicon transporter genes in rice. Plant Physiol. 145 (3), 919–924. https://doi.org/10.1104/pp.107.107599.

Mandlik, R., Thakral, V., Raturi, G., Shinde, S., Nikolić, M., Tripathi, D.K., Sonah, H., Deshmukh, R., 2020. Significance of silicon uptake, transport, and deposition in plants. J. Exp. Bot. 71 (21), 6703–6718. https://doi.org/10.1093/jxb/eraa301.

Mitani, N., Chiba, Y., Yamaji, N., Ma, J.F., 2009. Identification and characterization of maize and barley Lsi2-like silicon efflux transporters reveals a distinct silicon uptake system from that in rice. Plant Cell 21 (7), 2133–2142. https://doi.org/10.1105/tpc.109.067884.

Oraee, M.R., Tehranifar, A., 2024. Silicon in crop protection: effects on virus and pest management in agriculture. J. Plant Protect. Res. 64 (1), 18–29.

Pendias, A., Pendias, H., 2001. Trace Elements in Soils and Plants.

Ramírez, E., Rodríguez, N., de la Fuente, V., Moq, A., 2024. Unlocking opportunities for biosaline agriculture and improved human nutrition. Plants 13 (4), 496. https://doi.org/10.3390/plants13040496.

Reeves, R.R., Baker, A.J.M., 2000. Phytoremediation of Toxic Metals. Wiley Metal-accumulatind plants, pp. 193–229.

Rodríguez, N., Menéndez, N., Tornero, J., Amils, R., De La Fuente, V., 2005. Internal iron biomineralization in Imperata cylindrica , a perennial grass: chemical composition, speciation and plant localization. New Phytol. 165 (3), 781–789. https://doi.org/10.1111/j.1469-8137.2004.01264.x.

Rodríguez, N., Amils, R., Jiménez-Ballesta, R., Rufo, L., De La Fuente, V., 2007. Heavy metal content in Erica andevalensis: an endemic plant from the extreme acidic environment of tinto river and its soils. Arid Land Res. Manag. 21 (1), 51–65. https://doi.org/10.1080/15324980601074578.

Rufo, L., 2009. Caracterización de las fitocenosis del Río Tinto: Biodiversidad y aplicaciones biotecnológicas.

Rufo, L., Rodríguez, N., Amils, R., 2010. Fuente, Chemical and Mineralogical Characterization of the Soils of the Main Plant Communities of the 'Río Tinto' Basin, pp. 5–12.

Rufo, L., Rodríguez, N., Amils, R., de la Fuente, V., Jiménez-Ballesta, R., 2007. Surface geochemistry of soils associated to the Tinto River (Huelva, Spain). Sci. Total Environ. 378 (1–2), 223–227. https://doi.org/10.1016/j.scitotenv.2007.01.051.

Rufo, L., Franco, A., de la Fuente, V., 2014. Silicon in Imperata cylindrica (L.) P. Beauv: content, distribution, and ultrastructure. Protoplasma 251 (4), 921–930. https://doi.org/10.1007/s00709-013-0594-8.

Sánchez-Gavilán, I., Rufo, L., Rodríguez, N., de la Fuente, V., 2021. On the elemental composition of the Mediterranean euhalophyte Salicornia patula Duval-Jouve (Chenopodiaceae) from saline habitats in Spain (Huelva, Toledo and Zamora). Environ. Sci. Pollut. Control Ser. 28 (3), 2719–2727. https://doi.org/10.1007/s11356-020-10663-w.

Tornos, F., 2006. Environment of formation and styles of volcanogenic massive sulfides: the Iberian Pyrite Belt. Ore Geol. Rev. 28 (3), 259–307. https://doi.org/10.1016/j.oregeorev.2004.12.005.

Trembath-Reichert, E., Wilson, J.P., McGlynn, S.E., Fischer, W.W., 2015. Four hundred million years of silica biomineralization in land plants. Proc. Natl. Acad. Sci. U. S. A. 112 (17), 5449–5454. https://doi.org/10.1073/pnas.1500289112.

van Es, H., 2017. A new definition of soil. CSA News 62 (10), 20–21. https://doi.org/10.2134/csa2017.62.1016.

Zuluaga, J., Rodríguez, N., Rivas-Ramirez, I., De La Fuente, V., Rufo, L., Amils, R., 2011. An improved semiquantitative method for elemental analysis of plants using inductive coupled plasma mass spectrometry. Biol. Trace Elem. Res. 144 (1–3), 1302–1317. https://doi.org/10.1007/s12011-011-9140-8.

SECTION 3

Río Tinto as a Mars terrestrial analog

CHAPTER

Mineralogical and geochemical characterization of Río Tinto using combined spectroscopic methods: Implications for exploration of Mars

12

Fernando Rull[1], Aurelio Sanz[1], Guillermo Lopez-Reyes[1], Jorge L. Vago[2]
[1]*Research Group ERICA and LaDIS, Universidad de Valladolid, Edificio Alfonso VIII, Valladolid, Spain;* [2]*European Space Agency-ESTEC, Noordwijk, Netherlands*

Introduction

In March 2004, NASA announced an extraordinary discovery by the Opportunity rover (Mars Exploration Rover mission) at Meridiani Planum: the identification of the jarosite mineral, an iron sulfate whose presence provided unequivocal proof of past liquid water on the Martian surface (Squyres et al., 2004; Klingelhöfer et al., 2004). Jarosite (ide al formula $Fe^{3+}{}_3(SO_4)_2(OH)_6$) forms under aqueous and typically acidic conditions on Earth (Fernández-Remolar et al., 2005; Martinez-Frias et al., 2006). It was first discovered on Earth in 1852 by August Breithaupt in the Jaroso Ravine, Spain – the mineral's *type locality*. The detection of jarosite on Mars, together with other secondary minerals like iron oxides, phyllosilicates, and carbonates, has been regarded as strong evidence that wet (and likely acidic) environmental conditions occurred during Mars' geological history. In addition to jarosite, a variety of other hydrated sulfates such as gypsum ($CaSO_4{\cdot}2H_2O$), bassanite ($CaSO_4{\cdot}½H_2O$), kieserite ($MgSO_4{\cdot}H_2O$), epsomite ($MgSO_4{\cdot}7H_2O$), hexahydrite ($MgSO_4{\cdot}6H_2O$), pentahydrite ($MgSO_4{\cdot}5H_2O$), and starkeyite ($MgSO_4{\cdot}4H_2O$) have been detected on the Martian surface in recent years. These sulfate deposits, identified through orbital spectroscopy and in-situ rover analyses, indicate episodes of evaporation and mineral precipitation from acidic brines on ancient Mars (Bibring et al., 2005; Wang et al., 2016). Ongoing missions continue to reveal sulfates in diverse settings; for example, the Curiosity rover's instruments confirmed jarosite

Geomicrobiology of Río Tinto. https://doi.org/10.1016/B978-0-443-32948-7.00005-9

within Gale Crater sediments (Bristow et al., 2018) and the Perseverance rover's Raman spectrometers have detected sulfates with SuperCam (Lopez-Reyes et al., 2025) and SHERLOC (Corpolongo et al., 2023; Wang et al., 2024) in Jezero Crater. Furthermore, the SHERLOC instrument identified varying hydration states, underscoring that hydrated sulfates are widespread on Mars and retain clues to past water activity.

Given the significance of sulfates on Mars, investigations of terrestrial environments where sulfates form under analogous acidic, water-limited conditions are of great interest for planetary exploration. Studying such Mars analog sites on Earth can inform our interpretations of Martian mineralogy and geochemistry, as well as guide the development and test of rover instruments (Rull et al., 2022). One of the most well-known analogs is the Río Tinto mining district in Huelva, southwestern Spain, which has long been proposed as a geochemical and mineralogical analog of Meridiani Planum on Mars (Fairén et al., 2004; Fernández-Remolar et al., 2005). Río Tinto's acidic river system, fed by oxidative weathering of sulfidic ore deposits, produces abundant sulfate minerals similar to those on Mars. To date, approximately 15–20 different hydrated sulfate species have been identified in the Río Tinto system (Buckby et al., 2003; Edwards et al., 2007; Frost et al., 2005; Sobron et al., 2009a), including many of the same minerals detected on Mars (Carter et al., 2013). Another relevant analog site is the Jaroso Ravine in Almería, Spain – the type locality of jarosite. Jaroso has also been proposed as an analog for Martian acid sulfate terrains (Martinez-Frías et al., 2004) because it hosts assemblages of jarosite, natrojarosite ($NaFe^{3+}{}_3(SO_4)_2(OH)_6$), alunite ($KAl_3(SO_4)_2(OH)_6$), gypsum, and other sulfates similar to those found on Mars (Fernández-Remolar et al., 2005; Frost et al., 2005). An important difference, however, is that Río Tinto's sulfates precipitate in an environment teeming with extremophile microbial activity, whereas Jaroso Ravine's sulfates appear to form through purely abiotic processes in an older, dry hydrothermal setting. Comparative studies of such sites allow scientists to distinguish mineralogical signatures of biotic influence versus purely inorganic formation, which is relevant to astrobiology and the search for biosignatures on Mars.

Mars analog field sites are indispensable for developing and validating scientific approaches for Mars exploration. They enable testing of instruments under realistic field conditions, calibration of analytical protocols, and refining of strategies to detect organics or other biosignatures in Martian samples. In the context of recent and upcoming missions – including the Curiosity rover (MSL, 2012), Perseverance rover (Mars, 2020), the ExoMars Rosalind Franklin rover (scheduled for 2028), and a prospective Mars Sample Return – studies at analog sites like Río Tinto help predict what mineral assemblages and geochemical indicators these missions might encounter. Ultimately, insights gained on Earth can inform where and how to search for evidence of past life on the red planet.

Objectives

In this chapter, we review 2 decades of research at Río Tinto, with a focus on mineralogical and geochemical characterization of its iron-sulfate rich environment using a combination of spectroscopic techniques. The goal is to establish models of mineral precipitation and transformation in an extreme acidic system, and to draw comparisons to Martian environments. In particular, we devote special attention to the jarosite mineral: its occurrence at Río Tinto (where biological processes contribute to its formation) versus at Jaroso Ravine (where it forms abiotically). By comparing the spectroscopic signatures of jarosite from these sites, we explore potential criteria to differentiate biogenically influenced minerals from purely inorganic ones. A variety of analytical techniques have been employed in our investigations – including X-ray diffraction (XRD), Fourier-transform infrared spectroscopy (FTIR), laser-induced breakdown spectroscopy (LIBS), and Raman microscopy – both in the laboratory and in the field using portable prototypes of instruments developed for Mars missions. For brevity, the results presented here will emphasize the Raman spectroscopy findings, as Raman proved especially powerful in identifying the diverse minerals present.

Laser Raman spectroscopy has been recognized for decades as a powerful tool for in-situ mineralogy, capable of identifying both crystalline and amorphous materials in solid, liquid, or gaseous states. Its suitability for planetary exploration was noted early in the Mars rover era (Wang et al., 2003). Over the years, numerous Raman or combined Raman-LIBS instruments have been proposed for planetary missions. These include Mars-focused instruments (e.g., the initial ExoMars rover's Raman spectrometer concept (Courréges-Lacoste, B. and F., 2007), some of which are now operational on Mars, such as SHERLOC and SuperCam on NASA's Perseverance rover (Bhartia et al., 2021; Wiens et al., 2021). Others are in development or awaiting launch, for example the Raman Laser Spectrometer (RLS) on the ESA ExoMars Rosalind Franklin rover slated for 2028 (Rull et al., 2017), and the RAX instrument for JAXA's Mars Moon Explorer (MMX) which will perform Raman analyses on Phobos (Cho et al., 2021). Variants of Raman spectrometers have also been designed for other bodies – a remote Raman system was tested for potential Venus exploration (Sharma et al., 2003, 2011), a combined Raman/fluorescence sensor for organic detection on Mars and icy moons has been proposed (Fairén et al., 2020), and a versatile Raman instrument has even been developed for lunar exploration (Cloutis et al., 2022). Some of these devices couple Raman with LIBS or fluorescence spectroscopies to enhance their analytical capabilities. The successful deployment of Raman instruments on Perseverance, and the inclusion of RLS on ExoMars, demonstrate the value of Raman spectroscopy in planetary science. The work presented in this chapter not only yields scientific insights into the Río Tinto system as a Mars analog, but also provides a test-bed for these emerging technologies by challenging them with complex natural samples and mixtures analogous to those a rover might analyze on Mars.

Description of the study sites

Río Tinto (Huelva, SW Spain)

Río Tinto (meaning "Red River") is a river draining a historically active mining district in southwestern Spain (See Fig. 12.1). The river's basin lies within the Iberian Pyrite Belt (IPB), a 250-km long, 25–70 km wide metallogenic province renowned as one of the world's largest deposits of volcanogenic massive sulfides. The IPB bedrock consists of a Paleozoic volcano-sedimentary complex hosting abundant polymetallic sulfide ore bodies (dominantly pyrite, with copper, zinc, lead, etc.), which formed during Hercynian (Variscan) tectonism (Leistel et al., 1998). Mining activity in the Río Tinto area dates back to at least the Bronze Age, making it one of the longest continuously mined regions in the world. Extensive mining and exposure of sulfide minerals to oxygen and water have led to pervasive acid mine drainage (AMD) in the watershed. The oxidative dissolution of pyrite (FeS_2) and other sulfides generates sulfuric acid and releases large concentrations of metals into solution. As a result, the waters of the Río Tinto are extremely acidic (pH typically ~2) and enriched in sulfate and heavy metals. Dissolved sulfate levels can reach on the order of 10^4–10^5 mg/L (up to ~18,000 ppm as SO42−) in the river water. The striking red coloration of the river is caused by high concentrations of ferric iron, a product of pyrite oxidation (see Fig. 12.2).

Microorganisms—in particular, chemolithotrophic bacteria and archaea—play a key role in mediating the oxidation of sulfides and the generation of AMD at Río Tinto. The metabolic activities of acidophilic iron- and sulfur-oxidizing microbes greatly accelerate the breakdown of sulfide minerals, maintaining the high acidity and ferric iron levels in the water (López-Archtlla, Marín and Amils, 1993; Amaral Zettler et al., 2003; Gonzáles-Toril et al., 2005). This biologically enhanced leaching creates what has been described as an underground "bioreactor" of microbes driving sulfide oxidation (Amils et al., 2014). The acidic, iron-rich waters eventually emerge at the surface and flow along the river, where evaporation and mixing cause the precipitation of various sulfate minerals along the banks and floodplain. Hydrated sulfate salts (often brightly colored efflorescences or crusts) form as the acid brines evaporate, especially during drier seasons. The mineral precipitation sequence is influenced by seasonal climatic variations (temperature and rainfall) and local micro-environments (e.g., evaporative pools, wet vs. dry areas). Over 15 distinct sulfate minerals have been documented in the riverbank sediments and precipitates of Río Tinto. These include iron sulfates such as copiapite-group minerals, coquimbite, kornelite, and rozenite; magnesium sulfates like epsomite, hexahydrite, and starkeyite; calcium sulfate (gypsum); as well as the aforementioned jarosite and related K–Na–H3O iron sulfates (Buckby et al., 2003; Frost et al., 2005; Sobron et al., 2009a). Notably, many of these minerals are also found on Mars, making Río Tinto's mineral assemblage an excellent analog for acid-sulfate deposits on that planet (Carter et al., 2013). In addition to sulfates, secondary iron oxide/hydroxide minerals

FIGURE 12.1 Sampling sites.

(1) Sampling sites at the Peña del Hierro mine setting and Tinto river source. Efflorescence and coating samples were collected from the riverbanks of the river (P1–P7, northern branch; RP1-RP8, eastern branch; EG-1 to EG-13, western branch). The pH of the river branches at sampling has been indicated. (2) Liquid samples collected at the same places. RTNRE-eastern branch, RTNRN-Northern branch, RTNCO and RTNCOb two samples at the junction of these two branches with the western branch, RTNO-Western branch. (3) Examples of materials studied in Rio Tinto and conceptual diagram of mineral precipitation in this system. A. Primary materials (quartz, pyrite and chlorite); B. Typical reddish water in Rio Tinto; C. Ochre coatings on the riverbed (jarosite, schwertmannite, iron oxyhydroxides); D and E. Efflorescence on the riverbanks (hydrated sulfate minerals); F. Coatings comprising iron oxyhydroxides.

- Adapted from D.C. Fernández-Remolar, R.V. Morris, J.E. Gruener, R. Amils, A.H. Knoll, The Río Tinto Basin, Spain: mineralogy, sedimentary geobiology, and implications for interpretation of outcrop rocks at Meridiani Planum, Mars. Earth Planet Sci. Lett. 240 (1) (2005), 149–167, https://doi.org/10.1016/j.epsl.2005.09.043.

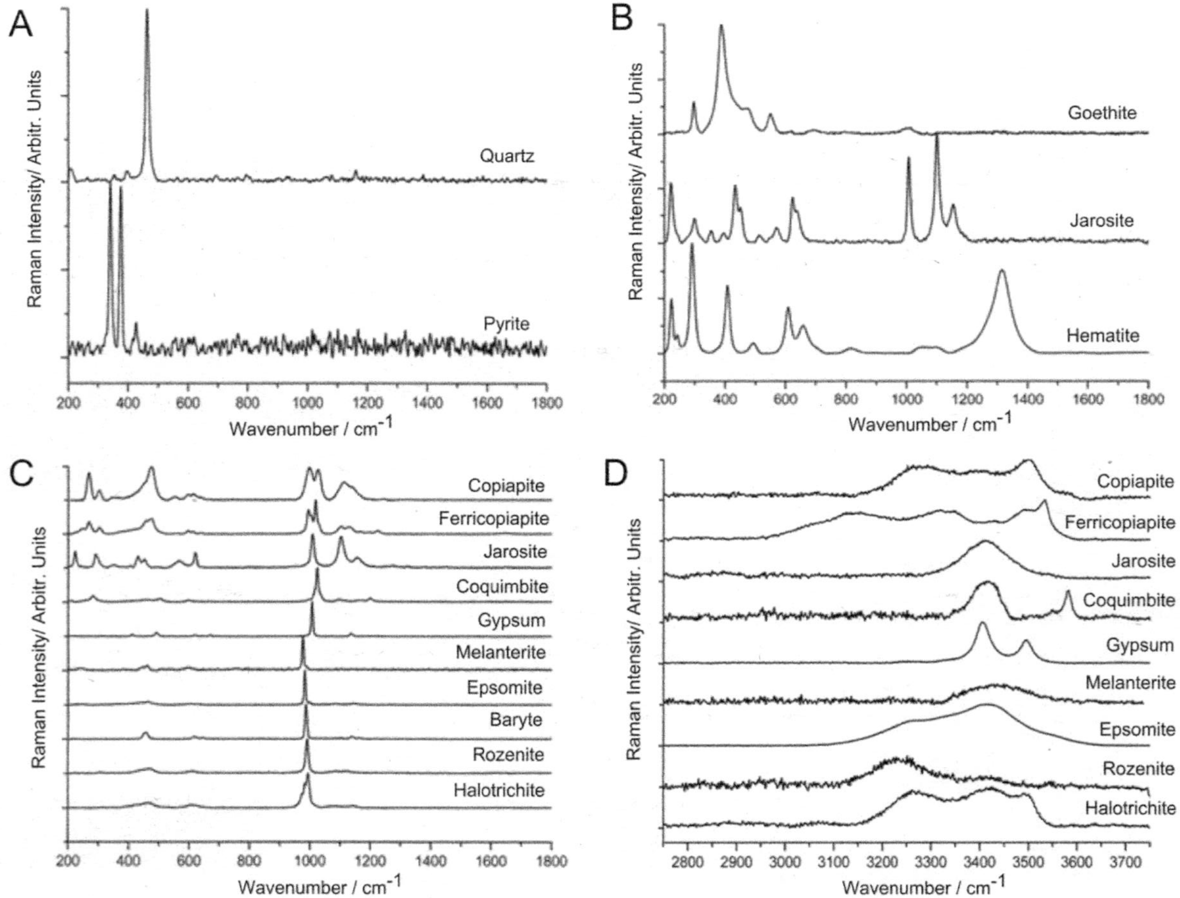

FIGURE 12.2 Raman spectra of minerals found in Río Tinto.

A. Primary materials (basement); B. Alteration minerals and coating on the riverbed; C. Hydrated sulfate mineral in the range 200–1800 cm^{-1}; D. Hydrated sulfate minerals in the v(OH) range 2800–3700 cm^{-1}.

(such as goethite and hematite) form in the sediments as products of iron oxidation and hydrolysis.

Beyond serving as a mineralogical analog, Río Tinto is also considered a potential astrobiological analog for Mars. Despite its extreme acidity (pH 1.5–3) and metal-rich waters, the system supports a robust ecosystem of acidophiles, including bacteria, archaea, and eukaryotic microorganisms. Their metabolic by-products contribute organic matter and can influence mineral formation (e.g., through biosorption or by providing nucleation sites). The question of whether biosignatures of such life could be detected remotely or in returned samples is highly relevant to Mars exploration. In the case of Río Tinto, the interplay of biological and inorganic processes offers a unique opportunity to study how microbial life might alter mineralogy and geochemistry in a Mars-like environment.

Jaroso Ravine (Almería, SE Spain)

The Jaroso Ravine (Barranco del Jaroso) is located in the Sierra Almagrera range of southeastern Spain. It is geologically distinct from Río Tinto and represents an older hydrothermal system. Jaroso is famously known as the "world type locality" of jarosite—it is the site where jarosite was first identified and described in the 19th century. The Jaroso Hydrothermal System (JHS) is interpreted as a late- or postvolcanic hydrothermal episode associated with Tertiary volcanism in the region (Martínez-Frías et al., 2004). The ravine cuts through a sequence of graphitic phyllites, quartz-rich schists, and quartzites of Paleozoic age, which serve as the host rocks for the hydrothermal mineralization. The alteration paragenesis in these rocks includes quartz, muscovite, graphite, and minor biotite and garnet (Martínez Frías, 1991; López-Gutiérrez et al., 1993). Within the ravine, metallic ore minerals (primarily Fe–Pb–Ag sulfides and sulfosalts) occur in veins. Although most of the sulfide ore is now weathered or removed, remnants of narrow polymetallic veins (10–30 cm wide) can still be observed on the ravine walls.

Jaroso can be considered an ancient analog of acid sulfate formation under abiotic conditions. The climate in this area is semiarid, with low annual rainfall (~210 mm/yr) that is irregularly distributed, and an average temperature of around 19.5°C. Water flow in the ravine is ephemeral – it remains dry for most of the year, with brief wet periods following occasional rain events. Consequently, sulfates in Jaroso form through three identified pathways: (1) Hydrothermal alteration products—relics of the original JHS activity, where hot, sulfur-rich fluids altered primary minerals (phyllosilicates and sulfides) and precipitated minerals like jarosite and alunite at depth or under subaqueous conditions (Fernández-Remolar et al., 2005). (2) Supergene alteration products—later (post-hydrothermal) minerals formed by weathering of primary sulfides in the presence of oxygen and water percolating through the rocks. These include secondary jarosite and related sulfates produced by oxidative acid leaching of the ore deposits, but on geologic timescales with no active biology. (3) Modern efflorescences—transient sulfate crusts that appear on the outcrop surfaces after rare rainfall events, when meteoric water dissolves soluble salts and redeposits them upon evaporation. These efflorescences (e.g., melanterite, epsomite) are seasonal and reflect present-day climate variability.

In contrast to Río Tinto, **no evidence of biological activity has been found associated with the sulfate formation in Jaroso Ravine** (Fernández-Remolar et al., 2005; Frost et al., 2007). The Jaroso system is essentially controlled by abiotic physico-chemical processes. This makes it an informative end-member analog for Mars—a scenario where sulfates form without any contribution from life, perhaps more comparable to current Martian surface conditions. The absence of organics or microbial influences in Jaroso's mineralogy means that any differences observed between Jaroso minerals and those from a biologically active site like Río Tinto could hint at biosignatures or biogenic effects. In particular, comparing jarosite from Jaroso (abiotic, hydrothermal origin) to jarosite from Río Tinto (potentially biomediated, supergene origin) allows us to investigate whether biological intervention leaves a detectable imprint on the mineral's properties (such as crystal chemistry, morphology, or spectroscopic features). Such comparisons are valuable for developing strategies to detect past life on Mars: if subtle spectral or structural differences can be correlated with biogenic processes, future rover missions might use those as additional criteria when analyzing Martian samples.

Analytical methodology

Sample collection

A wide range of representative samples was collected from both sites (Río Tinto and Jaroso) to encompass different minerals and formation conditions (See Fig. 12.3). These included sulfate salt crusts and sediments from various locations along the Río Tinto river (covering its western, northern, and eastern branches), as well as jarosite specimens from Jaroso Ravine of both hydrothermal origin (well-crystallized jarosite from an old mine tunnel in Jaroso) and supergene origin (microcrystalline jarosite from weathered outcrops). Samples were handled and stored to preserve their hydration state as much as possible prior to analysis.

Spectroscopic analyses

A combination of laboratory and field-portable spectroscopic instruments was used to analyze the mineral samples. In particular, extensive analyses were performed using Raman microscopy, both with a high-performance laboratory Raman system and a prototype of a rover-mounted Raman instrument:

- *Laboratory Micro-Raman System:* We employed a Kaiser Optical Systems HoloSpec f/1.8i Raman spectrometer, coupled via fiber optic to a Nikon Eclipse E600 microscope for microspot analyses. This system provides a spectral range of 180–3800 cm−1 with a spectral resolution of ~5 cm^{-1}. The detector is a 1024×256 pixel CCD (Andor Model DV420A-OE-130) cooled for low noise. The excitation source is a He–Ne laser (632.8 nm) directed through the microscope to a ~30 μm diameter spot (using a 50× objective) on the sample.

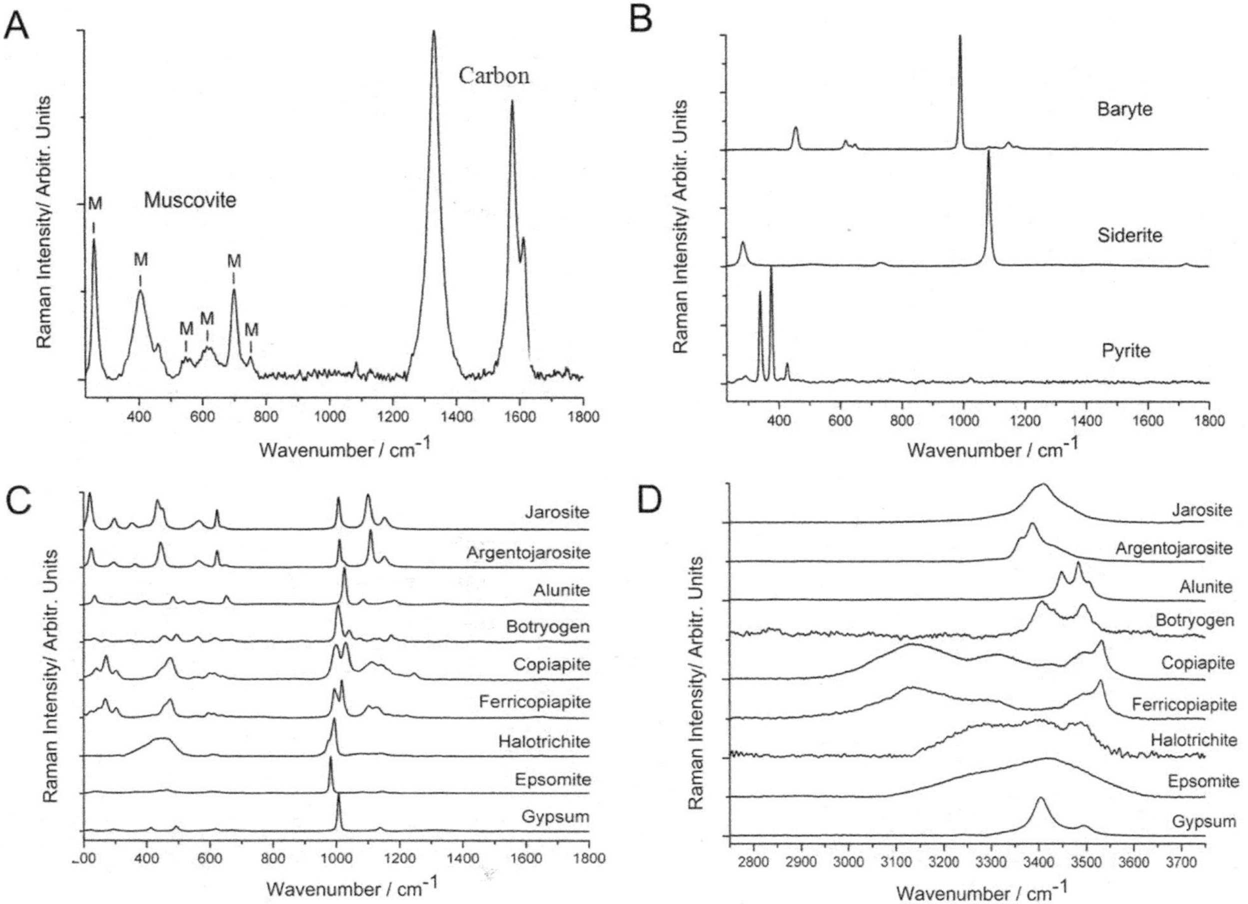

FIGURE 12.3 Raman spectra of minerals found in Jaroso ravine.

A. Primary materials (basement); B. Ore bodies; C. Hydrated sulfate mineral in the range 200–1800 cm^{-1}; D. Hydrated sulfate minerals in the range 2800–3700 cm^{-1}.

The laser power at the sample was maintained at ~16 mW, corresponding to an irradiance on the order of 5 kW/cm^2. This power level was conveniently reduced when necessary to avoid thermal damage to the sulfates, as evidenced by the absence of any black-body radiation continuum in the Raman spectra (McCreery, 2001). For more laser-sensitive minerals like iron oxides and hydroxides, the power was further reduced until no sample overheating was observed (monitored by checking that no broad thermal emission background appeared in the spectrum). Typical acquisition parameters for Raman spectra were 10 s integration time and 10 accumulations per spot, which provided a good signal-to-noise ratio for most minerals. Calibration of the Raman shift was checked using standard references (e.g., silicon wafer at 520.5 cm^{-1}).

- *ExoMars RLS SIM:* To compare results and evaluate instrumentation destined for Mars, we also analyzed samples with the RLS SIM (Lopez-Reyes et al., 2022) a scientific prototype of the Raman Laser Spectrometer (RLS) under development for ESA's ExoMars *Rosalind Franklin* rover mission. The RLS prototype operates with a 532 nm frequency-doubled Nd:YAG laser as its excitation source (green laser). The laser power at the sample was set around 20 mW with this system. The Raman optical head (designed to mimic the actual flight instrument) is fiber-coupled to a microscope or probe, yielding a larger analytical spot of ~50 μm diameter on the sample (thus an irradiance of ~1 kW/cm^2). The spectral range covered by the RLS spans approximately 150–4000 cm^{-1}, which includes the principal vibrational bands of interest (from lattice modes through OH stretching bands). The RLS spectrometer features a compact transmission grating and CCD detector optimized for the rover's conditions (Rull et al., 2017; Lopez-Reyes et al., 2013). Although the prototype's resolution is modest (~six to eight cm^{-1}) compared to the lab system, it is sufficient to resolve the key Raman peaks of the sulfate and oxide minerals in our study. We used the RLS prototype both in the lab and during field campaigns at Río Tinto to assess its performance on raw outcrop samples. Data from the RLS were processed with onboard-equivalent software for cosmic ray removal and fluorescence/background subtraction to simulate how the rover would handle spectra.

In addition to Raman spectroscopy, complementary analyses were performed: XRD was used to verify mineral phases (especially for bulk powder samples); FTIR spectroscopy (in ATR mode) provided additional identification of hydrated phases and molecular water in samples; and LIBS measurements were done on selected samples to determine elemental composition (not detailed here). Water chemistry of the Río Tinto river was analyzed in situ (pH, redox potential) and via laboratory assays (ICP-OES/ICP-MS for dissolved metals, ion chromatography for sulfate) to support geochemical modeling. In particular, Raman spectroscopy of liquid samples was used to investigate the speciation of Fe–SO_4 ions in the acidic river water, allowing us to model chemical equilibria in the $FeSO_4$–H_2SO_4–H_2O system (Venegas, 2014). These geochemical models help predict which mineral salts will precipitate under given conditions of concentration, pH and evaporation.

Results and discussion

Raman identification of sulfate minerals at Río Tinto

Field and laboratory Raman spectroscopy revealed a rich assemblage of secondary alteration phases consisting of sulfate minerals precipitating in the Río Tinto environment, in addition to primary minerals such as quartz and pyrite. Figure 12.2 shows the Raman spectra of these minerals. The ochre-colored sediments and crusts on the riverbed are composed predominantly of iron oxides/hydroxides and sulfates. Raman spectra confirmed the presence of jarosite, goethite (α-FeOOH), and hematite (Fe_2O_3) as major components of the solid precipitates coating the riverbed. Jarosite was detected both by in-situ measurements on outcrops and in laboratory spectra of collected samples, often in association with nanocrystalline goethite and hematite. The jarosite typically occurs as fine-grained coatings or disseminated particles, rather than large, well-formed crystals, consistent with rapid precipitation in the acidic stream. Additionally, Mössbauer spectroscopy measurements by other researchers had identified schwertmannite ($Fe^{3+}{}_{16}O_{16}(OH)_{12}(SO_4)_2{\cdot}nH_2O$), a metastable iron oxyhydroxysulfate, in the river sediments (Fernández-Remolar et al., 2005; Fleischer et al., 2010). Our Raman observations did not clearly distinguish schwertmannite (which has broad, poorly defined bands) from goethite in the mixtures, but the presence of schwertmannite is plausible in the freshest precipitates.

Micro-Raman analysis of the efflorescent sulfate salts along Río Tinto allowed us to identify at least 10 different hydrated sulfate minerals (see Table 12.1).

Table 12.1 Hydrated sulfate minerals found at Rio Tinto by Raman spectroscopy.

		Rio Tinto's branches		
Minerals	Formula	Western branch	Northern branch	Eastern branch
Copiapite	$Fe^{2+}(Fe^{3+})_4(SO_4)_6(OH)_2{\cdot}20H_2O$	X	X	X
Coquimbite	$Fe^{3+}{}_2(SO_4)_3{\cdot}9(H_2O)$	X		
Epsomite	$MgSO_4{\cdot}7H_2O$			X
Ferricopiapite	$Fe^{3+}{}_{2/3}Fe^{3+}{}_4(SO_4)_6(OH)_2{\cdot}20H_2O$	X	X	
Gypsum	$CaSO_4{\cdot}2H_2O$		X	X
Halotrichite	$Fe^{2+}Al_2(SO_4)_4{\cdot}22(H_2O)$	X	X	
Hexahydrite	$MgSO_4{\cdot}6H_2O$			X
Jarosite	$KFe^{3+}{}_3(SO_4)_2(OH)_6$		X	X
Melanterite	$FeSO_4{\cdot}7H_2O$		X	
Rozenite	$FeSO_4{\cdot}4H_2O$		X	X

The minerals detected include members of the copiapite group, coquimbite, ferricopiapite, halotrichite, rozenite, melanterite, gypsum, epsomite, hexahydrite, and jarosite. Many of these were identified by their distinctive Raman peak patterns in the sulfate stretching (ν1, ν3) and bending (ν2, ν4) regions. For example, copiapite [$(Fe^{2+})(Fe^{3+})_4(SO_4)_6(OH)_2{\cdot}20H_2O$] was recognized by a strong Raman band near ~1010 cm^{-1} (symmetric SO_4 stretch) and characteristic OH/H_2O bands, while coquimbite [$Fe^{3+}{}_2(SO_4)_3{\cdot}9H_2O$] showed its signature sulfate stretches around 1130–1150 cm^{-1}. Halotrichite [$Fe^{2+}Al_2(SO_4)_4{\cdot}22H_2O$] was identified by a pair of $\nu1(SO_4)$ bands at ~984 and 1004 cm^{-1} (indicative of mixed cation sulfates). Magnesium sulfates were also present: epsomite ($MgSO_4{\cdot}7H_2O$) and hexahydrite ($MgSO_4{\cdot}6H_2O$) were detected in samples from the eastern branch of the river, as evidenced by their Raman peaks in the 980–1040 cm^{-1} range and strong water band at ~3400 cm^{-1} (from crystal water). Gypsum ($CaSO_4{\cdot}2H_2O$) was less common but found in some eastern branch deposits (characteristic 1008 cm^{-1} sulfate stretch and 415 cm^{-1} bending mode). Iron(II) sulfates like melanterite ($FeSO_4{\cdot}7H_2O$), and its dehydrated form rozenite ($FeSO_4{\cdot}4H_2O$) were observed in some samples (notably in more sheltered niches that retain moisture). Melanterite shows a Raman ν1 at ~1022 cm^{-1} and rozenite at ~1018 cm^{-1}, with differences attributable to hydration state.

Despite the complex mixtures often present, the ~10–20 μm laser spot of the microRaman system was usually small enough to hit single mineral phases within a polycrystalline crust. In some cases, however, our spectra revealed admixtures of minerals in the same spot. For instance, jarosite was sometimes found intimately mixed with hematite and goethite (all products of pyrite oxidation). Likewise, copiapite and coquimbite frequently cooccurred, and we occasionally saw overlapping bands of hexahydrite with epsomite (indicating partial dehydration or coexistence). Using a larger laser spot (50 μm) or the unfocused RLS SIM accentuated this issue, yielding composite spectra from adjacent grains. Nonetheless, by careful selection of analysis spots and comparing against reference spectra, we successfully cataloged the Raman fingerprints of the major sulfates in Río Tinto.

The Raman analyses of the metamorphic materials that form the host rock of Jaroso Ravine reveal the presence of muscovite and carbon (Fig. 12.3A), in addition to quartz, anatase and rutile, hematite, magnetite, goethite, lepidocrocite, zircon, orthoclase and anorthoclase. The analyses of ore minerals of veins indicate the presence of pyrite and spharelite (Fig. 12.3B). Earlier works also found other sulphides in the mineral veins, including bournonite ($PbCuSbS_3$), boulangerite ($Pb_5Sb_4S_{11}$), tetrahedrite ($(Cu,Fe)_{12}Sb_4S_{13}$), galena (PbS), chalcopyrite ($CuFeS_2$) and arsenopyrite (FeAsS). Gangue minerals include barite, siderite and quartz (Fig. 12.3B)

Regarding minerals resulting from supergenic alteration of the veins and efflorescences (Fig. 12.3C and D), mainly minerals of the jarosite and alunite groups have been identified by using micro-Raman spectroscopy. Up to 14 different hydrated sulfates have been detected by this technique (See Table 12.2).

Table 12.2 Hydrated sulfate minerals found at Jaroso Ravine by Raman spectroscopy.

		Jaroso ravine	
Minerals	Formula	Supergenic alteration	Efflorescences
Aluminite	$Al_2SO_4(OH)_4{\cdot}7H_2O$	X	
Alunite	$KAl_3(SO_4)_2(OH)_6$	X	X
Botryogen	$MgFe^{3+}(SO_4)_2(OH){\cdot}7H_2O$		X
Copiapite	$Fe^{2+}(Fe^{3+})_4(SO_4)_6(OH)_2{\cdot}20H_2O$	X	
Epsomite	$MgSO_4{\cdot}7H_2O$	X	X
Fibroferrite	$Fe^{3+}(SO_4)(OH){\cdot}5H_2O$		X
Ferricopiapite	$Fe^{3+}{}_{2/3}Fe^{3+}{}_4(SO_4)_6(OH)_2{\cdot}20H_2O$		X
Gypsum	$CaSO_4{\cdot}2H_2O$	X	X
Halotrichite	$Fe^{2+}Al_2(SO_4)_4{\cdot}22(H_2O)$	X	X
Hexahydrite	$MgSO_4{\cdot}6H_2O$	X	X
Jarosite	$KFe^{3+}{}_3(SO_4)_2(OH)_6$	X	X
Natrojarosite	$NaFe^{3+}{}_3(SO_4)_2(OH)_6$	X	
Rozenite	$FeSO_4{\cdot}4H_2O$	X	X
Szomolnokite	$Fe^{2+}SO_4{\cdot}H_2O$	X	

Spatial mineral distribution at Rio Tinto

The occurrence of sulfate species was found to vary between different branches of the Río Tinto, reflecting local geochemical conditions. As summarized in Fig. 12.2 and Table 12.1, the western branch of the river (upstream, closer to certain mine sites) showed a more limited suite of hydrated Fe-sulfates – mainly copiapite, coquimbite, ferricopiapite, and halotrichite. In contrast, the northern and eastern branches (downstream, possibly receiving inputs from different host lithologies or experiencing different evaporation regimes) contained a wider variety of salts, including Mg-sulfates (epsomite/hexahydrite) and Fe^{2+} sulfates (melanterite/rozenite) in addition to the Fe^{3+} sulfates. This suggests that factors like groundwater composition, flow path, and evaporation rate differ across the river system, leading to locally distinct mineral assemblages. For example, the presence of Mg-sulfates in the eastern branch might indicate inputs of Mg (perhaps from bedrock or a tributary) and higher evaporation, whereas the western branch's dominance of ferric sulfates could imply more extensive oxidation (producing Fe^{3+}) and a relative paucity of Mg or Ca.

Our findings are consistent with previous mineralogical surveys of Río Tinto. In-situ Raman analyses reported by Fleischer et al., (2010) likewise confirmed copiapite, coquimbite, and rozenite as common efflorescences in parts of the river. An earlier study by Guerrero et al., (2010) had additionally identified rhomboclase

($H_5O_2Fe^{3+}(SO_4)_2{\cdot}2H_2O$) and szomolnokite ($FeSO_4{\cdot}H_2O$) in some precipitates. We did not explicitly observe these two in our Raman survey, perhaps due to their low abundance or overlapping features with other sulfates. However, an experimental study by (Guerrero, 2012) replicated evaporation of Río Tinto water under controlled conditions and detected magnesiocopiapite ($MgFe^{3+}{}_4(SO_4)_6(OH)_2{\cdot}20H_2O$), metavoltine ($K_4Na_4Fe^{2+}{}_{0.5}Zn_{0.5}Fe^{3+}{}_6(SO_4)12O_2{\cdot}20H_2O$), and aluminocopiapite ($Al^2/3Fe^{3+}{}_4(SO_4)_6(OH)_2{\cdot}20H_2O$) forming in addition to the minerals mentioned above. This indicates that the natural system likely contains an even greater complexity of sulfate phases at trace levels or in transient states. Overall, the Raman spectroscopic approach has proven adept at distinguishing these various sulfate minerals, even when they occur as fine-grained mixtures. This validates the capability of rover-mounted Raman instruments to unravel mineral assemblages in Martian samples. Minerals like copiapite and coquimbite have distinct vibrational signatures that can be recognized by the ExoMars RLS or Mars 2020 SHERLOC/SuperCam, thus aiding in paleoenvironmental interpretations on Mars (e.g., identifying episodes of evaporation or fluctuating water chemistry).

From a geochemical standpoint, the diversity of sulfates at Río Tinto reflects sequential evaporation and neutralization processes in an acid sulfate system. As pH and concentration thresholds are crossed, different sulfates precipitate: Fe^{3+} sulfates tend to form at lower pH and higher Fe/S ratios (jarosite, then copiapite group), whereas Mg and Fe^{2+} sulfates can form when some neutralization or mixing raises pH or when Fe^{2+} remains stable (melanterite in more reducing niches). The detection of these minerals on Mars in various locations likewise points to episodes of evaporative concentration of acidic, sulfate-rich waters in Mars' past (Gendrin et al., 2005; Farley et al., 2022). Thus, Río Tinto offers a tangible model: an acidic river depositing layer-cake sequences of sulfates that might resemble those in Martian sedimentary rocks (e.g., the Burns formation at Meridiani).

Raman spectroscopic characterization of Río Tinto acidic waters and analysis of mineral precipitation sequences

The physicochemical characterization of the acidic aqueous solutions of Río Tinto is of great importance for understanding its hydrogeochemistry and modeling the sequence of mineral precipitation. This characterization was carried out using in situ Raman spectroscopic analysis both directly on the river and on selected liquid samples collected from different branches at the river's source. In the latter case, after spectroscopic analysis, the samples were evaporated under laboratory conditions using a custom-designed evaporation simulator, as well as in the form of small droplets placed under a microscope for time-resolved, online monitoring (Rull et al., 2014).

In addition, the diffusion properties of sulfuric acid in water were investigated using dedicated diffusion chambers combined with in situ Raman spectroscopy (Sobrón et al., 2009b). These experimental results were complemented by computational modeling of thermodynamic equilibria in the acidic solutions using data from the $FeSO_4$–H_2SO_4–H_2O system (Sobron et al., 2007a).

Prior to analyzing the natural acidic solutions, the ionic speciation of aqueous sulfuric acid was investigated as a function of acid concentration. Sulfuric acid dissociates in two steps, producing bisulfate and sulfate ions as follows (Burriel et al., 1052):

$$H_2SO_4 + H_2O \leftrightarrow HSO_4^- + H_3O^+ \quad (12.1)$$

$$HSO_4^- + H_2O \leftrightarrow SO_4^{2-} + H_3O^+ \quad (12.2)$$

The degrees of dissociation corresponding to Eqs. (12.1) and (12.2) are denoted as α_1 and α_2, respectively. The first dissociation is essentially complete ($\alpha_1 \approx 1$) across most concentration ranges (Young et al., 1959). The second dissociation can be expressed in terms of ion molalities as:

$$\alpha_2 = m\left(SO_4^{2-}\right) / \left[m\left(SO_4^{2-}\right) + m\left(HSO_4^-\right)\right] \quad (12.3)$$

The molal concentrations of HSO_4^- and SO_4^{2-} can be estimated from the integrated intensities of the 1050 cm^{-1} and 980 cm^{-1} Raman bands, respectively.

The top shows Raman spectra of real samples collected from various source branches of Río Tinto and compares them with those from synthetic $FeSO_4$–H_2-SO_4–H_2O systems. The spectra, along with measured pH and colorimetric properties, indicate strong correspondence between the natural and synthetic acidic solutions (Fig. 12.4).

From these data, the ionic speciation and α_2 values were estimated, enabling the calculation of the equilibrium constant for the second dissociation, including the activity coefficient quotient as a function of sulfuric acid concentration (See, detail of the SO_4^{2-} and HSO_4^- internal vibrations) (Sobron et al., 2007b).

Bottom, displays in situ Raman spectra obtained at various depths in Río Tinto. Despite the inherent difficulty of these measurements, both SO_4^{2-} and HSO_4^- species were successfully detected and their relative intensities approximatively evaluated.

Another relevant aspect of this study was the analysis of the precipitation sequence from natural Río Tinto waters. To this end, a custom evaporation simulator was designed in which collected waters were placed on an inclined surface, simulating riverbank conditions (Guerrero et al., 2011; Venegas et al., 2012). The setup was illuminated using visible and near-infrared lamps to reproduce diurnal cycles. Complementary microscale experiments were performed using droplets of Río Tinto water deposited on various substrates (glass, aluminum, pyrite, and zinc) and analyzed by time-resolved micro-Raman spectroscopy at room temperature.

The experiments consisted of two stages: (i) a short-term observation phase lasting 90 min with spectral acquisition every 10 min to characterize the early stages of evaporation, and (ii) a long-term phase in which measurements were taken daily until complete evaporation (Rull et al., 2014). Nearly all mineral species previously identified in Río Tinto through in situ and laboratory techniques were also detected

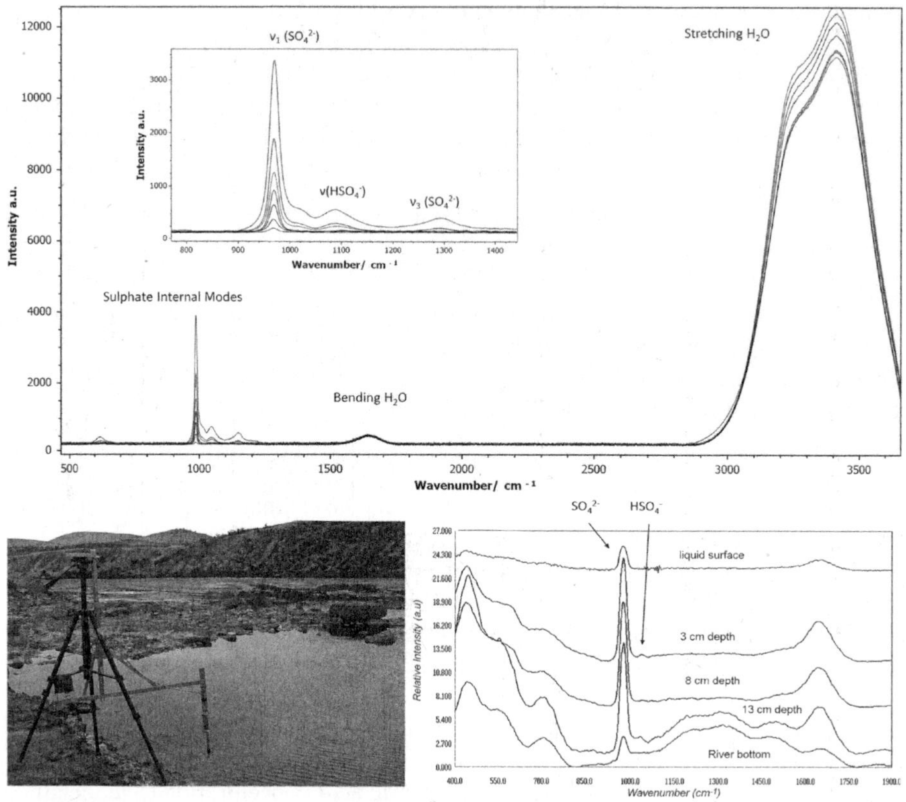

FIGURE 12.4

Top: Raman spectra obtained from the liquid samples at the Rio Tinto source branches and identification of the main chemical species on solution. Bottom: In-situ Raman spectra obtained at the river using a specific Raman probe for acidic and corrosive liquids. In both cases SO_4^{2-} and HSO_4^- chemical species were clearly identified and their relative concentration estimated.

in these simulation experiments. A summary of the results is provided in Table 12.3 (Fleischer et al., 2010; Rull et al., 2005; Rull et al., 2008).

Jarosite spectroscopy: Comparing biotic versus abiotic formation

Jarosite ($KFe^{3+}{}_3(SO_4)_2(OH)_6$) is a key mineral at both Río Tinto and Jaroso Ravine, and it serves as an insightful case study for how formation conditions (biotic vs. abiotic) might affect a mineral's characteristics. We performed detailed Raman spectroscopic analysis of jarosite samples from: (1) Jaroso Ravine hydrothermal veins (well-crystallized jarosite of abiotic, hydrothermal origin), (2) Jaroso Ravine supergene weathering products (jarosite of abiotic, but low-temperature secondary

Table 12.3 Mineral precipitation sequence obtained from droplets of Rio Tinto water samples along 30 days and identified by micro-Raman spectroscopy.

Time/Days	Mineral identified
1	Ferricopiapite
3	Coquimbite
5	Copiapite
7	Magnesiocopiapite
13	Hematite
16	Szomolnokite
18	Romboclase
30	Metavoltine

See (F. Rull, F. Sobrón, J. Guerrero, J. Medina, G. Venegas, F. Gázquez, J. Martínez-Frías, In-situ Raman analysis of the precipitation sequence of sulphate minerals using small droplets: application to Rio Tinto (Spain). Lecture Notes in Earth System Sciences. 0 (2014), 801–805. https://doi.org/10.1007/978-3-642-32408-6_173).

origin), and (3) Río Tinto riverbed sediments (jarosite precipitated with microbial involvement in an AMD environment). All these samples are the same mineral in terms of ideal composition, but they differ in crystallinity, grain size, and possibly minor substitutions or defects. Our aim was to see if any spectral differences correlate with the presence or absence of biological mediation during formation.

Raman spectral features of jarosite

In general, jarosite exhibits a characteristic set of Raman bands corresponding to the sulfate anion internal vibrations, the Fe–O and O–H lattice vibrations, and the OH stretching vibrations (See). A full factor-group analysis of jarosite's crystal symmetry ($R\overline{3}m$) is beyond the scope of this work (see Spratt et al., 2013, for a detailed treatment). We focus on the main observable bands. The $(SO_4)^{2-}$ tetrahedron in jarosite has slightly lowered symmetry ($C3v$) compared to free sulfate (Td), which causes some splitting of degenerate modes. In jarosite's Raman spectrum, the most intense band is typically the ν_1 symmetric stretch of SO_4, appearing near 1005 cm^{-1}. We indeed observe a very strong band at 1005 cm^{-1} in all our jarosite samples, which we assign to the ν_1(A1g) mode of sulfate. The ν_2 symmetric bend of sulfate (E mode splitting into Eg) manifests as a pair of medium bands around ~435 and 455 cm^{-1} in our spectra. These were present in each sample (we measured peaks at 433 and 452 cm^{-1} for the hydrothermal jarosite). The ν_3 antisymmetric stretch of sulfate splits into A1g and Eg components in jarosite; correspondingly, we detected an intense band at ~1100 cm^{-1} (1100–1102 cm^{-1}, varying slightly by sample, assigned to ν_3(A1g)) and a weaker feature around 1150–1160 cm^{-1} (assigned to ν_3(Eg)). For example, the Jaroso hydrothermal jarosite showed a sharp band at 1099 cm^{-1} and minor bands at 1150 and 1317 cm^{-1} (the latter might be an overtone or combination involving sulfate modes). The ν_4 antisymmetric bend of sulfate similarly splits; we observed a strong band at 623 cm^{-1} (A1g component) and a couple of weaker bands at ~570 and 640 cm^{-1} (Eg components). These assignments match well with prior Raman studies of natural jarosites (Frost et al., 2006) (Fig. 12.5).

At lower frequencies (below 400 cm^{-1}), jarosite displays lattice modes involving Fe–O vibrations and translations/librations of the sulfate groups. In our spectra, three main bands appeared at approximately 220, 300, and 350 cm^{-1}. Based on literature e.g (Frost et al., 2006), the band near 220 cm^{-1} can be attributed to a complex mode involving translations of Fe^{3+} cations against the sulfate framework (and possibly OH librations), while the band near 298 cm^{-1} may involve coupled Fe–O–S lattice vibrations. The feature around 352 cm^{-1} is assigned to an Fe–OH stretching mode within the $FeO_3(OH)_3$ octahedra of the jarosite structure. All three jarosite samples (biotic and abiotic) exhibited these lattice bands in roughly the same positions, with only subtle shifts (<5 cm^{-1}) and intensity differences. Thus, **for the internal sulfate modes and lattice modes, we found no significant differences between jarosites of different origin** – the peak positions and relative intensities were essentially the same within experimental precision. This indicates that the fundamental SO_4 structure and Fe–O framework of jarosite are preserved regardless of whether it formed quickly with microbial assistance or slowly in a hydrothermal setting. In other words, the primary vibrational "fingerprint" of the jarosite sulfate lattice is robust to formation conditions, which is an important point if one hopes to detect biosignatures: one cannot simply look at the sulfate stretch region and immediately tell a biotic from an abiotic jarosite.

However, a notable difference emerged in the O–H stretching region of the Raman spectra (approximately 3100–3600 cm^{-1}), which probes the vibrational modes of hydroxyl groups in the jarosite structure (right). Jarosite contains structural OH groups that are hydrogen-bonded to neighboring sulfate oxygen atoms or to other hydroxyls. The pattern and width of the OH-stretching bands can be sensitive to hydrogen bonding and disorder. In the hydrothermal jarosite from Jaroso (which is well-crystallized and relatively coarse-grained), the Raman spectrum showed multiple resolved OH bands: specifically, distinct peaks at 3319, 3389, 3409, and 3460 cm^{-1} were observed, superimposed on an OH envelope centered around 3410 cm^{-1}. These correspond to different types of OH environments in the crystal (each band likely represents OH groups with slightly different hydrogen-bond strengths). The combined OH band profile for the hydrothermal sample was quite narrow – the measured full width at half maximum (integrated over the whole OH region) was only about 56 cm^{-1} in this sample. By contrast, the supergene jarosite (microcrystalline) from Jaroso Ravine displayed a broader OH band feature. In that sample, the individual OH bands were not well resolved; instead, they merged into a broad envelope with an average integrated width of ~120 cm^{-1}. This broadening suggests a greater structural disorder or a range of H-bonding environments, likely due to poorer crystallinity or smaller crystallite size. Finally, the jarosite from Río Tinto showed the broadest OH stretching profile of all. Its OH band appeared as a single very broad feature (virtually no distinct subpeaks) spanning roughly

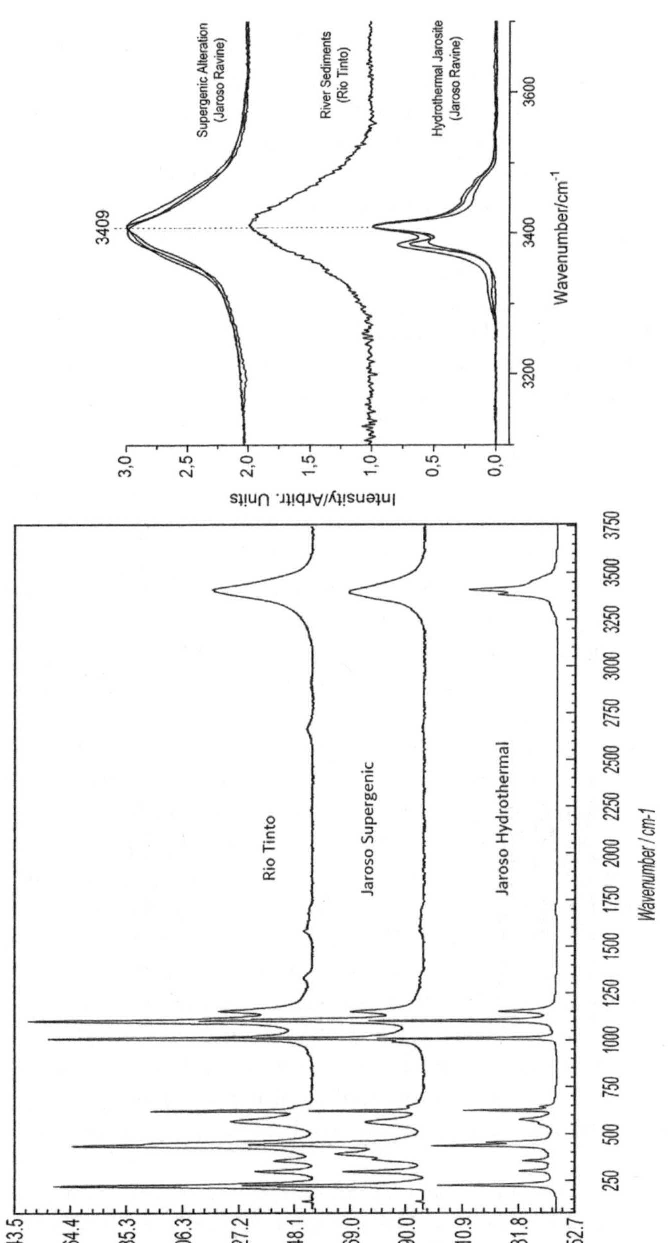

FIGURE 12.5

Left: Comparison of Raman spectra of jarosite samples from the sediments at the riverbed of Rio Tinto (top); jarosite of supergenic origin at the Jaroso ravine (medium) and hydrothermal jarosite from the Arteal tunnel also at Jaroso ravine (bottom). Right: Detail of the ν(OH) region of these samples showing the differences associated with their formation. The narrowest Raman bands in this region correspond to the hydrothermal jarosite, the broader bands in the OH-stretching region correspond to Rio Tinto jarosites.

3300–3500 cm^{-1}, with an integrated width of about 132 cm^{-1}. The center of the OH band envelope in all cases remained near 3400–3410 cm^{-1} (the average O–H stretch energy did not shift much), but **the bandwidth of the O–H stretching region increased from the hydrothermal to the biomediated jarosite**.

This systematic broadening of the OH stretching bands correlates with the degree of crystallinity and formation mechanism. Broader Raman bands generally indicate more structural disorder—meaning the OH bonds experience a wider range of bond strengths or environments in the disordered structure. The Rio Tinto jarosite, having formed rapidly at low pH with likely copious impurities and nano-size crystallites (and possibly proton substitutions in K sites, etc.), is indeed expected to have significant disorder. The hydrothermal jarosite, grown slowly from hot fluids, achieved a more ordered crystalline lattice, yielding sharp OH bands. The supergene Jaroso jarosite is intermediate. We interpret these results to mean that the **biogenic formation conditions at Río Tinto induce structural disorder in jarosite**, perhaps through rapid precipitation, incorporation of foreign ions (like H_3O^+, Na^+) or defects, and extremely fine grain size. Microbial activity might influence this by, for example, affecting nucleation rate or by binding certain crystal faces, but even purely abiotic rapid precipitation can cause similar effects. The broader OH band could be due to a multitude of slightly different O–H bond lengths caused by distortions in the jarosite lattice. In jarosite, each Fe^{3+} is octahedrally coordinated to four OH^- and two O (from sulfate). Some OH groups bridge to K^+ (or H_3O^+) in the large cavities. If the Fe–OH–K hydrogen bonding network is disrupted or variable, the OH stretching modes will vary more (Frost et al., 2006) noted that synthetic jarosites often show broader OH bands due to such disorder.

Importantly, while this spectral signature of disorder is consistent with biotic influence, it is not uniquely biotic. Structural disorder can arise from many factors unrelated to biology (e.g., rapid precipitation, low temperature, metastability). As a cautionary example, we cite the case of magnetite: biogenic magnetite crystals produced by bacteria can differ in crystal size and defects from abiotic magnetite, but **disorder alone is insufficient evidence of life since abiotic processes can yield similar features** (Perez-Gonzalez et al., 2010). In our study, the Rio Tinto jarosite's broad OH band could be taken as a potential biosignature, or at least a bio-indicator, since it occurs in a known biological context, but one must corroborate it with other lines of evidence. The broader OH band in Rio Tinto jarosite likely reflects its nanocrystalline, chemically heterogeneous nature caused by rapid biogeochemical precipitation. This could "assist in supporting additional clues" of biogenic activity when combined with other observations, but by itself, it does not prove biological origin.

The finding is in line with other work attempting to distinguish biotic versus abiotic jarosites. A comparative study by Loiselle et al., (2018) examined synthetic "abiotic" jarosite versus jarosite produced by iron-oxidizing bacteria (similar to Rio Tinto cultures) using multiple spectroscopic techniques. They found only minor differences between the two in Raman, IR, Mössbauer and VNIR spectra—largely consistent with our observation that the sulfate vibrational modes are unchanged, and only subtle features like organics or crystal morphology differed (Loiselle et al., 2018). That study concluded that identifying biosignatures in jarosite with in-situ rover instruments would be challenging, likely requiring sample return for definitive evidence, since organic molecules or textural biosignatures might not be detectable by spectroscopic means alone (Loiselle et al., 2018). Our results reinforce this cautious view: spectroscopic differentiation of biotic versus abiotic jarosite is subtle. The primary indicator we see (OH band broadening) suggests greater disorder consistent with biotic rapid formation, but one could imagine a scenario where Martian jarosite might also be disordered due to rapid inorganic precipitation.

Nonetheless, the comparative approach is still valuable. If a Mars rover encounters jarosite and finds it to be highly disordered or with unusual hydrogen-bond signatures, it could flag that sample for further scrutiny (perhaps that sample formed in an environment conducive to life). The broader context—such as detecting associated organic compounds, microbial fossils, or isotope anomalies—would then be needed to confirm any biogenic hypothesis. In summary, **Raman spectroscopy can detect differences in mineral quality and bonding that correlate with formation conditions**. In the case of jarosite, abiotic hydrothermal jarosite yields sharp, well-defined vibrational bands, whereas biomediated supergene jarosite shows broadened features. This observation, combined with complementary analyses (XRD showing peak broadening, electron microscopy showing crystal habit, etc.), can provide a holistic fingerprint of biogenic influence. While not a smoking gun for life, these mineralogical clues contribute to a weight-of-evidence strategy in astrobiology. On Mars, where any biosignature will likely be faint, the ability to compare mineral spectra to Earth analogs (biotic vs. abiotic) will be an important tool.

Conclusions

Río Tinto and Jaroso Ravine exemplify two end-member pathways of acid sulfate mineral formation on Earth—one strongly influenced by microbial activity in an active system, the other purely abiotic in an ancient hydrothermal context. Studying these sites with a suite of Raman and other spectroscopic methods yields insights directly transferable to Mars exploration. At Río Tinto, we demonstrated that a rover-like Raman spectrometer can unambiguously identify a wide range of hydrated sulfate minerals in situ, even amidst complex mixtures. Minerals such as jarosite, copiapite, coquimbite, halotrichite, and others were detected by their unique vibrational signatures. This validates the capability of instruments like ExoMars RLS or Mars2020 SHERLOC/SuperCam to perform mineral diagnostics on Mars, where similar sulfates have been observed from orbit and in situ. The mineral diversity and distribution at Río Tinto also provide a geochemical narrative analogous to Martian sites like Meridiani Planum: the sequence of sulfates points to evaporation of acidic, iron-rich waters, in line with the sulfate-rich sedimentary rocks that the Opportunity rover found on Mars (Squyres et al., 2004). By modeling the chemical equilibria of the Río Tinto water and comparing them with the precipitated minerals, we can better infer the conditions under which Martian sulfates formed (pH, water activity, episodicity of wetting/drying).

The comparative jarosite study between Río Tinto and Jaroso highlights the potential and the limitations of spectroscopic biosignature detection. On one hand, we found that certain spectral characteristics – notably the broadening of OH stretching bands – correlate with the jarosite that formed rapidly in a biologically active environment. This suggests that Raman spectroscopy might detect indirect signs of biogenic influence (increased structural disorder) in minerals. On the other hand, the overall spectral differences between biotic versus abiotic jarosite was subtle; the primary fingerprint of the sulfate anion remained the same. This implies that a standalone Raman or IR measurement would likely not be able to conclusively tell a "biotic" jarosite from an abiotic one on Mars. Instead, a combination of observations (mineral context, morphology, multi-technique data, and perhaps detection of organics or isotopic ratios) will be required to build a compelling case for past life. Our results support the approach of missions like ExoMars that integrate Raman with other instruments (e.g., organic molecule analyzers, X-ray diffraction) to cross-check for biosignatures (Vago et al., 2017; Westall et al., 2015).

In conclusion, the mineralogical and geochemical characterization of the Río Tinto system underscores the value of Earth analogs in preparing for Mars exploration. We have shown that combined spectroscopic methods can unravel the complexity of an extreme environment, yielding data on mineral identities, hydration states, and even formation processes (biotic vs. abiotic) that are directly relevant to Mars. Studies at Río Tinto have already influenced the design and testing of flight instruments—for instance, field trials of the Raman spectrometers and calibration of their detection limits for various sulfates were conducted here. Looking ahead, continued research on terrestrial analogs, including those without biological activity (like Jaroso) and those with potential biosignatures (like Río Tinto), will be crucial for interpreting upcoming Martian data. As Mars rovers keep sending back spectra of ancient rocks, we will be better poised to answer whether unusual features in those data hint at a once-living planet, or are merely the fingerprints of abiotic chemistry mimicking life. The work presented in this chapter illustrates a step toward that goal: translating the spectroscopic clues from Earth's deep past and extreme habitats into scientific wisdom for the exploration of Mars.

Acknowledgments

This work is supported through grants "PID2022-142490OB-C32" and "PCI2023-146000-2", funded by MICIU/AEI/10.13039/501100011033 and co-funded by FEDER, European Union. It is also supported by Consejeria de Educación (JCyL), grant VA026G24 and CLU-2023-21-05. Authors thank the different contributions of the team members and collaborators to the spectroscopic study of Rio Tinto and Jaroso along many years with particular acknowledgement to Jesus Medina, Gloria Venegas, Julia Guerrero, Fernando Gazquez, Jesus Martinez-Frias and Prof. Goestar Klinglerhofer (in memorian).

References

Amaral Zettler, L.A., Messerli, M.A., Laatsch, A.D., Smith, P.J.S., Sogin, M.L., 2003. From genes to genomes: beyond biodiversity in Spain's Rio Tinto. Biol. Bull. 204 (2), 205–209. https://doi.org/10.2307/1543560.

Amils, R., Fernández-Remolar, D., Team, T.I., 2014. Río tinto: a geochemical and mineralogical terrestrial analogue of mars. Life 4 (3), 511–534. https://doi.org/10.3390/life4030511.

Bhartia, R., Beegle, L.W., DeFlores, L., Abbey, W., Razzell Hollis, J., Uckert, K., Monacelli, B., Edgett, K.S., Kennedy, M.R., Sylvia, M., Aldrich, D., Anderson, M., Asher, S.A., Bailey, Z., Boyd, K., Burton, A.S., Caffrey, M., Calaway, M.J., Calvet, R., Cameron, B., Caplinger, M.A., Carrier, B.L., Chen, N., Chen, A., Clark, M.J., Clegg, S., Conrad, P.G., Cooper, M., Davis, K.N., Ehlmann, B., Facto, L., Fries, M.D., Garrison, D.H., Gasway, D., Ghaemi, F.T., Graff, T.G., Hand, K.P., Harris, C., Hein, J. D., Heinz, N., Herzog, H., Hochberg, E., Houck, A., Hug, W.F., Jensen, E.H., Kah, L. C., Kennedy, J., Krylo, R., Lam, J., Lindeman, M., McGlown, J., Michel, J., Miller, E., Mills, Z., Minitti, M.E., Mok, F., Moore, J., Nealson, K.H., Nelson, A., Newell, R., Nixon, B.E., Nordman, D.A., Nuding, D., Orellana, S., Pauken, M., Peterson, G., Pollock, R., Quinn, H., Quinto, C., Ravine, M.A., Reid, R.D., Riendeau, J., Ross, A.J., Sackos, J., Schaffner, J.A., Schwochert, M., O Shelton, M., Simon, R., Smith, C.L., Sobron, P., Steadman, K., Steele, A., Thiessen, D., Tran, V.D., Tsai, T., Tuite, M., Tung, E., Wehbe, R., Weinberg, R., Weiner, R.H., Wiens, R.C., Williford, K., Wollonciej, C., Wu, Y.H., Yingst, R.A., Zan, J., 2021. Perseverance's scanning habitable environments with Raman and luminescence for organics and chemicals (SHERLOC) investigation. Space Sci. Rev. 217 (4). https://doi.org/10.1007/s11214-021-00812-z.

Bibring, J.P., Langevin, Y., Gendrin, A., Gondet, B., Poulet, F., Berthé, M., Soufflot, A., Arvidson, R., Mangold, N., Mustard, J., Drossart, P., Erard, S., Fomi, O., Combes, M., Encrenaz, T., Fouchet, T., Merchiorri, R., Belluci, G.C., Altieri, F., Formisano, V., Bonello, G., Capaccioni, F., Cerroni, P., Coradini, A., Fonti, S., Kottsov, V., Ignatiev, N., Moroz, V., Titov, D., Zasova, L., Pinet, P., Douté, S., Schmitt, B., Sotin, C., Hauber, E., Hoffmann, H., Jaumann, R., Keller, U., Duxbury, T., Forget, F., 2005. Mars surface diversity as revealed by the OMEGA/Mars express observations. Science 307 (5715), 1576–1581. https://doi.org/10.1126/science.1108806.

Bristow, T.F., Rampe, E.B., Achilles, C.N., Blake, D.F., Chipera, S.J., Craig, P., Crisp, J.A., Des Marais, D.J., Downs, R.T., Gellert, R., Grotzinger, J.P., Gupta, S., Hazen, R.M., Horgan, B., Hogancamp, J.V., Mangold, N., Mahaffy, P.R., McAdam, A.C., Ming, D. W., Morookian, J.M., Morris, R.V., Morrison, S.M., Treiman, A.H., Vaniman, D.T., Vasavada, A.R., Yen, A.S., 2018. Clay mineral diversity and abundance in sedimentary rocks of Gale crater, Mars. Sci. Adv. 4 (6). https://doi.org/10.1126/sciadv.aar3330.

Buckby, T., Black, S., Coleman, M.L., Hodson, M.E., 2003. Fe-sulphate-rich evaporative mineral precipitates from the Río Tinto, southwest Spain. Mineral. Mag. 67 (2), 263–278. https://doi.org/10.1180/0026461036720104.

Burriel, F., Lucena, F., Arribas, Hernández, J., Química analítica cualitativa, 13a edición. (1052),

Carter, J., Poulet, F., Bibring, J.-P., Mangold, N., Murchie, S., 2013. Hydrous minerals on Mars as seen by the CRISM and OMEGA imaging spectrometers: updated global view. J. Geophys. Res.: Planets 118 (4), 831–858. https://doi.org/10.1029/2012je004145.

Cho, Y., Böttger, U., Rull, F., Hübers, H.-W., Belenguer, T., Börner, A., Buder, M., Bunduki, Y., Dietz, E., Hagelschuer, T., Kameda, S., Kopp, E., Lieder, M., Lopez-Reyes, G., Inza, A.G.M., Mori, S., Ogura, J.A., Paproth, C., Canora, C.P., Pertenais, M., Peter, G., Prieto-Ballesteros, O., Rockstein, S., Rodd-Routley, S., Perez, P.R., Ryan, C., Santamaria, P., Säuberlich, T., Schrandt, F., Schröder, S., Stangarone, C., Ulamec, S., Usui, T., Weber, I., Westerdorff, K., Yumoto, K., 2021. In situ science on Phobos with the Raman spectrometer for MMX (RAX): preliminary design and feasibility of Raman measurements. Earth Planets Space 73 (1). https://doi.org/10.1186/s40623-021-01496-z.

Cloutis, E.A., Caudill, C., Lalla, E.A., Newman, J., Daly, M., Lymer, E., Freemantle, J., Kruzelecky, R., Applin, D., Chen, H., Connell, S., Fernandes, D., Giusto, F., Hawke, J., Lamamry, J., Murzionak, P., Parkinson, A., Peng, Q.-Y., Turenne, N., Wolf, Z.U., 2022. LunaR: overview of a versatile Raman spectrometer for lunar exploration. Front. Astron. Space Sci. 9. https://doi.org/10.3389/fspas.2022.1016359.

Corpolongo, A., Jakubek, R.S., Burton, A.S., Brown, A.J., Yanchilina, A., Czaja, A.D., Steele, A., Wogsland, B.V., Lee, C., Flannery, D., Baker, D., Cloutis, E.A., Cardarelli, E., Scheller, E.L., Berger, E.L., McCubbin, F.M., Hollis, J.R., Hickman-Lewis, K., Steadman, K., Uckert, K., DeFlores, L., Kah, L., Beegle, L.W., Fries, M., Minitti, M., Haney, N.C., Conrad, P., Morris, R.V., Bhartia, R., Roppel, R., Siljeström, S., Asher, S.A., Bykov, S.V., Sharma, S., Shkolyar, S., Fornaro, T., Abbey, W., 2023. SHERLOC Raman mineral class detections of the mars 2020 crater floor campaign. J. Geophys. Res.: Planets 128 (3). https://doi.org/10.1029/2022JE007455.

Courréges-Lacoste, G.B., 2007. Combined Raman spectrometer/laser-induced breakdown spectrometer for the next ESA mission to Mars. Spectrochim. Acta, Part A 68 (1023).

Edwards, H.G.M., Vandenabeele, P., Jorge-Villar, S.E., Carter, E.A., Perez, F.R., Hargreaves, M.D., 2007. The Rio Tinto Mars Analogue site: an extremophilic Raman spectroscopic study. Spectrochim. Acta, Part A 68 (4), 1133–1137. https://doi.org/10.1016/j.saa.2006.12.080.

Fairén, A.G., Fernández-Remolar, D., Dohm, J.M., Baker, V.R., Amils, R., 2004. Inhibition of carbonate synthesis in acidic oceans on early Mars. Nature 431 (7007), 423–426. https://doi.org/10.1038/nature02911.

Fairén, A.G., Gómez-Elvira, J., Briones, C., Prieto-Ballesteros, O., Rodríguez-Manfredi, J.A., López Heredero, R., Belenguer, T., Moral, A.G., Moreno-Paz, M., Parro, V., 2020. The complex molecules detector (CMOLD): a fluidic-based instrument suite to search for (Bio)chemical complexity on mars and icy moons. Astrobiology 20 (9), 1076–1096. https://doi.org/10.1089/ast.2019.2167.

Farley, K.A., Stack, K.M., Shuster, D.L., Aqueously, D.L., 2022. Altered igneous rocks sampled on the floor of Jezero crater, Mars. Science 377 (6614).

Fernández-Remolar, D.C., Morris, R.V., Gruener, J.E., Amils, R., Knoll, A.H., 2005. The Río Tinto Basin, Spain: mineralogy, sedimentary geobiology, and implications for interpretation of outcrop rocks at Meridiani Planum, Mars. Earth Planet Sci. Lett. 240 (1), 149–167. https://doi.org/10.1016/j.epsl.2005.09.043.

Fleischer, I., Klingelhöfer, G., Rull, F., Wehrheim, S., Ebert, S., Panthöfer, M., Blumers, M., Schmanke, D., Maul, J., Schröder, C., 2010. In-situ mössbauer spectroscopy with MIMOS II at Rio tinto, Spain. J. Phys. Conf. 217 (1), 012062. https://doi.org/10.1088/1742-6596/217/1/012062.

Frost, R.L., Wills, R.A., Weier, M.L., Martens, W., Mills, S., 2006. A Raman spectroscopic study of selected natural jarosites. Spectrochim. Acta, Part A 63 (1), 1–8. https://doi.org/10.1016/j.saa.2005.03.034.

Frost, R.L., Weier, M., Martinez-Frias, J., Rull, F., Jagannadha Reddy, B., 2007. Sulphate efflorescent minerals from El Jaroso ravine, Sierra almagrera-an SEM and Raman spectroscopic study. Spectrochim. Acta, Part A 66 (1), 177–183. https://doi.org/10.1016/j.saa.2006.01.054.

Frost, R.L., Weier, M.L., Kloprogge, J.T., 2005. Raman spectroscopy of jarosites from Río Tinto, Spain: jarosite's significance in the geological record of Mars. Spectrochim. Acta Mol. Biomol. Spectrosc. 62, 166–171.

Gendrin, A., Mangold, N., Bibring, J.P., Langevin, Y., Gondet, B., Poulet, F., Bonello, G., Quantin, C., Mustard, J., Arvidson, R., LeMouélic, S., 2005. Sulfates in Martian layered terrains: the OMEGA/Mars express view. Science. 307 (5715), 1587–1591. https://doi.org/10.1126/science.1109087.

Gonzáles-Toril, E., Martínez-Frías, J., Gómez Gómez, J.M., Rull, F., Amils, R., 2005. Iron meteorites can support the growth of acidophilic chemolithoautotrophic microorganisms. Astrobiology 5 (3), 406–414. https://doi.org/10.1089/ast.2005.5.406.

Guerrero, J., 2012. Estudio espectroscópico y mineralógico de los sulfatos precipitados en Río Tinto a partir de aguas ácidas. Aplicaciones medioambientales.

Guerrero, J., Navarro, R., Medina, J., 2010. Caracterización mediante Espectroscopía Raman y LIBS de la Composición Geoquímica del Nacimiento del Rio Tinto. Macla, revista de la Sociedad Española de Mineralogía 13, 119–120.

Guerrero, J., Navarro, R., Sansano, A., 2011. Síntesis y caracterización de eflorescentes presentes en Río Tinto por espectroscopía Raman y difracción de rayos X. Macla, revista de la Sociedad Española de Mineralogía 15, 109–110.

Klingelhöfer, G., Morris, R.V., Bernhardt, B., Schröder, C., Rodionov, D.S., de Souza, P.A., Yen, A., Gellert, R., Evlanov, E.N., Zubkov, B., Foh, J., Bonnes, U., Kankeleit, E., Gütlich, P., Ming, D.W., Renz, F., Wdowiak, T., Squyres, S.W., Arvidson, R.E., 2004. Jarosite and hematite at Meridiani Planum from opportunity's mössbauer spectrometer. Science 306 (5702), 1740–1745. https://doi.org/10.1126/science.1104653.

Leistel, J.M., Marcoux, E., Thiéblemont, D., 1998. The volcanic-hosted massive sulfide deposits of the Iberian Pyrite Belt. Miner. Deposita 33, 2–30.

Loiselle, L., McCraig, M.A., Dyar, M.D., Léveillé, R., Shieh, S.R., Southam, G., 2018. A spectral comparison of jarosites using techniques relevant to the robotic exploration of biosignatures on mars. Life 8 (4). https://doi.org/10.3390/life8040061.

López-Archtlla, A.I., Marín, I., Amils, R., 1993. Bioleaching and interrelated acidophilic microorganisms from río tinto, Spain. Geomicrobiol. J. 11 (3–4), 223–233. https://doi.org/10.1080/01490459309377953.

López-Gutiérrez, J., Martínez-Frías, J., Lunar, R., 1993. Relationships between tectonics and base- and precious-metal mineralization in the Vera-Garrucha area (SE Spain). Current Research in Geology Applied to Ore Deposits. Society of Geology Applied to Mineral Deposits. 1, 735–739.

Lopez-Reyes, G., Rull, F., Venegas, G., Westall, F., Foucher, F., Bost, N., Sanz, A., Catalá-Espí, A., Vegas, A., Hermosilla, I., Sansano, A., Medina, J., 2013. Analysis of the scientific capabilities of the ExoMars Raman laser spectrometer instrument. Eur. J. Mineral 25 (5), 721–733. https://doi.org/10.1127/0935-1221/2013/0025-2317.

Lopez-Reyes, G., Veneranda, M., Manrique, J.A., González Martín, Á., Moral, A., Perez-Canora, C., Rodríguez Prieto, J.A., Sanz Arranz, A., Saiz Cano, J., Lalla, E., Konstantinidis, M., Prieto-Ballesteros, O., Medina, J., González, M.Á., Charro, E., Lopez, J.M., Rull, F., 2022. The Raman laser spectrometer ExoMars simulator (RLS Sim): a heavy-duty Raman tool for ground testing on ExoMars. J. Raman Spectrosc. 53 (3), 382–395. https://doi.org/10.1002/jrs.6281.

Lopez-Reyes, G., Manrique, J.A., Clave, E., 2025. Supercam Raman Activities at Jezero Crater, Mars: Observational Strategies, Data Processing, and Mineral Detections during the First 1000 Sols. ESS Open Archive.

Martínez Frías, J., 1991. Sulphide and sulphosalt mineralogy and paragenesis from the Sierra Almagrera veins, betic cordillera (SE Spain). Estud. Geol. 47 (5–6), 271–279. https://doi.org/10.3989/egeol.91475-6423.

Martinez-Frías, J., Lunar, R., Rodríguez-Losada, J.A., Delgado, A., Rull, F., 2004. The volcanism-related multistage hydrothermal system of El Jaroso (SE Spain): implications for the exploration of Mars. Earth Planets Space 56 (7), v. https://doi.org/10.1186/bf03352523.

Martinez-Frias, J., Amaral, G., Vázquez, L., 2006. Astrobiological significance of minerals on Mars surface environment. Rev. Environ. Sci. Biotechnol. 5 (2–3), 219–231. https://doi.org/10.1007/s11157-006-0008-x.

McCreery, R.L., 2001. Raman spectroscopy for chemical analysis. Meas. Sci. Technol. 12 (5), 653–654. https://doi.org/10.1088/0957-0233/12/5/704.

Perez-Gonzalez, T., Jimenez-Lopez, C., Neal, A.L., Rull-Perez, F., Rodriguez-Navarro, A., Fernandez-Vivas, A., Iañez-Pareja, E., 2010. Magnetite biomineralization induced by Shewanella oneidensis. Geochem. Cosmochim. Acta 74 (3), 967–979. https://doi.org/10.1016/j.gca.2009.10.035.

Rull, F., Sobrón, F., Guerrero, J., Medina, J., Venegas, G., Gázquez, F., Martínez-Frías, J., 2014. In-situ Raman analysis of the precipitation sequence of sulphate minerals using small droplets: application to Rio Tinto (Spain). Lecture Notes in Earth System Sciences 0, 801–805. https://doi.org/10.1007/978-3-642-32408-6_173.

Rull, F., Fleischer, I., Martinez-Frias, J., et al., 2008. Raman and Mössbauer spectroscopic characterisation of sulphate minerals from the Mars analogue sites at Rio Tinto and Jaroso Ravine, Spain. 39th Lunar and Planetary Science Conference. LPSC 2008, League City, Texas (U.S.A.), 10-14 March, 2008, Abstract #1616.

Rull, F., Martinez-Frias, J., Medina, J., 2005. Surface mineral analysis from two possible Martian analogs (Rio Tinto and Jaroso Ravine, Spain) using micro-, macro-, and remote laser Raman spectroscopy. European Geosciences Union. Geophysical Research Abstracts 7.

Rull, F., Veneranda, M., Manrique-Martinez, J.A., Sanz-Arranz, A., Saiz, J., Medina, J., Moral, A., Perez, C., Seoane, L., Lalla, E., Charro, E., Lopez, J.M., Nieto, L.M., Lopez-Reyes, G., 2022. Spectroscopic study of terrestrial analogues to support rover missions to Mars – a Raman-centred review. Anal. Chim. Acta 1209, 339003. https://doi.org/10.1016/j.aca.2021.339003.

Rull, F., Maurice, S., Hutchinson, I., Moral, A., Perez, C., Diaz, C., Colombo, M., Belenguer, T., Lopez-Reyes, G., Sansano, A., Forni, O., Parot, Y., Striebig, N., Woodward, S., Howe, C., Tarcea, N., Rodriguez, P., Seoane, L., Santiago, A., Rodriguez-Prieto, J.A., Medina, J., Gallego, P., Canchal, R., Santamaría, P., Ramos, G., Vago, J.L., 2017. On behalf of the RLS team, the Raman laser spectrometer for the

ExoMars rover mission to mars. Astrobiology 17 (6–7), 627–654. https://doi.org/10.1089/ast.2016.1567.

Sharma, S.K., Misra, A.K., Clegg, S.M., Barefield, J.E., Wiens, R.C., Acosta, T.E., Bates, D. E., 2011. Remote-Raman spectroscopic study of minerals under supercritical CO 2 relevant to Venus exploration. Spectrochim. Acta, Part A 80 (1), 75–81. https://doi.org/10.1016/j.saa.2011.01.033.

Sharma, S.K., Lucey, P.G., Ghosh, M., Hubble, H.W., Horton, K.A., 2003. Stand-off Raman spectroscopic detection of minerals on planetary surfaces. Spectrochim. Acta, Part A 59 (10), 2391–2407. https://doi.org/10.1016/S1386-1425(03)00080-5.

Sobron, P., Rull, F., Sobron, F., Sanz, A., Medina, J., Nielsen, C.J., 2007a. Modeling the physico-chemistry of acid sulfate waters through Raman spectroscopy of the system FeSO4 H2 SO4 H2O. J. Raman Spectrosc. 38 (9), 1127–1132. https://doi.org/10.1002/jrs.1729.

Sobron, P., Rull, F., Sobron, F., Sanz, A., Medina, J., Nielsen, C.J., 2007b. Raman spectroscopy of the system iron(III)–sulfuric acid–water: an approach to Tinto River's (Spain) hydrogeochemistry. Spectrochim. Acta Mol. Biomol. Spectrosc. 68 (4), 1138–1142. https://doi.org/10.1016/j.saa.2007.06.044.

Sobron, P., Sanz, A., Acosta, T., Rull, F., 2009a. A Raman spectral study of stream waters and efflorescent salts in Rio Tinto, Spain. Spectrochim. Acta, Part A 71 (5), 1678–1682. https://doi.org/10.1016/j.saa.2008.06.035.

Sobrón, P., Sobrón, F., Eide, U.M., et al., 2009b. Model-based measurements of diffusion of sulfuric acid into water using Raman spectroscopy. Appl. Spectrosc. 63 (12), 1382–1388.

Spratt, H.J., Rintoul, L., Avdeev, M., Martens, W.N., 2013. The crystal structure and vibrational spectroscopy of jarosite and alunite minerals. Am. Mineral. 98 (10), 1633–1643. https://doi.org/10.2138/am.2013.4486.

Squyres, S.W., Grotzinger, J.P., Arvidson, R.E., Bell, J.F., Calvin, W., Christensen, P.R., Clark, B.C., Crisp, J.A., Farrand, W.H., Herkenhoff, K.E., Johnson, J.R., Klingelhöfer, G., Knoll, A.H., McLennan, S.M., McSween, H.Y., Morris, R.V., Rice, J. W., Rieder, R., Soderblom, L.A., 2004. In situ evidence for an ancient aqueous environment at Meridiani Planum, mars. Science 306 (5702), 1709–1714. https://doi.org/10.1126/science.1104559.

Vago, J.L., Westall, F., Coates, A.J., Jaumann, R., Korablev, O., Ciarletti, V., Mitrofanov, I., Josset, J.L., De Sanctis, M.C., Bibring, J.P., Rull, F., Goesmann, F., Steininger, H., Goetz, W., Brinckerhoff, W., Szopa, C., Raulin, F., Edwards, H.G.M., Whyte, L.G., Fairén, A.G., Bridges, J., Hauber, E., Ori, G.G., Werner, S., Loizeau, D., Kuzmin, R.O., Williams, R.M.E., Flahaut, J., Forget, F., Rodionov, D., Svedhem, H., Sefton-Nash, E., Kminek, G., Lorenzoni, L., Joudrier, L., Mikhailov, V., Zashchirinskiy, A., Alexashkin, S., Calantropio, F., Merlo, A., Poulakis, P., Witasse, O., Bayle, O., Bayón, S., Meierhenrich, U., Carter, J., García-Ruiz, J.M., Baglioni, P., Haldemann, A., Ball, A.J., Debus, A., Lindner, R., Haessig, F., Monteiro, D., Trautner, R., Voland, C., Rebeyre, P., Goulty, D., Didot, F., Durrant, S., Zekri, E., Koschny, D., Toni, A., Visentin, G., Zwick, M., Van Winnendael, M., Azkarate, M., Carreau, C., 2017. Habitability on early mars and the search for biosignatures with the ExoMars rover. Astrobiology 17 (6–7), 471–510. https://doi.org/10.1089/ast.2016.1533.

Venegas, G., 2014. Raman study of sulphates formed by hydrothermal, evaporitic and weathering processes in the south east of spain.

Venegas, G., Guerrero, J., Sansano, A., 2012. Raman study of mineralogical precipitation sequence of Rio Tinto Mars analog. European Planetary Science Congress Abstracts 7.

Wang, A., Haskin, L.A., Lane, A.L., Wdowiak, T.J., Squyres, S.S., Wilson, R.J., Hovland, L. E., Manatt, K.S., Raouf, N., Smith, C.D., 2003. Development of the Mars microbeam Raman spectrometer (MMRS). J. Geophys. Res.: Planets 108 (1), 1–18. https://doi.org/10.1029/2002je001902.

Wang, A., Jolliff, B.L., Liu, Y., Connor, K., 2016. Setting constraints on the nature and origin of the two major hydrous sulfates on Mars: monohydrated and polyhydrated sulfates. J. Geophys. Res.: Planets 121 (4), 678–694. https://doi.org/10.1002/2015JE004889.

Wang, Y., Fang, P., Wang, H., Xin, Y., Yan, X., Duan, M., He, Z., Wan, X., 2024. Discovery of calcium sulfate at different hydration states on Mars - based on perseverance SHERLOC analysis. Vib. Spectrosc. 135, 103745. https://doi.org/10.1016/j.vibspec.2024.103745.

Westall, F., Foucher, F., Bost, N., Bertrand, M., Loizeau, D., Vago, J.L., Kminek, G., Gaboyer, F., Campbell, K.A., Bréhéret, J.-G., Gautret, P., Cockell, C.S., 2015. Biosignatures on mars: what, where, and how? Implications for the search for martian life. Astrobiology 15 (11), 998–1029. https://doi.org/10.1089/ast.2015.1374.

Wiens, R.C., Maurice, S., Robinson, S.H., Nelson, A.E., Cais, P., Bernardi, P., Newell, R.T., Clegg, S., Sharma, S.K., Storms, S., Deming, J., Beckman, D., Ollila, A.M., Gasnault, O., Anderson, R.B., André, Y., Michael Angel, S., Arana, G., Auden, E., Beck, P., Becker, J., Benzerara, K., Bernard, S., Beyssac, O., Borges, L., Bousquet, B., Boyd, K., Caffrey, M., Carlson, J., Castro, K., Celis, J., Chide, B., Clark, K., Cloutis, E., Cordoba, E.C., Cousin, A., Dale, M., Deflores, L., Delapp, D., Deleuze, M., Dirmyer, M., Donny, C., Dromart, G., George Duran, M., Egan, M., Ervin, J., Fabre, C., Fau, A., Fischer, W., Forni, O., Fouchet, T., Fresquez, R., Frydenvang, J., Gasway, D., Gontijo, I., Grotzinger, J., Jacob, X., Jacquinod, S., Johnson, J.R., Klisiewicz, R.A., Lake, J., Lanza, N., Laserna, J., Lasue, J., Le Mouélic, S., Legett, C., Leveille, R., Lewin, E., Lopez-Reyes, G., Lorenz, R., Lorigny, E., Love, S.P., Lucero, B., Madariaga, J.M., Madsen, M., Madsen, S., Mangold, N., Manrique, J.A., Martinez, J.P., Martinez-Frias, J., McCabe, K.P., McConnochie, T.H., McGlown, J.M., McLennan, S.M., Melikechi, N., Meslin, P.Y., Michel, J.M., Mimoun, D., Misra, A., Montagnac, G., Montmessin, F., Mousset, V., Murdoch, N., Newsom, H., Ott, L.A., Ousnamer, Z.R., Pares, L., Parot, Y., Pawluczyk, R., Glen Peterson, C., Pilleri, P., Pinet, P., Pont, G., Poulet, F., Provost, C., Quertier, B., Quinn, H., Rapin, W., Reess, J.M., Regan, A.H., Reyes-Newell, A.L., Romano, P.J., Royer, C., Rull, F., Sandoval, B., Sarrao, J.H., Sautter, V., Schoppers, M.J., Schröder, S., Seitz, D., Shepherd, T., Sobron, P., Dubois, B., Sridhar, V., Toplis, M.J., Torre-Fdez, I., Trettel, I.A., Underwood, M., Valdez, A., Valdez, J., Venhaus, D., Willis, P., 2021. The SuperCam instrument suite on the NASA mars 2020 rover: body unit and combined system tests. Space Sci. Rev. 217 (1). https://doi.org/10.1007/s11214-020-00777-5.

Young, T.F., Maranville, L.F., Smith, H.M., 1959. The Structure of Electrolyte Solutions. Chapman & Hall, Ltd.

CHAPTER

Methanogenesis in the Río Tinto basin and its implications regarding methane detection on Mars

13

Jose Luis Sanz[1], **Nuria Rodríguez**[2]
[1]*Universidad Autónoma de Madrid, Madrid, Spain;* [2]*Centro de Astrobiología (INTA-CSIC), Madrid, Spain*

Introduction

Río Tinto is an extreme environment produced by the biooxidation of metal sulfides in the subsurface of the Iberian Pyritic Belt. Río Tinto microbiology requires that a distinction be made between life in the water column and life in the sediments. The physicochemical characteristics of the water column: iron (~2.3 g L^{-1}), sulfate (~6 g L^{-1}), pH (2/2.5) and redox potential (+280/+650 mV), similar to acid rock drainage, impose very restrictive conditions on the prokaryotic community, so their prokaryotic diversity is very low, mostly members of the *Acidithiobacillus*, *Leptospirillum* and *Acidiphilium* genus (González-Toril et al., 2003). However, the physicochemical characteristics of sediments are, in general, less extreme allowing many other prokaryotes with different metabolic capacities to thrive in them. Absence of oxygen is necessary for the activity of iron-reducing bacteria, sulfate-reducing bacteria and methanogenic archaea, but pH and concentrations of iron and sulfate favor one activity or another. A geomicrobiological model of the Río Tinto sediments, indicating the microorganisms involved in the iron, sulfur and carbon cycles, can be found in Sánchez-Andrea et al. (2011).

In this chapter we are going to focus on methanogenesis, which constitutes the final stage of the mineralization of organic matter in the absence of oxygen. In it, the organic Carbon is converted into CH_4 and CO_2, usually 60:40, but proportions vary depending on the redox state of the starting material. There are three types of methanogenesis: hydrogenotrophic, acetoclastic and methylotrophic. Hydrogenotrophic methanogenesis involves the oxidation of H_2 by CO_2, which is why it is also known as carbonate respiration. It is the most widespread and, with one exception, all genera of methanogenic archaea carry it out. On Earth, hydrogenotrophic methanogenesis is considered one of the first metabolisms. In acetoclastic methanogenesis, acetate is split into CH_4 and CO_2. Although only two genera, *Methanosarcina*

Geomicrobiology of Río Tinto. https://doi.org/10.1016/B978-0-443-32948-7.00019-9

and *Methanosaeta* carry it out, 2/3 of all methane generated in environments rich in organic matter comes from acetate. The third type, methylotrophic methanogenesis, is highly restricted both taxonomically and environmentally. Only some genera belonging to the *Methanosarcinaceae* family carry it out, and it is limited to hypersaline environments, since this methanogenic pathway is not competitive at low salt contents (Oren, 2014), since it uses methyl compounds derived from their compatible solutes. However, in some terrestrial or freshwater environments with high sulfate content, sulfate can act as a final electron acceptor (sulfate respiration), displacing methanogenesis as the final stage of the mineralization of organic matter.

Methanogenesis is of economic, environmental and astrobiological interest. It generates a biofuel from both solid waste (domestic, agricultural or activated sludge digestion) and liquid waste (anaerobic wastewater treatment) in a relatively simple process. It is the dominant final stage of mineralization of organic matter in the absence of oxygen except in marine sediments and some sulfate-rich terrestrial or freshwater environments, where sulfate can act as a final electron acceptor (sulfate respiration), displacing methanogenesis as the final stage of the mineralization of organic matter. As a chemical biosignature methane can provide scientific evidence of life, which is why the origin of methane emissions detected on Mars is being thoroughly studied.

Methanogenesis in the Río Tinto sediments

Río Tinto sediments show a high degree of biodiversity, although the distribution of microorganisms depends on the physicochemical conditions. Thus, in the most oxidative zones of the sediments facultative acidophilic Fe(III)-reducing bacteria are predominant (e.g., *Acidithiobacillus*, *Sulfobacillus*, *Ferroplasma*), while as the pH increases and the solubility of iron decreases, sulfate-reducing bacteria become dominant. (e.g., *Desulfurella*, *Desulfosporosinus*) (Sánchez-Andrea et al., 2011). Several fermentative and sulfate-reducing bacteria have been isolated and identified in the Tinto sediments (see chapter 5 of this book).

In an essentially acidic environment dominated by iron and sulfur, the role of methanogenic archaea is limited. Low pH is inadequate for the growth of methanogenic archaea, which show an optimal pH close to neutrality and, the vast majority of methanogens, low tolerance to pHs below 5.5–6. At high sulfate concentrations sulfate-reducing bacteria (SRB) outcompete methanogenic archaea due to the Gibs' free energy for hydrogen and acetate for SRB bacteria -153.5 and -47.1 kJ mol^{-1}, respectively, versus -135 and -28.4 kJ mol^{-1} for methanogenic archaea.

However, enrichment cultures have shown that methanogens are remarkably widespread in sediments along the river (Fig. 13.1). The bed of Río Tinto is rocky, so, with the exception of a few man-made dams used for mining activities, the sediments are scarce and rarely exceed 10 cm in depth. Observations of bubbles emerging from the sediments, considered a sign of possible methane production, led to the collection of sediment samples directly into sterile tubes for further study

FIGURE 13.1

Sampling sites.

- *From left to right and from top to bottom: Campo de Galdierias (CG, 37.725362N, 6.561853W), Izquierda (IZQ, 37.72891N, 6.550255W), Jose Luis (JL, 37.691207N, 6.560587W), 2.9 (37.7242N, 6.56207W), Salinas (SAL, 37.669622N, 6.550212W), and Berrocal (BER, 37.601346 N, 6.549933W). For more detail, see Sánchez-Andrea, I., Rojas-Ojeda, P., Amils, R., Sanz, J.L. 2012. Screening of anaerobic activities in sediments of an acidic environment: Tinto River.* Extremophiles. *16(6), 829–839, https://doi.org/10.1007/s00792-012-0478-4.*

in the laboratory using anaerobic enrichment cultures. These were prepared anaerobically in glass serum bottles capped with butyl rubber stoppers. To ensure anaerobic conditions, the headspace was degassed with an N_2/CO_2 (80:20) mixture, and Na_2S added to the liquid. Resazurine was used as redox indicator. Enrichments were incubated at 30 °C in the dark. A basal medium previously described (Sanz et al., 1997) was used at pH 6. To stimulate specific methanogenic groups, six different carbon sources were used. Table 13.1 shows different sampling points along the river in

Table 13.1 Río Tinto sites in which methanogenesis has been detected in enrichment cultures. The pH and redox potential of sediments used as inocula and the used substrates are shown.

Site	pH	E_h(mV)	Substrate[3]	References	Identified methanogens
BER[a]	2.4	+482.8	H_2/CO_2, VFA, sodium acetate, sodium formate, methanol	Sanchez-Andrea et al. (2012)	*Methanosarcina* (*M. barkeri*)
IZQ[b]	3.9/4.0	+180/−20	Sodium formate, methanol	Sanchez-Andrea et al. (2012)	*Methanobrevibacter* (*M. arboriphilus*)
SAL[c]	2.3	+418.2	Mixture M	Sanchez-Andrea et al. (2012)	none
2.9[d]	2.9/3.7	+390	Methanol	Sanchez-Andrea et al. (2012)	none
GAL[b]	3.5/4.5[c] 5.9[d]	+200/+440[c] −205[d]	Non-stimulate, formate, acetate, VFA, lactate, formate + methanol, methanol, VFA + methanol	Sanz et al. (2011)	none
JL[b]	4/4.8[e] 5.3/5.9[f]	+20/+140[e] −30/−280[f]	Mixture M, H_2/CO_2, methanol	Sanchez-Andrea et al. (2012); Sanz et al. (2011)	*Methanosaeta concilii* *Methanobacterium bryantii*, *M. barkeri* *M. concilii*[g]

[a] *pH and redox potential (Eh) measured in water column.*
[b] *pH and Eh measured in sediment or core samples.*
[c] *Bulk.*
[d] *Macroniches.*
[e] *Reddish-brown bands.*
[f] *Black bands.*
[g] *In natural sediment without biostimulating.*

whose sediments methanogenic activity has been detected through the use of anaerobic enrichment cultures.

IZQ (*Izquierda* site), near the source of the river in Peña de Hierro, and BER (*Berrocal* site), in the middle course of the river with the formation of numerous stromatolites and varnishes on its banks, are two sampling points where sediments rarely accumulate. The analysis by Denaturing Gradient Gel Electrophoresis (DGGE) of the fragments of 16S rRNA amplified with the primers 622F-1100R, demonstrated the presence of methanogenic archaea in enrichment cultures (i) inoculated with sediments from IZQ and bioaugmented with formate, and (ii) inoculated with sludge from BER and acetate added as carbon source. The excision and subsequent sequencing of the majority bands of the DGGE allowed us to identify the presence of *Methanobrevibacter* (96%–100% similarity with *M. arboriphilus*) in the IZQ sampling site and *Methanosarcina* (94%–99% similarity with *M. barkeri*) in the BER sampling site (Sánchez-Andrea et al., 2012). Although formate is a common substrate for *Methanobrevibacter* (Rea et al., 2007), its presence is surprising as it has been described as a neutrophilic organism. Less unusual is the presence of *Methanosarcina*, given (i) its high growth rate (μmax $2.5{\cdot}10^{-2}\ h^{-1}$) at high acetate concentrations, such as those imposed in an enrichment culture; and (ii) its ability to tolerate pH as low as 4.2 has been observed (Florencio et al., 1993; Horn et al., 2003; Koschorreck et al., 2008).

CG (*Campo de Galdierias* site) is a small upwelling close to the course of the origin of the river. Negative redox potential values of −205 mV were measured at a specific location while, to our surprise, in the surrounding sediment, just 1 cm away, the redox potential values were close to +400 mV. Tubes were entirely filled in situ with sediment and water, to prevent oxygen diffusion, for subsequent microcosm studies in the laboratory (for more details see Sanz et al. (2011)). Methane production was measured in non-stimulated samples (0.7 μmols-CH_4/g-wet_sediment per day, endogenous methanogenic activity). The addition of substrates (formate, acetate, mixture of volatile fatty acids –VFA-, lactate) stimulated the methane production in the microcosms up to tenfold. When methanol was spiked together with VFA, the methanogenic activity was 30 times higher than basal methanogenesis. Methane production was associated with a decrease in the redox potentials (from +390 mV to −40/−110 mV), and an increase in the pH (from 3.3/3.8 to 5.4/6.6). Decrease in redox potential and increase in pH was also observed in all enrichment cultures where methane was produced.

JL (*José Luis site*) deserves a separate comment. JL is one of the few places in the Río Tinto where, thanks to the presence of a dam, sediment accumulates in an accessible way and, as far as we know, it is the only location with a well stratified sediment. Using a core sampler (7 ID cm and 45 cm long) several cores were extracted. These cores showed well-defined black and reddish-brown bands alternatively (Fig. 13.2). Redox potential measured in situ showed negative values for black bands and positive values for reddish ones. After extracting total DNA from the pooled black sediments, amplifiying with GenomiPhi and primers for the Archaea domain, and sequencing the 16S RNA amplicons, *Methanosaeta concilii* was identified. The

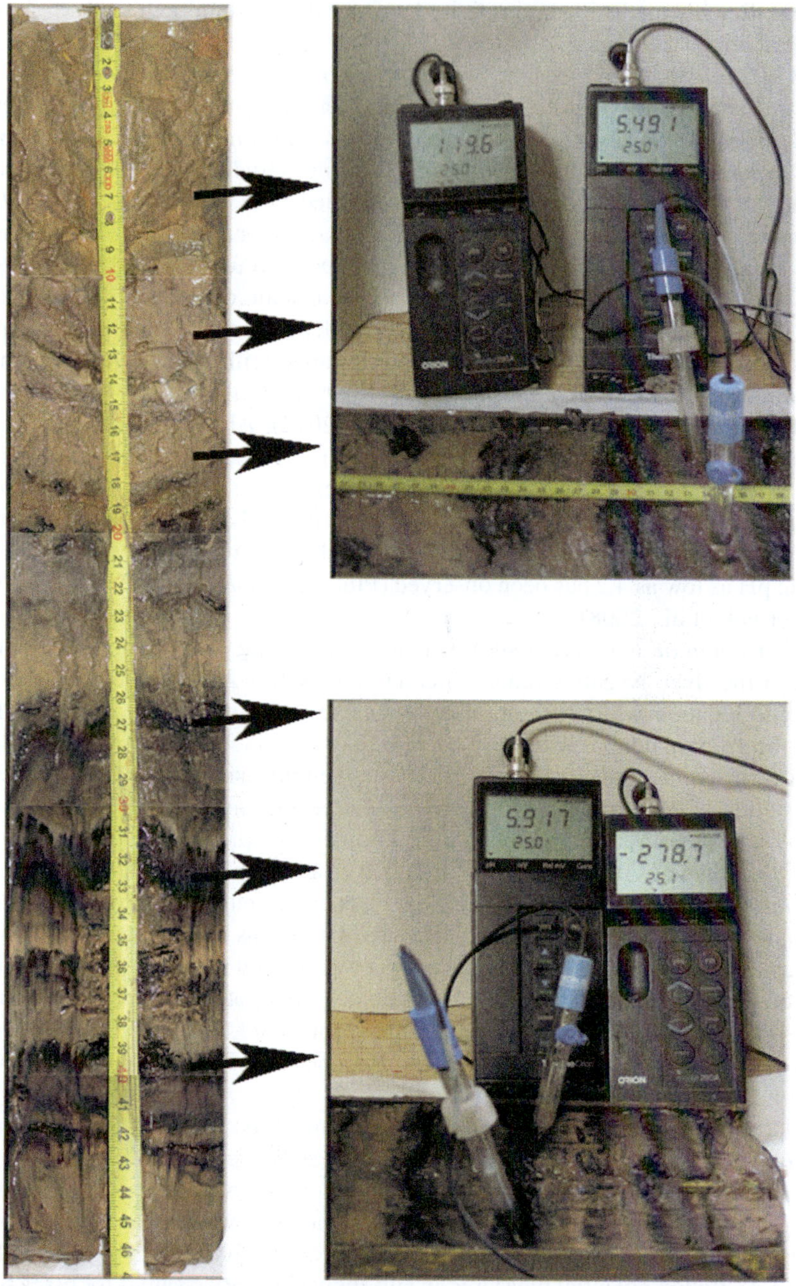

FIGURE 13.2 Sediment Core from JL Dam Site.

The column has characteristic black and reddish bands. Note the negative redox potential and higher pH of the blackish bands compared to the reddish-brown bands with characteristics closer to those typical of the Rio Tinto water column.

acetate content of the interstitial water of the reddish-brown bands was 0.3 mg L^{-1}, similar to the 0.6 mg L^{-1} measured in sediments from an acid mining lake by Koschorreck et al. (2003). Surprisingly, the acetate content of the interstitial water of the black bands was up to 79 mg L^{-1}, while formate was only 0.5 mg L^{-1} and the other VFAs were negligible. This high acetate concentration could explain the presence of the acetoclastic methanogen *M. concilii*.

Enrichment cultures with different methanogenic substrates were prepared. In the microcosm stimulated with a mixture of formate-acetate-propionate-butyrate-lactate-methanol-sucrose, the presence of *M. concilii*, and in much smaller proportion *Methanosarcina barkeri*, were identified. In the H_2/CO_2 or methanol stimulated cultures the only identified sequences were affiliated with *Methanobacterium bryantii* and *M. barkeri* species. *M. concilli* seems to be the predominant methanogen in this sample. *Methanosarcina* is the other acetoclastic methanogenic genus, but *Methanosaeta* (affinity constant, Ks, of 30 mg-acetate L^{-1}) outcompete with *Methanosarcina* (affinity constant of 300 mg-acetate L^{-1}) at the high acetate concentrations detected in the black bands of the sediments.

In many of the enrichment cultures studied, the reduction of ferric ion was observed prior to methane production. It has been shown that iron-reducing bacteria can outcompete methanogenic archaea for acetate and hydrogen (Lovley and Phillips, 1987; Van Bodegom et al., 2004). As the ferric ion is reduced in the anoxic sediments, the redox potential decreases, the pH increases and the concentrations of the inhibitor (Fe^{3+}) decrease, creating conditions favorable to methanogens.

The examples shown suggest that methanogenesis, although a residual activity compared to the biogeochemical cycles of iron or sulfur, is a generalized phenomenon in the Tinto basin.

Methanogenesis in the deep subsurface of the Iberian Pyrite Belt

The deep subsurface is one of the least known habitats on Earth. In fact, the extreme characteristics that allow microorganisms to survive there, e.g., depth and temperature or pressure, reduced physical space, limited nutrients, or water availability, can help us understand the ability of life to survive and thrive in extreme environments other than Earth and be useful in the search for life on extraterrestrial objects in future space missions (Purkamo, 2023).

Studying its geomicrobiology requires expensive drilling, so most studies have been carried out in offshore drilling related to oil platforms. Various reasons have drawn attention to the exploration of continental subsoil: origin of life; astrobiology, since the proliferation of underground life on our planet suggests the need to search for life underground in other extraterrestrial bodies; control of groundwater contamination by human activities; economic reasons, e.g., Corrosion in oil wells or production of surfactants for oil extraction, etc (Pedersen, 2000). Two devoted drillings were performed in the Río Tinto basin: Mars Analog Research and Technology Experiment (MARTE) project and the Iberian Pyritic Belt Subsurface Life Detection (IPBSL) project.

The MARTE project, designed to search for life in the non-porous volcanically hosted massive sulfide deposits of the Rio Tinto basement at *Peña de Hierro*, was carried out between 2003 and 2004. Three wells: BH1 (59 m of dark shale), BH4, and BH8 (both around 190 of pyrite stockwork), were drilled and the extracted samples were analyzed by complementary techniques to evaluate multiple parameters (Stoker et al., 2004; for more details see; Fernández-Remolar et al., 2004, 2008). Cores were placed in an anaerobic chamber and samples were obtained by splitting with a rotary hammer drill. Samples of 2–5 g of small pebbles were used as inoculums for enrichment cultures. The procedure was similar to that described in section "Methanogenesis in sediments". We focused our interest on methanogens due to the import role methanogenesis plays in the subsurface. In an H_2-driven biosphere in a deep granitic subsurface, hydrogenotrophic autotrophic methanogens and homoacetogens would form the base of the food chain, generating methane and acetate from H_2 and CO_2. Moreover, disproportionation and anaerobic respiration of acetate may then support acetoclastic methanogens, sulfate-reducers prokaryotes, and Fe(III)-reducers (Pedersen, 1997).

BH1. Due to alteration of its most superficial area (Fernández-Remolar et al., 2004), samples were taken from a depth of 30 m or more. In microcosms bio-stimulated with a mixture of VFA and methanol, methane production was detected in almost all of them, reaching a maximum with the sample obtained at 49 m below surface (mbs).

BH4. Due to the geological characteristics of the borehole, only samples obtained at depths below 130 m were analyzed. The production of methane was detected in the microcosms inoculated with samples from 133, 152, 156 y 168 mbs. Fernández-Remolar et al. (2008) analyzed dissolved gases (CH_4, H_2, CO_2, O_2) into borehole fluid of the BH4 and BH8: methane and hydrogen concentrations were inversely correlated. The hypothesis of the authors of the study is that, "under anoxic conditions, the reduction of ferric to ferrous iron through pyrite alteration initiates a process that produces hydrogen as a final by-product, which is then available for methanogenesis". Furthermore, at the depth at which the greatest methane production occurred, the measured pH was also the highest (5.64).

The IPBSL project was carried out from 2012 to 2015. Two boreholes were drilled (BH10: 613 m; BH11: 330 m, Fig. 13.3). From BH10, 47 core samples were extracted and subjected to exhaustive and complementary analysis (for more detailed information, see Amils et al. (2023)). Enrichment cultures were prepared using 1 g of powered rock samples from different depths, mineral medium (Sanz et al., 1997), and substrates for methanogenesis: 2 mM acetate (MG1), H_2:CO_2 (80:20) in the head space (MG2), and a mixture of propionate:butyrate:methanol (5 mM each) (MG3). In the 39 microcosms analyzed, methane production was detected in all of them for the three substrates tested (Table 13.2). In some of them the production was especially notable, particularly for the bio-stimulated

FIGURE 13.3 BH10 Drilling Site.

Typical 3 m long/60 mm diameter core.

samples with acetate (MG1) and H_2:CO_2 (MG2). The highest methanogenesis activity was recorded at 228.7 mbs. It is worth highlighting the production of methane at 414 mbs, even without added substrates ("natural activity"). At 10 of the depths studied, release of methane was observed in closed bottles filled with small fragments and rock splinters (Table 13.2, CH_4 occluded).

As a complement to the enrichment studies, 31 samples from selected depths were analyzed by catalyzed reported deposition fluorescence in situ hybridization (CARD-FISH). Probes for *Methanobacteriales*, *Methanosarcinales*, *Methanococcaceae*, and *Methanomicrobiales* were assayed. Although the number of samples hybridized with the last two probes was scarce (3 and 2, respectively), positive

Table 13.2 Methane production in enrichment cultures with MG1, MG2 and MG3 substrates along the borehole BH10. Release of occluded methane and produced methane without bioestimulation ("natural activity") are also shown. Hybridization with ***Methanosarcinales*** and ***Methanobacteriales*** probes is indicated.

mbs	Stimulated			No. estimulated	CH_4 occluded	CARD-FISH	
	MG1	MG2	MG3			*Methanosarcinales*	*Methanobacteriales*
90	+	+	+/–	+/–	+/–		
102.6	+	++	+/–	+/–	++		
121.8	+	+	+/–	+/–	+		
130.8	+	++	+/–	+/–	+/–		
139.4	++	++	+/–	+/–	+/–		
206.6	+	++	+/–	+/–	++	+	+
228.7	+++	+++	+++	+/–	–		
249.8	+++	+++	+	+/–	–		+
266.3	+	++	+	+/–	++	+	+
284	+++	++	+	+/–	–	+	+
304	+	++	+	+/–	–		
311.1	+++	++	+	+/–	–		
336.5	+++	++	+	+/–	–		
352.7	+	+++	+	+/–	+++		
353.2	+	+	+	–	+++	+	+
355.7	+	++	+	+/–	+/–		
393	+	+	+	+/–	+/–	+	
409.7	++	+	+	+/–	+/–	+	
411.2	+	+	+	–	+/–		
413.3	+	+	+	–	+/–		
414	+++	+++	+	+++	+/–	+	+

415.3	+	+	+	–	+/–		
416.6	++	+	+	+/–	+/–		
420	+	+	+	+/–	+/–		
433.3	+	+	+	+/–	+/–		
450.3	+	+	+++	+/–	+/–		
466.8	+++	++	+++	+	+/–		
487.2	+	++	+	+/–	+/–	+	
488	+	+	+	–	+/–	+	
492.8	+	+++	+	+	+	+	
496.8	+	++	+	+/–	+	+	
519.1	+	++	+	+/–	+	+	+
520	+++	++	+++	+/–	+		
538.5	+	++	+	+/–	+/–		
568.6	+	+++	+	+	+/–		+
607	+	+	+	+/–	+/–		+
612.9	++	+	+	+/–	+/–	+	

hybridization signals were observed for the *Methanosarcinales* (13 samples) and *Methanobacteriales* (9 samples) orders (Table 13.2). Moreover, *Methanobacterium* was detected using the immunological LDChip300 (Amils et al., 2023).

Three highly methane-producing enrichment cultures were selected: 414 plus MG1, 450 plus MG3 and 492 plus MG2 for 454 pyrosequencing. 5970 clean and reliable reads were considered for phylogenetic analysis. All the sequences were affiliated to only one taxon: *Methanomethylovorans* (100% similarity). Strangely, from the cloning and sequencing of nine samples from BH10 and BH11, only *Methanosarcina* was identified.

Methane is considered an important component of the deep subsurface ecosystems, since it has been detected in most drilling. With the objective of evaluating its biotic or abiotic origin, three depths were selected (414, 497 and 520 mbs) due to its high methanogenic activity observed in enrichment cultures. As in the occluded gas experiments discussed above, serum bottles were filled with rock splinters, and sealed. All treatments were carried out in an anaerobic chamber (for more details, see Sanz et al. (2021)). The experimental setup can be divided into two phases. Stage 1: Sterilized (control) and untreated (biotic production) samples were incubated in the dark for 100 days. Stage 2: Sterile milliQ water was spiked into the samples and the bottles were incubated for another 15 days. All procedures were performed maintaining sterile and anaerobic conditions. Sampling was carried out periodically and the methane concentration measured by GC. The results strongly support that at 414 and 520 mbs biotic methane production took place. Furthermore, carbon isotopic data measured in the well gases ($\delta^{13}C_{CH4}$ −43.0 to −31.2‰) confirmed the biological origin of the methane. In addition, archaea were detected by CARD-FISH in all the samples using the universal probe for the *Archaea* domain, but when specific methanogenic probes were tested, only positive hybridization was obtained with the *Methanobacteriales* probe at 520 mbs. This depth shows the highest H_2 production after water addition (3.4 μmols in our experimental conditions).

Our results regarding the gases released under the experimental conditions analyzed strongly suggest that the release of H_2 has a biological origin. Furthermore, some bacteria detected in the selected samples, i.e., *Shewanella*, *Rhodoplanes* and *Desulfosporosinus*, possess genes related to H_2 production. The continuous production of H_2 in this sample could justify the positive hybridization with the probe for *Methanobacteriales*, the most numerous group of hydrogenotrophic methanogenic archaea.

In summary, release of methane occluded in stone fragments, production of this gas in enrichment cultures, its biotic origin, and detection of methanogenic archaea by CARD-FISH in rock samples extracted directly from the cores along the length of the borehole, demonstrate beyond a shadow of a doubt that methanogenesis is practically ubiquitous in the deep subsurface of the Iberian Pyritic Belt. The presence of H_2 and CO_2 along the borehole, substrates for hydrogenotrophic methanogenesis (Amils et al., 2023; Sanz et al., 2021), support the existence of methanogenic activity in the ecosystem.

Methanogenesis on Mars

The study of biological activities (e.g., methanogenesis) in the deep subsurface has a marked astrobiological interest as it is concerned with the search for life on other planetary bodies. Methane is an abundant compound in the outer bodies of the solar system, both in solid form (e.g., clathrates in the Saturn Moon's Enceladus), liquid (e.g., oceans and lakes in the Saturn's Moon Titan), or gaseous form (e.g., atmospheres of Jupiter, Uranus or Neptune). However, its presence is scarce on the inner planets of the solar system. It is estimated that most of the 1.9 ppm of methane in the Earth's atmosphere is of biological origin while the amount and origin of methane on Mars remains controversial.

The first confirmed detections of methane on Mars, based on the analysis of spectroscopic observations made from Earth by Krasnopolsky et al. (2004) and from the Planetary Fourier Spectrometer (PFS) on board ESA's Mars Express orbiter by Formisano et al. (2004), were published in the year 2004. Throughout the multiple observations made since then, no repetitive seasonal, temporal or spatial trends were observed, but strong seasonal variations have been detected (Webster et al., 2018). In contrast, highly variable methane concentrations have been repeatedly reported (Luo et al., 2021). The Tunable Laser Spectrometer (TLS) and the Sample Analysis at Mars (SAM) on board NASA's Curiosity rover revealed a baseline of 0.41 ppbv (parts per billion by volume), with episodic spikes up to 21 ppbv, measurements made at Gale crater. Spectroscopic observations showed a plume of methane produced in northern soils during the summer of 2003 reached a peak of 45 $\pm$ 10 ppbv.

ACS and NOMAD instruments onboard ESA-Roscosmos ExoMars Trace Gas Orbiter (TGO) did not detect methane in 2018, contradicting SAM measurements, after a 6-month campaign in both hemispheres, despite the fact that its detection limit was 10–100 times lower than the previously used instruments (Korablev et al., 2019). One wonders: Did the MSL rover land next to one of the few places on Mars where methane is emitted, like Gale Crater? Or could the seasonal variability and occasional spikes of methane here be due to its release from soil gas pockets caused by the Curiosity rover movements or drilling? If so, the released methane appears to disappear quickly in the Martian atmosphere before it spreads through it. The reason for the short lifespan of methane on Mars is unknown.

As numerous studies show, the origin of methane on Mars may be multiple: serpentinization of olivine, ultraviolet degradation of organic compounds delivered by meteorites, release from clathrates, geothermical processes, and several more (Webster et al., 2018). Alternatively, it could be generated by methanogenic microorganisms, so the presence of atmospheric methane is a potential biosignature, and hence its great astrobiological interest.

Serpentinization may be at the origin of life on Earth, producing H_2 from H_2O. Fe^{3+} minerals generated in the crust by the oxidation of Fe^{2+} can act as catalysts for the reduction of CO_2. H_2 diffuses and reduces CO_2 to CH_4 and other organic compounds. Methanogens and acetogens, strictly anaerobic chemolithoautotrophic prokaryotes that use the acetyl-CoA pathway for CO_2 fixation, inhabit serpentinizing environments today. There is evidence that serpentinization once occurred on Mars and probably still occurs on Enceladus, Saturn's icy moon (Schwander et al., 2023).

It has also been proposed that Clathrates to play an important role in the volatile compound content of planets. Although on Earth the methane from clathrates is essentially of biological origin, on Mars it could come from hydrothermal reactions with olivine-rich material or have formed in the interstellar medium and subsequently embedded in clathrates together with water ice and incorporated in this form into comets, asteroids and planets. In this context, the release of methane from clathrates in the Martian cryosphere could be behind the variations in their seasonal abundance (Mousis et al., 2015).

Life as we know it uses oxidation-reduction processes to obtain energy, which involves electron donors and acceptors. Mars is primarily an oxidizing environment. Yung et al. (2018) in an excellent review ask "Where can we find redox gradients as energy sources for life on Mars?". The presence of a CH_4, a reduced compound, suggests the possibility that such a redox gradient could support life on Mars. In this context, identifying the origin of methane on Mars and distinguishing between biotic and abiotic sources of methane, e.g., by stable isotopic signature of CH_4 or/and increasing knowledge about serpentinization, is essential given the vast implications for geology and astrobiology. Do geochemical or biotic methane-producing activities persist today? During the Noachian period, the crust of Mars may have provided a favorable environment for microbial life. Regoliths could have played a key role. Sauterey et al. (2022) have examined numerous factors, e.g., subsurface habitability, biomass productivity, atmospheric composition in their search for possible locations of remains of this past life. The last question that we microbiologists must ask ourselves is, was CH_4 produced by extinct methanogenic microorganisms or is it still being produced by existing organisms?

Although the surface of Mars is today an inhospitable environment for life, deep subsurface might harbor most habitable conditions. Amils et al. (2011, 2023) have described multiple and widespread subsurface microbial activities in the Iberian Pyrite Belt. Among other reasons, the large deposits of iron sulfates and oxides on Mars and the iron sulfides in Tinto River, make the latter considered a terrestrial analog of Mars. In this chapter we have seen how, both in sediments and in boreholes, methanogenesis is very widespread in an environment so unsuitable for methanogenic archaea: low pH, oxidative redox potential, and high concentration of toxic heavy metals. The possibility that methanogenic organisms inhabit or have thrived in the subsurface of Mars cannot be ruled out. Studying it represents a challenge for future missions to Mars.

References

Amils, R., González-Toril, E., Aguilera, A., Rodríguez, N., Fernández-Remolar, D., Gómez, F., García-Moyano, A., Malki, M., Oggerin, M., Sánchez-Andrea, I., Sanz, J. L., 2011. From Río Tinto to Mars. The terrestrial and extraterrestrial ecology of acidophiles. Adv. Appl. Microbiol. 77, 41–70. https://doi.org/10.1016/B978-0-12-387044-5.00002-9.

Amils, R., Escudero, C., Oggerin, M., Sánchez, F.P., Arce Rodríguez, A., Remolar, D.F., Rodríguez, N., Villadangos, M.G., Sanz, J.L., Briones, C., Román, M.S.-, Gómez, F., Leandro, T., Moreno-Paz, M., Ballesteros, O.P.-, Molina, A., Tornos, F., Andrea, I.S.-, Timmis, K., Pieper, D.H., Parro, V., 2023. Coupled C, H, N, S and Fe biogeochemical cycles operating in the continental deep subsurface of the Iberian Pyrite Belt. Environ. Microbiol. 25 (2), 428–453. https://doi.org/10.1111/1462-2920.16291.

Fernández-Remolar, D.C., Prieto-Ballesteros, O., Rodríguez, N., Gómez, F., Amils, R., Gómez-Elvira, J., Stoker, C.R., 2008. Underground habitats in the Río Tinto basin: a model for subsurface life habitats on mars. Astrobiology 8 (5), 1023–1047. https://doi.org/10.1089/ast.2006.0104.

Fernández-Remolar, D.C., Prieto-Ballesteros, O., Stoker, C.R., 2004. Searching for an acidic aquifer in the Rio Tinto basin: first geobiology results of MARTE project. Lunar Planet. Sci. 1766. XXXV. https://www.researchgate.net/publication/23921334. XXXV.

Florencio, L., Nozhevnikova, A., van Langerak, A., Stams, A.J.M., Field, J.A., Lettinga, G., 1993. Acidophilic degradation of methanol by a methanogenic enrichment culture. FEMS Microbiol. Lett. 109 (1), 1–6. https://doi.org/10.1111/j.1574-6968.1993.tb06134.x.

Formisano, V., Atreya, S., Encrenaz, T., Ignatiev, N., Giuranna, M., 2004. Detection of methane in the atmosphere of Mars. Science 306 (5702), 1758–1761. https://doi.org/10.1126/science.1101732.

González-Toril, E., Llobet-Brossa, E., Casamayor, E.O., Amann, R., Amils, R., 2003. Microbial ecology of an extreme acidic environment, the Tinto river. Appl. Environ. Microbiol. 69 (8), 4853–4865. https://doi.org/10.1128/aem.69.8.4853-4865.2003.

Horn, M.A., Matthies, C., Küsel, K., Schramm, A., Drake, H.L., 2003. Hydrogenotrophic methanogenesis by moderately acid-tolerant methanogens of a methane-emitting acidic peat. Appl. Environ. Microbiol. 69 (1), 74–83. https://doi.org/10.1128/AEM.69.1.74-83.2003.

Korablev, O., Vandaele, A.C., Montmessin, F., Fedorova, A.A., Trokhimovskiy, A., Forget, F., Lefèvre, F., Daerden, F., Thomas, I.R., Trompet, L., Erwin, J.T., Aoki, S., Robert, S., Neary, L., Viscardy, S., Grigoriev, A.V., Ignatiev, N.I., Shakun, A., Patrakeev, A., Belyaev, D.A., Bertaux, J.L., Olsen, K.S., Baggio, L., Alday, J., Ivanov, Y.S., Ristic, B., Mason, J., Willame, Y., Depiesse, C., Hetey, L., Berkenbosch, S., Clairquin, R., Queirolo, C., Beeckman, B., Neefs, E., Patel, M.R., Bellucci, G., López-Moreno, J.J., Wilson, C.F., Etiope, G., Zelenyi, L., Svedhem, H., Vago, J.L., Alonso-Rodrigo, G., Altieri, F., Anufreychik, K., Arnold, G., Bauduin, S., Bolsée, D., Carrozzo, G., Clancy, R.T., Cloutis, E., Crismani, M., Da Pieve, F., D'Aversa, E., Duxbury, N., Encrenaz, T., Fouchet, T., Funke, B., Fussen, D., Garcia-Comas, M., Gérard, J.C., Giuranna, M., Gkouvelis, L., Gonzalez-Galindo, F., Grassi, D., Guerlet, S., Hartogh, P., Holmes, J., Hubert, B., Kaminski, J., Karatekin, O., Kasaba, Y., Kass, D., Khatuntsev, I., Kleinböhl, A., Kokonkov, N., Krasnopolsky, V., Kuzmin, R., Lacombe, G., Lanciano, O., Lellouch, E., Lewis, S., Luginin, M., Liuzzi, G., López-Puertas, M., López-Valverde, M., Määttänen, A., Mahieux, A., Marcq, E., Martin-Torres, J., Maslov, I., Medvedev, A., Millour, E., Moshkin, B., Mumma, M.J., Nakagawa, H., Novak, R.E., Oliva, F., Patsaev, D., Piccialli, A., Quantin-Nataf, C., Renotte, E., Ritter, B., Rodin, A., Schmidt, F., Schneider, N., Shematovich, V., Smith, M.D., Teanby, N.A., Thiemann, E., Thomas, N., Vander Auwera, J., Vazquez, L., Villanueva, G., Vincendon, M., Whiteway, J., Wilquet, V., Wolff, M.J.,

Wolkenberg, P., Yelle, R., Young, R., Zasova, L., Zorzano, M.P., 2019. No detection of methane on Mars from early ExoMars Trace gas orbiter observations. Nature 568 (7753), 517–520. https://doi.org/10.1038/s41586-019-1096-4.

Koschorreck, M., Wendt-Potthoff, K., Scharf, B., Richnow, H.H., 2008. Methanogenesis in the sediment of the acidic lake Caviahue in Argentina. J. Volcanol. Geoth. Res. 178 (2), 197–204. https://doi.org/10.1016/j.jvolgeores.2008.06.017.

Koschorreck, M., Wendt-Potthoff, K., Geller, W., 2003. Microbial sulfate reduction at low pH in sediments of an acidic lake in Argentina. Environ. Sci. Technol. 37 (6), 1159–1162. https://doi.org/10.1021/es0259584.

Krasnopolsky, V.A., Maillard, J.P., Owen, T.C., 2004. Detection of methane in the Martian atmosphere: evidence for life? Icarus 172 (2), 537–547. https://doi.org/10.1016/j.icarus.2004.07.004.

Lovley, D.R., Phillips, E.J.P., 1987. Competitive mechanisms for inhibition of sulfate reduction and methane production in the zone of ferric iron reduction in sediments. Appl. Environ. Microbiol. 53 (11), 2636–2641. https://doi.org/10.1128/aem.53.11.2636-2641.

Luo, Y., Mischna, M.A., Lin, J.C., Fasoli, B., Cai, X., Yung, Y.L., 2021. Mars methane sources in Northwestern Gale Crater inferred from back trajectory modeling. Earth Space Sci. 8 (11). https://doi.org/10.1029/2021ea001915.

Mousis, O., Chassefière, E., Holm, N.G., Bouquet, A., Waite, J.H., Geppert, W.D., Picaud, S., Aikawa, Y., Ali-Dib, M., Charlou, J.L., Rousselot, P., 2015. Methane clathrates in the solar system. Astrobiology 15 (4), 308–326. https://doi.org/10.1089/ast.2014.1189.

Oren, A., 2014. Taxonomy of halophilic Archaea: current status and future challenges. Extremophiles 18 (5), 825–834. https://doi.org/10.1007/s00792-014-0654-9.

Pedersen, K., 2000. Exploration of deep intraterrestrial microbial life: current perspectives. FEMS Microbiol. Lett. 185 (1), 9–16. https://doi.org/10.1016/s0378-1097(00)00061-6.

Pedersen, K., 1997. Microbial life in deep granitic rock. FEMS Microbiol. Rev. 20 (3–4), 399–414. https://doi.org/10.1016/s0168-6445(97)00022-3.

Purkamo, L., 2023. Life in the Deep Subsurface and its Implications for Astrobiology Oxford Research Encyclopedia of Planetary Science. Oxford University Press. https://doi.org/10.1093/acrefore/9780190647926.013.274.

Rea, S., Bowman, J.P., Popovski, S., Pimm, C., Wright, A.D.G., 2007. Methanobrevibacter millerae sp. nov. and Methanobrevibacter olleyae sp. nov., methanogens from the ovine and bovine rumen that can utilize formate for growth. Int. J. Syst. Evol. Microbiol. 57 (3), 450–456. https://doi.org/10.1099/ijs.0.63984-0.

Sánchez-Andrea, I., Rodríguez, N., Amils, R., Sanz, J.L., 2011. Microbial diversity in anaerobic sediments at Río Tinto, a naturally acidic environment with a high heavy metal content. Appl. Environ. Microbiol. 77 (17), 6085–6093. https://doi.org/10.1128/AEM.00654-11.

Sánchez-Andrea, I., Rojas-Ojeda, P., Amils, R., Sanz, J.L., 2012. Screening of anaerobic activities in sediments of an acidic environment: Tinto River. Extremophiles 16 (6), 829–839. https://doi.org/10.1007/s00792-012-0478-4.

Sanz, J.L., Rodríguez, N., Amils, R., 1997. Effect of chlorinated aliphatic hydrocarbons on the acetoclastic methanogenic activity of granular sludge. Appl. Microbiol. Biotechnol. 47 (3), 324–328. https://doi.org/10.1007/s002530050935.

Sanz, J.L., Rodriguez, N., Escudero, C., Carrizo, D., Amils, R., 2021. Biological production of H2, CH4 and CO2 in the deep subsurface of the Iberian Pyrite Belt. Environ. Microbiol. 23 (7), 3913–3922. https://doi.org/10.1111/1462-2920.15561.

Sanz, J.L., Rodríguez, N., Díaz, E.E., Amils, R., 2011. Methanogenesis in the sediments of Rio Tinto, an extreme acidic river. Environ. Microbiol. 13 (8), 2336–2341. https://doi.org/10.1111/j.1462-2920.2011.02504.x.

Sauterey, B., Charnay, B., Affholder, A., Mazevet, S., Ferrière, R., 2022. Early Mars habitability and global cooling by H2-based methanogens. Nat. Astron. 6 (11), 1263–1271. https://doi.org/10.1038/s41550-022-01786-w.

Schwander, L., Brabender, M., Mrnjavac, N., Wimmer, J.L.E., Preiner, M., Martin, W.F., 2023. Serpentinization as the source of energy, electrons, organics, catalysts, nutrients and pH gradients for the origin of LUCA and life. Front. Microbiol. 14. https://doi.org/10.3389/fmicb.2023.1257597.

Stoker, C.R., Dunagan, S., Stevens, T., Amils, R., Gómez-Elvira, J., Fernández, D., et al., 2004. Mars analog Rio Tinto Experiment (MARTE): 2003 drilling campaign to search for a subsurface biosphere at Rio Tinto Spain. Lunar Planet. Sci. 1967. XXXV. https://www.researchgate.net/publication/234227877. XXXV.

Van Bodegom, P.M., Scholten, J.C.M., Stams, A.J.M., 2004. Direct inhibition of methanogenesis by ferric iron. FEMS Microbiol. Ecol. 49 (2), 261–268. https://doi.org/10.1016/j.femsec.2004.03.017.

Webster, C.R., Mahaffy, P.R., Atreya, S.K., Moores, J.E., Flesch, G.J., Malespin, C., McKay, C.P., Martinez, G., Smith, C.L., Martin-Torres, J., Gomez-Elvira, J., Zorzano, M.P., Wong, M.H., Trainer, M.G., Steele, A., Archer, D., Sutter, B., Coll, P.J., Freissinet, C., Meslin, P.Y., Gough, R.V., House, C.H., Pavlov, A., Eigenbrode, J.L., Glavin, D.P., Pearson, J.C., Keymeulen, D., Christensen, L.E., Schwenzer, S.P., Navarro-Gonzalez, R., Pla-García, J., Rafkin, S.C.R., Vicente-Retortillo, Á., Kahanpää, H., Viudez-Moreiras, D., Smith, M.D., Harri, A.M., Genzer, M., Hassler, D. M., Lemmon, M., Crisp, J., Sander, S.P., Zurek, R.W., Vasavada, A.R., 2018. Background levels of methane in Mars' atmosphere show strong seasonal variations. Science 360 (6393), 1093–1096. https://doi.org/10.1126/science.aaq0131.

Yung, Y.L., Chen, P., Nealson, K., Atreya, S., Beckett, P., Blank, J.G., Ehlmann, B., Eiler, J., Etiope, G., Ferry, J.G., Forget, F., Gao, P., Hu, R., Kleinböhl, A., Klusman, R., Lefèvre, F., Miller, C., Mischna, M., Mumma, M., Newman, S., Oehler, D., Okumura, M., Oremland, R., Orphan, V., Popa, R., Russell, M., Shen, L., Sherwood Lollar, B., Staehle, R., Stamenković, V., Stolper, D., Templeton, A., Vandaele, A.C., Viscardy, S., Webster, C.R., Wennberg, P.O., Wong, M.L., Worden, J., 2018. Methane on mars and habitability: challenges and responses. Astrobiology 18 (10), 1221–1242. https://doi.org/10.1089/ast.2018.1917.

CHAPTER

14

From Río Tinto to Mars: Exploring potential microbial metabolisms and their biogeochemical signatures

Sergey Abramov

Stuttgart University, Environmental Microbiology, Stuttgart, Germany

Introduction

The surface of Mars, with its extreme environmental conditions, is currently inhospitable to life as we know it (Seto et al., 2019). Water, a fundamental requirement for life, is extremely scarce, and the planet's environment presents other challenges, such as oxidizing soils, low atmospheric pressure, drastic temperature fluctuations (−153 °C–20 °C), and a CO_2-rich atmosphere (96%). Additional stressors include intense ultraviolet and cosmic radiation, toxic heavy metals, nutrient deficiencies, and elevated pH levels (Berry et al., 2010; Mahaffy et al., 2013; Seto et al., 2019).

However, Mars is not always the desolate world that we see today. During its 4.5 billion year history, the planet has undergone significant climatic and atmospheric change. According to more conservative and pessimistic hypotheses, early Mars remained predominantly cold and dry, with only transient warming events likely triggered by volcanic or impact-related activity, resulting in brief episodes of liquid water stability (Wordsworth, 2016). By contrast, more optimistic models suggest that Mars may have supported stable bodies of liquid water in a denser, CO_2-rich atmosphere during the Noachian, Hesperian and early Amazonian periods, creating more favorable conditions for habitability (Fairén et al., 2004; Di Achille and Hynek, 2010; Andrews-Hanna and Lewis, 2011; Ramirez et al., 2014; Molina et al., 2017; Ramirez and Craddock, 2018; Kite et al., 2019; Mangold, 2021). Understanding this transition from a potentially habitable world to its current arid and frozen state is key to unraveling Mars' geological evolution and assessing its potential to host life, past, or present (Seto et al., 2019).

If life ever arose on Mars, it most likely emerged during the pre-Noachian to Noachian periods, when abundant surface water formed lakes and possibly an ocean in the northern hemisphere. Even in later periods, such as the Hesperian and

Geomicrobiology of Río Tinto. https://doi.org/10.1016/B978-0-443-32948-7.00003-5

Amazonian, transient surface habitats or stable subsurface environments may support microbial life (Fig. 14.1) (Paul and Mormile, 2020; Westall et al., 2015).

The presence of hydrated minerals is indicative of a variety of ancient environments, some of which may have facilitated life within the known terrestrial limits (Nixon et al., 2013). For example, Fe- and Mg-rich phyllosilicates (clays), which require the presence of liquid water and near-neutral pH conditions, are widespread over Noachian terrain (Wordsworth, 2016). Clay-containing sediments in the Gale Crater suggest the presence of a temperate, pH-neutral lake, which would have been conducive to long-term habitability during the Noachian and Hesperian periods. Conversely, the presence of SO_4^{2-} minerals, such as jarosite, indicates episodic acidic conditions and localized environmental variability, primarily on Hesperian terrain (Fairén et al., 2004; Grotzinger et al., 2014; Wordsworth, 2016; Rampe et al., 2017; Price et al., 2018). This is consistent with global models suggesting a transition from neutral to increasingly acidic surface conditions after the Noachian period, reducing the habitability of near-surface environments (Fig. 14.1) (Bibring et al., 2006; Nixon et al., 2013).

The mineralogy of Mars also reflects the chemical inventory of the planet, particularly the abundances of iron and sulfur. These elements are more concentrated on Mars than on Earth because of extensive Noachian-Hesperian volcanism, which released SO_2, and the high iron content of Martian basalts (Halevy et al., 2007; McSween et al., 2009; King and McLennan, 2010; Nixon et al., 2013).

In addition, sulfate minerals, including monohydrated and polyhydrated Mg sulfates, Ca sulfates, and less common Fe sulfates, are widespread on the Martian surface (Langevin et al., 2005; Ming et al., 2006; Bishop et al., 2009; Nixon et al., 2013). Mars is also covered by a global layer of nanophase Fe(III) oxide and oxyhydroxide dust (Nixon et al., 2013). The identified Fe-bearing minerals include hematite, which is often associated with evaporitic processes in stagnant water, and goethite, which is typically formed by hydrothermal alteration of basaltic crust (Tosca et al., 2005; Morris et al., 2008). Additional minerals, such as jarosite, ferrihydrite, and schwertmannite, indicate the influence of acidic hydrothermal systems on Martian geochemistry (Fig. 14.1) (Farrand et al., 2007; Price et al., 2018).

The widespread availability of iron and sulfur, especially in sulfate and oxide assemblages, suggests that if microbial life ever existed on Mars, it may have evolved iron- and sulfur-based metabolic pathways (Morris et al., 2008; Bishop et al., 2009; Nixon et al., 2013). These metabolisms could have used atmospheric gases such as CH_4, H_2, CO_2, H_2S, SO_2, and O_2 which were present in an atmosphere with an estimated upper pressure limit of about 0.9–3.8 bar about 3.6 Ga (Jakosky et al., 2017; Paul and Mormile, 2020; Ramirez et al., 2014; Wordsworth, 2016). The discovery of kerogen-like organic matter (OM) and other complex organic compounds in the Gale Crater mudstones and Martian meteorites supports the idea that organic compounds may have fueled redox reactions involving iron and sulfur (Steele et al., 2012; Ming et al., 2014; Eigenbrode et al., 2018; Ten Kate, 2018). If these reactions are biologically mediated, they may have left behind geochemical biosignatures.

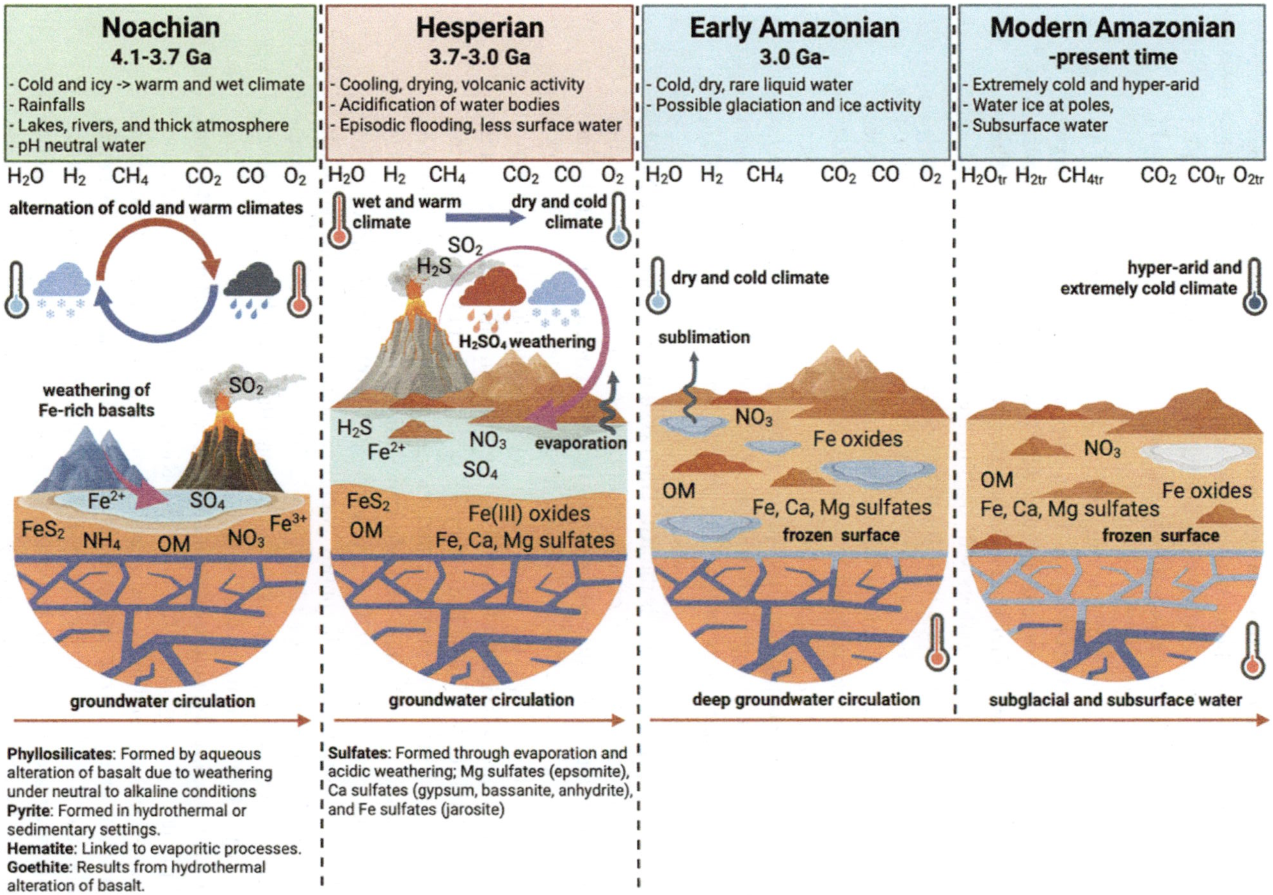

FIGURE 14.1 Environmental Evolution of Mars During the Noachian, Hesperian, and Amazonian Periods and Potential Metabolic Niches.

During the Noachian (~4.1–3.7 Ga), Mars likely hosted stable bodies of liquid water under a denser CO_2-rich atmosphere, supporting near-neutral pH conditions and the formation of Fe/Mg phyllosilicates. These conditions may have allowed microbial life to exploit electron donors such as H_2, CH_4, and Fe^{2+} and acceptors such as CO_2, O_2, Fe^{3+}, and/Fe(III) minerals. In the Hesperian (~3.7–3.0 Ga), decreasing atmospheric pressure and increasing volcanic activity led to episodic acidic conditions and widespread sulfate deposition (e.g., jarosite), potentially enabling acidophilic sulfur-based metabolisms. During the Amazonian period (~3.0 Ga to present), Mars became increasingly cold and arid, limiting surface habitability. However, subsurface environments with Fe(III), SO_4^{2-}, and residual organic matter may have supported chemolithotrophic life. tr - trace.

- Created in BioRender. Abramov, S. (2025) https://BioRender.com/wvr4i0t.

Consequently, the ancient geological record of Mars is a prime target for the search for biosignatures (Shen et al., 2022).

Such biosignatures, including mineral alteration patterns, gas fluxes, isotopic fractionation, and preserved metabolic by-products, are potential indicators of biological activities. As many of these signatures are pathway-specific, their detection could provide compelling evidence of ancient or extant life on Mars (Nixon et al., 2013).

Given the challenges of direct exploration, terrestrial Martian analogs play a critical role in astrobiology. These terrestrial and often extreme environments share key geological, geochemical, and ecological features with Mars and allow biosignature detection methods to be tested under realistic conditions (Gómez et al., 2011; Shen et al., 2022). The study of these analogs will help in the interpretation of Martian data and in the development of future mission strategies.

The ability of extremophiles to survive in the most inhospitable environments on Earth highlights the resilience of microbial life and its potential to persist under extreme conditions, both in the past and even today on Mars (Junge et al., 2004; Berry et al., 2010; Nicholson et al., 2013; Moissl-Eichinger et al., 2016; Seto et al., 2019). This resilience emphasizes the value of studying terrestrial Martian analogs, which provides critical insight into the potential habitability of the planet and guides the search for biosignatures in the Martian geochemical record (Tan et al., 2018).

Among the most valuable analogs is the Río Tinto system in southwestern Spain, which is considered representative of the Noachian (~4.1–3.7 Ga), Hesperian (~3.7–3.0 Ga), and even Amazonian Mars (~3.0-present) (Fairén et al., 2004; Carr and Head, 2010; Wordsworth, 2016). The river's acidic waters (pH 2.0–2.5), rich in Fe(III) and SO_4^{2-}, result from the oxidation of sulfide minerals and mimic processes inferred at Martian sites such as Meridiani Planum (Fernández-Remolar et al., 2005, 2008; Belilla et al., 2019; Shen et al., 2022; Finkel et al., 2023; Amils et al., 2024). The Río Tinto hosts minerals such as jarosite, which are also found on Mars at sites such as Terra Meridiani and Mawrth Vallis (Amils et al., 2007, 2024; Amils and Fernández-Remolar, 2020; Farrand et al., 2009; Klingelhöfer et al., 2004). Its sediments, dominated by hematite, goethite, and nanophase Fe(III) oxides, resemble the altered Martian surface (Amils et al., 2014, 2024), while SO_4^{2-}-containing evaporites such as gypsum provide further parallels with Martian sedimentary systems (Fig. 14.2) (Bibring et al., 2006).

In addition to the highly acidic sites, there are also moderately acidic (pH 5.0–6.0) and pH-neutral tributaries that enter the Río Tinto at its source and along its course. These sites, although less acidic, are still influenced by the sulfidic origin of the river and are characterized by relatively high concentrations of Fe(III) and SO_4^{2-}. Finally, the Río Tinto flows into the Atlantic Ocean through an estuary that extends approximately 15 km to the Niebla area. In estuary, both pH and salinity (e.g., NaCl concentrations) fluctuate in response to the diurnal tidal cycle. Waters with neutral to moderately acidic pH, but still enriched in Fe^{3+}, Fe^{2+}, and SO_4^{2-}

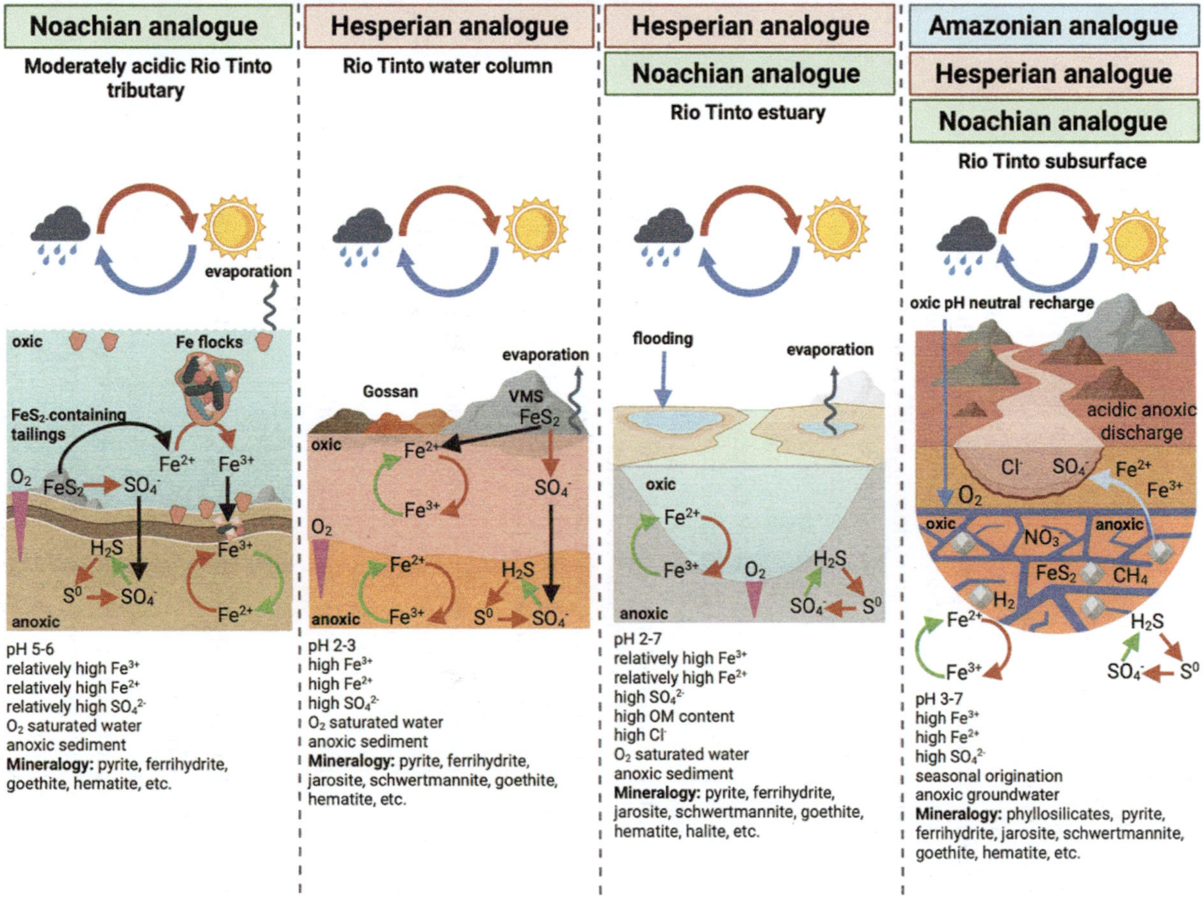

FIGURE 14.2 Terrestrial Analogs of Martian Environmental Conditions Across the Noachian, Hesperian, and Amazonian Periods Using Sites Along the Río Tinto System.

Four different Río Tinto environments illustrate how redox gradients, mineral assemblages, and geochemical conditions can reflect those that may existed on Mars. Noachian analog (first panel): A moderately acidic Río Tinto tributary represents a circumneutral to slightly acidic, Fe- and SO_4^{2-}-rich aquatic environment under fluctuating oxic-anoxic conditions, supporting iron and sulfur cycles. Hesperian analog (second panel): The Río Tinto water column reflects highly acidic and oxidizing conditions, which are driven by the weathering of volcanogenic massive sulfide (VMS), resulting in the formation of Fe(III) hydroxysulfates and Fe(III) (hydr)oxides. Noachian-Hesperian transition analog (third panel): The Río Tinto estuary shows the effect of episodic flooding and evaporation under variable pH and redox conditions, resulting in dynamic Fe and S cycling and chloride accumulation. Martian subsurface analog (fourth panel): The Río Tinto subsurface system mimics predominantly anoxic groundwater environments on Mars, where O_2-containing recharge water can temporally oxygenate anoxic groundwater establishing subsurface redox gradients. These conditions could support microbial processes such as nitrate reduction, methanogenesis, and Fe/S cycling, etc.

- Created in BioRender. Abramov, S. (2025) https://BioRender.com/pdofjle.

(Abramov et al., 2020), resemble ancient Noachian water bodies on Mars, such as the lake that once existed in the Gale Crater (Fig. 14.2).

Although life may have gradually disappeared from the Martian surface during the Late Noachian to Early Hesperian as the climate became increasingly cold and arid, it may have persisted in subsurface habitats or transient surface niches. Therefore, in addition to surface water, microorganisms may have colonized the Martian subsurface aquifers. The recent discovery of subglacial and subsurface liquid water on Mars supports this hypothesis, and highlights the potential for microbial life to persist in the modern Amazonian subsurface (Fig. 14.1) (Orosei et al., 2018).

The deep subsurface of the Iberian Pyrite Belt (IPB) near Peña de Hierro in the headwaters of the Río Tinto is characterized by active iron and sulfur biogeochemical cycles. It serves as a compelling terrestrial analog for the Martian subsurface during cold and dry periods when surface conditions were likely inhospitable, particularly during the Late Noachian and Amazonian periods. In addition, the mineralogical and geochemical conditions of the IPB subsurface also reflect environmental conditions thought to have existed in the earlier Martian history, including the Noachian and early Hesperian periods (Fig. 14.2) (Amils et al., 2023).

The Río Tinto also demonstrates potential for the preservation of biosignatures. Microbial structures in Fe(III) oxide and sulfate deposits dating back 25 million years remain well-preserved (Fernández-Remolar et al., 2021a). Even after mineralogical transitions, such as the transformation of jarosite into more stable phases, such as goethite, these deposits retain organic molecules. The encrustation of microbial structures by Fe(III) minerals (e.g., jarosite and goethite) and the preservation of DNA, peptides, and lipids emphasize the possibility of long-term encapsulation of biosignatures under Mars-like conditions (Fernández-Remolar et al., 2005, 2021a; Amils et al., 2024).

These results, together with the strong geochemical and mineralogical similarities between the Río Tinto and Martian environments, highlight its value as a terrestrial analog. Its accessibility, mineral diversity, and persistent biosignatures make the Río Tinto one of the most informative sites for investigating potential Martian metabolism and preserving their geological imprints.

Microbial Fe and sulfur metabolisms at the Río Tinto and their implications for Mars

In addition to carbon sources and liquid water, microorganisms require access to suitable electron donors and acceptors to sustain redox-driven metabolism (Westall et al., 2015). On Earth, many microorganisms generate energy through electron transport chains, in which electrons are transferred from electron donors to acceptors, ultimately driving ATP synthesis. These pathways are particularly important in environments such as the Río Tinto, where the abundance of iron and sulfur compounds provides the redox couples necessary for maintaining diverse microbial communities.

Key redox couples involving iron, sulfur, and carbon support diverse microbial metabolisms, with these elements often acting as both electron donors (e.g., Fe^{2+}, S^{2-}, $(CH_2O)_n$) and acceptors (e.g., Fe^{3+} and SO_4^{2-}) (Nixon et al., 2013). For example, acidophilic Fe(II)-oxidizing bacteria, such as *Acidithiobacillus ferrooxidans* use O_2 to oxidize Fe^{2+} to Fe^{3+}, whereas Fe(III)-reducing bacteria, such as *Acidiphilium* use organic substrates (e.g., $(CH_2O)_n$) to reduce Fe^{3+} to Fe^{2+} under both oxic and anoxic conditions. *A. ferrooxidans* can also contribute to the regeneration of Fe^{2+} from Fe^{3+} under anoxic conditions by coupling the reduction of Fe^{3+} with the oxidation of reduced sulfur compounds (RSCs) such as S^{2-}. Under oxic conditions, *A. ferrooxidans* can oxidize sulfide directly, with both pathways leading to the production of SO_4^{2-}. Sulfate can then serve as an electron acceptor for acidophilic sulfate-reducing bacteria (SRB) (Fig. 14.3) (González-Toril et al., 2003; Sánchez-Andrea et al., 2015; Abramov et al., 2022).

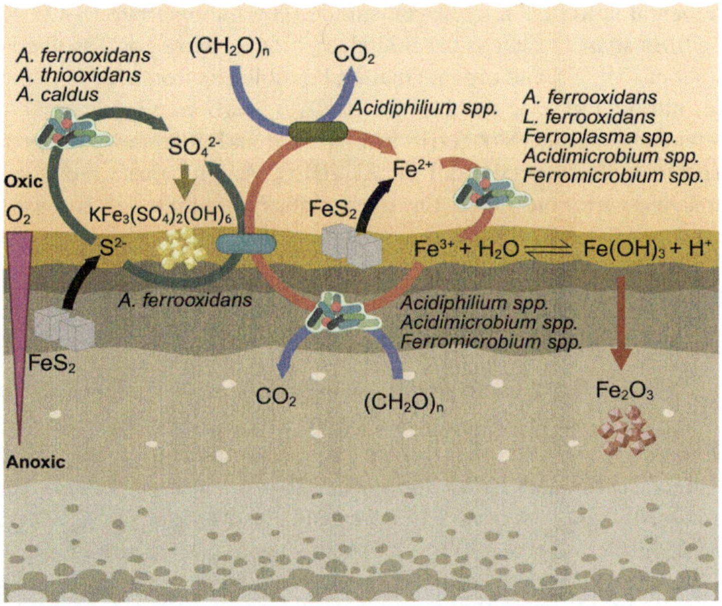

FIGURE 14.3 Schematic Representation of Iron and Sulfur Biogeochemical Cycles in the Acidic Water Column and Sediment of the Río Tinto (SW Spain).

Microbial processes driving redox transformations of Fe and S are shown in relation to environmental conditions. Figure adapted from González-Toril, E., Llobet-Brossa, E., Casamayor, E.O., Amann, R. Amils, R. 2003. Microbial ecology of an extreme acidic environment, the Tinto River. *Appl. Environ. Microbiol.* 69(8), 4853–4865. https://doi.org/10.1128/aem.69.8.4853-4865.2003; Abramov, S.M., Straub, D., Tejada, J., Grimm, L., Schädler, F., Bulaev, A., Thorwarth, H., Amils, R., Kappler, A., Kleindienst, S. 2022. Biogeochemical niches of Fe-cycling communities influencing heavy metal transport along the Río Tinto, Spain. *Appl. Environ. Microbiol.* 88(4), e0229021. https://doi.org/10.1128/aem.02290-21.

- Created in BioRender. Abramov, S. (2025) https://BioRender.com/t9lcaqz.

In addition, the supersaturation of river water with SO_4^{2-} and Fe^{3+} promotes the precipitation of SO_4^{2-}- and Fe(III)-containing minerals, such as jarosite, schwertmannite, and ferrihydrite. The balance between dissolved Fe^{3+} and ferrihydrite played a key role in maintaining the low pH of the river. When water is neutralized, this equilibrium shifts toward the formation of $Fe(OH)_3$, which causes a further decrease in pH. The resulting $Fe(OH)_3$ is deposited in the sediment where it recrystallizes into more stable Fe(III) oxides such as goethite and hematite (Fig. 14.3). These transformations drive the microbial iron and sulfur cycles in the Río Tinto, playing a key role in wider biogeochemical processes and maintaining the river's characteristically low pH.

On Mars, a range of potential electron donors may be available to support microbial metabolism. These include hydrogen (H_2), methane (CH_4), carbon monoxide (CO), reduced nitrogen compounds such as ammonium (NH_4^+) and nitrite (NO_2^-), reduced sulfur compounds such as sulfide (S^{2-}), bisulfide (HS^- in aqueous phase), elemental sulfur (S^0), hydrogen sulfide (H_2S), thiosulfate ($S_2O_3^{2-}$), ferrous iron (Fe^{2+}) from sources such as basaltic glass, biotite, pyrite, or dissolved in water, magnesium ions (Mg^{2+}), and organic matter. Possible electron acceptors include oxygen (O_2), nitrate (NO_3^-), perchlorate (ClO_4^-), carbon dioxide (CO_2), sulfate (SO_4^{2-}), elemental sulfur (S^0), ferric iron (Fe^{3+}), and potentially water (H_2O) or certain organic molecules (Westall et al., 2015). Among these redox pairs, iron and sulfur cycles are considered the most plausible sources of energy for early Martian life because of the predominantly anoxic conditions and distinctive mineralogy of Mars.

Microbial Fe(II) oxidation pathways

The basaltic Martian crust is rich in Fe-bearing silicates, such as olivine and pyroxene, in which Fe(II) predominates (McSween et al., 2009; Nixon et al., 2013). NASA's Curiosity rover identified olivine, pyroxene, pyrrhotite, and magnetite in the Gale Crater centimeters below the surface (Vaniman et al., 2014). Amorphous Fe(II)-rich basaltic glass is also widespread in the Martian crust (Morris et al., 2006a, 2006b; McSween et al., 2009), supporting the hypothesis that Fe(II) is readily available in subsurface and near-surface environments. Pyrite has also been identified in the Martian regolith, including in situ detection by the Mars Science Laboratory (MSL) at the Gale Crater, specifically in the Yellowknife Bay formation (Gil-Lozano et al., 2025). Pyrite, which is possibly formed by serpentinization, hydrothermal activity, radiolysis, or microbially mediated alteration, may also contribute to the Fe(II) pool. Thus, various Fe(II) minerals found in the Martian basalts and sediments would have provided a steady supply of Fe(II) as an electron donor.

Fe(II)-oxidizing microorganisms commonly use O_2 or NO_3^- as terminal electron acceptors (Nixon et al., 2013; Kappler et al., 2021). Evidence suggests that O_2 may have been present in the early Martian atmosphere (Tuff et al., 2013), while NO_3^- has been detected in Martian sediments (Price et al., 2018). Therefore, as on Earth, microbial Fe(II) oxidation could have been sustained by available electron acceptors such as O_2 or NO_3^- in past Martian environments.

Aerobic Fe(II) oxidation: oxygen as an electron acceptor

Although the current Martian atmospheric O_2 concentrations are only ~0.16%, they may still be sufficient to support aerobic microbial life in the past (Mahaffy et al., 2013; King, 2015). However, limited O_2 diffusion into sediments restricts its metabolism to surface or microoxic niches (Price et al., 2018). In ancient Martian lakes, such as the Gale Crater, redox stratification likely resulted in oxygenated surface waters and anoxic deeper layers (Hurowitz et al., 2017). These redox interfaces may support microaerophilic microorganisms, as observed in terrestrial lakes.

Despite these limitations, aerobic microorganisms, particularly microaerophiles, can persist in oxic-anoxic transition zones. In ancient Martian lakes, such as those in the Gale Crater, these zones would have been favorable for microaerophilic life due to redox stratification (Hurowitz et al., 2017). Microaerophilic Fe(II) oxidizers provide an alternative to aerobic acidophilic Fe(II) oxidizers and can outcompete abiotic oxidation under certain conditions (Price et al., 2018). These microorganisms thrive in oxic-anoxic boundary zones, typically at near-neutral pH, where abiotic Fe (II) oxidation is slow (Roden et al., 2004). Similarly, microaerophilic Fe (II)-oxidizing microorganisms successfully colonized several niches in the Fe(II)-rich, moderately acidic environments of the Río Tinto ecosystem. Unlike organisms restricted to thin microoxic zones in sediments, they can effectively compete for O_2 and oxidize the abundant dissolved Fe(II) (Fig. 14.2). These adaptations suggest that if similar environmental conditions existed on Mars, microaerophilic Fe(II) oxidizers could have played a role in the Martian iron cycle.

Microaerophilic Fe(II) oxidizers, such as *Marinobacter* and *Mariprofundus*, have been identified in the Río Tinto Estuary, an environment with important similarities to the sedimentary cycles that were active in Gale Crater during the Late Noachian and Early Hesperian periods. In this estuarine system, microaerophilic Fe(II) oxidizers are found in two distinct habitats: moderately acidic, Fe-rich sediments, and the water column. The latter is characterized by high O_2 saturation (11.4 mg/L), low mean pH (3.7), elevated Fe(II) concentrations (2.8 mg/L) and high salinity (27.7 g/L) (Abramov et al., 2022). These conditions create an ecological niche in which microaerophiles can outcompete abiotic Fe(II) oxidation, similar to what may have occurred in ancient Martian lakes.

Another intriguing niche for microaerophilic Fe(II)-oxidizing microorganisms is located in the headwaters of the Río Tinto near Peña de Hierro, where several tributaries with varying acidities converge. One of these tributaries, the "Right Stream," maintains a pH of 5–6 throughout the year. This stream hosts members of the microaerophilic family *Gallionellaceae*, which forms characteristic flocs throughout the water column (Fig. 14.2) (Bottaro et al., 2025a,b).

Interestingly, *Gallionella* and *Thiomonas* species have also been detected in the deep subsurface of the IPB near the Río Tinto. As part of the Mars Astrobiology Research and Technology Experiment (MARTE project; 2003–06), a Prediction

Analysis of Microarrays (PAM) combined with whole metagenomic amplification helped to identify *Gallionella* at a depth above 115 m below the surface (mbs) within a microoxic zone (84–116 mbs) between aerobic and anaerobic conditions, where they may play a key role in Fe(II) oxidation and biogeochemical cycling of iron (Puente-Sánchez et al., 2014).

These findings have important implications for Mars, where the subsurface may provide a more stable environment for microaerophilic life than the harsh oxidizing conditions at the surface. Higher partial pressures of O_2 in the Martian subsurface can create redox interfaces that support microaerophiles, making it a more favorable habitat for potential life (Fisk and Giovannoni, 1999).

Detailed geochemical analysis of samples taken from a 160 mbs deep borehole at the Peña de Hierro mining complex revealed increasing SO_4^{2-} concentrations below the water table. This zone is characterized by a relatively low pH (3–7) and high redox potential (400 mV), which are conditions that favor sulfide oxidation. Further analysis confirmed the presence of SO_4^{2-}, probably resulting from the oxidation of pyrite and other metallic sulfides. In addition, secondary minerals, such as jarosite, hematite, and goethite, were found within the fractures and crevices of the pyrite deposits.

Remarkably, molecular O_2 was found locally within the pyrite deposits despite the surrounding environment being predominantly anoxic (Harris et al., 2022). This phenomenon can be explained by the formation of H_2O_2 via reactions between pyrite and anoxic groundwater. The subsequent spontaneous decomposition of H_2O_2 is likely to generate localized O_2, creating microenvironments in which O_2-dependent processes can occur. In particular, O_2 can support microaerophilic Fe(II) oxidation even in an otherwise anoxic subsurface (Davila et al., 2008). The ability of microaerophiles to exploit these O_2-rich microenvironments in the Río Tinto subsurface suggests a compelling analog for potential habitable niches on Mars. If similar redox interfaces exist within Martian subsurface deposits, they could provide refuges for microaerophilic life, making them key targets in the search for extraterrestrial habitability (Fisk and Giovannoni, 1999).

Anaerobic Fe(II) oxidation: nitrate as electron acceptor

Many microorganisms on Earth use NO_3^- as an alternative electron acceptor in O_2-deficient environments because of its high redox potential. During denitrification, NO_3^- is sequentially reduced to produce molecular N_2 as the final product.

Nitrate has been detected in the sediments of the Gale Crater on Mars, at concentrations ranging from 70 to 1100 ppm (Stern et al., 2015, 2017). In addition, data from NASA's Mars Atmosphere and Volatile Evolution (MAVEN) mission obtained using an imaging ultraviolet spectrometer (IUVS) confirmed the presence of molecular N_2 (2.85%) in the Martian upper atmosphere (Stevens et al., 2015). On Earth, N_2 is mainly produced by denitrifying microorganisms (Fowler et al., 2013). In addition, NO_3^- reduction can be coupled with Fe(II) oxidation by autotrophic and heterotrophic NO_3^--reducing Fe(II)-oxidizing (NRFeOx) microorganisms (Stern et al., 2015). This suggests that Fe(II) and NO_3^- deposits on Mars may have served, or may still serve, as energy sources for such microbial life. If NRFeOx microorganisms existed on Mars, they could have thrived in ancient lakes and rivers and potentially persisted in subsurface aquifers long after the disappearance of surface water (Price et al., 2018).

Nitrate concentrations up to 2.5 ppm have been detected in the IPB subsurface (Amils et al., 2023). Catalyzed reporter deposition fluorescence in situ hybridization (CARD-FISH) analyses of rock samples from this environment revealed the presence of potential NRFeOx microorganisms, including *Acidovorax* and *Tessaracoccus*. Furthermore, NRFeOx activity was confirmed in the isolated strains of *Acidovorax*, *Tessaracoccus*, and *Pseudomonas*. Cultivation experiments combined with Raman spectroscopy demonstrated that *Acidovorax* can solubilize Fe from pyrite using acetate as a carbon and energy source and NO_3^- as an electron acceptor (Escudero et al., 2020; Amils et al., 2023). Iron(II) oxidation in these systems has been attributed to indirect chemical oxidation driven by reactive nitrogen intermediates such as NO_2^- and NO, both of which are produced during the stepwise reduction of NO_3^- (Amils et al., 2023).

Further insights come from enrichment cultures isolated from the "Right Stream," a moderately acidic tributary of the Río Tinto which enters the river near the Peña de Hierro. Mixotrophic NRFeOx cultures were obtained from this stream (Bottaro et al., 2025a,b). These results indicate that both microaerophilic and NRFeOx microorganisms can thrive in acidic, Fe- and S-rich environments and actively produce Fe(III) (oxyhydr)oxides and hydroxysulfates. These mineral by-products are considered promising geochemical biosignatures in Fe- and SO_4^{2-} rich sediments, such as those identified in Gale Crater.

Microbial Fe(III) reduction pathways

On Earth, Fe(III)-reducing microorganisms primarily use small organic molecules, such as acetate and lactate, as electron donors. As organic carbon has been detected on Mars, it is plausible that similar microbial processes could occur or have occurred there. Alternatively, H_2 may act as an electron donor, with CO_2 acting as a carbon source. On Mars, H_2 may originate from several sources, including photochemical production, serpentinization, pyritization, and microbial activity (Fernández-Remolar et al., 2008).

Another potential energy source is CO, which is an inorganic electron donor used by certain Fe(III)-reducing microorganisms. Carbon monoxide is produced photochemically in the Martian atmosphere, and some CO-oxidizing microorganisms can obtain both energy and carbon from CO alone, producing H_2 and CO_2 as metabolic by-products (Nixon et al., 2013).

Reduced nitrogen compounds, such as NH_4^+ and hydrogen cyanide (HCN), have not been detected on Mars, probably because of their instability under current surface conditions. However, several abiotic pathways for reduced nitrogen have been proposed for the Hadean Earth. These include NO_x reduction by metallic iron (Fe^0), reduction in aqueous environments, and the photochemical reduction of atmospheric N_2. Comparable processes may have occurred in the early Mars, particularly during the Noachian period, when less oxidizing conditions may have favored NH_4^+ stability (Koike et al., 2020).

These alternative electron donors suggest that if Fe(III)-reducing microorganisms existed on Mars, they could have relied on a variety of organic and inorganic substrates to sustain their metabolism (Nixon et al., 2013). In the Río Tinto, the high solubility of iron under acidic conditions results in Fe^{3+}-rich waters, where Fe^{3+} concentrations can exceed O_2 levels by several orders of magnitude, even in the surface layers. The abundance of Fe(III) makes it a widely available electron acceptor for microbial metabolism (Sánchez-Andrea et al., 2011).

Hydrogen-driven Fe(III) reduction

Atmospheric H_2 is scarce on Mars, and its presence in subsurface environments remains unconfirmed (Nixon et al., 2013). However, microbial H_2 production, together with processes such as pyritization, serpentinization, and photochemically driven oxidant generation, can contribute to H_2 availability (Paul and Mormile, 2020). In addition, cathodic depolarization of Fe^0 may further support H_2-dependent microbial metabolism in anoxic subsurface habitats, similar to that found in the Río Tinto subsurface (Fernández-Remolar et al., 2008).

Indeed, fluids extracted from the Río Tinto subsurface have been shown to contain dissolved gases, such as H_2, N_2O, and CH_4. Furthermore, biological production of H_2, CO_2, and CH_4 has been demonstrated in the IPB subsurface (Sanz et al., 2021). While initial hypotheses attributed H_2 formation to water-rock interactions, the discovery of H_2-producing bacteria suggests that microbial fermentation also contributes to H_2 production. Fermenters, such as *Paludibacter*, *Veillonella*, *Staphylococcus,* and *Clostridium* have been identified in river sediments. They can produce a complex mixture of intermediates. Some of these, such as propionate, are used by syntrophic bacteria such as *Syntrophobacter* to produce H_2 and acetate (Sánchez-Andrea et al., 2011). Hydrogen serves as an electron donor for Fe(III) reduction in acidophilic microorganisms, such as *A. ferrooxidans*, enabling Fe (III)-based metabolism in acidic environments (Johnson et al., 2012). Therefore, it may play a critical role as an electron donor in microbial ecosystems, supporting multiple metabolic pathways relevant to both terrestrial and potentially Martian subsurface environments (Amils, 2016).

Methane-driven Fe(III) reduction

On Earth, most methanogens produce CH_4 by coupling CO_2 reduction with H_2 oxidation, as observed in species such as *Methanosarcina barkeri*. Alternatively, some methanogens such as *Methanobacterium thermoautotrophicum* can produce CH_4 by reducing CO. These microorganisms typically inhabit low redox potential and neutral pH environments, conditions that contrast sharply with the oxidizing and acidic nature of the Río Tinto Basin.

The first detection of CH_4 in the Río Tinto occurred at Campo de Galdierias near the headwaters of the river (Amils et al., 2024). At this site, CH_4 was identified in sediment zones with negative redox potentials, and microcosm experiments using reduced sediments confirmed CH_4 production under anoxic conditions. Similarly, at JL Dam, CH_4 production was elevated in sediment cores that contained distinctive black bands often associated with reducing conditions. DNA analysis of these sediments identified *Methanosaeta concilii* as the dominant methanogen (Amils et al., 2014).

The presence of methanogens in apparently inhospitable environments can be explained by the formation of microniches in sediments, where the local physicochemical conditions differ from the surrounding environment. These reduced microenvironments are often formed by Fe(III)-reducing bacteria, which consume electron acceptors and decrease the redox potential downwards (Sanz et al., 2011). The ability of sediments to create such protective niches is even more pronounced in hard rock substrates of the deep subsurface of the IPB (Amils et al., 2024).

Microbial community analysis along the 166 m deep IPB borehole (project MARTE) revealed a consistent distribution pattern between methanogens and Fe (III) reducers (e.g., *Ferrimicrobium*), suggesting close ecological interactions. Interestingly, while methanogens were detected throughout the borehole, CH_4 was only found below 140 mbs. This discrepancy probably reflects the presence of methanotrophs in the overlying rock layers, which may prevent the accumulation of CH_4 in shallow zones (Puente-Sánchez et al., 2014).

As previously discussed, biological CH_4 production in the IPB subsurface has been well-documented (Sanz et al., 2021). Methanogenic activity in the Río Tinto region has been confirmed by MARTE and Iberian Pyrite Belt Subsurface Life Detection (IPBSL) drilling projects. In addition, enrichment cultures from subsurface samples have demonstrated the viability of methanogens in extreme environments (Amils et al., 2014).

The detection of methanogens in Fe-rich, acidic, and oxidized subsurface environments has significant astrobiological implications. This challenged the prevailing assumption that Mars is unsuitable for methanogenesis. Given the geochemical and mineralogical similarities between the Martian crust and the Río Tinto's IPB systems, the possibility of biological CH_4 production on Mars becomes a plausible and testable hypothesis (Amils and Fernández-Remolar, 2020).

Ammonium-driven Fe(III) reduction

Geochemical analyses of water-soluble cations from the IPB subsurface have revealed remarkable concentrations of NH_4^+ (up to 53.1 ppm), Fe^{2+} (up to 705 ppm), and Fe^{3+} (up to 311 ppm) (Amils et al., 2023). These conditions create a favorable environment for microbial redox processes, where NH_4^+ can act as an electron donor and dissolved Fe^{3+} or Fe(III)-bearing minerals act as electron acceptors. Such microbial interactions are thought to contribute to the formation of distinct mineralized structures within the subsurface environment.

The molecular composition of these structures provides valuable information regarding the metabolic pathways involved in their formation, effectively capturing and preserving microbial activity over time. Core samples obtained during the MARTE drilling project, analyzed by Time-of-Flight Secondary Ion Mass Spectrometry (ToF-SIMS) and Scanning Electron Microscopy (SEM) coupled with Energy-Dispersive X-ray Spectroscopy (EDS), revealed the presence of microfilament structures attributed to microorganisms capable of oxidizing NH_4^+ and NO_2^-, with NO_3^- identified as the final metabolic product (Fernández-Remolar et al., 2021b).

Interestingly, some of these NO_3^--containing structures appeared to consist of sheathed colonies of smaller microbial bodies aligned along the axis of microfilaments on the surface lamina. The formation of these microfilaments may have been driven by the dissolution of hydroxysulfate and oxyhydroxide surfaces, which releases Fe^{3+} and SO_4^{2-} into the surrounding microenvironment. These compounds can then serve as electron acceptors for the microbial oxidation of various organic substrates, CH_4, and NH_4^+.

In particular, the oxidation of NH_4^+ in this system is likely mediated by Fe(III)-dependent anaerobic ammonium oxidizing (Feammox) microorganisms. These microorganisms use Fe^{3+} as an electron acceptor to oxidize NH_4^+, producing N_2, NO_2^-, NO_3^- as by-products (Fernández-Remolar et al., 2021b). This suggests that Feammox microorganisms may be involved in Fe cycling in the IPB subsurface ecosystem.

Organic carbon as an electron donor for Fe(III) reduction

Most of the biomass in the Río Tinto system is concentrated along the riverbed and exposed bedrock surfaces where filamentous algae and fungi form dense biofilms. These biofilms provide a structured habitat for diverse prokaryotic communities that support a wide range of fermentation and respiratory metabolic pathways (Sánchez-Andrea et al., 2011). Within these communities, fermentative bacteria, such as *Paludibacter*, *Veillonella*, *Staphylococcus*, and *Clostridium* initiate the degradation of complex organic matter, producing a variety of metabolic intermediates that can be used by Fe(III) reducers.

In addition to fermentative and syntrophic interactions, heterotrophic Fe(III)-reducing bacteria play an important role in the microbial network of the Río Tinto. Genera such as *Acidiphilium*, *Sulfobacillus*, and *Acidobacterium* are capable of respiring Fe(III) by oxidizing organic substrates. These microorganisms colonize biofilms formed by fungi and algae, and are integrated into a wider microbial network. Surprisingly, they can also precipitate Fe(II) carbonates, such as siderite, in proximity to the cells (Sánchez-Román et al., 2014).

While heterotrophic Fe(III)-reducing bacteria dominate biofilm-associated communities, the Río Tinto subsurface also supports chemolithotrophic Fe(II)- and sulfur-oxidizing microorganisms that thrive in sulfide deposits. These environments are typically low in organic matter, however, smectite phyllosilicates contain organic carbon and preserve a wide range of biomarkers. In addition, ironstones in headwater terraces contain well-preserved macroscopic and microscopic biosignatures within Fe(III) oxide precipitates (Harris et al., 2022).

Despite these localized reservoirs, the distribution of organic carbon within the Río Tinto system is highly variable. For example, pyrite deposits contain extremely low levels of total organic carbon (TOC) (~0.05 wt%), which contrasts sharply with the higher organic carbon content of the surface soil cover (~3–10 wt%) and phyllosilicate-rich layers (~0.23 wt%) embedded in oxidized near-surface rocks. These results suggest that organic carbon does not accumulate within pyrite deposits (Davila et al., 2008).

Nevertheless, organic compounds, including organic acids (e.g., acetate up to 200.6 ppm, formate up to 6.5 ppm, propionate up to 5.2 ppm, and formate up to 6.7 ppm), carbohydrates (up to 72.1 ppm), and proteins (up to 50.3 ppm), have been detected in the subsurface fluids (Amils et al., 2023, 2024). These compounds provide crucial carbon and energy sources for heterotrophic microorganisms, supporting the viability of microbial life in subsurface ecosystems (Sanz et al., 2021). However, the origin of this organic matter, whether sedimentary, biological, or introduced from external sources, remains unclear (Amils et al., 2024).

Sulfur compounds as alternative electron donors for Fe(III) reduction

In the anoxic and acidic sediments of the Río Tinto, microbial communities are capable of coupling the oxidation of reduced sulfur compounds with the reduction of Fe(III). Species such as *A. ferrooxidans* and *Sulfobacillus* are known to facilitate this process using RSCs (e.g., S^{2-}, S^0) as an electron donor while reducing Fe(III) to Fe(II). This microbial redox interaction played a key role in the biogeochemical cycling of iron and sulfur within the Río Tinto system (Fig. 14.3). The microbial reduction of Fe(III) contributes to the continuous regeneration of Fe(II), which can subsequently be abiotically or biologically re-oxidized, thereby maintaining the Fe redox cycle under prevailing environmental conditions (Sánchez-Andrea et al., 2011).

Microbial sulfur oxidation pathways

Sulfur-rich and saline waters were likely widespread on early Mars, particularly during the Noachian-Hesperian transition (~4.1–3.0 Ga), and played a critical role in shaping the planet's geochemical environment (Fairén et al., 2004). Today, sulfur-bearing formations, including sulfides and sulfates, are widespread across the Martian surface. For example, sulfide minerals, such as pyrrhotite, have been identified in Martian meteorites (particularly in shergottites) and have been directly detected by lander missions. These minerals can potentially support microbial sulfur oxidation, a metabolism known to occur in extreme terrestrial environments (Nixon et al., 2013; Macey et al., 2020).

Pyrite may have formed on Mars in hydrothermal or sedimentary environments analogous to volcanogenic massive sulfide (VMS) deposits on Earth. Although prolonged surface oxidation may have degraded evidence of such ancient deposits, sulfides may have contributed to the formation of sulfate-rich formations observed throughout the Martian crust (Davila et al., 2008).

On Earth, sulfur oxidation can occur under both aerobic and anaerobic conditions. In anoxic environments, alternative electron acceptors such as NO_3^- or Fe (III) allow sulfur oxidation even at low pH (Sánchez-Andrea et al., 2011; Nixon et al., 2013). The oxidative weathering of pyrite under acidic conditions (pH 3–4) generates high redox potentials (~1000 mV), resulting in the precipitation of Fe (III) oxides and the release of SO_4^{2-} (Davila et al., 2008). The presence of SO_4^{2-}-bearing deposits on Mars further supports the importance of sulfur in the Martian geochemical cycle, influencing the interactions between the atmosphere, hydrosphere, and lithosphere.

Sulfate minerals, including Mg sulfates (epsomite), Ca sulfates (anhydrite, bassanite, gypsum), and Fe sulfates (e.g., jarosite), have been identified in Martian dust, soils, light-colored deposits (present within canyons, in chaos terrain), and within sedimentary rocks of Meridiani Planum. Sulfates occur as evaporitic crusts, mineral precipitates, and fillings of mineralized veins, and their hydrated phases can incorporate water into their crystal structures or retain them as fluid inclusions (Siljeström et al., 2024). Consequently, sulfate minerals may serve as important water reservoirs and play a central role in past geochemical cycles on Mars.

The Río Tinto system provides a compelling terrestrial analog for the study of sulfur cycling and sulfate formation on Mars. In this system, the sulfur cycle is driven by the microbial weathering of sulfides within the VMS deposits of the IPB. Acidic river waters originate from springs that transport sulfur and iron to the surface in the form of SO_4^{2-} and Fe(II)/Fe(III) complexes. These fluids are periodically replenished by meteoric water, which introduces oxygenated neutral water into subsurface fractures and sulfide-rich aquifers (Fernandez-Remolar et al., 2007).

Seasonal redox shifts and pH variations in the Río Tinto subsurface illustrate how oxidation of sulfide minerals, triggered by rainfall-driven recharge, generates SO_4^{2-}-enriched, acidic waters. A similar mechanism may have occurred on Mars, where long-term subsurface sulfur storage dating back to the Noachian period may have been later mobilized and oxidized during climatic or hydrological shifts (Fernandez-Remolar et al., 2007).

On Earth, the biogeochemical sulfur cycle involves diverse microbial pathways that mediate sulfur transformations across different redox states. The presence of sulfur species in different redox states on Mars, together with potential electron donors and acceptors (e.g., O_2, NO_3^-), supports the plausibility of sulfur-based microbial metabolisms. This is particularly relevant to Gale Crater, where mineralogical evidence suggests transitions between oxidizing and reducing conditions at the Noachian-Hesperian boundary (Horvath and Andrews-Hanna, 2021).

Oxygen-dependent sulfur oxidation

The sulfur cycling processes observed in the Río Tinto system provide valuable information for understanding the formation of extensive sulfate deposits on Mars during the Late Noachian-Hesperian transition (Bibring et al., 2006). In sulfur-rich terrestrial environments, such as the Río Tinto, primary microbial productivity is often driven by the aerobic oxidation of reduced sulfur species. For example,

A. ferrooxidans can oxidize both Fe(II) and RSCs under oxic conditions. Oxygenated, near-neutral meteoric waters introduced via fractures into sulfide-rich aquifers promote oxidative weathering, facilitating both the biotic and abiotic degradation of sulfide minerals. This leads to the formation of sulfuric acid and mobilization of the associated metal ions (e.g., Fe(II) and Fe(III), Cu, and Al).

Seasonal redox and pH shifts in Río Tinto groundwater illustrate the dynamic interplay between hydrology and microbial activity. During late summer, groundwater exhibits low redox potential, near-neutral pH, and elevated sulfide and Fe (II) concentrations, indicative of reducing conditions. In contrast, autumn conditions are characterized by lower pH, higher redox potential, increased SO_4^{2-} concentrations, and fluctuating Fe(II)/Fe(III) ratios. These changes reflect the onset of microbially mediated sulfide oxidation, where sulfide minerals react with O_2 and Fe(III), releasing protons and acidifying groundwater.

These observations support the hypothesis that the Martian sulfur cycle does not require high atmospheric sulfur flux from volcanic sources. Instead, Martian sulfate deposits may have been formed by the oxidation and remobilization of long-term subsurface sulfur reservoirs, which were probably stored as sulfide minerals or dissolved sulfur species since the early Noachian times (Fernandez-Remolar et al., 2007).

The oxidizing and acidifying conditions that promoted SO_4^{2-} formation on Mars may have been triggered by a climatic shift during the Late Noachian, possibly induced by renewed volcanic episodes. Such volcanism may have released large quantities of volatile compounds into the atmosphere, intensifying greenhouse warming, enhancing groundwater recharge, and promoting subsurface fluid circulation. Subsequent photochemical decomposition of volatiles, such as H_2O, CO_2, sulfurous gases, and CH_4, can produce oxidants and acidic species, facilitating the weathering of sulfide-rich subsurface deposits (Fernandez-Remolar et al., 2007).

Microbial SO_4^{2-} reduction pathways

The oxidation of sulfide minerals is believed to have contributed significantly to the widespread formation of sulfate minerals on Mars during the Late Noachian-Hesperian transition (Bibring et al., 2006). Sulfate deposits on Mars suggest the presence of a sulfur source most likely linked to volcanic activity (e.g., SO_2, H_2S) or long-term subsurface sulfur reservoirs (e.g., sulfide minerals), and often indicate acidic and/or saline formation conditions (Fernandez-Remolar et al., 2007; Wordsworth, 2016). While sulfate minerals are predominantly found in Hesperian-aged terrains, they are not exclusive to this period and may be associated with volcanic processes responsible for the formation of basaltic ridged plains (Wordsworth, 2016). These sulfates, now found in Martian soil and near-surface crust, represent a potentially important resource for microbial SO_4^{2-} reduction. Beyond sulfate, other oxidized sulfur species, such as SO_3^{2-} and $S_2O_3^{2-}$, can also serve as terminal electron acceptors for microbial metabolism. The detection of organic carbon, CH_4, and H_2 on Mars indicates the presence of potential electron donors that could support SRB, particularly in subsurface or groundwater-bearing environments (Nixon et al., 2013; Nachon et al., 2014).

In the Río Tinto system, Fe(III) is the dominant electron acceptor used for the oxidation of organic matter, primarily by *Acidiphilium*, *Acidobacterium*, and *Sulfobacillus*. As a more electropositive electron acceptor, Fe(III) suppresses both SO_4^{2-} reduction and methanogenesis. Indeed, modeling studies have demonstrated that the competitive advantage of Fe(III) reducers over SRB is pH dependent. Therefore, Fe (III) reducers can compete against SRB under acidic conditions, but might lose competition under alkaline conditions (Jin and Kirk, 2018). Indeed, with increasing pH, Fe(III) becomes less soluble, reducing its bioavailability and allowing SO_4^{2-} reduction to emerge as the primary pathway for organic matter degradation. The increase in pH can also be modulated by SRB via bicarbonate production in the microenvironment. Indeed, acidophilic SRB capable of dissimilatory SO_4^{2-} reduction are frequently detected in such extreme environments (Valdez-Nuñez et al., 2024). Sulfate-reducing bacteria, including *Desulfosporosinus*, *Desulfurella*, *Thermodesulfobium*, *Syntrophobacter*, *Desulfitobacterium*, and *Desulfotomaculum*, have been identified in anoxic sediments of the Río Tinto and at JL and SN dam sites located in the headwaters upstream of Nerva City (Sánchez-Andrea et al., 2012, 2013; Amils et al., 2014). Their presence under low-pH conditions confirms the role of microbial SO_4^{2-} reduction in completing the sulfur cycle in this acidic ecosystem.

In the sediments of the Río Tinto upper estuary (near the bridge in San Juan del Puerto), high organic matter, reducing conditions, and available SO_4^{2-} support the growth of SRB in the class *Deltaproteobacteria*, including *Desulfobacteraceae*, *Desulfobulbaceae*, and *Syntrophobacteraceae*. In this moderately acidic environment, SRB play a key role in the iron cycle through iron sequestration. The hydrogen sulfide they produce reacts with Fe(III) minerals to form secondary minerals, such as vivianite, mackinawite, pyrite, and greigite. In addition, SRB may directly use SO_4^{2-} from Fe(III) hydroxysulfates as electron acceptors, paralleling potential microbial SO_4^{2-} reduction in SO_4^{2-}-rich Martian environments (Abramov et al., 2022). In contrast, methanogens play a minor ecological role in Río Tinto. In SO_4^{2-}-rich environments, SRB outcompete methanogens for shared electron donors such as H_2 and acetate (Sánchez-Andrea et al., 2011).

In the subsurface, infiltrating water introduces dissolved O_2, which is rapidly consumed by pyrite-oxidizing microbes. This resulted in the formation of anoxic and acidic fluids enriched in SO_4^{2-}, Fe^{2+}, and Fe^{3+}. The oxidants (SO_4^{2-} and Fe^{3+}), in turn support the anaerobic microbial community, which includes both SO_4^{2-}- and Fe(III)-reducing microorganisms (Puente-Sánchez et al., 2014). Such microorganisms have been detected in subsurface samples, containing up to 634.5 ppm of SO_4^{2-}, collected during MARTE and IPBSL drilling projects (Amils et al., 2014, 2023). Analogous redox transitions and microbial processes may have occurred during several periods of Martian history, including the Early Noachian, the Late Noachian to Early Hesperian (Di Achille and Hynek, 2010), and the Late Hesperian to Early Amazonian (Molina et al., 2017), when repeated hydrological cycles allowed extensive water-rock interactions within the Martian crust. Therefore, the role of SRB in biosignature formation and the preservation of organic matter may be relevant for studying the potential life of early Mars (Valdez-Nuñez et al., 2024).

Organic carbon-driven SO_4^{2-} reduction

At the Río Tinto, a variety of metabolic intermediates produced by fermenting bacteria (e.g., propionate) serves as a key substrate for syntrophic bacteria such as *Syntrophobacter* spp., which further metabolize it to produce H_2 and acetate, two essential electron donors for sulfate reducers (Sánchez-Andrea et al., 2011). Oxidation of organic acids presents additional challenges. Within the cytoplasm, a higher pH environment causes these acids to dissociate, release protons, and lower the internal pH. However, some acidophilic SRB are capable of completely oxidizing organic acids to CO_2 at pH 3.9 and 5.0 (Sánchez-Andrea et al., 2022). In addition to simple organic compounds, SRB can degrade complex organic polymers (Valdez-Nuñez et al., 2024).

Microcosm experiments conducted with Río Tinto sediments from both the headwaters (Marismilla) and estuary (Moguer) have provided direct evidence of SO_4^{2-}-reducing activity under extreme conditions. These experiments were carried out at 30 °C using an artificial SO_4^{2-}-rich solution, with natural organic matter serving as the electron donor, confirming active SO_4^{2-} reduction in this Fe-rich, acidic environment (Velasco et al., 2010). In addition, heterotrophic SRB have been successfully enriched at pH 4 using alternative organic substrates, such as glycerol and methanol (Sánchez-Andrea et al., 2013).

Hydrogen-driven SO_4^{2-} reduction

As previously discussed, H_2 is a key product of fermentative metabolism and serves as a vital electron donor for SRB in acidic ecosystems such as the Río Tinto (Sánchez-Andrea et al., 2011). Hydrogen oxidation generates protons that must enter the cell for ATP production via ATP synthase, exacerbating proton stress under low pH conditions. However, acidophiles have evolved mechanisms to cope with the pH stress. Therefore, despite the potential challenges to ATP production that SRB are likely to face during electron transport phosphorylation when using as an electron donor, they are able to maintain energy production (Valdez-Nuñez et al., 2024). Therefore, acidotolerant H_2-oxidizing SRB have been enriched from sediments of the JL and SN dam systems, emphasizing their role in sustaining SO_4^{2-} reduction under low-pH, Fe-rich conditions (Sánchez-Andrea et al., 2013).

Methane-driven SO_4^{2-} reduction

Methane was detected both in the Río Tinto sediments and in the deep subsurface of the Peña de Hierro mining complex at 140 mbs (project MARTE). In shallower layers, CH_4 can be actively consumed by methanotrophic microorganisms, including those capable of forming syntrophic associations with SRB. In particular, close spatial associations between anaerobic methanotrophic archaea (ANMEs) and

SRB were observed at depths shallower than 149 mbs. These results suggest that anaerobic oxidation of methane (AOM) coupled with SO_4^{2-} reduction may serve as a significant CH_4 sink in these environments (Puente-Sánchez et al., 2014).

Preservation of geomicrobial biosignatures in the Río Tinto system

The search for extra-terrestrial life begins with the fundamental challenge of distinguishing biological signatures from abiotic features. These biosignatures cover a broad spectrum, including cellular and extracellular morphologies, biogenic rock fabrics, molecular structures (e.g., DNA, proteins, and lipids), chiral distributions, biogenic minerals formed by biomineralization, stable isotope fractionation in minerals and organic compounds, atmospheric gases, remotely detectable surface features, and temporal changes in planetary characteristics (Division on Engineering, Physical Sciences, Space Studies Board and Committee on Astrobiology Science Strategy for the Search for Life in the Universe, 2019; Shen et al., 2022).

The Río Tinto environment serves as a compelling terrestrial analog for Mars, particularly in its ability to preserve biosignatures formed under extreme geochemical conditions. The geological history of the river is preserved in diagenetically stabilized Fe(III) oxide/hydroxide terraces, ranging in age from 2.1 million to 10 thousand years. The subsurface mineral record at the Río Tinto reflects millions of years of Fe- and S-based chemolithotrophic microbial activity, providing valuable insights into ancient microbial processes (Fernández-Remolar and Knoll, 2008; Fernández-Remolar et al., 2008, 2018, 2021b). The acidic waters of the Río Tinto are often saturated with Fe compounds, such as Fe(III) oxyhydroxides (e.g., ferrihydrite) and Fe(III) hydroxysulfates (e.g., schwertmannite and jarosite), which facilitate the passive burial of microbial cells and microbial communities (e.g., biofilms and flocks), along with plant and invertebrate structures (Fernández-Remolar and Knoll, 2008; Amils et al., 2024). These Fe minerals fill voids, encapsulate microbial morphologies such as cell encrustation, act as protective cements, effectively mineralize cells, and protect them from abiotic and biotic degradation (Amils et al., 2024).

Biomineralization further enhances the preservation of biosignatures. Microorganisms in the Río Tinto system promote the localized precipitation of sulfate salts and Fe(III) oxyhydroxides both on their surfaces and within biofilms (Fernández-Remolar and Knoll, 2008; Amils et al., 2024). This microbial-induced mineralization results in the encrustation of microbial filaments and mat structures, promoting early stage mineral stabilization and the long-term preservation of geomicrobial features (Amils et al., 2024).

Although the preservation of organic matter in Río Tinto sediments is generally poor because of the oxidative and acidic conditions in the water column, various morphological and textural biosignatures remain detectable. These include

Fe-coated microbial streamers and microfossils embedded within Fe(III) oxides (e.g., goethite and hematite), which are considered high-priority targets for astrobiological exploration (Fernández-Remolar et al., 2005; Amils et al., 2014).

Biomineralization and Fe-based biosignatures

Biogenic mineralization plays a critical role in preserving microbial cells and their biosignatures over geological timescales (Amils et al., 2024). Contrary to the hypothesis that mineralization prevents microbial survival, rapid mineralization processes may create a protective environment that enhances long-term preservation of biological structures. These processes are particularly relevant in the context of early Mars, where similar mineralizing conditions may have facilitated the preservation of biosignatures, aiding their potential detection and analysis by future astrobiological missions (Price et al., 2018).

Iron(III) hydroxysulfate minerals, such as jarosite and schwertmannite, and Fe (III) (hydr)oxides (e.g., ferrihydrite, goethite, and hematite) are known to encrust organic materials, including plant debris (e.g., sticks and leaves), and microbial life forms such as filamentous fungi, bacteria, and algae. This encrustation process produces stable, long-lived biosignatures that can persist in the sedimentary record over geological timescales (Fernández-Remolar et al., 2005). These mineralized biological structures provide critical evidence of microbial activity in ancient environments.

In the Río Tinto system, microaerophilic Fe(II)-oxidizing microorganisms that inhabit moderately acidic headwaters produce distinctive extracellular stalks. These stalks, composed of organic fibers coated with Fe(III) oxides, were formed during the oxidation of Fe(II). In addition, NRFeOx microorganisms isolated from the same environment can encrust the cell walls with Fe(III) oxides. Such encrustation processes occur at high Fe(II) concentrations, often in the millimolar range, and are thought to have been prevalent in a variety of early Martian environments including hydrothermal systems, stratified lakes, and evaporitic basins formed during surface desiccation (Price et al., 2018).

Compelling evidence for long-term fossil preservation is provided by studies of a 10 m thick ferruginous outcrop located in the old terrace deposits of the Alto de la Mesa. These deposits confirm that mineral precipitation processes in the Río Tinto preserved biosignatures over time. However, hematite formation has obliterated many fossil fabrics and makes it difficult to identify these fabrics using standard optical and SEM techniques (Preston et al., 2011). Nevertheless, analytical techniques such as SEM-EDS and gas chromatography-mass spectrometry (GC-MS) can still help determine the biological origin of mineral textures in Río Tinto sediments (Souza-Egipsy et al., 2006).

Thus, if microbial cells encrusted with Fe(III) (hydr)oxides and hydroxysulfates are deposited in sediments and subsequently lithified, they could form visible macroscopic features such as flocs or laminated bands in an otherwise reducing matrix. These features may be preserved in sedimentary outcrops and may be accessible by erosion or mechanical excavation, thus providing ideal targets for astrobiological exploration on Mars (Price et al., 2018).

Iron minerals as records of ancient microbial activity on Earth and Mars

Primary sedimentary processes in the acidic river and groundwater of the Río Tinto system result in seasonal precipitation and mineralization of Fe compounds, primarily in the form of Fe(III) (hydr)oxides and hydroxysulfates. These include the nanophase goethite and jarosite. During the summer months, these compounds undergo recrystallization and partial transformation into goethite (Abramov et al., 2020). As diagenesis progressed over time, the goethite content of the sedimentary rocks increased, while continued SO_4^{2-} loss and mineral dehydration promoted the formation of hematite (Fig. 14.4).

Another scenario is that microorganisms can colonize deposits or penetrate seasonal micro-laminations and become embedded in these mineral phases. During the later stages of diagenesis, this matrix evolves into a composite of hematite and goethite, often retaining residual traces of sulfates. These processes result in the formation of microbial communities within fine-grained or cryptocrystalline matrices of Fe(III) minerals.

A representative example is found in the Fe(III) (hydr)oxide deposits of the Río Tinto Upper Terrace at the Alto de La Mesa. Rock samples from this site contained a cryptogranular matrix composed of Fe minerals (mainly goethite and hematite) with a high density of voids and fenestrated structures intersected by filamentous textures. These morphological features suggest a high input of plant detritus and other biological materials that are subsequently decomposed by fungi and bacteria. Such textures provide compelling evidence for the potential preservation of biosignatures through biologically mediated mineralization.

Advanced imaging and spectroscopic techniques have improved the understanding of these biosignatures. Spatial mapping of morphotypes using ptychographic X-ray computed laminography and elemental analysis using nano-X-ray fluorescence allows high-resolution visualization of microbial-mineral interactions. These techniques also help to elucidate fossilization processes over geological time scales (Maldanis et al., 2024).

In addition, ToF-SIMS analyses of subsurface rock samples from a depth of 40 mbs at the Peña de Hierro mining complex (part of the MARTE project) revealed the presence of inorganic complexes containing (PO_3^-, $SO_{x(2\text{ to }4)}^-$, $NO_{x(2,3)}^-$, $FeO_{x(1,2)}^-$, SiO_2^-, and Cl^-). The distribution of these ions was correlated with a number of organic molecules tentatively identified as saturated and monounsaturated fatty acids, polyunsaturated fatty acids, saccharides, phospholipids, sphingolipids, and possible peptide fragments. These molecular patterns help distinguish microbial structures involved in elemental cycling, such as branched macrofilaments, ovoid nodules with PO_4 lipids, mineralized nodules of chemoheterotrophic bacteria, and microfilaments of nitrogenous bacteria (Fernández-Remolar et al., 2021b). Taken together, these findings emphasize the strong potential of iron mineralization to preserve biosignatures, both morphologically and molecularly.

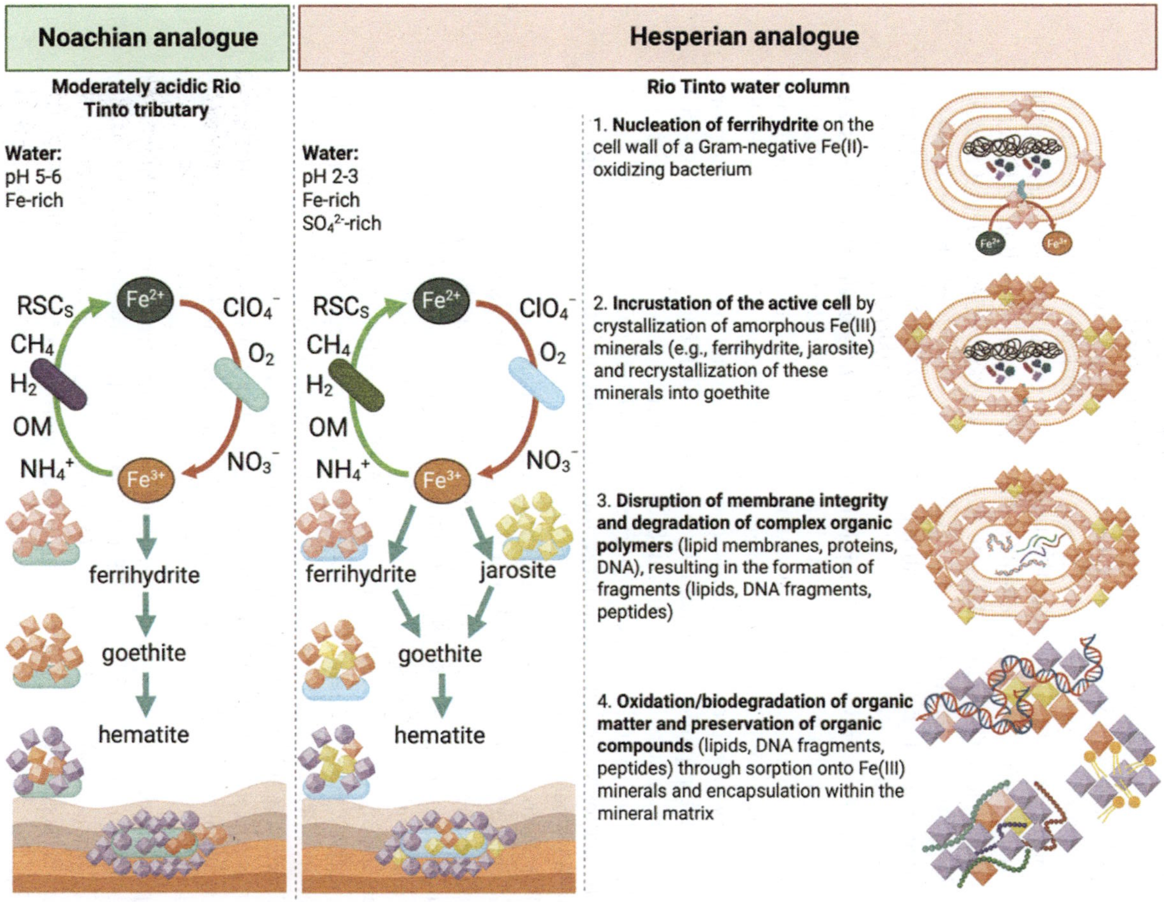

FIGURE 14.4 Iron Cycling and Biomolecule Preservation in the Río Tinto System.

Simplified iron biogeochemical cycles at the Río Tinto sites (moderately acidic tributary and acidic water of the Río Tinto) simulating the Noachian and Hesperian periods on Mars, together with a schematic representation of the preservation potential of key biomolecules (DNA, lipids, amino acids) in acidic Fe- and S-rich environments analogous to Hesperian conditions.

- Created in BioRender. Abramov, S. (2025) https://BioRender.com/w379irp.

Sulfur minerals as potential biosignatures on Earth and Mars

As previously noted, sulfide and sulfate minerals are widespread on the Martian surface. Although it should be taken into account that sulfide minerals may undergo degradation through prolonged surface oxidation (Davila et al., 2008; Nixon et al., 2013; Macey et al., 2020). Under acidic conditions, SRB can promote the precipitation of biominerals such as metal sulfides and Al hydroxides or hydroxysulfates. These biominerals, especially those closely associated with cell-derived organic carbon, can serve as valuable biosignatures. In acidic pit lakes, sulfide production by acidophilic SRB has been shown to results in the formation of minerals such as wurtzite (hexagonal ZnS), digenite ($Cu_{1.8}S$), djurleite ($Cu_{1.96}S$), and chalcocite (Cu_2S). Studies on neutrophilic SRB have also shown that biogenic metal sulfides often exhibit higher crystallinity than their abiogenic counterparts (Valdez-Nuñez et al., 2024).

In addition, the development and application of geochemical proxies are essential for identifying potential signs of early life on Mars, particularly in the context of future missions aimed at returning Martian samples to Earth. Among the most promising of these proxies are SO_4^{2-} minerals, particularly jarosite, because of their ability to preserve the biosignatures associated with sulfur-metabolizing microorganisms. Both neutrophilic and potentially acidophilic SRB can produce substantial sulfur isotope fractionation ($\delta^{34}S$) during SO_4^{2-} reduction, with sulfide depleted by up to 66‰ relative to SO_4^{2-}. When preserved in coexisting sulfide and SO_4^{2-} minerals in the geological record, such large fractionations are considered strong evidence for biosignatures (Valdez-Nuñez et al., 2024).

The Río Tinto Estuary near Moguer has revealed extreme sulfur isotope (^{34}S) fractionation, with an inverse correlation between SO_4^{2-} reduction rates and isotope fractionation. These values exceeded the 46‰ threshold predicted by the Rees model, suggesting the possibility of additional factors or processes operating in specific environments (Velasco et al., 2010).

Understanding these anomalous isotopic patterns is critical not only for reconstructing Earth's biogeochemical history, but also for interpreting potential biosignatures in extraterrestrial environments. The presence of jarosite in the estuary, a site of extreme sulfur isotope fractionation, suggests that analogous processes may have occurred on Mars, raising the possibility that similar isotopic signatures in Martian jarosite could serve as indicators of ancient microbial life (Velasco et al., 2010).

In support of this idea, stable isotope analyses of pyrite from terrestrial borehole sediments have revealed significant ^{34}S fractionation at various depths, consistent with SO_4^{2-}-reducing microbial activity. In addition, ^{13}C fractionation in the same samples provides further evidence of subsurface microbial processes (Amils et al., 2014). Together, sulfur and carbon isotope signatures provide a robust suite of geochemical biosignatures for detecting both past and present life, whether on Earth or Mars.

Exploring carbonate minerals as potential biosignatures in terrestrial and Martian environments

Carbonate minerals discovered in ancient Martian rocks have often been interpreted as indicators of past pH-neutral water conditions that are favorable to carbonate precipitation. However, carbonate minerals are scarce on the Martian surface. This is important because, under warm and wet conditions with a basaltic crust and a CO_2-rich atmosphere, carbonate formation is typically expected. One possible explanation for their low abundance is that acidic surface conditions, potentially caused by the dissolution of SO_2 in water, may inhibit carbonate precipitation (Wordsworth, 2016).

Significant neutralization of Río Tinto waters, creating favorable conditions for carbonate precipitation, occurs primarily in areas where sediments are rich in organic matter and Fe(III), or at confluence with pH-neutral tributaries. However, the identification of carbonates in both surface and subsurface samples at the Río Tinto challenges this assumption (Fernández-Remolar et al., 2018). These results show that carbonate minerals can be formed even under acidic conditions, particularly by microbially mediated processes that are not in equilibrium with the surrounding environment and could therefore serve as potential biosignatures (Valdez-Nuñez et al., 2024).

At the Río Tinto, micron-scale carbonate precipitates, such as siderite ($FeCO_3$), are commonly associated with the interface of bacterial biofilms, mineralized subsurface colonies, and biogenic microstructures such as globules, platelets, and dumbbell-like formations. However, the formation of siderite in the acidic Río Tinto system is unexpected based on the prevailing water geochemistry (Valdez-Nuñez et al., 2024). These carbonate structures are likely to be derived from microbial metabolism, in which the reduction of Fe(III) is coupled with the oxidation of organic matter by chemoheterotrophs, such as *Acidiphillium*. This activity creates localized circumneutral microenvironments at the scale of microbial colonies or even single cells despite the persistently low pH of the surrounding environment (Fig. 14.5).

Acidophilic SRB can also create microenvironments with pH, CO_2 activity, and H_2S concentrations that differ significantly from those of the bulk solution, facilitating siderite formation (Valdez-Nuñez et al., 2024). Over millions of years, siderite has precipitated through the activity of acidophilic Fe(III) reducers, and SRB can be preserved as a potential biosignature.

For example, SEM-EDS of rock samples from four boreholes at the Peña de Hierro mining complex (166–610 mbs; MARTE and IPBSL projects) revealed the microbial structures. These include hydrothermal phases, such as pyrite and barite ($BaSO_4$; dated to ~345 Ma), as well as biologically induced minerals less than 7 million years old. Hydrothermal minerals, including pyrite and barite, are covered by bacterial or filamentous microbial structures and secondary Fe carbonates such as siderite, with laminar to spherical shapes. Siderite spheroids were observed as

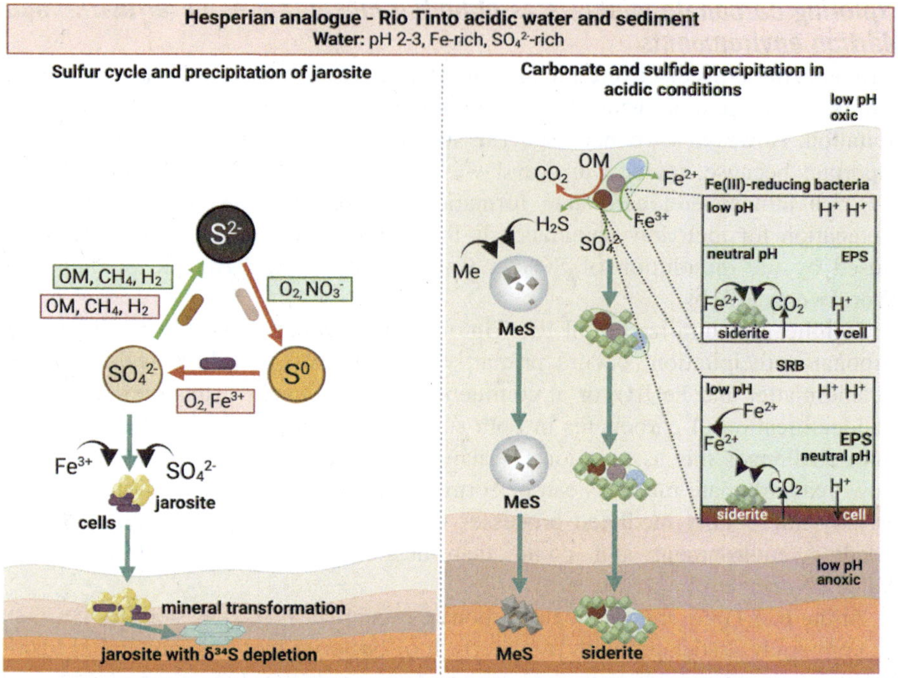

FIGURE 14.5 Sulfur Cycling and Mineral Formation at Sediment of the Río Tinto.

Simplified sulfur biogeochemical cycle at the Río Tinto environments simulating the Hesperian period on Mars, together with a schematic representation of formation of metal sulfides (MeS) and carbonate minerals in acidic sediment of the Río Tinto.

- Created in BioRender. Abramov, S. (2025) https://BioRender.com/2jfars4.

complex aggregates, forming rosette structures that completely enclosed the pyrite crystals. The co-occurrence of spheroidal siderite and pyrite, together with obvious signs of pyrite corrosion, suggests that pyrite degradation may result from the oxidation of organic matter coupled with SO_4^{2-} reduction. In this context, siderite can be formed anaerobically by the release of Fe^{2+} during pyrite oxidation or degradation and the concomitant heterotrophic production of HCO_3^- from the microbial oxidation of organic compounds (e.g., oxalates), with sulfate from barite acting as an electron acceptor for SRB. In addition, HCO_3^- can be released by Fe(III)-reducing chemoheterotrophs, such as *Acidiphillium*.

In this model, the association between sulfur-bearing minerals, dissolution scars, and secondary minerals of biological origin is a complex process involving microbial attack on mineral surfaces by SRB, followed by precipitation of Fe-rich carbonates. Thus, Fe sulfide compounds, such as pyrite, act as electron donors during microbial oxidation, while SO_4^{2-} minerals, such as barite, act as electron acceptors through SO_4^{2-} reduction. The formation of siderite would have resulted from carbonate biomineralization by SRB, Fe(III)-reducing chemoheterotrophs, or other microorganisms that concentrate carbonate through metabolic pathways (Fernández-Remolar et al., 2018).

This finding suggests that if microbial life existed on Mars, it could have facilitated carbonate precipitation even in acidic habitats by creating similarly localized circumneutral niches. The long-term preservation of carbonates at the Río Tinto also supports the possibility that analogous carbonate deposits on Mars could have been preserved over geological timescales (Fernández-Remolar et al., 2012).

Preservation of organic biosignatures in Fe-rich sediments

Over thousands to millions of years, diagenetic processes have gradually transformed primary minerals such as jarosite and ferrihydrite into more stable Fe(III) oxides such as goethite and hematite (Amils et al., 2024). Despite this mineralogical evolution, traces of sulfates and organic compounds often persist, suggesting that the transformation is incomplete. This partial alteration has played a crucial role in the preservation of organic compounds and biomolecules over geological timescales (Fernández-Remolar and Knoll, 2008; Amils et al., 2024). In particular, evidence of exceptional preservation in the ancient acidic deposits of the Río Tinto includes the discovery of peptide sequences in terrace deposits over 2 million years of age (Amils et al., 2024).

However, the preservation of organic matter in Fe(III)-rich sediments presents significant challenges because of the inherently oxidizing conditions (Tan et al., 2018). Such acidic Fe(III)-rich conditions, known for their strong oxidative capacity, have also been documented in the subsurface environment (Davila et al., 2008). Similarly, on Mars, subsurface migration of groundwater interacting with VMS deposits would have mobilized acidic and oxidizing fluids through large parts of the crust, resulting in widespread deposition of sulfates and Fe(III) oxides. This groundwater could have leached significant volumes of aquifer material and crustal rock, potentially erasing any organic compounds to depths of hundreds of meters. Therefore, the preservation of organic biosignatures must have been severely limited in parts of the ancient Martian crust that were exposed to aqueous processes (Davila et al., 2008).

Nevertheless, there is compelling evidence that molecular organic compounds can persist in Fe-bearing sediments in both recent and ancient deposits worldwide, including those at the Río Tinto. For example, up to 21% of the total organic carbon in some sediments is bound to Fe mineral phases in excess of what would be expected based on typical rates of biochemical degradation (Tan et al., 2018). A high total organic carbon (TOC) content, indicating relatively good preservation of the organic fraction, has been observed in several Río Tinto terraces representing different depositional episodes ranging from the late Pliocene to the Holocene, as well as in a regionally developed Gossan dating to the end of the Miocene. This preserved organic matter is associated with organic matter sorbed onto Fe(III) oxides, organo-Fe precipitates, and/or ternary complexes involving Fe(III) oxides, organic matter, and clay minerals (Sánchez-García et al., 2017). This suggests that the unique geochemical environment at Río Tinto facilitates molecular stabilization by mitigating chemical and biological degradation pathways (Amils et al., 2024).

The exceptional preservation potential at Río Tinto is primarily attributed to mineral encapsulation, where Fe minerals encapsulate organic matter and form diffusion barriers that protect it from oxidants. Encapsulation creates closed microenvironments that limit microbial activity and oxidative damage. In addition, high concentrations of metal ions inhibit microbial metabolism, whereas low water activity limits hydrolysis and enzymatic degradation. These combined effects prevent conventional taphonomic degradation pathways, allowing Fe minerals to protect organic compounds and biomolecules from both physical and chemical transformations over long geological timescales (Fig. 14.4) (Fernández-Remolar et al., 2008; Amils et al., 2024).

Fatty acids and lipids as molecular fossils

In contrast to thermophiles and halophiles, acidophiles have no distinct lipid biomarker. However, in polyextreme Fe sulfate systems, different mineral matrices may preserve lipids to different extents, potentially aiding the detection of past life in such environments (Fig. 14.4). Lipid biomarkers, particularly hydrocarbon cores of cell membranes, are highly resistant to degradation and are often preserved in sedimentary deposits. The oldest known biomarkers on Earth, lipids such as hopanoids and isoprenoids, have been detected in 1.64–1.73 billion year old Australian rocks (Finkel et al., 2023).

Research on Fe-rich environments, such as hot springs, suggests that Fe(III) oxide encrustation may protect lipids from enzymatic degradation by heterotrophs, thereby increasing their long-term stability. If life existed on Early Mars, its organic record may include preserved microbial lipids, making it an important reservoir of fossil carbon. Lipids, particularly alkanoic (fatty) acids, are useful for distinguishing between the different domains of life. The highest concentrations of lipids were found in the goethite layers, indicating high water-rock ratios and a greater potential for habitability. Goethite recrystallizes into hematite, which is abundant in Mars. Martian missions should prioritize the detection of fatty acids or their diagenetic products in Fe(III) (hydr)oxides, especially in sulfur-rich environments (Tan et al., 2018).

Studies from the Río Tinto system have shown that both microbial diversity and lipid complexity increase with depth, probably because of reduced surface hydrological degradation and less extreme acidity. Identifying lipid biomarkers in such environments requires a multianalyte approach to achieve greater biosource specificity. For example, a 2020 drilling mission at the Río Tinto combined 16S rRNA gene sequencing, immunoassays, and molecular lipid analysis to characterize the subsurface biosphere (Finkel et al., 2023).

In an evaporitic esplanade near the Peña de Hierro mining complex, a 2 m depth profile revealed a higher proportion of odd-numbered long-chain *n*-alkanes compared to even-numbered chains down to 1 m depth. This finding is consistent with the presence of increasing concentrations of iso- and anteiso-branched C15 and C17 fatty acids, indicating the presence of SRB (Sánchez-García et al., 2020). In addition, unsaturated fatty acids associated with cyanobacteria, such as 18:1ω9 and 18:2ω6, were found predominantly at depth, suggesting colonization by subsurface microniches with higher pH values (>4), as confirmed by RNA sequencing and immunoassays (Cirés et al., 2017; Sánchez-García et al., 2020; Finkel et al., 2023).

Other monounsaturated fatty acids, such as 18:1ω8, were attributed to methanotrophs, whereas dicarboxylic acids (C6–C15) were linked to thermophiles. Phytosterols such as β-sitosterol and stigmastanol have been identified as inputs from land plants (Sánchez-García et al., 2020; Finkel et al., 2023).

In addition, microbial structures and molecular biosignatures have been exceptionally well preserved in the Río Tinto Gossan deposit, which consists of ancient Fe(III) oxide and sulfate minerals dating back to approximately 25 million years. Fatty acids ranging from C11 to C30 were identified using ToF-SIMS, and their distribution indicated a contribution from multiple biological sources, including microbial communities and higher plants (Fernández-Remolar et al., 2021a).

Additionally, different glycerolipids, monoacylglycerides, diacylglycerides, and glycerophospholipids, were detected in the same samples. The diglyceride structures were further confirmed by the detection of carbonyl-bearing fragment ions (Fernández-Remolar et al., 2021a).

Peptides as indicators of microbial life

Proteins are of great taxonomic value, however, their structural instability limits their long-term preservation. Proteins have been shown to persist for up to 3.8 million years. In contrast, amino acids are relatively stable in subsurface environments, and can persist for billions of years. However, their extensive degradation over time complicates the assessment of their biogenic origins (Finkel et al., 2023). Notably, short-chain peptides have been identified in the oldest terrace of ferruginous sedimentary deposits in the Río Tinto, which is over 2 million years old. These peptides were produced under the acidic and oxidizing conditions of the Río Tinto, providing compelling evidence for the long-term stability of biological molecules in such extreme environments (Colín-García et al., 2011). The preserved peptide fragments, attributable to specific organisms from the ancient paleo-fluvial system, demonstrate that identifiable biomolecules can persist over geological timescales (Fig. 14.4).

Further evidence of exceptional biomolecular preservation comes from the Río Tinto Gossan deposit (Fernández-Remolar et al., 2021a). This deposit contained well-preserved microbial structures and biosignatures. Advanced analytical techniques, such as ToF-SIMS, have revealed filamentous microbial morphologies encrusted by jarosite as well as the co-occurrence of fatty acids and peptides.

Gossan was formed as a result of a complex chemolithotrophic alteration of pyrite, involving both direct and indirect microbial processes. Such complex diagenesis, involving an interplay between dramatic pH and redox changes, hydrolysis, and dehydration, should have prevented the preservation of molecular compounds of biological origin. Nevertheless, the presence of diverse and abundant biomolecules, including peptides, in Gossan indicates that alternative preservation mechanisms have played a crucial role in the fossilization of microbial organic matter in the subsurface of the Peña de Hierro mining complex.

DNA as a marker of recent microbial activity

Like proteins, DNA has important taxonomic value, but is unlikely to be preserved over long geological timescales. DNA has been shown to persist for up to one million years. These timeframes are relatively short in the context of Martian surface conditions, where oxidative soils and intense UV radiation prevail for billions of years, greatly reducing the likelihood of DNA preservation (Fig. 14.4) (Finkel et al., 2023).

Fourier Transform Infrared (FTIR) spectroscopy has proven effective in identifying the organic components in the Río Tinto samples. For example, spectra from the lower unit of the Las Zarandas Holocene terrace (maximum age of ~11,000 years), composed of orange/red crystalline masses with visible filamentous structures, show numerous absorption bands influenced by minerals such as jarosite and goethite. In addition, some of these bands may correspond to ribose-phosphate backbone vibrations, which are possibly indicative of DNA.

Additional FTIR analysis of an intermediate-age terrace near the Huerto del Loco, composed mainly of goethite with hematite and minor quartz grains, identified absorption bands potentially attributable to ribose in DNA or RNA. These samples were collected from areas showing morphological evidence of filamentous bacteria (Preston et al., 2011).

Consequently, genetic biosignatures (i.e., DNA sequences) may not be the primary means for detecting traces of life on Mars. Instead, DNA may serve as a biomarker of current or recent microbial activity, for example in the form of environmental DNA (eDNA) (Sánchez-García et al., 2020).

Other organic compounds and their detection at the Río Tinto

In addition, aliphatic hydrocarbon vibrational modes and signatures of organic matter were detected via FTIR spectroscopy in samples from both the young terrace at Las Zarandas and the intermediate-age terrace near the Huerto del Loco. Similar signatures were also found in the oldest deposits at the Alto de la Mesa outcrop, dated to the late Pliocene (~2.0 ± 0.1 Ma). These highly stratified sediments consist mainly of goethite with minor hematite and quartz and contain preserved debris of rocks, minerals, insects, and plants (Preston et al., 2011).

ToF-SIMS analyses of Fe-bearing deposits from the Upper Gossan unit in the Peña de Hierro mining complex revealed a diverse range of organic molecules. These include n-alkanes, straight-chain hydrocarbons with low unsaturation, branched hydrocarbons, isoprenoids, polycyclic aromatic hydrocarbons (PAHs), steroids, hopanoids, polyphenols, lignin derivatives, phytochemicals, and potentially larger biomolecular fragments, such as polysaccharides or polyketides (Fernández-Remolar et al., 2021a). Raman spectroscopy has been used to identify carotenoids, scytonemin, and mycosporine-like amino acids in samples of efflorescent mineral deposits and epiliths collected along the edge of water at several sites in the Jaroso Ravine (Edwards et al., 2007).

The presence of large molecular species (e.g., peptides and lipids) indicates that acidic ferruginous materials have a significant preservation potential. In this regard, preservation is favored in the acidic deposits of the Río Tinto, despite the high water availability for mineral transformation. In contrast, the absence of such preservation in sediments formed under low-pH conditions on Mars with limited water availability suggests that high preservation should be expected in Hesperian materials. This supports the hypothesis that ancient Hesperian ferruginous terrains on Mars formed under acidic and oxidizing conditions are prime targets in the search for ancient biosignatures, assuming that life once existed on the planet (Fernández-Remolar et al., 2021a).

Conclusion

Studying extreme environments such as the Río Tinto provides essential insights into how microbial life can persist under geochemical and physical conditions that are hostile to life. This improves our understanding of Martian past and potential habitability. Mars were once potentially home to stable bodies of liquid water and chemically diverse environments, but underwent a transition toward increasing acidity, oxidation, aridity, and decreasing temperature. These changes likely led to the extinction of life or its possible escape into the subsurface, making terrestrial analogs critical for guiding the search for biosignatures.

The Río Tinto, with its acidic, Fe- and SO_4^{2-}-rich waters, mineral diversity, and stratified redox gradients, mimics Martian environments ranging from the Noachian to Amazonian. Its mineral assemblages, including jarosite, goethite, and hematite, closely resemble those observed on Mars, particularly in areas such as Meridiani Planum and Gale Crater. Crucially, the Río Tinto system not only supports diverse extremophilic microbial life, but also demonstrates how biosignatures can be preserved under Mars-like conditions. These include macroscopic microbial mats, microscopic cellular structures (e.g., biofilms and single cells), and organic molecules (e.g., DNA, lipids, and peptides) encapsulated in Fe-rich minerals, despite the overall poor preservation of organic material.

Furthermore, the active iron and sulfur biogeochemical cycles in the deep subsurface of the Iberian Pyrite Belt highlight the potential of similar subsurface ecosystems on Mars, especially during later colder epochs. The preservation of biosignatures in ancient Río Tinto deposits despite mineralogical transformation suggests that such evidence could persist over geological timescales on Mars.

Therefore, the Río Tinto serves as a robust terrestrial analog for ancient and modern Martian environments, supporting the idea that if life existed on Mars, it may have left detectable biosignatures in Fe- and S-rich deposits. A study of the biogeochemical cycles in the Río Tinto system not only enhances our understanding of the adaptability of life on Earth but also provides strategic information for mission planning, biosignature detection, and interpretation of Martian astrobiological data.

References

Abramov, S.M., Straub, D., Tejada, J., Grimm, L., Schädler, F., Bulaev, A., Thorwarth, H., Amils, R., Kappler, A., Kleindienst, S., 2022. Biogeochemical niches of Fe-cycling communities influencing heavy metal transport along the Rio Tinto, Spain. Appl. Environ. Microbiol. 88 (4). https://doi.org/10.1128/aem.02290-21.

Abramov, S.M., Tejada, J., Grimm, L., Schädler, F., Bulaev, Tomaszewski, A., et al., 2020. Role of biogenic Fe (III) minerals as a sink and carrier of heavy metals in the Rio Tinto. Sci. Total Environ. 718, 137294.

Amils, R., Escudero, C., Oggerin, M., Puente Sánchez, F., Arce Rodríguez, A., Fernández Remolar, D., Rodríguez, N., García Villadangos, M., Sanz, J.L., Briones, C., Sánchez-Román, M., Gómez, F., Leandro, T., Moreno-Paz, M., Prieto-Ballesteros, O., Molina, A., Tornos, F., Sánchez-Andrea, I., Timmis, K., Pieper, D.H., Parro, V., 2023. Coupled C, H, N, S and Fe biogeochemical cycles operating in the continental deep subsurface of the Iberian Pyrite Belt. Environ. Microbiol. 25 (2), 428–453. https://doi.org/10.1111/1462-2920.16291.

Amils, R., Escudero, C., Huang, T., Fernádez-Remolar, D., 2024. Geomicrobiology of Río Tinto (Iberian Pyrite Belt): a geological and mineralogical Mars analogue. In: Geomicrobiology: Natural and Anthropogenic Settings. Springer Nature, Spain, pp. 123–150. https://doi.org/10.1007/978-3-031-54306-7_7.

Amils, R., González-Toril, E., Fernández-Remolar, D., Gómez, F., Aguilera, Á., Rodríguez, N., Malki, M., García-Moyano, A., Fairén, A.G., de la Fuente, V., Sanz, J. L., 2007. Extreme environments as Mars terrestrial analogs: the Rio Tinto case. Planet. Space Sci. 55 (3), 370–381. https://doi.org/10.1016/j.pss.2006.02.006.

Amils, R., 2016. Lessons learned from thirty years of geomicrobiological studies of Río Tinto. Res. Microbiol. 167 (7), 539–545. https://doi.org/10.1016/j.resmic.2016.06.001.

Amils, R., Fernández-Remolar, D., 2020. Rio Tinto: An extreme acidic environmental model of astrobiological interest. Extremophiles as Astrobiological Models, pp. 21–44.

Amils, R., Fernández-Remolar, D., The IPBSL team, 2014. Río Tinto: a geochemical and mineralogical terrestrial analogue of mars. Life 4 (3), 511–534. https://doi.org/10.3390/life4030511.

Andrews-Hanna, J.C., Lewis, K.W., 2011. Early Mars hydrology: 2. Hydrological evolution in the Noachian and Hesperian epochs. J. Geophys. Res.: Planets 116 (2). https://doi.org/10.1029/2010JE003709.

Belilla, J., Moreira, D., Jardillier, L., Reboul, G., Karim, B., López-García, J.M., Bertolino, P., López-Archilla, A.I., 2019. Purificación López-García, Hyperdiverse archaea near life limits at the polyextreme geothermal Dallol area. Nat. Ecol. Evol. 3 (11), 1552–1561. https://doi.org/10.1038/s41559-019-1005-0.

Berry, B.J., Jenkins, D.G., Schuerger, A.C., 2010. Effects of simulated mars conditions on the survival and growth of *Escherichia coli* and *Serratia liquefaciens*. Appl. Environ. Microbiol. 76 (8), 2377–2386. https://doi.org/10.1128/AEM.02147-09.

Bibring, J.P., Langevin, Y., Mustard, J.F., Poulet, F., Arvidson, R., Gendrin, A., Gondet, B., Mangold, N., Pinet, P., Forget, F., Berthe, M., Gomez, C., Jouglet, D., Soufflot, A., Vincendon, M., Combes, M., Drossart, P., Encrenaz, T., Fouchet, T., Merchiorri, R., Belluci, G.C., Altieri, F., Formisano, V., Capaccioni, F., Cerroni, P., Coradini, A., Fonti, S., Korablev, O., Kottsov, V., Ignatiev, N., Moroz, V., Titov, D., Zasova, L., Loiseau, D., Pinet, P., Douté, S., Schmitt, B., Sotin, C., Hauber, E., Hoffmann, H., Jaumann, R., Keller, U., Arvidson, R., Duxbury, T., Forget, F., Neukum, G., 2006. Global mineralogical and aqueous Mars history derived from OMEGA/Mars express data. Science 312 (5772), 400–404. https://doi.org/10.1126/science.1122659.

Bishop, J.L., Parente, M., Weitz, C.M., Noe Dobrea, E.Z., Roach, L.H., Murchie, S.L., McGuire, P.C., McKeown, N.K., Rossi, C.M., Brown, A.J., Calvin, W.M., Milliken, R., Mustard, J.F., 2009. Mineralogy of Juventae Chasma: sulfates in the light-toned mounds, mafic minerals in the bedrock, and hydrated silica and hydroxylated ferric sulfate on the plateau. J. Geophys. Res.: Planets 114 (11). https://doi.org/10.1029/2009JE003352.

Bottaro, M, Abramov, S, Amils, R, Martinez-Bonilla, A, ThomasArrigo, L, Straub, D, et al., 2025a. Coupling of Fe and N cycles by nitrate-reducing Fe(II)-oxidizing microorganisms in the tidal sediments of an extreme acidic river (Río Tinto, Spain). Appl. Geochem. 190, 106488. https://doi.org/10.1016/j.apgeochem.2025.106488.

Bottaro, M, Abramov, S, Amils, R, Straub, D, Kühnel, S, Mollenkopf, M, et al., 2025b. Impact of pH and salinity fluctuations on oxidation of Fe(II) by nitrate-reducing microorganisms enriched from the reduced tidal sediment of an extreme acidic river (Río Tinto, Spain). FEMS Microbiol. Ecol. 101, fiaf083. https://doi.org/10.1093/femsec/fiaf083.

Carr, M.H., Head, J.W., 2010. Geologic history of Mars. Earth Planet Sci. Lett. 294 (3–4), 185–203. https://doi.org/10.1016/j.epsl.2009.06.042.

Cirés, S., Casero, M., Quesada, A., 2017. Toxicity at the edge of life: a review on cyanobacterial toxins from extreme environments. Mar. Drugs 15 (7), 233. https://doi.org/10.3390/md15070233.

Colín-García, M., Kanawati, B., Harir, M., Schmitt-Kopplin, P., Amils, R., Parro, V., García, M., Fernández-Remolar, D., 2011. Detection of peptidic sequences in the ancient acidic sediments of Río Tinto, Spain. Orig. Life Evol. Biosph. 41 (6), 523–527. https://doi.org/10.1007/s11084-011-9258-x.

Davila, A.F., Fairén, A.G., Gago-Duport, L., Stoker, C., Amils, R., Bonaccorsi, R., Zavaleta, J., Lim, D., Schulze-Makuch, D., McKay, C.P., 2008. Subsurface formation of oxidants on Mars and implications for the preservation of organic biosignatures. Earth Planet Sci. Lett. 272 (1–2), 456–463. https://doi.org/10.1016/j.epsl.2008.05.015.

Di Achille, G., Hynek, B.M., 2010. Ancient ocean on Mars supported by global distribution of deltas and valleys. Nat. Geosci. 3 (7), 459–463. https://doi.org/10.1038/ngeo891.

Division on Engineering, Physical Sciences, Space Studies Board and Committee on astrobiology science strategy for the search for life in the universe. In: An Astrobiology Strategy for the Search for Life in the Universe, 2019. National Academies Press.

Edwards, H.G.M., Vandenabeele, P., Jorge-Villar, S.E., Carter, E.A., Perez, F.R., Hargreaves, M.D., 2007. The Rio Tinto Mars Analogue site: an extremophilic Raman spectroscopic study. Spectrochim. Acta, Part A Mol. Biomol. Spectrosc. 68 (4), 1133–1137. https://doi.org/10.1016/j.saa.2006.12.080.

Eigenbrode, J.L., Summons, R.E., Steele, A., Freissinet, C., Millan, M., Navarro-González, R., Sutter, B., McAdam, A.C., Franz, H.B., Glavin, D.P., Archer, P.D.,

Mahaffy, P.R., Conrad, P.G., Hurowitz, J.A., Grotzinger, J.P., Gupta, S., Ming, D.W., Sumner, D.Y., Szopa, C., Malespin, C., Buch, A., Coll, P., 2018. Organic matter preserved in 3-billion-year-old mudstones at Gale crater, Mars. Science 360 (6393), 1096–1110. https://doi.org/10.1126/science.aas9185.

Escudero, C., del Campo, A., Ares, J.R., Sánchez, C., Martínez, J.M., Gómez, F., Amils, R., 2020. Visualizing microorganism-mineral interaction in the Iberian Pyrite Belt subsurface: the acidovorax case. Front. Microbiol. 11. https://doi.org/10.3389/fmicb.2020.572104.

Fairén, A.G., Fernández-Remolar, D., Dohm, J.M., Baker, V.R., Amils, R., 2004. Inhibition of carbonate synthesis in acidic oceans on early Mars. Nature 431 (7007), 423–426. https://doi.org/10.1038/nature02911.

Farrand, W.H., Glotch, T.D., Rice, J.W., Hurowitz, J.A., Swayze, G.A., 2009. Discovery of jarosite within the Mawrth Vallis region of Mars: implications for the geologic history of the region. Icarus 204 (2), 478–488. https://doi.org/10.1016/j.icarus.2009.07.014.

Farrand, W.H., Bell, J.F., Johnson, J.R., Jolliff, B.L., Knoll, A.H., McLennan, S.M., Squyres, S.W., Calvin, W.M., Grotzinger, J.P., Morris, R.V., Soderblom, J., Thompson, S.D., Watters, W.A., Yen, A.S., 2007. Visible and near-infrared multispectral analysis of rocks at Meridiani Planum, Mars, by the Mars exploration rover opportunity. J. Geophys. Res.: Planets 112 (6). https://doi.org/10.1029/2006JE002773.

Fernández-Remolar, D.C., Banerjee, N., Gómez-Ortiz, D., Izawa, M., Amils, R., 2018. A mineralogical archive of the biogeochemical sulfur cycle preserved in the subsurface of the Río Tinto system. Am. Mineral. 103 (3), 394–411. https://doi.org/10.2138/am-2018-6288.

Fernández-Remolar, D.C., Preston, L.J., Sánchez-Román, M., Izawa, M.R.M., Huang, L., Southam, G., Banerjee, N.R., Osinski, G.R., Flemming, R., Gómez-Ortíz, D., Prieto Ballesteros, O., Rodríguez, N., Amils, R., Darby Dyar, M., 2012. Carbonate precipitation under bulk acidic conditions as a potential biosignature for searching life on Mars. Earth Planet Sci. Lett. 351–352, 13–26. https://doi.org/10.1016/j.epsl.2012.07.015.

Fernandez-Remolar, D.C., Prieto-Ballesteros, O., Osburn, M.R., Gomez-Ortiz, D., Arvidson, R.E., Morris, R.V., Ming, D., Amils, R., Friendlander, L.R., 2007. Mars sulfate formation sourced in sulfide-enriched subsurface fluids: the Rio Tinto model. Lunar Planet. Sci. Conf. XXXVIII, 1580.

Fernández-Remolar, D.C., Gomez-Ortiz, D., Huang, T., Anglés, A., Shen, Y., Hu, Q., Amils, R., Rodríguez, N., Escudero, C., Banerjee, N.R., 2021b. The molecular record of metabolic activity in the subsurface of the Río Tinto Mars analog. Astrobiology 21 (11), 1387–1405. https://doi.org/10.1089/ast.2020.2431.

Fernández-Remolar, D.C., Morris, R.V., Gruener, J.E., Amils, R., Knoll, A.H., 2005. The Río Tinto Basin, Spain: mineralogy, sedimentary geobiology, and implications for interpretation of outcrop rocks at Meridiani Planum, Mars. Earth Planet Sci. Lett. 240 (1), 149–167. https://doi.org/10.1016/j.epsl.2005.09.043.

Fernández-Remolar, D.C., Prieto-Ballesteros, O., Rodríguez, N., Gómez, F., Amils, R., Gómez-Elvira, J., Stoker, C.R., 2008. Underground habitats in the Río Tinto basin: a model for subsurface life habitats on mars. Astrobiology 8 (5), 1023–1047. https://doi.org/10.1089/ast.2006.0104.

Fernández-Remolar, D.C., Carrizo, D., Harir, M., Huang, T., Amils, R., Schmitt-Kopplin, P., Sánchez-García, L., Gomez-Ortiz, D., Malmberg, P., 2021a. Unveiling microbial preservation under hyperacidic and oxidizing conditions in the Oligocene Rio Tinto deposit. Sci. Rep. 11 (1). https://doi.org/10.1038/s41598-021-00730-8.

Fernández-Remolar, D.C., Knoll, A.H., 2008. Fossilization potential of iron-bearing minerals in acidic environments of Rio Tinto, Spain: implications for Mars exploration. Icarus 194 (1), 72–85. https://doi.org/10.1016/j.icarus.2007.10.009.

Finkel, P.L., Carrizo, D., Parro, V., Sánchez-García, L., 2023. An overview of lipid biomarkers in terrestrial extreme environments with relevance for Mars exploration. Astrobiology 23 (5), 563–604. https://doi.org/10.1089/ast.2022.0083.

Fisk, M.R., Giovannoni, S.J., 1999. Sources of nutrients and energy for a deep biosphere on Mars. J. Geophys. Res.: Planets 104 (5), 11805–11815. https://doi.org/10.1029/1999JE900010.

Fowler, D., Coyle, M., Skiba, U., Sutton, M.A., Cape, J.N., Reis, S., Sheppard, L.J., Jenkins, A., Grizzetti, B., Galloway, J.N., Vitousek, P., Leach, A., Bouwman, A.F., Butterbach-Bahl, K., Dentener, F., Stevenson, D., Amann, M., Voss, M., 2013. The global nitrogen cycle in the twenty first century. Phil. Trans. Biol. Sci. 368 (1621). https://doi.org/10.1098/rstb.2013.0164.

Gil-Lozano, C., Mateo-Martí, E., Gago-Duport, L., Losa-Adams, E., Sampedro, M.F., Bishop, J.L., Chevrier, V., Fairén, A.G., 2025. Evaluating the reactivity of pyrite on Mars under current and ancient geochemical environments. Front. Astron. Space Sci. 12. https://doi.org/10.3389/fspas.2025.1504288.

Gómez, F., Berenguer, J., Gargaud, M., Amils, R., Quintanilla, J.C., Cleaves, H.J., Irvine, W. M., Pinti, D.L., 2011. Encyclopedia of Astrobiology. Springer Terrestrial Analog.

González-Toril, E., Llobet-Brossa, E., Casamayor, E.O., Amann, R., Amils, R., 2003. Microbial ecology of an extreme acidic environment, the Tinto River. Appl. Environ. Microbiol. 69 (8), 4853–4865. https://doi.org/10.1128/aem.69.8.4853-4865.2003.

Grotzinger, J.P., Sumner, D.Y., Kah, L.C., Stack, K., Gupta, S., Edgar, L., Rubin, D., Lewis, K., Schieber, J., Mangold, N., Milliken, R., Conrad, P.G., DesMarais, D., Farmer, J., Siebach, K., Calef, F., Hurowitz, J., McLennan, S.M., Ming, D., Vaniman, D., Crisp, J., Vasavada, A., Edgett, K.S., Malin, M., Blake, D., Gellert, R., Mahaffy, P., Wiens, R.C., Maurice, S., Grant, J.A., Wilson, S., Anderson, R.C., Beegle, L., Arvidson, R., Hallet, B., Sletten, R.S., Rice, M., Bell, J., Griffes, J., Ehlmann, B., Anderson, R.B., Bristow, T.F., Dietrich, W.E., Dromart, G., Eigenbrode, J., Fraeman, A., Hardgrove, C., Herkenhoff, K., Jandura, L., Kocurek, G., Lee, S., Leshin, L.A., Leveille, R., Limonadi, D., Maki, J., McCloskey, S., Meyer, M., Minitti, M., Newsom, H., Oehler, D., Okon, A., Palucis, M., Parker, T., Rowland, S., Schmidt, M., Squyres, S., Steele, A., Stolper, E., Summons, R., Treiman, A., Williams, R., Yingst, A., Science Team, M.S.L., Kemppinen, O., Bridges, N., Johnson, J.R., Cremers, D., Godber, A., Wadhwa, M., Wellington, D., McEwan, I., Newman, C., Richardson, M., Charpentier, A., Peret, L., King, P., Blank, J., Weigle, G., Li, S., Robertson, K., Sun, V., Baker, M., Edwards, C., Farley, K., Miller, H., Newcombe, M., Pilorget, C., Brunet, C., Hipkin, V., Léveillé, R., Marchand, G., Sobrón Sánchez, P., Favot, L., Cody, G., Flückiger, L., Lees, D., Nefian, A., Martin, M., Gailhanou, M., Westall, F., Israël, G., Agard, C., Baroukh, J., Donny, C., Gaboriaud, A., Guillemot, P., Lafaille, V., Lorigny, E., Paillet, A., Pérez, R., Saccoccio, M., Yana, C., Armiens-Aparicio, C., Caride Rodríguez, J., Carrasco Blázquez, I., Gómez, F.G., Gómez-Elvira, J., Hettrich, S., Malvitte, A.L., Jiménez, M. M., Martínez-Frías, J., Martín-Soler, J., Javier Martín-Torres, F., Jurado, A.M., Mora-Sotomayor, L., Caro, G.M., Navarro López, S., Peinado-González, V., Pla-García, J., Manfredi, J.A.R., Romeral-Planelló, J.J., Sans Fuentes, S.A., Martinez, E.S., Torres Redondo, J., Urqui-O'Callaghan, R., Mier, M.-P.Z., Chipera, S., Lacour, J.-L.,

Mauchien, P., Sirven, J.-B., Manning, H., Fairén, A., Hayes, A., Joseph, J., Sullivan, R., Thomas, P., Dupont, A., Lundberg, A., Melikechi, N., Mezzacappa, A., DeMarines, J., Grinspoon, D., Reitz, G., Prats, B., Atlaskin, E., Genzer, M., Harri, A.-M., Haukka, H., Kahanpää, H., Kauhanen, J., Paton, M., Polkko, J., Schmidt, W., Siili, T., Fabre, C., Wray, J., Wilhelm, M.B., Poitrasson, F., Patel, K., Gorevan, S., Indyk, S., Paulsen, G., Bish, D., Gondet, B., Langevin, Y., Geffroy, C., Baratoux, D., Berger, G., Cros, A., d'Uston, C., Forni, O., Gasnault, O., Lasue, J., Lee, Q.-M., Meslin, P.-Y., Pallier, E., Parot, Y., Pinet, P., Schröder, S., Toplis, M., Lewin, É., Brunner, W., Heydari, E., Achilles, C., Sutter, B., Cabane, M., Coscia, D., Szopa, C., Robert, F., Sautter, V., Le Mouélic, S., Nachon, M., Buch, A., Stalport, F., Coll, P., François, P., Raulin, F., Teinturier, S., Cameron, J., Clegg, S., Cousin, A., DeLapp, D., Dingler, R., Jackson, R. S., Johnstone, S., Lanza, N., Little, C., Nelson, T., Williams, R.B., Jones, A., Kirkland, L., Baker, B., Cantor, B., Caplinger, M., Davis, S., Duston, B., Fay, D., Harker, D., Herrera, P., Jensen, E., Kennedy, M.R., Krezoski, G., Krysak, D., Lipkaman, L., McCartney, E., McNair, S., Nixon, B., Posiolova, L., Ravine, M., Salamon, A., Lee, S., Stoiber, K., Supulver, K., Van Beek, J., Van Beek, T., Zimdar, R., Louise French, K., Iagnemma, K., Miller, K., Goesmann, F., Goetz, W., Hviid, S., Johnson, M., Lefavor, M., Lyness, E., Breves, E., Darby Dyar, M., Fassett, C., Edwards, L., Haberle, R., Hoehler, T., Hollingsworth, J., Kahre, M., Keely, L., McKay, C., Bleacher, L., Brinckerhoff, W., Choi, D., Dworkin, J.P., Floyd, M., Freissinet, C., Garvin, J., Glavin, D., Harpold, D., Martin, D.K., McAdam, A., Pavlov, A., Raaen, E., Smith, M.D., Stern, J., Tan, F., Trainer, M., Posner, A., Voytek, M., Aubrey, A., Behar, A., Blaney, D., Brinza, D., Christensen, L., DeFlores, L., Feldman, J., Feldman, S., Flesch, G., Jun, I., Keymeulen, D., Mischna, M., Morookian, J.M., Pavri, B., Schoppers, M., Sengstacken, A., Simmonds, J.J., Spanovich, N., de la Torre Juarez, M., Webster, C.R., Yen, A., Archer, P.D., Cucinotta, F., Jones, J.H., Morris, R.V., Niles, P., Rampe, E., Nolan, T., Fisk, M., Radziemski, L., Barraclough, B., Bender, S., Berman, D., Dobrea, E.N., Tokar, R., Cleghorn, T., Huntress, W., Manhès, G., Hudgins, J., Olson, T., Stewart, N., Sarrazin, P., Vicenzi, E., Bullock, M., Ehresmann, B., Hamilton, V., Hassler, D., Peterson, J., Rafkin, S., Zeitlin, C., Fedosov, F., Golovin, D., Karpushkina, N., Alexander, K., Litvak, M., Malakhov, A., Mitrofanov, I., Mokrousov, M., Nikiforov, S., Prokhorov, V., Sanin, A., Tretyakov, V., Varenikov, A., Vostrukhin, A., Kuzmin, R., Clark, B., Wolff, M., Oliver, B., Drake, D., Bean, K., Lemmon, M., Schwenzer, S.P., Lee, E.M., Sucharski, R., Hernández, M.Á.de P., José Blanco Ávalos, J., Ramos, M., Kim, M.-H., Malespin, C., Plante, I., Muller, J.-P., Navarro-González, R., Ewing, R., Boynton, W., Downs, R., Fitzgibbon, M., Harshman, K., Morrison, S., Kortmann, O., Williams, A., Lugmair, G., Wilson, M.A., Jakosky, B., Balic-Zunic, T., Frydenvang, J., Kløvgaard Jensen, J., Kinch, K., Koefoed, A., Madsen, M.B., Stipp, S.L.S., Boyd, N., Campbell, J.L., Perrett, G., Pradler, I., VanBommel, S., Jacob, S., Owen, T., Savijärvi, H., Boehm, E., Böttcher, S., Burmeister, S., Guo, J., Köhler, J., Martín García, C., Mueller-Mellin, R., Wimmer-Schweingruber, R., Bridges, J.C., McConnochie, T., Benna, M., Franz, H., Bower, H., Brunner, A., Blau, H., Boucher, T., Carmosino, M., Atreya, S., Elliott, H., Halleaux, D., Rennó, N., Wong, M., Pepin, R., Elliott, B., Spray, J., Thompson, L., Gordon, S., Ollila, A., Williams, J., Vasconcelos, P., Bentz, J., Nealson, K., Popa, R., Moersch, J., Tate, C., Day, M., Francis, R., McCullough, E., Cloutis, I.L.ten K., Scholes, D., Slavney, S., Stein, T., Ward, J., Berger, J., Moores, J.E., 2014. A habitable fluvio-lacustrine environment at

Yellowknife Bay, Gale Crater, Mars. Science 343 (6169). https://doi.org/10.1126/science.1242777.

Halevy, I., Zuber, M.T., Schrag, D.P., 2007. A sulfur dioxide climate feedback on early Mars. Science 318 (5858), 1903–1907. https://doi.org/10.1126/science.1147039.

Harris, C.M., Maclay, M.T., Lutz, K.A., Nathan, V., Ortega Dominguez, N.A., Leavitt, W.D., Palucis, M.C., 2022. Remote and in-situ characterization of mars analogs: coupling scales to improve the search for microbial signatures on Mars. Front. Astron. Space Sci. 9. https://doi.org/10.3389/fspas.2022.849078.

Horvath, D.G., Andrews-Hanna, J.C., 2021. The hydrology and climate of Mars during the sedimentary infilling of Gale crater. Earth Planet Sci. Lett. 568. https://doi.org/10.1016/j.epsl.2021.117032.

Hurowitz, J.A., Grotzinger, J.P., Fischer, W.W., McLennan, S.M., Milliken, R.E., Stein, N., Vasavada, A.R., Blake, D.F., Dehouck, E., Eigenbrode, J.L., Fairén, A.G., Frydenvang, J., Gellert, R., Grant, J.A., Gupta, S., Herkenhoff, K.E., Ming, D.W., Rampe, E.B., Schmidt, M.E., Siebach, K.L., Stack-Morgan, K., Sumner, D.Y., Wiens, R.C., 2017. Redox stratification of an ancient lake in Gale crater, Mars. Science 356 (6341). https://doi.org/10.1126/science.aah6849.

Jakosky, B.M., Slipski, M., Benna, M., Mahaffy, P., Elrod, M., Yelle, R., Stone, S., Alsaeed, N., 2017. Mars' atmospheric history derived from upper-atmosphere measurements of 38 Ar/36 Ar. Science 355 (6332), 1408–1410. https://doi.org/10.1126/science.aai7721.

Jin, Q., Kirk, M.F., 2018. pH as a primary control in environmental microbiology: 1. thermodynamic perspective. Front. Environ. Sci. 6 (May). https://doi.org/10.3389/fenvs.2018.00021.

Johnson, D.B., Kanao, T., Hedrich, S., 2012. Redox transformations of iron at extremely low pH: fundamental and applied aspects. Front. Microbiol. 3. https://doi.org/10.3389/fmicb.2012.00096.

Junge, K., Eicken, H., Deming, J.W., 2004. Bacterial activity at -2 to -20°C in Arctic wintertime sea ice. Appl. Environ. Microbiol. 70 (1), 550–557. https://doi.org/10.1128/AEM.70.1.550-557.2004.

Kappler, A., Bryce, C., Mansor, M., Lueder, U., Byrne, J.M., Swanner, E.D., 2021. An evolving view on biogeochemical cycling of iron. Nat. Rev. Microbiol. 19 (6), 360–374. https://doi.org/10.1038/s41579-020-00502-7.

King, G.M., 2015. Carbon monoxide as a metabolic energy source for extremely halophilic microbes: implications for microbial activity in Mars regolith. Proc. Natl. Acad. Sci. 112 (14), 4465–4470. https://doi.org/10.1073/pnas.1424989112.

King, P.L., McLennan, S.M., 2010. Sulfur on Mars. Elements 6 (2), 107–112. https://doi.org/10.2113/gselements.6.2.107.

Kite, E.S., Mayer, D.P., Wilson, S.A., Davis, J.M., Lucas, A.S., de Quay, G.S., 2019. Persistence of intense, climate-driven runoff late in Mars history. Sci. Adv. 5 (3). https://doi.org/10.1126/sciadv.aav7710.

Klingelhöfer, G., Morris, R.V., Bernhardt, B., Schröder, C., Rodionov, D.S., De Souza, P.A., Yen, A., Gellert, R., Evlanov, E.N., Zubkov, B., Foh, J., Bonnes, U., Kankeleit, E., Gütlich, P., Ming, D.W., Renz, F., Wdowiak, T., Squyres, S.W., Arvidson, R.E., 2004. Jarosite and hematite at Meridiani Planum from opportunity's Mössbauer spectrometer. Science 306 (5702), 1740–1745. https://doi.org/10.1126/science.1104653.

Koike, M., Nakada, R., Kajitani, I., Usui, T., Tamenori, Y., Sugahara, H., Kobayashi, A., 2020. In-situ preservation of nitrogen-bearing organics in Noachian Martian carbonates. Nat. Commun. 11 (1). https://doi.org/10.1038/s41467-020-15931-4.

Langevin, Y., Poulet, F., Bibring, J.P., Gondet, B., 2005. Sulfates in the north polar region of Mars detected by OMEGA/Mars express. Science 307 (5715), 1584–1586. https://doi.org/10.1126/science.1109091.

Macey, M.C., Fox-Powell, M., Ramkissoon, N.K., Stephens, B.P., Barton, T., Schwenzer, S.P., Pearson, V.K., Cousins, C.R., Olsson-Francis, K., 2020. The identification of sulfide oxidation as a potential metabolism driving primary production on late Noachian Mars. Sci. Rep. 10 (1). https://doi.org/10.1038/s41598-020-67815-8.

Mahaffy, P.R., Webster, C.R., Atreya, S.K., Franz, H., Wong, M., Conrad, P.G., Harpold, D., Jones, J.J., Leshin, L.A., Manning, H., Owen, T., Pepin, R.O., Squyres, S., Trainer, M., Kemppinen, O., Bridges, N., Johnson, J.R., Minitti, M., Cremers, D., Bell, J.F., Edgar, L., Farmer, J., Godber, A., Wadhwa, M., Wellington, D., McEwan, I., Newman, C., Richardson, M., Charpentier, A., Peret, L., King, P., Blank, J., Weigle, G., Schmidt, M., Li, S., Milliken, R., Robertson, K., Sun, V., Baker, M., Edwards, C., Ehlmann, B., Farley, K., Griffes, J., Grotzinger, J., Miller, H., Newcombe, M., Pilorget, C., Rice, M., Siebach, K., Stack, K., Stolper, E., Brunet, C., Hipkin, V., Lveill, R., Marchand, G., Snchez, P.S., Favot, L., Cody, G., Steele, A., Flckiger, L., Lees, D., Nefian, A., Martin, M., Gailhanou, M., Westall, F., Isral, G., Agard, C., Baroukh, J., Donny, C., Gaboriaud, A., Guillemot, P., Lafaille, V., Lorigny, E., Paillet, A., Prez, R., Saccoccio, M., Yana, C., ArmiensAparicio, C., Rodrguez, J.C., Blzquez, I.C., Gmez, F.G., Gmez-Elvira, J., Hettrich, S., Malvitte, A.L., Jimnez, M.M., Martnez-Fras, J., Martn-Soler, J., Martn-Torres, F.J., Jurado, A.M., Mora-Sotomayor, L., Caro, G.M., Lpez, S.N., Peinado-Gonzlez, V., Pla-Garca, J., Manfredi, J.A.R., Romeral-Planell, J.J., Fuentes, S.A.S., Martinez, E.S., Redondo, J.T., Urqui-O'Callaghan, R., Mier, M.P.Z., Chipera, S., Lacour, J.L., Mauchien, P., Sirven, J. B., Fairn, A., Hayes, A., Joseph, J., Sullivan, R., Thomas, P., Dupont, A., Lundberg, A., Melikechi, N., Mezzacappa, A., DeMarines, J., Grinspoon, D., Reitz, G., Prats, B., Atlaskin, E., Genzer, M., Harri, A.M., Haukka, H., Kahanp, H., Kauhanen, J., Paton, M., Polkko, J., Schmidt, W., Siili, T., Fabre, C., Wray, J., Wilhelm, M.B., Poitrasson, F., Patel, K., Gorevan, S., Indyk, S., Paulsen, G., Gupta, S., Bish, D., Schieber, J., Gondet, B., Langevin, Y., Geffroy, C., Baratoux, D., Berger, G., Cros, A., Duston, C., Forni, O., Gasnault, O., Lasue, J., Lee, Q.M., Maurice, S., Meslin, P.Y., Pallier, E., Parot, Y., Pinet, P., Schrder, S., Toplis, M., Lewin, R., Brunner, W., Heydari, E., Achilles, C., Oehler, D., Sutter, B., Cabane, M., Coscia, D., Szopa, C., Dromart, G., Robert, F., Sautter, V., Le Moulic, S., Mangold, N., Nachon, M., Buch, A., Stalport, F., Coll, P., Franois, P., Raulin, F., Teinturier, S., Cameron, J., Clegg, S., Cousin, A., DeLapp, D., Dingler, R., Jackson, R.S., Johnstone, S., Lanza, N., Little, C., Nelson, T., Wiens, R.C., Williams, R.B., Jones, A., Kirkland, L., Treiman, A., Baker, B., Cantor, B., Caplinger, M., Davis, S., Duston, B., Edgett, K., Fay, D., Hardgrove, C., Harker, D., Herrera, P., Jensen, E., Kennedy, M.R., Krezoski, G., Krysak, D., Lipkaman, L., Malin, M., McCartney, E., McNair, S., Nixon, B., Posiolova, L., Ravine, M., Salamon, A., Saper, L., Stoiber, K., Supulver, K., Van Beek, J., Van Beek, T., Zimdar, R., French, K.L., Iagnemma, K., Miller, K., Summons, R., Goesmann, F., Goetz, W., Hviid, S., Johnson, M., Lefavor, M., Lyness, E., Breves, E., Dyar, M.D., Fassett, C., Blake, D.F., Bristow, T., DesMarais, D., Edwards, L., Haberle, R., Hoehler, T., Hollingsworth, J., Kahre, M., Keely, L.,

McKay, C., Bleacher, L., Brinckerhoff, W., Choi, D., Dworkin, J.P., Eigenbrode, J., Floyd, M., Freissinet, C., Garvin, J., Glavin, D., Martin, D.K., McAdam, A., Pavlov, A., Raaen, E., Smith, M.D., Stern, J., Tan, F., Meyer, M., Posner, A., Voytek, M., Anderson, R.C., Aubrey, A., Beegle, L.W., Behar, A., Blaney, D., Brinza, D., Calef, F., Christensen, L., Crisp, J.A., DeFlores, L., Feldman, J., Feldman, S., Flesch, G., Hurowitz, J., Jun, I., Keymeulen, D., Maki, J., Mischna, M., Morookian, J.M., Parker, T., Pavri, B., Schoppers, M., Sengstacken, A., Simmonds, J.J., Spanovich, N., De La Torre Juarez, M., Vasavada, A.R., Yen, A., Archer, P.D., Cucinotta, F., Ming, D., Morris, R.V., Niles, P., Rampe, E., Nolan, T., Fisk, M., Radziemski, L., Barraclough, B., Bender, S., Berman, D., Dobrea, E.N., Tokar, R., Vaniman, D., Williams, R.M.E., Yingst, A., Lewis, K., Cleghorn, T., Huntress, W., Manhs, G., Hudgins, J., Olson, T., Stewart, N., Sarrazin, P., Grant, J., Vicenzi, E., Wilson, S.A., Bullock, M., Ehresmann, B., Hamilton, V., Hassler, D., Peterson, J., Rafkin, S., Zeitlin, C., Fedosov, F., Golovin, D., Karpushkina, N., Kozyrev, A., Litvak, M., Malakhov, A., Mitrofanov, I., Mokrousov, M., Nikiforov, S., Prokhorov, V., Sanin, A., Tretyakov, V., Varenikov, A., Vostrukhin, A., Kuzmin, R., Clark, B., Wolff, M., McLennan, S., Botta, O., Drake, D., Bean, K., Lemmon, M., Schwenzer, S. P., Anderson, R.B., Herkenhoff, K., Lee, E.M., Sucharski, R., De Pablo Hernndez, M., Blanco Valos, J.J., Ramos, M., Kim, M.H., Malespin, C., Plante, I., Muller, J.P., Navarro-Gonzlez, R., Ewing, R., Boynton, W., Downs, R., Fitzgibbon, M., Harshman, K., Morrison, S., Dietrich, W., Kortmann, O., Palucis, M., Sumner, D.Y., Williams, A., Lugmair, G., Wilson, M.A., Rubin, D., Jakosky, B., Balic-Zunic, T., Frydenvang, J., Jensen, J.K., Kinch, K., Koefoed, A., Madsen, M.B., Stipp, S.L.S., Boyd, N., Campbell, J.L., Gellert, R., Perrett, G., Pradler, I., VanBommel, S., Jacob, S., Rowland, S., Savijrvi, H., Boehm, E., Bttcher, S., Burmeister, S., Guo, J., Khler, J., Garca, C.M., Mueller-Mellin, R., Wimmer-Schweingruber, R., Bridges, J.C., McConnochie, T., Benna, M., Bower, H., Brunner, A., Blau, H., Boucher, T., Carmosino, M., Elliott, H., Halleaux, D., Renn, N., Elliott, B., Spray, J., Thompson, L., Gordon, S., Newsom, H., Ollila, A., Williams, J., Vasconcelos, P., Bentz, J., Nealson, K., Popa, R., Kah, L.C., Moersch, J., Tate, C., Day, M., Kocurek, G., Hallet, B., Sletten, R., Francis, R., McCullough, E., Cloutis, E., Ten Kate, I.L., Arvidson, R., Fraeman, A., Scholes, D., Slavney, S., Stein, T., Ward, J., Berger, J., Moores, J.E., 2013. Abundance and isotopic composition of gases in the martian atmosphere from the Curiosity rover. Science 341 (6143), 263–266. https://doi.org/10.1126/science.1237966.

Maldanis, L., Fernandez-Remolar, D., Lemelle, L., Knoll, A.H., Guizar-Sicairos, M., Holler, M., da Silva, F.M.C., Magnin, V., Mermoux, M., Simionovici, A., 2024. Unveiling challenging microbial fossil biosignatures from Rio Tinto with micro-to-nanoscale chemical and ultrastructural imaging. Astrobiology 24 (7), 721–733. https://doi.org/10.1089/ast.2023.0127.

Mangold, N., 2021. Intermittent warmth on young Mars. Nat. Geosci. 14 (3), 112–113. https://doi.org/10.1038/s41561-021-00700-9.

McSween, H.Y., Jeffrey Taylor, G., Wyatt, M.B., 2009. Elemental composition of the Martian crust. Science 324 (5928), 736–739. https://doi.org/10.1126/science.1165871.

Ming, D.W., Mittlefehldt, D.W., Morris, R.V., Golden, D.C., Gellert, R., Yen, A., Clark, B.C., Squyres, S.W., Farrand, W.H., Ruff, S.W., Arvidson, R.E., Klingelhöfer, G., McSween, H. Y., Rodionov, D.S., Schröder, C., de Souza, J.A., Wang, A., 2006. Geochemical and mineralogical indicators for aqueous processes in the Columbia Hills of Gusev crater, Mars. J. Geophys. Res.: Planets 111 (2). https://doi.org/10.1029/2005JE002560.

Ming, D.W., Archer, P.D., Glavin, D.P., Eigenbrode, J.L., Franz, H.B., Sutter, B., Brunner, A. E., Stern, J.C., Freissinet, C., McAdam, A.C., Mahaffy, P.R., Cabane, M., Coll, P., Campbell, J.L., Atreya, S.K., Niles, P.B., Bell, J.F., Bish, D.L., Brinckerhoff, W.B., Buch, A., Conrad, P.G., Des Marais, D.J., Ehlmann, B.L., Fairén, A.G., Farley, K., Flesch, G.J., Francois, P., Gellert, R., Grant, J.A., Grotzinger, J.P., Gupta, S., Herkenhoff, K.E., Hurowitz, J.A., Leshin, L.A., Lewis, K.W., McLennan, S.M., Miller, K.E., Moersch, J., Morris, R.V., Navarro-González, R., Pavlov, A.A., Perrett, G. M., Pradler, I., Squyres, S.W., Summons, R.E., Steele, A., Stolper, E.M., Sumner, D.Y., Szopa, C., Teinturier, S., Trainer, M.G., Treiman, A.H., Vaniman, D.T., Vasavada, A.R., Webster, C.R., Wray, J.J., Kemppinen, O., Bridges, N., Johnson, J.R., Minitti, M., Cremers, D., Edgar, L., Farmer, J., Godber, A., Wadhwa, M., Wellington, D., McEwan, I., Newman, C., Richardson, M., Charpentier, A., Peret, L., King, P., Blank, J., Weigle, G., Schmidt, M., Li, S., Milliken, R., Robertson, K., Sun, V., Baker, M., Edwards, C., Griffes, J., Miller, H., Newcombe, M., Pilorget, C., Rice, M., Siebach, K., Stack, K., Brunet, C., Hipkin, V., Léveillé, R., Marchand, G., Sánchez, P. S., Favot, L., Cody, G., Flückiger, L., Lees, D., Nefian, A., Martin, M., Gailhanou, M., Westall, F., Israël, G., Agard, C., Baroukh, J., Donny, C., Gaboriaud, A., Guillemot, P., Lafaille, V., Lorigny, E., Paillet, A., Pérez, R., Saccoccio, M., Yana, C., Armiens-Aparicio, C., Caride Rodríguez, J., Carrasco Blázquez, I., Gómez Gómez, F., Gómez-Elvira, J., Hettrich, S., Malvitte, A.L., Jiménez, M.M., Martínez-Frías, J., Martín-Soler, J., Martín-Torres, F.J., Jurado, A.M., Mora-Sotomayor, L., Caro, G.M., López, S. N., Peinado-González, V., Pla-García, J., Manfredi, J.A.R., Romeral-Planelló, J.J., Fuentes, S.A.S., Martinez, E.S., Redondo, J.T., Urqui-O'Callaghan, R., Zorzano, M.P., Chipera, S., Lacour, J.L., Mauchien, P., Sirven, J.B., Manning, H., Hayes, A., Joseph, J., Sullivan, R., Thomas, P., Dupont, A., Lundberg, A., Melikechi, N., Mezzacappa, A., DeMarines, J., Grinspoon, D., Reitz, G., Prats, B., Atlaskin, E., Genzer, M., Harri, A.M., Haukka, H., Kahanpää, H., Kauhanen, J., Kemppinen, O., Paton, M., Polkko, J., Schmidt, W., Siili, T., Fabre, C., Wilhelm, M.B., Poitrasson, F., Patel, K., Gorevan, S., Indyk, S., Paulsen, G., Schieber, J., Gondet, B., Langevin, Y., Geffroy, C., Baratoux, D., Berger, G., Cros, A., d'Uston, C., Forni, O., Gasnault, O., Lasue, J., Lee, Q.M., Maurice, S., Meslin, P.Y., Pallier, E., Parot, Y., Pinet, P., Schröder, S., Toplis, M., Lewin, É., Brunner, W., Heydari, E., Achilles, C., Oehler, D., Coscia, D., Israël, G., Dromart, G., Robert, F., Sautter, V., Le Mouélic, S., Mangold, N., Nachon, M., Stalport, F., François, P., Raulin, F., Teinturier, S., Cameron, J., Clegg, S., Cousin, A., DeLapp, D., Dingler, R., Jackson, R.S., Johnstone, S., Lanza, N., Little, C., Nelson, T., Wiens, R.C., Williams, R.B., Jones, A., Kirkland, L., Baker, B., Cantor, B., Caplinger, M., Davis, S., Duston, B., Edgett, K., Fay, D., Hardgrove, C., Harker, D., Herrera, P., Jensen, E., Kennedy, M.R., Krezoski, G., Krysak, D., Lipkaman, L., Malin, M., McCartney, E., McNair, S., Nixon, B., Posiolova, L., Ravine, M., Salamon, A., Saper, L., Stoiber, K., Supulver, K., Van Beek, J., Van Beek, T., Zimdar, R., French, K.L., Iagnemma, K., Goesmann, F., Goetz, W., Hviid, S., Johnson, M., Lefavor, M., Lyness, E., Breves, E., Dyar, M.D., Fassett, C., Blake, D.F., Bristow, T., Edwards, L., Haberle, R., Hoehler, T., Hollingsworth, J., Kahre, M., Keely, L., McKay, C., Wilhelm, M.B., Bleacher, L., Choi, D., Dworkin, J.P., Floyd, M., Garvin, J., Harpold, D., Jones, A., Martin, D.K., Raaen, E., Smith, M.D., Tan, F., Meyer, M., Posner, A., Voytek, M., Anderson, R.C.,

Aubrey, A., Beegle, L.W., Behar, A., Blaney, D., Brinza, D., Calef, F., Christensen, L., Crisp, J.A., DeFlores, L., Feldman, J., Feldman, S., Jun, I., Keymeulen, D., Maki, J., Mischna, M., Morookian, J.M., Parker, T., Pavri, B., Schoppers, M., Sengstacken, A., Simmonds, J.J., Spanovich, N., de la Torre Juarez, M., Yen, A., Cucinotta, F., Jones, J. H., Rampe, E., Nolan, T., Fisk, M., Radziemski, L., Barraclough, B., Bender, S., Berman, D., Dobrea, E.N., Tokar, R., Williams, R.M.E., Cleghorn, T., Huntress, W., Manhès, G., Hudgins, J., Olson, T., Stewart, N., Sarrazin, P., Vicenzi, E., Wilson, S.A., Bullock, M., Ehresmann, B., Hamilton, V., Hassler, D., Peterson, J., Rafkin, S., Zeitlin, C., Fedosov, F., Golovin, D., Karpushkina, N., Kozyrev, A., Litvak, M., Malakhov, A., Mitrofanov, I., Mokrousov, M., Nikiforov, S., Prokhorov, V., Sanin, A., Tretyakov, V., Varenikov, A., Vostrukhin, A., Kuzmin, R., Clark, B., Wolff, M., Botta, O., Drake, D., Bean, K., Lemmon, M., Schwenzer, S.P., Anderson, R.B., Lee, E. M., Sucharski, R., de Pablo Hernández, M.Á., Ávalos, J.J.B., Ramos, M., Kim, M.H., Malespin, C., Plante, I., Muller, J.P., Ewing, R., Boynton, W., Downs, R., Fitzgibbon, M., Harshman, K., Morrison, S., Dietrich, W., Kortmann, O., Palucis, M., Williams, A., Lugmair, G., Wilson, M.A., Rubin, D., Jakosky, B., Balic-Zunic, T., Frydenvang, J., Jensen, J.K., Kinch, K., Koefoed, A., Bo Madsen, M., Stipp, S.L.S., Boyd, N., VanBommel, S., Jacob, S., Owen, T., Rowland, S., Atlaskin, E., Savijärvi, H., Boehm, E., Böttcher, S., Burmeister, S., Guo, J., Köhler, J., García, C.M., Mueller-Mellin, R., Wimmer-Schweingruber, R., Bridges, J.C., McConnochie, T., Benna, M., Bower, H., Blau, H., Boucher, T., Carmosino, M., Elliott, H., Halleaux, D., Rennó, N., Wong, M., Pepin, R., Elliott, B., Spray, J., Thompson, L., Gordon, S., Newsom, H., Ollila, A., Williams, J., Vasconcelos, P., Bentz, J., Nealson, K., Popa, R., Kah, L.C., Tate, C., Day, M., Kocurek, G., Hallet, B., Sletten, R., Francis, R., McCullough, E., Cloutis, E., ten Kate, I.L., Kuzmin, R., Arvidson, R., Fraeman, A., Scholes, D., Slavney, S., Stein, T., Ward, J., Berger, J., Moores, J.E., 2014. Volatile and organic compositions of sedimentary rocks in Yellowknife Bay, Gale Crater, Mars. Science 343 (6169). https://doi.org/10.1126/science.1245267.

Moissl-Eichinger, C., Cockell, C., Rettberg, P., 2016. Venturing into new realms? Microorganisms in spacea. FEMS Microbiol. Rev. 40 (5), 722–737. https://doi.org/10.1093/femsre/fuw015.

Molina, A., López, I., Prieto-Ballesteros, O., Fernández-Remolar, D., de Pablo, M.Á., Gómez, F., 2017. Coogoon Valles, western Arabia Terra: hydrological evolution of a complex Martian channel system. Icarus 293, 27–44. https://doi.org/10.1016/j.icarus.2017.04.002.

Morris, R.V., Klingelhöfer, G., Schröder, C., Rodionov, D.S., Yen, A., Ming, D.W., de Souza, P.A., Fleischer, I., Wdowiak, T., Gellert, R., Bernhardt, B., Evlanov, E.N., Zubkov, B., Foh, J., Bonnes, U., Kankeleit, E., Gütlich, P., Renz, F., Squyres, S.W., Arvidson, R.E., 2006a. Mössbauer mineralogy of rock, soil, and dust at Gusev crater, Mars: spirit's journey through weakly altered olivine basalt on the plains and pervasively altered basalt in the Columbia Hills. J. Geophys. Res.: Planets 111 (E2). https://doi.org/10.1029/2005je002584.

Morris, R.V., Klingelhöfer, G., Schröer, C., Fleischer, I., Ming, D.W., Yen, A.S., Gellert, R., Arvidson, R.E., Rodionov, D.S., Crumpler, L.S., Clark, B.C., Cohen, B.A., McCoy, T.J., Mittlefehldt, D.W., Schmidt, M.E., De Souza, J.A., Squyres, S.W., 2008. Iron mineralogy and aqueous alteration from Husband Hill through Home Plate at Gusev Crater, Mars: results from the Mössbauer instrument on the Spirit Mars Exploration Rover. J. Geophys. Res.: Planets 113 (12). https://doi.org/10.1029/2008JE003201.

Morris, R.V., Klingelhöfer, G., Schröder, C., Rodionov, D.S., Yen, A., Ming, D.W., de Souza Jr., P.A., Wdowiak, T., Fleischer, I., Gellert, R., Bernhardt, B., Bonnes, U., Cohen, B.A., Evlanov, E.N., Foh, J., Gütlich, P., Kankeleit, E., McCoy, T., Mittlefehldt, D.W., Renz, F., Schmidt, M.E., Zubkov, B., Squyres, S.W., Arvidson, R. E., 2006b. Mössbauer mineralogy of rock, soil, and dust at Meridiani Planum, Mars: opportunity's journey across sulfate-rich outcrop, basaltic sand and dust, and hematite lag deposits. J. Geophys. Res.: Planets 111 (E12). https://doi.org/10.1029/2006JE002791.

Nachon, M., Clegg, S.M., Mangold, N., Schröder, S., Kah, L.C., Dromart, G., Ollila, A., Johnson, J.R., Oehler, D.Z., Bridges, J.C., Le Mouélic, S., Forni, O., Wiens, R.C., Anderson, R.B., Blaney, D.L., Bell, J.F., Clark, B., Cousin, A., Dyar, M.D., Ehlmann, B., Fabre, C., Gasnault, O., Grotzinger, J., Lasue, J., Lewin, E., Léveillé, R., McLennan, S., Maurice, S., Meslin, P.-Y., Rapin, W., Rice, M., Squyres, S.W., Stack, K., Sumner, D.Y., Vaniman, D., Wellington, D., 2014. Calcium sulfate veins characterized by ChemCam/Curiosity at Gale crater, Mars. J. Geophys. Res.: Planets 119 (9), 1991–2016. https://doi.org/10.1002/2013je004588.

Nicholson, W.L., Krivushin, K., Gilichinsky, D., Schuerger, A.C., 2013. Growth of Carnobacterium spp. from permafrost under low pressure, temperature, and anoxic atmosphere has implications for Earth microbes on Mars. Proc. Natl. Acad. Sci. 110 (2), 666–671. https://doi.org/10.1073/pnas.1209793110.

Nixon, S.L., Cousins, C.R., Cockell, C.S., 2013. Plausible microbial metabolisms on Mars. Astron. Geophys. 54 (1), 1–13. https://doi.org/10.1093/astrogeo/ats034.

Orosei, R., Lauro, S.E., Pettinelli, E., Cicchetti, A., Coradini, M., Cosciotti, B., Di Paolo, F., Flamini, E., Mattei, E., Pajola, M., Soldovieri, F., Cartacci, M., Cassenti, F., Frigeri, A., Giuppi, S., Martufi, R., Masdea, A., Mitri, G., Nenna, C., Noschese, R., Restano, M., Seu, R., 2018. Radar evidence of subglacial liquid water on Mars. Science 361 (6401), 490–493. https://doi.org/10.1126/science.aar7268.

Paul, V.G., Mormile, M.R., 2020. Potential energy production and utilization pathways of the Martian subsurface: clues from extremophilic microorganisms on Earth. Extremophiles as Astrobiological Models. Scrivener Publishing LLC, pp. 291–315.

Preston, L.J., Shuster, J., Fernández-Remolar, D., Banerjee, N.R., Osinski, G.R., Southam, G., 2011. The preservation and degradation of filamentous bacteria and biomolecules within iron oxide deposits at Rio Tinto, Spain. Geobiology 9 (3), 233–249. https://doi.org/10.1111/j.1472-4669.2011.00275.x.

Price, A., Pearson, V.K., Schwenzer, S.P., Miot, J., Olsson-Francis, K., 2018. Nitrate-dependent iron oxidation: a potential Mars metabolism. Front. Microbiol. 9. https://doi.org/10.3389/fmicb.2018.00513.

Puente-Sánchez, F., Moreno-Paz, M., Rivas, L.A., Cruz-Gil, P., García-Villadangos, M., Gómez, M.J., Postigo, M., Garrido, P., González-Toril, E., Briones, C., Fernández-Remolar, D., Stoker, C., Amils, R., Parro, V., 2014. Deep subsurface sulfate reduction and methanogenesis in the Iberian Pyrite Belt revealed through geochemistry and molecular biomarkers. Geobiology 12 (1), 34–47. https://doi.org/10.1111/gbi.12065.

Ramirez, R.M., Kopparapu, R., Zugger, M.E., Robinson, T.D., Freedman, R., Kasting, J.F., 2014. Warming early Mars with CO2 and H2. Nat. Geosci. 7 (1), 59–63. https://doi.org/10.1038/ngeo2000.

Ramirez, R.M., Craddock, R.A., 2018. The geological and climatological case for a warmer and wetter early Mars. Nat. Geosci. 11 (4), 230–237. https://doi.org/10.1038/s41561-018-0093-9.

Rampe, E.B., Ming, D.W., Blake, D.F., Bristow, T.F., Chipera, S.J., Grotzinger, J.P., Morris, R. V., Morrison, S.M., Vaniman, D.T., Yen, A.S., Achilles, C.N., Craig, P.I., Des Marais, D.J., Downs, R.T., Farmer, J.D., Fendrich, K.V., Gellert, R., Hazen, R.M., Kah, L.C., Morookian, J.M., Peretyazhko, T.S., Sarrazin, P., Treiman, A.H., Berger, J.A., Eigenbrode, J., Fairén, A.G., Forni, O., Gupta, S., Hurowitz, J.A., Lanza, N.L., Schmidt, M.E., Siebach, K., Sutter, B., Thompson, L.M., 2017. Mineralogy of an ancient lacustrine mudstone succession from the Murray formation, Gale crater, Mars. Earth Planet Sci. Lett. 471, 172–185. https://doi.org/10.1016/j.epsl.2017.04.021.

Roden, E.E., Sobolev, D., Glazer, B., Luther, G.W., 2004. Potential for microscale bacterial Fe redox cycling at the aerobic-anaerobic interface. Geomicrobiol. J. 21 (6), 379–391. https://doi.org/10.1080/01490450490485872.

Sánchez-Andrea, I., van der Graaf, C.M., Hornung, B., Bale, N.J., Jarzembowska, M., Sousa, D.Z., Rijpstra, W.I.C., Sinninghe Damsté, J.S., Stams, A.J.M., 2022. Acetate degradation at low pH by the moderately acidophilic sulfate reducer acididesulfobacillus acetoxydans gen. nov. sp. nov. Front. Microbiol. 13. https://doi.org/10.3389/fmicb.2022.816605.

Sánchez-Andrea, I., Stams, A.J.M., Hedrich, S., Ňancucheo, I., Johnson, D.B., 2015. Desulfosporosinus acididurans sp. nov.: an acidophilic sulfate-reducing bacterium isolated from acidic sediments. Extremophiles 19 (1), 39–47. https://doi.org/10.1007/s00792-014-0701-6.

Sánchez-Andrea, I., Stams, A.J.M., Amils, R., Sanz, J.L., 2013. Enrichment and isolation of acidophilic sulfate-reducing bacteria from Tinto River sediments. Environ. Microbiol. Rep. 5 (5), 672–678. https://doi.org/10.1111/1758-2229.12066.

Sánchez-Andrea, I., Rodríguez, N., Amils, R., Sanz, J.L., 2011. Microbial diversity in anaerobic sediments at Río Tinto, a naturally acidic environment with a high heavy metal content. Appl. Environ. Microbiol. 77 (17), 6085–6093. https://doi.org/10.1128/AEM.00654-11Spain.

Sánchez-Andrea, I., Rojas-Ojeda, P., Amils, R., Sanz, J.L., 2012. Screening of anaerobic activities in sediments of an acidic environment: Tinto River. Extremophiles 16 (6), 829–839. https://doi.org/10.1007/s00792-012-0478-4.

Sánchez-García, L., Carrizo, D., Fernández-Remolar, D., Parro, V., 2017. In: Lipidic Biosignatures in Diagenetically Stabilized Ironstones Terraces of Rio Tinto, an Acidic Environment with Analogies to Mars European Planetary Science Congress.

Sánchez-García, L., Fernández-Martínez, M.A., Moreno-Paz, M., Carrizo, D., García-Villadangos, M., Manchado, J.M., Stoker, C.R., Glass, B., Parro, V., 2020. Simulating Mars drilling mission for searching for life: ground-truthing lipids and other complex microbial biomarkers in the iron-sulfur rich Río Tinto analog. Astrobiology 20 (9), 1029–1047. https://doi.org/10.1089/ast.2019.2101.

Sánchez-Román, M., Fernández-Remolar, D., Amils, R., Sánchez-Navas, A., Schmid, T., Martin-Uriz, P.S., Rodríguez, N., McKenzie, J.A., Vasconcelos, C., 2014. Microbial mediated formation of Fe-carbonate minerals under extreme acidic conditions. Sci. Rep. 4 (1). https://doi.org/10.1038/srep04767.

Sanz, J.L., Rodriguez, N., Escudero, C., Carrizo, D., Amils, R., 2021. Biological production of H2, CH4 and CO2 in the deep subsurface of the Iberian Pyrite Belt. Environ. Microbiol. 23 (7), 3913–3922. https://doi.org/10.1111/1462-2920.15561.

Sanz, J.L., Rodríguez, N., Díaz, E.E., Amils, R., 2011. Methanogenesis in the sediments of Rio Tinto, an extreme acidic river. Environ. Microbiol. 13 (8), 2336–2341. https://doi.org/10.1111/j.1462-2920.2011.02504.x.

Seto, M., Noguchi, K., Cappellen, P.V., 2019. Potential for aerobic methanotrophic metabolism on Mars. Astrobiology 19 (10), 1187–1195. https://doi.org/10.1089/ast.2018.1943.
Shen, J.X., Chen, Y., Sun, Y., Liu, L., Pan, Y.X., Lin, W., 2022. Detection of biosignatures in Terrestrial analogs of Martian regions: strategical and technical assessments. Earth Planet. Phys. 6 (5), 431–450. https://doi.org/10.26464/epp2022042.
Siljeström, S., Czaja, A.D., Corpolongo, A., Berger, E.L., Li, A.Y., Cardarelli, E., Abbey, W., Asher, S.A., Beegle, L.W., Benison, K.C., Bhartia, R., Bleefeld, B.L., Burton, A.S., Bykov, S.V., Clark, B., DeFlores, L., Ehlmann, B.L., Fornaro, T., Fox, A., Gómez, F., Hand, K., Haney, N.C., Hickman-Lewis, K., Hug, W.F., Imbeah, S., Jakubek, R.S., Kah, L.C., Kivrak, L., Lee, C., Liu, Y., Martínez-Frías, J., McCubbin, F.M., Minitti, M., Moore, K., Morris, R.V., Núñez, J.I., Osterhout, J.T., Phua, Y.Y., Randazzo, N., Hollis, J.R., Rodriguez, C., Roppel, R., Scheller, E.L., Sephton, M., Sharma, S.K., Sharma, S., Steadman, K., Steele, A., Tice, M., Uckert, K., VanBommel, S., Williams, A.J., Williford, K.H., Winchell, K., Wu, M.K., Yanchilina, A., Zorzano, M.P., 2024. Evidence of sulfate-rich fluid alteration in Jezero Crater Floor. J. Geophys. Res.: Planets 129 (1). https://doi.org/10.1029/2023JE007989.
Souza-Egipsy, V., Ormö, J., Bowen, B.B., Chan, M.A., Komatsu, G., 2006. Ultrastructural study of iron oxide precipitates: implications for the search for biosignatures in the Meridiani hematite concretions, Mars. Astrobiology 6 (4), 527–545. https://doi.org/10.1089/ast.2006.6.527.
Steele, A., McCubbin, F.M., Fries, M., Kater, L., Boctor, N.Z., Fogel, M.L., Conrad, P.G., Glamoclija, M., Spencer, M., Morrow, A.L., Hammond, M.R., Zare, R.N., Vicenzi, E. P., Siljeström, S., Bowden, R., Herd, C.D.K., Mysen, B.O., Shirey, S.B., Amundsen, H. E.F., Treiman, A.H., Bullock, E.S., Jull, A.J.T., 2012. A reduced organic carbon component in Martian basalts. Science 337 (6091), 212–215. https://doi.org/10.1126/science.1220715.
Stern, J.C., Sutter, B., Jackson, W.A., Navarro-González, R., McKay, C.P., Ming, D.W., Archer, P.D., Mahaffy, P.R., 2017. The nitrate/(per)chlorate relationship on Mars. Geophys. Res. Lett. 44 (6), 2643–2651. https://doi.org/10.1002/2016GL072199.
Stern, J.C., Sutter, B., Freissinet, C., Navarro-González, R., McKay, C.P., Douglas Archer, P., Buch, A., Brunner, A.E., Coll, P., Eigenbrode, J.L., Fairen, A.G., Franz, H.B., Glavin, D. P., Kashyap, S., McAdam, A.C., Ming, D.W., Steele, A., Szopa, C., Wray, J.J., Javier Martín-Torres, F., Zorzano, M.-P., Conrad, P.G., Mahaffy, P.R., Kemppinen, O., Bridges, N., Johnson, J.R., Minitti, M., Cremers, D., Bell, J.F., Edgar, L., Farmer, J., Godber, A., Wadhwa, M., Wellington, D., McEwan, I., Newman, C., Richardson, M., Charpentier, A., Peret, L., King, P., Blank, J., Weigle, G., Schmidt, M., Li, S., Milliken, R., Robertson, K., Sun, V., Baker, M., Edwards, C., Ehlmann, B., Farley, K., Griffes Shechet, J., Grotzinger, J., Miller, H., Newcombe, M., Pilorget, C., Rice, M., Siebach, K., Stack, K., Stolper, E., Brunet, C., Hipkin, V., Léveillé, R., Marchand, G., Sobrón Sánchez, P., Favot, L., Cody, G., Steele, A., Flückiger, L., Lees, D., Nefian, A., Martin, M., Westall, F., Israël, G., Agard, C., Baroukh, J., Donny, C., Guillemot, P., Lafaille, V., Lorigny, E., Paillet, A., Pérez, R., Saccoccio, M., Yana, C., Aparicio, C.A., Rodríguez, J.C., Carrasco Blázquez, I., Gómez, F.G., Gómez-Elvira, J., Hettrich, S., Malvitte, A.L., Jiménez, M.M., Martínez-Frías, J., Martín-Soler, J., Javier Martín - Torres, F., Molina Jurado, A., Mora-Sotomayor, L., Caro, G.M., Navarro López, S., Peinado-González, V., Pla-García, J., Rodriguez Manfredi, J.A., Romeral-Planelló, J.J., Fuentes, S.A.S., Martinez, E.S., Torres Redondo, J., Urqui-O'Callaghan, R., Mier, M.-P.Z., Chipera, S., Lacour, J.-L., Mauchien, P., Sirven, J.-B., Manning, H., Fairén, A.,

Hayes, A., Joseph, J., Squyres, S., Sullivan, R., Thomas, P., Dupont, A., Lundberg, A., Melikechi, N., Mezzacappa, A., DeMarines, J., Grinspoon, D., Reitz, G., Prats, B., Atlaskin, E., Genzer, M., Harri, A.-M., Haukka, H., Kahanpää, H., Kauhanen, J., Kemppinen, O., Paton, M., Polkko, J., Schmidt, W., Siili, T., Fabre, C., Wray, J., Wilhelm, M.B., Poitrasson, F., Patel, K., Gorevan, S., Indyk, S., Paulsen, G., Gupta, S., Bish, D., Schieber, J., Gondet, B., Langevin, Y., Geffroy, C., Baratoux, D., Berger, G., Cros, A., d'Uston, C., Forni, O., Gasnault, O., Lasue, J., Lee, Q.-M., Maurice, S., Meslin, P.-Y., Pallier, E., Parot, Y., Pinet, P., Schröder, S., Toplis, M., Lewin, É., Brunner, W., Heydari, E., Achilles, C., Oehler, D., Sutter, B., Cabane, M., Coscia, D., Israël, G., Szopa, C., Dromart, G., Robert, F., Sautter, V., Le Mouélic, S., Mangold, N., Nachon, M., Buch, A., Stalport, F., Coll, P., François, P., Raulin, F., Teinturier, S., Cameron, J., Clegg, S., Cousin, A., DeLapp, D., Dingler, R., Jackson, R.S., Johnstone, S., Lanza, N., Little, C., Nelson, T., Wiens, R.C., Williams, R.B., Jones, A., Kirkland, L., Treiman, A., Baker, B., Cantor, B., Caplinger, M., Davis, S., Duston, B., Edgett, K., Fay, D., Craig, H., Harker, D., Herrera, P., Jensen, E., Kennedy, M.R., Krezoski, G., Krysak, D., Lipkaman, L., Malin, M., McCartney, E., McNair, S., Nixon, B., Posiolova, L., Ravine, M., Salamon, A., Lee, S., Stoiber, K., Supulver, K., Van Beek, J., Van Beek, T., Zimdar, R., Louise French, K., Iagnemma, K., Miller, K., Summons, R., Goesmann, F., Goetz, W., Hviid, S., Johnson, M., Lefavor, M., Lyness, E., Breves, E., Darby Dyar, M., Fassett, C., Blake, D.F., Bristow, T., DesMarais, D., Edwards, L., Haberle, R., Hoehler, T., Hollingsworth, J., Kahre, M., Keely, L., McKay, C., Wilhelm, M.B., Bleacher, L., Brinckerhoff, W., Choi, D., Conrad, P., Dworkin, J.P., Eigenbrode, J., Floyd, M., Freissinet, C., Garvin, J., Glavin, D., Harpold, D., Jones, A., Mahaffy, P., Martin, D.K., McAdam, A., Pavlov, A., Raaen, E., Smith, M.D., Stern, J., Tan, F., Trainer, M., Meyer, M., Posner, A., Voytek, M., Anderson, R.C., Aubrey, A., Beegle, L.W., Behar, A., Blaney, D., Brinza, D., Calef, F., Christensen, L., Crisp, J.A., DeFlores, L., Ehlmann, B., Feldman, J., Flesch, G., Hurowitz, J., Jun, I., Keymeulen, D., Maki, J., Mischna, M., Morookian, J.M., Parker, T., Pavri, B., Schoppers, M., Sengstacken, A., Simmonds, J.J., Spanovich, N., Juarez, M.de la T., Vasavada, A.R., Webster, C.R., Yen, A., Archer, P. D., Cucinotta, F., Jones, J.H., Ming, D., Morris, R.V., Niles, P., Rampe, E., Nolan, T., Fisk, M., Radziemski, L., Barraclough, B., Bender, S., Berman, D., Noe Dobrea, E., Tokar, R., Vaniman, D., Rebecca, M.E., Williams, Yingst, A., Lewis, K., Leshin, L., Cleghorn, T., Wesley, H., Manhès, G., Hudgins, J., Olson, T., Stewart, N., Sarrazin, P., Grant, J., Vicenzi, E., Wilson, S.A., Bullock, M., Ehresmann, B., Hamilton, V., Hassler, D., Peterson, J., Rafkin, S., Zeitlin, C., Fedosov, F., Golovin, D., Karpushkina, N., Kozyrev, A., Litvak, M., Malakhov, A., Mitrofanov, I., Mokrousov, M., Nikiforov, S., Prokhorov, V., Sanin, A., Tretyakov, V., Varenikov, A., Vostrukhin, A., Kuzmin, R., Clark, B., Wolff, M., McLennan, S., Oliver, B., Drake, D., Bean, K., Lemmon, M., Schwenzer, S.P., Anderson, R.B., Herkenhoff, K., Lee, E.M., Sucharski, R., Hernández, M.Á.de P., José Blanco Ávalos, J., Ramos, M., Kim, M.-H., Malespin, C., Plante, I., Muller, J.-P., Navarro-González, R., Ewing, R., Boynton, W., Downs, R., Fitzgibbon, M., Harshman, K., Morrison, S., Dietrich, W., Kortmann, O., Palucis, M., Sumner, D.Y., Williams, A., Lugmair, G., Wilson, M.A., Rubin, D., Jakosky, B., Balic-Zunic, T., Frydenvang, J., Jensen, J.K., Kinch, K., Koefoed, A., Madsen, M.B., Stipp, S.L.S., Boyd, N., Campbell, J.L., Gellert, R., Perrett, G., Pradler, I., Scott, V.B., Jacob, S., Owen, T., Rowland, S., Atlaskin, E., Savijärvi, H., Boehm, E., Böttcher, S., Burmeister, S., Guo, J., Köhler, J., García, C.M., Mueller-

Mellin, R., Wimmer-Schweingruber, R., Bridges, J.C., McConnochie, T., Benna, M., Franz, H., Bower, H., Brunner, A., Blau, H., Boucher, T., Carmosino, M., Atreya, S., Elliott, H., Douglas, H., Rennó, N., Wong, M., Pepin, R., Elliott, B., Spray, J., Thompson, L., Gordon, S., Newsom, H., Ollila, A., Williams, J., Vasconcelos, P., Bentz, J., Nealson, K., Popa, R., Kah, L.C., Moersch, J., Tate, C., Day, M., Kocurek, G., Hallet, B., Sletten, R., Francis, R., McCullough, E., Cloutis, I.L.ten K., Kuzmin, R., Arvidson, R., Fraeman, A., Scholes, D., Slavney, S., Stein, T., Ward, J., Berger, J., Moores, J.E., Mitchell, J., Jean-Rigaud, L., Lee, R., Nikitczuk, M., Thomas, A., Kaplan, H., Buz, J., Lapôtre, M., Nastan, A., Anderson, D., Said, D., Larif, M.-F., Mousset, V., Verdasca, J.A.A., Martín-Sauceda Martín, J., Moreno, G., Valentín-Serrano, P., Ullán-Nieto, A., Cal, T.M.de, Azeez, S., Ren, J., Mrigakshi, A., Ritter, B., Berger, T., Matthiä, D., Genzer, M., Mäkinen, T., Merikallio, S., Spring, J., Daydou, Y., Rapin, W., Beck, P., Douglas Archer, P., Abbey, W., Blacksberg, J., Lichtenberg, K., Milkovich, S., Schroeder, J., Scodary, A., Williford, K., Bonnet, J.-Y., Millan, M., Thuillier, F., Joulin, M., Le Deit, L., Dequaire, T., Hardy, K., Pagel, H., Parra, A., Beach, M., Magee, A., McBride, M., Kotrc, B., Berlanga, G., Ozanne, M., Lepore, K., Arevalo, R., Domagal-Goldman, S., Getty, S., Graham, H., Cropper, K., Lisov, D., Shterts, R., Dehouck, E., Battalio, M., Mason, E., McLain, H., Starr, R., Kirk, R., Rubalcaba, Á.de V.-R., Guzewich, S., Fendrich, K., Appel, J.K., Kharytonov, A., Lohf, H., Orthen, T., Edwards, P., Kashyap, S., Pinnick, V., Carey, C., Borlina, C.S., Desrousseaux, R., Fischer, E., Martínez, G.M., Becker, R., Liu, L., Stewart, W., Coman, E.I., Fox, V., McCullough, E., Francis, R., Matthews, T., Moore, C., Jacob, K., 2015. Evidence for indigenous nitrogen in sedimentary and aeolian deposits from the curiosity rover investigations at Gale crater, Mars. Proc. Natl. Acad. Sci. 112 (14), 4245–4250. https://doi.org/10.1073/pnas.1420932112.

Stevens, M.H., Evans, J.S., Schneider, N.M., Stewart, A.I.F., Deighan, J., Jain, S.K., Crismani, M., Stiepen, A., Chaffin, M.S., McClintock, W.E., Holsclaw, G.M., Lefèvre, F., Lo, D.Y., Clarke, J.T., Montmessin, F., Bougher, S.W., Jakosky, B.M., 2015. New observations of molecular nitrogen in the Martian upper atmosphere by IUVS on MAVEN. Geophys. Res. Lett. 42 (21), 9050–9056. https://doi.org/10.1002/2015gl065319.

Tan, J., Lewis, J.M.T., Sephton, M.A., 2018. The fate of lipid biosignatures in a Mars-analogue sulfur stream. Sci. Rep. 8 (1). https://doi.org/10.1038/s41598-018-25752-7.

Ten Kate, I.L., 2018. Organic molecules on Mars. Science 360 (6393), 1068–1069. https://doi.org/10.1126/science.aat2662.

Tosca, N.J., McLennan, S.M., Clark, B.C., Grotzinger, J.P., Hurowitz, J.A., Knoll, A.H., Schröder, C., Squyres, S.W., 2005. Geochemical modeling of evaporation processes on Mars: insight from the sedimentary record at Meridiani Planum. Earth Planet Sci. Lett. 240 (1), 122–148. https://doi.org/10.1016/j.epsl.2005.09.042.

Tuff, J., Wade, J., Wood, B.J., 2013. Volcanism on Mars controlled by early oxidation of the upper mantle. Nature 498 (7454), 342–345. https://doi.org/10.1038/nature12225.

Valdez-Nuñez, L.F., Kappler, A., Ayala-Muñoz, D., Chávez, I.J., Mansor, M., 2024. Acidophilic sulphate-reducing bacteria: diversity, ecophysiology, and applications. Environ. Microbiol. Rep. 16 (5). https://doi.org/10.1111/1758-2229.70019.

Vaniman, D.T., Bish, D.L., Ming, D.W., Bristow, T.F., Morris, R.V., Blake, D.F., Chipera, S.J., Morrison, S.M., Treiman, A.H., Rampe, E.B., Rice, M., Achilles, C.N., Grotzinger, J.P., McLennan, S.M., Williams, J., Bell, J.F., Newsom, H.E., Downs, R.T., Maurice, S., Sarrazin, P., Yen, A.S., Morookian, J.M., Farmer, J.D., Stack, K., Milliken, R.E.,

Ehlmann, B.L., Sumner, D.Y., Berger, G., Crisp, J.A., Hurowitz, J.A., Anderson, R., Des Marais, D.J., Stolper, E.M., Edgett, K.S., Gupta, S., Spanovich, N., Agard, C., Verdasca, J. A.A., Anderson, R., Archer, D., Armiens-Aparicio, C., Arvidson, R., Atlaskin, E., Atreya, S., Aubrey, A., Baker, B., Baker, M., Balic-Zunic, T., Baratoux, D., Baroukh, J., Barraclough, B., Bean, K., Beegle, L., Behar, A., Bender, S., Benna, M., Bentz, J., Berger, J., Berman, D., Blanco Avalos, J.J., Blaney, D., Blank, J., Blau, H., Bleacher, L., Boehm, E., Botta, O., Böttcher, S., Boucher, T., Bower, H., Boyd, N., Boynton, B., Breves, E., Bridges, J., Bridges, N., Brinckerhoff, W., Brinza, D., Brunet, C., Brunner, A., Brunner, W., Buch, A., Bullock, M., Burmeister, S., Cabane, M., Calef, F., Cameron, J., Iain, J., Campbell, Cantor, B., Caplinger, M., Rodríguez, J.C., Carmosino, M., Carrasco Blázquez, I., Charpentier, A., Choi, D., Clark, B., Clegg, S., Cleghorn, T., Cloutis, E., George, C., Coll, P., Conrad, P., Coscia, D., Cousin, A., Cremers, D., Cros, A., Cucinotta, F., d'Uston, C., Davis, S., Day, M.K., Juarez, M.de la T., DeFlores, L., DeLapp, D., DeMarines, J., Dietrich, W., Dingler, R., Donny, C., Drake, D., Dromart, G., Dupont, A., Duston, B., Dworkin, J., Darby Dyar, M., Edgar, L., Edwards, C., Edwards, L., Ehresmann, B., Eigenbrode, J., Elliott, B., Elliott, H., Ewing, R., Fabre, C., Fairén, A., Farley, K., Fassett, C., Favot, L., Fay, D., Fedosov, F., Feldman, J., Feldman, S., Fisk, M., Fitzgibbon, M., Flesch, G., Floyd, M., Flückiger, L., Forni, O., Fraeman, A., Francis, R., François, P., Franz, H., Freissinet, C., Louise French, K., Frydenvang, J., Gaboriaud, A., Gailhanou, M., Garvin, J., Gasnault, O., Geffroy, C., Gellert, R., Genzer, M., Glavin, D., Godber, A., Goesmann, F., Goetz, W., Golovin, D., Gómez, F.G., Gómez-Elvira, J., Gondet, B., Gordon, S., Gorevan, S., Grant, J., Griffes, J., Grinspoon, D., Guillemot, P., Guo, J., Guzewich, S., Haberle, R., Halleaux, D., Hallet, B., Hamilton, V., Hardgrove, C., Harker, D., Harpold, D., Harri, A.-M., Harshman, K., Hassler, D., Haukka, H., Hayes, A., Herkenhoff, K., Herrera, P., Hettrich, S., Heydari, E., Hipkin, V., Hoehler, T., Hollingsworth, J., Hudgins, J., Huntress, W., Hviid, S., Iagnemma, K., Indyk, S., Israël, G., Jackson, R., Jacob, S., Jakosky, B., Jensen, E., Kløvgaard Jensen, J., Johnson, J., Johnson, M., Johnstone, S., Jones, A., Jones, J., Joseph, J., Jun, I., Kah, L., Kahanpää, H., Kahre, M., Karpushkina, N., Kasprzak, W., Kauhanen, J., Keely, L., Kemppinen, O., Keymeulen, D., Kim, M.-H., Kinch, K., King, P., Kirkland, L., Kocurek, G., Koefoed, A., Köhler, J., Kortmann, O., Kozyrev, A., Krezoski, J., Krysak, D., Kuzmin, R., Lacour, J.L., Lafaille, V., Langevin, Y., Lanza, N., Lasue, J., Le Mouélic, S., Lee, E.M., Lee, Q.-M., Lees, D., Lefavor, M., Lemmon, M., Lepinette Malvitte, A., Leshin, L., Léveillé, R., Lewin-Carpintier, É., Lewis, K., Shuai, L., Lipkaman, L., Little, C., Litvak, M., Lorigny, E., Lugmair, G., Lundberg, A., Lyness, E., Madsen, M., Mahaffy, P., Maki, J., Malakhov, A., Malespin, C., Malin, M., Mangold, N., Manhes, G., Manning, H., Marchand, G., Marín Jiménez, M., García, C.M., Martin, D., Martin, M., Martínez-Frías, J., Martín-Soler, J., Javier Martín-Torres, F., Mauchien, P., McAdam, A., McCartney, E., McConnochie, T., McCullough, E., McEwan, I., McKay, C., McNair, S., Melikechi, N., Meslin, P.-Y., Meyer, M., Mezzacappa, A., Miller, H., Miller, K., Minitti, M., Mischna, M., Mitrofanov, I., Moersch, J., Mokrousov, M., Molina Jurado, A., Moores, J., Mora-Sotomayor, L., Mueller-Mellin, R., Muller, J.-P., Muñoz Caro, G., Nachon, M., Navarro López, S., Navarro-González, R., Nealson, K., Nefian, A., Nelson, T., Newcombe, M., Newman, C., Nikiforov, S., Niles, P., Nixon, B., Noe Dobrea, E., Nolan, T., Oehler, D., Ollila, A., Olson, T., Owen, T., Hernández, M.Á.de P., Paillet, A., Pallier, E., Palucis, M., Parker, T., Parot, Y.,

Patel, K., Paton, M., Paulsen, G., Pavlov, A., Pavri, B., Peinado-González, V., Pepin, R., Peret, L., Perez, R., Perrett, G., Peterson, J., Pilorget, C., Pinet, P., Pla-García, J., Plante, I., Poitrasson, F., Polkko, J., Popa, R., Posiolova, L., Posner, A., Pradler, I., Prats, B., Prokhorov, V., Wilson Purdy, S., Raaen, E., Radziemski, L., Rafkin, S., Ramos, M., Raulin, F., Ravine, M., Reitz, G., Rennó, N., Richardson, M., Robert, F., Robertson, K., Manfredi, J.A.R., Romeral-Planelló, J.J., Rowland, S., Rubin, D., Saccoccio, M., Salamon, A., Sandoval, J., Sanin, A., Fuentes, S.A.S., Lee, S., Sautter, V., Savijärvi, H., Schieber, J., Schmidt, M., Schmidt, W., Dan, D., Scholes, Schoppers, M., Schröder, S., Schwenzer, S., Martinez, E.S., Sengstacken, A., Shterts, R., Siebach, K., Siili, T., Simmonds, J., Sirven, J.-B., Slavney, S., Sletten, R., Smith, M., Sánchez, P.S., Spray, J., Squyres, S., Stalport, F., Steele, A., Stein, T., Stern, J., Stewart, N., Svane Stipp, S.L., Stoiber, K., Sucharski, B., Sullivan, R., Summons, R., Sun, V., Supulver, K., Sutter, B., Szopa, C., Tan, F., Tate, C., Teinturier, S., ten Kate, I., Thomas, P., Thompson, L., Tokar, R., Toplis, M., Torres Redondo, J., Trainer, M., Tretyakov, V., Urqui-O'Callaghan, R., Van Beek, J., Van Beek, T., VanBommel, S., Varenikov, A., Vasavada, A., Vasconcelos, P., Vicenzi, E., Vostrukhin, A., Voytek, M., Wadhwa, M., Ward, J., Webster, C., Weigle, E., Wellington, D., Westall, F., Craig Wiens, R., Wilhelm, M.B., Williams, A., Williams, R., Mouser, R.B., Williams, Wilson, M., Wimmer-Schweingruber, R., Wolff, M., Wong, M., Wray, J., Wu, M., Yana, C., Yingst, A., Zeitlin, C., Zimdar, R., Zorzano Mier, M.-P., 2014. Mineralogy of a mudstone at Yellowknife Bay, Gale Crater, Mars. Science 343 (6169). https://doi.org/10.1126/science.1243480.

Velasco, E., Mason, Gonzalez-Toril, E., Zegers, T., Davies, G., Amils, R., 2010. Microbial sulphur isotope fractionation in a Mars analogue environment at Rio Tinto. In: EGU General Assembly Conference Abstracts.

Westall, F., Foucher, F., Bost, N., Bertrand, M., Loizeau, D., Vago, J.L., Kminek, G., Gaboyer, F., Campbell, K.A., Bréhéret, J.G., Gautret, P., Cockell, C.S., 2015. Biosignatures on Mars: what, where, and how? Implications for the search for Martian life. Astrobiology 15 (11), 998–1029. https://doi.org/10.1089/ast.2015.1374.

Wordsworth, R.D., 2016. The climate of early Mars. Annu. Rev. Earth Planet Sci. 44, 381–408. https://doi.org/10.1146/annurev-earth-060115-012355.

CHAPTER

Technological developments for the characterization of the environment and subsurface of the Iberian pyrite belt and Mars

15

Jose A. Rodriguez-Manfredi[1], Sara Navarro[1], Victor Parro[2], Eduardo Sebastián[1], Germán Martínez[1], Josefina Torres[3], Javier Gómez-Elvira[1,3]

[1]*Department of Instrumentation and Space Exploration, Centro de Astrobiologia (CAB), INTA-CSIC, Madrid, Spain;* [2]*Department of Molecular Evolution, Centro de Astrobiologia (CAB), INTA-CSIC, Torrejón de Ardoz, Madrid, Spain;* [3]*Instituto Nacional de Técnica Aeroespacial, Department of Payloads, Madrid, Spain*

Introduction: bridging two worlds through technology

Exploring and trying to understand planetary environments is one of the most important scientific endeavors of our time. Among the planets in our Solar System, Mars stands out as a prime target due to its geological complexity, its potential to host past or present life, its relevance for comparative planetology with Earth, and perhaps as a possible future destination for humanity. However, the tools and strategies needed to explore such distant and hostile terrains cannot emerge in a vacuum; these require long and complex design processes. Terrestrial analogs provide a critical bridge between conceptual designs and actual planetary missions. In this context, the Iberian Pyrite Belt (IPB), and in particular the Río Tinto region in southwestern Spain, has emerged as a site of unparalleled value (Amils et al., 2007).

The Río Tinto area hosts a unique geobiological environment shaped by intense hydrothermal activity, extreme acidity, and abundant iron and sulfur cycles. These conditions have led to the formation of an ecosystem that is not only resilient to harsh environments but also biogeochemically complex (Fernández-Remolar et al., 2005). Such features closely mirror key characteristics of ancient Martian environments inferred from data gathered by orbiters and in situ missions (Squyres et al., 2004; Grotzinger et al., 2014). The oxidation states, mineral assemblages, and microbe-mineral interactions observed at Río Tinto provide clues about the kinds of biosignatures that might be found on Mars, while also allowing us to test the durability and performance of the technologies needed to detect them.

Geomicrobiology of Río Tinto. https://doi.org/10.1016/B978-0-443-32948-7.00007-2

Over the years, extensive field campaigns and controlled experiments in Río Tinto have demonstrated the site's value, not only for science validation, but also for the development of technology and engineering processes. Aspects such as real-time data acquisition, environmental monitoring under variable geochemical conditions, deployment of autonomous robotic systems, and remote operations have all been tested in this setting. Consequently, Río Tinto has served both as a testing ground for technologies intended for Mars and as a laboratory for refining our scientific questions about the Red Planet. It has become a cornerstone for iterating the loop between conceptual planetary science, engineering design, and practical implementation in the field.

In this chapter, we adopt the Río Tinto system, one of the most scientifically valuable Mars analogs on Earth, not only as a geological and astrobiological reference, but as an operational and conceptual framework for understanding the challenges of Martian subsurface exploration. Drawing on the unique geochemical and microbial characteristics of the IPB, we examine the rationale for drilling and environmental monitoring on Mars, and trace the development of the technologies required to access, characterize, and interpret buried environments. Through detailed analysis of drilling systems, autonomous platforms, and atmospheric payloads, we show how analogue field campaigns at Río Tinto have helped validate mission architectures and instrumentation. The chapter concludes with a forward-looking assessment of future exploration strategies, highlighting the role of small robotic systems, the potential of In-Situ Resource Utilization (ISRU), and the integration of environmental science into mission planning, critical elements for ensuring the sustainability and scientific return of upcoming Mars missions.

The Iberian Pyrite Belt and Mars: A tale of two analogs

The Iberian Pyrite Belt, extending across southern Spain and Portugal, is among the largest and most mineralogically diverse sulfide regions on Earth. Its geological history has been shaped by prolonged hydrothermal activity, leading to the formation of massive polymetallic deposits (Tornos, 2006). In the Río Tinto area, acidic surface waters and subsurface fluids have interacted with pyritic substrates for millions of years, producing highly oxidized, iron-rich terrains and chemically extreme environments. These conditions offer strong parallels with those inferred for early Mars, particularly during the Noachian and early Hesperian periods (Bibring et al., 2006).

Remote sensing and in situ exploration of Mars have revealed the presence of iron oxides, sulfates, and silica-rich deposits, minerals typically associated with aqueous alteration (Squyres et al., 2008). In addition, the detection of sedimentary layers and hydrothermal veins suggests that Mars once hosted active water-rock interactions in localized settings. These environments, perhaps hydrothermal or groundwater-fed, could have sustained microbial life, assuming sufficient time and favorable chemical gradients were present.

The redox gradients and acidic conditions found in the Río Tinto system make it an excellent analogue for these Martian contexts. From a biological standpoint, the microbial communities thriving in Río Tinto's subsurface and surface waters, many of which are iron- and sulfur-oxidizing extremophiles, provide a living model for hypothetical life-supporting metabolisms in early Mars (Amils et al., 2011). These communities provide insight into how life can flourish under conditions that are both chemically challenging and in the absence of sunlight. They also serve as a reference for calibrating instruments designed to detect biosignatures.

This analogy, however, is not merely conceptual, going beyond theory. Over the past 2 decades, Río Tinto has been used as a site for testing planetary instruments and operations, from spectrometers and subsurface probes to autonomous drilling systems (Stoker et al., 2008). These field campaigns have validated instrument performance under field conditions, improved protocols for life-detection missions, and contributed to refining mission architectures for Mars exploration. In this sense, Río Tinto functions not only as a science analogue but as a systems integration environment.

Scientific imperatives for subsurface characterization

Subsurface exploration has become a cornerstone of planetary science, driven by a growing recognition that many of the most compelling scientific targets on Mars lie beneath its surface (Vago et al., 2017). Unlike the heavily irradiated and oxidized surface, the subsurface may preserve biosignatures, unaltered geochemical records, and even transient water activity shielded from extreme conditions. The Martian surface has been exposed to cosmic radiation and oxidative agents for billions of years; in contrast, subsurface niches may offer more favorable conditions for long-term preservation.

Understanding the stratigraphy, geochemistry, and potential habitability of these subsurface domains is key to reconstructing Mars's past and assessing its astrobiological potential. Subsurface environments may contain not only mineralogical records of past aqueous alteration but also direct evidence of past or extant microbial life, should it exist. This is particularly true in the presence of hydrated minerals or amorphous phases capable of trapping organic compounds.

On Earth, subsurface studies in extreme environments have revolutionized our understanding of deep microbial ecosystems, mineral-weathering processes, and fluid-rock interactions (Onstott et al., 1999). These environments are often anoxic, energy-limited, and rich in reduced chemical species, features that make them similar to possible Martian subsurface environments. In Río Tinto, drilling campaigns have revealed redox interfaces, mineralogical gradients, and biological niches that are inaccessible from surface studies alone (Amils et al., 2008). These findings provide a valuable interpretative framework for designing dedicated instrumentation and shaping future Mars missions that prioritize deep drilling and in situ analysis.

Furthermore, the potential use of local resources for ISRU adds another layer of motivation for subsurface exploration, particularly for future crewed missions. Water, brines, and certain mineral phases could be harnessed for life support and fuel production (Boynton et al., 2002; Gayen et al., 2020). Assessing the availability and accessibility of these resources will require detailed subsurface mapping and characterization. In this way, scientific goals and operational needs converge, reinforcing the importance of investing in robust exploration technologies.

The role of technology in unlocking geological and astrobiological mysteries

Technological innovation is central to subsurface exploration. From the earliest core-sampling drills to the latest autonomous robotic systems, each step forward has opened new frontiers in our ability to investigate otherwise inaccessible environments. Historically, planetary exploration relied on surface-based observations, but the development of miniaturized, robust instruments has made subsurface access feasible within the constraints of space missions (Vago et al., 2006).

In the case of Río Tinto, numerous technologies have been developed and tested through collaborative international projects. The IPBSL (Iberian Pyrite Belt Subsurface Life) project, for example, deployed advanced drilling systems, integrated sensor packages, and continuous monitoring platforms to study biological and geochemical processes in situ (Amils et al., 2013). These efforts laid the groundwork for analogous operations on Mars, such as the ExoMars Rosalind Franklin rover, which has been designed to deploy a drill capable of reaching 2 m beneath the Martian surface (Vago et al., 2017).

Equally important has been the integration of autonomous systems capable of performing scientific operations with minimal human intervention. This includes onboard decision-making algorithms, adaptive sampling strategies, and real-time data processing. These capabilities are essential for missions operating under communication delays and limited bandwidth, as is the case on Mars. The progressive incorporation of artificial intelligence tools and machine learning classifiers is now enabling rovers to prioritize sample collection, perform spectral interpretation onboard, and dynamically adjust navigation plans.

Technological development is also about robustness and resilience. Drilling on Mars or in Río Tinto is not just a matter of penetrating rock. It involves mechanical challenges, contamination control, instrument miniaturization, and real-time analytics. The iterative field testing of these systems in Río Tinto plays a key role in reducing risk in future space missions, and often uncovers subtle design flaws not evident in laboratory settings.

In sum, the interplay between terrestrial analogs, scientific objectives, and technological innovation forms a triad that drives planetary exploration forward. The Río Tinto experience exemplifies how Earth-based research can derisk mission architectures, validate instrument design, and refine scientific objectives.

Río Tinto as a technological testing ground for planetary exploration

Río Tinto beyond a martian analogue

The region has long been recognized for its value as a terrestrial analog of Mars. However, limiting its relevance only to astrobiology would not only be incomplete but also a missed scientific and technological opportunity. Río Tinto is a unique geological, geochemical, and biological environment that serves as a multifaceted platform for a wide range of research: from precision mining to advanced instrumentation (Fernández-Remolar et al., 2005; Amils et al., 2011).

One of the most striking features of the area is the coexistence of geological and biological processes under extreme conditions: high acidity, elevated concentrations of heavy metals, and intense environmental oxidation. These characteristics have given rise to a specialized microbial ecosystem composed of acidophilic archaea and bacteria actively involved in iron and sulfur biogeochemical cycling (Amils et al., 2007). Far from being just an astrobiological testing ground, this interplay between abiotic and biotic processes makes Río Tinto a natural laboratory for organic geochemistry, mineral biosynthesis, and ecosystem resilience, offering opportunities to study these phenomena using dedicated technologies.

Another key feature that expands the site's interest beyond astrobiology is the uniqueness of its hydrological system. Surface and subsurface waters maintain a consistently acidic regime (pH near 1), largely driven by microbial activity on sulfide minerals. This dynamic equilibrium has persisted for millennia, making it exceptionally stable from a geochemical perspective. In addition, the mineral stratigraphy and subsurface geophysics allow detailed correlations between mineral alterations, ancient hydrothermal processes, and modern seismic signals, providing unique opportunities to validate remote sensing and geophysical prospecting methodologies.

In this context, the site has also served as a reference environment for evaluating emerging technologies applicable not only to planetary exploration but also to sectors like sustainable mining or autonomous environmental monitoring. For example, continuous chemical sensors, hyperspectral spectroscopy, miniaturized geophysical systems, and terrestrial robotic platforms have all been tested there. The interaction of these tools with a dynamic and chemically challenging environment has provided real-world data for calibrating, validating, and optimizing control algorithms, data acquisition methods, and decision-making protocols (Stoker et al., 2008).

Moreover, the heritage and historical significance of the region add another layer of socioeconomic complexity. The coexistence of a long-standing mining tradition, an extreme ecosystem, and growing scientific activity calls for a holistic approach, one that takes into account the ethical, social, and cultural dimensions of technological intervention. In this sense, the area becomes a space where cutting-edge science meets the challenges of sustainable development and land-use legacy management.

In short, Río Tinto is more than an excellent analog for Mars. It is a singular environment for developing and testing advanced geoscientific and environmental technologies. Its multidimensional richness positions it as a privileged setting to explore the integration of disciplines, objectives, and technologies that, while initially designed for planetary exploration, find here a testbed with immediate relevance to Earth-based applications.

The MARTE project: Lessons from an analog deep drilling mission

The Mars Astrobiology Research and Technology Experiment (MARTE) project was a foundational research initiative in the development of technologies for Martian subsurface exploration (Stoker et al., 2008). Conducted between 2002 and 2005 under NASA's ASTEP (Astrobiology Science and Technology for Exploring Planets) program, this project aimed to demonstrate the technical and scientific feasibility of searching for life beneath the Martian surface (Cannon et al., 2007). Led by NASA Ames Research Center, the project also involved collaboration from the Centro de Astrobiología (CSIC-INTA), Honeybee Robotics, the University of Oklahoma, and other international partners. It represented one of the most ambitious field tests carried out to date.

The selection of Río Tinto as the site for this simulation was based on the area's extreme geochemical characteristics (its acidic waters, high concentrations of heavy metals, and specialized microbial communities, as described), all of which provide an exceptional terrestrial analog for ancient Martian hydrothermal environments.

Strategic objectives and methodological approach

MARTE was designed with a dual experimental architecture that reflected the inherent complexity of modern planetary exploration. The project aimed to scientifically characterize the anaerobic, chemolithoautotrophic subsurface biosphere present in the Río Tinto region, powered largely by iron and sulfide minerals. At the same time, it aimed to develop and demonstrate an integrated technical approach to automated drilling for subsurface exploration on Mars (Glass et al., 2005), including sample handling and in situ life detection (Cannon et al., 2007).

The experimental strategy included three primary objectives (Stoker et al., 2008). The first was to search for and characterize subsurface life in Río Tinto, along with the physical and chemical properties of its environment and the energy sources that sustain it. The second objective involved conducting a high-fidelity simulation of a robotic Martian drilling mission to characterize the subsurface and search for subsurface life. The third focused on demonstrating drilling, sample handling, and instrumentation technologies relevant to the search for life on Mars.

Technological development and systems integration

The MARTE robotic system represented one of the most sophisticated drilling and sample analysis platforms developed for planetary applications (Cannon et al., 2007). The integrated platform, weighing approximately 900 kg and standing about 3 m tall with a diameter of 2.4 m, incorporated six primary subsystems operating in coordinated fashion under automated control. Although not optimized for flight in terms of mass, the system was sized to fit under the fairing of a Delta II rocket, setting realistic parameters for future missions (Stoker et al., 2008).

The drilling subsystem, developed by Honeybee Robotics, was the technological core of the project (Glass et al., 2005). This 10-axis, 150 W average power drill was designed to recover subsurface samples up to 10 m deep without using drilling fluids. The system produced cores 27 mm in diameter and 250 mm long, creating boreholes 48 mm wide through dry rotary coring techniques involving carbide and monocrystalline diamond cutters.

The Core Sample Handling System (CSHS), developed by the University of Oklahoma, provided automated capabilities for sample manipulation and preparation (Cannon et al., 2007). This subsystem included 24-degree-of-freedom core grippers, rotary storage units for up to nine cores, faceting and subsampling saws, and systems for crushing and transferring samples. The integration of these components enabled full sample processing from extraction to analysis, while maintaining strict contamination control protocols.

Scientific instrumentation and analytical systems

The remote science instrument suite incorporated cutting-edge technology for mineralogical characterization and sampling location selection (Cannon et al., 2007). The Core Context Imager captured color images of full cores at a resolution of 125 μm/pixel, while the Microscopic Imager provided color microscopic images at 7 μm/pixel over a $2.5 \times 1.7\ cm^2$ field of view. The VNIR spectrometers, both point and imaging, enabled mineral identification through spectral analysis in the 400–1050 nm range.

The SOLID2 (Signs Of LIfe Detector, version 2) instrument, developed by the Centro de Astrobiología (Parro et al., 2005, 2007), marked a significant innovation in life detection using antibody-based microarrays. This automated device used 157 different antibodies to detect biochemical compounds, including nucleic acids, proteins, and polysaccharides, with parts-per-billion sensitivity. The ability of SOLID2 to automatically process 0.5 g of powdered sample and perform in situ immunological analyses sets new standards for planetary life-detection instrumentation.

Borehole inspection and operational architecture

The Borehole Inspection System (BHIS), also developed by the Centro de Astrobiología, provided in situ borehole wall inspection capabilities using panoramic cameras and Raman spectroscopy (Stoker et al., 2008). The BHIT (Borehole Inspection Tool) included two color CCD cameras and a Raman spectrometer equipped with a 5 mW He–Ne laser, enabling the acquisition of 360 degrees panoramic images and georeferenced Raman spectra with 10 cm^{-1} spectral resolution. The system also incorporated a second probe, the BHIST (Borehole Inspection Standard Tool), which featured a magnetic susceptibility sensor and provided invaluable data for characterizing the magnetic properties of the subsurface around the drilled hole.

The remote operations subsystem used the Contingent Rover Language (CRL) to execute automated mission plans (Cannon et al., 2007). This architecture, previously validated on NASA Ames rovers, allowed all subsystems to be operated via high-level commands sent from remote science teams. Satellite communication facilitated daily transmission of scientific data and receipt of operational plans, simulating realistic conditions for a Martian mission.

Operational results and technical achievements

The mission simulation, conducted in September 2005 at Peña de Hierro, in Río Tinto, met and exceeded all predefined success criteria (Stoker et al., 2008). Over 30 days of operation, the system reached depths of up to 6.1 m, recovering 29 cores for scientific analysis. Despite the challenging geological conditions of poorly consolidated gossan, which limited core recovery to an average of 32%, the system demonstrated robust operational performance.

Remote science teams operating from Madrid, Spain, and California, USA, sequentially managed to correctly interpret the geological nature of the drilled deposits compared to field truth (Stoker et al., 2008). Analysis of microscopic images and spectral data allowed the identification of diagnostic features of gossan, including botryoidal textures, mineral dissolution cavities, and pseudomorphs of oxidized sulfides. VNIR spectroscopy confirmed the presence of goethite and hematite, corroborating interpretations based on color imagery.

Biosignature detection and scientific validation

The results from the SOLID2 instrument constituted one of the most significant achievements of the project (Stoker et al., 2008). Out of 15 samples analyzed from depths ranging from 0.65 to 5.57 m, biosignatures were detected in 12 samples. These included compounds such as cyclic AMP, cortisol-related steroids, and specific signatures of bacteria like *Leptospirillum ferrooxidans* and species of Acidithiobacillus. Control experiments using sterilized samples validated the biological origin of the detected signals.

Identified technical challenges

The MARTE project revealed significant technical limitations that provided valuable insights for the development of future planetary drilling technologies (Cannon et al., 2007). Low core recovery rates in poorly consolidated material highlighted the need for more effective capture mechanisms. The system relied on small jaws and friction to retain cores, which proved inadequate for fractured materials.

The 77 drilling system failures during operations underscored the importance of automated diagnostics and autonomous recovery capabilities (Cannon et al., 2007). Cutting transport emerged as a critical issue, with obstructions causing torque spikes and potentially mission-compromising jams. Experimental use of compressed air to remove cuttings improved penetration rates but compromised core recovery in unconsolidated materials. Furthermore, although the system was designed to operate on limited power (~150 W), the combined energy load profile, including active cooling and repetitive multi-axis movements, produced unexpected demand peaks, which could pose risks in future solar-powered missions with strict energy constraints.

Impact on mission architecture development

The lessons learned from the MARTE project have significantly influenced the design of subsequent Martian missions (Stoker et al., 2008). The successful demonstration of automated drilling, sample handling, and remote life detection established the technical viability of subsurface life-search missions. The contamination control protocols developed during the project provided reference frameworks for future planetary missions (Vago et al., 2017).

In addition, the IceBreaker mission concept proposed by NASA, focused on studying ice and organic materials in Martian polar latitudes, explicitly incorporates several subsystems validated in MARTE, including the modular approach, SOLID-based analysis, and coordination between drilling and spectroscopic analysis (McKay et al., 2013).

The successful integration of multiple technologies into a single coherent platform demonstrated the feasibility of complex subsurface exploration systems (Cannon et al., 2007). The project showed that a 10-m drilling mission on Mars is technically achievable within approximately 3 months of mission time, though further technological development is required for fully reliable systems. Experience gained in remote operations and automated decision-making has served as the basis for the development of control architectures for current and future Martian rovers.

Legacy and future prospects

The legacy of the MARTE project extends beyond its immediate technical achievements, establishing a paradigm for integrating cutting-edge science, technological development, and validation in terrestrial analog environments (Stoker et al., 2008). Its contribution to understanding the operational and technological requirements for subsurface exploration of Mars continues to serve as a basis for the development of future planetary astrobiology missions.

The project demonstrated that it is possible to build complex systems capable of functioning autonomously in extreme environments, that life can be detected in the subsurface using robotic technology, and that these operations can be performed from thousands of kilometers away (Cannon et al., 2007). These capabilities represent exactly the fundamental requirements needed for the successful exploration of Mars.

In the context of future crewed missions, the knowledge gained from MARTE provided a useful foundation for planning the logistics and reliability of autonomous drilling systems that could operate alongside astronauts, reducing their exposure to hazardous environments. Systems like the one tested at Peña de Hierro could, after adaptations, help establish geological prospecting routines or astrobiological assessments near human settlements on Mars.

Moreover, the current strategic context of Mars exploration, with missions such as Mars Sample Return and proposals like Enceladus Orbilander or Europa Clipper, which also explore remote drilling, highlights the value of trials like MARTE. Such simulations offer a realistic framework in which to evaluate not only technologies but also operational architecture design, resource allocation, and risk mitigation strategies.

In this sense, the MARTE model could be replicated with tailored variations in other terrestrial locations, including cold or polar environments such as Arctic permafrost, which would be even closer to the astrobiological targets of regions with surface or subsurface ice on Mars.

Lessons from Río Tinto applied to Mars

Río Tinto has served as a unique terrestrial analogue for Mars, enabling the testing of hypotheses, technologies, and operations in a setting that replicates key Martian conditions. One of the primary lessons is the value of conducting research in natural environments that combine extreme geochemistry, microbial life, and operational challenges. This complexity allows for a more realistic preparation for planetary exploration.

Scientifically, Río Tinto has shown that biosignatures can be preserved in minerals like jarosite and goethite, supporting the idea that similar targets on Mars may hold evidence of past life. The coexistence of abiotic and biotic processes also helped refine strategies to distinguish between biological and non-biological signals.

On the technological side, the MARTE project highlighted challenges in sample recovery, contamination control, and system coordination. These lessons informed the design of future drilling systems, emphasizing the need for modularity, reliability, and clean operation. Río Tinto also demonstrated the importance of autonomy: delayed communications and environmental unpredictability showed that systems must be able to operate independently with robust onboard decision-making.

Operationally, the experience improved understanding of how to prioritize science goals while managing limited time, energy, and resources. It also helped refine workflows between scientific planning and engineering execution, which is vital for real mission scenarios.

Finally, Río Tinto's success has encouraged the identification of other analogue sites, especially cold or icy environments, to simulate Mars' polar regions. The knowledge gained from Río Tinto continues to inform current and future missions like ExoMars and IceBreaker, strengthening the foundations for robotic and human exploration of the Martian subsurface.

Cutting-edge technologies for martian subsurface and environmental characterization

Current technologies in subsurface exploration

Mars has long been a central target for planetary exploration, driven by the search for signs of past or present life and the desire to reconstruct its geological and climatic evolution. Subsurface exploration plays a key role in this endeavor, as the underground layers of the planet are more likely to preserve biosignatures and environmental records shielded from the harsh surface conditions. Several missions have contributed significantly to this field by deploying a variety of technologies to access and investigate the shallow subsurface.

NASA's Mars Science Laboratory mission, led by the Curiosity rover, has been operating since 2012 in Gale Crater. While primarily focused on surface exploration, the mission has nonetheless contributed invaluable insights into the shallow subsurface. Curiosity is equipped with a rotary-percussion drill capable of sampling rock material from a few centimeters below the surface. These powdered samples are delivered to onboard analytical instruments, SAM (Sample Analysis at Mars) and CheMin (Chemistry and Mineralogy X-ray Diffraction), which have revealed the presence of organic molecules and minerals such as clays, suggesting past aqueous environments and habitability (Grotzinger et al., 2014). In addition, the DAN (Dynamic Albedo of Neutrons) instrument has mapped variations in hydrogen content below the surface to depths of about 1 m, indirectly indicating the presence of bound water or hydrated minerals (Mitrofanov et al., 2012). Although its reach is limited, Curiosity has helped lay the groundwork for deeper and more targeted subsurface missions.

Launched in 2018, the InSight lander focused on geophysical exploration. Its HP^3 instrument (Flow and Physical Properties Package) was designed to penetrate up to 5 m below the surface to measure thermal gradients, but unexpectedly cohesive soil conditions prevented the "mole" from achieving its target depth. Despite this limitation, the mission yielded important data on regolith properties and mechanical resistance. More notably, the SEIS seismometer provided the first direct evidence of seismic activity on Mars, enabling a clearer view of its internal structure and layering (Lognonné et al., 2020).

The Perseverance rover, part of NASA's Mars 2020 mission, has introduced substantial advancements in subsurface exploration. Its RIMFAX instrument (Radar Imager for Mars Subsurface Experiment), a ground-penetrating radar, delivers high-resolution vertical profiles of the subsurface to depths of up to 10 m, allowing researchers to identify sedimentary structures and potential ice-rich layers (Farley et al., 2020). Additionally, Perseverance is collecting and caching core samples for future retrieval, linking in situ science with upcoming Earth-based analyses.

Looking ahead, the Rosalind Franklin rover from ESA's ExoMars program is designed to drill down to 2 m, significantly deeper than any previous rover. The objective is to retrieve samples that may have been shielded from radiation and oxidation since their deposition. Onboard instruments such as MOMA (Mars Organic Molecule Analyzer) will search for molecular biosignatures in these well-preserved materials (Vago et al., 2017).

Altogether, these missions represent a cumulative effort to access, characterize, and understand the Martian subsurface. While technical challenges (such as unexpected soil behavior) have emerged, each mission has expanded our capabilities and informed the design of the next. As technologies evolve, future missions may probe even deeper, bringing us closer to resolving one of Mars exploration's most compelling questions: whether life ever emerged beneath its surface.

Beyond drilling: Autonomous robotic exploration of the subsurface

Drilling on Mars presents significant challenges due to the planet's harsh environment, including ubiquitous dust, broad temperature fluctuations, heterogenous soil composition, and the possible presence of subsurface ice. The development of more robust, low-mass and autonomous drilling systems, along with core retrieval and sample handling capabilities, represents a critical step forward for upcoming planetary missions. Field campaigns and analogue research in terrestrial sites such as Río Tinto (Stoker et al., 2008) have proven essential for testing and developing future technologies for Martian subsurface exploration.

Planetary autonomous drilling missions require systems that can operate without human intervention. This requires sophisticated architectures that address power limitations, mass constraints, and the need for robust fault detection and recovery mechanisms. The design involves developing automated drilling systems with intelligent software for operation, fault diagnosis, and corrective actions, all while minimizing power consumption and mass. Power is a precious resource on Mars, requiring drilling systems to be designed for low power consumption to maximize operational time and mission lifespan; solar power or RTGs (Radioactive Thermal Generators) are likely to be the energy source used, which will require efficient drilling methods. Low mass is also crucial due to launch and operational limitations.

Robotic exploration of Mars' subsurface can be broadly categorized into two main types: drilling and shallow in-situ analysis. Drilling involves physically penetrating the Martian surface to access deeper layers, ranging from centimeters to several meters. Deeper access allows the recovery of materials potentially shielded from radiation and oxidation, while shallower, laser-based techniques are generally applied for in situ geochemical assessments, offering limited penetration but rapid analytical feedback. In all cases, drilling depth strongly constrains the selection and engineering of the drilling mechanism.

Several terrestrial technologies are currently being adapted for space applications. Mechanical drilling systems employing diamond-impregnated bits, discrete cutter heads, or rotary-percussive approaches are being refined for use under Martian conditions. Plasma-based drilling, a novel approach under development, uses high-voltage electrical discharges to vapourize rock instead of fracturing it mechanically, offering a potential advantage in icy or consolidated substrates (Tang et al., 2022).

Laser drilling systems, exemplified by instruments like SuperCam instrument aboard Perseverance, employ focused laser pulses to vapourize surface material and analyze its composition via spectrometry. Although limited in penetration, laser techniques are highly effective for noncontact, localized geochemical analyses, and are well suited for rapid surface characterization.

Mechanical drilling platforms, such as those employed by the Curiosity and Perseverance rovers, have demonstrated the feasibility of retrieving powdered or cored samples for onboard analysis. These operations require a suite of ancillary systems for sample acquisition, manipulation, and delivery to scientific instruments. Critical to such operations is the implementation of contamination control protocols, particularly sterilization of any components contacting the subsurface, to comply with planetary protection guidelines.

An emerging paradigm involves embedding analytical capabilities directly into the drilling architecture. This could enable real-time scientific decision-making, improve operational efficiency, and potentially reduce reliance on sample return strategies (Glass et al., 2014). In the context of future missions, a combined approach, integrating both mechanical and laser-based techniques, may offer the optimal trade-off between penetration depth, analytical precision, and resource consumption.

Mechanical systems excel in achieving depth and bulk sampling, offering greater control and scalability for ISRU scenarios, while laser drilling provides rapid, targeted, and minimally invasive analysis. By leveraging the respective strengths of these approaches, mission architectures can be tailored to address diverse scientific goals, ranging from biosignature detection to resource mapping.

Advanced geophysical methods

The application of geophysical methods to Mars has enabled a deeper understanding of the planet's subsurface. Supporting this effort, a variety of instruments have been developed and, in some cases, already deployed either on the surface or in orbit.

Ground Penetrating Radars (GPRs) have played a key role in subsurface exploration, enabling the mapping of subsurface ice, sediment layers and, potentially in future missions, lava tubes and subsurface water deposits. These instruments have been deployed aboard orbiters: i.e., SHARAD -Mars SHAllow RADar sounder on the Mars Reconnaissance Orbiter-reaching a ~1 km depth approximately, and MARSIS -Mars Advanced Radar for Subsurface and Ionosphere Sounding on Mars Express-to a depth of ~5 km. GPRs have been proposed for several rover and lander missions, with RIMFAX being the first GPR deployed on the Martian surface aboard the Perseverance rover (Hamran et al., 2022).

Seismology is another field of great interest, encompassing the study of Mars quakes and the characterization of meteoroid impacts. NASA's InSight mission (Banerdt et al., 2020) was designed to investigate Martian seismic activity, deploying the SEIS (Seismic Experiment for Interior Structure) instrument, which enabled a detailed study and provided unprecedented insight into the planet's inner structure (crustal thickness and inner core size). This mission also carried a heat flow measurement probe (HP3), designed to measure heat flow and thermal conductivity by drilling a hole to insert the probe.

Similarly, electromagnetic induction probes, though not yet deployed on Mars, would allow monitoring of the electrical conductivity of subsurface materials, enabling the detection of water reservoirs, stratigraphy and lithography in the subsurface.

Magnetic field studies also contribute to our understanding of Mars' geologic history. Mars Global Surveyor orbiter mapped local magnetic fields, originated by the planet's crustal anomalies (Acuña et al., 1998). Magnetometers are frequently included in rover and lander payload proposals to extend magnetic field investigations to the surface. The IFG instrument (InSight Fluxgate Magnetometer) on InSight exemplifies this approach, providing enhanced resolution of surface magnetic field measurements (Mittelholz et al., 2022).

The planet's gravitational field and its anomalies, which serve to identify crustal thickness variations, have been subsequently mapped with orbiter radiometer payloads of both Mars Reconnaisance Orbiter (MRO) and Mars Odyssey (Genova et al., 2016). Local measurements have also been performed using the navigation instruments of the Mars rovers.

The detection of water or hydrated minerals is also of paramount scientific interest. This search was carried out in the past using neutron spectrometry (Dynamic Albedo of Neutrons instrument -DAN- of the MSL mission), and future developments of nuclear magnetic resonance instruments will enable more thorough and improved hydrogen measurements to detect water. Dielectric spectroscopy can be employed, either independently or in combination with neutron spectroscopy, to detect all phases of water within the first few centimeters of the subsurface.

Gamma-ray spectrometers, such as the GRS aboard Mars Odyssey (Boynton et al., 2004), have been successfully used to detect Mars' natural radioactivity, thus determining the composition of subsurface rocks and minerals, giving further insight into the planet's geologic history.

Environmental monitoring systems for Mars planetary exploration

Environmental monitoring has played a central role in Mars exploration, offering critical insights into the planet's present and past climate, its habitability potential, and the operational challenges for robotic and future crewed missions. These objectives have progressively shaped mission architectures, aligning scientific goals with technological innovation in atmospheric instrumentation.

NASA's long-term Mars Exploration Program (structured around four goals: Life, Climate, Geology, and Preparation for Human Exploration), explicitly integrates atmospheric science across each theme. While these goals have been articulated through the Mars Exploration Program Analysis Group (MEPAG) since the early 2000s, environmental monitoring has been integral to planetary exploration since its earliest days. Even before MEPAG, missions like Viking recognized the importance of characterizing the atmosphere for both scientific understanding and the safe design of landing systems.

The Viking missions (1976) had broad scientific objectives, including the search for life, geologic characterization, and atmospheric analysis. Specifically, the Viking mission sought to characterize mesoscale and synoptic systems, detect small- and medium-scale weather phenomena, and probe the structure of the boundary layer, providing foundational data to model Mars' climate and surface conditions. To achieve these environmental objectives, the Viking landers featured relatively simple but robust meteorological packages measuring temperature, pressure, and wind (Chamberlain et al., 1976). These instruments provided the first long-term weather records on Mars and established a baseline for atmospheric variability (Hess et al., 1977).

Pathfinder (1997) shared similar goals, with an emphasis on demonstrating new technologies and characterizing the surface environment in support of future missions (Golombek et al., 1997). Atmospheric science played a contextual role, aimed at understanding the diurnal variability of pressure, temperature, and wind near the surface, and refining predictions for entry, descent, and landing (EDL) scenarios. Pathfinder's Atmosphere Structure Instrument/Meteorology package (ASI/MET) experiment had similar meteorological sensors to those of the Viking landers, but operated with improved temporal resolution (Schofield et al., 1997).

The Mars Exploration Rovers (MERs) Spirit and Opportunity (2004) were primarily tasked with assessing past water activity and geological history (Squyres et al., 2003). Atmospheric measurements supported these goals indirectly by helping characterize the present-day surface environment and dust cycle, critical for interpreting surface weathering and erosional processes. The MER rovers lacked a dedicated atmospheric suite but leveraged their Miniature Thermal Emission Spectrometer (Mini-TES) instrument to derive thermal profiles and atmospheric dust opacity from infrared spectra (Smith et al., 2004).

Phoenix (2008), designed to study subsurface ice and assess potential habitability in Mars' northern plains, integrated atmospheric science more centrally (Smith et al., 2008). Its goals included characterizing local weather, investigating vapor exchange between the atmosphere and ice-rich regolith, and determining isotopic ratios of key atmospheric constituents to probe climate evolution. These objectives directly supported assessments of present climate and surface conditions relevant to astrobiology. Phoenix represented a leap forward, incorporating the first lidar to profile dust and cloud layers, as well as instruments for humidity, isotopic analysis, and subsurface–atmosphere interactions (Whiteway et al., 2009; Fischer et al., 2019). This marked a shift toward an integrated perspective on surface and atmospheric processes.

With Curiosity (2012), atmospheric science was explicitly embedded in the mission's planetary process objectives, which included assessing the long-term evolution of Mars' atmosphere and characterizing present-day water and CO_2 cycles (Grotzinger et al., 2012). The meteorological monitoring also supported habitability assessments by providing data on surface conditions, including radiation, dust, and humidity, key factors for biosignature preservation. Compared to previous environmental stations, Curiosity's Rover Environmental Monitoring Station (REMS) system added UV sensors and refined wind, temperature, relative humidity, and pressure sensors with continuous, long-duration observations (Gómez-Elvira et al., 2012). Its continuous, high-resolution data enabled the study of microscale dynamics, including turbulence and convection.

InSight (2018) was the first dedicated geophysical station on Mars, designed to investigate the planet's internal structure and thermal evolution through seismic, heat flow, and rotational measurements (Banerdt et al., 2020). Although primarily geophysical, the mission included supporting atmospheric monitoring via the Auxiliary Payload Sensor Suite (APSS), which comprised a high-sensitivity pressure sensor, a fluxgate magnetometer (IFG), and the Temperature and Winds for InSight (TWINS) sensors, derived from the REMS suite on Curiosity (Spiga et al., 2018; Banfield et al., 2019). A separate radiometer (RAD) measured ground temperature. The continuous, high-sensitivity measurements provided by APSS enabled discoveries such as infrasound, soliton-like bores, baroclinic waves, and vortex activity (Banfield et al., 2020).

Perseverance (2021) continued MSL's trend by targeting both scientific and operational goals (Farley et al., 2020). Atmospheric observations supported model validation and dust characterization to inform future sample return efforts and human exploration scenarios. Furthermore, they helped define the environmental context for sample caching and Ingenuity's flights. To achieve these objectives, the Mars Environmental Dynamics Analyzer (MEDA) onboard the Mars 2020 Perseverance rover represents the most sophisticated and comprehensive environmental monitoring system ever sent to the Martian surface (Rodriguez-Manfredi et al., 2021). In addition to upgrading REMS sensors, MEDA introduced new capabilities: a Thermal Infrared Sensor (TIRS) for IR radiative fluxes, a Radiation and Dust Sensor (RDS) for solar fluxes and optical depth and aerosol properties, and a camera (SkyCam) dedicated to observing the Martian sky.

Finally, China's Tianwen-1 (2021) broadened international participation in Mars surface exploration (Zou et al., 2021). One of its key science goals was to characterize the Martian climate to support future missions. Atmospheric measurements were designed to enable predictive modeling of seasonal and regional meteorological patterns. Specifically, its Mars Climate Station (MCS) was equipped with four core environmental sensors: a temperature sensor, a pressure sensor, a wind speed and direction sensor, and a microphone capable of capturing ambient acoustic data (Jiang et al., 2023). Despite its modest capabilities, MCS demonstrates China's capacity for autonomous environmental monitoring and highlights a scalable model for future meteorological networks.

From Viking to Perseverance and Tianwen-1, atmospheric payloads have become progressively more integrated, autonomous, and science-driven, mirroring the evolution of Mars exploration itself.

Instrumentation for the detection and characterization of life

The search for traces of life in planetary environments represents one of the most complex technological challenges in modern astrobiology. Developing instrumentation capable of detecting and characterizing microbial life in extreme settings requires the integration of multiple analytical techniques: from traditional approaches like gas chromatography and mass spectrometry to groundbreaking innovations such as biosensors based on immuno-microarrays (Moreno-Paz et al., 2018). The diversity of available strategies reflects both the inherent complexity of biosignature detection and the need to address different types of biological evidence, ranging from simple organic molecules to complex biopolymers (Parro et al., 2011).

Spectroscopic techniques have proven particularly effective in identifying biomolecules. Raman spectroscopy, recently validated through space experiments lasting 469 days outside the International Space Station, has confirmed its ability to detect biomolecules shielded from ultraviolet radiation. This nondestructive technique allows for the identification of photosynthetic pigments, bacterial hopanoids, and other organic compounds that could persist for millions of years (Parro et al., 2018). At the same time, laser-induced fluorescence spectroscopy has shown promising results in detecting amino acids and other biological compounds, with applications not only in planetary missions but also in the analysis of meteoritic samples.

Planetary mass spectrometry remains a fundamental technique for life detection, especially through the development of molecular assembly indices that can distinguish between biological and abiotic compounds based on characteristic fragmentation patterns (Sellam et al., 2025). Gas chromatography coupled with mass spectrometry has evolved significantly, incorporating derivatization techniques that allow the analysis of amino acids and other thermolabile compounds under the extreme conditions of space (Moreno-Paz et al., 2023). These improvements have been tested in missions like Curiosity, where complex organic compounds have been identified in Martian soil.

Next-generation instrumentation includes innovative techniques such as capillary electrophoresis, liquid chromatography, and nanopore sensors, offering low-mass and low-power alternatives for future planetary missions (Parro et al., 2005). The development of biosensors based on immunoassays represents a particularly promising approach, capable of detecting a wide range of molecular biomarkers with extraordinary sensitivity.

In this context, the achievements of the SOLID instrument take on particular significance for the future of planetary exploration. Beyond its successful demonstration in the MARTE project, SOLID technology has established a new paradigm in automated biosignature detection, proving the feasibility of immunoassays under extreme planetary conditions (Stoker et al., 2008). The results obtained have shown that it is possible to detect specific biomarkers of microorganisms such as *Leptospirillum ferrooxidans* and species of *Acidithiobacillus* in subsurface samples, confirming the survival of immunological reagents in space (Parro et al., 2011) and establishing contamination control protocols essential for future life-detection missions (Moreno-Paz et al., 2023).

The continued development of SOLID technology, now in version 3.1 with capacity for more than 300 different antibodies, marks a significant step toward fully automated life-detection missions. Its potential impact on future mission architectures is considerable, especially for initiatives like ocean worlds missions, where in situ biosignature detection will be critical for scientific interpretation.

Among the most promising upcoming missions is the Mars Life Explorer (MLE), recommended by the latest planetary science decadal survey as the next priority medium-class mission for the Mars exploration program (NASA, 2022). Planned for launch in the 2030s, MLE would be capable of drilling up to 2 m into ice at mid-latitudes on Mars and analyzing samples in situ for signs of life. NASA's Dragonfly mission to Titan, scheduled for launch in 2028, will also incorporate advanced biosignature detection instruments to study prebiotic chemistry on Saturn's moon.

Additionally, the Europa Clipper mission (launched in 2024) and the proposed Orbilander concept for Enceladus represent complementary approaches to the search for life in the ocean worlds of the outer solar system (Mackenzie et al., 2020). These initiatives illustrate how technologies developed and validated in terrestrial analogs like Río Tinto continue to be relevant in the design of the next generation of planetary life-detection instruments.

The role of small robots and drones in planetary prospecting

Robotics in planetary prospecting has undergone substantial advances over the last few decades. In many contexts, robotic systems can perform operations that are difficult, hazardous, or outright unfeasible for human operators. Planetary exploration is a good example of such an environment, where robots are essential to execute scientific and operational tasks under extreme conditions. Recent missions such as Mars 2020 have shown the growing potential and autonomy of robotic platforms on Mars (Balaram et al., 2018). Traditionally, planetary mobility has been dominated by static landers or wheeled rovers, whose exploration range is constrained by energy consumption, terrain limitations, and operational speed.

Small robotic systems are now emerging as key assets in planetary prospecting, offering enhanced capabilities for accessing complex terrains, conducting autonomous science, and supporting infrastructure development. These platforms can reach remote or hazardous zones, conduct reconnaissance, and perform tasks not easily achievable by larger vehicles. Their roles include scouting for science targets or landing sites, conducting localized geological surveys, and prospecting for resources such as subsurface water ice.

Examples of such small robotic systems include:

- Aerial drones and quadcopters, which are increasingly considered for aerial reconnaissance and terrain mapping. These systems can fly over broad areas, generating high-resolution imagery and topographic data unreachable by ground-based rovers (Sharma et al., 2022). They are especially valuable for scouting potential water ice deposits or identifying geologically interesting formations, thus guiding surface operations more effectively.
- Legged or articulated robots, designed to traverse rugged and topographically complex terrains. These systems, equipped with adaptive limbs or tracks, can access craters, escarpments, and stratified outcrops that are inaccessible to traditional wheel-based platforms, expanding the operational envelope of robotic missions (Haluška et al., 2022).
- Swarms of small landers or cooperative rovers, which can be deployed either from a central platform or via independent entry points to increase spatial coverage. Working collectively, these systems can carry complementary instrumentation to different locations, enabling simultaneous data acquisition and multi-site characterization. Some units may include integrated analytical instruments for in situ analysis, capable of identifying mineralogical compositions, detecting biosignature candidates, or mapping local resource distribution.

These robots could also support future ISRU by identifying and extracting materials of interest or by contributing to the construction of infrastructure, such as habitats or landing pads, through cooperative behavior. In the longer term, such capabilities may become essential for prehuman mission architectures, enabling automated site preparation and risk mitigation.

The use of small robots and drones in planetary prospecting is progressing rapidly, offering a multitude of benefits for future space exploration missions. Their versatility in accessing difficult environments, collecting diverse data, and performing tasks with minimal risk makes them invaluable tools for expanding our understanding of the Solar System and preparing us for future endeavors beyond Earth.

The role of Río Tinto in the next generation of Mars exploration missions

A glimpse into the future of Mars exploration

The future of Mars exploration stands at a pivotal crossroads. While past and current missions have largely focused on Mars' surface and atmosphere, future exploration will pivot increasingly toward the subsurface, the search for biosignatures, and developing the environmental understanding required for eventual human presence.

In parallel, growing interest in ISRU, the rise of low-cost and commercial missions, and the uncertain fate of Mars Sample Return (MSR) are reshaping the strategic landscape.

At the heart of these developments lies a sustained interest in the habitability and potential for life on Mars, both past and present. The MEPAG Goals document and the most recent Planetary Science Decadal Survey (2023–32) place strong emphasis on understanding the environmental context of life: where it might have arisen, how long it could have persisted, and whether biosignatures might be detectable today. These questions will increasingly require access to environments shielded from radiation, chemically preserved, and minimally altered over geological time, attributes typically associated with deep or protected subsurface settings.

Indeed, the subsurface has emerged as the new frontier of Mars exploration. Current data suggest that beneath the surface lies a more stable thermal and chemical environment, with possible brine activity, preserved organics, and reduced exposure to destructive cosmic radiation. While earlier missions have scratched the surface, future efforts will reach several meters deep, targeting stratigraphic layers or buried deposits that predate the harsh surface conditions seen today. For instance, missions like the proposed MLE concept aim to deliver such capabilities, employing advanced drilling systems, ground-penetrating radar, and precision landing strategies to reach scientifically rich but challenging terrains.

Accessing and characterizing these buried environments is not only important for science but also for engineering future human activities on Mars. Subsurface ice represents a crucial resource for ISRU: it could support drinking water, agriculture, radiation shielding, and be converted into fuel via electrolysis. ESA's long-term Exploration Architecture (*ESA Terrae Novae Exploration Strategy 2040*, 2020) calls for the predeployment of infrastructure capable of utilizing these resources. In this context, understanding how water ice is distributed, how it changes with seasons and latitude, and how dust or regolith layers affect its accessibility becomes a shared priority for both science and exploration.

Although the long-anticipated MSR campaign remains a cornerstone of NASA's scientific vision, recent developments have cast doubt on its feasibility under current budgetary and institutional frameworks. The program's high cost, complex architecture, and delays have spurred interest in alternative or scaled-back approaches, including sample caching followed by lightweight retrieval missions, potentially leveraging international or commercial partnerships. Meanwhile, international players and commercial ventures are reshaping the pace and diversity of Mars exploration. China's Tianwen-1 mission, followed by the planned Tianwen-3 sample return campaign, represents a significant investment in long-term Mars science. In the U.S., private companies like SpaceX are pushing for end-to-end Mars architectures with the potential for large payloads, frequent launches, and eventually crewed missions. These efforts are still in early stages, but their ambition and capital are already influencing programmatic decisions across agencies.

A key trend to watch is the rise of low-cost, high-frequency missions, a paradigm enabled by miniaturization, modular design, and piggybacking on launch opportunities. Small orbiters, landers, and even aerial platforms like balloons or rotorcraft could offer focused scientific returns at a fraction of traditional mission costs. These could be used to target specific hypotheses, serve as environmental scouts for more complex operations, or act as nodes in a broader sensor network.

In this changing landscape, the path forward will require strategic alignment between science goals, technical capabilities, and international cooperation. Efforts to understand the Martian subsurface, climate, and potential for life must be tightly integrated with considerations of mission risk, cost, and long-term sustainability. Environmental monitoring, once a secondary objective, will now be a core pillar supporting this integration. It not only helps interpret geological and chemical data but also enables safe operations and provides critical inputs for human system design.

Río Tinto as a test field for Mars instrumentation

As planetary exploration evolves toward increasingly ambitious missions, there is a growing need for analogue environments that not only resemble Martian conditions but also provide operational flexibility, geological diversity, and biological constraints against which new technologies can be tested. In this context, Río Tinto remains a uniquely valuable terrestrial laboratory, not only for its capacity to perfectly mimic the Martian environment, but for its ability to simulate and challenge critical aspects of scientific and engineering workflows that will define the next decade of Mars exploration.

One of the areas where Río Tinto continues to offer significant promise is in the testing of new spectroscopic instruments. Technologies such as advanced Raman, deep UV, laser-induced breakdown spectroscopy (LIBS), and integrated hyperspectral systems will play an increasingly important role in in situ characterization of mineralogy and organic molecules on Mars. Río Tinto's abundance of iron oxides, sulfates, and redox gradients presents a useful analog for evaluating the detection limits and specificity of these tools under realistic geological noise conditions. Future iterations of miniaturized spectrometers, some already being considered for missions targeting Mars, Europa, and Enceladus, can benefit from calibration campaigns and blind field testing in Río Tinto's stratified mineral zones (Maurice et al., 2021).

In parallel, the site offers compelling opportunities to validate next-generation drilling systems, particularly those focused on automated, contamination-free subsurface access. The challenges of accessing pristine material from depths greater than 2 m, critical for detecting preserved biosignatures, require mechanical robustness and autonomous control under variable soil cohesion, mineral crusting, and pore-space heterogeneity. Previous tests conducted at Río Tinto, such as those associated with the aforementioned MARTE project, have demonstrated the viability of early stage systems. Building on this foundation, new platforms intended for missions such as IceBreaker, Enceladus Orbilander, or crew-assisted ISRU scenarios could use Río Tinto to trial robotic arm coordination, sterile core transfer, and on-board decision-making algorithms (McKay et al., 2013).

Moreover, autonomous navigation and robotic mobility will be central to future exploration missions, especially in terrains where human support is delayed or absent. Río Tinto's heterogeneous topography, presence of loose detritus, crusted surfaces, and microponds offer a safe yet challenging analogue for training navigation systems, including visual odometry, slip estimation, and hazard detection in partially structured environments. These tests could inform both surface mobility architectures and coordinated rover-drone reconnaissance scenarios, as envisioned in NASA's Dragonfly mission or in ESA's exploration roadmaps.

Of particular relevance is the potential of Río Tinto to support instrumentation for biosignature detection under controlled but biologically relevant conditions. While the formation of biofilms and mineral precipitates by microbial action has already been studied in the area, the next step involves evaluating how emerging instruments, including lab-on-a-chip systems, microfluidic antibody arrays, and real-time sequencing tools, perform in detecting such biosignatures in situ. These efforts are especially critical in the context of future missions where triage protocols must decide which materials are worthy of retrieval based on partial or preliminary astrobiological evidence. Río Tinto could become a testbed not only for performance validation but also for developing detection protocols under noisy backgrounds, assessing false-positive risk, and training AI-based interpretation frameworks.

While it is important to acknowledge that Río Tinto is not a perfect match for Mars, its deviation from Martian conditions can be reframed as a strength. The availability of acidic liquid water, the presence of active biology, and the possibility to perform repetitive, nondestructive trials make it an ideal environment for iterative instrument development and calibration under partially analog but scientifically constrained conditions. As noted in reviews of planetary analogue research (Preston and Dartnell, 2014), the versatility of terrestrial test sites must be understood not as literal surrogates, but as controlled environments where specific planetary hypotheses and technical constraints can be realistically explored.

Finally, Río Tinto offers a valuable interface between planetary science and mission operations. The region has already been used for training astronaut crews, conducting simulated mission campaigns, and hosting interdisciplinary field schools. Going forward, its role may expand into a training ground for integrated field campaigns, involving coordinated geology, biology, robotics, and data systems—an essential rehearsal space for crewed missions and science autonomy scenarios.

In conclusion, the future of Mars exploration will increasingly depend on analogue sites that enable rigorous, interdisciplinary, and scalable instrument validation. Río Tinto, with its unique combination of geological fidelity, biological relevance, and operational flexibility, is well-positioned to continue playing a critical role in shaping the missions that will define the next era of planetary exploration.

Conclusions and perspectives

Research conducted in Río Tinto has left a significant legacy in the field of planetary exploration. It has not only enabled the validation of life-detection strategies under extreme conditions but has also decisively contributed to the design of technologies tailored to the challenges of Martian subsurface exploration. Over the past 2 decades, the geological and geochemical setting of the Río Tinto system has functioned as a highly effective testbed for evaluating scientific hypotheses, validating instruments, and simulating operations under remote and autonomous conditions.

Among the most relevant lessons learned is the confirmation that biomarkers can be preserved in highly oxidizing environments like those prevalent on Mars, provided there are specific mineralogical settings capable of protecting them. Likewise, the experience has highlighted the need to integrate deep drilling strategies with systems combining mechanical robustness, operational autonomy, and analytical sensitivity. The MARTE project clearly exemplified this convergence between science and technology, establishing an operational platform that influenced missions such as ExoMars and IceBreaker.

Regarding the evolution of subsurface exploration instrumentation, the field has progressed from merely exploratory, low-resolution approaches to complex, modular systems capable of operating autonomously for weeks or even months. Technologies such as dry rotary drilling, automated core-handling systems, and onboard spectroscopic or immunological instruments have demonstrated their feasibility and operational value. This progression suggests that within the next decade, subsurface exploration systems will incorporate embedded artificial intelligence, reconfigurable operational capabilities, and context-driven decision-making strategies.

From a scientific perspective, the coming years will be marked by renewed interest in Martian subsurface exploration. The current roadmap includes missions like MLE and IceBreaker, all of which have explicit astrobiological goals. These initiatives share a common logic: if life ever existed on Mars, its traces are more likely to be preserved underground, shielded from cosmic radiation, oxidizing agents, and extreme surface temperature gradients.

For this reason, future strategies must include hybrid architectures that combine remote geophysical sensing, selective drilling, autonomous in situ analysis, and strict contamination control protocols. The Río Tinto experience has shown that these capabilities can be integrated and deployed in real-world conditions, an outcome that reinforces their feasibility for application on Mars.

In addition, the evolution of the planetary exploration landscape, marked by increased international competition, commercial engagement, and a shift toward more flexible, low-cost missions, calls for analog testing environments that are adaptable and mission-representative. In this regard, Río Tinto not only remains relevant but is increasingly strategic. Looking ahead, the integration of small robotic systems, advanced environmental monitoring, and ISRU into planetary architectures will require sustained technological validation. Río Tinto offers a realistic, scalable platform for rehearsing these mission components under conditions that simulate many of Mars' operational constraints.

In sum, Río Tinto has been more than just a scientific analog for Mars; it has served as a true design laboratory for the next generation of planetary missions. Its contributions to science, technology, and exploration engineering will remain relevant for years to come, not only for Mars but also for other astrobiologically promising bodies across the Solar System.

AI disclosure

During the preparation of this work, the author(s) used ChatGPT in order to search for references; homogenization of style in English. After using this tool/service, the author(s) reviewed and edited the content as needed and take(s) full responsibility for the content of the publication.

References

Acuña, M.H., Connerney, J.E.P., Wasilewski, P., Lin, R.P., Anderson, K.A., Carlson, C.W., McFadden, J., Curtis, D.W., Mitchell, D., Reme, H., Mazelle, C., Sauvaud, J.A., d'Uston, C., Cros, A., Medale, J.L., Bauer, S.J., Cloutier, P., Mayhew, M., Winterhalter, D., Ness, N.F., 1998. Magnetic field and plasma observations at mars: initial results of the mars global surveyor mission. Science 279 (5357), 1676–1680. https://doi.org/10.1126/science.279.5357.1676.

Amils, R., Fernández-Remolar, D.C., Parro, V., Sánchez, A., Rodríguez, N., Rodríguez, J.M., et al., 2013. Detection of biomarkers in Río Tinto analog environments using integrated micro–analytical techniques. BioMed Res. Int. 907278. https://doi.org/10.1155/2013/907278.

Amils, R., Fernández-Remolar, D.C., Parro, V., Sánchez, A., Rodríguez, N., Rodríguez, J.M., 2011. Detection of biosignatures in acid mine drainage environments: implications for Martian life detection. Astrobiology 11 (10), 1013–1021. https://doi.org/10.1089/ast.2010.0545.

Amils, R., Fernández-Remolar, D.C., Parro, V., Sánchez, A., Rodríguez, N., Rodríguez, J.M., 2007. Extremophiles in Río Tinto: the contribution of microbial biomineralization and acidophilic microbial processes to acidic drainage. Nat. Rev. Microbiol. 5 (11), 902–912. https://doi.org/10.1038/nrmicro1741.

Amils, R., Fernández-Remolar, D.C., Parro, V., Sánchez, A., Rodríguez, N., Rodríguez, J.M., et al., 2008. The mars analog rio Tinto experiment (MARTE): searching for a subsurface biosphere at Río Tinto, Spain. Planet. Space Sci. 56 (9), 1324–1333. https://doi.org/10.1016/j.pss.2008.06.017.

Balaram, J.B., Canham, T., Duncan, C.G.M., Grip, H.F., Johnson, W.M.J., Quon, A., Stern, R., Zhu, D., 2018. Mars helicopter technology demonstrator. In: AIAA Atmospheric Flight Mechanics Conference, 2018. American Institute of Aeronautics and Astronautics Inc, AIAA, United States. https://doi.org/10.2514/6.2018-0023.

Banerdt, W.B., Smrekar, S.E., Banfield, D., Giardini, D., Golombek, M., Johnson, C.L., Lognonné, P., Spiga, A., Spohn, T., Perrin, C., Stähler, S.C., Antonangeli, D., Asmar, S., Beghein, C., Bowles, N., Bozdag, E., Chi, P., Christensen, U., Clinton, J., Collins, G.S., Daubar, I., Dehant, V., Drilleau, M., Fillingim, M., Folkner, W., Garcia, R.F., Garvin, J., Grant, J., Grott, M., Grygorczuk, J., Hudson, T., Irving, J.C.E., Kargl, G., Kawamura, T., Kedar, S., King, S., Knapmeyer-Endrun, B., Knapmeyer, M., Lemmon, M., Lorenz, R., Maki, J.N., Margerin, L., McLennan, S.M., Michaut, C., Mimoun, D., Mittelholz, A., Mocquet, A., Morgan, P., Mueller, N.T., Murdoch, N., Nagihara, S., Newman, C., Nimmo, F., Panning, M., Pike, W.T., Plesa, A.C., Rodriguez, S., Rodriguez-Manfredi, J.A., Russell, C.T., Schmerr, N., Siegler, M., Stanley, S., Stutzmann, E., Teanby, N., Tromp, J., van Driel, M., Warner, N., Weber, R., Wieczorek, M., 2020. Initial results from the InSight mission on Mars. Nat. Geosci. 13 (3), 183–189. https://doi.org/10.1038/s41561-020-0544-y.

Banfield, Rodriguez-Manfredi, Russell, Rowe, Leneman, Lai, Cruce, Means, Johnson, Mittelholz, Joy, Chi, Mikellides, Carpenter, Navarro, Sebastian, Gomez-Elvira, Torres, Mora, Peinado, Lepinette, Hurst, Lognonné, Smrekar, Banerdt, 2019. InSight auxiliary payload sensor suite (APSS). Space Sci. Rev. 215 (1). https://doi.org/10.1007/s11214-018-0570-x.

Banfield, D., Spiga, A., Newman, C., Forget, F., Lemmon, M., Lorenz, R., Murdoch, N., Viudez-Moreiras, D., Pla-Garcia, J., Garcia, R.F., Lognonné, P., Karatekin, Ö., Perrin, C., Martire, L., Teanby, N., Van Hove, B., Maki, J.N., Kenda, B., Mueller, N.T., Rodriguez, S., Kawamura, T., McClean, J.B., Stott, A.E., Charalambous, C., Millour, E., Johnson, C.L., Mittelholz, A., Määttänen, A., Lewis, S.R., Clinton, J., Stähler, S.C., Ceylan, S., Giardini, D., Warren, T., Pike, W.T., Daubar, I., Golombek, M., Rolland, L., Widmer-Schnidrig, R., Mimoun, D., Beucler, É., Jacob, A., Lucas, A., Baker, M., Ansan, V., Hurst, K., Mora-Sotomayor, L., Navarro, S., Torres, J., Lepinette, A., Molina, A., Marin-Jimenez, M., Gomez-Elvira, J., Peinado, V., Rodriguez-Manfredi, J.-A., Carcich, B.T., Sackett, S., Russell, C.T., Spohn, T., Smrekar, S.E., Banerdt, W.B., 2020. The atmosphere of Mars as observed by InSight. Nat. Geosci. 13 (3), 190–198. https://doi.org/10.1038/s41561-020-0534-0.

Bibring, J.P., Langevin, Y., Mustard, J.F., Poulet, F., Arvidson, R., Gendrin, A., Gondet, B., Mangold, N., Pinet, P., Forget, F., Berthe, M., Gomez, C., Jouglet, D., Soufflot, A., Vincendon, M., Combes, M., Drossart, P., Encrenaz, T., Fouchet, T., Merchiorri, R., Belluci, G.C., Altieri, F., Formisano, V., Capaccioni, F., Cerroni, P., Coradini, A., Fonti, S., Korablev, O., Kottsov, V., Ignatiev, N., Moroz, V., Titov, D., Zasova, L., Loiseau, D., Pinet, P., Douté, S., Schmitt, B., Sotin, C., Hauber, E., Hoffmann, H., Jaumann, R., Keller, U., Arvidson, R., Duxbury, T., Forget, F., Neukum, G., 2006. Global mineralogical and aqueous Mars history derived from OMEGA/Mars express data. Science 312 (5772), 400–404. https://doi.org/10.1126/science.1122659.

Boynton, W.V., Feldman, W.C., Squyres, S.W., Prettyman, T.H., Brückner, J., Evans, L.G., Reedy, R.C., Starr, R., Arnold, J.R., Drake, D.M., Englert, P.A.J., Metzger, A.E., Mitrofanov, I., Trombka, J.I., D'Uston, C., Wänke, H., Gasnault, O., Hamara, D.K., Janes, D.M., Marcialis, R.L., Maurice, S., Mikheeva, I., Taylor, G.J., Tokar, R., Shinohara, C., 2002. Distribution of hydrogen in the near surface of Mars: evidence for subsurface ice deposits. Science 297 (5578), 81–85. https://doi.org/10.1126/science.1073722.

Boynton, W.V., Feldman, W.C., Mitrofanov, I.G., Evans, L.G., Reedy, R.C., Squyres, S.W., Starr, R., Trombka, J.I., d'Uston, C., Arnold, J.R., Englert, P.A.J., Metzger, A.E., Wänke, H., Brückner, J., Drake, D.M., Shinohara, C., Fellows, C., Hamara, D.K.,

Harshman, K., Kerry, K., Turner, C., Ward, M., Barthe, H., Fuller, K.R., Storms, S.A., Thornton, G.W., Longmire, J.L., Litvak, M.L., Ton'chev, A.K., 2004. The mars Odyssey gamma-ray spectrometer instrument suite. Space Sci. Rev. 110 (1/2), 37–83. https://doi.org/10.1023/B:SPAC.0000021007.76126.15.

Cannon, H.N., Stocker, C.R., Dunagan, S.E., Davis, K., Gómez-Elvira, J., Glass, B.J., Lemke, L.G., Miller, D., Bonaccorsi, R., Branson, M., Christa, S., Rodríguez-Manfredi, J.A., Mumm, E., Paulsen, G., Roman, M., Winterholler, A., Zavaleta, J.R., 2007. MARTE: technology development and lessons learned from a Mars drilling mission simulation. J. Field Robot. 24 (10), 877–905. https://doi.org/10.1002/rob.20224.

Chamberlain, T.E., Cole, H.L., Dutton, R.G., Greene, G.C., Tillman, J.E., 1976. Atmospheric measurements on Mars: the Viking meteorology experiment. Bull. Am. Meteorol. Soc. 57 (9), 1094–1105. https://doi.org/10.1175/1520-0477(1976)057<1094:AMOMTV>2.0.CO;2.

ESA Terrae Novae Exploration Strategy 2040. European Space Agency, (2020).

Farley, K.A., Williford, K.H., Stack, K.M., Bhartia, R., Chen, A., de la Torre, M., Hand, K., Goreva, Y., Herd, C.D.K., Hueso, R., Liu, Y., Maki, J.N., Martinez, G., Moeller, R.C., Nelessen, A., Newman, C.E., Nunes, D., Ponce, A., Spanovich, N., Willis, P.A., Beegle, L.W., Bell, J.F., Brown, A.J., Hamran, S.E., Hurowitz, J.A., Maurice, S., Paige, D.A., Rodriguez-Manfredi, J.A., Schulte, M., Wiens, R.C., 2020. Mars 2020 mission overview. Space Sci. Rev. 216 (8). https://doi.org/10.1007/s11214-020-00762-y.

Fernández-Remolar, D.C., Morris, R.V., Gruener, J.E., Amils, R., Knoll, A.H., 2005. The Río Tinto Basin, Spain: mineralogy, sedimentary geobiology, and implications for interpretation of outcrop rocks at Meridiani Planum, Mars. Earth Planet Sci. Lett. 240 (1), 149–167. https://doi.org/10.1016/j.epsl.2005.09.043.

Fischer, E., Martínez, G.M., Rennó, N.O., Tamppari, L.K., Zent, A.P., 2019. Relative humidity on mars: new results from the Phoenix TECP sensor. J. Geophys. Res.: Planets 124 (11), 2780–2792. https://doi.org/10.1029/2019je006080.

Gayen, P., Sankarasubramanian, S., Ramani, V.K., 2020. Fuel and oxygen harvesting from Martian regolithic brine. Proc. Natl. Acad. Sci. U. S. A 117 (50), 31685–31689. https://doi.org/10.1073/pnas.2008613117.

Genova, A., Goossens, S., Lemoine, F.G., Mazarico, E., Neumann, G.A., Smith, D.E., Zuber, M.T., 2016. Seasonal and static gravity field of Mars from MGS, Mars Odyssey and MRO radio science. Icarus 272, 228–245. https://doi.org/10.1016/j.icarus.2016.02.050.

Glass, B.J., Cannon, H., Stoker, C., Davis, K., 2005. Robotic and human-tended collaborative drilling automation for subsurface exploration. International Astronautical Federation - 56th International Astronautical Congress 2005 2, 1169–1175.

Glass, B.J., Dave, A., McKay, C.P., Paulsen, G., 2014. Robotics and automation for "icebreaker. J. Field Robot. 31 (1), 192–205. https://doi.org/10.1002/rob.21487.

Golombek, M.P., Cook, R.A., Economou, T., Folkner, W.M., Haldemann, A.F.C., Kallemeyn, P.H., Knudsen, J.M., Manning, R.M., Moore, H.J., Parker, T.J., Rieder, R., Schofield, J.T., Smith, P.H., Vaughan, R.M., 1997. Overview of the mars pathfinder mission and assessment of landing site predictions. Science 278 (5344), 1743–1748. https://doi.org/10.1126/science.278.5344.1743.

Gómez-Elvira, J., Armiens, C., Castañer, L., Domínguez, M., Genzer, M., Gómez, F., Haberle, R., Harri, A.M., Jiménez, V., Kahanpää, H., Kowalski, L., Lepinette, A., Martín, J., Martínez-Frías, J., McEwan, I., Mora, L., Moreno, J., Navarro, S., De Pablo, M.A., Peinado, V., Peña, A., Polkko, J., Ramos, M., Renno, N.O., Ricart, J.,

Richardson, M., Rodríguez-Manfredi, J., Romeral, J., Sebastián, E., Serrano, J., De La Torre Juárez, M., Torres, J., Torrero, F., Urquí, R., Vázquez, L., Velasco, T., Verdasca, J., Zorzano, M.P., Martín-Torres, J., 2012. REMS: the environmental sensor suite for the Mars Science Laboratory rover. Space Sci. Rev. 170 (1–4), 583–640. https://doi.org/10.1007/s11214-012-9921-1.

Grotzinger, J.P., Sumner, D.Y., Kah, L.C., Stack, K., Gupta, S., Edgar, L., Rubin, D., Lewis, K., Schieber, J., Mangold, N., Milliken, R., Conrad, P.G., DesMarais, D., Farmer, J., Siebach, K., Calef, F., Hurowitz, J., McLennan, S.M., Ming, D., Vaniman, D., Crisp, J., Vasavada, A., Edgett, K.S., Malin, M., Blake, D., Gellert, R., Mahaffy, P., Wiens, R.C., Maurice, S., Grant, J.A., Wilson, S., Anderson, R.C., Beegle, L., Arvidson, R., Hallet, B., Sletten, R.S., Rice, M., Bell, J., Griffes, J., Ehlmann, B., Anderson, R.B., Bristow, T.F., Dietrich, W.E., Dromart, G., Eigenbrode, J., Fraeman, A., Hardgrove, C., Herkenhoff, K., Jandura, L., Kocurek, G., Lee, S., Leshin, L.A., Leveille, R., Limonadi, D., Maki, J., McCloskey, S., Meyer, M., Minitti, M., Newsom, H., Oehler, D., Okon, A., Palucis, M., Parker, T., Rowland, S., Schmidt, M., Squyres, S., Steele, A., Stolper, E., Summons, R., Treiman, A., Williams, R., Yingst, A., , MSL Science Team, Kemppinen, O., Bridges, N., Johnson, J.R., Cremers, D., Godber, A., Wadhwa, M., Wellington, D., McEwan, I., Newman, C., Richardson, M., Charpentier, A., Peret, L., King, P., Blank, J., Weigle, G., Shuai, L., Robertson, K., Sun, V., Baker, M., Edwards, C., Farley, K., Miller, H., Newcombe, M., Pilorget, C., Brunet, C., Hipkin, V., Léveillé, R., Marchand, G., Sobrón Sánchez, P., Favot, L., Cody, G., Flückiger, L., Lees, D., Nefian, A., Martin, M., Gailhanou, M., Westall, F., Israël, G., Agard, C., Baroukh, J., Donny, C., Gaboriaud, A., Guillemot, P., Lafaille, V., Lorigny, E., Paillet, A., Pérez, R., Saccoccio, M., Yana, C., Armiens-Aparicio, C., Rodríguez, J.C., Blázquez, I.C., Gómez, F.G., Gómez-Elvira, J., Hettrich, S., Malvitte, A.L., Jiménez, M.M., Martínez-Frías, J., Martín-Soler, J., Martín-Torres, F.J., Jurado, A.M., Mora-Sotomayor, L., Caro, G.M., Navarro López, S., Peinado-González, V., Pla-García, J., Manfredi, J.A.R., Romeral-Planelló, J.J., Fuentes, S.A.S., Martinez, E.S., Torres Redondo, J., Urqui-O'Callaghan, R., Mier, M.-P.Z., Chipera, S., Lacour, J.-L., Mauchien, P., Sirven, J.-B., Manning, H., Fairén, A., Hayes, A., Joseph, J., Sullivan, R., Thomas, P., Dupont, A., Lundberg, A., Melikechi, N., Mezzacappa, A., DeMarines, J., Grinspoon, D., Reitz, G., Prats, B., Atlaskin, E., Genzer, M., Harri, A.-M., Haukka, H., Kahanpää, H., Kauhanen, J., Paton, M., Polkko, J., Schmidt, W., Siili, T., Fabre, C., Wray, J., Wilhelm, M.B., Poitrasson, F., Patel, K., Gorevan, S., Indyk, S., Paulsen, G., Bish, D., Gondet, B., Langevin, Y., Geffroy, C., Baratoux, D., Berger, G., Cros, A., d'Uston, C., Forni, O., Gasnault, O., Lasue, J., Lee, Q.-M., Meslin, P.-Y., Pallier, E., Parot, Y., Pinet, P., Schröder, S., Toplis, M., Lewin, É., Brunner, W., Heydari, E., Achilles, C., Sutter, B., Cabane, M., Coscia, D., Szopa, C., Robert, F., Sautter, V., Mouélic, S.Le, Nachon, M., Buch, A., Stalport, F., Coll, P., François, P., Raulin, F., Teinturier, S., Cameron, J., Clegg, S., Cousin, A., DeLapp, D., Dingler, R., Jackson, R.S., Johnstone, S., Lanza, N., Little, C., Nelson, T., Williams, R.B., Jones, A., Kirkland, L., Baker, B., Cantor, B., Caplinger, M., Davis, S., Duston, B., Fay, D., Harker, D., Herrera, P., Jensen, E., Kennedy, M.R., Krezoski, G., Krysak, D., Lipkaman, L., McCartney, E., McNair, S., Nixon, B., Posiolova, L., Ravine, M., Salamon, A., Saper, L., Stoiber, K., Supulver, K., Beek, J.Van, Beek, T.Van, Zimdar, R., Louise French, K., Iagnemma, K., Miller, K., Goesmann, F., Goetz, W., Hviid, S., Johnson, M., Lefavor, M., Lyness, E., Breves, E., Darby Dyar, M., Fassett, C.,

Edwards, L., Haberle, R., Hoehler, T., Hollingsworth, J., Kahre, M., Keely, L., McKay, C., Bleacher, L., Brinckerhoff, W., Choi, D., Dworkin, J.P., Floyd, M., Freissinet, C., Garvin, J., Glavin, D., Harpold, D., Martin, D.K., McAdam, A., Pavlov, A., Raaen, E., Smith, M.D., Stern, J., Tan, F., Trainer, M., Posner, A., Voytek, M., Aubrey, A., Behar, A., Blaney, D., Brinza, D., Christensen, L., DeFlores, L., Feldman, J., Feldman, S., Flesch, G., Jun, I., Keymeulen, D., Mischna, M., Morookian, J.M., Pavri, B., Schoppers, M., Sengstacken, A., Simmonds, J.J., Spanovich, N., Juarez, M.de la T., Webster, C.R., Yen, A., Paul, D.A., Cucinotta, F., Jones, J.H., Morris, R.V., Paul, N., Rampe, E., Nolan, T., Fisk, M., Radziemski, L., Barraclough, B., Bender, S., Berman, D., Dobrea, E.N., Tokar, R., Cleghorn, T., Huntress, W., Manhès, G., Hudgins, J., Olson, T., Stewart, N., Sarrazin, P., Vicenzi, E., Bullock, M., Ehresmann, B., Hamilton, V., Hassler, D., Peterson, J., Rafkin, S., Zeitlin, C., Fedosov, F., Golovin, D., Karpushkina, N., Kozyrev, A., Litvak, M., Malakhov, A., Mitrofanov, I., Mokrousov, M., Nikiforov, S., Prokhorov, V., Sanin, A., Tretyakov, V., Varenikov, A., Vostrukhin, A., Kuzmin, R., Clark, B., Wolff, M., Oliver, B., Drake, D., Bean, K., Lemmon, M., Schwenzer, S.P., Lee, E.M., Sucharski, R., Hernández, M.Á.de P., Ávalos, J.J.B., Ramos, M., Kim, M.-H., Malespin, C., Plante, I., Muller, J.-P., Navarro-González, R., Ewing, R., Boynton, W., Downs, R., Fitzgibbon, M., Harshman, K., Morrison, S., Kortmann, O., Williams, A., Lugmair, G., Wilson, M.A., Jakosky, B., Balic-Zunic, T., Frydenvang, J., Jensen, J.K., Kinch, K., Koefoed, A., Madsen, M.B., Stipp, S.L.S., Boyd, N., Campbell, J.L., Perrett, G., Pradler, I., VanBommel, S., Jacob, S., Owen, T., Savijärvi, H., Boehm, E., Böttcher, S., Burmeister, S., Guo, J., Köhler, J., García, C.M., Mueller-Mellin, R., Wimmer-Schweingruber, R., Bridges, J.C., McConnochie, T., Benna, M., Franz, H., Bower, H., Brunner, A., Blau, H., Boucher, T., Carmosino, M., Atreya, S., Elliott, H., Halleaux, D., Rennó, N., Wong, M., Pepin, R., Elliott, B., Spray, J., Thompson, L., Gordon, S., Ollila, A., Williams, J., Vasconcelos, P., Bentz, J., Nealson, K., Popa, R., Moersch, J., Tate, C., Day, M., Raymond, F., McCullough, E., Cloutis, E., ten Kate, I.L., Scholes, D., Slavney, S., Stein, T., Ward, J., Berger, J., Moores, J.E., 2014. A habitable fluvio-lacustrine environment at yellowknife bay, Gale crater, mars. Science 343 (6169). https://doi.org/10.1126/science.1242777.

Grotzinger, J.P., Crisp, J., Vasavada, A.R., Anderson, R.C., Baker, C.J., Barry, R., Blake, D.F., Conrad, P., Edgett, K.S., Ferdowski, B., Gellert, R., Gilbert, J.B., Golombek, M., Gómez-Elvira, J., Hassler, D.M., Jandura, L., Litvak, M., Mahaffy, P., Maki, J., Meyer, M., Malin, M.C., Mitrofanov, I., Simmonds, J.J., Vaniman, D., Welch, R.V., Wiens, R.C., 2012. Mars Science Laboratory mission and science investigation. Space Sci. Rev. 170 (1–4), 5–56. https://doi.org/10.1007/s11214-012-9892-2.

Haluška, J., Koval, A., Nikolakopoulos, G., 2022. On the unification of legged and aerial robots for planetary exploration missions. Appl. Sci. 12 (8), 3983. https://doi.org/10.3390/app12083983.

Hamran, S.E., Paige, D.A., Allwood, A., Amundsen, H.E.F., Berger, T., Brovoll, S., Carter, L., Casademont, T.M., Damsgård, L., Dypvik, H., Eide, S., Fairén, A.G., Ghent, R., Kohler, J., Mellon, M.T., Nunes, D.C., Plettemeier, D., Russell, P., Siegler, M., Øyan, M.J., 2022. Ground penetrating radar observations of subsurface structures in the floor of Jezero crater, Mars. Sci. Adv. 8 (34). https://doi.org/10.1126/sciadv.abp8564.

Hess, S.L., Henry, R.M., Leovy, C.B., Ryan, J.A., Tillman, J.E., 1977. Meteorological results from the surface of mars: viking 1 and 2. J. Geophys. Res. 82 (28), 4559–4574. https://doi.org/10.1029/js082i028p04559.

Jiang, C., Jiang, Y., Li, H., Du, S., 2023. Initial results of the meteorological data from the first 325 sols of the Tianwen-1 mission. Sci. Rep. 13 (1). https://doi.org/10.1038/s41598-023-30513-2.

Lognonné, P., Banerdt, W.B., Pike, W.T., Giardini, D., Christensen, U., Garcia, R.F., Kawamura, T., Kedar, S., Knapmeyer-Endrun, B., Margerin, L., Nimmo, F., Panning, M., Tauzin, B., Scholz, J.-R., Antonangeli, D., Barkaoui, S., Beucler, E., Bissig, F., Brinkman, N., Calvet, M., Ceylan, S., Charalambous, C., Davis, P., van Driel, M., Drilleau, M., Fayon, L., Joshi, R., Kenda, B., Khan, A., Knapmeyer, M., Lekic, V., McClean, J., Mimoun, D., Murdoch, N., Pan, L., Perrin, C., Pinot, B., Pou, L., Menina, S., Rodriguez, S., Schmelzbach, C., Schmerr, N., Sollberger, D., Spiga, A., Stähler, S., Stott, A., Stutzmann, E., Tharimena, S., Widmer-Schnidrig, R., Andersson, F., Ansan, V., Beghein, C., Böse, M., Bozdag, E., Clinton, J., Daubar, I., Delage, P., Fuji, N., Golombek, M., Grott, M., Horleston, A., Hurst, K., Irving, J., Jacob, A., Knollenberg, J., Krasner, S., Krause, C., Lorenz, R., Michaut, C., Myhill, R., Nissen-Meyer, T., ten Pierick, J., Plesa, A.-C., Quantin-Nataf, C., Robertsson, J., Rochas, L., Schimmel, M., Smrekar, S., Spohn, T., Teanby, N., Tromp, J., Vallade, J., Verdier, N., Vrettos, C., Weber, R., Banfield, D., Barrett, E., Bierwirth, M., Calcutt, S., Compaire, N., Johnson, C.L., Mance, D., Euchner, F., Kerjean, L., Mainsant, G., Mocquet, A., Rodriguez Manfredi, J.A., Pont, G., Laudet, P., Nebut, T., de Raucourt, S., Robert, O., Russell, C.T., Sylvestre-Baron, A., Tillier, S., Warren, T., Wieczorek, M., Yana, C., Zweifel, P., 2020. Constraints on the shallow elastic and anelastic structure of Mars from InSight seismic data. Nat. Geosci. 13 (3), 213–220. https://doi.org/10.1038/s41561-020-0536-y.

Mackenzie, S.M., Caswell, T.E., Phillips-Lander, C.M., Stavros, E.N., Hofgartner, J.D., Sun, V.Z., 2020. Science objectives for flagship-class mission concepts for the search for evidence of life at Enceladus. Astrobiology 20 (11), 1315–1351. https://doi.org/10.1089/ast.2020.2268.

Maurice, S., Wiens, R.C., Bernardi, P., Caïs, P., Robinson, S., Nelson, T., Gasnault, O., Reess, J.M., Deleuze, M., Rull, F., Manrique, J.A., Abbaki, S., Anderson, R.B., André, Y., Angel, S.M., Arana, G., Battault, T., Beck, P., Benzerara, K., Bernard, S., Berthias, J.P., Beyssac, O., Bonafous, M., Bousquet, B., Boutillier, M., Cadu, A., Castro, K., Chapron, F., Chide, B., Clark, K., Clavé, E., Clegg, S., Cloutis, E., Collin, C., Cordoba, E.C., Cousin, A., Dameury, J.C., D'Anna, W., Daydou, Y., Debus, A., Deflores, L., Dehouck, E., Delapp, D., De Los Santos, G., Donny, C., Doressoundiram, A., Dromart, G., Dubois, B., Dufour, A., Dupieux, M., Egan, M., Ervin, J., Fabre, C., Fau, A., Fischer, W., Forni, O., Fouchet, T., Frydenvang, J., Gauffre, S., Gauthier, M., Gharakanian, V., Gilard, O., Gontijo, I., Gonzalez, R., Granena, D., Grotzinger, J., Hassen-Khodja, R., Heim, M., Hello, Y., Hervet, G., Humeau, O., Jacob, X., Jacquinod, S., Johnson, J.R., Kouach, D., Lacombe, G., Lanza, N., Lapauw, L., Laserna, J., Lasue, J., Le Deit, L., Le Mouélic, S., Le Comte, E., Lee, Q.M., Legett, C., Leveille, R., Lewin, E., Leyrat, C., Lopez-Reyes, G., Lorenz, R., Lucero, B., Madariaga, J.M., Madsen, S., Madsen, M., Mangold, N., Manni, F., Mariscal, J.F., Martinez-Frias, J., Mathieu, K., Mathon, R., McCabe, K.P., McConnochie, T., McLennan, S.M., Mekki, J., Melikechi, N., Meslin, P.Y., Micheau, Y., Michel, Y., Michel, J.M., Mimoun, D., Misra, A., Montagnac, G., Montaron, C., Montmessin, F., Moros, J., Mousset, V., Morizet, Y., Murdoch, N., Newell, R.T., Newsom, H., Nguyen Tuong, N., Ollila, A.M., Orttner, G., Oudda, L., Pares, L., Parisot, J., Parot, Y., Pérez, R., Pheav, D., Picot, L., Pilleri, P., Pilorget, C.,

Pinet, P., Pont, G., Poulet, F., Quantin-Nataf, C., Quertier, B., Rambaud, D., Rapin, W., Romano, P., Roucayrol, L., Royer, C., Ruellan, M., Sandoval, B.F., Sautter, V., Schoppers, M.J., Schröder, S., Seran, H.C., Sharma, S.K., Sobron, P., Sodki, M., Sournac, A., Sridhar, V., Standarovsky, D., Storms, S., Striebig, N., Tatat, M., Toplis, M., Torre-Fdez, I., Toulemont, N., Velasco, C., Veneranda, M., Venhaus, D., Virmontois, C., Viso, M., Willis, P., Wong, K.W., 2021. The SuperCam instrument suite on the mars 2020 rover: science objectives and mast-unit description. Space Sci. Rev. 217 (3). https://doi.org/10.1007/s11214-021-00807-w.

McKay, C.P., Stoker, C.R., Glass, G.J., Davé, A.I., Davila, A.F., Heldmann, J.L., Marinova, M. M., Fairen, A.G., Quinn, R.C., Zacny, K.A., Paulsen, G., Smith, P.H., Parro, V., Andersen, D.T., Hecht, M.H., Lacelle, D., Pollard, W.H., 2013. The icebreaker life mission to Mars: a search for biomolecular evidence for life. Astrobiology 13 (4), 334–353. https://doi.org/10.1089/ast.2012.0878.

Mitrofanov, I.G., Litvak, M.L., Varenikov, A.B., Barmakov, Y.N., Behar, A., Bobrovnitsky, Y. I., Bogolubov, E.P., Boynton, W.V., Harshman, K., Kan, E., Kozyrev, A.S., Kuzmin, R.O., Malakhov, A.V., Mokrousov, M.I., Ponomareva, S.N., Ryzhkov, V.I., Sanin, A.B., Smirnov, G.A., Shvetsov, V.N., Timoshenko, G.N., Tomilina, T.M., Tret'yakov, V.I., Vostrukhin, A.A., 2012. Dynamic Albedo of neutrons (DAN) experiment onboard NASA's mars science laboratory. Space Sci. Rev. 170 (1–4), 559–582. https://doi.org/10.1007/s11214-012-9924-y.

Mittelholz, A., Johnson, C.L., Fillingim, M.O., Joy, S., Langlais, B., Thorne, S.N., et al., 2022. The surface magnetic field environment from InSight. EPSC Abstracts 16. https://doi.org/10.5194/epsc2022-988.

Moreno-Paz, M., Gómez-Cifuentes, A., Ruiz-Bermejo, M., Hofstetter, O., Maquieira, Á., Manchado, J.M., Morais, S., Sephton, M.A., Niessner, R., Knopp, D., Parro, V., 2018. Detecting nonvolatile life- and nonlife-derived organics in a carbonaceous chondrite analogue with a new multiplex immunoassay and its relevance for planetary exploration. Astrobiology 18 (8), 1041–1056. https://doi.org/10.1089/ast.2017.1747.

Moreno-Paz, M., dos Santos Severino, R.S., Sánchez-García, L., Manchado, J.M., García-Villadangos, M., Aguirre, J., Fernández-Martínez, M.A., Carrizo, D., Kobayashi, L., Dave, A., Warren-Rhodes, K., Davila, A., Stoker, C.R., Glass, B., Parro, V., 2023. Life detection and microbial biomarker profiling with signs of life detector-life detector chip during a mars drilling simulation campaign in the hyperarid core of the atacama desert. Astrobiology 23 (12), 1259–1283. https://doi.org/10.1089/ast.2021.0174.

NASA, 2022. Mars life explorer: a proposed mission to search for life in the martian subsurface. National Academies Planetary Science and Astrobiology Decadal Survey.

Onstott, T.C., Colwell, F.S., Bach, W., 1999. Evidence for microbial activity deep within the Earth's crust. Science 284 (5423), 479–482. https://doi.org/10.1126/science.284.5423.479.

Parro, V., De Diego-Castilla, G., Moreno-Paz, M., Blanco, Y., Cruz-Gil, P., Rodríguez-Manfredi, J.A., Fernández-Remolar, D., Gómez, F., Gómez, M.J., Rivas, L.A., Demergasso, C., Echeverría, A., Urtuvia, V.N., Ruiz-Bermejo, M., García-Villadangos, M., Postigo, M., Sánchez-Román, M., Chong-Díaz, G., Gómez-Elvira, J., 2011. A microbial oasis in the hypersaline atacama subsurface discovered by a life detector chip: implications for the search for life on mars. Astrobiology 11 (10), 969–996. https://doi.org/10.1089/ast.2011.0654.

Parro, V., Blanco, Y., Puente-Sánchez, F., Rivas, L.A., Moreno-Paz, M., Echeverría, A., Chong-Díaz, G., Demergasso, C., Cabrol, N.A., 2018. Biomarkers and metabolic patterns in the sediments of evolving glacial lakes as a proxy for planetary lake exploration. Astrobiology 18 (5), 586–606. https://doi.org/10.1089/ast.2015.1342.

Parro, V., Rodríguez-Manfredi, J.A., Briones, C., Compostizo, C., Herrero, P.L., Vez, E., Sebastián, E., Moreno-Paz, M., García-Villadangos, M., Fernández-Calvo, P., González-Toril, E., Pérez-Mercader, J., Fernández-Remolar, D., Gómez-Elvira, J., 2005. Instrument development to search for biomarkers on mars: terrestrial acidophile, iron-powered chemolithoautotrophic communities as model systems. Planet. Space Sci. 53 (7), 729–737. https://doi.org/10.1016/j.pss.2005.02.003.

Parro, V., Rodríguez-Nava, N., Rivas, L.A., Fernández-Remolar, D.C., Gómez-Elvira, J., Gómez, F., 2007. Performance of the SOLID instrument in terrestrial Mars analogs. Astrobiology 7 (3), 600–622. https://doi.org/10.1089/ast.2006.0121.

Preston, L.J., Dartnell, L.R., 2014. Planetary habitability: lessons learned from terrestrial analogues. Int. J. Astrobiol. 13 (1), 81–98. https://doi.org/10.1017/S1473550413000396.

Rodriguez-Manfredi, J.A., de la Torre Juárez, M., Alonso, A., Apéstigue, V., Arruego, I., Atienza, T., Banfield, D., Boland, J., Carrera, M.A., Castañer, L., Ceballos, J., Chen-Chen, H., Cobos, A., Conrad, P.G., Cordoba, E., del Río-Gaztelurrutia, T., de Vicente-Retortillo, A., Domínguez-Pumar, M., Espejo, S., Fairen, A.G., Fernández-Palma, A., Ferrándiz, R., Ferri, F., Fischer, E., García-Manchado, A., García-Villadangos, M., Genzer, M., Giménez, S., Gómez-Elvira, J., Gómez, F., Guzewich, S.D., Harri, A.M., Hernández, C.D., Hieta, M., Hueso, R., Jaakonaho, I., Jiménez, J.J., Jiménez, V., Larman, A., Leiter, R., Lepinette, A., Lemmon, M.T., López, G., Madsen, S.N., Mäkinen, T., Marín, M., Martín-Soler, J., Martínez, G., Molina, A., Mora-Sotomayor, L., Moreno-Álvarez, J.F., Navarro, S., Newman, C.E., Ortega, C., Parrondo, M.C., Peinado, V., Peña, A., Pérez-Grande, I., Pérez-Hoyos, S., Pla-García, J., Polkko, J., Postigo, M., Prieto-Ballesteros, O., Rafkin, S.C.R., Ramos, M., Richardson, M.I., Romeral, J., Romero, C., Runyon, K.D., Saiz-Lopez, A., Sánchez-Lavega, A., Sard, I., Schofield, J.T., Sebastian, E., Smith, M.D., Sullivan, R.J., Tamppari, L.K., Thompson, A.D., Toledo, D., Torrero, F., Torres, J., Urquí, R., Velasco, T., Viúdez-Moreiras, D., Zurita, S., 2021. The mars environmental dynamics analyzer, MEDA. A suite of environmental sensors for the mars 2020 mission. Space Sci. Rev. 217 (3). https://doi.org/10.1007/s11214-021-00816-9.

Schofield, J.T., Barnes, J.R., Crisp, D., Haberle, R.M., Larsen, S., Magalhães, J.A., Murphy, J. R., Seiff, A., Wilson, G., 1997. The mars pathfinder atmospheric structure investigation/meteorology (ASI/MET) experiment. Science 278 (5344), 1752–1758. https://doi.org/10.1126/science.278.5344.1752.

Sellam, Y., Riedo, A., Tulej, M., Neubeck, A., Ivarsson, M., Wiesendanger, R., et al., 2025. Utilizing laser ablation mass spectrometry to identify ancient life in Earth's gypsum deposits. Front. Astron. Space Sci. 12, 1–15. https://doi.org/10.3389/fspas.2025.1247823.

Sharma, M., Gupta, A., Gupta, S.K., Alsamhi, S.H., Shvetsov, A.V., 2022. Survey on unmanned aerial vehicle for mars exploration: deployment use case. Drones 6 (1). https://doi.org/10.3390/drones6010004.

Smith, M.D., Wolff, M.J., Lemmon, M.T., Spanovich, N., Banfield, D., Budney, C.J., Clancy, R.T., Ghosh, A., Landis, G.A., Smith, P., Whitney, B., Christensen, P.R., Squyres, S.W., 2004. First atmospheric science results from the Mars exploration rovers Mini-TES. Science 306 (5702), 1750–1753. https://doi.org/10.1126/science.1104257.

Smith, P.H., Tamppari, L., Arvidson, R.E., Bass, D., Blaney, D., Boynton, W., Carswell, A., Catling, D., Clark, B., Duck, T., DeJong, E., Fisher, D., Goetz, W., Gunnlaugsson, P., Hecht, M., Hipkin, V., Hoffman, J., Hviid, S., Keller, H., Kounaves, S., Lange, C.F.,

Lemmon, M., Madsen, M., Malin, M., Markiewicz, W., Marshall, J., McKay, C., Mellon, M., Michelangeli, D., Ming, D., Morris, R., Renno, N., Pike, W.T., Staufer, U., Stoker, C., Taylor, P., Whiteway, J., Young, S., Zent, A., 2008. Introduction to special section on the Phoenix mission: landing site characterization experiments, mission overviews, and expected science. J. Geophys. Res.: Planets 113 (E3). https://doi.org/10.1029/2008je003083.

Spiga, A., Banfield, D., Teanby, N.A., Forget, F., Lucas, A., Kenda, B., Manfredi, J.A.R., Widmer-Schnidrig, R., Murdoch, N., Lemmon, M.T., Garcia, R.F., Martire, L., Karatekin, Ö., Le Maistre, S., Van Hove, B., Dehant, V., Lognonné, P., Mueller, N., Lorenz, R., Mimoun, D., Rodriguez, S., Beucler, É., Daubar, I., Golombek, M.P., Bertrand, T., Nishikawa, Y., Millour, E., Rolland, L., Brissaud, Q., Kawamura, T., Mocquet, A., Martin, R., Clinton, J., Éléonore, S., Spohn, T., Smrekar, S., Banerdt, W. B., 2018. Atmospheric science with InSight. Space Sci. Rev. 214 (7). https://doi.org/10.1007/s11214-018-0543-0.

Squyres, S.W., Arvidson, R.E., Ruff, S., Gellert, R., Morris, R.V., Ming, D.W., Crumpler, L., Farmer, J.D., Des Marais, D.J., Yen, A., McLennan, S.M., Calvin, W., Bell, J.F., Clark, B. C., Wang, A., McCoy, T.J., Schmidt, M.E., de Souza, P.A., 2008. Detection of silica-rich deposits on mars. Science 320 (5879), 1063–1067. https://doi.org/10.1126/science.1155429.

Squyres, S.W., Grotzinger, J.P., Arvidson, R.E., Bell, J.F., Calvin, W., Christensen, P.R., Clark, B.C., Crisp, J.A., Farrand, W.H., Herkenhoff, K.E., Johnson, J.R., Klingelhöfer, G., Knoll, A.H., McLennan, S.M., McSween, H.Y., Morris, R.V., Rice, J. W., Rieder, R., Soderblom, L.A., 2004. In situ evidence for an ancient aqueous environment at meridiani planum, mars. Science 306 (5702), 1709–1714. https://doi.org/10.1126/science.1104559.

Squyres, S.W., Arvidson, R.E., Baumgartner, E.T., Bell, J.F., Christensen, P.R., Gorevan, S., Herkenhoff, K.E., Klingelhöfer, G., Madsen, M.B., Morris, R.V., Rieder, R., Romero, R. A., 2003. Athena Mars rover science investigation. J. Geophys. Res.: Planets 108 (12). https://doi.org/10.1029/2003je002121.

Stoker, C.R., Cannon, H.N., Dunagan, S.E., Lemke, L.G., Glass, B.J., Miller, D., Gomez-Elvira, J., Davis, K., Zavaleta, J., Winterholler, A., Roman, M., Rodriguez-Manfredi, J. A., Bonaccorsi, R., Bell, M.S., Brown, A., Battler, M., Chen, B., Cooper, G., Davidson, M., Fernández-Remolar, D., Gonzales-Pastor, E., Heldmann, J.L., Martínez-Frías, J., Parro, V., Prieto-Ballesteros, O., Sutter, B., Schuerger, A.C., Schutt, J., Rull, F., 2008. The 2005 MARTE robotic drilling experiment in Río Tinto, Spain: objectives, approach, and results of a simulated mission to search for life in the martian subsurface. Astrobiology 8 (5), 921–945. https://doi.org/10.1089/ast.2007.0217.

Tang, X., Adkins, C., Mallams, J., Burnette, M., Barnes, T., Mier-Hicks, F., Meirion-Griffith, G., Staack, D., 2022. Plasma drilling on Martian ice: enabling efficient deep subsurface access to Mars' polar layered deposits. Planet. Space Sci. 223, 105578. https://doi.org/10.1016/j.pss.2022.105578.

Tornos, F., 2006. Environment of formation and styles of volcanogenic massive sulfides: the Iberian Pyrite Belt. Ore Geology Reviews 28 (3), 259–307. https://doi.org/10.1016/j.oregeorev.2004.12.005.

Vago, J.L., Westall, F., Coates, A.J., Jaumann, R., Korablev, O., Ciarletti, V., Mitrofanov, I., Josset, J.L., De Sanctis, M.C., Bibring, J.P., Rull, F., Goesmann, F., Steininger, H., Goetz, W., Brinckerhoff, W., Szopa, C., Raulin, F., Edwards, H.G.M., Whyte, L.G., Fairén, A.G., Bridges, J., Hauber, E., Ori, G.G., Werner, S., Loizeau, D., Kuzmin, R.O.,

Williams, R.M.E., Flahaut, J., Forget, F., Rodionov, D., Svedhem, H., Sefton-Nash, E., Kminek, G., Lorenzoni, L., Joudrier, L., Mikhailov, V., Zashchirinskiy, A., Alexashkin, S., Calantropio, F., Merlo, A., Poulakis, P., Witasse, O., Bayle, O., Bayón, S., Meierhenrich, U., Carter, J., García-Ruiz, J.M., Baglioni, P., Haldemann, A., Ball, A.J., Debus, A., Lindner, R., Haessig, F., Monteiro, D., Trautner, R., Voland, C., Rebeyre, P., Goulty, D., Didot, F., Durrant, S., Zekri, E., Koschny, D., Toni, A., Visentin, G., Zwick, M., Van Winnendael, M., Azkarate, M., Carreau, C., 2017. Habitability on early mars and the search for biosignatures with the ExoMars rover. Astrobiology 17 (6–7), 471–510. https://doi.org/10.1089/ast.2016.1533.

Vago, J.L., Westall, F., Horneck, G., Foucher, F., Pullan, D., Kminek, G., 2006. Habitability on early Mars and the search for biosignatures with the ExoMars rover. Adv. Space Res. 38 (4), 709–720. https://doi.org/10.1016/j.asr.2005.05.119.

Whiteway, J.A., Komguem, L., Dickinson, C., Cook, C., Illnicki, M., Seabrook, J., Popovici, V., Duck, T.J., Davy, R., Taylor, P.A., Pathak, J., Fisher, D., Carswell, A.I., Daly, M., Hipkin, V., Zent, A.P., Hecht, M.H., Wood, S.E., Tamppari, L.K., Renno, N., Moores, J.E., Lemmon, M.T., Daerden, F., Smith, P.H., 2009. Mars water-ice clouds and precipitation. Science 325 (5936), 68–70. https://doi.org/10.1126/science.1172344.

Zou, Y., Zhu, Y., Bai, Y., Wang, L., Jia, Y., Shen, W., Fan, Y., Liu, Y., Wang, C., Zhang, A., Yu, G., Dong, J., Shu, R., He, Z., Zhang, T., Du, A., Fan, M., Yang, J., Zhou, B., Wang, Y., Peng, Y., 2021. Scientific objectives and payloads of Tianwen-1, China's first Mars exploration mission. Adv. Space Res. 67 (2), 812–823. https://doi.org/10.1016/j.asr.2020.11.005.

SECTION

4

Geomicrobiology: Fundamentals and applications

CHAPTER

From rock to river: Subsurface microbial processes shaping the chemistry of Río Tinto

16

Adrián Martínez-Bonilla[1], Ignacio Carrasco-Ropero[1], Cristina Escudero[2]

[1]*Centro de Biología Molecular Severo Ochoa (CSIC-UAM), Interactions with the Environment, Madrid, Spain;* [2]*Eberhard-Karls-University Tuebingen, Department of Geosciences, Tübingen, Germany*

Introduction

Río Tinto, located in the province of Huelva, southwestern Spain, is one of the most extreme and unique ecosystems of the Iberian Peninsula. The main characteristic of the river is its constant acidic pH, a direct consequence of the strong buffering capacity of the ferric ion (Fe^{3+}), which also is responsible for the river's characteristic red color. The concentration of dissolved ferric iron in the river is at least one or two orders of magnitude higher than that of other acidic rivers found in the area, such as the Odiel and Agrio rivers (Amils et al., 2002).

Fe^{3+} plays a key role in buffering the system: at high pH values, dissolved Fe^{3+} precipitates in the form of ferric hydroxide ($Fe(OH)_3$), releasing protons and thus acidifying the water; at low pH, ferric hydroxide dissolves, consuming protons and compensating the acidification (Amils et al., 2002) (Eq. 16.1). This dynamic maintains a remarkably stable pH around 2.3 throughout the year.

$$Fe^{3+} + 3H_2O \leftrightarrow Fe(OH)_3 + 3H^+ \tag{16.1}$$

Iron, therefore, is the primary driver behind the extreme physicochemical conditions of the Río Tinto ecosystem. But what is the source of such elevated Fe concentrations in the river? Historically, the prevailing explanation has been that these high Fe^{3+} concentrations result from intense mining activity in the region.

Anthropogenic influence: Five millennia of mining in the IPB

Río Tinto rises in Peña de Hierro, in the core of the Iberian Pyrite Belt (IPB) (Amaral Zettler et al., 2002; González-Toril et al., 2003; López-Archilla et al., 2001). The IPB is an extensive geological formation stretching 250 km in length and 25–70 km in width across the South Portuguese Zone of the Iberian Peninsula. It

Geomicrobiology of Río Tinto. https://doi.org/10.1016/B978-0-443-32948-7.00020-5

contains 80 known deposits and more than 1.6 Mt of sulfide ore, making it one of the largest reserves of metallic sulfides worldwide (Tornos, 2006). The regional geology is predominantly composed of pyrite (FeS_2) and chalcopyrite ($CuFeS_2$), with minor amounts of other sulfides such as sphalerite (ZnS) and galena (PbS) (Tornos, 2006). These geological features have allowed extensive mining of its massive sulfide deposits for approximately 5000 years (Davis Jr. et al., 2000; Leblanc et al., 2000).

Mining activity in the IPB began in the third millennium BCE (Before Common Era), with the existence in the region of communities dedicated to the extraction of copper from carbonates, oxides and certain sulfides (Macías and Domínguez, 2011). Initially limited to shallow deposits (Nocete et al., 2005), mining methods improved with the arrival of Tartessian communities, whose economy revolved around metal production (Martiañez, 2000). During the Iron Age, new tools enabled deeper excavation, and evidence of underground mining at Río Tinto dates back to the seventh century BCE (Macías and Domínguez, 2011). This activity waned with the fall of Tartessos and the expansion of Carthage and was not revitalized until the Roman Age (Vara, 1963).

With the arrival of the Roman Empire (second century BCE), mining in the IPB experienced a great boost. Technological advances in extraction, drainage and metallurgy were introduced, allowing large-scale exploitation, unknown until that time. Romans developed mainly subsurface mining, focused on silver ore, reaching considerable depths. Subsequently, they began to extract copper, with the maximum activity in the first century CE (Common Era) (Macías and Domínguez, 2011). Afterward, activity gradually declined and largely ceased following the Visigoth conquest in 405 CE (Olías and Nieto, 2015).

Medieval mining was traditionally considered limited, although some activities attributed to the Romans may date from this period, as Roman techniques persisted until the end of the 15th century. In the 16th century, mining focused mainly on the collection of seasonally formed evaporite salts, used for medicinal and industrial purposes (Macías and Domínguez, 2011). Systematic mining resumed in 1725, targeting copper. The global copper demand of the 19th century further spurred mining, although operations were disrupted during the Spanish War of Independence (1808–14) (Caballero, 1981; Ortíz Mateo and Romero Macías, 2004). In 1839, new technology allowed the extraction of low-grade copper ores, although limited to superficial subsurface operations reaching 64 m, lower than Roman depths (del Bayo, 1852). Until 1850, Río Tinto was the only active mine in the IPB, but soon after there was a boom in the exploitation of metallic sulfides, with several mines, including Peña de Hierro, active in 1855 (Olías and Nieto, 2015).

The Spanish government reactivated the Río Tinto mine in 1850 but sold it in 1873 to British investors, who founded The Rio Tinto Company Limited (RTC). This triggered a major industrial phase, shifting from underground to open-pit mining. Operations such as Veta Sur (1876), Veta Norte (1890–1900), and Corta Atalaya (1910) marked the peak of sulfide extraction. Although underground mining continued, the Veta Norte and Veta Sur open pit mines remained active until

1930–40, and Corta Atalaya operated until the end of the 20th century (Olías and Nieto, 2015).

In 1954, control of the mine reverted to Spanish investors and, although pyrite mining declined in the 1970s due to competition from industrial sulfur, increased world demand for copper led to the start of hydrometallurgical recovery at Cerro Colorado mine in 1970, marking a new production peak. Environmental concerns and market shifts in the 1980s led to mine closure in 2001 (Martiañez, 2000; Olías and Nieto, 2015). Río Tinto reopened in 2015 under new international ownership, focusing on open-pit copper extraction at Cerro Colorado and the production of copper concentrates with silver as a by-product. Exploration for additional resources, including zinc, lead, gold, and silver, is ongoing (Mining, 2026).

Given this long and intense mining history, the distinctive chemical characteristics of Río Tinto waters have traditionally been attributed to acid mine drainage (AMD) resulting from centuries of mineral exploitation. As a result, the river has long been regarded as a contaminated system requiring remediation (Davis Jr. et al., 2000; Leblanc et al., 2000; Nieto et al., 2007; Van Geen et al., 1997).

Is Río Tinto a case of acid mine drainage?

AMD is widely recognized as one of the most severe environmental consequences associated with the exploitation of sulfide-rich ore deposits. It is primarily responsible for the acidification and metal contamination of the surface and groundwater near mining areas (Johnson and Hallberg, 2003). AMD originates from the oxidative dissolution of sulfide minerals, primarily pyrite (FeS_2) but also sphalerite (ZnS), galena (PbS), chalcopyrite ($CuFeS_2$), and arsenopyrite (FeAsS), when these are exposed to oxygen and water (Amils et al., 2002, 2004; Johnson and Hallberg, 2003). This process involves a complex interplay of abiotic reactions and microbial activity, regulated by physical, chemical, and biological factors (Fig. 16.1).

The generation of AMD begins under circumneutral pH conditions, typically during the initial exposure of pyrite to the atmosphere, causing the so-called weathering processes. In this early stage, oxygen reacts with the pyrite releasing ferrous iron (Fe^{2+}), protons (H^+) and sulfur species such as thiosulfate ($S_2O_3^{2-}$) (Eq. 16.2). These sulfur species and Fe^{2+} are oxidized abiotically by dissolved oxygen producing sulfate and Fe^{3+} respectively (Eqs. 16.3-16.4).

$$FeS_2 + 7/2\ O_2 + H_2O \rightarrow Fe^{2+} + 2\ H^+ + 2\ S_2O_3^{2-} \tag{16.2}$$

$$S_2O_3^{2-} + 2\ O_2 + H_2O \rightarrow 2\ SO_4^{2-} + 2\ H^+ \tag{16.3}$$

$$2\ Fe^{2+} + 1/2\ O_2 + 2\ H^+ \rightarrow 2\ Fe^{3+} + H_2O \tag{16.4}$$

Under neutral to mildly acidic pH, Fe^{3+} is not stable in solution and precipitates as ferric hydroxide, releasing additional H^+ (Eq. 16.1). The H^+ released eventually causes the acidification of the environment, increasing the solubility of Fe^{3+}. Once

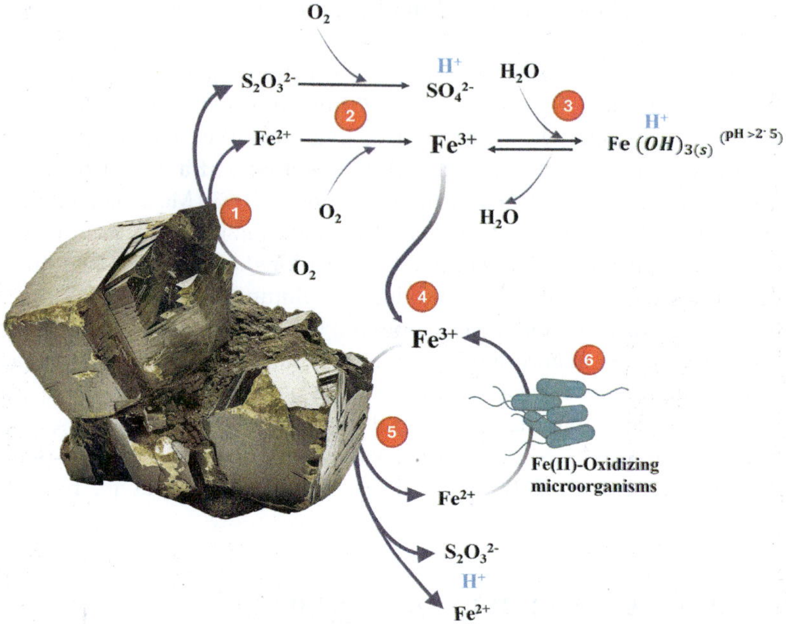

FIGURE 16.1 Conceptual model of the stepwise oxidation of pyrite during the onset and progression of acid mine drainage (AMD).

The process begins with the oxidation of pyrite (1) by dissolved oxygen under circumneutral pH conditions, releasing ferrous iron (Fe^{2+}) and intermediate sulfur species such as thiosulfate ($S_2O_3^{2-}$). Thiosulfate is subsequently oxidized abiotically by oxygen (2) to produce sulfate (SO_4^{2-}) and protons (H^+), contributing to acidification. The Fe^{2+} released is also oxidized by oxygen (3) to Fe^{3+}, which, under pH > 2.5, precipitates as ferric hydroxide [$Fe(OH)_3(s)$], releasing additional protons. As the pH decreases, Fe^{3+} remains soluble and becomes the dominant oxidant of pyrite (4), leading to further release of Fe^{2+} and acidity (5). Under these acidic conditions, Fe(II)-oxidizing acidophilic microorganisms (6) catalyze the regeneration of Fe^{3+}, greatly enhancing the rate and extent of the AMD process.

Fe^{3+} is solubilized (around pH 3), it becomes the principal oxidant in the ongoing dissolution of pyrite under acidic conditions (Eqs. 16.5-16.6) (Sand et al., 2001).

$$FeS_2 + 6\,Fe^{3+} + 3\,H_2O \rightarrow 7\,Fe^{2+} + S_2O_3^{2-} + 6\,H^+ \tag{16.5}$$

$$S_2O_3^{2-} + 8\,Fe^{3+} + 5\,H_2O \rightarrow 8\,Fe^{2+} + 2\,SO_4^{2-} + 10\,H^+ \tag{16.6}$$

The Fe^{2+} produced can be oxidized again by O_2, regenerating Fe^{3+} and sustaining the cycle (Eq. 16.2). However, at low pH, the solubility of O_2 decreases, and its oxidation capacity becomes limited. This is where acidophilic Fe(II)-oxidizing microorganisms play a crucial role. These microbes can oxidize Fe^{2+} several orders of magnitude faster than O_2, maintaining the supply of Fe^{3+} and playing a central role in sustaining and accelerating the oxidation of sulfide minerals.

Microbial diversity in AMD systems is generally low due to the extreme conditions, especially the low pH, high levels of toxic metals and the limited range of electron donors and acceptors (Denef et al., 2010). These communities are dominated by members of the phyla Pseudomonadota, Nitrospirota, and Euryarchaeota (Kuang et al., 2013). Key acidophilic Fe(II)-oxidizing microorganisms are *Leptospirillum ferrooxidans, Acidithiobacillus ferrooxidans, Acidimicrobium ferrooxidans, Sulfolobus metallicus*, all of which regenerate Fe^{3+} through aerobic respiration (Sand et al., 2001). Less abundant taxa, such as *Ferrovum* spp., are also relevant, particularly in moderately acidic conditions (Hallberg et al., 2006; Kuang et al., 2013).

Under natural, undisturbed conditions, most sulfide deposits remain buried in water-saturated, anoxic sediments, limiting their interaction with oxygen and thus suppressing AMD formation. Mining activities break this balance by exposing sulfides to the atmosphere and to oxygenated water, triggering AMD processes. After mining ceases, the rise of the water table and the dissolution of previously formed salts can cause temporary spikes in contaminant concentrations. While these levels gradually decrease, they often remain elevated for centuries or even thousands of years (Younger, 1997).

Despite the strong geochemical resemblance between Río Tinto and typical AMD systems, multiple lines of geological and mineralogical evidence suggest a different origin. Fernández-Remolar and collaborators demonstrated that the iron-rich sedimentary terraces along the Río Tinto, although differing in age, exhibit a remarkably consistent mineralogical, geochemical profile and evolutionary process (Fernández-Remolar et al., 2003, 2005). Their results support that the oldest terraces, dated to the late Miocene, were formed under environmental conditions similar to those currently found in Río Tinto, characterized by high concentration of Fe^{3+} and low pH. Therefore, it was concluded that, although the mining activity can influence the properties of Río Tinto, its extreme conditions would have originated about seven million years ago. This origin is long before the oldest recorded mining activity, strongly supporting the hypothesis that this acidic river Río Tinto is not the result of AMD processes alone, but rather a naturally acidic ecosystem. In fact, the river rises from the subsurface through natural springs, showing its extreme characteristics. Currently, Río Tinto is recognized as one of the largest natural acidic ecosystems in the world (Fernández-Remolar et al., 2005) and, since December 2004, it has been considered as a Protected Natural Landscape by the Junta de Andalucía (Boja, 2005).

Based on these findings, it was proposed that Río Tinto is the result of an underground bioreactor in which the sulfide bodies, mainly pyrite, would be dissolved by chemolithotrophic microorganisms (Gómez-Ortiz et al., 2014). This hypothesis was explored in detail through the MARTE (Mars Astrobiology Research and Technology Experiment, 2003–06) and IPBSL (Iberian Pyrite Belt Subsurface Life Detection, 2011–15) drilling projects, both of which aimed to detect microbial life and investigate iron-sulfide oxidation processes in the deep, oxygen-limited subsurface of the IPB (Amils et al., 2014; Amils et al., 2023).

To experimentally validate whether native microbial communities from the IPB subsurface can promote iron release under such natural anoxic conditions, during the IPBSL project long-term microcosm experiments were performed using powdered rock samples from three depths (102.6, 139.4, and 228.6 m), selected for their high pyrite content (10.6%, 8.5%, and 8.5%, respectively). Both sterile and nonsterile replicates were incubated under nitrogen atmosphere, in the dark, with only sterile, anoxic MilliQ water without the addition of any external electron donors, acceptors, or inoculum. Over a period of 10 months, total iron concentrations were periodically measured (Amils et al., 2023; Escudero, 2018). The results were clear and compelling. Significant iron release was observed in all nonsterile incubations, with final total Fe concentrations reaching 4.6 g/L and 6.2 g/L at 102.6 and 139.4 m, respectively, compared to less than 1 g/L in the sterile controls. These findings are conceptually significant because they demonstrate that iron mobilization from pyritic rocks can occur under strictly anoxic conditions, driven solely by native microbial communities. This reveals the existence of a biologically mediated, anaerobic form of acid rock drainage (ARD), distinct from the classical, oxygen-dependent AMD systems described to date. While traditional AMD models rely on the exposure of sulfides to oxygen and surface water, the results from these microcosms show that endogenous microbial metabolisms in deep, oxygen-depleted environments can independently trigger iron cycling and pyrite alteration.

The results of this experiment expand the current definition of AMD/ARD to include anaerobic processes and supports the interpretation of Río Tinto as a surface expression of deep, microbially driven acid generation. These incubations, representing the closest simulation to natural IPB subsurface conditions, provide direct experimental evidence that Fe release can occur in the absence of oxygen, validating the subsurface bioreactor hypothesis and reinforcing the view that Río Tinto is a naturally acidic ecosystem powered by geobiological processes occurring underground.

Anaerobic pyrite oxidation: From hypothesis to microbial evidence

The oxidation of pyrite (FeS_2) has been extensively studied under aerobic conditions, primarily due to its role in acid mine drainage formation and its application in industrial bioleaching processes (Brierley and Brierley, 2013). In these systems, the process is enhanced by acidophilic iron- and sulfur-oxidizing microorganisms and has been exploited for the extraction of valuable metals such as copper, gold, lead, and zinc from sulfide ores (Rawlings et al., 2003). In contrast, pyrite oxidation under anoxic conditions has historically received far less attention and was long considered improbable. The prevailing assumption was that pyrite remains stable in oxygen-depleted sediments, such as marine, lacustrine, or deep aquifer environments, due to the lack of effective oxidants under those conditions (Schippers and Jørgensen, 2002).

This view was further supported by the known geochemistry of Fe^{3+}, which is a more potent oxidant of pyrite than O_2 but is extremely insoluble at circumneutral pH. As such, pyrite biooxidation at pH > 2.5 has traditionally been overlooked, given that Fe^{3+} rapidly precipitates and becomes unavailable for redox reactions (Nordstrom, 1982). Consequently, most of the data on microbial pyrite oxidation derive from acidophilic systems where oxygen is used as the terminal electron acceptor and Fe^{3+} remains soluble (Rohwerder et al., 2003; Vera et al., 2013). However, chemical studies have shown that pyrite oxidation by Fe^{3+} is in fact possible at neutral pH, and the rate of oxidation under these conditions can even exceed that observed with molecular oxygen (Moses and Herman, 1991; Moses et al., 1987).

Support for this scenario comes from a range of geochemical and microbiological studies conducted over the past decades. Field studies in nitrate-contaminated aquifers have shown that nitrate-reducing microbial activity under anoxic, circumneutral conditions is frequently associated with increased concentrations of dissolved sulfate and iron when pyrite is present in the system (Postma et al., 1991; Schwientek et al., 2008; Zhang et al., 2009). These systems, often impacted by agricultural fertilizer runoff, have also shown enrichment in trace metals whose concentrations correlate with the pyrite content of the local sediment (Houben et al., 2017). Due to the limited availability of labile organic matter in these environments, it has been proposed that denitrifying chemolithoautotrophic microorganisms may use pyrite as electron donor (Postma et al., 1991). The dissolution of pyrite in this context would release sulfate, iron, and trace metals, accompanied by a decrease in nitrate concentration. This hypothesis is further supported by isotopic signatures of nitrogen and sulfur compounds in groundwater, which indicate biologically mediated redox transformations consistent with nitrate reduction and sulfur oxidation (Schwientek et al., 2008; Zhang et al., 2009).

Nonetheless, the exact mechanism of pyrite oxidation in these systems remains contested. One major point of controversy is whether chemolithotrophic denitrifiers, or any other microorganisms, can use pyrite directly as an electron donor. Several studies over the past decades have yielded conflicting results. For example, Schippers and Jøgensen concluded, based on work in marine sediments, that pyrite oxidation under nitrate-reducing conditions is abiotic and driven by manganese oxides (Mn(IV)), rather than by microbial activity (Schippers and Jørgensen, 2001, 2002). In their experiments, only poorly crystalline iron sulfide (FeS) could be oxidized biologically, via the so-called polysulfide mechanism, and bioreactor assays further supported this finding: sulfate production occurred only when FeS, but not pyrite, was added (Haaijer et al., 2007).

On the other hand, other studies have provided experimental support for a biological influence on pyrite dissolution. Several studies have shown that adding pyrite to sediment incubations stimulates the growth of nitrate-reducing microbial communities and leads to increased sulfate concentrations in solution (Jørgensen et al.,

2009; Torrentó et al., 2011). Moreover, enrichment cultures of *Thiobacillus*, particularly *Thiobacillus denitrificans*, have demonstrated growth using pyrite as the sole electron donor (Bosch et al., 2012; Vaclavkova et al., 2014; Torrentó et al., 2011). Yet, subsequent analyses cast doubt on these findings. It was later reported that *T. denitrificans* was not truly oxidizing pyrite, but instead relying on intracellular sulfur stores or sulfur species contaminating the mineral surface (Yan et al., 2019), highlighting the importance of verifying the purity and nature of the electron donor in such experiments. In addition, despite the stoichiometry of Eq. (16.7) being used as a theoretical basis, many incubation studies fail to reproduce it in full, with sulfate levels often too low relative to nitrate consumption. Some authors suggest this discrepancy could be due to the formation of intermediate sulfur species (e.g., thiosulfate, tetrathionate) (Bosch et al., 2012) or due to the presence of additional nitrate-reducing metabolisms (Torrentó et al., 2011; Vaclavkova et al., 2014), which would decouple pyrite oxidation from full denitrification.

$$5\,FeS_2 + 14\,NO_3^- + 4\,H^+ \rightarrow 7\,N_2 + 10\,SO_4^{2-} + Fe^{2+} + 2\,H_2O \qquad (16.7)$$

To clarify the role of Fe^{3+} and explore whether it could mediate pyrite oxidation under anoxic conditions, controlled laboratory incubations have been conducted using different iron and nitrogen species. These experiments demonstrated that soluble Fe^{3+} complexes, such as ferric citrate, can effectively oxidize pyrite at circumneutral pH in the absence of oxygen. In contrast, poorly soluble Fe^{3+} salts (like $FeCl_3$) and reactive nitrogen species such as nitrite (NO_2^-) did not result in significant pyrite dissolution (Martínez-Bonilla, 2025). These findings underscore the importance of Fe^{3+} solubility and suggest that microbial processes which increase Fe^{3+} availability, either through acidification or the production of chelators like siderophores (Gomes et al., 2024), could be critical in enabling pyrite oxidation under anoxic conditions.

To date, two microbial systems have been conclusively shown to promote pyrite oxidation under anoxic conditions. The first is the autotrophic nitrate-reducing Fe (II)-oxidizing (NRFeOx) enrichment culture AG, which is capable of directly oxidizing synthetic pyrite while reducing nitrate (Jakus et al., 2021). The second is the mixotrophic NRFeOx bacterium *Acidovorax* sp. BoFeN1. In contrast to culture AG, *Acidovorax* does not seem to oxidize pyrite directly. Instead, it generates Fe^{3+} under anoxic conditions through mixotrophic NRFeOx. This biologically produced Fe^{3+} then acts as an effective oxidant, promoting indirect pyrite dissolution even at neutral pH (Amils et al., 2023; Escudero, 2018). This distinction becomes important when considered in the context of the IPB subsurface. While *Acidovorax* has been consistently detected in core samples from borehole BH10, no microbial community resembling the enrichment culture AG has been identified to date. This absence strongly suggests that, in the subsurface of the IPB, pyrite oxidation is not driven by direct microbial attack but by an indirect mechanism: the microbial production of Fe^{3+} under anoxic, neutral conditions. As such, all current evidence points to Fe^{3+} as the key oxidant responsible for pyrite dissolution in the IPB subsurface, a process initiated not by oxygen or manganese oxides, but by nitrate-reducing microorganisms that maintain Fe cycling in this unique geological setting.

Iron cycling and nitrate-dependent Fe(II) oxidation in the IPB subsurface

Recent studies have revealed that the oxidation of Fe(II) coupled to nitrate reduction is a widespread and mechanistically diverse process in anoxic environments. Microorganisms capable of this metabolism are commonly grouped into three functional categories (Bryce et al., 2018; Jakus et al., 2021). The first group includes autotrophic enrichment cultures such as KS, BP, AG, and HP, which oxidize Fe^{2+} enzymatically while fixing CO_2 (Fig. 16.2A) (Grimm et al., 2025; Huang et al., 2022). These cultures are considered to perform direct enzymatic oxidation of Fe(II), although the enzymatic mechanisms are still under investigation.

The second group is composed of mixotrophic microorganisms, which use both Fe^{2+} and organic matter as energy and carbon source respectively (Fig. 16.2B). Among this group, *Acidovorax* is by far the most studied and best characterized species. Multiple strains, including *Acidovorax* sp. BoFeN1 (Kappler et al., 2005), BrG1 (Straub et al., 2004), 2AN (Chakraborty and Picardal, 2013), and TPSY (Carlson et al., 2013), have shown a remarkable ability to oxidize high concentrations of Fe^{2+} under anoxic, neutral pH conditions in the presence of nitrate.

The third group, chemodenitrifiers, reduces nitrate without enzymatically oxidizing Fe(II); instead, iron is abiotically oxidized by reactive nitrogen species (RNS) produced during denitrification, such as nitrite (NO_2^-) and nitric oxide (NO) (Fig. 16.2C). These compounds are strong oxidants under anoxic and circumneutral conditions (Van Cleemput and Baert, 1983). Experimental studies have shown that both NO_2^- and NO can rapidly oxidize Fe^{2+} in aqueous solution, with nitrous oxide (N_2O) typically forming as the main product (Moraghan and Buresh, 1977; Yan et al., 2015). Klueglein and Kappler (2013) demonstrated that 14% of a 1 mM Fe^{2+} solution was oxidized by 1 mM NO_2^- within the first minute of incubation, with nearly complete oxidation after just 20 minutes. These findings highlight the high reactivity of RNS and their potential to drive abiotic Fe(II) oxidation in microbial systems lacking dedicated iron-oxidizing enzymes. Consequently, the contribution of biologically mediated Fe(II) oxidation by NRFeOx microorganisms, including *Acidovorax*, may be overestimated in some studies. The abiotic oxidation of Fe^{2+} by RNS produced during nitrate respiration can mimic or enhance microbial Fe(II) oxidation, complicating the distinction between enzymatic and chemically driven processes.

Some authors have therefore proposed that all nitrate-reducing microorganisms may possess an inherent capacity to oxidize Fe^{2+} indirectly through RNS production (Carlson et al., 2013). This is further supported by observations that organisms not typically associated with iron metabolism, such as *Escherichia coli* and

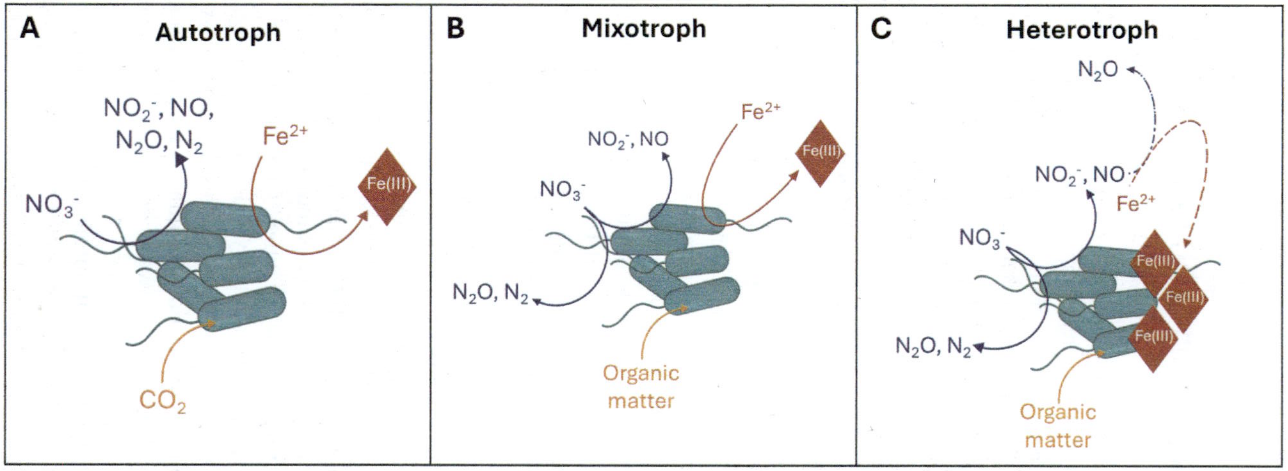

FIGURE 16.2 Metabolic pathways of nitrate-reducing Fe(II) oxidation.

Schematic representation of the three main metabolic strategies by which Fe(II) can be oxidized in the absence of oxygen through nitrate reduction. (A) Fe(II) oxidation is performed by autotrophic microorganisms via enzymatic activity, (B) Fe(II) is oxidized by mixotrophic bacteria that use Fe(II) as electron donors and organic matter as carbon source, (C) heterotrophic microorganisms do not oxidize Fe(II) directly but produce reactive nitrogen species (e.g., NO_2^-, NO) that abiotically oxidize Fe(II). Solid arrows indicate biologically catalyzed reactions, while dashed arrows denote abiotic transformations. In the three panels, the red diamonds represent insoluble and precipitated Fe(III).

- This figure is based on Bryce, C., Blackwell, N., Schmidt, C., Otte, J., Huang, Y.-M., Kleindienst, S., Tomaszewski, E., Schad, M., Warter, V., Peng, C., Byrne, J.M., Kappler, A., 2018. Microbial anaerobic Fe(II) oxidation—ecology, mechanisms and environmental implications. Environ. Microbiol. 20 (10), 3462–3483, https://doi.org/10.1111/1462-2920.14328.

Pseudomonas spp., can oxidize Fe^{2+} when using nitrate as an electron acceptor (Li et al., 2018; Brons et al., 1991), suggesting that RNS-mediated iron oxidation may be a widespread incidental outcome of nitrate respiration.

In the IPB, all the conditions necessary for NRFeOx are present. Both MARTE and IPBSL drilling projects reported nitrate concentrations above 100 ppm across multiple depths (Amils et al., 2011, 2023). In addition, a range of dissolved organic compounds—including acetate, formate, propionate, oxalate, carbohydrates, and proteins—was detected in borehole fluids and core samples (Puente-Sánchez et al., 2014). These compounds can serve as electron donors for heterotrophic NRFeOx activity.

Iron geochemistry also supports the existence of an active iron cycle in the deep subsurface. MARTE reported Fe^{2+} concentrations averaging 95 ppm and Fe^{3+} at 22 ppm, while IPBSL detected 41 ppm of Fe^{2+} and 63 ppm of Fe^{3+} in BH10 (Amils et al., 2023; González-Toril et al., 2010). Microbial enrichments and genomic analyses confirmed the presence of Fe(III) reducers such as *Acidiphilium* and *Shewanella*, as well as *Acidithiobacillus*, which is known to reduce Fe^{3+} under anoxic conditions (Quatrini and Johnson, 2019).

Although no *Acidovorax* strains have yet been isolated from the IPB, CARD-FISH and sequencing have confirmed their presence in rock core samples (Amils et al., 2023; Escudero, 2018). Actually, 82% of the microorganisms detected in the IPB subsurface have the potential for nitrate reduction and, thus, for chemodenitrification. This observation suggests that nitrate-reducing processes, either through direct enzymatic oxidation of Fe(II) or via the production of reactive nitrogen species capable of abiotic oxidation, may play a central role in IPB subsurface iron cycling.

Diversity and ecological role of NRFeOx microorganisms in the IPB

Building upon the detection of *Acidovorax* and other putative Fe(II)-oxidizing taxa in the IPB subsurface, efforts have focused on expanding the known diversity of microorganisms capable of NRFeOx. As part of the IPBSL project, 38 isolates representing 23 different genera were recovered from anaerobic enrichment cultures derived from boreholes BH10 and BH11 (Martínez-Bonilla, 2025). These isolates were subjected to a sequential screening strategy designed to determine their ability to oxidize Fe(II) using nitrate as the terminal electron acceptor under anoxic conditions.

Each isolate was first subjected to whole-genome sequencing and functional annotation to identify genes associated with nitrate reduction, including *narGHI*, *napAB*, *nirK*, *norBC*, and *nosZ* (Martínez-Bonilla, 2025). Based on these genomic profiles, a subset of strains with nitrate-reducing potential was selected for further physiological testing. Their ability to oxidize Fe(II) was evaluated under strictly anaerobic conditions, using Fe(II) and nitrate as substrates in controlled incubations. In total, 15 isolates were confirmed to oxidize Fe^{2+} in the presence of nitrate, despite lacking canonical Fe(II)-oxidizing genes such as *mtoAB* or *cyc2*. This strongly suggests that Fe(II) oxidation in these strains proceeds via an indirect mechanism (Bryce et al., 2018; Martínez-Bonilla, 2025). In fact, the stoichiometry of Fe(II) oxidized to nitrate reduced observed in these experiments closely aligned with that expected from chemodenitrification reactions (Klueglein and Kappler, 2013; Bryce et al., 2018).

Among the tested isolates, representatives of *Pseudomonas*, *Shewanella*, *Stutzerimonas*, and *Tessaracoccus* achieved complete Fe(II) oxidation within the duration of the experiments. The isolation of *Tessaracoccus lapidicaptus* was particularly significant, as it is one of the most abundant microorganisms detected in the BH10 (Amils et al., 2023; Puente-Sánchez et al., 2014). Although the chemodenitrification by *Tessaracoccus* was notably slower, requiring over 45 days for near-complete oxidation, the oxidation of Fe(II) was unambiguous and was supported by the detection of ferric iron via fluorescence microscopy using Fe(III)-sensitive probes (Amils et al., 2023).

The remaining 11 isolates demonstrated only partial Fe(II) oxidation. While physiological limitations cannot be fully excluded, several environmental and metabolic factors likely contribute to this incomplete activity. Indeed, NRFeOx efficiency is highly sensitive to both environmental and physiological constraints, with Fe(II) toxicity and Fe(III) encrustation emerging as major limiting factors in experimental systems. Although less studied under anaerobic conditions, Fe(II) toxicity is well documented: micromolar concentrations can inhibit microbial growth, as shown in *Rhodobacter* (5 μM) (Poulain and Newman, 2009), and millimolar levels can be lethal, as observed in *Streptococcus* ($>$10 mM) (Dunning et al., 1998). Several IPB isolates displaying incomplete Fe(II) oxidation may be experiencing similar stress.

Fe(III) production poses another significant challenge as, at neutral pH, Fe(III) is extremely insoluble ($\sim 10^{-9}$–10^{-10} M), leading to the rapid formation of $Fe(OH)_3$. These precipitates can encrust cell surfaces, physically isolating them from their surroundings, limiting nutrient and electron exchange, and ultimately causing metabolic arrest or death (Miot et al., 2015). This phenomenon has been observed via SEM and EDX analyses, particularly in strains belonging to *Cellulomonas*, *Ciceribacter*, *Propionicimonas*, and *Shewanella* (Fig. 16.3). Importantly, some microorganisms might mitigate these encrustation-induced stresses. For example, the production of extracellular polymeric substances (EPS) by *Acidovorax* may help sequester Fe(III) within the biofilm matrix (Escudero, 2018), which could help maintain redox activity in environments prone to ferric precipitation.

Additionally, some microorganisms may be sensitive to localized acidification caused by the precipitation of Fe(III) produced during NRFeOx, a process that releases protons and can lower the pH in the immediate microenvironment (Eq. 16.1). Beyond microbial strategies, the mineralogical context of the IPB may also play a buffering role. Carbonate phases such as siderite ($FeCO_3$) and ankerite

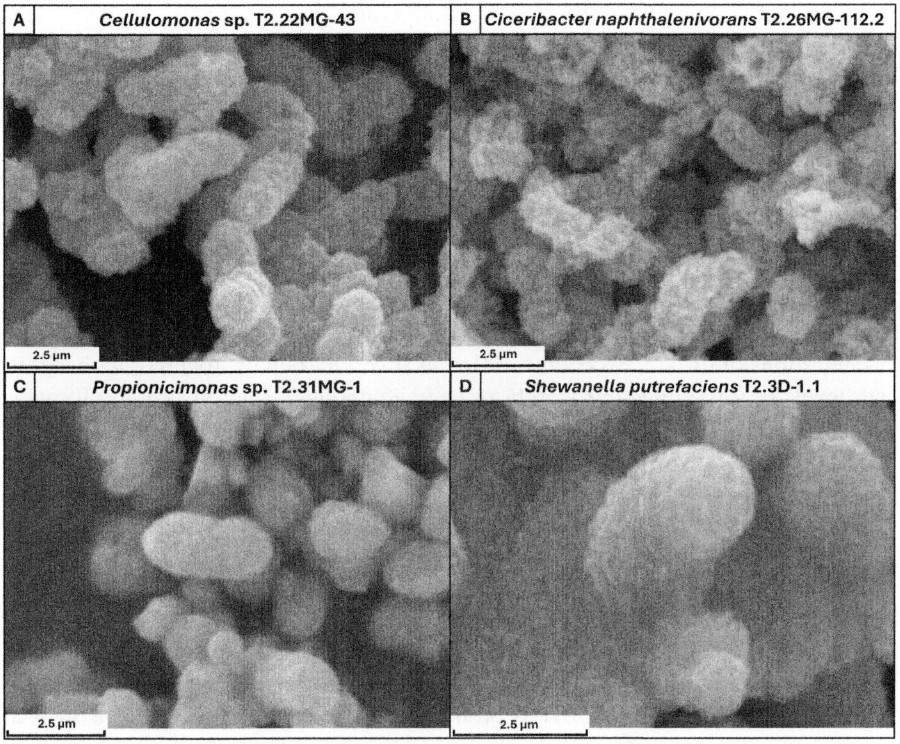

FIGURE 16.3 Iron encrustations detected in isolates from the IPB.

Scanning electron microscopy (SEM) images showing Fe encrustations on the cell surfaces of four isolates obtained from the Iberian pyrite belt. Panel (A) shows *Cellulomonas* sp. T2.22 MG-43, panel (B) *Ciceribacter naphthalenivorans* T2.26 MG-112.2, panel (C) *Propionicimonas* sp. T2.31MG-1, and panel (D) *Shewanella putrefaciens* T2.3D-1.1. These precipitates, predominantly composed of ferric iron, were formed during nitrate-dependent Fe(II) oxidation (NRFeOx) incubations and are consistent with Fe(III) mineral deposition on cellular membranes. Scale bars, 2.5 µm.

($CaFe(CO_3)_2$), which have been detected in rock cores, can neutralize protons generated during Fe(II) oxidation, dampening local pH drops. These conditions may enhance microbial persistence and activity in situ, even if such constraints limit performance under laboratory conditions (Amils et al., 2023; Martínez-Bonilla, 2025).

The presence of these NRFeOx-capable organisms across boreholes BH10 and BH11 suggests a broad spatial distribution in the IPB subsurface. Notably, these boreholes differ significantly in their physicochemical conditions. BH10 exhibits near-neutral pH, while BH11 displays more acidic conditions (pH 3.5–4), likely due to extensive sulfide oxidation (Amils et al., 2023). Despite this contrast, NRFeOx microorganisms were detected in both locations, suggesting that this metabolism can occur across a range of pH values. This tolerance may be enhanced by the presence of mineral buffering or by microbially generated biofilms that create more favorable microscale conditions (Martínez-Bonilla, 2025).

Mineralogical context was also shown to affect Fe(II) oxidation efficiency. Experiments testing the influence of different substrates, including hematite (Fe_2O_3), pyrite (FeS_2), and quartz (SiO_2), demonstrated that hematite significantly enhanced Fe(II) oxidation rates in strains such as *Lelliottia* and *Stutzerimonas*, whereas pyrite and quartz reduced oxidation efficiency. This was interpreted as a mineral-catalyzed effect, in which reactive Fe(III) species on hematite surfaces promote further Fe(II) oxidation (Martínez-Bonilla, 2025).

Notably, the IPB provides the key chemical and physical conditions needed to sustain NRFeOx metabolism. Both the MARTE and IPBSL projects reported high concentrations of nitrate, exceeding 100 ppm at certain depths, together with significant levels of low-molecular-weight organic compounds such as acetate, formate, propionate, oxalate, and various carbohydrates and peptides (Amils et al., 2023; Puente-Sánchez et al., 2014). These compounds can serve as electron donors or carbon sources for heterotrophic microorganisms capable of coupling nitrate reduction to Fe(II) oxidation. This potential is further supported by data from borehole BH10 (Fig. 16.4), where nitrate and nitrite were consistently detected at multiple depths, indicating the persistence of nitrogen-based electron acceptors. Enrichment cultures from the same core confirmed the presence of active nitrate-reducing communities, including several strains able to oxidize Fe(II) under anoxic conditions. Moreover, CARD-FISH analyses revealed the widespread distribution of NRFeOx-capable microorganisms throughout the BH10 profile. Detected taxa include genera such as *Acidovorax*, *Pseudomonas*, and *Tessaracoccus* spp., all of which are known or have been demonstrated to carry out nitrate-dependent Fe(II) oxidation (Fig. 16.4).

Collectively, the data show that the IPB subsurface harbors a taxonomically and metabolically diverse community of microorganisms capable of performing nitrate-dependent Fe(II) oxidation through both direct and indirect mechanisms. Their widespread presence, broad pH tolerance, and functional capacity under anoxic conditions support the hypothesis that NRFeOx is a key driver of iron cycling in the deep subsurface. By generating Fe(III), these microorganisms not only influence the redox state of iron but may also facilitate the sustained anaerobic oxidation of pyrite, contributing to the geochemical and acidic character of the Río Tinto system.

Evidence of pyrite oxidation by NRFeOx microorganisms in the IPB subsurface

The hypothesis that nitrate-reducing microorganisms actively contribute to pyrite dissolution in the deep subsurface of the IPB finds its most compelling support not in theoretical models or laboratory simulations, but in direct, in situ observations of microbial-mineral interactions. Recent applications of advanced imaging and spectroscopy techniques, particularly the combination of confocal Raman

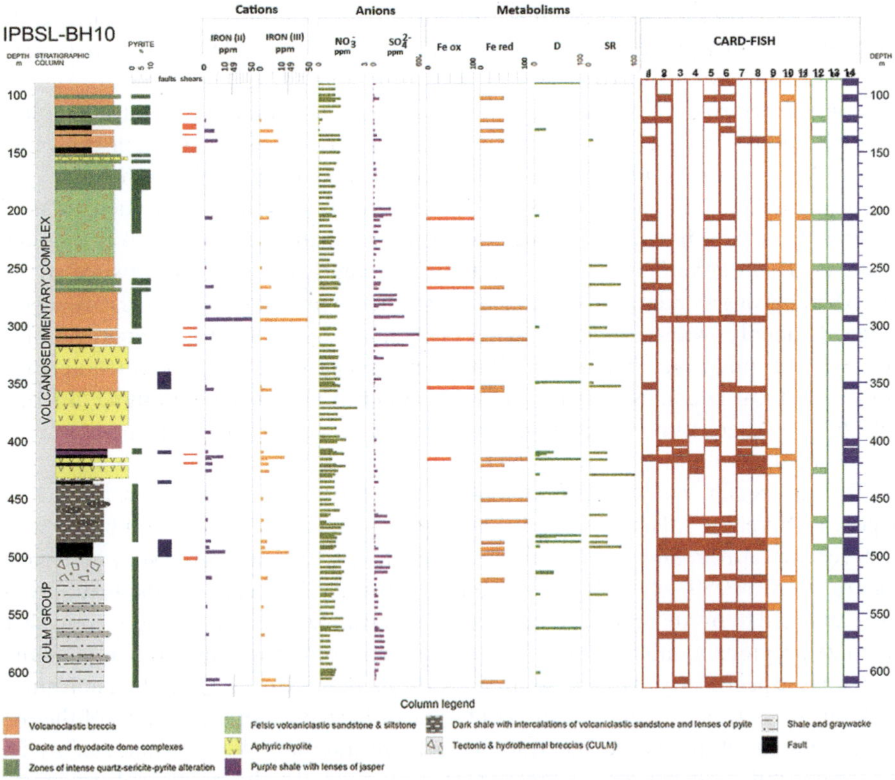

FIGURE 16.4 Integrated biogeochemical and microbiological profile of borehole BH10 in the IPB subsurface.

From left to right: mineralogical composition of the core, highlighting the depth distribution of pyrite; vertical profiles of Fe(II), Fe(III), nitrate (NO_3^-), and sulfate (SO_4^{2-}); depth-resolved results of anaerobic enrichment cultures showing microbial activities associated with iron oxidation (Fe ox), iron reduction (Fe red), denitrification (D) and sulfate reduction (SR); CARD–FISH–based detection of microbial taxa grouped by functional role. Mixotrophic and chemodenitrifying Fe(II)-oxidizing bacteria (shown in red) include [1–2] *Acidovorax* spp., [3] *Rhizobium* spp., [4] *Brevundimonas* spp., [5] *Rhodoplanes* spp., [6] *Pseudomonas* spp., [7] *Tessaracoccus* spp., and [8] *Tessaracoccus lapidicaptus*. Fe(III)-reducing bacteria (in orange): [9] *Acidiphilium* spp., [10–11] *Acidithiobacillus* spp., sulfate-reducing bacteria (SRB, in green): [12] Deltaproteobacteria (general SRB group), [13] *Desulfosarcina/Desulfococcus* group. Sulfur-oxidizing bacteria (in blue): [14] *Sulfobacillus* spp.

microscopy (CRM) with fluorescence in situ hybridization (FISH), known as Raman-FISH (Huang et al., 2007), have provided unprecedented resolution in tracking microbial colonization of mineral surfaces and their associated chemical transformations. These tools have revealed, with high spatial precision, that key

NRFeOx bacteria such as *Acidovorax* are not only present in the IPB subsurface but are also physically associated with pyrite and likely responsible for its active dissolution (Escudero et al., 2020).

Raman spectroscopy is a nondestructive analytical technique that detects molecular vibrations through the inelastic scattering of monochromatic light, typically from a laser source (Dieing et al., 2011; Smith and Dent, 2013). That's to say, when applied to minerals, it allows for their identification and the detection of compositional and molecular changes with high sensitivity and spatial resolution. In the context of the IPB, Raman spectroscopy can detect not only pristine pyrite (FeS_2), but also secondary iron oxides that are known products of its oxidative dissolution. Crucially, when Raman imaging is correlated with FISH, an approach that tags specific microbial taxa using fluorescent rRNA-targeted probes (Moter and Göbel, 2000), it becomes possible to overlay microbial identity with mineral structure, thereby identifying not just who is there, but what is happening chemically in their immediate vicinity.

This approach was applied to powdered natural rock samples from various depths in borehole BH10, where FISH targeting *Acidovorax* was combined with high-resolution Raman mapping (Escudero et al., 2020). The results showed a clear pattern: in over 95% of analyzed mineral fragments, *Acidovorax* cells were physically attached to pyrite grains. More importantly, Raman spectra acquired from these colonized regions exhibited systematic deviations from those collected in uncolonized areas. Specifically, a decrease in the Ag/Eg band ratio and an upshift in Raman peak positions were observed, consistent with surface-level modifications of the pyrite crystal structure. These spectral changes are interpreted as evidence of pyrite oxidation and restructuring, triggered or enhanced by the microbial presence. Such shifts were never observed in abiotic controls, nor in mineral areas devoid of *Acidovorax*, pointing to a direct causal relationship between microbial activity and mineral alteration.

In addition to Raman evidence of mineral modification, fluorescence-based iron mapping revealed colocalization of Fe^{3+} with *Acidovorax* and *Tessaracoccus* cells. Using highly specific Fe^{3+}-sensitive fluorescent probes, a strong ferric iron signal was observed in direct association with microbial colonies. This provides functional evidence that these microorganisms are not only capable of oxidizing Fe^{2+} under nitrate-reducing conditions (as previously demonstrated in laboratory settings), but that they are doing so in situ (Amils et al., 2023; Escudero, 2018). In the case of *Acidovorax*, Fe^{3+} was often concentrated around the cells, indicative of active oxidation occurring at or near the cell membrane. In some colonies, Raman also detected iron oxide minerals such as hematite (Escudero et al., 2020), secondary products expected from Fe^{2+} oxidation and pyrite weathering (Caldeira et al., 2003).

Beyond individual cell-mineral associations, a remarkable feature observed in these samples was the formation of extensive microbial biofilms composed of *Acidovorax* or other NRFeOx organisms such as *Tessaracoccus*. These biofilms were visualized using both lectin staining (to detect extracellular polysaccharides) and Raman imaging (to detect organic compounds like proteins and sugars) (Amils et al., 2023; Escudero, 2018; Escudero et al., 2020). Their presence is significant for several reasons. First, biofilms facilitate stable adhesion to mineral surfaces (Flemming, 2011; Flemming et al., 2016), anchoring the cells in optimal positions to access electron donors such as Fe^{2+} from pyrite. Second, they might create physicochemical microenvironments distinct from the surrounding bulk medium (Flemming, 2011; Flemming et al., 2016). Within these matrices, protons released from Fe^{2+} oxidation can accumulate, locally lowering pH and enhancing mineral dissolution (Rohwerder et al., 2003; Sand and Gehrke, 2006). Third, and perhaps most critically, the EPS matrix can help chelate and stabilize Fe^{3+} (Yu et al., 2016), preventing it from precipitating immediately as $Fe(OH)_3$. This buffering effect is essential in neutral to mildly acidic environments like the IPB subsurface, where Fe^{3+} is otherwise sparingly soluble and prone to forming passivating mineral crusts.

The role of EPS in preventing encrustation is not trivial. In many NRFeOx systems, microbial inactivation due to Fe(III) precipitation is a limiting factor as discussed before. Cells become coated in amorphous iron oxides, which block nutrient exchange and inhibit further activity. However, in the IPB subsurface, this problem appears to be mitigated. Raman-FISH data show that microbial colonies with active Fe^{3+} production are viable, uncoated, and metabolically intact. This strongly suggests that EPS production and the hydraulic dynamics of the fractured rock matrix together allow for continuous ferric iron turnover without the formation of inhibitory precipitates.

When these data are taken together, the picture becomes difficult to ignore: we are not simply observing microorganisms that are *capable* of driving pyrite oxidation—we are witnessing them doing it in situ. *Acidovorax* is present, metabolically active, physically associated with pyrite, embedded in biofilms that concentrate Fe^{3+}, and coinciding with Raman-detectable changes in the mineral consistent with oxidative attack. The presence of secondary iron oxides like hematite in colonized zones adds another layer of confirmation, as these are classic products of Fe^{2+} oxidation under circumneutral conditions.

This evidence represents the strongest support to date for the microbial origin of Río Tinto's geochemical peculiarities. Unlike classical acid mine drainage systems, where oxygen is the primary oxidant and microbial diversity is limited to acidophiles, the IPB subsurface hosts a metabolically diverse, facultatively anaerobic microbial community that drives a self-sustaining redox cycle. Nitrate-reducing organisms, including both mixotrophic NRFeOx taxa and chemodenitrifiers, oxidize Fe^{2+} to Fe^{3+}, which then abiotically or biologically dissolves pyrite, regenerating Fe^{2+} and protons in the process. Embedded within biofilms, these processes are physically and chemically coupled, enabling mineral attack in the absence of oxygen, light, or human disturbance.

Although many mechanistic details remain to be clarified—including the enzymatic pathways involved, the relative contribution of chemodenitrification versus direct Fe(II) oxidation, and the long-term kinetics of pyrite transformation—the evidence overwhelmingly points to nitrate-reducing microorganisms as key agents in the genesis and maintenance of Río Tinto's unique geochemistry. Their presence, activity, and direct association with mineral alteration provide a mechanistic foundation for the subsurface bioreactor hypothesis. In this view, Río Tinto is not simply a river "stained" by mining, but the visible outlet of a deep, slow, and biologically orchestrated geochemical engine.

Pyrite-centered syntrophic networks and microbial mineral interaction

Beyond the activity of individual nitrate-reducing Fe(II)-oxidizing (NRFeOx) microorganisms, increasing evidence suggests that microbial life in the deep subsurface of the IPB is structured around complex syntrophic networks. These consortia involve functionally distinct microorganisms that interact metabolically, spatially, and chemically, forming stable, self-sustaining systems capable of driving biogeochemical cycling under the energy-limited and anoxic conditions of the deep biosphere.

CARD-FISH analyses and enrichment studies have consistently shown that *Acidovorax* is not acting alone in the subsurface, but is spatially colocalized with other microbial taxa, including *Sulfobacillus*, *Acidiphilium*, and sulfate-reducing bacteria (SRB) (Amils et al., 2023) (Fig. 16.5). These coassociations seems to be particularly frequent around pyrite grains, suggesting that this mineral serves as a shared resource and structural anchor for cooperative microbial communities. The

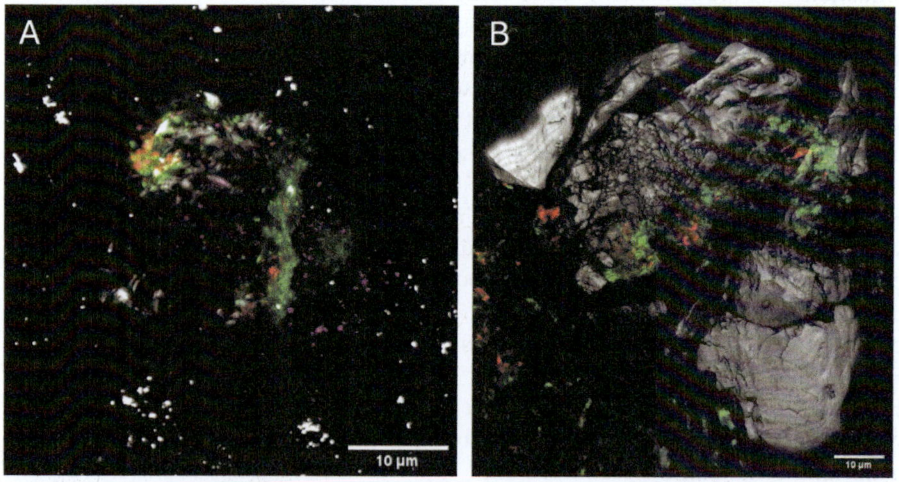

FIGURE 16.5 Visualization of microbial interactions in the IPB subsurface by CARD-FISH.

(A) Putative coupling between the iron and sulfur cycles mediated by *Acidovorax* (green), *Sulfobacillus* (red) and *Acidiphilium* (purple). (B) Putative sulfur cycling involving *Sulfobacillus* and sulfate-reducing bacteria (SRB). Scale bars: 10 μm.

proposed model of syntrophic interaction begins with *Acidovorax* oxidizing Fe(II) to Fe(III) and potentially producing intermediate sulfur species such as thiosulfate ($S_2O_3^{2-}$) or tetrathionate ($S_4O_6^{2-}$) during the process of pyrite oxidation. These sulfur species can then be utilized by *Sulfobacillus*, which is known to oxidize reduced sulfur compounds while reducing Fe(III), effectively closing a redox loop that regenerates substrates for *Acidovorax*. *Acidiphilium* and SRB may utilize Fe(III) or sulfate as terminal electron acceptors respectively. SRBs reduce sulfate to sulfide, while *Acidiphilium* can reduce Fe(III) and use organic compounds excreted by other members of the community. The mutualistic interactions among these organisms would ensure the recycling of key redox-active species (Fe and S) (Escudero, 2018). The existence of these microbial consortia has important implications for subsurface geochemistry. Pyrite grains, which dominate the mineralogy of the IPB, not only serve as a source of Fe and S but also as a physical scaffold for microbial colonization. The interactions among NRFeOx bacteria, sulfur oxidizers, and metal reducers would enable the complete transformation of pyrite and sustain geochemical gradients over long timescales and spatial scales.

Taken together, the data support a model in which pyrite acts as the energetic and structural backbone of the IPB subsurface ecosystem. It fuels a cascade of redox transformations that are driven and stabilized by microbial syntrophy. This systemic view emphasizes the importance of mineral-microbe interfaces and cross-feeding interactions in shaping biogeochemical processes in extreme environments like Río Tinto.

Final remarks

The study of the origin of Río Tinto and the Iberian Pyrite Belt subsurface has fundamentally reshaped our understanding of biogeochemical processes in extreme subsurface environments. Once regarded primarily as an example of anthropogenically induced acid mine drainage, the Río Tinto system is now recognized as one of the most prominent natural acidic ecosystems on Earth. Multiple lines of evidence, ranging from geochronological studies of iron terraces to direct microbiological and geochemical investigations, support the conclusion that the river's acidity and iron load are not solely the product of historical mining activities. Instead, they are sustained by a deep biosphere hosted within the IPB, where microbial processes play a central role in generating and maintaining the river's extreme geochemical properties.

At the core of this hypothesis is the concept of a subsurface microbial bioreactor, in which buried pyrite is continuously oxidized under anoxic conditions, releasing Fe^{2+}, sulfur species, and protons into rising groundwater. This groundwater emerges at the surface already enriched in ferric iron and characterized by low pH, consistent with observations made along the entire course of the Río Tinto. The viability of this model has been corroborated through extensive drilling campaigns (MARTE and

IPBSL), which have provided direct access to deep rock cores, water, and microbial communities residing hundreds of meters below the surface.

Central to the functioning of this bioreactor is the metabolic process of nitrate-reducing Fe(II) oxidation (NRFeOx), the only known mechanism by which microorganisms can oxidize Fe^{2+} in the absence of oxygen. The presence of nitrate, Fe^{2+}, and diverse organic substrates in the IPB subsurface creates ideal conditions for this metabolism to thrive. Genomic and cultivation-based studies have identified a wide array of potential NRFeOx organisms in this environment. Among them, *Acidovorax* stands out due to its metabolic efficiency, colonization of pyrite surfaces, and ability to form EPS-rich biofilms that facilitate Fe(III) accumulation and mineral dissolution.

The ecosystem is not limited to single-species metabolism. Rather, it is shaped by intricate syntrophic interactions among NRFeOx bacteria, sulfur oxidizers such as *Sulfobacillus*, metal reducers like *Acidiphilium*, and sulfate-reducing bacteria (SRB). These consortia colocalize on pyrite grains, forming spatially organized biofilms that allow for the recycling of redox-active compounds and the stabilization of microenvironments necessary for metabolic cooperation. This syntrophic web enables the full oxidation of pyrite and the maintenance of geochemical gradients, ensuring system stability even under oligotrophic and anoxic conditions.

Despite the robustness of these processes, NRFeOx efficiency is constrained by factors such as Fe(II) toxicity and Fe(III) encrustation. Experimental evidence has shown that high concentrations of ferrous iron can inhibit microbial growth, while the precipitation of Fe(III) can physically encase cells, leading to metabolic arrest or death. These limitations can be partly mitigated by EPS production, mineral buffering, and spatial segregation within the biofilms. Nevertheless, they represent critical variables that must be considered when assessing the ecological performance and resilience of subsurface microbial populations.

The case of Río Tinto and the IPB challenges classical paradigms in environmental microbiology and geochemistry. It exemplifies how microbial life can shape large-scale geological and hydrological features, not as a consequence of human disturbance, but as a persistent natural process sustained over millions of years. It also expands the known limits of life by highlighting microbial strategies that operate in metal-rich, acidic, and anoxic conditions once thought too extreme to support sustained biological activity.

In addition to its Earth-based significance, the Río Tinto system provides a valuable analog for astrobiological exploration. The capacity of microbial consortia to drive iron and sulfur cycling in the absence of oxygen has direct implications for the search for life in subsurface environments on Mars and other planetary bodies.

Ultimately, the Río Tinto system stands as a testament to the power of microbial life to transform its environment, creating chemically extreme yet biologically dynamic niches in the most unlikely places. The continued study of this ecosystem not only enriches our understanding of microbial ecology and biogeochemistry but also challenges us to reconsider the boundaries of life on Earth and beyond.

AI disclosure

During the preparation of this work the author(s) used ChatGPT in order to English proofreading. After using this tool/service, the author(s) reviewed and edited the content as needed and take(s) full responsibility for the content of the publication.

References

Amaral Zettler, L.A., Gómez, F., Zettler, E., Keenan, B.G., Amils, R., Sogin, M.L., 2002. Eukaryotic diversity in Spain's river of fire. Nature 417 (6885). https://doi.org/10.1038/417137a, 137–137.

Amils, R., Escudero, C., Oggerin, M., Puente Sánchez, F., Arce Rodríguez, A., Fernández Remolar, D., Rodríguez, N., García Villadangos, M., Sanz, J.L., Briones, C., Sánchez-Román, M., Gómez, F., Leandro, T., Moreno-Paz, M., Prieto-Ballesteros, O., Molina, A., Tornos, F., Sánchez-Andrea, I., Timmis, K., Pieper, D.H., Parro, V., 2023. Coupled C, H, N, S and Fe biogeochemical cycles operating in the continental deep subsurface of the Iberian Pyrite Belt. Environ. Microbiol. 25 (2), 428–453. https://doi.org/10.1111/1462-2920.16291. Available from: http://onlinelibrary.wiley.com/journal/10.1111/(ISSN)1462-2920.

Amils, R., González-Toril, E., Aguilera, A., Rodríguez, N., Fernández-Remolar, D., Gómez, F., García-Moyano, A., Malki, M., Oggerin, M., Sánchez-Andrea, I., Sanz, J.L., 2011. From Río Tinto to Mars. The terrestrial and extraterrestrial ecology of acidophiles. Adv. Appl. Microbiol. 77, 41–70. https://doi.org/10.1016/B978-0-12-387044-5.00002-9.

Amils, R., González-Toril, E., Gómez, F., Fernández-Remolar, D., Rodríguez, N., Malki, M., Zuluaga, J., Aguilera, A., Amaral-Zettler, L.A., 2004. Importance of Chemolithotrophy for Early Life on Earth: The Tinto River (Iberian Pyritic Belt) Case. Springer Science and Business Media LLC, pp. 463–480. https://doi.org/10.1007/1-4020-2522-x_28.

Amils, R., González-Toril, E., Fernández-Remolar, D., Gómez, F., Rodríguez, N., Durán, C., 2002. Interaction of the sulfur and iron cycles in the Tinto River ecosystem. Rev. Environ. Sci. Biotechnol. 1 (4), 299–309. https://doi.org/10.1023/a:1023232002312.

Amils, R., Fernández-Remolar, D., T.IPBSL Team, 2014. Río Tinto: a geochemical and mineralogical terrestrial analogue of mars. Life 4 (3), 511–534. https://doi.org/10.3390/life4030511.

Boja, 2005. DECRETO 558/2004, de 14 de diciembre, por el que se declara el Paisaje Protegido de Río Tinto. In: Junta de Andalucía: Boletín Oficial de la. Junta de Andalucía, pp. 24–55.

Bosch, J., Lee, K.Y., Jordan, G., Kim, K.W., Meckenstock, R.U., 2012. Anaerobic, nitrate-dependent oxidation of pyrite nanoparticles by thiobacillus denitrificans. Environ. Sci. Technol. 46 (4), 2095–2101. https://doi.org/10.1021/es2022329.

Brierley, C.L., Brierley, J.A., 2013. Progress in bioleaching: Part B: applications of microbial processes by the minerals industries. Appl. Microbiol. Biotechnol. 97 (17), 7543–7552. https://doi.org/10.1007/s00253-013-5095-3.

Brons, H.J., Hagen, W.R., Zehnder, A.J.B., 1991. Ferrous iron dependent nitric oxide production in nitrate reducing cultures of *Escherichia coli*. Arch. Microbiol. 155 (4), 341–347. https://doi.org/10.1007/BF00243453.

Bryce, C., Blackwell, N., Schmidt, C., Otte, J., Huang, Y.-M., Kleindienst, S., Tomaszewski, E., Schad, M., Warter, V., Peng, C., Byrne, J.M., Kappler, A., 2018. Microbial anaerobic Fe(II) oxidation—ecology, mechanisms and environmental implications. Environ. Microbiol. 20 (10), 3462–3483. https://doi.org/10.1111/1462-2920.14328.

Caballero, 1981. Las antiguas explotaciones de Río Tinto. Publication EXCMA.

Caldeira, C.L., Ciminelli, V.S.T., Dias, A., Osseo-Asare, K., 2003. Pyrite oxidation in alkaline solutions: nature of the product layer. Int. J. Miner. Process. 72 (1–4), 373–386. https://doi.org/10.1016/s0301-7516(03)00112-1.

Carlson, H.K., Clark, I.C., Blazewicz, S.J., Iavarone, A.T., Coates, J.D., 2013. Fe(II) oxidation is an innate capability of nitrate-reducing bacteria that involves abiotic and biotic reactions. J. Bacteriol. 195 (14), 3260–3268. https://doi.org/10.1128/jb.00058-13.

Chakraborty, A., Picardal, F., 2013. Induction of nitrate-dependent Fe(II) oxidation by Fe(II) in Dechloromonas sp. strain UWNR4 and Acidovorax sp. strain 2AN. Appl. Environ. Microbiol. 79 (2), 748–752. https://doi.org/10.1128/aem.02709-12.

Davis Jr., R.A., Welty, A.T., Borrego, J., Morales, J.A., Pendon, J.G., Ryan, J.G., 2000. Rio Tinto estuary (Spain): 5000 years of pollution. Environ. Geol. 39 (10), 1107–1116. https://doi.org/10.1007/s002549900096.

del Bayo, E.J., 1852. Memorias sobre las minas nacionales de Rio-Tinto. Imprenta de la Viuda de D. Antonio Yenes, Madrid.

Denef, V.J., Mueller, R.S., Banfield, J.F., 2010. AMD biofilms: using model communities to study microbial evolution and ecological complexity in nature. ISME J. 4 (5), 599–610. https://doi.org/10.1038/ismej.2009.158.

Dieing, T., Hollricher, O., Toporski, J., 2011. Confocal Raman microscopy. Ser. Opt. Sci. 158. https://doi.org/10.1007/978-3-642-12522-5.

Dunning, J.C., Ma, Y., Marquis, R.E., 1998. Anaerobic killing of oral streptococci by reduced, transition metal cations. Appl. Environ. Microbiol. 64 (1), 27–33. https://doi.org/10.1128/aem.64.1.27-33.1998.

Escudero, C., del Campo, A., Ares, J.R., Sánchez, C., Martínez, J.M., Gómez, F., Amils, R., 2020. Visualizing microorganism-mineral interaction in the Iberian pyrite belt subsurface: the Acidovorax case. Front. Microbiol. 11. https://doi.org/10.3389/fmicb.2020.572104. Available from: https://www.frontiersin.org/journals/microbiology#.

Escudero, C., 2018. Fluorescence Microscopy for the in Situ Study of the Iberian Pyrite Belt Subsurface Geomicrobiology.

Fernández-Remolar, D.C., Rodriguez, N., Gómez, F., Amils, R., 2003. Geological record of an acidic environment driven by iron hydrochemistry: the Tinto river system. J. Geophys. Res.: Planets 108 (7), 1–16. https://doi.org/10.1029/2002je001918. Available from: http://agupubs.onlinelibrary.wiley.com/agu/jgr/journal/10.1002/(ISSN)2169-9100/.

Fernández-Remolar, D.C., Morris, R.V., Gruener, J.E., Amils, R., Knoll, A.H., 2005. The Río Tinto Basin, Spain: mineralogy, sedimentary geobiology, and implications for interpretation of outcrop rocks at Meridiani Planum, Mars. Earth Planet Sci. Lett. 240 (1), 149–167. https://doi.org/10.1016/j.epsl.2005.09.043.

Flemming, H.C., Wingender, J., Szewzyk, U., Steinberg, P., Rice, S.A., Kjelleberg, S., 2016. Biofilms: an emergent form of bacterial life. Nat. Rev. Microbiol. 14 (9), 563–575. https://doi.org/10.1038/nrmicro.2016.94.

Flemming, H.C., 2011. The perfect slime. Colloids Surf. B Biointerfaces 86 (2), 251–259. https://doi.org/10.1016/j.colsurfb.2011.04.025.

Gomes, A.F.R., Almeida, M.C., Sousa, E., Resende, D.I.S.P., 2024. Siderophores and metallophores: metal complexation weapons to fight environmental pollution. Sci. Total Environ. 932, 173044. https://doi.org/10.1016/j.scitotenv.2024.173044.

Gómez-Ortiz, D., Fernández-Remolar, D.C., Granda, A., Quesada, C., Granda, T., Prieto-Ballesteros, O., Molina, A., Amils, R., 2014. Identification of the subsurface sulfide bodies responsible for acidity in Río Tinto source water, Spain. Earth Planet Sci. Lett. 391, 36–41. https://doi.org/10.1016/j.epsl.2014.01.022.

González-Toril, E., Llobet-Brossa, E., Casamayor, E.O., Amann, R., Amils, R., 2003. Microbial ecology of an extreme acidic environment, the Tinto river. Appl. Environ. Microbiol. 69 (8), 4853–4865. https://doi.org/10.1128/aem.69.8.4853-4865.2003.

González-Toril, E., Aguilera, A., Rodriguez, N., Fernández-Remolar, D., Gómez, F., Diaz, E., García-Moyano, A., Sanz, J.L., Amils, R., 2010. Microbial ecology of Río Tinto, a natural extreme acidic environment of biohydrometallurgical interest. Hydrometallurgy 104 (3–4), 329–333. https://doi.org/10.1016/j.hydromet.2010.01.011.

Grimm, H., Lorenz, J., Straub, D., Joshi, P., Shuster, J., Zarfl, C., Muehe, E.M., Kappler, A., Bose, A., 2025. Nitrous oxide is the main product during nitrate reduction by a novel lithoautotrophic iron(II)-oxidizing culture from an organic-rich paddy soil. Appl. Environ. Microbiol. 91 (1). https://doi.org/10.1128/aem.01262-24.

Haaijer, S.C.M., Lamers, L.P.M., Smolders, A.J.P., Jetten, M.S.M., Op den Camp, H.J.M., 2007. Iron sulfide and pyrite as potential electron donors for microbial nitrate reduction in freshwater wetlands. Geomicrobiol. J. 24 (5), 391–401. https://doi.org/10.1080/01490450701436489.

Hallberg, K.B., Coupland, K., Kimura, S., Johnson, D.B., 2006. Macroscopic streamer growths in acidic, metal-rich mine waters in North Wales consist of novel and remarkably simple bacterial communities. Appl. Environ. Microbiol. 72 (3), 2022–2030. https://doi.org/10.1128/AEM.72.3.2022-2030.2006.

Houben, G.J., Sitnikova, M.A., Post, V.E.A., 2017. Terrestrial sedimentary pyrites as a potential source of trace metal release to groundwater—a case study from the Emsland, Germany. Appl. Geochem. 76, 99–111. https://doi.org/10.1016/j.apgeochem.2016.11.019.

Huang, W.E., Stoecker, K., Griffiths, R., Newbold, L., Daims, H., Whiteley, A.S., Wagner, M., 2007. Raman-FISH: combining stable-isotope Raman spectroscopy and fluorescence in situ hybridization for the single cell analysis of identity and function. Environ. Microbiol. 9 (8), 1878–1889. https://doi.org/10.1111/j.1462-2920.2007.01352.x.

Huang, Y., Jakus, N., straub, D., Konstantinidis, K.T., Blackwell, N., Kappler, A., Kleindienst, S., 2022. Candidatus ferrigenium altingense'sp. nov., are autotrophic Fe (II)-oxidizing bacteria of the family Gallionellaceae. Syst. Appl. Microbiol. 45 (3).

Jakus, N., Mellage, A., Höschen, C., Maisch, M., Byrne, J.M., Mueller, C.W., Grathwohl, P., Kappler, A., 2021. Anaerobic neutrophilic pyrite oxidation by a chemolithoautotrophic nitrate-reducing iron(II)-Oxidizing culture enriched from a fractured aquifer. Environ. Sci. Technol. 55 (14), 9876–9884. https://doi.org/10.1021/acs.est.1c02049.

Johnson, D.B., Hallberg, K.B., 2003. The microbiology of acidic mine waters. Res. Microbiol. 154 (7), 466–473. https://doi.org/10.1016/S0923-2508(03)00114-1.

Jørgensen, C.J., Jacobsen, O.S., Elberling, B., Aamand, J., 2009. Microbial oxidation of pyrite coupled to nitrate reduction in anoxic groundwater sediment. Environ. Sci. Technol. 43 (13), 4851–4857. https://doi.org/10.1021/es803417sDenmark. Available from: http://pubs.acs.org/doi/pdfplus/10.1021/es803417s.

Kappler, A., Schink, B., Newman, D.K., 2005. Fe(III) mineral formation and cell encrustation by the nitrate-dependent Fe(II)-oxidizer strain BoFeN1. Geobiology 3 (4), 235–245. https://doi.org/10.1111/j.1472-4669.2006.00056.x.

Klueglein, N., Kappler, A., 2013. Abiotic oxidation of Fe(II) by reactive nitrogen species in cultures of the nitrate-reducing Fe(II) oxidizer Acidovorax sp. BoFeN1—questioning the existence of enzymatic Fe(II) oxidation. Geobiology 11 (2), 180–190. https://doi.org/10.1111/gbi.12019.

Kuang, J.L., Huang, L.N., Chen, L.X., Hua, Z.S., Li, S.J., Hu, M., Li, J.T., Shu, W.S., 2013. Contemporary environmental variation determines microbial diversity patterns in acid mine drainage. ISME J. 7 (5), 1038–1050. https://doi.org/10.1038/ismej.2012.139.

Leblanc, M., Morales, J.A., Borrego, J., Elbaz-Poulichet, F., 2000. 4,500-year-old mining pollution in southwestern Spain: long-term implications for modern mining pollution. Econ. Geol. 95 (3), 655–662. https://doi.org/10.2113/gsecongeo.95.3.655.

Li, S., Li, X., Li, F., 2018. Fe(II) oxidation and nitrate reduction by a denitrifying bacterium, Pseudomonas stutzeri LS-2, isolated from paddy soil. J. Soils Sediments 18 (4), 1668–1678. https://doi.org/10.1007/s11368-017-1883-1.

López-Archilla, A.I., Marin, I., Amils, R., 2001. Microbial community composition and ecology of an acidic aquatic environment: the Tinto river, Spain. Microb. Ecol. 41 (1), 20–35. https://doi.org/10.1007/s002480000044.

Macías, J.P., Domínguez, A.D., 2011. Ingeniería minera antigua y medieval en el suroeste ibérico. Bol. Geol. Min. 122.

Martiañez, C., 2000. Historia de la Faja Pirítica Ibérica, pp. 10–49.

Martínez-Bonilla, A., 2025. Anaerobic Iron Oxidation by Microorganisms Isolated from the Deep Subsurface of the Iberian Pyrite Belt. Doctoral dissertation.

Mining, Atalaya. Atalaya Mining. https://atalayamining.com/es/.

Miot, J., Remusat, L., Duprat, E., Gonzalez, A., Pont, S., Poinsot, M., 2015. Fe biomineralization mirrors individual metabolic activity in a nitrate-dependent Fe(II)- oxidizer. Front. Microbiol. 6.

Moraghan, J.T., Buresh, R.J., 1977. Chemical reduction of nitrite and nitrous oxide by ferrous iron. Soil Sci. Soc. Am. J. 41 (1), 47–50. https://doi.org/10.2136/sssaj1977.03615995004100010017x.

Moses, C.O., Kirk Nordstrom, D., Herman, J.S., Mills, A.L., 1987. Aqueous pyrite oxidation by dissolved oxygen and by ferric iron. Geochem. Cosmochim. Acta 51 (6), 1561–1571. https://doi.org/10.1016/0016-7037(87)90337-1.

Moses, C.O., Herman, J.S., 1991. Pyrite oxidation at circumneutral pH. Geochem. Cosmochim. Acta 55 (2), 471–482. https://doi.org/10.1016/0016-7037(91)90005-P.

Moter, A., Göbel, U.B., 2000. Fluorescence in situ hybridization (FISH) for direct visualization of microorganisms. J. Microbiol. Methods 41 (2), 85–112. https://doi.org/10.1016/S0167-7012(00)00152-4.

Nieto, J.M., Sarmiento, A.M., Olías, M., Canovas, C.R., Riba, I., Kalman, J., Delvalls, T.A., 2007. Acid mine drainage pollution in the Tinto and Odiel rivers (Iberian pyrite belt, SW Spain) and bioavailability of the transported metals to the Huelva estuary. Environ. Int. 33 (4), 445–455. https://doi.org/10.1016/j.envint.2006.11.010.

Nocete, F., Álex, E., Nieto, J.M., Sáez, R., Bayona, M.R., 2005. An archaeological approach to regional environmental pollution in the south-western Iberian Peninsula related to Third millennium BC mining and metallurgy. J. Archaeol. Sci. 32 (10), 1566–1576. https://doi.org/10.1016/j.jas.2005.04.012.

Nordstrom, D.K., 1982. Aqueous Pyrite Oxidation and the Consequent Formation of Secondary Iron Minerals Acid Sulfate Weathering. Wiley, United States, pp. 37–56. https://doi.org/10.2136/sssaspecpub10.c3. Available from: https://onlinelibrary.wiley.com/doi/book/10.2136/sssaspecpub10.

Olías, M., Nieto, J., 2015. Background conditions and mining pollution throughout history in the Río Tinto (SW Spain). Environments 2 (3), 295–316. https://doi.org/10.3390/environments2030295.

Ortíz Mateo, M., Romero Macías, E., 2004. La metalurgia en las Minas de Riotinto, desde su rehabilitación al alquiler del Marqués de Remisa (1725–1849) y obtención de indicadores ambientales del consumo de combustible en los procesos metalúrgicos. Bol. Geol. Min. 115 (1), 103–114.

Postma, D., Boesen, C., Kristiansen, H., Larsen, F., 1991. Nitrate reduction in an unconfined sandy aquifer: water chemistry, reduction processes, and geochemical modeling. Water Resour. Res. 27 (8), 2027–2045. https://doi.org/10.1029/91wr00989.

Poulain, A.J., Newman, D.K., 2009. Rhodobacter capsulatus catalyzes light-dependent Fe(II) oxidation under anaerobic conditions as a potential detoxification mechanism. Appl. Environ. Microbiol. 75 (21), 6639–6646. https://doi.org/10.1128/AEM.00054-09.

Puente-Sánchez, F., Moreno-Paz, M., Rivas, L.A., Cruz-Gil, P., García-Villadangos, M., Gómez, M.J., Postigo, M., Garrido, P., González-Toril, E., Briones, C., Fernández-Remolar, D., Stoker, C., Amils, R., Parro, V., 2014. Deep subsurface sulfate reduction and methanogenesis in the Iberian Pyrite Belt revealed through geochemistry and molecular biomarkers. Geobiology 12 (1), 34–47. https://doi.org/10.1111/gbi.12065.

Quatrini, R., Johnson, D.B., 2019. Acidithiobacillus ferrooxidans. Trends Microbiol. 27 (3), 282–283. https://doi.org/10.1016/j.tim.2018.11.009.

Rawlings, D.E., Dew, D., Du Plessis, C., 2003. Biomineralization of metal-containing ores and concentrates. Trends Biotechnol. 21 (1), 38–44. https://doi.org/10.1016/S0167-7799(02)00004-5.

Rohwerder, T., Gehrke, T., Kinzler, K., Sand, W., 2003. Bioleaching review part A: progress in bioleaching: fundamentals and mechanisms of bacterial metal sulfide oxidation. Appl. Microbiol. Biotechnol. 63 (3), 239–248. https://doi.org/10.1007/s00253-003-1448-7.

Sand, W., Gehrke, T., Jozsa, P.-G., Schippers, A., 2001. (Bio)chemistry of bacterial leaching—direct vs. indirect bioleaching. Hydrometallurgy 59 (2–3), 159–175. https://doi.org/10.1016/s0304-386x(00)00180-8.

Sand, W., Gehrke, T., 2006. Extracellular polymeric substances mediate bioleaching/biocorrosion via interfacial processes involving iron(III) ions and acidophilic bacteria. Res. Microbiol. 157 (1), 49–56. https://doi.org/10.1016/j.resmic.2005.07.012.

Schippers, A., Jørgensen, B.B., 2001. Oxidation of pyrite and iron sulfide by manganese dioxide in marine sediments. Geochem. Cosmochim. Acta 65 (6), 915–922. https://doi.org/10.1016/s0016-7037(00)00589-5.

Schippers, A., Barker Jørgensen, B., 2002. Biogeochemistry of pyrite and iron sulfide oxidation in marine sediments. Geochem. Cosmochim. Acta 66 (1), 85–92. https://doi.org/10.1016/s0016-7037(01)00745-1.

Schwientek, M., Einsiedl, F., Stichler, W., Stögbauer, A., Strauss, H., Maloszewski, P., 2008. Evidence for denitrification regulated by pyrite oxidation in a heterogeneous porous groundwater system. Chem. Geol. 255 (1–2), 60–67. https://doi.org/10.1016/j.chemgeo.2008.06.005.

Smith, E., Dent, G., 2013. Modern Raman Spectroscopy: A Practical Approach. John Wiley and Sons.

Straub, K.L., Schönhuber, W.A., Buchholz-Cleven, B.E.E., Schink, B., 2004. Diversity of ferrous iron-oxidizing, nitrate-reducing bacteria and their involvement in oxygen-independent iron cycling. Geomicrobiol. J. 21 (6), 371–378. https://doi.org/10.1080/01490450490485854.

Tornos, F., 2006. Environment of formation and styles of volcanogenic massive sulfides: the Iberian Pyrite Belt. Ore Geology Reviews 28 (3), 259–307. https://doi.org/10.1016/j.oregeorev.2004.12.005.

Torrentó, C., Urmeneta, J., Otero, N., Soler, A., Viñas, M., Cama, J., 2011. Enhanced denitrification in groundwater and sediments from a nitrate-contaminated aquifer after addition of pyrite. Chem. Geol. 287 (1–2), 90–101. https://doi.org/10.1016/j.chemgeo.2011.06.002.

Vaclavkova, S., Jørgensen, C.J., Jacobsen, O.S., Aamand, J., Elberling, B., 2014. The importance of microbial iron sulfide oxidation for nitrate depletion in anoxic Danish sediments. Aquat. Geochem. 20 (4), 419–435. https://doi.org/10.1007/s10498-014-9227-x.

Van Cleemput, O., Baert, L., 1983. Nitrite stability influenced by iron compounds. Soil Biol. Biochem. 15 (2), 137–140. https://doi.org/10.1016/0038-0717(83)90093-7.

Van Geen, A., Adkins, J.F., Boyle, E.A., Nelson, C.H., Palanques, A., 1997. A120 yr record of widespread contamination from mining of the Iberian pyrite belt. Geology 25 (4), 291–294. https://doi.org/10.1130/0091-7613(1997)025<0291:AYROWC>2.3.CO;2.

Vara, I.P., 1963. Piritas de Huelva: su historia, minería y aprovechamiento. Suc. de Rivadeneyra. Suc. de Rivadeneyra.

Vera, M., Schippers, A., Sand, W., 2013. Progress in bioleaching: fundamentals and mechanisms of bacterial metal sulfide oxidation—part A. Appl. Microbiol. Biotechnol. 97 (17), 7529–7541. https://doi.org/10.1007/s00253-013-4954-2.

Yan, R., Kappler, A., Muehe, E.M., Knorr, K.H., Horn, M.A., Poser, A., Lohmayer, R., Peiffer, S., 2019. Effect of reduced sulfur species on chemolithoautotrophic pyrite oxidation with nitrate. Geomicrobiol. J. 36 (1), 19–29. https://doi.org/10.1080/01490451.2018.1489915.

Yan, R., Kappler, A., Peiffer, S., 2015. Interference of nitrite with pyrite under acidic conditions: implications for studies of chemolithotrophic denitrification. Environ. Sci. Technol. 49 (19), 11403–11410. https://doi.org/10.1021/acs.est.5b02981.

Younger, P.L., 1997. The longevity of mine water pollution: a basis for decision-making. Sci. Total Environ. 194–195, 457–466. https://doi.org/10.1016/S0048-9697(96)05383-1.

Yu, S., Wei, Q., Zhao, T., Guo, Y., Ma, L.Z., 2016. A survival strategy for *Pseudomonas aeruginosa* that uses exopolysaccharides to sequester and store iron to stimulate psl-dependent biofilm formation. Appl. Environ. Microbiol. 82 (21), 6403–6413. https://doi.org/10.1128/AEM.01307-16.

Zhang, Y.C., Slomp, C.P., Broers, H.P., Passier, H.F., Cappellen, P.V., 2009. Denitrification coupled to pyrite oxidation and changes in groundwater quality in a shallow sandy aquifer. Geochem. Cosmochim. Acta 73 (22), 6716–6726. https://doi.org/10.1016/j.gca.2009.08.026.

CHAPTER

Biogeochemical cycles operating in the deep subsurface of the Iberian Pyrite Belt

17

Ricardo Amils[1,2]

[1]*Universidad Autónoma de Madrid, Centro de Biología Molecular Severo Ochoa, Department of Molecular Biology, Madrid, Spain;* [2]*Centro de Astrobiología, Department of Planetology and Habitability, Madrid, Spain*

Introduction

Although Charles Darwin predicted the existence of life independent of solar radiation in the deep subsurface (Darwin, 1839), continental subsurface geomicrobiology has only recently emerged as a significant area of biological research. This delay is largely due to the initial skepticism of science toward early drilling results, which were often dismissed as contamination, as well as the technological challenges associated with overcoming this problem (Bastin et al., 1926; Zobell, 1938). Recent discoveries have revealed that a substantial portion of prokaryotic biomass and diversity resides in the so call dark biosphere (Colman et al., 2017; Magnabosco et al., 2018). As a result more precise information about deep subsurface geomicrobiology has become critical to the advance of our understanding of the fundamental biogeochemical cycles operation. Additionally, studying life in the deep subsurface serves as a valuable model for exploring the origins of life on Earth (Pedersen, 2000) and for its detection beyond our planet (Purkamo et al., 2020).

Most deep subsurface studies to date have relied on groundwater samples collected from artesian wells. While these studies have yielded some insights into microbial diversity, they cannot provide crucial information about the interactions between microorganisms and the mineralogical structures they inhabit (Nuppunen-Puputti et al., 2022; Purkamo et al., 2018; Stevens and McKinley, 1995). To better understand the diversity, abundance, and functional roles of microbiomes in continental subsurface rock systems, dedicated drilling efforts are required to obtain cores at different depths from which to generate samples for their analysis using diverse and complementary methodologies. The first report describing the results of dedicated continental rock matrix drilling was published 2 decades ago (Zhang et al., 2005).

In general, the few devoted hard rock continental drilling projects conducted so far have primarily relied on a single methodology, sequencing, focused on vastly

Geomicrobiology of Río Tinto. https://doi.org/10.1016/B978-0-443-32948-7.00015-1

different locations and depths. This approach has provided limited insights into how biogeochemical cycles operate in the deep subsurface (Lehman et al., 2004; Zhang et al., 2005; Fry et al., 2009; Breuker et al., 2011; Momper et al., 2017; Dutta et al., 2018; Liu et al., 2020; Cockell et al., 2021). To address these limitations the Iberian Pyrite Belt drilling projects, the Mars Astrobiology Research and Technology Experiment Analogue (MARTE, 2003–05) and the Iberian Pyrite Belt Subsurface Life Detection (IPBSL, 2011–16), were designed to demonstrate the existence of an underground bioreactor that generate the extreme conditions detected along the Río Tinto basin. These projects provided core samples from various depths which were then analyzed by multiple complementary methodologies, enabling a more thorough analysis of the biogeochemical cycles operating in the deep subsurface of the Iberian Pyrite Belt.

The Iberian Pyrite Belt

The Iberian Pyrite Belt (IPB), ranges for nearly 250 km across the southwestern Iberian Peninsula. It holds the largest concentration of massive metal sulfide deposits on Earth. These deposits originated through active hydrothermalism during the Hercynian orogeny, forming the planet's most extensive sulfide anomaly (Tornos, 2006).

Río Tinto, which rises in the core of the IPB in Peña de Hierro (Iron Mountain), is a nearly 100 km long river flowing into the Atlantic Ocean near Huelva. Río Tinto is an unusual extreme environment characterized by constant acidity, a high concentration of heavy metals, and a remarkable microbial diversity, predominantly eukaryotic (Amaral-Zettler et al., 2002; González-Toril et al., 2003; Aguilera et al., 2006; Oggerin et al., 2024). Historically, the extreme conditions in the Tinto basin were attributed to 5000 years of mining in the IPB (Leblanc et al., 2000). However, recent geomicrobiological studies propose that the IPB subsurface functions as a vast microbial bioreactor, where metal sulfides fuel metabolic activities, and the resulting products drain into the river (Gómez-Ortiz et al., 2014; Allman et al., 2021). For a detailed description of the geology, mineralogy and microbial diversity of Río Tinto see Chapters 1 to 9 of this volume.

The MARTE drilling project

The aim of the MARTE project was to search for microbial life in the deep subsurface of the IPB and to test new developed drilling technology to search for signs of life on the Mars subsurface. The MARTE project was a joint adventure between NASA Ames Research Center (USA) and the recently founded Centro de Astrobiología (Spain), associated to the NASA Astrobiology Institute (USA).

During 2003 and 2004 three boreholes were drilled, BH1 59 m below surface (mbs), BH4 166 mbs, and BH8 166 mbs. The drilling was carried out using a wire-line drilling system that produced 60-mm-diameter cores. During drilling, cores were encased in plastic liners to avoid excessive contact with the drilling fluid, required to refrigerate the bit and to remove drilling rubble. To control the level of sample contamination during the drilling three methodologies were tested: (a) fluorescence beats; (b) a microbiological marker, *Lactobacillus* sp., not detected in the subsurface of the IPB; and (c) a chemical marker, NaBr, selected due to the negligible concentration of Br detected in a preliminary survey of outcrop samples from the drilling area. All were added to the drilling fluid, the greatest source of contamination. Once the cores were retrieved from the drilling rig, they were cut into manageable pieces, placed in plastic bags, gassed with N_2 to preserve the anoxic conditions existing in the subsurface, and transported to the laboratory nearby at the Museo Minero de Ríotinto for further processing and analysis. In the laboratory, cores were placed in a sterile anaerobic chamber, logged, split at selected sites with a hydraulic press and powdered samples were drilled out the central part of the core with a rotary hammer drill using sterile bits. These powdered samples were used for biological, geological, and geochemical analysis (Puente-Sánchez et al., 2014a).

After drilling, Multi-Level Diffusion Samplers were installed in the boreholes to assess the presence and evolution of dissolved gases and ions in the groundwater. Dissolved H_2, CH_4 and CO_2 were detected in the three boreholes. H_2 and CH_4 could be of use as source of energy for the microorganisms of the ecosystem, and CO_2 as a source of carbon for autotrophic microorganisms. The highest concentration of H_2 was detected between the aerobic and anaerobic interface at 90 mbs, the level of the water table, as a result of its gaseous nature. The ion chromatographic analysis reported the presence of nitrate, nitrite, sulfate and ferric iron, all of them potential electron acceptors for anaerobic respiration and sulfide and ferrous iron as possible inorganic electron donors. Organic acids like acetate and formate, which could also be used as a source of carbon and as electron donors were also identified.

Microbial diversity was evaluated using several methodologies. The use of an immunosensor Life Detector Chip (LDChip200), developed at the Centro de Astrobiología for the detection of signs of life in future space missions, holding 200 specific antibodies raised to detect different microorganisms and metabolic products (Parro et al., 2011) identified the presence of Gram-positive bacteria, which was further confirmed by hybridization with an oligonucleotide microarray (PAM). PAM was specially designed for the detection of a broad group of prokaryotic microorganisms (Garrido et al., 2008), which also detected the presence of members of the Actinobacteriota and Bacillota phyla, bacteria involved in the Fe and S cycles, sulfate reducing bacteria (SRB), and methanogenic Archaea. Finally, PCR-amplified 16S rRNA gene from samples that gave positive immunological and hybridization signals were cloned and sequenced. Most of the obtained sequences corresponded to members of the Beta- and Gamma-proteobacteria and the

Acidobacteria division. Fluorescence in situ hybridization (FISH) and DNA staining with 4′,6-diamino-2-phenylindole (DAPI) were used to detect the presence of microorganisms using epifluorescent microscopy. Few microorganisms in a variable distribution along the column were observed, as expected for the heterogeneous mineral matrix in which they grow. The highest cell density was observed in samples corresponding to faults. High levels of epifluorescence due to the presence of diverse minerals in the core samples challenged the identification of real hybridization signals. Some of the observed clusters of microorganisms were efficiently hybridized with FISH probes, a strong indication that they were metabolically active (Puente-Sánchez et al., 2014a).

Enrichment cultures to detect expected metabolisms such as iron oxidizers, sulfate reducers, denitrifiers and methanogens gave positive results in several uncontaminated core samples. Unfortunately, some of the analyzed samples had to be discarded due to the detection of bromide during the ionic chromatography analysis, meaning that they were in contact with the drilling fluid and susceptible to contamination. However, many important questions remained unanswered or required further confirmation. These preliminary results supported the hypothesis of the existence of a subsurface bioreactor with metabolically active microorganisms able to use different detected electron donors and acceptors and which, thanks to their metabolism, could dissolve the metallic sulfides from the subsurface of the IPB, giving rise to the extreme conditions detected in the Tinto River.

The main lessons learned during the development of the MARTE project were: (a) NaBr was the best drilling fluid contamination marker, (b) ion chromatography analysis had to be done right after the drilling in order to select the most interesting uncontaminated samples for a further deeper analysis, (c) the use of Multi-Level Diffusion Samplers was not the most appropriate methodology to measure gases related to the different sample metabolisms due to the complexity of the operation and the possibility of contamination generating artifacts, (d) enrichment cultures were a fast way to detect active microbial activities operating in the subsurface, (e) immunodetection and oligonucleotide hybridization were suitable methodologies for the analysis of deep subsurface rock diversity, (f) fluorescence in situ hybridization was a powerful methodology to consider in future drilling operations due to the small amount of sample required to obtain suitable microbial diversity information, and (g) the use of different and complementary microbial diversity identification methodologies was an appropriate way to overcome the bias and limitations introduced by each of them.

The IPBSL drilling project

Following the preliminary results of the MARTE project, we applied for an ERC Advance Grant to perform an in-depth analysis of the subsurface microbial bioreactor operating in the IPB. The subsequent IPBSL project, aimed at addressing many of the question produced during the development of the MARTE project, assess the geomicrobiological cycles driving the deep subsurface ecosystem, and evaluate their role in generating the extreme conditions, notably low pH and high concentration of ferric iron, detected in the Río Tinto basin. The drilling conditions and the generation of the core samples were the same as those described in the MARTE project section.

Geophysical, geochemical, mineralogical and biochemical analysis

A systematic geophysical analysis of the Peña de Hierro (Iron Mountain) area employing Time-domain Electromagnetic Sounding (TDEM) and Electrical Resistivity Tomography (ERT) provided sufficient data to identify zones where groundwater and reduced minerals intersect (for additional information see Chapter 3). This facilitated the selection of optimal drilling sites (Gómez-Ortiz et al., 2014). The selected drilling sites, BH10 and BH11, are located on the north flank of the Río Tinto anticline. Both boreholes were drilled through the upper part of the volcano-sedimentary complex of the area, 614 mbs for BH10, and 340 mbs for BH11. These boreholes intersect a significant hydrothermal altered zone characterized by widespread mineral replacement including quartz, clinochlore, illite, pyrite, sericite and siderite. With the exception of quartz, most of these minerals contain iron, consistent with the elemental composition detected by ICP-MS in the rock samples and their soluble extraction fraction (Fig. 17.1). In this chapter only results from borehole BH10 are reported, the shorter BH11 borehole remains under study to validate the results advanced with the analysis of borehole BH10.

Analyzing the soluble ion content of retrieved core samples overnight after drilling facilitated the selection of the most interesting samples for further in-depth analysis. This selection relied on the absence of bromide (Br), which served as a drilling fluid contamination control as well as on the presence of molecules indicative of microbial activity (Fig. 17.1). The detection of short-chain organic acids, such as formate, acetate, propionate and oxalate, at various depths along the borehole strongly suggested active microbial processes. Additionally, the stochastic distribution of sulfate along the borehole (Fig. 17.1) was a signal of bioleaching activities on the metal sulfides existing in the IPB, and the possible existence of sulfate reducing bacteria (SRB), a microbial metabolism previously identified in the MARTE project and other drilling operations (Itävaara et al., 2011; Bell et al., 2020).

The presence of a rather constant concentration of soluble nitrate along the column, despite the lack of a clear mineralogical explanation, coupled with that of nitrite and NH_4^+, strongly suggests the existence of an active N cycle in the deep subsurface (Fig. 17.1), which had already been advanced by the MARTE project and suggested by other authors (Lau et al., 2014; Rempfert et al., 2017). The detection of proteins and carbohydrates together with the identification of metabolically important organic acids and anions throughout the borehole provides

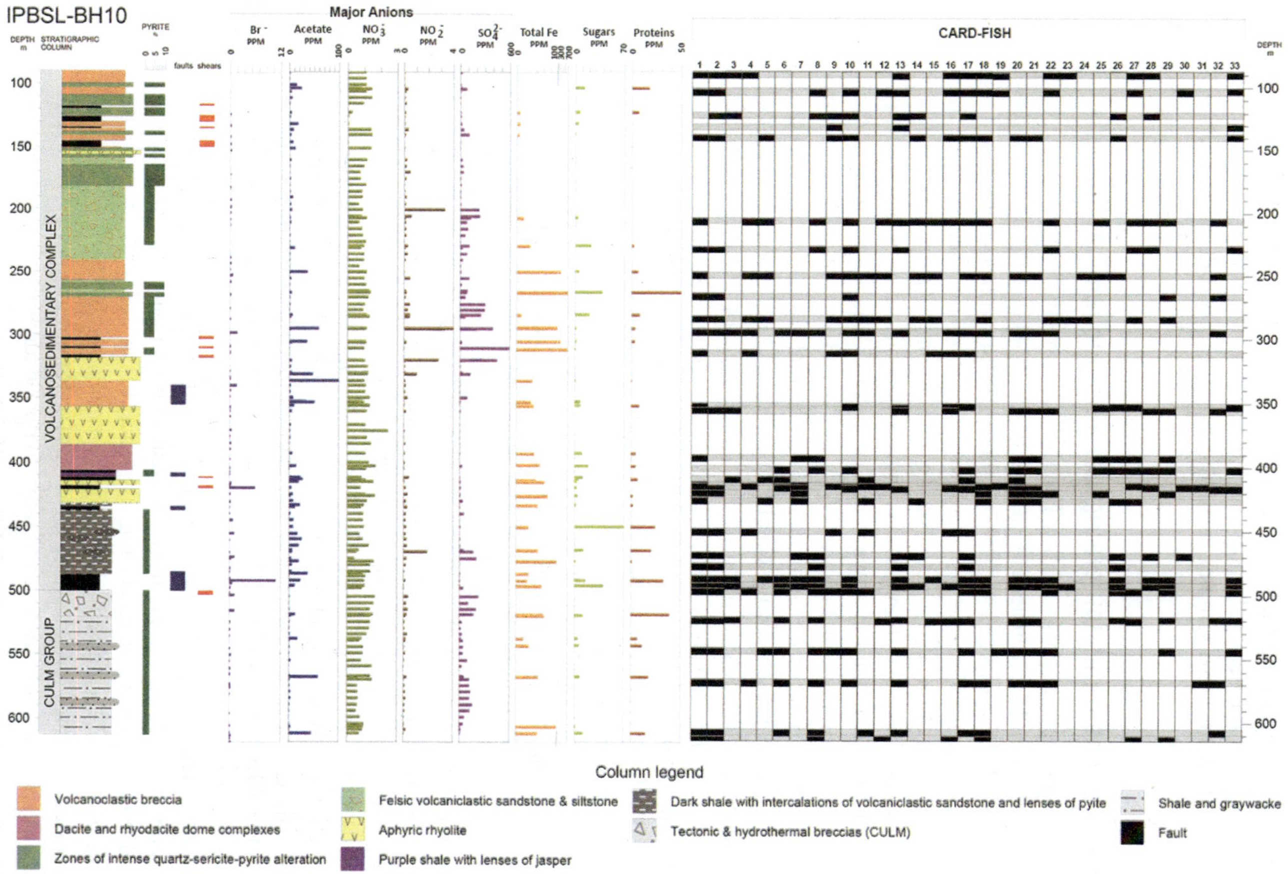

FIGURE 17.1 Mineralogy, physicochemical variables and microbial diversity detected by hybridization along borehole BH10.

Specificity of the hybridization probes: 1: Bacteria (I and III); 2: Planctomycetales; 3: ANAMMOX, 4: α-Proteobacteria; 5: *Acidiphilium* spp.; 6: *Rhizobium* spp.; 7: *Brevundimonas* spp.; 8: *Rhodoplanes* spp.; 9: β-Proteobacteria; 10: *Acidovorax* spp. (I and II); 11: γ-Proteobacteria; 12: *Acidithiobacillus* spp. (I + II); 13: *Pseudomonas* spp.; 14: SRB/δ-Proteobacteria; 15: *Desulfosarcina/Desulfococcus*; 16: Firmicutes (a and b); 17: *Sulfobacillus* spp.; 18: Actinobacteria; 19: *Propionibacterium* spp.; 20: *Tessaracoccus* spp.; 21: *T. lapidicaptus*; 22: Acidobacteria; 23: Bacteroidetes; 24: *Leptospirillum* spp.; 25: Cyanobacteria; 26: Chloroflexi (I and II); 27: Archaea; 28: Euryarchaeota; 29: Methanosarcinales; 30: Metahanococcaceae; 31: Methanomicrobiales; 32: Methanobacteriales; 33: Crenarchaeota.

compelling evidence for the existence of geomicrobiological activity in the deep subsurface of the IPB (Fig. 17.1).

From the MARTE project we learned that measurements of the gases produced by different metabolic processes operating along the borehole using a Multi-Level Diffusion Sampler were unreliable. In contrast, the IPBSL project measured occluded H_2, CO_2 and CH_4 in the core samples using small vials (27 mL) filled with chips and powdered samples, leaving minimal head space. These vials were prepared in the anaerobic chamber under strict anaerobic conditions, sealed with gas-tight rubber septa and aluminum caps, and flushed with N_2. Incubations were conducted statically, in the absence of light, in a thermostatic chamber at 30°C. Released or consumed gases were analyzed periodically via gas chromatography. Similarly, production or consumption of H_2, CH_4 and CO_2 were measured after microbial metabolic activation of powdered samples by adding a mineral solution. Measurements followed the same protocol as occluded gases. The highest concentrations of gases were detected at 284 and 414 mbs (Table 17.1).

As demonstrated in the MARTE drilling project the use of complementary methodologies is an appropriate way to overcome the bias introduced by the use of only one methodology, usually sequencing, to describe the existing microbial diversity and identifying microbial contamination (Sheik et al., 2018). In the IPBSL project we employed five independent methodologies, some of them already tested in the earlier MARTE drilling project: immunological detection, DNA sequencing, hybridization, enrichment cultures, and isolation from active enrichment cultures.

Immunological and DNA analysis

An improved version of the Immunosensor Life Detector Chip (LDChip300), containing 300 specific antibodies targeting microorganisms and metabolic products, were used to analyze 21 core samples from borehole BH10. Immunological detection identified *Acidithiobacillus*, *Desulfosporosinus*, *Methanobacterium*, *Nostoc*, *Shewanella*, *Sulfobacillus* and *Streptomyces* as the most abundant bacterial and archeal genera (Fig. 17.2).

Although the efficiency of nucleic acids extraction from rock cores is very low, requiring the use of large amounts of sample, we were able to recover sufficient DNA from 16 core samples from borehole BH10 to assess their microbial diversity. Despite using diverse archeal PCR primers and amplification protocols we were unable to amplify 16S rDNA archeal sequences, even though we knew through other methodologies (geochemical, immunological, fluorescence in situ hybridization and enrichment cultures) that members of this Domain were present along the borehole. Most probably insufficient extracted DNA or its quality are the most reasonable causes for not obtaining information on this important Domain using this methodology. This is a nice example of the need for diverse complementary methodologies to insure an accurate analysis of the microbial diversity exiting in the hard rock deep subsurface.

Table 17.1 Metabolic activities and production of H_2, CO_2 and CH_4 at different depths.

Depth mbs	MT	MTP	ACT	DNT	SBR	FeOx	FeRd	H_2	CO_2	CH_4
90	+	–	–	–	++	nd	Nd	+/–	+++	+/–
121.8	+	–	–	–	+/–	–	+	+	++	+/–
139.4	++	+	+	–	–	–	+	+++	++	+/–
206.6	+	–	–	+/–	+/–	+	–	+/–	++	+/–
228.7	+++	–	–	–	+	–	+	+/–	+	+/–
249.8	+++	–	++	–	++	++	–	+/–	+	+/–
266.3	+	–	++	–	–	+	+	+/–	+++	+/–
284	++	–	+	–	+++	–	++	+++	+++	+/–
304	+	–	+	–	+/–	–	++	+++	+	+/–
311.1	++	–	+++	–	–	++	++	++	+	+/–
336.5	++	–	–	–	–	++	++	+/–	++	+/–
352.7	++	+++	–	+++	+++	+	+	++	+++	+/–
393	+	–	–	–	+/–	+	+	++	++	+/–
409.7	+	+/–	–	–	+/–	nd	nd	+++	++	+/–
411.2	+	–	–	++	+	–	++	++	++	–
414	+++	–	+	–	+/–	++	++	+	+++	+++
415.3	+	++	–	+/–	+/–	nd	nd	++	++	–
420	+	+/–	+	–	–	–	+	+/–	++	+/–
433.3	+	–	+++	–	–	–	+	++	+	+/–
450.3	++	–	–	+	++	–	++	+	++	+/–
466.8	+++	–	+	–	–	–	+	+/–	++	+
488	+	++	–	+/–	–	–	+	+	++	–
496.8	+	–	++	–	++	–	+	++	+	+/–
520	+++	–	–	–	+++	–	+	+	+	+/–
538.5	+	–	+	+++	–	–	+	+/–	+	+/–
568.6	++	–	++	+++	++	+	++	+	+++	+
607	+	–	–	–	–	–	+	++	++	+/–
612.9	+	–	–	+++	–	nd	nd	++	+	+/–

MT, *methanogenesis;* MTP, *methanotrophy;* ACT, *acetogenesis;* DNT, *denitrification;* SRB, *sulfate reduction;* FeOx, *iron oxidation;* FeRd, *iron reduction;* nd, *non determined. For medium components, growth conditions and gas chromatography conditions see Amils et al. (2023).*

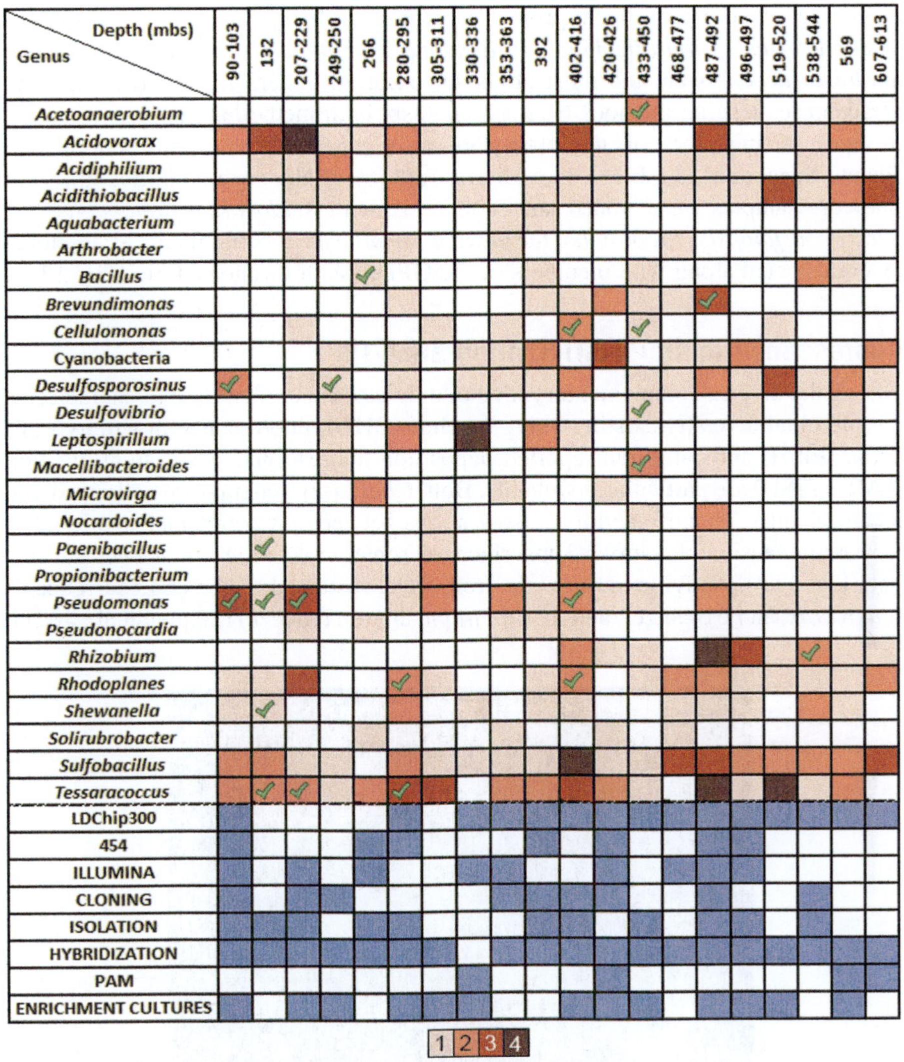

FIGURE 17.2 Methodologies and detected genera in borehole BH10.

Distribution of the genera detected using different methodologies. Hot tones correspond to the number of times that a given genus has been detected in a depth interval; in blue: number of independent methodologies that identified the selected genera along the column; √: members of the genus isolated from BH10 enrichment cultures at a given depth interval.

PCR amplicons from these 16 samples were analyzed through hybridization with the Prokaryotic Acidophile Microarray (PAM), cloning and massive 16S rRNA gene sequencing with Ultra-Deep Sequencing (UDS) using Roche 454 and Illumina MiSeq technologies, in our case as UDS methodologies had evolved during the development of the project. *Acidovorax*, *Aquabacterium*, *Arthrobacter*, *Bifidobacterium*, *Bradirhizobium*, *Psudomonas*, *Propionibacterium*, *Rhodoblastus* and *Tessaracocus* were the most abundant bacterial genera identified using both ultradeep sequencing methods (Fig. 17.2). Also, both methodologies identified the presence of members of the class GpI of Cyanobacteria at different depths (Puente-Sánchez et al., 2018). In addition, full 16S rRNA gene amplification from nine core samples were cloned and sequenced, with *Acidovorax*, *Aquabacterium*, *Chrocococcidiopsis*, *Propionibacterium* and *Simplicispira* being the most abundant genera detected along with members of the GPI class of Cyanobacteria (Fig. 17.2).

Fluorescence in situ hybridization analysis

Among the diverse complementary methodologies used to identify and characterize the microbial diversity existing along borehole BH10, fluorescence in situ hybridization (FISH) and catalyzed reported deposition fluorescence in situ hybridization (CARD-FISH) are particularly valuable. Both techniques generate spatially resolved data to identify, quantify and map associations between microorganisms, critical to our comprehension of complex subsurface microbial interactions in the deep subsurface (Fig. 17.3). Activities such as the iron cycle, sustained by the co-occurrence of *Acidovorax*, an Fe oxidizer, and *Acidiphilium*, an iron reducer (Kappler et al., 2021);

FIGURE 17.3 Microbial consortia in the IPB subsurface detected by CARD-FISH.

DOPE-FISH detection of a mixed biofilm of bacteria (blue) and archaea (red) at 139.4 mbs using EU 338(I-III), ARC915 hybridization probes and a mixture (ConA, AAL and PNA) lectin dyes (green).

the sulfur cycle driven by the close contact of *Sulfobacillus*, an S oxidizer, and Sulfate Reducing Bacteria (Bell et al., 2020) (Figs. 17.1 and 17.2); or novel subsurface microbial activities such as ANAMMOX, responsible for the anaerobic oxidation of NH_4^+ along the borehole (Strous et al., 2006); or anaerobic methane oxidation (methanotrophy) mediated by close contact between archeal ANME-2 and Sulfate Reducing Bacteria (Knittel and Boetius, 2009) were detected along borehole BH10 (Fig. 17.1).

Furthermore, a major advantage of FISH and CARD-FISH is their minimal sample requirement enabling microscopic-scale spatial resolution, unlike the rest of the methodologies that require 5–10 g of core sample and yielding diversity results impossible to correlate at microbial level interactions. In addition, FISH and CARD-FISH are non-destructive methodologies, allowing repeated hybridizations with complementary probes to identify the diverse microorganisms present in the sample and the evaluation of any possible interaction between them and the mineral substrate of their habitat (Fig. 17.3).

The positive FISH hybridizations obtained for different microorganisms at different depths strongly suggest that they are metabolically active, a critical question in subsurface geomicrobiology (Escudero et al., 2018a, b), and references therein). Microbial diversity in borehole B10 was primarily identified using CARD-FISH, which amplifies the signal, facilitating the identification of microorganisms and the existence of possible artifacts (Escudero et al., 2018a,b). The most abundant bacterial genera detected corresponded to *Acidovorax*, *Pseudomonas*, *Rhizobium*, *Rhodoplanes,* and *Tessaracoccus* (Figs. 17.1 and 17.2), while archeal members of the orders Merhanobacteriales and Methanosarcinales exhibited the strongest hybridization signals. Notably methanogens were detected within these archeal groups, underscoring their presence in the deep subsurface ecosystem.

Contrary to a commonly accepted idea, fluorescence hybridizations using lectin-specific dyes revealed a robust association between detected microorganisms and exopolysaccharides (EPS), demonstrating, the existence of biofilms in the oligotrophic conditions prevailing in the deep subsurface (Fig. 17.3). These biofilms likely facilitate the generation of microniches with distinct environmental conditions as well as their interconnectivity (Escudero et al., 2018a). Our results challenge the notion that biofilm formation is resource-prohibitive in the deep subsurface. Instead, microorganisms strategically utilize limited resources to create biofilms, leveraging advantages such as desiccation resistance, metabolic substrates and products diffusion control and microniche networking, among others (Coyte et al., 2017; Vera et al., 2013).

Consistent with other continental drillings, Archaea in the IPB deep subsurface are less abundant than Bacteria (Fig. 17.1) (Zhang et al., 2005; Amils et al., 2023, Dutta et al., 2018). The distribution of identified microorganisms along the 612 m borehole BH10 was relatively uniform, unlike the gradients reported in other drilling operations (Fig. 17.1).

Despite challenges in quantifying cells in low-porosity, mineralogically heterogeneous rock, hybridization data from varying depths yielded estimates of 10^4–10^5

cells/g of rock sample, consistent with cell densities reported for other hard rock drillings (Fry et al., 2009; Cockell et al., 2021). Extrapolating these values, we estimate that the IPB harbors approximately 4×10^{24} microorganisms within its first kilometer of depth, equivalent to 90 Mt tons of carbon.

Activated microbial activities and enrichment cultures

Activated microbial activities refer to processes observed in core samples following the addition of a mineral solution. This method provides the closest approximation to native microbial activity in the deep subsurface, as it minimizes alterations to the in situ sample conditions. As mentioned, this approach has been employed to stimulate microbial gas production and consumption. Metabolic activated activities generally increased the production of H_2, CO_2 and CH_4 when compared to the occluded ones. However, gas consumption was also observed, reflecting utilization in diverse microbial metabolic processes, such as H_2 and CO_2 consumption by methanogenic and acetogenic pathways (Amils et al., 2023). This approach has demonstrated that over 60% of subsurface H_2 and CO_2 in the subsurface of the IPB originates from biological activity (Sanz et al., 2021), rather than geochemical reactions as has often been assumed in other drilling studies (Apps and Van De Kamp, 1993). A similar approach was used to activate ferrous iron anaerobic oxidizing activities in BH10 core samples (122, 139 and 229 mbs), in this case adding only sterile water. In all cases, ferric iron generation was detected after several months of incubation at 30°C (Amils et al., 2023).

Enrichment cultures are an effective method for identifying microbial metabolisms in environmental samples. However, this methodology has the important bias that the in vitro conditions are quite different than those existing in the native in situ habitats, especially those related with substrate concentrations and temperature, which might facilitate the development of opportunistic microorganisms existing in the sample. For the analysis of metabolic activities in borehole BH10, we focused on metabolisms either expected, such as iron and sulfide oxidation (Fernández-Remolar et al., 2008), or previously detected in the MARTE project and other drilling operations, such as sulfate reduction and methanogenesis. Aditionally, we included expected activities related with the geochemical conditions detected in the borehole, such as nitrate reduction (nitrate), acetogenesis (acetate) and methanotrophy (methane).

Although archeal 16S rDNA could not be amplified from extracted DNA, methanogenic activity along the borehole was confirmed through complementary methodologies, such as immunological detection of *Methanobacterium,* CARD-FISH identification of Methanobacteriales and Methanosarcinales, detection of occluded CH_4 in core samples and CH_4 generation after activating the core samples with mineral solutions, presence of H_2 and CO2 substrates for hydrogenotrophic methanogens, and CH_4 production in enrichment cultures with diverse substrates (Fig. 17.1 and Table 17.1). Methane production in enrichment cultures varied depending on the use of different hydrogenotrophic, acetoclastic or methylotrophic substrates. Methanogenesis is among the earliest metabolic activities detected in the continental deep subsurface and one of the most widespread, as reported in numerous drilling operations (Kotelnikova, 2002).

The presence of methane along the borehole prompted investigation of methanotrophic activity, which was confirmed via methane consumption in enrichment cultures at different depths (Table 17.1). As in other anaerobic systems, this involved syntrophy between anaerobic methane oxidizing archaea (ANME 2) and sulfate reducing bacteria (SRB), visualized by hybridization using specific fluorescence probes (Fig. 17.1). Although sulfate reducing bacteria and ANME archaea have been identified previously in the continental deep subsurface (Motamedi and Pedersen, 1998; Fry et al., 2009), this is the first report of their syntrophic activity in such environments. Although direct evidence is lacking, the co-occurrence of iron reducing bacteria and ANME 2 archaea along the borehole raises the possibility of their syntrophic interaction oxidizing methane anaerobically in the deep subsurface (Schnakenberg et al., 2021).

In addition to methanogenic and methanotrophic activities (methane cycle), other notable deep subsurface microbial activities were detected with variable intensities at different depths using enrichment cultures, including sulfate reduction (SRB), Fe oxidation Fe and reduction (Fe cycle), autotrophic denitrification and acetogenesis (Table 17.1). Sulfate reducing activity (SRB), commonly observed in deep subsurface environments was anticipated given the detection of sulfate throughout the borehole (Fig. 17.1). In SRB enrichment cultures H_2 and a mixture of lactate, glycerol and methanol were used as electron donors for anaerobic respiration. The resulting H_2S production from this activity may explain the framboidal pyrite observed in some of the core samples (Fernández-Remolar et al., 2018).

Anaerobic Fe oxidation, already detected by activation of some core samples by addition of sterile water, and Fe reduction activities required for an active Fe cycle, were assessed along the column (Table 17.1). For anaerobic Fe oxidation enrichment cultures acetate was selected as electron donor and nitrate as electron acceptor, both detected in the geochemical analysis of the column (Fig. 17.1). For anaerobic Fe reduction glucose was selected as electron donor as carbohydrates have been also detected along the column (Fig. 17.1). Co-occurrence of both activities in 25% of samples, strongly suggest an active Fe cycle in the IPB subsurface (Kappler et al., 2021). This is further supported by the already mentioned hybridization using specific fluorescence probes for microorganisms known by their capability to oxidize Fe, like *Acidovorax*, and to reduce it, like *Acidiphilium*, both detected in close proximity (Fig. 17.1).

The detection of nitrate and nitrite in the geochemical analysis and the generation of reduced S compounds due to SRB activities, suggested the interest of testing the presence of autotrophic denitrification activities along the borehole (Lau et al., 2016). Similarly, high concentration of acetate, along with H_2 and CO_2, both substrates for acetogenic activity, and the reports describing this activity in the continental deep subsurface (Magnabosco et al., 2016; Rempfert et al., 2017) justified testing this activity. As shown in Table 17.1 both activities, denitrification and acetogenesis were detected at different depths with varying efficiencies. The high number of microbial activities identified at 353 mbs could be due to the existence of a fault facilitating the flow of water and the availability of substrates at this depth (Fig. 17.1 and Table 17.1).

Ilumina MiSeq sequence analysis of the 16S rRNA gene in the most active methanogenic, sulfate reducing and acetogenic enrichment cultures revealed *Acetobacterium, Desulfobibrio, Desulfurispora, Pseudomonas, Propionibacterium* and *Tessaracoccus* as the most abundant identified genera.

Twenty-nine bacteria belonging to the *Acetoanaerobium, Aestuarimicrobium, Bacillus, Brevundimonas, Cellulomonas, Desulfovibrio, Lelliotia, Macellibacteroides, Microbacterium, Nocardoides, Paenibacillus, Propionicimonas, Pseudomonas, Rhizobium, Rhodoplanes, Shewanella, Tessaracoccus* and *Xanthobacter* genera were isolated under strict anaerobic conditions using the most active enrichment cultures for hydrogenotrophic methanogens and autotrophic denitrifiers (Leandro, 2018; Leandro et al., 2018). Members of most of these bacterial genera have been reported in other continental deep subsurface drilling operations (Amils et al., 2023, and references therein). However, it is worth noting that members of the *Acetoanaerobium, Aestuarimicrobium, Lelliotia, Macellibacteroides* and *Xanthobacter* genera were identified and isolated from continental deep subsurface core samples for the first time. Three of these, *Acetoanaerobium, Aestuarimicrobium* and *Lelliotia* have been further characterized genotypically to investigate their potential roles in the deep subsurface of the IPB (Martínez, 2025).

Using the methodologies described above we identified the presence of 670 distinct prokaryotic microorganisms along borehole BH10 (Amils et al., 2023). A key question is whether this diversity is unique to the IPB, given its distinctive mineralogy, or reflects broader trends observed in other continental drilling operations with differing mineralogical and geographical contexts. A phylum level comparison between the IPBSL drilling project and 18 other continental drilling operations across diverse locations is shown in Fig. 17.4. This comparison reveals that the subsurface microbial diversity (dark biosphere) is remarkably homogeneous at the phylum level, regardless of mineralogy and location, and distinct from the patchier distribution observed in sunlit surface ecosystems (Ruff et al., 2024).

The most representative genera along borehole BH10

To overcome the previously mentioned bias resulting from relying on a single methodology to describe the microbial diversity in the deep subsurface of the IPB, we selected the most representative genera among those identified using complementary methodologies following these strict criteria: to be identified at least by two independent methods and at five depth intervals of the 20 analyzed. Sixteen genera belonging to the Pesudomonadota, Actinomycetota, Bacillota and Nitrospirota phyla met this selection criteria: *Acidiphilium, Acidovorax, Acidithiobacillus, Arthrobacter, Bacillus, Brevundimonas, Desulfosporosinus, Desulfovibrio,*

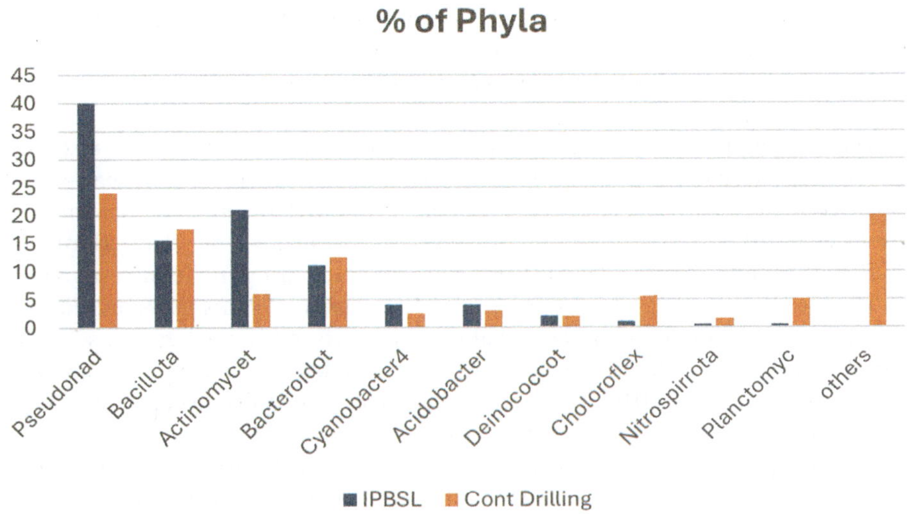

FIGURE 17.4

Comparison of the microbial diversity at the phylum level detected in the deep subsurface of the Iberian Pyrite belt with the diversity reported in 18 different drilling operations.

Leptospirillum, Propionibacterium, Pseudomonas, Rhizobium, Rhodoplanes, Shewanella, Sulfobacillus and *Tessaracoccus*. The class Cyanobacteria was also included. Not only did it meet the selection criteria, but also exhibited the highest frequency of identification using different methodologies along the borehole. However, most used methodologies could not resolve its members to the genus level. Due to the ecological significance of detecting Cyanobacteria in a system deprived of light, they were retained for further analysis. Members of these representative genera identified in borehole BH10 were previously reported as members of the continental deep subsurface (Amils et al., 2023) and the references therein.

The only representative genus identified in the deep subsurface of IPB that had not been previously reported was *Rhodoplanes*. The annotation of genes encoding for nitrate ammonification, sulfur oxidation, thiosulfate reduction and denitrification in the sequenced genome of the IPB isolate *Rhodoplanes* sp. T2.6MG-98 supports its development in the deep subsurface of the IPB (Mariñán et al., 2019).

Members of the genera *Acidithiobacillus, Acidovorax, Desulfosporosinus, Pseudomonas, Sulfobacillus* and *Tessaracoccus*, along with the class Cyanobacteria were detected in 73% of the analyzed intervals using at least three independent methodologies (Fig. 17.2). While Cyanobacteria had been reported in prior drilling studies (Bose et al., 2020; Dutta et al., 2018; Gihring et al., 2006; Momper et al., 2017), their presence in light-independent ecosystems was often overlooked due to the challenge of explaining their occurrence in the absence of sunlight. In 2018, our metagenomic analysis of borehole BH10 (Puente-Sánchez et al., 2018) demonstrated that Cyanobacteria, at various depths, utilize alternative metabolisms independent of light, such as H_2 oxidation, for its development. This finding enabled subsequent identification of Cyanobacteria in the deep subsurface and raised questions about their evolutionary origin (for further information see chapter 7).

The isolation and characterization of *Tessaracoccus lapidicaptus* by (Puente-Sánchez et al., 2014b), an endemic heterotrophic nitrate reducing bacterium identified in most enrichment cultures generated from BH10 borehole core samples (Amils et al., 2023) and isolated from some of them (Leandro et al., 2017; Martínez, 2025) highlight its critical role in the IPB subsurface requiring further investigation.

Three of the most representative genera, *Acidovorax*, *Pseudomonas* and *Rhizobium* belong to a core of eight terrestrial deep subsurface bacterial genera selected after the analysis of eight drilling operations through diverse mineral substrates and locations (Soares et al., 2019). In the deep subsurface of the IPB, members of one of these genera, *Rhizobium*, were identified by four independent methodologies across 10 depth intervals, while members of the other two selected genera, *Acidovorax* and *Pseudomonas*, were detected in 13 depth intervals using three independent methodologies. These findings underscore the ecological significance of these genera in the deep subsurface.

The identification of the most representative genera of the IPB subsurface allowed a principal component analysis (PCA) to identify correlations between these microorganisms and the detected metals and mineral substrates. Members of the genera *Arthrobacter*, *Acidovorax*, *Pseudomonas*, *Rhodoplanes*, and *Sulfobacillus* exhibited positive correlations with pyrite and iron. Notably, *Acidovorax*, *Pseudomonas* and *Rhodoplanes* are known to oxidize ferrous iron under anaerobic conditions. *Shewanella*, an Fe reducer, showed a positive correlation with siderite, a ferrous carbonate; *Tessaracoccus*, an iron oxidizer, and *Acidiphilium*, an iron reducer, with clinochlore, an iron phyllosilicate, and cooper; *Brevundimonas* with illite, another iron phyllosilicate, and zinc; and *Bacillus* with manganese. These correlations underscore the interplay between the most representative microorganisms and the mineral content detected along the borehole (Amils et al., 2023).

Coupled biogeochemical cycles operating in the deep subsurface of the IPB

The limited quantity and quality of DNA extracted from core samples hindered the generation of metanogenic data required for a comprehensive understanding of the IPB subsurface ecosystem. However, access to the genomic information of different representative isolates together with shotgun metagenomic data from Cyanobacteria (Puente-Sánchez et al., 2018), generate sufficient information on the enzymatic activities linked to the coupled operation of the most important biogeochemical cycles operating in the IPB deep subsurface (Amils et al., 2023). The complete set of genes able to functionally maintain the carbon cycle (CO_2 fixation, CO_2 production through respiration and fermentation), the H cycle (H_2 production and oxidation), the nitrogen cycle (N_2 fixation, dissimilatory nitrate reduction to ammonia, nitrate reduction, denitrification, and ANAMMOX), the sulfur cycle (S oxidation and reduction), and the iron cycle (iron reduction) have been detected in eight core samples obtained at different depths (392, 416, 4767, 487, 584, 569 and 607 mbs). The proposed roles of the most relevant identified and isolated microbial genera, their genes, and metabolic products in the maintenance of the basic C, H, N, S and Fe biogeochemical cycles operating in a coupled mode in the deep subsurface of the IPB is shown in Fig. 17.5. While the subsurface functioning of biogeochemical cycles has been proposed in prior studies using groundwater samples (Nyyssönen et al., 2014; Momper et al., 2017; Bell et al., 2020), with the important limitations already mentioned, the results of the IPBSL project not only provides the first direct evidence of their existence but also how these cycles function in a coupled manner to ensure the required efficiency given the extreme oligotrophic conditions in which they operate.

The existence of a significant number of relevant microorganisms identified at varying depths, whose genomes contain annotated genes actively involved in the nitrogen cycle underscores the critical role of this biogeochemical process in the deep subsurface (Fig. 17.5). While earlier studies suggested its importance (Nuppunen-Puputti et al., 2022; Lau et al., 2016; Rempfert et al., 2017), our findings explicitly reaffirm its central role in the continental deep subsurface.

The anaerobic deep subsurface of the IPB hosts microbial iron oxidation processes driven by nitrate reduction, where nitrite (NO_2^-) and nitric oxide (NO) act as indirect oxidants for Fe^{2+}. Diverse isolates from different depths demonstrated variable Fe^{2+} oxidation efficiencies supporting microbial involvement (Martínez, 2025). Genomic analysis of dominant subsurface genera revealed widespread nitrate reductase genes, yet no direct genetic markers for enzymatic Fe oxidation were identified. Fluorescence in situ hybridization (FISH) coupled with Raman microscopy linked the genus *Acvidovorax* to pyrite dissolution under nitrate-reducing conditions, implicating nitrate-dependent Fe oxidizers (NDFeO) in subsurface iron cycling (Escudero et al., 2021). These processes generate soluble Fe^{3+}, secondary Fe minerals, and sulfide dissolution, contributing to the extreme geochemistry of the Río Tinto basin. While nitrate reduction by products is proposed as the primary Fe oxidation pathway, the potential for direct enzymatic activity remains unresolved. Experimental evidence, including anaerobic pyrite solubilization by *Acidovorx* underscores a critical biotic-abiotic interplay driving Fe biogeochemical cycling in anaerobic conditions, emphasizing the role of nitrate-reducing microorganisms in shaping iron mineralogy and acidic ecosystems. However, the absence of definitive genetic pathways for anaerobic Fe oxidation underscores persistent knowledge gaps in microbial-metal interactions in anaerobic habitats (for a deeper analysis see Chapter 16).

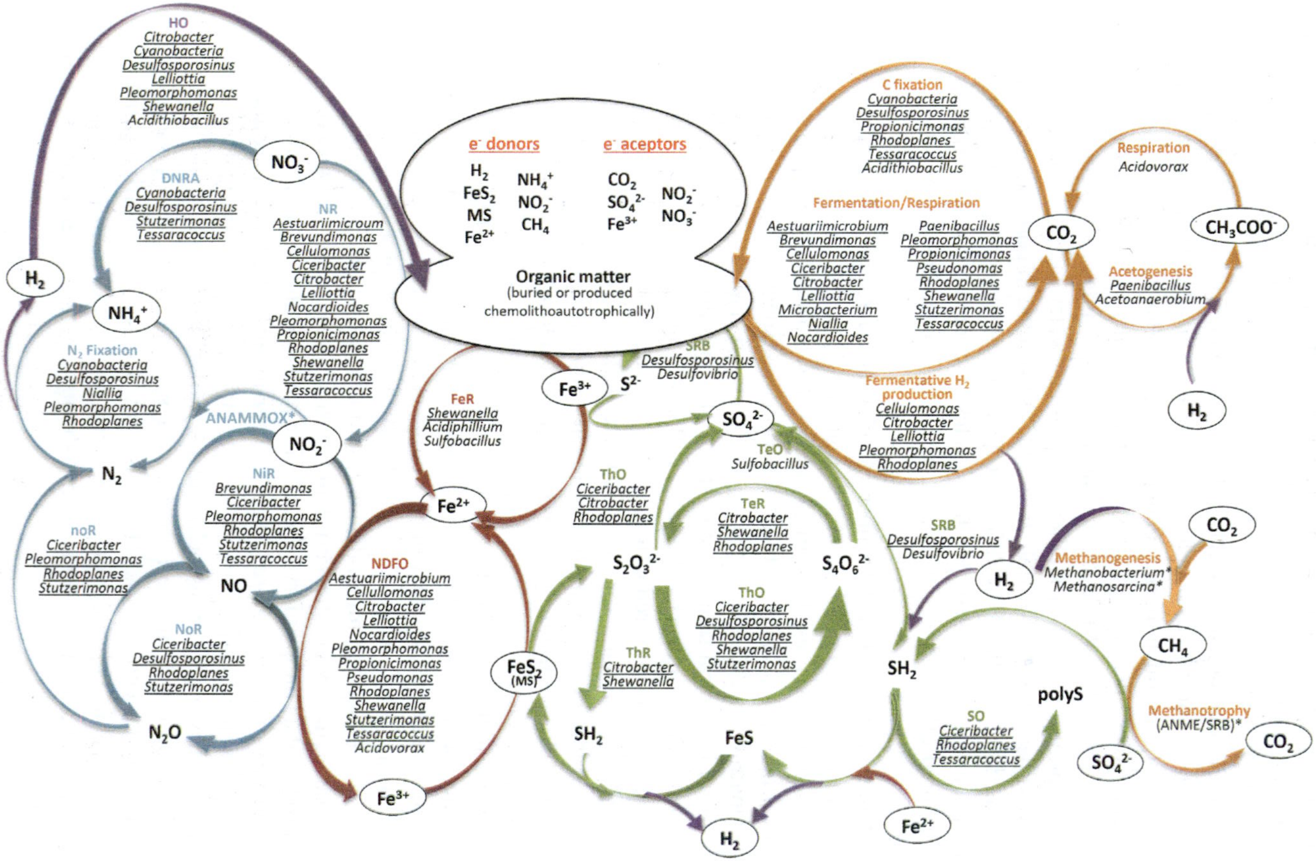

FIGURE 17.5 Geomicrobiological model of the main biogeochemical cycles operating in the deep subsurface of the IPB.

Purple: H cycle; blue: N cycle; red: Fe cycle; green: S cycle and brown: C cycle. MS, metal sulfides; F, fermentation; R, respiration. Genes: HO, hydrogen oxidase; NR, nitrate reduction; DNRA, dissimilatory nitrate reduction to ammonium; NiR, nitrite reductase; NoR, nitric oxide reductase; noR, nitrous oxide reductase; ANAMMOX, anaerobic ammonium oxidation; FeR, iron reduction; SRB, sulfate reducing bacteria; SO, sulfur oxidation; TeR, tetrathionate reductase; ThO, thiosulfate oxidase, ThR, thiosulfate reductase, ANME, anaerobic methane oxidizing archaea. NDFO, nitrate dependent Fe oxidizing bacteria; FeR, Fe reducing bacteria. Genera listed under the different genes: sequenced isolates in which the presence of the corresponding gene has been identified. Not underlined genera: those only identified using different methodologies. * identified by CARD-FISH.

Astrobiological interest of subsurface life detection in the IPB

As mentioned, among the astrobiology community the interest in continental subsurface life is gaining momentum due to its profound implications that could redefine habitability and identify conditions that could support life on diverse planetary bodies. The current concept of the habitable zone (HZ) considers a planet habitable only if the distance to the star and its environmental conditions permits the presence of liquid water on its surface (Cockell et al., 2016; Kasting and Catling, 2003). However, the discovery of Earth's deep subsurface biosphere, where microbial life thrives independently of surface conditions, calls for expanding the current concept of habitability to include subsurface habitability (SSHZ), especially for rocky exoplanets that due to environmental constrains do not meet the surface liquid water condition (McMahon, O'Malley-James and Parnell, 2013; Escudero and Amils, 2024).

A key limitation is that current technologies primarily analyze exoplanets surfaces and atmospheres, thus offering no insights into their subsurface. In contrast, solar system exploration missions can directly investigate the subsurface or bring to Earth subsurface samples for a thorough analytical evaluation. Mars is an interesting study case. According to the accumulated data from different missions Mars lost its surface liquid water and the actual existing extreme conditions of radiation, oxidizing conditions and low temperature make it very unlikely that life can develop on its surface (Margulis et al., 1979; Martínez et al., 2017). Recent experimental results from Mars Express (Orosei et al., 2018) and Insight (Wright et al., 2024) missions produced data that are consistent with the existence of liquid water in the deep subsurface of the planet. This is important because of the three conditions that life requires: energy, basic elements (C, H, O, N, P, S) and an adequate solvent, only the presence of solvent is the limiting factor, because energy and basic elements have been detected throughout the universe. Therefore, recent astrobiological missions are focusing on the search for water, which due to its outstanding properties facilitates the existence of life in our planet.

The demonstration of Darwin's prediction of life in the deep subsurface has revolutionized astrobiology, urging future Mars missions to prioritize subsurface exploration if we want to detect extant life on Mars. These days the use of rovers to explore Mars has revealed very interesting surface data of possible past habitability areas but they have proved not useful for the detection of extant life as they have no access to the deep subsurface. The discovery of microbial life sustaining the operation of the most important biogeochemical cycles in the subsurface of one of the best geochemical and mineralogical Mars analogs, the Iberian Pyrite Belt (Fernández-Remolar et al., 2005), underscores the potential for subsurface life on Mars (Boston, Ivanov and P. McKay, 1992; Price et al., 2018; Price et al., 2022; Koike et al., 2020; Purkamo et al., 2020). Future drilling on Mars missions, despite the risks and challenges involved, is the only way to see whether there actually is life on Mars.

Despite technical challenges, integrating subsurface exploration into mission design is critical. The IPB's deep subsurface geomicrobiology, which thrives in high sulfidic and low pH conditions, exemplifies how life could persist in Mars' subsurface through chemolithotrophic metabolism. These findings align with astrobiological studies advocating for advance life detection technologies, such as autonomous drilling systems and metabolomic sensors, to overcome current exploration limitations.

Final considerations

The identification and characterization of the most relevant microorganisms inhabiting the hard rock subsurface of the Iberian Pyrite Belt, along with the identification and quantification of their substrates and metabolic products, strongly suggest the operation of coupled C, H, N, S and Fe biogeochemical cycles in the absence of radiation. These results demonstrate that the deep subsurface of the IPB contains sufficient biologically accessible energy to sustain approximately 4×10^{24} microorganisms within the first kilometer of subsurface. These findings support the existence of a subsurface bioreactor in the IPB, that generate the extreme conditions detected in the Tinto basin. The subsurface biogeochemical cycles contribute significantly to the search for life in Earth's deep subsurface environments and to the future search for life in the subsurface of other planetary bodies.

References

Aguilera, A., Manrubia, S.C., Gómez, F., Rodríguez, N., Amils, R., 2006. Eukaryotic community distribution and its relationship to water physicochemical parameters in an extreme acidic environment, Río Tinto (Southwestern Spain). Appl. Environ. Microbiol. 72 (8), 5325–5330. https://doi.org/10.1128/AEM.00513-06.

Allman, C.J, Gómez-Ortiz, D., Burke, A., Amils, R., Rodríguez, N, Fernández-Remolar, D, 2021. Hydrogeochemical variability of the acidic springs in the Rio Tinto. Water 13 (20), 2861. https://doi.org/10.3390/W13202861.

Amaral-Zettler, L., Gómez, F., Zettler, E.R., Keenan, B.G., Amils, R., Sogin, M., 2002. Eukaryotic diversity in Spain's river of fire. Nature 417. https://doi.org/10.1038/417137a.

Amils, R., Escudero, C., Oggerin, M., Sánchez, F.P., Rodríguez, A.A., Remolar, D.F., Rodríguez, N., Villadangos, M.G., Sanz, J.L., Briones, C., Román, M.S.-, Gómez, F., Leandro, T., Moreno-Paz, M., Ballesteros, O.P.-, Molina, A., Tornos, F., Andrea, I.S.-, Timmis, K., Pieper, D.H., Parro, V., 2023. Coupled C, H, N, S and Fe biogeochemical cycles operating in the continental deep subsurface of the Iberian Pyrite Belt. Environ. Microbiol. 25 (2), 428–453. https://doi.org/10.1111/1462-2920.16291.

Apps, J.A., Van De Kamp, P.C., 1993. Energy gases of abiogenic origin in the Earth's crust. U. S. Geol. Surv. Prof. Pap. 1570, 81–132. https://doi.org/10.3133/pp1570.

Bastin, E.S., Greer, F.E., Merritt, C.A., Moulton, G., 1926. The presence of sulphate reducing bacteria in oil field waters. Science 63 (1618), 21–24. https://doi.org/10.1126/science.63.1618.21.

Bell, E., Lamminmäki, T., Alneberg, J., Andersson, A.F., Qian, C., Xiong, W., Hettich, R.L., Frutschi, M., Bernier-Latmani, R., 2020. Active sulfur cycling in the terrestrial deep subsurface. ISME J. 14 (5), 1260–1272. https://doi.org/10.1038/s41396-020-0602-x.

Bose, H., Dutta, A., Roy, A., Gupta, A., Mukhopadhyay, S., Mohapatra, B., Sarkar, J., Roy, S., Kazy, S.K., Sar, P., 2020. Microbial diversity of drilling fluids from 3000 m deep Koyna pilot borehole provides insights into the deep biosphere of continental earth crust. Sci. Drill. 27, 1–23. https://doi.org/10.5194/sd-27-1-2020.

Boston, P.J., Ivanov, M.V., McKay, C.P., 1992. On the possibility of chemosynthetic ecosystems in subsurface habitats on Mars. Icarus 95 (2), 300–308. https://doi.org/10.1016/0019-1035(92)90045-9.

Breuker, A., Köweker, G., Blazejak, A., Schippers, A., 2011. The deep biosphere in terrestrial sediments in the chesapeake bay area, Virginia, USA. Front. Microbiol. 2 (JULY). https://doi.org/10.3389/fmicb.2011.00156.

Cockell, C.S., Bush, T., Bryce, C., Direito, S., Fox-Powell, M., Harrison, J.P., Lammer, H., Landenmark, H., Martin-Torres, J., Nicholson, N., Noack, L., O'Malley-James, J., Payler, S.J., Rushby, A., Samuels, T., Schwendner, P., Wadsworth, J., Zorzano, M.P., 2016. Habitability: a review. Astrobiology 16 (1), 89–117. https://doi.org/10.1089/ast.2015.1295.

Cockell, C.S., Schaefer, B., Wuchter, C., Coolen, M.J.L., Grice, K., Schnieders, L., Morgan, J. V., Gulick, S.P.S., Wittmann, A., Lofi, J., Christeson, G.L., Kring, D.A., Whalen, M.T., Bralower, T.J., Osinski, G.R., Claeys, P., Kaskes, P., de Graaff, S.J., Déhais, T., Goderis, S., Hernandez Becerra, N., Nixon, S., 2021. Shaping of the present-day deep biosphere at chicxulub by the impact catastrophe that ended the cretaceous. Front. Microbiol. 12. https://doi.org/10.3389/fmicb.2021.668240. Available from:

Colman, D.R., Poudel, S., Stamps, B.W., Boyd, E.S., Spear, J.R., 2017. The deep, hot biosphere: twenty-five years of retrospection. Proc. Natl. Acad. Sci. U. S. A 114 (27), 6895–6903. https://doi.org/10.1073/pnas.1701266114.

Coyte, K.Z., Tabuteau, H., Gaffney, E.A., Fostera, K.R., Durham, W.M., 2017. Microbial competition in porous environments can select against rapid biofilm growth. Proc. Natl. Acad. Sci. U. S. A 114 (2), E161. https://doi.org/10.1073/pnas.1525228113.

Darwin, C., Voyages of the adventure and beagle. J. Remarks III (1839), 1832–1836.

Dutta, A., Gupta, S.D., Gupta, A., Sarkar, J., Roy, S., Mukherjee, A., Sar, P., 2018. Exploration of deep terrestrial subsurface microbiome in Late Cretaceous Deccan traps and underlying Archean basement, India. Sci. Rep. 8 (1). https://doi.org/10.1038/s41598-018-35940-0.

Escudero, C., Campo, Ares, J.R., Sánchez, C., Martínez, J.M., Gómez, F., et al., 2021. Visualizing microorganisms-mineral interaction in the iberian pyrite belt subsurface: the Acidovorax case. Front. Microbiol. 11, 572104. https://doi.org/10.3389/fmicb.2020.572104.

Escudero, C., Vera, M., Oggerin, M., Amils, R., 2018a. Active microbial biofilms in deep poor porous continental subsurface rocks. Sci. Rep. 8 (1). https://doi.org/10.1038/s41598-018-19903-z.

Escudero, C., Oggerin, M., Amils, R., 2018b. The deep continental subsurface: the dark biosphere. Int. Microbiol. 21 (1–2), 3–14. https://doi.org/10.1007/s10123-018-0009-y.

Escudero, C., Amils, R., 2024. Hard rock dark biosphere and habitability. Front. Astron. Space Sci. 10, 1203845. https://doi.org/10.3389/fspas.2023.1203845.

Fernández-Remolar, D.C., Banerjee, N., Gómez-Ortiz, D., Izawa, M., Amils, R., 2018. A mineralogical archive of the biogeochemical sulfur cycle preserved in the subsurface of the Río Tinto system. Am. Mineral. 103 (3), 394–411. https://doi.org/10.2138/am-2018-6288.

Fernández-Remolar, D.C., Morris, R.V., Gruener, J.E., Amils, R., Knoll, A.H., 2005. The Río Tinto Basin, Spain: mineralogy, sedimentary geobiology, and implications for interpretation of outcrop rocks at Meridiani Planum, Mars. Earth Planet Sci. Lett. 240 (1), 149–167. https://doi.org/10.1016/j.epsl.2005.09.043.

Fernández-Remolar, D.C., Prieto-Ballesteros, O., Rodríguez, N., Gómez, F., Amils, R., Gómez-Elvira, J., Stoker, C.R., 2008. Underground habitats in the Río Tinto basin: a model for subsurface life habitats on mars. Astrobiology 8 (5), 1023–1047. https://doi.org/10.1089/ast.2006.0104.

Fry, J.C., Horsfield, B., Sykes, R., Cragg, B.A., Heywood, C., Kim, G.T., Mangelsdorf, K., Mildenhall, D.C., Rinna, J., Vieth, A., Zink, K.G., Sass, H., Weightman, A.J., Parkes, R.J., 2009. Prokaryotic populations and activities in an interbedded coal deposit, including a previously deeply buried section (1.6-2.3 km) above ~150 Ma basement rock. Geomicrobiol. J. 26 (3), 163–178. https://doi.org/10.1080/01490450902724832.

Garrido, P., González-Toril, E., García-Moyano, A., Moreno-Paz, M., Amils, R., Parro, V., 2008. An oligonucleotide prokaryotic acidophile microarray: its validation and its use to monitor seasonal variations in extreme acidic environments with total environmental RNA. Environ. Microbiol. 10 (4), 836–850. https://doi.org/10.1111/j.1462-2920.2008.01477.x.

Gihring, T.M., Moser, D.P., Lin, L.H., Davidson, M., Onstott, T.C., Morgan, L., Milleson, M., Kieft, T.L., Trimarco, E., Balkwill, D.L., Dollhopf, M.E., 2006. The distribution of microbial taxa in the subsurface water of the kalahari shield, South Africa. Geomicrobiol. J. 23 (6), 415–430. https://doi.org/10.1080/01490450600875696.

Gómez-Ortiz, D., Fernández-Remolar, D.C., Granda, Á., Quesada, C., Granda, T., Prieto-Ballesteros, O., Molina, A., Amils, R., 2014. Identification of the subsurface sulfide bodies responsible for acidity in Río Tinto source water, Spain. Earth Planet Sci. Lett. 391, 36–41. https://doi.org/10.1016/j.epsl.2014.01.022.

González-Toril, E., Llobet-Brossa, E., Casamayor, E.O., Amann, R., Amils, R., 2003. Microbial ecology of an extreme acidic environment, the Tinto River. Appl. Environ. Microbiol. 69 (8), 4853–4865. https://doi.org/10.1128/aem.69.8.4853-4865.2003.

Itävaara, M., Nyyssönen, M., Kapanen, A., Nousiainen, A., Ahonen, L., Kukkonen, I., 2011. Characterization of bacterial diversity to a depth of 1500m in the Outokumpu deep borehole, Fennoscandian Shield. FEMS Microbiol. Ecol. 77 (2), 295–309. https://doi.org/10.1111/j.1574-6941.2011.01111.x.

Kappler, A., Bryce, C., Mansor, M., Lueder, U., Byrne, J.M., Swanner, E.D., 2021. An evolving view on biogeochemical cycling of iron. Nat. Rev. Microbiol. 19 (6), 360–374. https://doi.org/10.1038/s41579-020-00502-7.

Kasting, J.F., Catling, D., 2003. Evolution of a habitable planet. Annu. Rev. Astron. Astrophys. 41, 429–463. https://doi.org/10.1146/annurev.astro.41.071601.170049.

Knittel, K., Boetius, A., 2009. Anaerobic oxidation of methane: progress with an unknown process. Annu. Rev. Microbiol. 63 (1), 311–334. https://doi.org/10.1146/annurev.micro.61.080706.093130.

Koike, M., Nakada, R., Kajitani, I., Usui, T., Tamenori, Y., Sugahara, H., Kobayashi, A., 2020. In-situ preservation of nitrogen-bearing organics in Noachian Martian carbonates. Nat. Commun. 11 (1), 1988. https://doi.org/10.1038/s41467-020-15931-4.

Kotelnikova, S., 2002. Microbial production and oxidation of methane in deep subsurface. Earth Sci. Rev. 58 (3–4), 367–395. https://doi.org/10.1016/s0012-8252(01)00082-4.

Lau, M.C.Y., Cameron, C., Magnabosco, C., Titus Brown, C., Schilkey, F., Grim, S., Hendrickson, S., Pullin, M., Barbara Sherwood, L., Heerden, E.van, Kieft, T.L., Onstott, T.C., 2014. Phylogeny and phylogeography of functional genes shared among seven terrestrial subsurface metagenomes reveal N-cycling and microbial evolutionary relationships. Front. Microbiol. 5, 531. https://doi.org/10.3389/fmicb.2014.00531.

Lau, M.C.Y., Kieft, T.L., Kuloyo, O., Linage-Alvarez, B., Van Heerden, E., Lindsay, M.R., Magnabosco, C., Wang, W., Wiggins, J.B., Guo, L., Perlman, D.H., Kyin, S., Shwe, H. H., Harris, R.L., Oh, Y., Yi, M.J., Purtschert, R., Slater, G.F., Ono, S., Wei, S., Li, L., Lollar, B.S., Onstott, T.C., 2016. An oligotrophic deep-subsurface community dependent on syntrophy is dominated by sulfur-driven autotrophic denitrifiers. Proc. Natl. Acad. Sci. U. S. A 113 (49), E7927. https://doi.org/10.1073/pnas.1612244113.

Leandro, T., Rodríguez, N., Sanz, J.L., Costa, R., Amils, A., 2018. Study of Methanogenic Enrichment Cultures of Rock Cores from the Deep Subsurface of the Iberian Pyrite Belt.

Leandro, T., Costa, Sanz, J.L., Amils, R., 2017. Complete genome of Tessaracoccus sp. strain T2.5-30 isolated from 139.5 m deep on the subsurface of the Iberian Pyrite Belt. Genome Announc. 5. https://doi.org/10.1088/1361-6641/aa7295.57, 238–17.

Leandro, T., 2018. Microbial Diversity and Anaerobic Metabolisms in the Subsurface of the Iberian Pyrite Belt. Thesis. Coimbra University. Coimbra University, p. Thesis.

Leblanc, M., Morales, J.A., Borrego, J., Elbaz-Poulichet, F., 2000. 4,500-year-old mining pollution in southwestern Spain: long-term implications for modern mining pollution. Econ. Geol. 95 (3), 655–662. https://doi.org/10.2113/gsecongeo.95.3.655.

Lehman, R.M., O'Connell, S.P., Banta, A., Fredrickson, J.K., Reysenbach, A.L., Kieft, T.L., Colwell, F.S., 2004. Microbiological comparison of core and groundwater samples collected from a fractured basalt aquifer with that of dialysis chambers incubated in Situ. Geomicrobiol. J. 21 (3), 169–182. https://doi.org/10.1080/01490450490275848.

Liu, Q., Adler, K., Lipus, D., Kämpf, H., Bussert, R., Plessen, B., Schulz, H.-M., Krauze, P., Horn, F., Wagner, D., Mangelsdorf, K., Alawi, M., 2020. Microbial signatures in deep CO2-saturated miocene sediments of the active Hartoušov Mofette system (NW Czech Republic). Front. Microbiol. 11, 543260. https://doi.org/10.3389/fmicb.2020.543260.

Magnabosco, C., Ryan, K., Lau, M.C.Y., Kuloyo, O., Sherwood Lollar, B., Kieft, T.L., Van HeerDen, E., Onstott, T.C., 2016. A metagenomic window into carbon metabolism at 3 km depth in Precambrian continental crust. ISME J. 10 (3), 730–741. https://doi.org/10.1038/ismej.2015.150.

Magnabosco, C., Lin, L.-H., Dong, H., Bomberg, M., Ghiorse, W., Stan-Lotter, H., Pedersen, K., Kieft, T.L., van Heerden, E., Onstott, T.C., 2018. The biomass and biodiversity of the continental subsurface. Nat. Geosci. 11 (10), 707–717. https://doi.org/10.1038/s41561-018-0221-6.

Margulis, L., Mazur, P., Barghoorn, E.S., Halvorson, H.O., Jukes, T.H., Kaplan, I.R., 1979. The viking mission: implications for life on mars. J. Mol. Evol. 14 (1–3), 223–232. https://doi.org/10.1007/BF01732380.

Mariñán, N., Martínez, J.M., Leandro, T., Amils, R., 2019. Draft genome sequence of Rhodoplanes sp. strain T2.26MG-98, isolated from 492.6 meters deep on the subsurface of the Iberian pyrite belt. Microbiol. Resour. Announc. 8 (16), e00070–19. https://doi.org/10.1128/MRA.00070-19. Available from:

Martínez, A., 2025. Anaerobic Iron Oxidation by Microorganisms Isolated from the Deep Subsurface of the Iberian Pyrite Belt. Thesis. Universidad Autónoma de Madrid, p. Thesis.

Martínez, G.M., Newman, C.N., De Vicente-Retortillo, A., Fischer, E., Renno, N.O., Richardson, M.I., Fairén, A.G., Genzer, M., Guzewich, S.D., Haberle, R.M., Harri, A.-M., Kemppinen, O., Lemmon, M.T., Smith, M.D., de la Torre-Juárez, M., Vasavada, A. R., 2017. The modern near-surface Martian climate: a review of in-situ meteorological data from viking to curiosity. Space Sci. Rev. 212 (1–2), 295–338. https://doi.org/10.1007/s11214-017-0360-x.

McMahon, S., O'Malley-James, J., Parnell, J., 2013. Circumstellar habitable zones for deep terrestrial biospheres. Planet. Space Sci. 85, 312–318. https://doi.org/10.1016/j.pss.2013.07.002.

Momper, L., Kiel Reese, B., Zinke, L., Wanger, G., Osburn, M.R., Moser, D., Amend, J.P., 2017. Major phylum-level differences between porefluid and host rock bacterial communities in the terrestrial deep subsurface. Environ. Microbiol. Rep. 9 (5), 501–511. https://doi.org/10.1111/1758-2229.12563. Available from:

Motamedi, M., Pedersen, K., 1998. Note: Desulfovibrio aespoeensis sp. nov., a mesophilic sulfate-reducing bacterium from deep groundwater at aspo hard rock laboratory, Sweden. Int. J. Syst. Bacteriol. 48 (1), 311–315. https://doi.org/10.1099/00207713-48-1-311.

Nuppunen-Puputti, M., Kietäväinen, R., Raulio, M., Soro, A., Purkamo, L., Kukkonen, I., et al., 2022. Eplilithic microbial community functionality in deep oligotrophic continental bedrock. Front. Microbiol. 13, 826048. https://doi.org/10.3389/fmicb.2022.826048.

Nyyssönen, M., Hultman, J., Ahonen, L., Kukkonen, I., Paulin, L., Laine, P., Itävaara, M., Auvinen, P., 2014. Taxonomically and functionally diverse microbial communities in deep crystalline rocks of the Fennoscandian shield. ISME J. 8 (1), 126–138. https://doi.org/10.1038/ismej.2013.125.

Oggerin, M., del Moral, C., Rodriguez, N., Fernandez-Gonzalez, N., Martínez, J.M., Lorca, I., Amils, R., 2024. Metal tolerance of Río Tinto fungi. Front. Fungal Biol. 5, 1446674. https://doi.org/10.3389/ffunb.2024.1446674. Available from:

Orosei, R., Lauro, S.E., Pettinelli, E., Cicchetti, A., Coradini, M., Cosciotti, B., Di Paolo, F., Flamini, E., Mattei, E., Pajola, M., Soldovieri, F., Cartacci, M., Cassenti, F., Frigeri, A., Giuppi, S., Martufi, R., Masdea, A., Mitri, G., Nenna, C., Noschese, R., Restano, M., Seu, R., 2018. Radar evidence of subglacial liquid water on Mars. Science 361 (6401), 490–493. https://doi.org/10.1126/science.aar7268.

Parro, V., De Diego-Castilla, G., Moreno-Paz, M., Blanco, Y., Cruz-Gil, P., Rodríguez-Manfredi, J.A., Fernández-Remolar, D., Gómez, F., Gómez, M.J., Rivas, L.A., Demergasso, C., Echeverría, A., Urtuvia, V.N., Ruiz-Bermejo, M., García-Villadangos, M., Postigo, M., Sánchez-Román, M., Chong-Díaz, G., Gómez-Elvira, J., 2011. A microbial oasis in the hypersaline atacama subsurface discovered by a life detector chip: implications for the search for life on mars. Astrobiology 11 (10), 969–996. https://doi.org/10.1089/ast.2011.0654.

Pedersen, K., 2000. Exploration of deep intraterrestrial microbial life: current perspectives. FEMS Microbiol. Lett. 185 (1), 9–16. https://doi.org/10.1016/s0378-1097(00)00061-6.

Price, A., Pearson, V.K., Schwenzer, S.P., Miot, J., Olsson-Francis, K., 2018. Nitrate-dependent iron oxidation: a potential Mars metabolism. Front. Microbiol. 9, 513. https://doi.org/10.3389/fmicb.2018.00513. Available from:

Price, A., Macey, M.C., Pearson, V.K., Schwenzer, S.P., Ramkissoon, N.K., Olsson-Francis, K., 2022. Oligotrophic growth of nitrate-dependent Fe2+-oxidising microorganisms under simulated early Martian conditions. Front. Microbiol. 13, 800219. https://doi.org/10.3389/fmicb.2022.800219. Available from:

Puente-Sánchez, F., Moreno-Paz, M., Rivas, L., Cruz-Gil, P., García-Villadangos, M., Gómez, M.J., Postigo, et al., 2014a. Deep subsurface sulfate reducing and methanogenesis in the Iberian Pyrite Belt revealed through geochemistry and molecular biomarkers. Geobiology 12, 34–37.

Puente-Sánchez, F., Sánchez-Román, M., Amils, R., Parro, V., 2014b. Tessaracoccus lapidicaptus sp. nov., an actinobacterium isolated from the deep subsurface of the Iberian pyrite belt. Int. J. Syst. Evol. Microbiol. 64 (Pt_10), 3546–3552. https://doi.org/10.1099/ijs.0.060038-0.

Puente-Sánchez, F., Arce-Rodríguez, A., Oggerin, M., García-Villadangos, M., Moreno-Paz, M., Blanco, Y., Rodríguez, N., Bird, L., Lincoln, S.A., Tornos, F., Prieto-Ballesteros, O., Freeman, K.H., Pieper, D.H., Timmis, K.N., Amils, R., Parro, V., 2018. Viable cyanobacteria in the deep continental subsurface. Proc. Natl. Acad. Sci. 115 (42), 10702–10707. https://doi.org/10.1073/pnas.1808176115.

Purkamo, L., Kietäväinen, R., Nuppunen-Puputti, M., Bomberg, M., Cousins, C., 2020. Ultradeep microbial communities at 4.4 km within crystalline bedrock: implications for habitability in a planetary context. Life 10 (1), 2. https://doi.org/10.3390/life10010002. Available from:

Purkamo, L., Kietäväinen, R., Miettinen, H., Sohlberg, E., Kukkonen, I., Itävaara, M., Bomberg, M., 2018. Diversity and functionality of archaeal, bacterial and fungal communities in deep Archaean bedrock groundwater. FEMS (Fed. Eur. Microbiol. Soc.) Microbiol. Ecol. 94 (8). https://doi.org/10.1093/femsec/fiy116.

Rempfert, K.R., Miller, H.M., Bompard, N., Nothaft, D., Matter, J.M., Kelemen, P., Fierer, N., Templeton, A.S., 2017. Geological and geochemical controls on subsurface microbial life in the Samail Ophiolite, Oman. Front. Microbiol. 8, 56. https://doi.org/10.3389/fmicb.2017.00056. Available from:

Ruff, S.E., De Angelis, I.H., Mullis, M., Payet, J.P., Magnabosco, C., Lloyd, K.G., Sheik, C.S., Steen, A.D., Shipunova, A., Morozov, A., Reese, B.K., Bradley, J.A., Lemonnier, C., Schrenk, M.O., Joye, S.B., Huber, J.A., Probst, A.J., Morrison, H.G., Sogin, M.L., Ladau, J., Colwell, F., 2024. A global comparison of surface and subsurface microbiomes reveals large-scale biodiversity gradients, and a marine-terrestrial divide. Sci. Adv. 10 (51). https://doi.org/10.1126/sciadv.adq0645. Available from:

Sanz, J.L., Rodriguez, N., Escudero, C., Carrizo, D., Amils, R., 2021. Biological production of H2, CH4 and CO2 in the deep subsurface of the Iberian pyrite belt. Environ. Microbiol. 23 (7), 3913–3922. https://doi.org/10.1111/1462-2920.15561. Available from:

Schnakenberg, A., Aromokeye, D.A., Kulkarni, A., Maier, L., Wunder, L.C., Richter-Heitmann, T., Pape, T., Ristova, P.P., Bühring, S.I., Dohrmann, I., Bohrmann, G., Kasten, S., Friedrich, M.W., 2021. Electron acceptor availability shapes anaerobically methane oxidizing archaea (ANME) communities in south Georgia sediments. Front. Microbiol. 12, 617280. https://doi.org/10.3389/fmicb.2021.617280. Available from:

Sheik, C.S., Reese, B.K., Twing, K.I., Sylvan, J.B., Grim, S.L., Schrenk, M.O., Sogin, M.L., Colwell, F.S., 2018. Identification and removal of contaminant sequences from ribosomal gene databases: lessons from the census of deep life. Front. Microbiol. 9, 840. https://doi.org/10.3389/fmicb.2018.00840. Available from:

Soares, A., Edwards, A., An, D., Bagnoud, A., Bomberg, M., Budwill, K., Caffrey, S.M., Fields, M., Gralnick, J., Kadnikov, V., Momper, L., Osburn, M., Moreau, J.W., Moser, D., Mu, A., Purkamo, L., Rassner, S.M., Sheik, C.S., Sherwood Lollar, B., Toner, B.M., Voordouw, G., Wouters, K., Mitchell, A.C., 2019. A global perspective on microbial diversity in the terrestrial deep subsurface. bioRxiv. https://doi.org/10.1101/602672. Available from:

Stevens, T.O., McKinley, J.P., 1995. Lithoautotrophic microbial ecosystems in deep basalt aquifers. Science 270 (5235), 450–455. https://doi.org/10.1126/science.270.5235.450.

Strous, M., Pelletier, E., Mangenot, S., Rattei, T., Lehner, A., Taylor, M.W., Horn, M., Daims, H., Bartol-Mavel, D., Wincker, P., Barbe, V., Fonknechten, N., Vallenet, D., Segurens, B., Schenowitz-Truong, C., Médigue, C., Collingro, A., Snel, B., Dutilh, B. E., Op Den Camp, H.J.M., Van Der Drift, C., Cirpus, I., Van De Pas-Schoonen, K.T., Harhangi, H.R., Van Niftrik, L., Schmid, M., Keltjens, J., Van De Vossenberg, J., Kartal, B., Meier, H., Frishman, D., Huynen, M.A., Mewes, H.W., Weissenbach, J., Jetten, M.S.M., Wagner, M., Le Paslier, D., 2006. Deciphering the evolution and metabolism of an anammox bacterium from a community genome. Nature 440 (7085), 790–794. https://doi.org/10.1038/nature04647.

Tornos, F., 2006. Environment of formation and styles of volcanogenic massive sulfides: the Iberian Pyrite Belt. Ore Geol. Rev. 28 (3), 259–307. https://doi.org/10.1016/j.oregeorev.2004.12.005.

Vera, M., Schippers, A., Sand, W., 2013. Progress in bioleaching: fundamentals and mechanisms of bacterial metal sulfide oxidation—part A. Appl. Microbiol. Biotechnol. 97 (17), 7529–7541. https://doi.org/10.1007/s00253-013-4954-2.

Wright, V., Morzfeld, M., Manga, M., 2024. Liquid water in the Martian mid-crust. Proc. Natl. Acad. Sci. 121 (35). https://doi.org/10.1073/pnas.2409983121 e2409983121.

Zhang, G., Dong, H., Xu, Z., Zhao, D., Zhang, C., 2005. Microbial diversity in ultra-high-pressure rocks and fluids from the Chinese continental scientific drilling project in China. Appl. Environ. Microbiol. 71 (6), 3213–3227. https://doi.org/10.1128/aem.71.6.3213-3227.2005.

Zobell, C.E., 1938. Studies on the bacterial flora of marine bottom sediments. J. Sediment. Res. 8, 10–18.

CHAPTER

18

Deciphering microbial metabolism under polyextreme conditions in the Río Tinto through metatranscriptomics

Mercedes Moreno-Paz[1], Ángeles Aguilera[1], Elena González-Toril[2], Fernando Puente-Sánchez[3], Victor Parro[1]

[1]*Centro de Astrobiología (CAB), INTA-CSIC, Torrejón de Ardoz, Madrid, Spain;* [2]*Instituto de Técnica Aeroespacial Esteban Terradas (INTA), Torrejón de Ardoz, Madrid, Spain;* [3]*Department of Aquatic Sciences and Assessment, Swedish University of Agricultural Sciences, Uppsala, Sweden*

Introduction

Metatranscriptomics is the study of actively expressed genes in complex microbial communities. It helps us analyze microbial activity at a given point in time under a particular set of environmental conditions. Traditionally, microbiologists have used methods such as reverse transcriptase PCR (RT-PCR) and quantitative PCR (RT-qPCR) to quantify gene expression of a limited number of genes of interest. However, the advent of high-throughput technologies, such as microarray technology, and more recently, next-generation sequencing methods, such as RNA-seq, has significantly expanded our understanding of microbial activities in environmental conditions. Unlike metagenomics, which provides insights into the genetic potential of a microbiome, metatranscriptomics reveals which genes are actively transcribed, offering a real-time perspective on microbial functions and interactions. Furthermore, its integration with other high-throughput 'omics' technologies, including proteomics and metabolomics, has enabled a comprehensive and integrative analysis of microbial community functions, as well as their dynamic interactions with the surrounding environment. These advances have allowed the rapid and accurate measurement of expression levels of hundreds or thousands of transcripts both from individual organisms and entire microbial communities, and led us to new insights into the mechanisms of adaptation of microorganisms to extreme environments.

The main difficulty when performing transcriptomic and metatranscriptomic analyses is to achieve a good quality of RNA during its extraction, particularly in extreme environments where harsh conditions complicate sample collection, as is the case in acid mine drainage (AMD) environments. This is especially relevant

Geomicrobiology of Río Tinto. https://doi.org/10.1016/B978-0-443-32948-7.00014-X

when dealing with the Río Tinto (Spain) samples. The river is a polyextreme environment mainly characterized by its high metal concentrations and extremely low pH, which can degrade the RNA messengers during extraction. As the poly extreme conditions present in the river have been described in other chapters, hereafter, we will focus on the mechanisms of adaptation of the microbial diversity inhabiting it. Regarding prokaryotes, the low cell number frequently found in some places in the course of the river can be an additional drawback in obtaining enough quantities of RNA for downstream analysis. Another disadvantage of conducting this type of analysis in complex natural ecosystems is not only the broad biodiversity present but also the numerous environmental parameters that can influence gene expression. This could explain, in part, why there are relatively few studies based on transcriptomics or metatranscriptomics, as most research has focused on in vitro assays rather than on environmental analyses.

Adaptation to acidic environments: Iron and pH homeostasis

Most microorganisms maintain intracellular pH close to neutrality. In the particular case of acidophilic microorganisms, they are subjected to a large proton gradient that requires a large amount of energy. The adaptive mechanisms to environments with high proton concentration are based on the maintenance of a neutral pH in their cytoplasm, employing both unique transport systems and proton pumps that expel protons to the outside (Mirete et al., 2017). They also synthesize proteins with a high molecular weight, which increases the bonds between amino acids and makes the secondary structure more stable. This makes it more difficult to break the bonds that maintain the structure and keep the protein functional.

Adaptive mechanisms of *Acidithiobacillus ferrooxidans* to extreme pH

The mechanisms underlying pH homeostasis are not yet completely elucidated, and to date, only a limited number of transcriptomic assays, most of them based on quantitative RT-PCR, have been published. Two *A. ferroxidans* strains isolated from the bio-heap of Zijinshan copper mine are able to grow at extreme pH (Feng et al., 2015). As the pH decreases, the capsule of *A. Ferrooxidans* become thicker, accompanied by alterations in cell morphology. Another mechanical barrier for these species is to alter the fatty acid composition to increase the proportion of unsaturated ones as a measure to protect cell integrity. Furthermore, an increase in the levels of chaperone proteins (DnaK, DnaJ, GrpE) helps cells repair and properly process peptide chains, accompanied by enhanced H^+-ATPase activity to satisfy the energy requirements of each acid-tolerance component.

These results were partially corroborated by another recent and well-studied example in an *A. ferrooxidans* isolated from the Río Tinto basin (Huelva, Spain). In vitro studies performed using RT-PCR and proteomics have demonstrated that

A. ferrooxidans can grow across a broad pH range (0.5–5.5) (Ibáñez et al., 2023). Surprisingly, *A. ferrooxidans* is able to grow at near-neutral pHs (pH 6.0), possibly activating a response due to the expression of chaperone proteins. On the contrary, this bacterium is unable to survive at pH below 0.5, likely due to disruption and destabilization of the capsule and cell membranes (Feng et al., 2015).

Acidithiobacillus species play an important role in the biogeochemical cycle of iron, as well as in high metal tolerance. The genome of *A. ferrooxidans* reveals an extensive repertoire of genes involved in energy conservation through the oxidation of ferrous iron, reduced sulfur compounds, and hydrogen. These pathways are central to the organism's survival in low-pH, metal-rich environments (Valdés et al., 2008). This work highlights the modular organization of electron transfer chains, including the presence of multiple cytochromes, rusticyanin, and other redox proteins that facilitate electron flow from substrates to terminal electron acceptors. Moreover, the integration of genomic and proteomic analyses reveals the potential regulatory mechanisms that allow this bacterium to switch between different energy sources depending on environmental availability.

Concerning iron metabolism, metagenomic and protein translation studies have shown that *A. ferrooxidans* genomes harbor genes and proteins specifically involved in iron uptake and storage, as well as proteins related to the regulation of iron homeostasis. The comparative analysis of the genomes has exhibited an amino acid composition comparable to that of proteins from neutrophilic bacterial species (Quatrini et al., 2005). The energetic metabolism of iron and sulfur in the wild-type strain has been elucidated through transcriptomic analyses, providing insights into the molecular mechanisms underlying these processes (Quatrini et al., 2006). Unexpectedly, *A. ferrooxidans* also contains genes clustered within operon-like structures as putative siderophore systems for Fe(III) uptake. However, genes coding for the biosynthesis of those compounds have not been identified in its genome. The existence of these siderophores as part of multiple Fe(III) uptake systems can favor their survival in aerobic environments where iron is scarce. These results may elucidate the organism's inability to withstand elevated Fe(III) concentrations in bioleaching processes, where *Leptospirillum* species outcompete it. These findings have significant implications for bioleaching operations, especially in copper biomining, as this microorganism serves as a model organism.

Adaptive strategies of *Leptospirillum* in response to extreme physicochemical gradients in the Río Tinto ecosystem

The microbial diversity of the extremely acidic Río Tinto ecosystem is classified according to its ecological niche into planktonic and benthic communities. These include microorganisms inhabiting the water column or mining lakes, and those present in sediments or forming biofilms. In all cases, the spatial distribution of these communities is strongly influenced by prevailing environmental conditions. Water columns are predominantly composed of bacteria, with more than 80% of the cells belonging to *Acidithiobacillus*, *Leptospirillum,* and *Acidiphilium* genera, which are the most abundant, while Archaea represent only a small fraction of the microbial community (González-Toril and Aguilera, 2019). The water column in the middle course (pH 2.2) is also colonized by iron oxidizers affiliated with *Acidithiobacillus* and *Leptospirillum.* Although in a smaller proportion than the upper part of the river, they share this ecological niche with other taxa (Abramov et al., 2022). Accordingly, these three bacterial genera have been selected as models to understand the adaptation mechanisms to the environmental conditions present in Río Tinto waters.

Species of the *Leptospirillum* genus are chemolithotrophic, Gram-negative bacteria that obtain energy exclusively from the oxidation of Fe(II). They oxidize Fe(II) as an electron donor to Fe(III), with oxygen as the electron acceptor. The components of the electron transfer chain of *Leptospirillum* spp. are conserved among different leptospirilli groups, suggesting the existence of a common pathway for electron transport between Fe(II) and oxygen. A genome comparative study of four *Leptospirillum* genomes representing the three *Leptospirillum* groups revealed a high abundance of cytochrome *c* genes, encompassing at least 19 proteins. This finding highlights the crucial role of these genes in Fe(II) oxidation in *Leptospirillum* (Levicán et al., 2012).

It has been described that bacteria of the genus *Leptospirillum,* an aerobic and strict iron oxidizer, predominate in the upper reaches of the river and the vicinity of the old Peña de Hierro mine (García-Moyano et al., 2012), probably due to the more extreme conditions of these areas. The pH is consistently low, accompanied by a high concentration of ferric ion, which reaches up to 20 g/L near the river's spring (González-Toril et al., 2003). The water temperature in the source of the river is slightly higher than downstream and close to 16°C, which seems to remain constant with the seasons. These factors, together with the shallow depth of the water column—just a few centimeters—create the optimal conditions that appear to be particularly well-suited for the growth of *Leptospirillum,* which seems to be better adapted to them than other iron- and sulfur-oxidizing chemolithoautotrophs. The low prokaryotic diversity in the river, where *L. ferrooxidans* is one of the prevalent species, has enabled us to perform global gene expression analyses throughout transcriptomic profiling using shotgun DNA microarrays (Parro and Moreno-Paz, 2003). This approach has been a powerful tool in the last 2 decades, not only for gene expression studies but also for gene discovery, especially for environmental organisms whose genome sequences are unknown.

The study performed the analysis of RNA transcripts of the two sites of the water column: the river course (Site 1) and a dam located 2 kilometers away (Site 2). Site 1 was characterized by high salinity (mainly due to elevated concentrations of iron and sulfate) and lower oxygen concentrations compared to Site 2 (Fig. 18.1A).

The integration of gene expression data with physicochemical parameters led to the identification of specific genetic and metabolic adaptations that enable

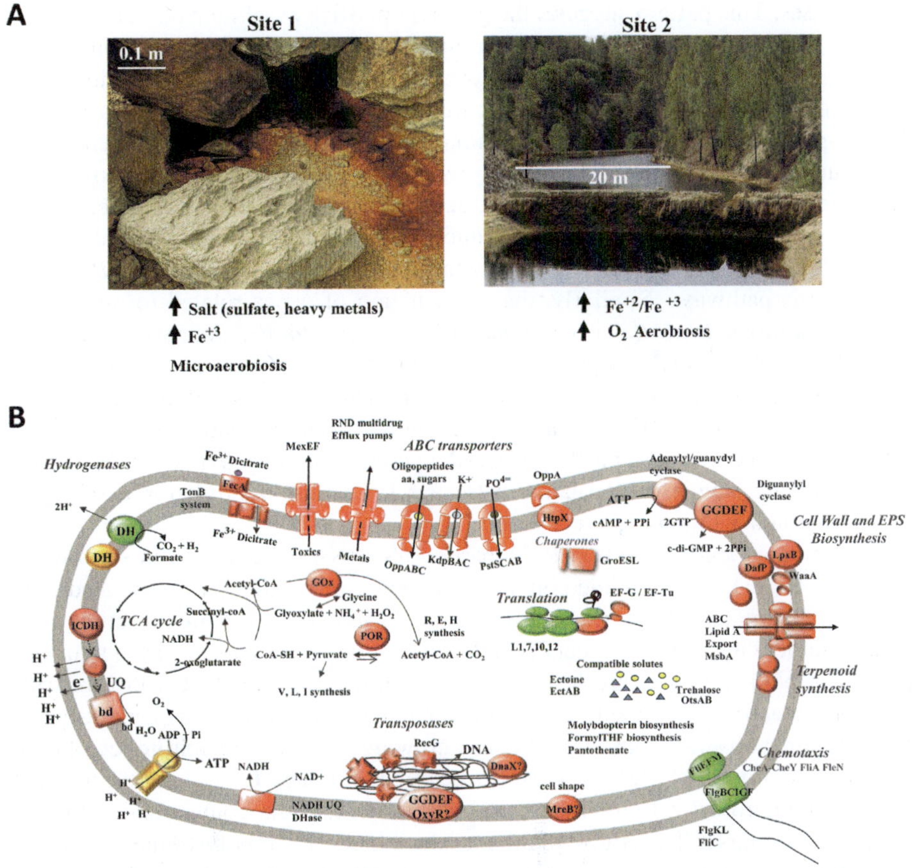

FIGURE 18.1 Schematic representation of the most relevant *Leptospirillum ferrooxidans* proteins whose genes are preferentially expressed in acidic headwaters of the Río Tinto under high-salt, high-metal and low-oxygen concentrations.

(A) Site 1 (S1) corresponds to a perennial spring that emerges from beneath a pile of mineral deposit, a remnant of historical mining activity. In contrast, Site 2 (S2) is an acidic impoundment located approximately 2 km away, formerly used as a water reservoir. (B) Proteins upregulated in S1 are depicted in red. Proteins shown in red indicate those proteins whose genes are either preferentially expressed at S2 or lacking a clear site-specific expression pattern (orange). Single-letter amino acid codes are as follows: R (arginine), E (glutamic acid), H (histidine), V (valine), L (leucine), I (isoleucine).

L. ferrooxidans to thrive in highly acidic, metal-rich, and low-oxygen environments, highlighting the complex, multilayered physiological strategies employed by *L. ferrooxidans* in situ (Fig. 18.1B).

The results indicate that the elevated iron-oxidizing metabolic activity at Site 1 is associated with high expression of genes encoding mobile genetic elements, such as transposases. This pattern suggests the genomic plasticity of L. ferrooxidans constitutes a key adaptive mechanism to environmental pressures. Notably, the upregulation of genes involved in cell envelope biogenesis, exopolysaccharide (EPS) production, and biofilm formation indicated surface attachment and a sessile lifestyle as protective strategies. The induction of carotenoid synthesis, DNA repair systems, and antioxidant regulators (*oxyR*, *recG genes)* evidenced a robust oxidative stress response. Additionally, specific genes are differentially expressed under distinct stressors: genes involved in osmotic stress tolerance are activated under high salinity conditions, including ectoine and trehalose (*ectAB*, *otsAB genes*) biosynthesis pathways. Similarly, the upregulation of genes related to microaerophilic conditions (*cydAB*), iron uptake (*fecA-exbBD-tonB*), potassium transport (*kdpBAC*), high-affinity phosphate transport system (*pstSCAB operon*), and various multidrug and metal efflux pumps reflects coordinated responses to nutrient limitation and metal toxicity. Finally, activating central metabolic pathways such as the glycine cleavage system and the porGAB operon, encoding pyruvate: ferredoxin oxidoreductase, suggests alternative pathways for carbon assimilation and energy production under low oxygen conditions.

One notable study based on RNA-based 16S rRNA gene amplicon sequencing has revealed that in the middle course, the planktonic fraction and associated minerals are dominated by *Leptospirillum* and *Acidithiobacillus,* both exhibiting high metabolic activity, while the Fe(III) reducers *Acidiphilium* and *Acetobacteraceae*, present in both fractions, are less active (Abramov et al., 2022). On the contrary, benthonic fractions (sediments) show higher functional diversity with *Leptospirillum, Acidiferrobacter*, and *Acidibacter*, growing in microaerophilic conditions (Sánchez-Andrea et al., 2011; Abramov et al., 2022). The RNA activity indicates a dynamic microbial activity, mainly based on the high iron content and pH in the main river's course. This study highlights the metabolic activity found is relevant for stabilizing pH and maintaining the iron cycle, the formation of Fe(III) minerals that can trap metals, as well as heavy metal transport (As, Cd, Cr, Pb) as dissolved biomineral aggregates along the RíoTinto.

Leptospirillum also dominates benthic environments in the river's middle course, where low oxygen concentrations give it an advantage over other iron-oxidizing taxa. The limitation of O_2 in these sediments could be explained by the overexpression of proteins involved in the last step of aerobic respiration, enabling the adaptation of this organism to low oxygen levels (Abramov et al., 2022).

These findings support the conclusion that specific gene expression patterns can be valuable indicators for the physiological conditions in a defined ecosystem. Moreover, *Leptospirillum ferrooxidans* exhibits adaptive responses to fluctuations in physicochemical parameters in its natural environment, and the upregulation of specific genes and operons provides valuable insights into prevailing environmental conditions, such as nutrient limitations and various stress factors, leading to the understanding of the correlations between global gene expression patterns and environmental parameters in natural ecosystems.

Adaptation to oligotrophic environments

Nitrogen fixation under nitrogen deprivation

Most extremely acidic environments exhibit low productivity, which is associated with relatively low concentrations of nutrients: ammonium, phosphate, or nitrate (Spijkerman, 2007). Given that nitrogen availability can limit microbial growth, atmospheric nitrogen fixation (diazotrophy) is the primary mechanism for assimilation of this element, and is widespread among Bacteria and Archaea. Acidophilic communities essentially consist of chemolithoautotrophic microorganisms that oxidize ferrous ions and reduced sulfur compounds as energy sources, obtaining most of their carbon and nitrogen from the atmospheric CO_2 and N_2, respectively (Tyson et al., 2005).

A pioneering work in transcriptomics in acidic environments was the study of the *nif* regulon in *L. ferrooxidans,* which dominates the column water along with *Acidithiobacillus* and *Acidiphilium* genera. While *A. ferrooxidans* is a well-known nitrogen fixer, species of *Acidiphilium* lack this capacity (Li et al., 2020; Martin-Uriz et al., 2011; Valdés et al., 2008). In vitro experiments using transcriptomic analyses by DNA microarrays addressed the genetic mechanisms behind the ability to fix nitrogen in *L. ferrooxidans* strain L3.2, particularly within the context of its iron-oxidizing metabolism. Using DNA microarrays, gene expression profiles were compared between cultures enriched with ammonium and those deprived of a nitrogen source. The study confirmed that *L. ferrooxidans* genome encodes a complete operon (*nifHDKENX*) for the molybdenum–iron (Mo–Fe) nitrogenase, the enzyme responsible for nitrogen fixation, and typically found in other acidophilic nitrogen-fixing bacteria (Mackintosh, 1978; Stevens et al., 1986), as well as other *nif* gene clusters involved in the assembly of Fe–S clusters, transport systems and some regulatory genes (Parro and Moreno-Paz, 2003) (Fig. 18.2).

Interestingly, the *nif* regulon is not constitutively expressed in *Leptospirillum.* Instead, its expression is modulated based on the availability of nitrogen in the medium. When nitrogen is scarce, there is a significant upregulation of *nif* gene expression, which indicates that *L. ferrooxidans* adjusts its metabolic processes to ensure nitrogen fixation in nutrient-limited environments. Moreover, the *nif* regulon expression seems to be iron-dependent and influenced by the bacterium's iron-oxidizing activity (Moreno-Paz and Parro, 2006). Under iron-limiting conditions, *nif* genes are downregulated, indicating a potential interaction between nitrogen fixation and iron oxidation pathways. This suggests that *L. ferrooxidans* prioritizes iron oxidation over nitrogen fixation when iron (FeII) availability is low (Moreno-Paz and Parro, 2006). On the contrary, putative operon hydrogenases (HyfBDE, HyfBAC) are preferentially expressed in ammonium-rich medium. Several potential regulatory proteins controlling the expression of the *nif* genes (e.g., *NifA* and *ntrC-like* genes), were also identified, along with proteins for nitrogen uptake regulation (*glnB* genes), and those involved in transport and chemotaxis (e.g., *modABC* and *amtB genes*) (Fig. 18.2). These proteins are thought to respond to environmental

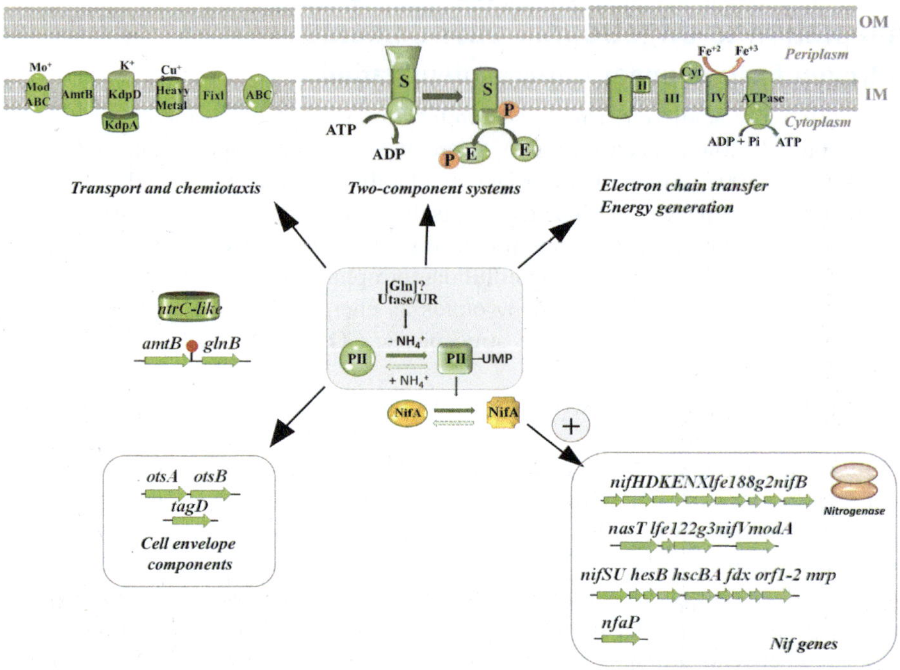

FIGURE 18.2 Schematic representation of the nitrogen fixation regulon and regulatory mechanisms *in Leptospirillum ferrooxidans* under laboratory conditions.

Arrows and figures represent genes and proteins, respectively, which are upregulated under nitrogen fixation conditions. Based on comparative analyses with other bacterial systems, the presence of an uridylyltransferase/uridylyl-removing (UTase/UR) enzyme is proposed. This enzyme is hypothesized to sense the cellular nitrogen status by monitoring the concentration of the PII signal transduction protein (Gln). The gray rectangle represents the central role that PII may play in regulating different cell functions: electron chain transfer, transport and chemotaxis, two-component systems, nitrogenase operon (*nifHDKENXlfe188g2nifB*). Components of the electron transport chain in *L. ferrooxidans* are represented by NADH oxidoreductases, ubiquinones, and cytochromes. *OM*, outer membrane; *IM*, inner membrane.

signals such as nitrogen availability, iron levels, and the overall metabolic state of the bacterium.

In contrast, transcriptomic data from environmental samples from the Río Tinto and an AMD biofilm show that there is no evidence of significant expression of the *nif* operon (Ram et al., 2005; Parro et al., 2007). This lack of nitrogen fixation implies the presence of a more readily available nitrogen source. In fact, the strong expression of ammonium transporter (*amt*) and the glnA gene, which encodes glutamine synthase, indicates that ammonium incorporation is a significant nitrogen source in situ (Chen et al., 2015; Parro et al., 2007). Thus, the *nif* regulon's expression is tightly controlled by environmental factors, which highlights the bacterium's metabolic flexibility and its ecological adaptation.

Despite inducing *nif* genes, other metabolic functions are coordinated with nitrogen fixation to force cells to adapt to a more energetically demanding nitrogen source: i) energy production, ii) cell envelope modification, iii) chemotactic and pumping mechanisms, iv) sensing and signal transduction systems, and v) sulfur metabolism (Parro and Moreno-Paz, 2004). These adaptations suggest that *L. ferrooxidans* has a sophisticated regulatory network that balances nitrogen fixation and iron oxidation, two critical processes for its survival and function in acidic environments.

Genes involved in nitrogen fixation have been identified in the *Leptospirillum* groups I and III (Parro and Moreno-Paz, 2003; Tyson et al., 2005). Although group II species (*L.ferriphilum*) can grow in media deprived of any nitrogen source, only the *nif* regulon has been found in some isolates (Christel et al., 2018; LIU et al., 2020). In fact, nitrogenase homologous genes in different Río Tinto *L. ferriphilum* isolates have not been detected (García-Moyano et al., 2008). Nonetheless, genes for ammonium transport (*amt* genes) are present in *L. ferriphilum* isolated from a Chilean mine (Levicán et al., 2008).

A DNA microarray specifically designed for *L. ferrooxidans* L3.2, was used to investigate whether isolates of *L. ferriphilum* were able to fix nitrogen (García-Moyano et al., 2008). The results confirmed the absence of a homologous gene encoding for the nitrogenase in different *L. ferriphilum* isolates from Río Tinto, while the upregulation of *amtB* and *glnB* under nitrogen-limited conditions supports the use of available dissolved nitrogen rather than atmospheric N_2 (Garrido et al., 2008). These observations suggest that the *L. ferriphilum* isolated from this environment is unable to fix atmospheric nitrogen in a similar way to *L. ferrooxidans,* and may instead depend on environmental nitrogen or the metabolic activity of coexisting diazotrophs.

Transcriptomic analyses of environmental samples confirmed that nitrogen fixers in the Río Tinto do not activate nitrogen fixation mechanisms (Parro et al., 2007). Considering that nitrogen fixation is an energetically costly process, and that the nitrogenase is very sensitive to oxygen, it is very probable that microorganisms utilize other alternative nitrogen sources present in the water (Ram et al., 2005; Parro et al., 2007). Although direct detection of nitrite, nitrate, or ammonium in the Río Tinto has proven difficult, the identification of the *glnA* gene suggests that there must be additional sources of nitrogen available in the river or alternative metabolic mechanisms to compensate for the limitation of this nutrient. Furthermore, some acidophilic chemolithotrophs growing in the absence of an external nitrogen source are believed to absorb nitrogen ions from the atmosphere (Holden, 1991). However, it is more likely that *L. ferriphilum* relies on alternative dissolved inorganic nitrogen sources present in the medium, or on the activity of coexisting diazotrophic microorganisms in the environment that support its efficient growth, and hence, explain the specific distribution of *Leptospirillum* spp. found in Río Tinto (García-Moyano et al., 2008).

Altogether, these studies have provided a deeper understanding of the genetic mechanisms behind nitrogen fixation in iron-oxidizing bacteria inhabiting the Río Tinto and other AMD environments. They underscore how environmental conditions shape metabolism and gene expression. This methodology is particularly relevant in biooxidation of metal sulfides, a crucial process in biohydrometallurgy and other biotechnological processes where *Leptospirillum* spp. are involved (Remonsellez et al., 2009; Remonsellez et al., 2009). Notably, gene expression patterns enabled the association of specific transcripts with environmental samples, regardless of their taxonomic origin, as demonstrated in an industrial bioleaching heap (Chile) (Remonsellez et al., 2009).

Regulatory adaptations to phosphate starvation

Phosphate limitation is a well-characterized environmental stress that induces high-affinity phosphate transport systems in bacteria to maintain cellular phosphate homeostasis. In *Escherichia coli*, for example, phosphate starvation triggers the upregulation of the *pstSCAB* operon, which encodes a high-affinity ABC-type phosphate transporter (Aguena et al., 2002). Consistently, phosphate concentrations are often below detection limits in extremely acidic environments, such as the Río Tinto (Parro et al., 2007) and other AMD systems, where phosphate remains largely undetectable (Chen et al., 2015). Previous studies developed in AMD microorganisms, and specifically on *Acidithiobacillus* spp., have found that the expression of the *pstSCAB* operon is ubiquitous and seems to be upregulated under phosphate starvation (Vera et al., 2003). Parro et al. (2007) demonstrated that in response to phosphate starvation, *L. ferrooxidans* induced the expression of the pstCAB operon. Notably, this operon is located immediately downstream of a putative phosphate regulon response regulator gene, phoB, suggesting a coordinated regulatory control. The scarcity of phosphate in the acidic Río Tinto waters is likely a consequence of its precipitation by trivalent cations, particularly Fe^{3+} and Al^{3+} (Grzmil and Wronkowski, 2006). A survey performed during two consecutive summers studied how temporal variations in the geochemical conditions alter metal concentrations at different locations in the river, leading to the upregulation of genes involved in phosphate uptake (Parro et al., 2007). Further metatranscriptomic analyses showed that the *pstSCAB* operon is also preferentially expressed in planktonic cells rather than in biofilms, as reported in *L. ferrooxidans* populations from the Río Tinto (Moreno-Paz et al., 2010). Under sustained phosphate limitation, high levels of PstS appear to modulate biofilm production, and once the biofilm reaches a mature state, the *pstSCAB* operon is repressed. A similar response has been documented in *Pseudomonas aeruginosa* following *in vitro* experiments, where phosphate depletion led to an upregulated expression of PstS, subsequently promoting enhanced biofilm production (Long et al., 2008). Therefore, the expression of the pstCAB operon in response to phosphate scarcity appears to be a conserved and widespread strategy among bacteria to cope with phosphate limitation, which also promotes cell aggregation.

Physiological and transcriptomic differences between planktonic and biofilm-associated *Leptospirillum* spp

Biofilms are cell aggregates embedded in an extracellular polysaccharide matrix whose formation is influenced by many factors, including temperature, pH, salt, relative humidity, oxygen availability, and nutrient concentration. Bacteria can be planktonic (free-living), but in most natural ecosystems, they prefer to grow as biofilms (sessile) at interfaces (Kilic, 2025). This preferential colonization is attributed to an adaptive response to adverse environmental conditions.

During the summer, when temperatures rise and the water flow decreases, prokaryotic macroscopic filaments are observed in specific locations of the Río Tinto. In particular, brown filaments are mainly found in the river's upper course, attached to solid surfaces or floating in the water stream. These filaments, rich in *Leptospirillum*, are usually accompanied by *Acidithiobacillus ferrooxidans* and *Acidiphilium* (García-Moyano et al., 2007).

Moreno-Paz et al. (2010) conducted environmental transcriptomic analyses to elucidate the physiological differences between planktonic (free-living) and sessile (biofilm-associated) lifestyles of Leptospirillum spp. within their natural microbial community. The microbial community composition was also studied using a prokaryotic acidophile microarray (PAM), which confirmed that *Leptospirillum* spp. and *A. ferrooxidans* were the dominant taxa across both environments and enabled the identification of the most metabolically active prokaryotes (García-Moyano et al., 2007; Garrido et al., 2008). Furthermore, gene expression analyses were performed using a *L. ferrooxidans* shotgun DNA microarray, which revealed a lifestyle-specific gene expression.

The physiological parameters indicated that planktonic cells grew in a high-salt, oxygen-rich environment, while biofilm cells thrived in microaerobic niches. Transcriptomic profiling showed that biofilms (sessile cells) exhibited gene induction related to biofilm formation and maintenance, which includes quorum sensing (*mqsR*, *rpfG*), chemotaxis (*cheAY*, *bdlA*), motility (*fliA*, *motA*), and production of extracellular matrix components and cell envelope and wall synthesis (*murA/murB*) (Fig. 18.3A).

Genes involved in stress and proteolysis, such as chaperones (clpB, groESL) and proteases (clpXP, lon), were also induced. Moreover, genes related to mixed acid fermentation (encoding pyruvate oxidase, phosphotransacetylase, and acetate kinase enzymes) were highly induced, which suggests acetate production and active fermentative metabolism under microaerophilic conditions. In fact, acetate concentrations in situ can reach up to 3.54 mM. Although its role is unclear, acetate can serve as an energy source for heterotrophic Fe^{3+} reducers and inhibit chemolithotrophic growth, thus shaping microbial interactions and, hence, biogeochemical cycling.

In contrast, planktonic (free-living) cells upregulate genes that are mainly involved in the biosynthesis of amino acids, fatty acids, and compatible solutes

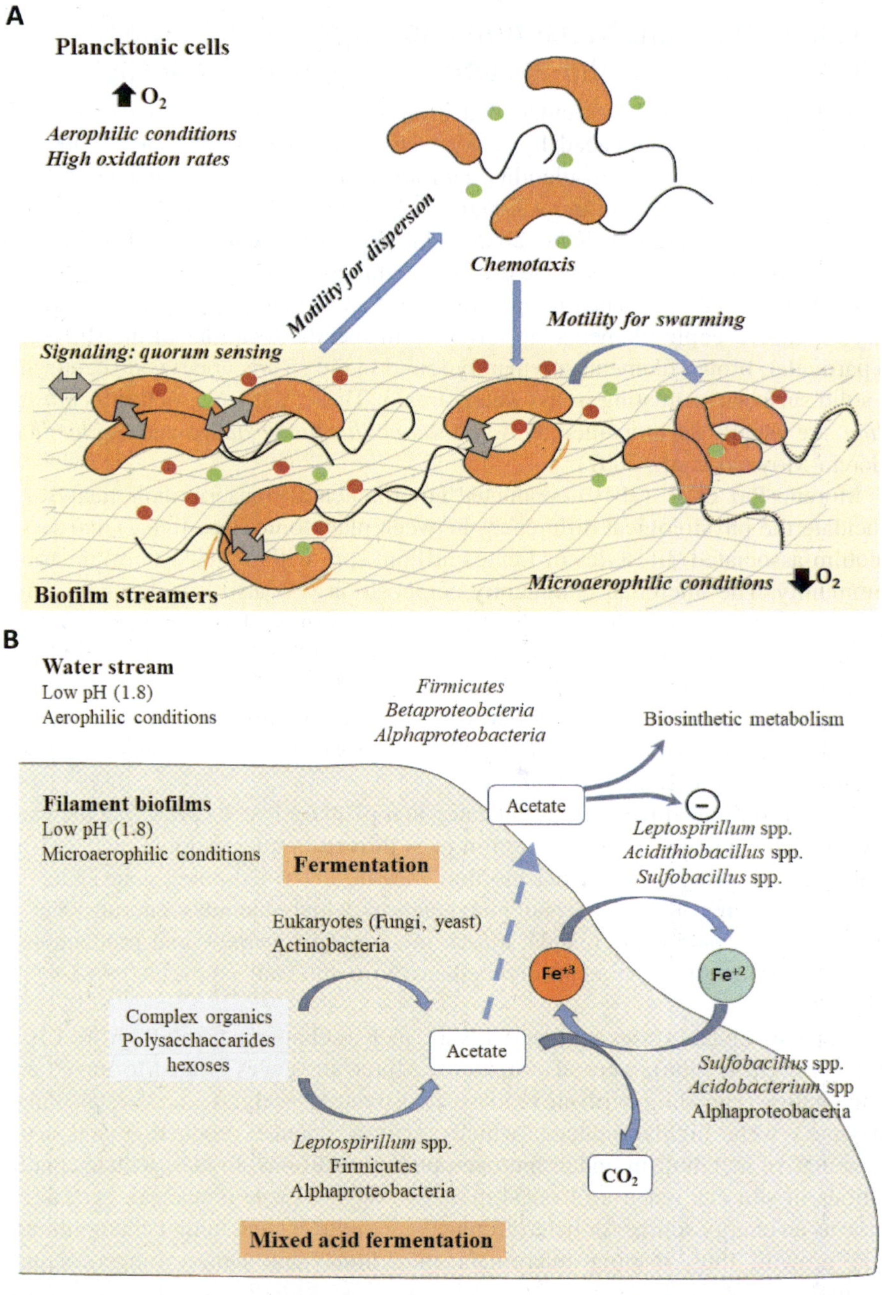

FIGURE 18.3 Schematic representation of the main functions related to biofilm formation, dispersion, and maintenance detected by environmental transcriptomic studies and the role of acetate in the extremely acidophilic microbial community.

(A) Two different metabolic states can be inferred: (1) within the biofilm, cells activate the fermentative pathways leading to net acetate production under microaerophilic conditions; and (2) free-living cells under aerophilic conditions exhibit a highly active

like trehalose (treS), which can act as an osmoprotective agent in cells. Some genes associated with transport were also upregulated, particularly those related to phosphate (*pstSCAB*) and potassium (*kdpBAC*), possibly involved in osmoregulatory or stress-adaptive responses. Finally, tricarboxylic acid (TCA) cycle genes, such as NAD^+-dependent isocitrate dehydrogenase (idh), are indicative of active oxidative metabolism. Therefore, differences in the expression of genes related to nutrient transport, metal homeostasis, and pH regulation reflect a dynamic and specific adaptation to microenvironments within the river.

The study reveals pronounced specific lifestyles that depend on physiological traits. Biofilm streamers are metabolically adapted to low oxygen and nutrient gradients, upregulating genes involved in cell structural maintenance, communication via quorum sensing, stress response, or the activation of fermentative pathways, producing small molecules like acetate (Fig. 18.3B). This metabolite, in turn, serves as a source of carbon to heterotrophic iron reducers (e.g., Sulfobacillus, Acidobacterium). This response indicates a greater ability to withstand adverse environmental conditions (low pH, heavy metals, and low oxygen availability). On the contrary, free-living cells (planktonic cells) maintain a classical chemolithoautotrophic lifestyle, that is, an aerobic metabolism for energy production and nutrient uptake fueled by the TCA cycle and iron oxidation, and prepare for stress conditions through robust transporter systems. This duality highlights the ecological plasticity of Leptospirillum and its central role in acidic metal-rich environments.

Adaptation to high metal concentrations

Nickel (Ni) tolerance and resistance in the *Acidiphilium* genus

Microorganisms inhabiting acidic environments are able to withstand unusually high metal concentrations, conditions that are unthinkable for neutrophilic species.

tricarboxylic acid (TCA) cycle for energetic metabolism. Dashed arrows indicate quorum-sensing activity, while dots represent different metabolites (e.g., acetate, formate). (B) The microorganisms identified through PAM analysis are located as a function of their key metabolic role. Sugars derived from the biofilm (indicated by the shaded and hatched area), along with other complex organic compounds, are degraded by heterotrophic organisms such as fungi and actinobacteria. Secondary fermenters could render acetate that can be used by iron reducers to produce Fe(II), subsequently serving as an energy source for iron oxidizers. Accumulated acetate may inhibit the growth of chemolithotrophs like *Leptospirillum* spp. and *Acidithiobacillus* spp. (−), while supporting the growth of *Firmicutes*, *Sulfobacillus* spp., and members of the Alpha- and Betaproteobacteria.

Although most of the heavy metals at low concentrations in the water are critical in many cellular functions, their lack can lead to dysfunction and even affect their survival. Conversely, at high concentrations, metals become toxic. Extremely acidophilic microorganisms have developed specialized mechanisms to cope with heavy metal-rich environments. A primary strategy involves the active extrusion of metals and metalloids from the cell, facilitated by genetically encoded efflux pump systems and typically activated in response to elevated concentrations of toxic metals. These systems play a crucial role in detoxification and survival under metal stress conditions (Aliyu et al., 2024).

The Río Tinto is characterized by its high metal content, ranging from micromolar to millimolar concentrations varying depending on the location (García-Moyano et al., 2012). The acidity of its water enhances their bioavailability and, consequently, their toxicity, as most metals typically occur as soluble cations readily accessible to microorganisms. In fact, the ability of microorganisms to extract metals during bioleaching processes partly depends on its capacity to withstand the heavy metal concentrations present in the environment.

Acidiphilium spp. are significant members in the iron and sulfur biogeochemical cycles and are regarded as some of the most metal-resistant microorganisms known to thrive in acidic environments. These bacteria are abundant throughout the entire course of the river, preferentially inhabiting the deep water column, sediments, and forming biofilms (García-Moyano et al., 2012).

Particularly, *Acidiphilium* sp. strain PM (DSM 24941), isolated from the Río Tinto waters, exhibits exceptional resistance to multiple metals, including aluminum (Al), nickel (Ni), and zinc (Zn). The strain shows moderate resistance to nickel (Ni, up to 10 mM) and other metals after prolonged incubation (>1 month). While extended incubations in high metal concentrations lead to cell aggregation, a small fraction bypasses or compensates for Ni toxicity, growing at concentrations approaching 1 M. Therefore, this ecological success in metal-contaminated habitats highlights their relevance as model organisms for studying acidophilic metal resistance mechanisms (Table 18.1).

To decipher the genetic basis of this phenotype, San Martín-Úriz et al. (2014) developed a shotgun library of *Acidiphilium* sp. and conducted a functional screening in *Escherichia coli*. The study revealed the presence of several open reading frames (ORFs) conferring Ni resistance (Table 18.1), also reported in other stress-tolerant bacteria (Guazzaroni et al., 2013). The most notable finding identified by metagenomics was the operon encoding the ATP-dependent protease complex HslVU (ClpQY), which is essential under oxidative stress conditions (Rohrwild et al., 1996). Its expression in *E. coli* confers enhanced resistance not only to Ni but also to cobalt (Co), which implies a role in broader metal stress responses. Phylogenetic analyses of this operon indicate vertical inheritance within *Acidiphilium*, supporting the hypothesis that key stress-related operons have been conserved and functionally adapted over evolutionary time in these extreme environments. This phenotype probably indicates that Ni resistance in *Acidiphilium* sp. PM is mediated

Table 18.1 Main proteins related to nickel (Ni) tolerance and resistance in *Acidiphilium* sp. PM identified by metagenomics (Martín-Úriz et al., 2014).

Protein	Function	Proposed Ni resistance effect	References
HslVU (ClpQY)	ATP-dependent protease complex is essential under oxidative stress	Enhances resistance to Ni and Co through involvement in stress response	Rohrwild et al. (1996)
Glycosyltransferase	Lipopolysaccharide (LPS) biosynthesis	Increased cellular Ni content, likely via outer membrane sequestration	Raetz and Whitfield (2002)
Nonspecific protein tyrosine kinase	Exopolysaccharide biosynthesis and membrane remodeling	Ni resistance through structural modifications	Whitfield (2006)
Polysaccharide export protein	Exopolysaccharide export	Extracellular barrier formation	Whitfield (2006)
Malonyl-CoA:ACP transacylase	Fatty acid biosynthesis	Membrane modification under metal stress	Whitfield (2006)
Dihydroxy-acid dehydratase	Branched-chain amino acid biosynthesis	Reduce damage to Fe–S cluster-containing enzymes under Ni stress	MacOmber and Hausinger (2011)

by different mechanisms, such as proteolytic degradation, cell envelope remodeling, and metabolic resilience.

The overall response of *Acidiphilium* sp. PM to Ni was finally deciphered by coupling whole-genome transcriptomics and proteomics methods (Martín-Úriz et al., 2014). RNA transcripts were analyzed following exposure to low Ni levels (10 mM) to decipher the Ni resistance mechanisms and evaluate Ni toxicity during the early stages of growth.

Exposure to nickel resulted in the downregulation of genes essential in DNA replication, such as *dnaA*, DNA gyrase, and DNA ligase, while *sirA*, a replication inhibitor that binds to DnaA, was up-regulated. The synthesis of DNA and RNA precursors was also affected, accompanied by the upregulation of enzymes involved in their catabolism. The reduction in DNA replication was also accompanied by a significant reduction in RNA synthesis, as evidenced by the downregulation of RNA polymerase subunit genes *rpoA*, *rpoB*, and *rpoC*. (genes encoding RNA polymerase subunits) resulting in a marked reduction of cellular transcription levels, which provoked delayed cell division. Protein synthesis was also severely reduced, not only due to an inefficient ribosome assembly and maturation (gene *rimM*) but also by the reduction of aminoacyl-tRNAs, which triggered a reduction of the transcription and translation levels in the cell elongation of the polypeptide chain during translation (EF-Tu, EF-Ts genes). Furthermore, transcriptomic analysis showed a downregulation of genes associated with peptidoglycan biosynthesis, which seems to indicate that cell wall assembly is suppressed as part of the growth inhibition mechanism. These results indicate a broad cellular response that prioritizes survival over proliferation when exposed to metal-induced stress.

Furthermore, changes in energy metabolism are mainly related to the expression of glycolysis and the tricarboxylic acid cycle (TCA), which seems to be repressed in the presence of Ni in favor of enzymes from the glyoxylate cycle, such as isocitrate lyase. Additionally, *Acidiphilium* sp. PM harbors two operons encoding distinct variants of respiratory chain Complex I, which are differentially repressed in response to Ni, suggesting a preferential use of these complexes depending on environmental conditions.

In contrast to the general repression of main cellular processes, genes involved in cation homeostasis, stress response, and the 2-methylcitrate cycle were overexpressed in the early stages of growth. Ni also induces a set of genes encoding metal homeostasis, such as regulators and transporters, like a family of metal-binding transcriptional regulators (MerR), and an arsenite-oxidizing operon (*arsR*), which is involved in metal ion detoxification, efflux, and sequestration of different metals. Similar results are obtained after exposing *Acidiphilium* sp. PM to Zn (Martín-Úriz, 2015).

Although Ni generates lower oxidative stress compared to Fe(III) and Cu(II), *Acidiphilium* mitigates this stress by producing rubrerythrin, a protein involved in oxidative stress protection found in other acidophile archaea (Neira et al., 2022). Furthermore, Ni exposure triggered significant upregulation of a plasmid-encoded toxin-antitoxin (TA) system, *mazEF*. In this system, the antitoxin MazE is degraded by proteases, while the toxin MazF is active and can cause potentially growth arrest or death. This response further supports the induction of a dormancy-like state under metal stress.

Taken together, these data indicate that Ni addition triggers a rapid transcriptomic response, leading to cell growth arrest by repressing several cellular processes: DNA replication, transcription, protein synthesis, and ATP production. These changes, likely mediated by the *mazEF* toxin-antitoxin system, ultimately result in a halt in cell growth. Furthermore, some enzymes involved in the tricarboxylic acid cycle, the 2-methylcitrate pathway, or in the degradation of itaconic acid may be the primary targets of Ni-induced toxicity.

Interestingly, *Acidiphilium* PM isolate from the Río Tinto has demonstrated the remarkable ability to couple glucose oxidation with electron transfer to graphite electrodes under aerobic conditions, an uncommon capability with promising biotechnological applications (Malki et al., 2008).

Eukaryotic adaptive mechanisms to metal stress

The eukaryotic community thriving in the Río Tinto dominates this ecological niche, surpassing prokaryotic diversity in terms of biomass. The most frequent species are members of the phylum Chlorophyta, such as *Chlamydomonas*, *Chlorella*, and *Euglena*, followed by *Klebsormidium* and *Zygnemopsis* (Aguilera et al., 2007; Aguilera, 2013). Indeed, the most acidic part of the river is inhabited by a eukaryotic community dominated by two species related to the genera *Dunaliella* and *Cyanidium* (Rhodophyta), both well known for their tolerance to heavy metals and acidity. Although the adaptive mechanisms of these microorganisms to survive such conditions are still poorly understood, initial transcriptomic studies are shedding light on key survival strategies across different green algae from different habitats.

Like prokaryotes, these organisms overcome high concentrations of metals by combining active (intracellular) and passive (extracellular) transport systems. Passively, acidophilic eukaryotes are commonly found forming biofilms and producing extracellular polymeric substances, which can sequester metals and significantly reduce the amount that enters the cells (Souza-Egipsy et al., 2011). Actively, the most frequently triggered intracellular defense mechanism in response to metal entry is the production of phytochelatins (Díaz et al., 2020) and metallothioneins, small peptides capable of covalently binding metals and transporting them into vacuoles, which are subsequently exocytosed (Zhu et al., 2024), thereby neutralizing metal toxicity. Although natural populations are typically exposed to multiple cations simultaneously, our current knowledge on heavy metals and detoxification mechanisms is primarily based on studies exposing isolates to a single stressor.

The genus *Dunaliella* is known for its exceptional tolerance to extreme conditions and is therefore an attractive model for studying the adaptive mechanisms that enable algae to survive and thrive under the acidic and polyextreme conditions found in the Río Tinto ecosystem (González-Toril and Aguilera, 2019). Within the genus, *D. salina* and *D. acidophila* are among the most described species to date. In particular, *D. acidophila* has been associated with the lowest pH and highest heavy metal concentrations in the Río Tinto (Aguilera et al., 2006). To better understand microbial adaptation to toxic metals, the stress responses of *D. acidophila* and *C. acidophila* strains have been investigated under in vitro conditions using comparative transcriptomic analyses that simulate environmentally relevant metal concentrations (Puente-Sánchez et al., 2016a, Puente-Sánchez et al., 2018).

Dunaliella cells were cultured both in natural high-metal water from the Río Tinto and in an artificial medium supplemented with equivalent cadmium (Cd) concentrations, revealing significant transcriptomic differences compared to a control without Cd exposure. Even in the absence of heavy metals in the medium, the cells continue to produce numerous transcripts associated with general and oxidative stress, suggesting the existence of a constitutive stress response in this species, even after prolonged cultivation.

These transcriptomic analyses suggest significantly higher basal, metabolic, and energetic activity when the cells were exposed to natural metal-rich waters compared to the controls, or when they were exposed to a single metal. This difference is likely due to the fact that the river water has a more complex metal composition and contains natural chelators and potentially other growth-promoting compounds. The transcription of genes related to protein catabolism, amino acid degradation and the transport of nitrogenous compounds was also increased in the natural water compared to the other two experimental conditions. The increased rate of protein turnover may help acidophilic algae in preserving the functional integrity of their proteome in such extreme environmental conditions, as heavy metals are known to disrupt protein functionality (Johnson, 2009).

Key detoxification enzymes, such as phytochelatin synthase—an enzyme important for heavy metal detoxification in plants and other organisms—and glutathione S-transferase, and metallothioneins (MTs), showed up to 10 times higher transcript abundances. These enzymes are directly involved in cellular responses to heavy metals and detoxification processes (Hirooka et al., 2017; Olsson et al., 2015), facilitating heavy metal binding and homeostasis (Zhu et al., 2024). The phytochelatin synthase gene (*CaPCS2),* metallothioneins, and transporter proteins of the ACR3 and AcrB types were also observed in *C. acidophila* in response to both cadmium and arsenic (As) exposure, suggesting that it is also involved in the detoxification process (Puente-Sánchez et al., 2018; Díaz et al., 2020). However, other neutrophilic green algae (such as *Chlamydomonas reinhardtii*) may lack these proteins or fail to express them with the same intensity or efficiency. Results based on differential gene expression revealed that exposure to cadmium significantly increases the expression of several efflux family proteins and genes involved in oil biosynthesis in *C. acidophila*, suggesting these may be key components in adaptation to acidic environments. Notably, genetic differences between *C. acidophila* and *C. reinhardtii* have been observed, including unique genes and distinctive amino acid usage patterns, which may indicate strong selective pressure in acidic conditions. Furthermore, following exposure to cadmium, there is a strong induction of transposable elements (retrotransposons), suggesting a regulatory role in the adaptation to metal-induced stress.

Metals also induced significant ultrastructural changes, such as increased vacuolization and accumulation of starch and lipid reserves. Transmission electron microscopy observations have shown the presence of electron-dense bodies within intracellular vacuoles in *C. acidophila* and *Cyanidium caldarium*, as well as in other microalgal species exposed to heavy metals (Barrie Johnson and Aguilera, 2019). These dense bodies contain heavy metals, confirming that vacuoles are also a mechanism contributing to metal tolerance by reducing cytoplasmic metal concentrations. Notably, cytoplasmic vacuoles and the extracellular capsule in *C. acidophila* contain Cd and As, along with high phosphorus content, indicating a role for polyphosphate in metal detoxification, probably through mechanisms of metal sequestration, bioaccumulation, and biosorption. Additionally, the alga excretes exopolymers, leading to capsule thickening and enhanced metal sequestration. These findings highlight the multiple strategies employed by *C. acidophila* to tolerate and adapt to high concentrations of toxic metals.

High Cd concentrations can also cause structural and functional alterations in the protein complexes involved in photosynthesis (photosystems PSI and PSII, and F-type H^+/Na^+-transporting ATPases (aptB)), which generally lead to growth retardation. In a recent study, researchers identified that more than half of the differentially expressed genes are downregulated in the presence of Cd in a *D. salina* strain, suggesting the existence of an inhibition effect in the photosynthetic machinery of this green algae (Zhu et al., 2024). Naturally resistant mutants to cadmium in this strain upregulate different genes involved in photosynthesis, carbon fixation, and cell wall remodeling that are absent in the wild type, which can indicate that photosynthetic proteins can be an important mechanism of resistance to Cd. Enhanced expression of glutathione S-transferase (GST) and superoxide dismutase (SOD) suggests a robust oxidative stress response, while overexpression of ABC transporters contributes to cadmium efflux and intracellular compartmentalization. Heat shock proteins (HSPs) were differentially regulated, supporting protein protection and stress adaptation. Therefore, cadmium stress triggers a widespread change in gene expression and reflects a metabolic reprogramming, involving oxidative stress pathways, metal transport, and energy metabolism (Puente-Sánchez et al., 2018).

Another element present at concentrations of up to 6 mM in some areas of the Río Tinto is copper (Cu). Transcriptomic studies performed in the microalgae *C. acidophila* at concentrations of 500 μM of copper (Cu) indicate that this species may have developed resistance mechanisms different from nonacidophilic organisms (Olsson et al., 2015). Several transporters, including copper-regulating ABC transporters and others of unknown function, are down-regulated by copper exposure. Despite the high levels of Cu used in the study, the analysis of the transcriptomic libraries revealed no detectable alterations in the expression of transcripts encoding proteins associated with antioxidant defense mechanisms. In contrast to previous findings reported for this microalga under cadmium (Cd) exposure, it is noteworthy that *C. acidophila* did not exhibit significant changes in photosynthetic activity, even when exposed to copper concentrations known to be toxic to other species.

Photosynthetic adaptations in the Rio Tinto environment

The photosynthetic activity in acidic environments is poorly understood, and most research has focused on studying the primary productivity of acidic lakes. Photosynthesis is particularly sensitive to environmental stressors such as high salinity, pH, or toxic metal concentrations. While pH alone does not considerably diminish photosynthetic activity, its combination with heavy metals or low concentrations of soluble phosphate in the water can significantly reduce it (Spijkerman et al., 2007; Tangahu et al., 2014). In fact, photosynthetic inhibition is a key cellular response to metal stress.

Most of the photosynthetic eukaryotes in the Tinto ecosystem are forming biofilms. As in the case of prokaryotes, the diversity of benthic eukaryotic microorganisms in highly acidic aquatic environments ($pH \leq 3.5$) is limited to species that have developed strategies to tolerate elevated H^+ concentrations, dissolved metals, and nutrient scarcity (Brake and Hasiotis, 2010). Consequently, the distribution of these photosynthetic communities seems to be more influenced by the presence of these environmental stressors.

Cid et al. (2010) compared global expression patterns of soluble cellular proteins from a *C. acidophila* strain isolated from the Río Tinto after culturing cells both in artificial media without metals (pH 2) and in natural Río Tinto water. The comparison of two conditions demonstrated that RuBisCo, the key enzyme of the Calvin cycle responsible for fixing CO_2 to organic carbon, was drastically repressed. Similar results have also been reported in its neutrophilic counterpart, *C. reinhardtii*, as well as other photosynthetic organisms when cadmium and copper are present in the medium. These results are closely linked to the high levels of heavy metals in natural acidic waters.

Light acts as another limiting factor of photosynthetic activity in acidic waters, probably due to the high concentration of soluble Fe(III) ions, which cause both qualitative and quantitative changes in its distribution throughout the water column. Unlike Rubisco, phytochromes were among the most overexpressed protein families in *Chlamydomonas* (Cid et al., 2010). In photosynthetic organisms, these proteins work as photoreceptors of red (600–700 nm) and far-red light (700–800 nm), which could explain the induction of these types of receptors in the Río Tinto acidic waters. This high level of phytochromes may represent an adaptive response, enabling photosynthetic microorganisms to optimize the use of the available light spectrum in these environmental conditions. Studies in acidic mining lakes show that only red light reaches the bottom sediments. Similarly, benthic diatom biofilms from acidic lakes have developed a strong adaptation to low light intensities. Despite the scarce light, these microalgae likely maintain photosynthesis through an efficient absorption of red light, the dominant wavelength in the intensely red acidic waters (Souza-Egipsy et al., 2011).

To investigate the underlying mechanisms of photosynthetic adaptation in autotrophic microalgae and autotrophic protists inhabiting the Río Tinto, oximetry and Pulse-Amplitude Modulated fluorometry (hereafter PAM-fluorometry) techniques have been used. The measurement of these parameters determined that maximal photosynthesis values (Pmax) are low and similar to those obtained in other acid mine environments, suggesting a limited availability of nutrients in the water. However, the high biomass of photosynthetic organisms in the Río Tinto—exceeding 65%—does not support such low productivity levels (Spijkerman et al., 2007). Furthermore, ultraviolet (UV) radiation can be extremely harmful to microorganisms by severely affecting DNA synthesis, nutrient uptake, biomass production, biochemical activity, and biodiversity within aquatic ecosystem communities.

Photosynthetic parameters vary significantly across species and sampling locations. Notably, microscopic analyses of *E. mutabilis* biofilms reveal that this phototrophic alga accounts for more than 95% of the total photosynthetic eukaryotic biomass. Metatranscriptomic studies based on RNA-seq and Illumina sequencing were conducted to delve deeper into the molecular mechanisms underlying this adaptive response (Puente-Sánchez et al., 2016b). The study was performed over a diel cycle (12:00, 20:00, and night) in an *E. mutabilis* biofilm to investigate its metabolic capacity for surviving in acidic waters. Although the overexpressed RNA transcripts at all sampling times were associated with several physiological processes, approximately 15% of the identified and annotated genes were involved in photosynthetic activity and general stress responses.

These results revealed extensive transcriptomic reprogramming in natural *E. mutabilis* biofilms during the central hours of the day, the period of highest light intensity and UV radiation (Puente-Sánchez et al., 2016b). At midday, there is a strong photoinhibition as a result of stress induced by excessive solar radiation. At this time, cells cease their photosynthetic activity to protect the photosynthetic apparatus from excess light and UV radiation. However, photosynthetic function is fully restored when solar radiation decreases markedly. When UV radiation was blocked using a filter, no photoinhibition was observed at any time of the day. This observation indicates the presence of a dynamic photoinhibition system in *Euglena* biofilms triggered by UV radiation rather than light intensity, as evidenced by PAM-fluorometryanalysis.

For example, the loss of Photosystem II activity triggers repair responses centered on the D1 protein, the primary target of oxidative damage, namely a rapid degradation of photo-damaged D1 protein by Deg proteases, followed by replacement with newly synthesized functional copies. In *E. mutabilis*, transcriptomic data revealed a strong overexpression of transcripts related to the D1 protein and Deg proteases, indicating their continuous synthesis throughout the diel cycle. This ensures a constant supply of new D1 protein, increasing its turnover rate, minimizing light-induced damage to Photosystem II, and reducing recovery time. Such a rapid response constitutes a photosynthetic protective strategy adapted to highly variable environments, such as those found in acidic waters, and represents a key repair mechanism that is crucial for the survival of photosynthetic organisms under light stress conditions.

To reduce the damage of solar radiation, these euglenophytes present a circadian rhythm with synchronized cell division cycles, where the majority of their photosynthetic activity occurs during early morning hours. During daylight hours, the majority of expressed transcripts are associated with genes involved in Photosystem II stability and repair, DNA repair from UV damage, dissipation of excess energy from the photosystems, and oxidative stress responses, all of which confirm the photoinhibition phenomena detected via PAM-fluorometry at midday. DNA replication and cell division occur primarily throughout the night, when irradiance is substantially lower, thereby reducing the risk of UV-induced mutagenesis.

Furthermore, transcriptomic analyses indicated a general oxidative stress response that upregulates genes associated with protein refolding (chaperones) and degradation (ubiquitin/proteasome system), which play a role in stabilizing proteins and membranes, thereby protecting cells from stress damage.

It is worth highlighting that, although *E. mutabilis* shares many stress responses with other organisms, a large proportion of unknown genes, many of which have not been previously detected, were found throughout the transcriptomic analyses. This underscores that many of the molecular mechanisms underlying environmental stress adaptation in these extremophiles remain undescribed.

Conclusions

The exploration of microbial metabolism in polyextreme environments such as the Río Tinto through metatranscriptomic approaches has revealed remarkable physiological and genetic plasticity among microbial communities capable of thriving under conditions of extreme acidity, high metal concentrations, low oxygen availability, and nutrient scarcity. The application of high-throughput transcriptomic technologies has been instrumental in uncovering the molecular mechanisms that underlie such resilience and adaptability.

Microorganisms like *A. ferrooxidans* and *L. ferrooxidans* exhibit complex strategies for maintaining homeostasis in the face of proton gradients, iron toxicity, and metal stress, including proton extrusion systems, chaperone protein expression, and modular pathways for alternative energy generation. In particular, *Leptospirillum* spp. demonstrate a highly regulated gene expression response to environmental stimuli, such as oxygen and nitrogen availability, and show a strong propensity to form biofilms as structural and metabolic adaptations to harsh conditions.

The presence and environmental regulation of nitrogen fixation pathways, particularly the *nif* operon, underscores the metabolic sophistication of these organisms. *L. ferrooxidans* activates this operon under nitrogen-depleted conditions, though field data suggest a preference for more readily available nitrogen sources when present. Similarly, phosphate limitation induces the expression of high-affinity transport systems such as the pstSCAB operon, indicating tightly controlled nutrient uptake mechanisms, often integrated with quorum sensing and biofilm development.

Transcriptomic analyses have also highlighted distinct physiological and gene expression profiles among planktonic and biofilm-associated lifestyles. Sessile cells display enhanced gene expression linked to biofilm formation, stress responses, and fermentative metabolism, whereas planktonic cells favor chemolithoautotrophic metabolism and exhibit strong stress-response systems for nutrient acquisition and osmotic regulation.

Metal tolerance in *Acidiphilium* spp. and extremophilic microalgae such as *Chlamydomonas acidophila* involves efflux systems, production of metallothioneins and phytochelatins, oxidative stress responses, and cell envelope remodeling. These adaptations allow survival and even proliferation in environments with toxic levels of metals.

Lastly, the photosynthetic adaptations observed in eukaryotic phototrophs, particularly in *Euglena mutabilis*, reveal a dynamic regulatory system responsive to UV stress, with diel transcriptomic shifts that protect and repair the photosynthetic apparatus.

Collectively, these findings offer crucial insights into resistance mechanisms against multiple stressors. This comprehensive understanding is vital not only for developing robust models of microbial life in extreme terrestrial environments, which supports the advancement of innovative biotechnological processes, but also provides valuable perspectives on the evolution of life and the potential for life beyond Earth.

Acknowledgments

This work was supported by the Spain Ministry of Science, Innovation and Universities/State Agency of Research MICIU/AEI/10.13039/501100011033 and by "ERDF A way of making Europe" under project no. PID2021-126746NB-I00.

AI disclosure

During the preparation of this work the author(s) used ChatGPT in order to check and improve English language writing. After using this tool/service, the author(s) reviewed and edited the content as needed and take(s) full responsibility for the content of the publication.

References

Abramov, S.M., Straub, D., Tejada, J., Grimm, L., Schädler, F., Bulaev, A., Thorwarth, H., Amils, R., Kappler, A., Kleindienst, S., 2022. Biogeochemical niches of Fe-cycling communities influencing heavy metal transport along the Rio Tinto, Spain. Appl. Environ. Microbiol. 88 (4). https://doi.org/10.1128/aem.02290-21.

Aguena, M., Yagil, E., Spira, B., 2002. Transcriptional analysis of the pst operon of *Escherichia coli*. Mol. Genet. Genom. 268 (4), 518–524. https://doi.org/10.1007/s00438-002-0764-4.

Aguilera, A., Gómez, F., Lospitao, E., Amils, R., 2006. A molecular approach to the characterization of the eukaryotic communities of an extreme acidic environment: methods for DNA extraction and denaturing gradient gel electrophoresis analysis. Syst. Appl. Microbiol. 29 (7), 593–605. https://doi.org/10.1016/j.syapm.2006.01.006.

Aguilera, A., Zettler, E., Gómez, F., Amaral-Zettler, L., Rodríguez, N., Amils, R., 2007. Distribution and seasonal variability in the benthic eukaryotic community of Río Tinto (SW, Spain), an acidic, high metal extreme environment. Syst. Appl. Microbiol. 30 (7), 531–546. https://doi.org/10.1016/j.syapm.2007.05.003.

Aguilera, A., 2013. Eukaryotic organisms in extreme acidic environments, the Río Tinto case. Life 3 (3), 363–374. https://doi.org/10.3390/life3030363.

Aliyu, G.O., Ezugworie, F.N., Onwosi, C.O., Nnamchi, C.I., Ekwealor, C.C., Igbokwe, V.C., Sani, R.K., 2024. Multi-stress adaptive lifestyle of acidophiles enhances their robustness for biotechnological and environmental applications. Sci. Total Environ. 954. https://doi.org/10.1016/j.scitotenv.2024.176190.

Barrie Johnson, D., Aguilera, A., 2019. Extremophiles and acidic environments. In: Encyclopedia of Microbiology. Elsevier, United Kingdom, pp. 206–227. https://doi.org/10.1016/B978-0-12-809633-8.90687-3.

Brake, S.S., Hasiotis, S.T., 2010. Eukaryote-dominated biofilms and their significance in acidic environments. Geomicrobiol. J. 27 (6–7), 534–558. https://doi.org/10.1080/01490451003702966.

Chen, L.X., Hu, M., Huang, L.N., Hua, Z.S., Kuang, J.L., Li, S.J., Shu, W.S., 2015. Comparative metagenomic and metatranscriptomic analyses of microbial communities in acid mine drainage. ISME J. 9 (7), 1579–1592. https://doi.org/10.1038/ismej.2014.245.

Christel, S., Herold, M., Bellenberg, S., El Hajjami, M., Buetti-Dinh, A., Pivkin, I.V., Sand, W., Wilmes, P., Poetsch, A., Dopson, M., Julia Pettinari, M., 2018. Multi-omics reveals the lifestyle of the acidophilic, mineral-oxidizing model species Leptospirillum ferriphilum T. Appl. Environ. Microbiol. 84 (3). https://doi.org/10.1128/aem.02091-17.

Cid, C., Garcia-Descalzo, L., Casado-Lafuente, V., Amils, R., Aguilera, A., 2010. Proteomic analysis of the response of an acidophilic strain of Chlamydomonas sp. (Chlorophyta) to natural metal-rich water. Proteomics 10 (10), 2026–2036. https://doi.org/10.1002/pmic.200900592.

Díaz, S., De Francisco, P., Olsson, S., Aguilera, Á., González-Toril, E., Martín-González, A., 2020. Toxicity, physiological, and ultrastructural effects of arsenic and cadmium on the extremophilic microalga Chlamydomonas acidophila. Int. J. Environ. Res. Publ. Health 17 (5), 1650. https://doi.org/10.3390/ijerph17051650.

Feng, S., Yang, H., Wang, W., 2015. System-level understanding of the potential acid-tolerance components of Acidithiobacillus thiooxidans ZJJN-3 under extreme acid stress. Extremophiles 19 (5), 1029–1039. https://doi.org/10.1007/s00792-015-0780-z.

García-Moyano, A., González-Toril, E., Aguilera, A., Amils, R., 2012. Comparative microbial ecology study of the sediments and the water column of the Río Tinto, an extreme acidic environment. FEMS (Fed. Eur. Microbiol. Soc.) Microbiol. Ecol. 81 (2), 303–314. https://doi.org/10.1111/j.1574-6941.2012.01346.x.

García-Moyano, A., González-Toril, E., Moreno-Paz, M., Parro, V., Amils, R., 2008. Evaluation of Leptospirillum spp. in the Río Tinto, a model of interest to biohydrometallurgy. Hydrometallurgy 94 (1–4), 155–161. https://doi.org/10.1016/j.hydromet.2008.05.046.

García-Moyano, A., González-Toril, E., Aguilera, A., Amils, R., 2007. Prokaryotic community composition and ecology of floating macroscopic filaments from an extreme acidic environment, Río Tinto (SW, Spain). Syst. Appl. Microbiol. 30 (8), 601–614. https://doi.org/10.1016/j.syapm.2007.08.002.

Garrido, P., Toril, E.G., Moyano, A.G., Moreno-Paz, M., Amils, R., Parro, V., 2008. An oligonucleotide prokaryotic acidophile microarray: its validation and its use to monitor seasonal variations in extreme acidic environments with total environmental RNA. Environ. Microbiol. 10 (4), 836–850. https://doi.org/10.1111/j.1462-2920.2008.01477.x.

González-Toril, E., Llobet-Brossa, E., Casamayor, E.O., Amann, R., Amils, R., 2003. Microbial ecology of an extreme acidic environment, the Tinto river. Appl. Environ. Microbiol. 69 (8), 4853–4865. https://doi.org/10.1128/aem.69.8.4853-4865.2003.

González-Toril, E., Aguilera, Á., 2019. Microbial ecology in extreme acidic environments. In: Microbial Diversity in the Genomic Era. Elsevier, Spain, pp. 227–238. https://doi.org/10.1016/B978-0-12-814849-5.00014-9.

Grzmil, B., Wronkowski, J., 2006. Removal of phosphates and fluorides from industrial wastewater. Desalination 189 (1–3), 261–268. https://doi.org/10.1016/j.desal.2005.07.008.

Guazzaroni, M.E., Morgante, V., Mirete, S., González-Pastor, J.E., 2013. Novel acid resistance genes from the metagenome of the Tinto River, an extremely acidic environment. Environ. Microbiol. 15 (4), 1088–1102. https://doi.org/10.1111/1462-2920.12021.

Hirooka, S., Hirose, Y., Kanesaki, Y., Higuchi, S., Fujiwara, T., Onuma, R., Era, A., Ohbayashi, R., Uzuka, A., Nozaki, H., Yoshikawa, H., Miyagishima, S.Y., 2017. Acidophilic green algal genome provides insights into adaptation to an acidic environment. Proc. Natl. Acad. Sci. U. S. A 114 (39), E8304. https://doi.org/10.1073/pnas.1707072114.

Holden, P., 1991. Identification of structural nif genes in an iron-oxidising moderate thermophile. FEMS (Fed. Eur. Microbiol. Soc.) Microbiol. Lett. 83 (1), 7–10. https://doi.org/10.1016/0378-1097(91)90434-c.

Ibáñez, A., Barreiro, C., Diez-Galán, A., Cobos, R., Calvo-Peña, C., Coque, J.J.R., 2023. Molecular identification and acid stress response of an Acidithiobacillus thiooxidans strain isolated from Rio Tinto (Spain). Int. J. Mol. Sci. 24 (17), 13391. https://doi.org/10.3390/ijms241713391.

Johnson, D.B., 2009. Extremophiles: Acidic Environments Encyclopedia of Microbiology, third ed. Elsevier, United Kingdom, pp. 107–126. https://doi.org/10.1016/B978-012373944-5.00278-9.

Kilic, T., 2025. Factors Affecting Biofilm Formation and the Effects of These Factors on Bacteria. IntechOpen, IntechOpen. https://doi.org/10.5772/intechopen.1008877.

Levicán, G., Ugalde, J.A., Ehrenfeld, N., Maass, A., Parada, P., 2008. Comparative genomic analysis of carbon and nitrogen assimilation mechanisms in three indigenous bioleaching bacteria: predictions and validations. BMC Genom. 9. https://doi.org/10.1186/1471-2164-9-581.

Levicán, G., Gómez, M.J., Chávez, R., Orellana, O., Moreno-Paz, M., Parro, V., 2012. Comparative genomic analysis reveals novel facts about ***Leptospirillum*** spp. cytochromes. Microb. Physiol. 22 (2), 94–104. https://doi.org/10.1159/000338105.

Li, L., Liu, Z., Zhang, M., Meng, D., Liu, X., Wang, P., Li, X., Jiang, Z., Zhong, S., Jiang, C., Yin, H., Summers, Z.M., 2020. Insights into the metabolism and evolution of the genus acidiphilium , a typical acidophile in acid mine drainage. mSystems 5 (6). https://doi.org/10.1128/msystems.00867-20.

Liu, H. Wei, Xu, L.Feng, Guo, X., Jiang, H. Dan, Liu, X. Duan, Liang, Y. Li, Yin, H. Qun, Liu, Y. Zi, 2020. Comparative genome analysis on intraspecific evolution and nitrogen fixation of Leptospirillum ferriphilum isolates. Trans. Nonferrous Met. Soc. China 30 (6), 1635–1646. https://doi.org/10.1016/s1003-6326(20)65326-2.

Long, J., Zaborina, O., Holbrook, C., Zaborin, A., Alverdy, J., 2008. Depletion of intestinal phosphate after operative injury activates the virulence of P aeruginosa causing lethal gut-derived sepsis. Surgery 144 (2), 189–197. https://doi.org/10.1016/j.surg.2008.03.045.

Mackintosh, M.E., 1978. Nitrogen fixation by Thiobacillus ferrooxidans. J. Gen. Microbiol. 105 (2), 215–218. https://doi.org/10.1099/00221287-105-2-215.

MacOmber, L., Hausinger, R.P., 2011. Mechanisms of nickel toxicity in microorganisms. Metallomics 3 (11), 1153–1162. https://doi.org/10.1039/c1mt00063b.

Malki, M., De Lacey, A.L., Rodríguez, N., Amils, R., Fernandez, V.M., 2008. Preferential use of an anode as an electron acceptor by an acidophilic bacterium in the presence of oxygen. Appl. Environ. Microbiol. 74 (14), 4472–4476. https://doi.org/10.1128/AEM.00209-08.

Martin-Uriz, P.S., Gomez, M.J., Arcas, A., Bargiela, R., Amils, R., 2011. Draft genome sequence of the electricigen Acidiphilium sp. strain PM (DSM 24941). J. Bacteriol. 193 (19), 5585–5586. https://doi.org/10.1128/JB.05386-11Spain.

Martín-Uriz, S., 2015. Isolation and Characterization of Acidiphilium Sp. PM, a Nickel-Resistant Electricigen from Río Tinto.

Martín-Úriz, P.S., Mirete, S., Alcolea, P., Gómez, M., Amils, R., González-Pastor, J.E., 2014. Nickel-resistance determinants in *Acidiphilium* sp. PM identified by genome-wide functional screening. Plos One 9 (4), e95041. https://doi.org/10.1371/journal.pone.0095041.

Mirete, S., Morgante, V., González-Pastor, J.E., 2017. Acidophiles: Diversity and Mechanisms of Adaptation to Acidic Environments Adaption of Microbial Life to Environmental Extremes: Novel Research Results and Application, second ed. Springer International Publishing, Spain, pp. 227–251. https://doi.org/10.1007/978-3-319-48327-6_9.

Moreno-Paz, M., Gómez, M.J., Arcas, A., Parro, V., 2010. Environmental transcriptome analysis reveals physiological differences between biofilm and planktonic modes of life of the iron oxidizing bacteria Leptospirillum spp. in their natural microbial community. BMC Genom. 11 (1). https://doi.org/10.1186/1471-2164-11-404Spain.

Moreno-Paz, M., Parro, V., 2006. Amplification of low quantity bacterial RNA for microarray studies: time-course analysis of Leptospirillum ferrooxidans under nitrogen-fixing conditions. Environ. Microbiol. 8 (6), 1064–1073. https://doi.org/10.1111/j.1462-2920.2006.00998.x.

Neira, G., Vergara, E., Cortez, D., Holmes, D.S., 2022. A large-scale multiple genome comparison of acidophilic archaea (pH $\leq$ 5.0) extends our understanding of oxidative stress responses in polyextreme environments. Antioxidants 11 (1), 59. https://doi.org/10.3390/antiox11010059.

Olsson, S., Puente-Sánchez, F., Gómez, M.J., Aguilera, A., 2015. Transcriptional response to copper excess and identification of genes involved in heavy metal tolerance in the extremophilic microalga Chlamydomonas acidophila. Extremophiles 19 (3), 657–672. https://doi.org/10.1007/s00792-015-0746-1.

Parro, V., Moreno-Paz, M., 2003. Gene function analysis in environmental isolates: the nif regulon of the strict iron oxidizing bacterium Leptospirillum ferrooxidans. Proc. Natl. Acad. Sci. 100 (13), 7883–7888. https://doi.org/10.1073/pnas.1230487100.

Parro, V., Moreno-Paz, M., González-Toril, E., 2007. Analysis of environmental transcriptomes by DNA microarrays. Environ. Microbiol. 9 (2), 453–464. https://doi.org/10.1111/j.1462-2920.2006.01162.x.

Parro, V., Moreno-Paz, M., 2004. Nitrogen fixation in acidophile iron-oxidizing bacteria: the nif regulon of Leptospirillum ferrooxidans. Res. Microbiol. 155 (9), 703–709. https://doi.org/10.1016/j.resmic.2004.05.010.

Puente-Sánchez, F., Olsson, S., Aguilera, A., 2016b. Comparative transcriptomic analysis of the response of Dunaliella acidophila (Chlorophyta) to short-term cadmium and chronic natural metal-rich water exposures. Microb. Ecol. 72 (3), 595–607. https://doi.org/10.1007/s00248-016-0824-7.

Puente-Sánchez, F., Olsson, S., Gómez-Rodriguez, M., Souza-Egipsy, V., Altamirano-Jeschke, M., Amils, R., Parro, V., Aguilera, A., 2016a. Solar radiation stress in natural acidophilic biofilms of Euglena mutabilis revealed by metatranscriptomics and PAM fluorometry. Protist 167 (1), 67–81. https://doi.org/10.1016/j.protis.2015.12.003.

Puente-Sánchez, F., Díaz, S., Penacho, V., Aguilera, A., Olsson, S., 2018. Basis of genetic adaptation to heavy metal stress in the acidophilic green alga Chlamydomonas acidophila. Aquat. Toxicol. 200, 62–72. https://doi.org/10.1016/j.aquatox.2018.04.020.

Quatrini, R., Jedlicki, E., Holmes, D.S., 2005. Genomic insights into the iron uptake mechanisms of the biomining microorganism Acidithiobacillus ferrooxidans. J. Ind. Microbiol. Biotechnol. 32 (11–12), 606–614. https://doi.org/10.1007/s10295-005-0233-2.

Quatrini, R., Appia-Ayme, C., Denis, Y., Ratouchniak, J., Veloso, F., Valdes, J., Lefimil, C., Silver, S., Roberto, F., Omar, O., Denizot, F., Jedlicki, E., Holmes, D., Bonnefoy, V., 2006. Insights into the iron and sulfur energetic metabolism of Acidithiobacillus ferrooxidans by microarray transcriptome profiling. Hydrometallurgy 83 (1–4), 263–272. https://doi.org/10.1016/j.hydromet.2006.03.030.

Raetz, C.R.H., Whitfield, C., 2002. Lipopolysaccharide endotoxins. Annu. Rev. Biochem. 71 (1), 635–700. https://doi.org/10.1146/annurev.biochem.71.110601.135414.

Ram, R.J., VerBerkmoes, N.C., Thelen, M.P., Tyson, G.W., Baker, B.J., Blake, R.C., Shah, M., Hettich, R.L., Banfield, J.F., 2005. Community proteomics of a natural microbial biofilm. Science 308 (5730), 1915–1920. https://doi.org/10.1126/science.1109070.

Remonsellez, F., Moreno-Paz, M., Parro, V., Demergasso, C., 2009a. Differential gene expression of industrial bioleaching transcriptomes by Leptospirillum ferrooxidans DNA microarray. Adv. Mater. Res. 71–73, 227–230. https://doi.org/10.4028/www.scientific.net/amr.71-73.227.

Remonsellez, F., Galleguillos, F., Moreno-Paz, M., Parro, V., Acosta, M., Demergasso, C., 2009b. Dynamic of active microorganisms inhabiting a bioleaching industrial heap of low-grade copper sulfide ore monitored by real-time PCR and oligonucleotide prokaryotic acidophile microarray. Microb. Biotechnol. 2 (6), 613–624. https://doi.org/10.1111/j.1751-7915.2009.00112.x.

Rohrwild, M., Coux, O., Huang, H.C., Moerschell, R.P., Yoo, S.J., Seol, J.H., Chung, C.H., Goldberg, A.L., 1996. HslV-HslU: a novel ATP-dependent protease complex in *Escherichia coli* related to the eukaryotic proteasome. Proc. Natl. Acad. Sci. U. S. A 93 (12), 5808–5813. https://doi.org/10.1073/pnas.93.12.5808.

Sánchez-Andrea, I., Rodríguez, N., Amils, R., Sanz, J.L., 2011. Microbial diversity in anaerobic sediments at Río Tinto, a naturally acidic environment with a high heavy metal content. Appl. Environ. Microbiol. 77 (17), 6085–6093. https://doi.org/10.1128/aem.00654-11.

Souza-Egipsy, V., Altamirano, M., Amils, R., Aguilera, A., 2011. Photosynthetic performance of phototrophic biofilms in extreme acidic environments. Environ. Microbiol. 13 (8), 2351–2358. https://doi.org/10.1111/j.1462-2920.2011.02506.x.

Spijkerman, E., Barua, D., Gerloff-Elias, A., Kern, J., Gaedke, U., Heckathorn, S.A., 2007. Stress responses and metal tolerance of Chlamydomonas acidophila in metal-enriched lake water and artificial medium. Extremophiles 11 (4), 551–562. https://doi.org/10.1007/s00792-007-0067-0.

Spijkerman, E., 2007. Phosphorus acquisition by Chlamydomonas acidophila under autotrophic and osmo-mixotrophic growth conditions. J. Exp. Bot. 58 (15–16), 4195–4202. https://doi.org/10.1093/jxb/erm276.

Stevens, C.J., Dugan, P.R., Tuovinen, O.H., 1986. Acetylene reduction (nitrogen fixation) by Thiobacillus ferrooxidans. Biotechnol. Appl. Biochem. 8, 351–359.

Tangahu, B.V., Saptarini, D., Warmadewanthi, I.D.A.A., Pudjiastuti, L., Agus Mardyanto Tardan, M., Luqman, A., 2014. Ferronickel slag toxicity tests on Chlorella vulgaris and Artemia sp. Plant Sci. Today 1 (3), 155–160. https://doi.org/10.14719/pst.2014.1.3.29.

Tyson, G.W., Lo, I., Baker, B.J., Allen, E.E., Hugenholtz, P., Banfield, J.F., 2005. Genome-directed isolation of the key nitrogen fixer Leptospirillum ferrodiazotrophum sp. nov. from an acidophilic microbial community. Appl. Environ. Microbiol. 71 (10), 6319–6324. https://doi.org/10.1128/AEM.71.10.6319-6324.2005.

Valdés, J., Pedroso, I., Quatrini, R., Dodson, R.J., Tettelin, H., Blake, R., Eisen, J.A., Holmes, D.S., 2008. Acidithiobacillus ferrooxidans metabolism: from genome sequence to industrial applications. BMC Genom. 9. https://doi.org/10.1186/1471-2164-9-597.

Vera, M., Guiliani, N., Jerez, C.A., 2003. Proteomic and genomic analysis of the phosphate starvation response of Acidithiobacillus ferrooxidans. Hydrometallurgy 71 (1–2), 125–132. https://doi.org/10.1016/s0304-386x(03)00148-8.

Whitfield, C., 2006. Biosynthesis and assembly of capsular polysaccharides in *Escherichia coli*. Annu. Rev. Biochem. 75 (1), 39–68. https://doi.org/10.1146/annurev.biochem.75.103004.142545.

Zhu, Q., Wu, M., Xiao, J., Gao, L., Song, R., Yang, Z., Xia, L., Zheng, J., Liu, J., 2024. Comparative transcriptome combined with physiological analyses revealed key genes and pathways for cadmium tolerance in wild-type and mutant microalgae Dunaliella salina. Algal Res. 79, 103475. https://doi.org/10.1016/j.algal.2024.103475.

CHAPTER

Biotechnological applications of the Río Tinto basin biodiversity

19

Irma Marín, Carlos Pernas-Pleite, José P. Abad
Department of Molecular Biology, Universidad Autónoma de Madrid, Madrid, Spain

Introduction

Extremophilic microorganisms

Extremophilic microorganisms are those that have adapted to extreme environmental conditions and include thermophiles, psychrophiles, acidophiles, alkalophiles, halophiles, piezophiles, metallophiles, xerophiles, radiophiles, barophiles, endolitics, and crytoendolitics (Wang et al., 2024) (Fig. 19.1). Polyextremophiles are those organisms that can thrive in more than one type of extreme environment (Gupta et al., 2014).

Since these microorganisms adapt to these extreme environments by means of biochemical and physiological changes that produce a wide range of products, such as extremozymes, biosurfactants and extremolytes, among others, they can be used in a number of applications, including sustainable agriculture, food, cosmetics, pharmaceuticals and textile industries (Rawat et al., 2024; Kochhar et al., 2022). Some extremophiles are also applied to biofuel or biopolymer production, to copper, gold and other metal extraction in industrial biomining, or in bioremediation by the use of different approaches to remove pollutants from contaminated environments. The use of extremophiles offers an eco-friendly alternative to traditional processes. Thus, for instance, in some cases, these microorganisms do not require sterile techniques since the extreme conditions in which they grow impede the presence of other microorganisms. Thus, operational costs are reduced and, in certain industrial processes, are more efficient and have a lower environmental impact. However, one of the drawbacks of using extremophiles is the difficulty of isolating and cultivating new species in the laboratory. This affects the accurate evaluation of the genetic diversity of these microorganisms. In recent years, advances and optimization in molecular biology, synthetic biology, and fermentation technologies have significantly enhanced screening and identification possibilities. This has led to the identification of new genes and metabolic pathways which will allow the development of new biotechnological applications (Rawat et al., 2024; Wang et al., 2024).

Extremophiles have also played an important role in the scope of Astrobiology as a science, since their discovery opened the possibility that life could exist on other planets (Thombre et al., 2020; Rothschild and Mancinelli, 2001). To support this

Geomicrobiology of Río Tinto. https://doi.org/10.1016/B978-0-443-32948-7.00010-2

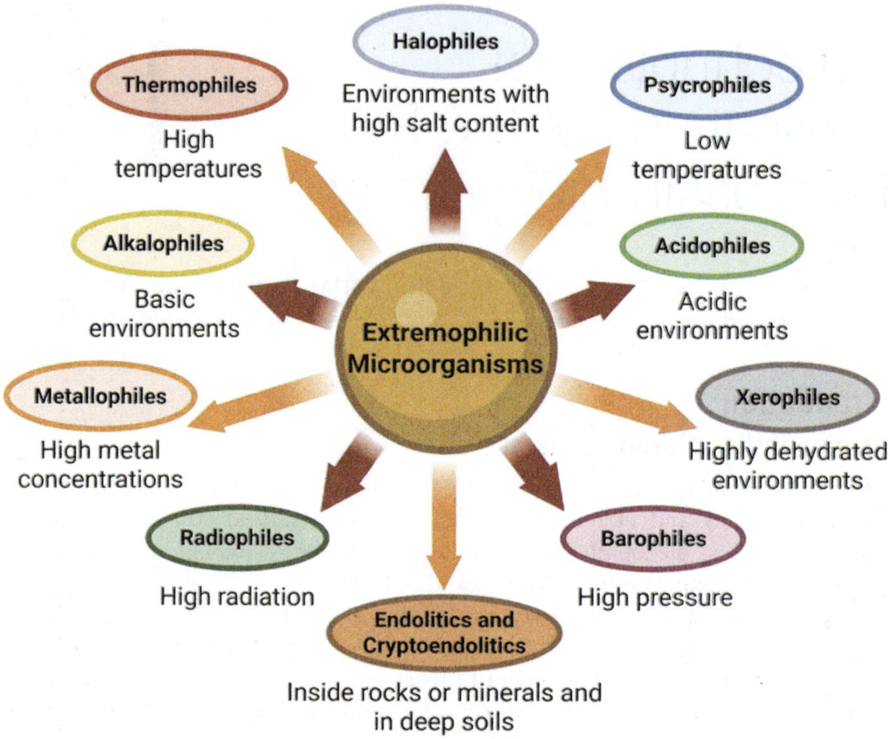

FIGURE 19.1

Types of extremophilic microorganisms.

hypothesis, it is essential to find Earth environments that could serve as analogies of planets, the so-called Planetary Fields. This allows us to study how microorganisms survive and adapt to extreme conditions in order to approximate how they might survive in other planetary environments. In this sense, the Río Tinto has been considered a Mars' analogue.

Acidophilic microorganisms are extremophiles that have developed strategies for growth in harsh, low pH conditions. They produce small organic molecules with unique biological activities and enzymes adapted for efficient functioning under extremely acidic conditions (Tapadar et al., 2021). They are also among the best candidates for detoxification of heavy-metal pollutants through bio-oxidation (bioleaching) (Jeong and Choi, 2020), and other biotechnological applications.

The Río Tinto ecosystem and its microbial diversity

The Río Tinto is an extreme environment because of the low pH ($pH < 3$ and $Eh > 400$ mV) along its almost 100 Km length and its high heavy-metal concentration (Amils et al., 2011; Fernández-Remolar et al., 2005). Early studies assumed that these conditions were associated with the mining activity conducted in the region by the Tartessians, Phoenicians, Greeks, Romans, British and Spanish over the last 5000 years (Davis et al., 2000; Leblanc et al., 2000). However, subsequent geological and microbiological works have shown that the high levels of heavy metals, ferric ions, and low pH are a consequence of the evolution of its microbial ecosystem and thus, that the microbiota of this environment is responsible for the characteristic extreme features of the Río Tinto. Moreno et al. (2003) carried out a series of geological studies related to the appearance of the oxidation processes that gave rise to the generation of the gossan found in the Río Tinto deposits. The conclusion was that this occurred during the Miocene age, long before the mining in the region began. It is now well established that the extreme conditions of the Río Tinto are the result of the metabolism of chemolithotrophic microorganisms, which thrive in the high concentrations of metal sulfides found in the Iberian Pyritic Belt (Gómez-Ortiz et al., 2014).

Since the beginning of the microbial ecology study of the river in the 1980s to the present day, microorganisms, including bacteria, archaea, and eukaryotes, have been isolated, which led to a detailed understanding of the river's physicochemical characteristics and to an ongoing evaluation of their possible biotechnological applications (Amils 2016). From the first studies carried out with both classical and molecular techniques, it was concluded that the most abundant bacteria in the water column corresponded to species of the *Acidithiobacillus*, *Leptospirillum,* and *Acidiphilium* genera, all of which are related to the iron cycle (Garcia-Moyano et al., 2012; López-Archilla et al., 2001). The biodiversity of the river is noticeably higher in its sediments, where representatives of the phyla Actinomycetota, Acidobacteriota, Pseudomonadota, Planctomycetota, Bacillota, and Chloroflexota have been detected (Garcia-Moyano et al., 2012; Sanchez-Andrea et al., 2011).

Given the low pH and high heavy metal concentrations in the Río Tinto, it is usually assumed that eukaryotic growth and diversity would be quite limited. However, an unexpectedly wide range of diverse eukaryotic organisms constitutes up to 65% of the total river waters' biomass (Fig. 19.2) (Aguilera et al., 2006). Thus, this river is a unique acidic extreme ecosystem in which members of the Bacillariophyta, Chlorophyta, Euglenozoa and Rotifera phyla, Hypotrichiaciliates, Heliozoa, cercomonads, amoebae, stramenopiles, fungi and yeast have been found (Sabater et al., 2003; Amaral et al., 2002; López-Archilla et al., 2001).

The greatest eukaryote diversity in the Río Tinto corresponds to fungi with an initial 400 isolates described in the earliest study of the rivers' biodiversity (López-Archilla et al., 2001, 2004). An additional set of 359 were isolated and identified later (Oggerin et al., 2024), from different samples and locations along the river. The last isolated strains were classified based on internal transcribed spacer DNA sequences, and found to belong principally to the phylum Ascomycota, with others assigned to Basidiomycota and Mucoromycota.

In recent years, some studies have focused on the river's intertidal zone, where the environmental and geochemical characteristics are modified twice a day according to the tidal cycle. This area is furthest from the river source, from the San Juan

FIGURE 19.2

Microalgae at various Río Tinto locations.

del Puerto village area to La Rábida at the river's estuary (Fig. 19.3) (Eduardo-Correia et al., 2020; Vizioli, 2017; Eduardo-Correia, 2015). Several microalgae belonging to the genera *Coelastrella*, *Parachlorella*, *Tetradesmus*, *Tetratostichococcus*, *Porphyridium*, and one from the family Trebouxyphiaceae have been isolated

FIGURE 19.3

The intertidal and estuary zones of the Río Tinto. ***Map lines delineate study areas and do not necessarily depict accepted national boundaries.***

from this zone (Vizioli, 2017), as well as 549 bacteria belonging to diverse phylogenetic groups, including 48 genera, diversified in more than 90 species (Eduardo-Correia et al., 2020; Eduardo-Correia, 2015).

Viral particles have also been detected in the Río Tinto. They seem to have had a role in iron mineralization at low pH, with viral capsids and tails owing to mineral precipitation (Kyle et al., 2008). These authors also found a correlation, albeit weak, between the virus and prokaryotic abundance, perhaps indicating that the detected viruses were mainly bacteriophages. The search for viral diversity in this environment is still at its beginning, but studies are underway (Alcamí et al., personal communication). Enzymes from these viruses will probably have interesting biotechnological applications.

Río Tinto biodiversity as a source for biotechnological applications

Biotechnological applications of Río Tinto's biodiversity have been explored in various studies, including metal bioleaching, metal extraction by bioaccumulation, bioremediation of metal-contaminated environments, production of compounds of interest (including antimicrobial agents) and enzymes, and synthesis of silver nanoparticles with different biotechnological applications. These applications are listed below, classified by the type of organism used.

Bacteria and archaea

Bioleaching

This technology makes use of chemolithotrophic microorganisms to extract metals from low-grade sulfide ores, solubilizing the metals contained in them by oxidation. It is considered an eco-friendly technology since the microorganisms involved in the mineral oxidation processes are autotrophs that function at lower temperatures and pressures than classical metallurgical processes (Ibañez et al., 2023). The first studies focusing on the bioleaching process using Río Tinto microorganisms dated back to the early 1980s, though bioleaching studies had been initiated in the late 1950s when a bacterium isolated from the waters of the Bingham Canyon mine (Kennecott, USA) was tested for this purpose (Bryner and Jameson, 1958). This bacterium was identified as *Thiobacillus ferrooxidans* (now *Acidithiobacillus ferrooxidans*).

Bioleaching mechanisms include the so-called direct enzymatic oxidation by microorganisms (1) in acidic pH, also produced by microorganisms (2), and the indirect chemical oxidation of the sulfide by biologically produced ferric ions (3) (Ehrlich et al., 1999).

$$2Fe^{2+} + {}^{1}\!/_{2}O_2 + 2H^+ \xrightarrow{Microorganisms} 2Fe^{3+} + H_2O \quad (19.1)$$

$$S + {}^{3}\!/_{2}O_2 + H_2O \xrightarrow{Microorganisms} H_2SO_4 \quad (19.2)$$

$$MS + 2Fe^{3+} \rightarrow M^{2+} + 2Fe^{2+} \tag{19.3}$$

With in situ mining operations, the most common and profitable microorganisms used for low-grade iron and sulfur minerals are sulfur-oxidizing bacteria belonging to the genera *Acidithiobacillus* and *Leptospirillum,* together with obligate chemoorganotrophic bacteria of the *Acidiphilium* genus. These microorganisms generate and maintain the extreme conditions found in their habitats (Ibañez et al., 2023; González-Toril et al., 1999). *Acidithiobacillus ferrooxidans* has been used as the model acidophile microorganism both in copper biomining and in genetic modification to improve its bioleaching capacity (Ibañez et al., 2023; Gumulya et al., 2018). Depending on the ore and the conditions of the bacterial biomining process, microorganisms belonging to the bacterial genera *Acidimicrobium, Alicyclobacillus*, and *Sulfobacillus*, and the archaeal genera *Acidianus, Ferroplasma, Metallosphaera,* and *Sulfolobus,* have also been used (Gumulya et al., 2018; Schippers, 2007).

Chalcopyrite is the most abundant copper mineral in Nature with a high commercial interest. It is found in association with other minerals, being highly resistant to chemical attack by reagents normally used in hydrometallurgy, since only strong oxidizing conditions can achieve any metal solubilization. Thermophiles show a greater effect than mesophiles in attacking minerals such as chalcopyrite, although temperature can hinder the diffusion of gases such as O_2 or CO_2, limiting the microorganisms' growth (Mier et al., 1995).

One of the microorganisms able to extract copper from chalcopyrite was isolated from the Riotinto mines and was named *Sulfolobus rivotincti* CECT 4489. It is a thermophilic archaeon, with optimum growth at 65–68°C and a pH of around 2. It is perfectly adapted to very acidic environments with high concentrations of metallic ions. The metabolism of this microorganism is strictly aerobic and chemolithotrophic, obtaining energy from the oxidation of Fe^{2+} to Fe^{3+} and several mineral sulfides. Due to its capacity to solubilize copper from chalcopyrite minerals, which is superior to that of other members of the *Sulfolobus* genus as *S. solfataricus* or *S. acidocaldarius*, this microorganism is of great technological interest. This microorganism was the first isolate patented in Spain and the first for biomining worldwide (Marín et al., 1996).

One of the main limitations of bioleaching on metallic concentrates or high-grade ore extraction is the time required to obtain an appropriate metal solubilization level. This time usually ranges from a few days to months. In attempts to reduce the bioleaching time of metal complexes, Carranza et al. (1997) developed the effect separation process (IBES process) that was evaluated with a chalcopyrite/sphalerite concentrate from the Río Tinto. This process uses indirect mechanisms to biologically produce Fe^{3+}, which is then used to oxidize the sulfide. The microorganism used to generate the Fe^{3+} was an *Acidithiobacillus ferrooxidans* isolate from drainage waters of the Riotinto mine, and the process was able to solubilize 95% of the copper and zinc from the sulfides' concentrate in only 8 h.

Biopolymers production

Exopolysaccharides (EPSs) are natural biopolymers produced by various microorganisms in which they perform a variety of roles. These include regulation of biofilm formation and cell growth, help in the transfer of genetic material between cells, and protection of the microbial cells from environmental conditions (Netrusov et al., 2023).

Polysaccharides derived from plants and algae have traditionally dominated the market. However, microbial EPSs present a large diversity in structural composition (xanthan, dextran, alginate, cellulose, hyaluronic acid) and their use in pharmaceuticals, nutraceuticals, functional foods, cosmetics, and insecticides has increased significantly due to their unique physical and chemical properties (Kopperi et al., 2021; Barcelos et al., 2020).

The ability of exopolysaccharides (EPSs) to scavenge reactive oxygen species (ROS) and the antioxidant activity of EPSs from *Bacillus* species have been studied. Additionally, EPSs are recognized as a source of metal-chelating compounds, which is significant given that metal ions can generate highly reactive oxygen-free radicals through Fenton or Haber–Weiss reactions (Karamać 2009).

Bacillus xiamenensis RT6 is a strain isolated from Río Tinto and an EPS producer (Huang-Lin et al., 2022). The polymer contains glucose (60%), mannose (20%), and galactose (20%). The production of this EPS was optimized and its possible applications evaluated, finding antioxidant, metal-chelating, and emulsifying properties. The highest exopolysaccharide yield occurred when glucose was used as the sole carbon source, achieving an EPS yield of up to 700 mg/mL. The in vitro antioxidant studies have demonstrated that it can effectively scavenge free radicals such as 1,1-diphenyl-2-picryl-hydrazyl (DPPH•) and hydroxyl (•OH). Additionally, this EPS is biocompatible at the cellular level, protecting cells from reactive oxygen species (ROS). This EPS also exhibited notable metal-chelating activities at high concentrations, which remained effective even at pH 2.5. Furthermore, it showed an excellent emulsifying capacity with certain natural oils, such as olive and sesame. These findings suggest that EPS is well-suited for industrial applications, including bioremediation for natural oils, food emulsifying processes, and cosmetic or pharmaceutical applications as an antioxidant agent.

Nanotechnology and nanosilver applications

Nanothechnology is the science that studies the production, characterization and applications of materials with at least one dimension sized $\leq$100 nm. These nanomaterials can be produced in various ways, starting with materials of higher dimensions that are later reduced to the nanometric scale (top-down approaches) or from materials with smaller sizes that can later be converted to nanomaterials of the appropriate sizes (bottom-up approaches) (Cele, 2020; Silva, 2004). Nanomaterials can have different structures, such as nanoparticles, nanotubes, nanowalls, and nanotriangles, among many others (Srivastava and Bhargava, 2021), depending on the synthesis method and the starting material. Among these materials, those based on metals in the elemental form or as oxides, sulfides or others are among the most frequently studied (Dhaka et al., 2023; Srivastava and Bhargava, 2021; Ding et al., 2015; Stark et al., 2015). A wide range of applications for nanomaterials has been described. Particularly, silver nanoparticles are considered useful as antibacterial agents, besides other applications in dye decolorization, electronics, to mention only a few (Moulahoum and Ghorbanizamani, 2024; Manjua et al., 2023). Silver nanomaterials are most frequently obtained by bottom-up approaches from ions in their soluble salts, e.g., $AgNO_3$. Methods for the synthesis of these nanomaterials are based on chemical (Nguyen et al., 2023), physical (Surender and Gahlot, 2021) or biological approaches (Pandit et al., 2022). The last ones use extracts of organisms or culture media containing reducing and protecting agents to generate silver nanomaterials. Unlike chemical and physical methods, these are cheaper and do not use toxic chemicals or sophisticated specific equipment for their synthesis, and are, therefore, considered green methods. The biological materials used are very diverse and come from different parts of plants (Thomas et al., 2024), bacteria (Rodrigues et al., 2024), algae (Jacob et al., 2021), fungi (Vijayakumar et al., 2024), lichens (Rattan et al., 2021), among others. In most cases, silver nanoparticles have a so called core-shell structure, the core being in most cases elemental silver, AgCl, Ag_2O or Ag_2S, which is covered by a corona composed, in the case of biological methods of synthesis, of organic compounds or biomolecules acting as capping agents to prevent aggregation among the nanoparticles, and relevant to their interaction with others compounds or cells (Fig. 19.4). The components present in the biological materials used for silver nanoparticles

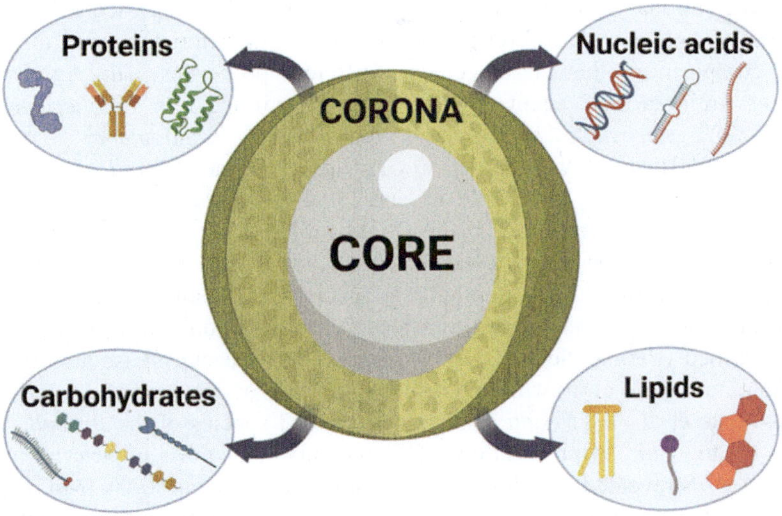

FIGURE 19.4

Core-shell structure of biogenic silver nanoparticles (AgNPs).

(AgNPs) synthesis should be capable of generating the cores, in the case of Ag^0 cores, by reducing silver ions, and binding to the cores to cap them.

One of the most serious global health problems in recent years has been the rise of antibiotic-resistant pathogenic bacteria. The inappropriate use of these drugs for human and animal treatment has diminished their effectiveness, facilitating the emergence of multi-resistant bacteria. According to experts, by 2050, 10 million people are projected to die from this cause, in what has come to be known as the 'silent pandemic'. The solutions are complex and involve the discovery of new antibiotics. However, pharmaceutical industries have largely abandoned the search for new antibiotic products due to the high costs associated with research and development (Murray et al., 2022; Rudramurthy et al., 2016).

Alternatives to the use of antibiotics are currently being explored. These include the recovery of known antibiotics that previously produced side effects, the combined use of two or more antibiotics, or the search for alternative methods of combating infections. Such methods include the use of phages, monoclonal antibodies, compounds derived from metals, such as silver nanoparticles, or mixtures of antibiotics with other molecules that can act synergistically with them, among others (Plotniece et al., 2023; Mateo and Jiménez, 2022).

Silver nanoparticles are postulated as a possible alternative to antibiotics, alone or combined with classical antibiotics (Ribeiro et al., 2022).

The article by Pernas-Pleite et al. (2022) described the biological synthesis of new types of silver nanoparticles (AgNPs) using a microorganism isolated from the Río Tinto estuary, which was identified as *Pseudomonas alloputida*. The study also details the AgNPs characteristics and properties as antimicrobial agents. The bacterium was cultured in two different media, and broths were taken at the exponential and stationary growth phases. The collected broths were used to synthesize four types of AgNPs with different physicochemical characteristics (size, dispersity, corona composition) that could generate different properties. The AgNPs were tested as antimicrobial agents against three Gram-negative (*Escherichia coli* ATCC 25922, *Klebsiella pneumoniae* ATCC 2966, *Pseudomonas aeruginosa* CECT 108, PA01, and PA14) and three Gram-positive ones (*Bacillus subtilis* 168, *Staphylococcus aureus* CECT 794, and *Staphylococcus epidermidis* ATCC 12228), showing differential and very high antibacterial activity levels against planktonic cells and the formation of biofilms, and synergistically in combination with the classical antibiotics streptomycin against *E. coli* and *S. aureus* and with ampicillin against *S. aureus*. The synthesis of eight additional types of AgNPs has been described (Pernas-Pleite, 2024) using two other bacterial isolates from the Río Tinto, a strain of a still undescribed species of the genus *Lysinibacillus* (Pernas-Pleite et al., 2025) and another undescribed species of the *Brucella* genus (Pernas-Pleite et al., in preparation). The assessment of the antibacterial activity of these AgNPs revealed very low minimal inhibitory concentrations (MIC) against the same bacterial species and strains indicated above, as well as synergy with additional classical antibiotics.

Besides the application of the AgNPs obtained from culture broths of Río Tinto bacterial isolates as antibacterial agents, other applications have been explored, including their catalytic activity on the reductive and oxidative decolorization of dyes, such as the recalcitrant carmine indigo, among others, and the detection of heavy metals in aqueous solutions, with positive results (Pernas-Pleite, 2024).

Microbial fuel cells

Several anaerobic metal-reducing bacteria can donate electrons to an electrode, which is of great interest for microbial fuel cells development, since this avoids the O_2 diffusion from the cathodic compartment to the sensitive anodic one when the microorganism colonizes the electrodes. *Acidiphilium* sp. strain 3.2 Sup 5, isolated from the Río Tinto, has been evaluated for this purpose (Malki et al., 2008). In this study, the authors demonstrated that the bacterium colonized carbon electrodes and that the electrocatalytic oxidation of glucose was unaffected by the presence of O_2 in the electrolyte solution, producing electrocatalytic currents up to 3 A/m^3 and a poised potential of +0.15 V. This is a singular behavior of this strain, which initially could be considered to be related to the ability of this bacterium to reduce ferric ion levels in the presence of O_2. However, other isolates of the same genus, also able to produce the ferric ion reduction in the same conditions, were not able to transfer electrons to the electrode as the Río Tinto strain had, indicating that this strain can adhere quite efficiently to the electrode in a unique form (Malki et al., 2013).

Fungi

Since fungi from extreme habitats are subjected to environmental conditions not met by mesophilic fungi, it can be expected that certain compounds they synthesize are specific to them, which might be useful in various applications (Hujslová et al., 2019).

Metal sequestration

The first studies on the Río Tinto fungi resulted in the collection of more than 400 isolates from all along the river (López-Archilla et al., 2001; Durán et al., 1999) that were assigned to Zygomycetes, Deuteromycetes, Basidiomycetes, and Ascomycetes Phyla. These microorganisms were evaluated for biotechnological purposes to find less expensive and environmentally safer methods to recover heavy metals from industrial waters. Certain isolates were able to remove metallic cations from diluted solutions and were proposed as an alternative to the differential precipitation and ion exchange techniques usually employed for that purpose (Durán et al., 1999). The potential mechanisms by which these microorganisms can remove metals include adsorption to the negative charges of cell envelopes, production of chelating agents, changes in metal redox potential, and facilitation of metal transport. Metal resistance could be related to metal sequestration, so the authors carried out a screening by choosing 60 taxonomically different acidophilic fungal isolates and eight metals (Ag^{+}, Hg^{2+}, Cu^{2+}, Zn^{2+}, As^{5+}, Cr^{3+}, Ni^{2+}, and Cd^{2+}), with a metal selection based on metal toxicity, economic interest and presence in the Río Tinto habitat. Among the tested isolates, the most effective ones were *Alternaria* sp. I14, able to remove 95% of the Hg^{2+} from a 0.1 mM $HgSO_4$ solution, *Penicillium* sp. I25, removing 93% of the Zn^{2+} from a 200 mM $ZnCl_2$ solution, and *Penicillium* sp. I28 and *Scytalidium* sp. P65, which removed 90% of the Ni^{2+} and Cd^{2+}, respectively, from solutions containing 1 mM $NiSO_4.6H_2O$ or 10 mM $CdSO_4.8/3H_2O$, respectively. *Penicillium* sp. P34 showed a high specificity and efficiency for copper sequestering from solutions of copper sulfate and with industrial bioleaching solutions from chalcopyrite concentrates, and the transmission electron microscopy and EDAX probe showed dense particles inside the fungi cells, suggesting that active transport and intracellular storage in vesicular structures might be responsible for specific copper sequestering.

In another study, the same mechanism was proposed for another Río Tinto water isolate, *Aspergillus* sp. P37, capable of removing arsenic from culture media with concentrations higher than 10 mM by forming vacuoles. In addition, this isolate was also resistant to metals such as copper and chromium (Canovas et al., 2003).

The metal-specific sequestration performed by Río Tinto environment isolates raises the possibility of its use from a biotechnological point of view as a way to extract and concentrate metals from solutions, as well as for bioremediation of the environment polluted with specific metals or multi-metal contaminated ones. In this last case, using less specific metal-sequestering fungi. The study of these possibilities is still quite limited, considering the large number and diversity of fungi isolated from the Río Tinto. Other studies using this river's fungi have shown the metal resistance of these microorganisms, which can be a starting point for further applied studies on their use for extracting metals and bioremediation (Oggerin et al., 2024).

Enzymes

Lignocellulosic residues are the most abundant materials on Earth (Zhang, 2008). In recent years, there has been an increasing interest in using lignocellulosic biomass in the so-called second generation (2G) technology for biofuels production, as is the case with bioethanol, and in biorefineries able to produce compounds of industrial interest (Castilla-Archilla et al., 2019; Valdivia et al., 2016). The significance of this technology lies in its exploitation of biomass derived from agricultural, forestry, and other nonfood plant resources (Zhang, 2013). This use requires a pretreatment of these materials to remove lignin and for the cellulose and hemicellulose breakdown to obtain monosaccharides for goods production (Godoy et al., 2021). There are commercial cellulolytic cocktails of enzymes capable of operating at high temperatures and pH conditions of around 5.5, in which undesirable microorganism contamination can occur during the manufacturing process. Because of this, the discovery of cellulolytic enzymes with activity at low pHs is of great interest. The potential for finding fungi from an acidic environment, such as the Río Tinto, with cellulolytic and hemicellulolytic activities at low pH has been analyzed (Daddaoua et al., 2022). A total of 182 species from a Río Tinto fungi collection were tested, finding several activities related to cellulose deconstruction (endoglucanase, cellobiohydrolase, and β-glucosidase) and two enzymatic activities related to hemicellulose breakdown (β-xylosidase and xyloglucanase) present in their secretomes, which can release monosaccharides from preindustrial cocktails containing complex carbohydrates. The authors concluded that the Río Tinto is a unique niche populated by fungi that show various enzymatic activities against plant materials that may be of use in biotechnological applications.

Algae

Metal sequestration

The tolerance of three *Chlamydomonas* sp. isolates from Río Tinto to the metals in its waters and to Cd^{2+} in artificial water solutions has been analyzed (Aguilera and Amils, 2005). In this report, a tolerance to 0.2 mM Cd^{2+} was detected, which is about two orders of magnitude higher than for isolates from other previously described environments. Nevertheless, at that concentration, certain effects on the structure and physiology of the microalgae cells were observed, including a lack of motility, an increase in the number and relative volume of starch granules, and a reduction of pyrenoid volume. Transmission electron microscopy showed the accumulation of an electron-dense material inside vacuoles that contain Cd, which was less abundant in the cytoplasm. This accumulation could be useful in the bioremediation of Cd-contaminated waters or the extraction of this metal.

Tolerance to various concentrations of Cd^{2+}, Cu^{2+}, Fe^{2+}, and Pb^{2+} was assessed at pHs 4 and 7 for other microalgae isolated from the Río Tinto intertidal zone (Vizioli, 2017). The highest tolerance measured by the IC_{50} was found for the following: a *Coelastrella* sp. to Pb^{2+}(322.5 mg/L of $Pb(NO_3)_2$), a *Parachlorella* sp. to Cu^{2+} (14.7 mg/L of $CuSO_4.5H_20$), a *Porphyridium* sp. to Cd^{2+} (5.9 mg/L of $CdSO_4.8/3H_2O$), a *Stichococcus bacillaris* to Fe^{2+} (21 mg/L of $FeSO_4$) and an alga from the Trebouxyphiaceae family to Cd^{2+} (6.5 mg/L of $CdSO_4.8/3H_2O$). A light microscopy study detected morphometric changes in the cells exposed to the metals, as well as the appearance of an extracellular mucilaginous structure producing cell aggregation, in some cases. The bioremediation capabilities of the same algae were tested with solutions of Cu^{2+} or Fe^{2+} at pHs 4 and 7 after a 15 day incubation, finding that all the algae were able to remove considerable percentages of the metals from the solution, with most of them eliminating more than 50% of the metal present in the solutions, and many of them over 90%, specially the Trebouxyphiaceae alga reached 93% removal of Cu^{2+} and 99% of Fe^{2+}, and the *Parachlorella* sp. almost 99% of the Fe^{2+}.

Edible biomass

Studies have been conducted on the nutritional value of *Coccomyxa onubensis* biomass (Navarro et al., 2016). This alga, isolated from the acidic waters of the Río Tinto, can grow at pHs in the range 2.5–9, producing high biomass yields with optimum growth between pHs 2.5 and 4.5 (Fuentes et al., 2016; Garbayo et al., 2012). From a nutritional point of view, this microalga biomass contained a high proportion of proteins (44.6% of its dry weight (dw)), with a moderate carbohydrate content (24.8% (dw)) and dietary fiber (15.7% (dw)), as well as a low lipid (5.4% (dw)) and nucleic acids (4.8% (dw)) content. In addition, a high level of polyunsaturated fatty acids was measured (65% of the total fatty acid mass), as well as carotenoids with a high proportion of lutein and β-carotene, which are of interest as animal food supplements as they have a significant hypocholesterolemic and hypotriglyceridemic effect in animals. The capability of this microalga to grow at low pHs and high salt concentrations is also advantageous for biomass production at a large scale without contamination by other microorganisms.

Lutein

Lutein ((3R,3′R,6′R)-β,ε-carotene-3,3′-diol) is a xanthophyll that may help prevent or mitigate the effects of degenerative human diseases, such as age-related macular degeneration (AMD) and cataracts. Additionally, lutein can contribute to skin health (Fernandez-Sevilla et al., 2010). Currently, marigold (*Tagetes erecta*) petals are the predominant commercial source of lutein, containing this compound in 0.02%–0.28% of their dry weight. Although lutein has also been produced synthetically, the cost is not competitive with that of marigold-derived lutein. Microalgae are considered an alternative source of lutein due to their potential for higher yield and lower resource requirements. Algae species of the genera *Chlorella*, *Dunaliella* and *Muriellopsis,* and the species *Haematococcus pluvialis, Scenedesmus almeriensis*, and *Neospongiococcus gelatinosum* have been reported to contain various levels of lutein (0.5%–1.2% dry weight), higher than that of the marigold petals. On the other hand, microalgae cultures can also produce additional valuable products, such as other pigments and lipids. Research has been published on optimizing microalgae cultivation for lutein production, extraction, and encapsulation (Muhammad et al., 2024).

C. onubensis, a microalga isolated from the Río Tinto, can accumulate a high concentration of lutein (9.7 mg g^{-1} (dw)) when grown at a laboratory scale. On a larger scale, using an outdoor 800 L tubular photobioreactor at a temperature range of 10–25°C and pHs in the range 2.5–3, a 10 mg g^{-1} (dw), lutein yield has been reported (Fuentes et al., 2020).

Antimicrobials

Navarro et al. tested the antimicrobial activity of extracts of *C. onubensis* performed with nonpolar solvents like hexane, diethyl ether, chloroform and dichloromethane, and additionally with polar solvents, against Gram-negative bacteria (*Pseudomonas aeruginosa* ATCC 27853, *Escherichia coli* ATCC 25922 (β-lactamase sensitive), *E. coli* ESBL ATCC BAA-2471 (β-lactamase resistant), *Salmonella enterica* ATCC 19430, and *Proteus mirabilis* ATCC 29906) and Gram-positive bacteria (*Staphylococcus aureus* ATCC 29213 and *S. aureus* MR ATCC BAA-2422 (methicillin-resistant)), as well as against the yeast *Candida albicans* ATCC 10231. Depending on the solvent used for the extraction, activity was found against different microorganisms, thus, hexane extract was active on *P. mirabilis*, *S. enterica* and the chloroform extract was additionally active on *E. coli*, while the dichloromethane extract was only active against *S. enterica*, probably indicating the presence in the various extracts of different active compounds. No antimicrobial activity was found against Gram-positive bacteria. Polar-solvent extracts showed no noticeable activity (Navarro et al., 2017).

Nanotechnology and nanosilver applications

In a preliminary study, various chlorophytes from the Río Tinto intertidal zone, including one strain each of *Coelastrella* sp., *Parachlorella* sp., *Tetradesmus obliquus*, *Stichococcus bacillaris*, and a microalga from the Trebouxyphiaceae family, were assessed for AgNPs production. Culture broths of the microalgae grown at pHs 4 or 7 were evaluated for AgNPs synthesis with positive results. These nanomaterials showed antibacterial activity against *E. coli*, *K. pneumoniae* and *S. epidermidis* (Vizioli, 2017).

Further in-depth studies on the AgNPs syntheses and possible applications by some of these microalgae were performed later, particularly of the *Parachlorella* sp. and the formerly assigned as *Stichococcus bacillaris*, based on 18S rRNA sequence, and better identified as *Tetratostichococcus* sp. based on the ITS 2 sequence (Pernas-Pleite, 2024; Pernas-Pleite et al., 2023). Broths were obtained from cultures at pHs 4 or 7 in two media, BG-11 and chloride-depleted BG-11, from the early and late growth phases. Eight types of AgNPs were obtained from each alga, their physicochemical characteristics were determined, and their applications were studied. The synthesis kinetics and production of the AgNPs varied depending on the broth used, reaching maximum yields in reaction times ranging from hours to various days. The study of their physicochemical properties was performed by transmission electron microscopy for the shape and core size distribution, dynamic light scattering for the hydrodynamic size, Fourier transform infrared spectroscopy for the corona composition, and UV-Vis spectroscopy for concentration determination in combination with total reflection X-ray fluorescence, using this last technique to determine AgNPs elemental composition. These parameters allowed the differentiation of the eight types of AgNPs and the establishment of relationships between these characteristics and their aptness for biotechnological applications.

Applications for silver nanoparticles in various fields have been reported (Fig. 19.5), one of the more relevant being as antimicrobial agents, as discussed

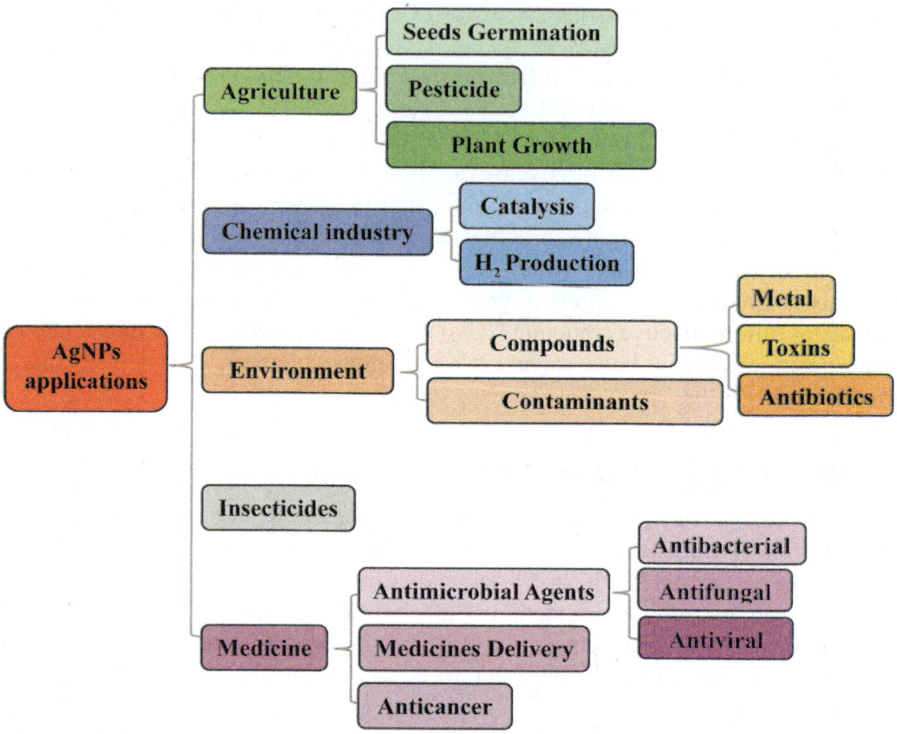

FIGURE 19.5

Biotechnological applications of silver nanoparticles in various fields.

above. Other applications of interest that have been explored using AgNPs originated from algae isolated from the Río Tinto (Pernas-Pleite, 2024) include their use in dye decolorization for bioremediation of dye-contaminated environments and for a simple and cheap method to detect and quantify heavy metals in aqueous solution.

As antibacterial agents, the AgNPs were evaluated against the three Gram-negative and three Gram-positive bacteria used in previous works with bacterial isolates from the Río Tinto (see above in the section on bacteria and archaea). All the AgNPs were active against all the test bacteria with different activity levels, depending on the origin of the biological material used in their synthesis.

Besides the direct antibacterial activity of the AgNPs, their activity as enhancers of classical antibiotics' activity was checked (Pernas-Pleite et al., 2023), and a synergistic antibacterial activity against *E. coli* and *S. aureus* with streptomycin and kanamycin, against *E. coli* with tetracycline, and against *S. aureus* with ampicillin was found.

Lipids production, biogas, and biodiesel generation

Two microalgae isolated from the Río Tinto, *Parachlorella* sp. and *Tetradesmus obliquus*, have been evaluated for lipid production for biogas and biodiesel generation (González Sierra, 2017). Different growth variables, including temperature, pH, and nitrogen presence in the culture media, were studied, finding that the highest yield was obtained in nitrogen-depleted media at 28°C and pH 4 for *Parachlorella* sp., and at the same temperature but at pH 7 for *T. obliquus*. The total lipid yields related to biomass were very high, 83% for *Parachlorella* sp. and 92% for *T. obliquus*. This represents theoretical percentages of saturated fatty acids of 61% and 65%, respectively. The study of the production of biomethane and biodiesel indicated maximum productions for the first of 0.8 and 0.1 t ha^{-1} yr^{-1}, and for biodiesel of 4.2 and 6.8 t ha^{-1} yr^{-1} for *Parachlorella* sp. and *T. obliquus*, respectively. These results indicate that *Parachlorella* sp. was better for methane generation than *T. obliquus*, and the reverse was true for biodiesel production.

Plants

Metal sequestration

The possibilities for metal sequestration using organisms adapted to growing in the Río Tinto and the surrounding habitats are not limited to microorganisms, as studies with plants have shown (Fuente et al., 2016). In this report, plants in the fields around the Río Tinto were shown to accumulate minerals such as ferrihydrite, jarosite, hematite, and maghemite/magnetite in their roots and rhizomes. The oxides were observed by TEM-EDX in the form of intra- and intercellular aggregates as dark deposits (200–600 nm long) covering the inside of the cell membrane and organelles. In roots and rhizomes, iron was found in oxidized form (Fe^{3+}), and ferrous form was not detected. The analysis of incinerated plant material allowed the researchers to detect monodisperse nanocrystals with sizes ranging from 5 to 10 nm. This study extends the previously known range of nanosized iron oxides inside *Imperata cylindrica* tissues. From a biotechnological point of view, these results show that plants from the Río Tinto habitat, such as *Imperata cylindrical* could be of great interest in bioremediation. of soils contaminated with toxic metals.

Prospects and future developments

The Río Tinto is a completely unique acidic extreme environment that includes metal-rich earths along most of the river course and saline marshes in the intertidal and estuary zones. This makes the whole region a source of diverse extreme organisms suitable for study from many points of view, the peculiarities of the biodiversity, biochemistry, and metabolism of the organisms in their various ecosystems, both from an academic point of view, as well as for the enormous biotechnological potential of their biota. Only a small part of the rich biodiversity has been studied up to now, and of these, only a few have been evaluated for biotechnological applications. Thus, the potential, both academic and biotechnological, of most of the eucaryotes, including fungi, microalgae, and some plants, along with prokaryotes present in the Rio Tinto environment have yet to be studied in depth, although some promising preliminary studies have raised high expectations for biotechnological applications. More in-depth work is needed to demonstrate the applicability of the already explored, but also to widen the search for new ones. This would have to include new methods for isolating species still to be discovered and evaluating their properties.

Minimal study has been made of the applications of organisms from the Río Tinto for metal sequestration, principally to assess the microorganisms' tolerance to metals, but the potential for effective bioremediation and metal extraction still has a long way to go to become practically useful. Moreover, only a few compounds and enzymes from the Río Tinto biota have been studied, and lutein biomolecule is the only one whose usefulness has been demonstrated on an industrial scale. Among the compounds of possible use, those with antimicrobial activity are the most important since the onset of the silent pandemic of bacterial antibiotic resistance. Sadly, there are virtually no studies for the screening of these types of substance reported. Also, very few of the species of algae have been studied for their use as edible biomass, or biogas or biodiesel generation, and only at a laboratory scale. Nanomaterials production, mainly of metal nanoparticles, has been addressed using some bacteria and microalgae, but only silver nanoparticles have been produced and, again, only at a laboratory scale, and only some of their putative applications have been evaluated. Synthesis of other metal or metalloid nanoparticles, such as those of gold, platinum, selenium, and magnetic iron oxides, among others, is still an option for biotechnological applications of the Río Tinto microbiota diversity. There are thus enormous possibilities for further exploration and development of biotechnological applications of the organisms inhabiting the Río Tinto and its surrounding ecosystems.

References

Aguilera, A., Manrubia, S.C., Gómez, F., Rodriguez, N., Amils, R., 2006. Eukaryotic community distribution and its relationship to water physicochemical parameters in an extreme acidic environment, Rio Tinto (Southwestern Spain). Appl. Environ. Microbiol. 72, 5325–5330. https://doi.org/10.1128/AEM.00513-06.

Aguilera, A., Amils, R., 2005. Tolerance to cadmium in sp. (Chlorophyta) strains isolated from an extreme acidic environment, the Tinto River (SW, Spain). Aquat. Toxicol. 75, 316–329. https://doi.org/10.1016/j.aquatox.2005.09.002.

Amaral, L.A., Gómez, F., Zettler, E., Keenan, B.G., Amils, R., Sogin, M.L., 2002. Eukaryotic diversity in Spain's river of fire. Nature 417, 137.

Amils, R., 2016. Lessons learned from thirty years of geomicrobiological studies of Río Tinto. Res. Microbiol. 167, 539–545. https://doi.org/10.1016/j.resmic.2016.06.001.

Amils, R., González-Toril, E., Aguilera, A., Rodríguez, N., Fernández-Remolar, D., Gómez, F., et al., 2011. From Río Tinto to Mars: the terrestrial and extraterrestrial ecology of acidophiles. Adv. Appl. Microbiol. 77, 41–70. https://doi.org/10.1016/B978-0-12-387044-5.00002-9.

Barcelos, M.C.S., Vespermann, K.A.C., Pelissari, F.M., Molina, G., 2020. Current status of biotechnological production and applications of microbial exopolysaccharides. Crit. Rev. Food Sci. Nutr. 60, 1475–1495. https://doi.org/10.1080/10408398.2019.1575791.

Bryner, L.C., Jameson, A.K., 1958. Microorganisms in leaching sulfide minerals. Appl. Microbiol. 6 (4), 281–287. https://doi.org/10.1128/am.6.4.281-287.1958.

Cánovas, D., Durán, C., Rodríguez, N., Amils, R., de Lorenzo, V., 2003. Testing the limits of biological tolerance to arsenic in a fungus isolated from the River Tinto. Environ. Microbiol. 5 (2), 133–138. https://doi.org/10.1046/j.1462-2920.2003.00386.x.

Carranza, F., Palencia, I., Romero, R., 1997. Silver catalyzed IBES process: application to a Spanish copper-zinc sulphide concentrate. Hydrometallurgy 44, 29–42.

Castilla-Archilla, J., O'Flaherty, V., Lens, P.N.L., 2019. Biorefineries: industrial innovation and tendencies. In: Bastidas-Oyanedel, J.R., Schmidt, J.E. (Eds.), Biorefinery. Integrated Sustainable Processes for Biomass Conversion to Biomaterials, Biofuels, and Fertilizers, vol 37. Springer Nature, Switzerland AG, pp. 3–36. https://doi.org/10.1007/978-3-030-10961-5.

Cele, T., 2020. Preparation of nanoparticles. In: Avramescu, S.M., Akhtar, K., Fierascu, I., Khan, S.B., Ali, F., Asiri, A.M. (Eds.), Engineered Nanomaterials - Health and Safety. IntechOpen. https://doi.org/10.5772/intechopen.90771.

Daddaoua, A., Álvarez, C., Oggerin, M., Rodriguez, N., Duque, E., Amils, R., et al., 2022. Rio Tinto as a niche for acidophilus enzymes of industrial relevance. Microb. Biotechnol. 16, 1069–1086. https://doi.org/10.1111/1751-7915.14192.

Davis, R.A., Welty, A.T., Borrego, J., Morales, J.A., Pendon, J.G., Ryan, J.G., 2000. Rio Tinto estuary (Spain): 5000 years of pollution. Environ. Geol. 39 (10), 1107–1116.

Dhaka, A., Mali, S.C., Sharma, S., Trivedi, R., 2023. A review on biological synthesis of silver nanoparticles and their potential applications. Results Chem. 6, 101108. https://doi.org/10.1016/j.rechem.2023.101108.

Ding, Q., Ding, F., Qian, T., Zhao, D., Wang, L., 2015. Reductive immobilization of rhenium in soil and groundwater using pyrite nanoparticles. Water Air Soil Pollut. 226, 409. https://doi.org/10.1007/s11270-015-2588-y.

Duran, C., Marin, I., Amils, R., 1999. Specific metal sequestering acidophilic fungi. In: Amils, R., Ballester, A. (Eds.), Biohydrometallurgy and the Environment towards the Mining of the 21st Century, Part B. Elsevier, Amsterdam, The Netherlands, pp. 521–530.

Eduardo-Correia, E.B., 2015. Abundancia, diversidad y perfiles de multirresistencia de bacterias cultivables resistentes a los antibióticos en la ría de Huelva y la chorrera de Despeñalagua (Guadalajara). Ph.D. Thesis. Universidad Autónoma de Madrid, Madrid, Spain.

Eduardo-Correia, B., Morales-Filloy, H., Abad, J.P., 2020. Bacteria from the multi-contaminated Tinto River estuary (SW, Spain) show high multi-resistance to antibiotics and point to *Paenibacillus* spp. as antibiotic-resistance-dissemination players. Front. Microbiol. 10, 3071. https://doi.org/10.3389/fmicb.2019.03071.

Ehrlich, H.I., 1999. Past, present and future o biohydrometallurgy. In: Amils, R., Ballester, A. (Eds.), Biohydrometallurgy and the Environment towards the Mining of the 21st Century. Part A. Elsevier, Amsterdam, The Netherlands, pp. 3–12.

Fernández-Remolar, D.C., Morris, R.V., Gruener, J.E., Amils, R., Knoll, A.H., 2005. The Río Tinto Basin, Spain: mineralogy, sedimentary geobiology, and implications for interpretation of outcrop rocks at Meridiani Planum, Mars. Earth Planet Sci. Lett. 240, 149–167. https://doi.org/10.1016/j.epsl.2014.01.022.

Fernández-Sevilla, J.M., Acién Fernández, F.G., Molina Grima, E., 2010. Biotechnological production of lutein and its applications. Appl. Microbiol. Biotechnol. 86, 27–40. https://doi.org/10.1007/s00253-009-2420-y.

Fuente, V., Rufo, L., Juárez, B.H., Menéndez, N., García-Hernández, M., Salas-Colera, E., et al., 2016. Formation of biomineral iron oxides compounds in a Fe hyperaccumulator plant: *Imperata cylindrica* (L.) P. Beauv. J. Struct. Biol. 193, 23–32. https://doi.org/10.1016/j.jsb.2015.11.005.

Fuentes, J.L., Montero, Z., Cuaresma, M., Ruiz-Domínguez, M.C., Mogedas, B., Garbayo Nores, I., et al., 2020. Outdoor large-scale cultivation of the acidophilic microalga *Coccomyxa onubensis* in a vertical close photobioreactor for lutein production. Processes 8, 324. https://doi.org/10.3390/pr8030324.

Fuentes, J.L., Huss, V.A.R., Montero, Z., Torronteras, R., Cuaresma, M., Garbayo, I., et al., 2016. Phylogenetic characterization and morphological and physiological aspects of a novel acidotolerant and halotolerant microalga *Coccomyxa onubensis* sp. nov. (Chlorophyta, Trebouxiophyceae). J. Appl. Phycol. 28, 3269–3279. https://doi.org/10.1007/s10811-016-0887-3.

Garbayo, I., Torronteras, R., Forján, E., Cuaresma, M., Casal, C., Mogedas, B., et al., 2012. Identification and physiological aspects of a novel carotenoid-enriched, metal-resistant microalga isolated from an acidic river in Huelva (Spain). J. Phycol. 48, 607–614. https://doi.org/10.1111/j.1529-8817.2012.01160.x.

García-Moyano, A., González-Toril, E., Aguilera, A., Amils, R., 2012. Comparative microbial ecology study of the sediments and water column of Río Tinto, an extreme acidic environment. FEMS Microbiol. Ecol. 81, 303–314. https://doi.org/10.1111/j.1574-6941.2012.01346.x.

Godoy, P., García-Franco, A., Recio, M., Ramos, J., Duque, E., 2021. Synthesis of aromatic amino acids from 2G lignocellulosic substrates. Microb. Biotechnol. 14, 1931–1943. https://doi.org/10.1111/1751-7915.13844.

Gómez-Ortiz, D., Fernández-Remolar, D.C., Granda, A., Quesada, C., Granda, T., Prieto-Ballesteros, O., et al., 2014. Identification of the subsurface sulfide bodies responsible for acidity in Río Tinto source water, Spain. Earth Planet Sci. Lett. 391, 36–41. https://doi.org/10.1016/j.epsl.2014.01.022.

González Sierra, J., 2017. Estudio de la posible aplicación de microalgas aisladas de Rio Tinto, para la producción de biodiesel y biogás. Undergraduate Thesis. Universidad Autónoma de Madrid, Madrid, Spain.

González-Toril, E., Llobet-Brossa, E., Casamayor, E.O., Amman, R., Amils, R., 2003. Microbial ecology of an extreme acidic environment, the Río Tinto. Appl. Environ. Microbiol. 69, 4853–4865. https://doi.org/10.1128/AEM.69.8.4853-4865.2003.

Gumulya, Y., Boxall, N.J., Khaleque, H.N., Santala, V., Carlson, R.P., Kaksonen, A.H., 2018. In a quest for engineering acidophiles for biomining applications: challenges and opportunities. Genes 9, 116. https://doi.org/10.3390/genes9020116.

Gupta, G.N., Srivastava, S., Khare, S.K., Prakash, V., 2014. Extremophiles: an overview of microorganism from extreme environment. Int. J. Agric. Environ. Biotechnol. 7 (2), 371–380. https://doi.org/10.5958/2230-732X.2014.00258.7.

Huang-Lin, E., Sánchez-León, E., Amils, R., Abrusci, C., 2022. Potential applications of an exopolysaccharide produced by *Bacillus xiamenensis* RT6 isolated from an acidic environment. Polymers 14, 3918. https://doi.org/10.3390/polym14183918.

Hujslová, M., Bystrianský, L., Benada, O., Gryndler, M., 2019. Fungi, a neglected component of acidophilic biofilms: do they have a potential for biotechnology? Extremophiles 23, 267–275. https://doi.org/10.1007/s00792-019-01085-9.

Ibáñez, A., Garrido-Chamorro, S., Coque, J.J.R., Barreiro, C., 2023. From genes to bioleaching: unraveling sulfur metabolism in *Acidithiobacillus* genus. Genes 14 (9), 1772. https://doi.org/10.3390/genes14091772.

Jacob, J.M., Ravindran, R., Narayanan, M., Samuel, S.M., Pugazhendhi, A., Kumar, G., 2021. Microalgae: a prospective low-cost green alternative for nanoparticle synthesis. Curr. Opin. Environ. Sci. Health 20, 100163. https://doi.org/10.1016/j.coesh.2019.12.005.

Jeong, S.W., Choi, Y.J., 2020. Extremophilic microorganisms for the treatment of toxic pollutants in the environment. Molecules 25 (21), 4916. https://doi.org/10.3390/molecules25214916.

Karamać, M., 2009. Chelation of Cu (II), Zn (II), and Fe (II) by tannin constituents of selected edible nuts. Int. J. Mol. Sci. 10, 5485–5497. https://doi.org/10.3390/ijms10125485.

Kochhar, N., Kavya, I..K., Shrivastava, S., Ghosh, G., Rawat, V.S., Sodhi, et al., 2022. Perspectives on the microorganism of extreme environments and their applications. Curr. Res. Microb. Sci. 3, 100134. https://doi.org/10.1016/j.crmicr.2022.100134.

Kopperi, H., Amulya, K., Mohan, S.V., 2021. Simultaneous biosynthesis of bacterial polyhydroxybutyrate (PHB) and extracellular polymeric substances (EPS): process optimization and scale-up. Bioresour. Technol. 341, 125735. https://doi.org/10.1016/j.biortech.2021.125735.

Kyle, J.E., Pedersen, K., Grant Ferris, F., 2008. Virus mineralization at low pH in the Río Tinto, Spain. Geomicrobiol. J. 25, 338–345. https://doi.org/10.1080/01490450802402703.

Leblanc, M., Morales, J.A., Borrego, J., Elbaz-Poulichez, F., 2000. 4500-year-old mining pollution in Southwestern Spain: long-Term implications for modern mining pollution. Econ. Geol. 95 (3), 655–662. https://doi.org/10.2113/gsecongeo.95.3.655.

López-Archilla, A.I., Marín, I., Amils, R., 2001. Microbial community composition and ecology of an acidic aquatic environment: the Tinto River, Spain. Microb. Ecol. 41, 20–35. https://doi.org/10.1007/s002480000044.

López-Archilla, A.I., González, A.E., Terrón, M.C., Amils, R., 2004. Ecological study of the fungal populations of the acidic Tinto River in southwestern Spain. Can. J. Microbiol. 50, 923–934. https://doi.org/10.1139/w04-089.

Malki, M., Casado, S., López, M.F., Caillard, R., Palomares, F.J., Martín Gago, J.A., et al., 2013. Physicochemical characterization of *Acidiphilium* sp. biofilms. ChemPhysChem 14 (6), 1237–1244. https://doi.org/10.1002/cphc.201201034.

Malki, M., De Lacey, A.L., Rodríguez, N., Amils, R., Fernandez, V.M., 2008. Preferential use of an anode as an electron acceptor by an acidophilic bacterium in the presence of oxygen. Appl. Environ. Microbiol. 74 (14), 4472–4476. https://doi.org/10.1128/AEM.00209-08.

Manjua, K., Ranjini, H.K., Niranjan Raja, S., Chandra Nayaka, S., Lavanya, S.N., Chouhan, R.S., et al., 2023. Nanoagrosomes: future prospects in the management of drug resistance for sustainable agricultura. Plant Nano Biol. 4, 100039. https://doi.org/10.1016/j.plana.2023.100039.

Marín, I., Amils, R., Ballester, A., Gómez, E., 1996. Nuevo microorganismo del género *Sulfolobus* y su uso en la biolixiviación de sulfuros metálicos para la recuperación de metales (OEPM, Patent ES 2087033).

Mateo, E.M., Jiménez, M., 2022. Silver nanoparticle-based therapy: can it be useful to combat multi-drug resistant bacteria? Antibiotics 11 (9), 1205. https://doi.org/10.3390/antibiotics11091205.

Mier, J.L., Ballester, A., Blázquez, M.L., 1995. Influence of metallic ions in the bioleaching of chalcopyrite by *Sulfolobus* BC: experiments using pneumatically stirred reactors and massive samples. Miner. Eng. 8 (9), 949–965.

Moreno, C., Capitán, M.A., Doyle, M., Nicro, J.M., Ruíz, F., Sáez, R., 2003. A minimum age for the Las Cruces gossan: inferences about the beginning of extreme ecosystems at the Iberian Pyryte Belt. Geogaceta 33, 67–70.

Moulahoum, H., Ghorbanizamani, F., 2024. Navigating the development of silver nanoparticles based food analysis through the power of artificial intelligence. Food Chem. 445, 138800. https://doi.org/10.1016/j.foodchem.2024.138800.

Muhammad, G., Butler, T.O., Chen, B., Lv, Y., Xiong, W., Zhao, X., et al., 2024. Sustainable production of lutein—an underexplored commercially relevant pigment from microalgae. Biomass Convers. Biorefinery 14, 7255–7276. https://doi.org/10.1007/s13399-022-03349-5.

Murray, C.J., Ikuta, K.S., Sharara, F., Swetschinski, L., Robles Aguilar, G., Gray, A., et al., 2022. Global burden of bacterial antimicrobial resistance in 2019: a systematic analysis. Lancet 399, 629–655. https://doi.org/10.1016/S0140-6736(21)02724-0.

Navarro, F.J., Forján, E., Vázquez, M., Montero, Z., Bermejo, E., Castaño, M.A., et al., 2016. Microalgae as a safe food source for animals: nutritional characteristics of the acidophilic microalga *Coccomyxa onubensis*. Food Nutr. Res. 60, 30472. https://doi.org/10.3402/fnr.v60.30472.

Navarro, F., Forján, E., Vázquez, M., Toimil, A., Montero, Z., Ruiz-Domínguez, M.C., et al., 2017. Antimicrobial activity of the acidophilic eukaryotic microalga *Coccomyxa onubensis*. Phycol. Res. 65, 38–43. https://doi.org/10.1111/pre.12158.

Netrusov, A.I., Liyaskina, E.V., Kurgaeva, I.V., Liyaskina, A.U., Yang, G., Revin, V.V., 2023. Exopolysaccharides producing bacteria: a review. Microorganisms 11, 1541. https://doi.org/10.3390/microorganisms11061541.

Nguyen, N.P.U., Dang, N.T., Doan, L., Nguyen, T.T.H., 2023. Synthesis of silver nanoparticles: from conventional to 'modern' methods: a review. Processes 11, 2617. https://doi.org/10.3390/pr11092617.

Oggerin, M., del Moral, C., Rodriguez, N., Fernandez-Gonzalez, N., Martínez, J.M., Lorca, I., et al., 2024. Metal tolerance of Río Tinto fungi. Front. Fungal Biol. 5, 1446674. https://doi.org/10.3389/ffunb.2024.1446674.

Pandit, C., Roy, A., Ghotekar, S., Khusro, A., Islam, M.N., Emran, T.B., Bradley, D.A., 2022. Biological agents for synthesis of nanoparticles and their applications. J. King Saud Univ. Sci. 34 (3), 101869. https://doi.org/10.1016/j.jksus.2022.101869.

Pernas-Pleite, C., Conejo-Martínez, A.M., Marín, I., Abad, J.P., 2022. Green extracellular synthesis of silver nanoparticles by *Pseudomonas alloputida*, their growth and biofilm-formation inhibitory activities and synergic behavior with three classical antibiotics. Molecules 27, 7589. https://doi.org/10.3390/molecules27217589.

Pernas-Pleite, C., Conejo-Martínez, A.M., Fernández Freire, P., Hazen, M.J., Marín, I., Abad, J.P., 2023. Microalga broths synthesize antibacterial and non-cytotoxic silver nanoparticles showing synergy with antibiotics and bacterial ROS induction and can be reused for successive AgNP batches. Int. J. Mol. Sci. 24, 16183. https://doi.org/10.3390/ijms242216183.

Pernas-Pleite, C., 2024. Síntesis biogénica extracelular de nanopartículas de plata (AgNPs) a partir de bacterias y microalgas del Río Tinto (Huelva). Aplicaciones antibacterianas y medioambientales. Ph.D. Thesis. Universidad Autónoma de Madrid, Madrid, Spain.

Pernas-Pleite, C., Conejo-Martínez, A.M., Marín, I., Abad, J.P., 2025. Silver nanoparticles (AgNPs) from *Lysinibacillus* sp. culture broths: antibacterial activity, mechanism insights and synergy with classical antibiotics. Biomolecules 15 (5), 731. https://doi.org/10.3390/biom15050731.

Plotniece, A., Sobolev, A., Supuran, C.T., Carta, F., Björkling, F., Franzyk, H., et al., 2023. Selected strategies to fight pathogenic bacteria. J. Enzym. Inhib. Med. Chem. 38 (1), 2155816. https://doi.org/10.1080/14756366.2022.2155816.

Rattan, R., Shukla, S., Sharma, B., Bhat, M., 2021. A mini-review on lichen-based nanoparticles and their applications as antimicrobial agents. Front. Microbiol. 12, 633090. https://doi.org/10.3389/fmicb.2021.633090.

Rawat, M., Chauhan, M., Pandey, A., 2024. Extremophiles and their expanding biotechnological applications. Arch. Microbiol. 206, 247. https://doi.org/10.1007/s00203-024-03981-x.

Ribeiro, A.I., Dias, A.M., Zille, A., 2022. Synergistic effects between metal nanoparticles and commercial antimicrobial agents: a review. ACS Appl. Nano Mater. 5 (3), 3030–3064. https://doi.org/10.1021/acsanm.1c03891.

Rodrigues, A.S., Batista, J.G.S., Rodrigues, M.Á.V., Thipe, V.C., Minarini, L.A.R., Lopes, P. S., 2024. Advances in silver nanoparticles: a comprehensive review on their potential as antimicrobial agents and their mechanisms of action elucidated by proteomics. Front. Microbiol. 15, 1440065. https://doi.org/10.3389/fmicb.2024.1440065.

Rothschild, L.J., Mancinelli, R.L., 2001. Life in extreme environments. Nature 409, 1092–1101. https://doi.org/10.1038/35059215.

Rudramurthy, G., Swamy, M.K., Sinniah, U.R., Ghasemzadeh, A., 2016. Nanoparticles: alternatives against drug-resistant pathogenic microbes. Molecules 21, 836. https://doi.org/10.3390/molecules21070836.

Sabater, S., Buchaca, T., Cambra, J., Catalan, J., Guasch, H., Ivorra, N., et al., 2003. Structure and function of benthic algal communities in an extremely acid river. J. Phycol. 39 (3), 481–489. https://doi.org/10.1046/j.1529-8817.2003.02104.x.

Sánchez-Andrea, I., Rodríguez, N., Amils, R., Sanz, J.L., 2011. Microbial diversity of anaerobic sediments of Río Tinto. a natural acid and high heavy metals content environment. Appl. Environ. Microbiol. 77, 17. https://doi.org/10.1128/AEM.00654-11.

Schippers, A., 2007. Microorganisms involved in bioleaching and nucleic acid-based molecular methods for their identification and quantification. In: Sand, W., Donati, E.R. (Eds.), Microbial Processing of Metal Sulfides. Springer, Heidelberg, Germany, pp. 3–33.

Silva, G.A., 2004. Introduction to nanotechnology and its applications to medicine. Surg. Neurol. 61 (3), 216–220. https://doi.org/10.1016/j.surneu.2003.09.036.

Srivastava, S., Bhargava, A., 2021. Green Nanoparticles: The Future of Nanobiotechnology. Springer, Berlin/Heidelberg, Germany. https://doi.org/10.1007/978-981-16-7106-7.

Stark, W.J., Stoessel, P.R., Wohlleben, W., Hafner, A., 2015. Industrial applications of nanoparticles. Chem. Soc. Rev. 44 (16), 5793–5805. https://doi.org/10.1039/c4cs00362d.

Surender, Gahlot, A., 2021. Physical methods of nanoparticles preparation: an overview. Int. J. Adv. Eng. Manag. 3 (12), 812–816. https://doi.org/10.35629/5252-0312812816.

Tapadar, S., Tripathi, D., Pandey, S., Goswami, K., Bhattacharjee, A., Das, K., et al., 2021. Role of extremophiles and extremophilic proteins in industrial waste treatment. In: Shah, M.P. (Ed.), Removal of Emerging Contaminants through Microbial Processes. Springer, Singapore, pp. 217–235. https://doi.org/10.1007/978-981-15-5901-3_11.

Thomas, S., Gonsalves, R.A., Jose, J., Zyoud, S.H., Prasad, A.R., Garvasis, J., 2024. Plant-based synthesis, characterization approaches, applications and toxicity of silver nanoparticles: a comprehensive review. J. Biotechnol. 394, 135–149. https://doi.org/10.1016/j.jbiotec.2024.08.009.

Thombre, R.S., Vaishampayan, P.A., Gomez, F., 2020. Applications of extremophiles in Astrobiology. In: Salwan, R., Sharma, V. (Eds.), Physiological and Biotechnological Aspects of Extremophiles. Academic Press, pp. 89–104. https://doi.org/10.1016/B978-0-12-818322-9.00007-1. Chapter 7.

Valdivia, M., Galan, J.L., Laffarga, J., Ramos, J., 2016. Biofuels 2020: biorefineries based on lignocellulosic materials. Microb. Biotechnol. 9, 585–594. https://doi.org/10.1111/1751-7915.12387.

Vijayakumar, G., Kim, H.J., Jo, J.W., Rangarajulu, S.K., 2024. Macrofungal mediated biosynthesis of silver nanoparticles and evaluation of its antibacterial and wound-healing efficacy. Int. J. Mol. Sci. 25 (2), 861. https://doi.org/10.3390/ijms25020861.

Vizioli, C., 2017. Diversidad bacteriana asociada a la zona intermareal del estuario del Río Tinto. Aislamiento y propiedades biotecnológicas de aislados microalgales ácido-tolerantes. Ph.D. Thesis. Universidad Autónoma de Madrid, Madrid, Spain.

Wang, Y., Qian, J., Shi, T., Wang, Y., Ding, Q., Ye, C., 2024. Application of extremophile cell factories in industrial biotechnology. Enzym. Microb. Technol. 175, 110407. https://doi.org/10.1016/j.enzmictec.2024.110407.

Zhang, Y.-H.P., 2008. Reviving the carbohydrate economy via multi-product lignocellulose biorefineries. J. Ind. Microbiol. Biotechnol. 35, 367–375. https://doi.org/10.1007/s10295-007-0293-6.

Zhang, Y.-H.P., 2013. Next generation biorefineries will solve the food, biofuels, and environmental trilemma in the energy-food-water nexus. Energy Sci. Eng. 1 (1), 27–41. https://doi.org/10.1002/ese3.2.

CHAPTER

Vitamin B_{12} in the Tinto basin and the deep subsurface of the Iberian Pyrite Belt

20

Guillermo Mateos[1,2], Adrián Martínez-Bonilla[1,2]

[1]*Universidad Autónoma de Madrid, Departamento de Biología Molecular, Madrid, Spain;* [2]*Centro de Biología Molecular Severo Ochoa (CSIC-UAM), Universidad Autónoma de Madrid, Madrid, Spain*

Introduction

Vitamin B_{12}, also known as cobalamin, consists of a corrin ring with a cobalt (Co) atom covalently linked in the center and two ligands above and below the corrin ring. The lower ligand contains a 5,6′ dimethylbenzimidazole (DMB) group, while the upper ligand can vary between a methyl group (methylcobalamin), a hydroxyl group (hydroxocobalamin), a cyanide (cyanocobalamin), or a 5′ deoxyadenosyl (adenosylcobalamin). Both methylcobalamin and adenosylcobalamin are of biological origin, while hydroxocobalamin and cyanocobalamin are synthetic forms for cyanide poisoning and as a dietary supplement, respectively. Vitamin B_{12} belongs to a larger group of molecules, cobamides, which all include the same scaffold but vary the lower ligand between benzimidazoles (like DMB), purines, or phenol groups (Sokolovskaya et al., 2020). Out of all known cobamides, vitamin B_{12} has been studied the most due to its relevance in human health.

Nevertheless, its relevance is widespread within all of life's domains. An example of this is the epoxyqueosine reductase QueG, involved in the synthesis of queosine, a modified nucleoside present in eukaryotes and bacteria that is present in the wobble position of tRNA for the synthesis of histidine, aspartate, tyrosine, and asparagine (Miles et al., 2011). Similar to this, the ribonucleotide reductase NrdJZ is also dependent on vitamin B_{12} and catalyzes the conversion of ribonucleotides to deoxyribonucleotides, necessary for the repair and replication of DNA (Borovok et al., 2004). Although some of these enzymes do have vitamin B_{12} independent counterparts, others, like the methylmalonyl-CoA mutase, which is widespread in bacteria and conserved in humans, to the best of our knowledge, have no known vitamin B_{12} independent counterpart. Vitamin B_{12} can also be relevant for the synthesis of secondary metabolites like antibiotics (gentamicin, carbapenem) (Marous et al., 2015; Kim et al., 2013), the posttranslational modification of proteins (Deobald et al., 2018; Gagsteiger et al., 2022), or the synthesis of ribosomal natural products, also

Geomicrobiology of Río Tinto. https://doi.org/10.1016/B978-0-443-32948-7.00004-7

known as RiPP (Parent et al., 2016). Even for processes like Mercury (Hg) methylation or reductive dehalogenation, vitamin B_{12} or similar corrinoids have been shown to be necessary (Zheng et al., 2024; Liao et al., 2016). Thus, vitamin B_{12} is involved in the regulation of a wide variety of metabolisms, possibly having a critical role in the structure and relationships in microbial communities from different environments.

The synthesis of vitamin B_{12} can take place through either an aerobic pathway or an anaerobic pathway. Depending on the organism and the substrate, this process can require over 30 different steps to complete, being an expensive process in both evolutionary and energetic terms (Bryant et al., 2020). To date, only prokaryotes have been reported to produce this cofactor, and as such, eukaryotes depend heavily on them to obtain it. Hence, sharing and obtaining this essential cofactor can shape ecosystems and the interactions between their components (Sokolovskaya et al., 2020; Romine et al., 2017). The most recent bulk genome analysis of bacterial production of vitamin B_{12} shows that its distribution is uneven within the different phyla. Out of 11,436 bacterial species, 37% can produce cobalamin de novo, with Pseudomonadota, Bacillota, and Actinomycota being enriched, while Bacteroidota (previously known as Proteobacteria, Firmicutes, Actinobacteria, and Bacteroidetes, respectively) (Oren and Garrity, 2021) show a depletion compared to the other phyla (Shelton et al., 2019). Phyla like Thermotogota have shown evidence of lateral gene transfer for many genes involved in the salvage and synthesis of vitamin B_{12} (Swithers et al., 2012). Interestingly enough, marine Cyanobacteria produce pseudocobalamin instead of cobalamin, which has an adenine as the lower ligand instead of DMB (Heal et al., 2017). Adding to the diversity of cofactors that can be found in the environment, it is also an opportunity for other microbes that could remodel other corrinoids into vitamin B_{12}, depending on their needs. Although Archaea can synthesize vitamin B_{12} as well, they are less studied in this regard. Regardless, Zhou et al. (2021) studied the prevalence of vitamin B_{12} synthesis genes in different environments, showing that in their study, most genes mapped in bacterial genomes (98.07%), while a small percentage mapped in archaeal genomes (1.93%). This study also shows the prevalence of phyla like Pseudomonadota, Actinobacteriota, and Bacillota as the top three phyla with more vitamin B_{12} synthesis-related genes. In the case of archaea, Euryarchaeota, Crenarchaeota, and Thaumarchaeota were the ones with more genes related to the synthesis of vitamin B_{12} (Zhou et al., 2021). Within the Euryarchaeota phylum, methanogenic archaea can synthesize vitamin B_{12} analogs that allow CO_2 fixation and methane production (Kanno et al., 2022; Wood et al., 1982).

Model organisms for the production of vitamin B_{12} include *Pseudomonas denitrificans*, *Propionibacterium shermanii*, and *Sinorhizobium meliloti*. Where *P. denitrificans* uses the aerobic pathways for B_{12} synthesis, while *P. shermanii* and *S. meliloti* use the anaerobic pathways (Moore and Warren, 2012; Rodionov et al., 2003). Yields between 200 and 300 mg/L of vitamin B_{12} are currently being achieved with a marketplace value of 339.8 million USD in 2020 (Tripathi et al., 2024). Genetic limitations due to the intrinsic regulation of the pathway, its inherent intricacy, and the lack of genetic tools for the model strains are holding back higher yields of this valuable cofactor (Fang et al., 2017). Other bacteria have been reported to produce vitamin B_{12}, like *Priestia megaterium* (previously known as *Bacillus megaterium*), *Salmonella typhimurium*, or *Thermosipho melanesiensis* (Mohammed et al., 2014; Jeter et al., 1984; Swithers et al., 2012). Some attempts have been made to increase the yield of vitamin B_{12} by bypassing riboswitches (Moore et al., 2014) or through the construction of heterologous pathways in model organisms like *Escherichia coli* (Fang et al., 2018; Ko et al., 2014).

The synthesis of adenosylcobalamin (Fig. 20.1) begins with the conversion of either glutamate or succinyl-CoA into 5-aminolevulinic acid (5 ALA), which is then transformed into porphobilinogen (PBG). Four PBG molecules are polymerized into hydroxymethylbilane (HMB), and then it is transformed into uroporphyrinogen III (Uro III). These first steps are shared among other tetrapyrrole synthesis pathways like heme groups, cofactor F_{430}, or chlorophyll. Uro III can enter either the aerobic pathway or the anaerobic pathway; aside from the need for O_2 or not, they have different enzymes, numbers of reactions, requirements, and timing for the insertion of the Co atom inside the corrin ring (Fig. 20.1). Nevertheless, both pathways require the addition of eight methyl groups, the insertion of the Co atom, its reduction to Co^{1+} and its adenylation (Bryant et al., 2020). For both routes, the end product is adenosylcobyrinic acid. Many of the enzymes involved in either the aerobic or anaerobic pathways have orthologues in the opposing one, which makes it very difficult to differentiate between aerobic and anaerobic pathways just based on sequence-based screenings (Shelton et al., 2019). The last steps attach the lower ligand, which first requires the addition of aminopropanol phosphate, followed by a GDP molecule, and finally the DMB. In this last step, the DMB molecule is first transformed into α-ribazole phosphate, which displaces the GDP molecule and finally renders adenosylcobalamin, also known as vitamin B_{12}. The synthesis of the lower ligand can also have different pathways. On the one hand, it can start from a reduced flavin like flavin mononucleotide (FMN), and through the addition of O_2, the enzyme BluB can yield DMB. On the other hand, there is an anaerobic pathway that starts from 5-aminoimidazole ribotide, an intermediary in purine biosynthesis, and that has four steps instead of just one (Bryant et al., 2020).

Considering the wide range of processes that require vitamin B_{12}, it is also very important to understand how cross-feeding takes place and what approaches exist to obtain it. Most of the genomes assayed in genomic studies do not contain the full pathway for the synthesis of vitamin B_{12}, but in many cases include a portion of it (Shelton et al., 2019; Lu et al., 2020; Zhou et al., 2021). It has been hypothesized that prokaryotes that need it but cannot produce it can remodel different cobamides, use intermediaries, and complete the final steps, or capture the final product. It has been reported that the ligands attached to cobamides can often vary depending on the prokaryote, yielding cobamides that are not vitamin B_{12} but can still be of use to other organisms (Helliwell et al., 2016). These different types of cobamides are thought to be a consequence of the availability of lower ligand bases, e.g., benzimidazole, phenolic, or purine bases, in the environment (Sokolovskaya et al., 2020). In

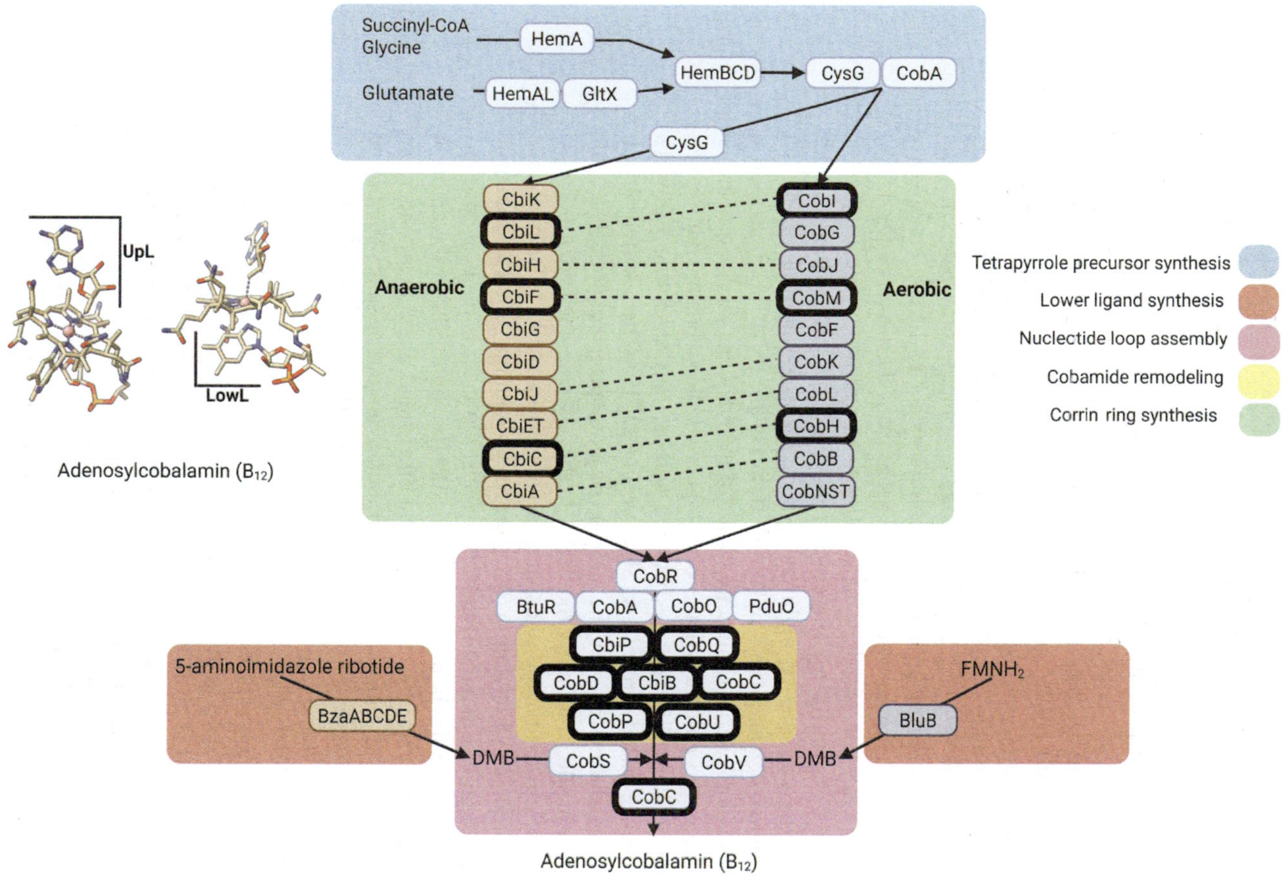

FIGURE 20.1 Adenosylcobalamin biosynthetic pathway.

Aerobic coring ring synthesis genes are highlighted in dark blue, while the anaerobic counter pathway genes are highlighted in orange. Based on the results by Shelton et al. (2019), proteins that are orthologs in the aerobic or anaerobic synthesis pathways are connected by dashed black lines. Proteins that have been determined as conserved for vitamin B_{12} production have been highlighted with thick black edges. Adenosylcobalamin 3D structure is represented on the left, with two positions, one where the upper ligand (UpL) is visible and the other one where the lower ligand (LowL) is more visible.

- Created in BioRender. Balsa, E. (2025) https://BioRender.com/ej2tydz

addition to previously mentioned pseudocobalamin production by cyanobacteria, for example, *Dehalococoides mccartyi* can salvage and remodel different cobamides into cobalamin by attaching DMB supplemented in the media (Yi et al., 2012). Environmental and genomic evidence support the presence of different cobamide molecules coexisting in marine environments (Beauvais et al., 2023; Heal et al., 2017). Therefore, the ability to salvage and remodel the available molecules to obtain the needed cobamide variant is of great importance since auxotrophs cannot synthesize it de novo. In this case, microbes rely on the proteins involved in the nucleotide loop assembly to remodel corrinoids into cobamide cofactors that suit their specific needs (Shelton et al., 2019). These genes are shown in yellow in Fig. 20.1. Starting from adenosylcobyrinic acid, these enzymes allow the synthesis of adenosylcobalamin (Villa and Escalante-Semerena, 2024). Other microbes like *Vibrio cholerae* and *Rhodobacter spharoides* have also shown remarkable capability to remodel available cobamides. *Vibrio cholerae* has a similar enzyme to CbiR, named *Vc*CobS, that can also cleave the lower ligand, but in addition, it can attach the DMB nucleotide to yield adenosylcobalamin (Ma et al., 2020). The CbiZ enzyme is present in both Archaea and Bacteria and can also release the lower ligand, but instead of acting on the phosphodiester bond, it cleaves the amide bond (Gray and Escalante-Semerena, 2009). CbiR from *Akkermansia muciniphila* aids in the conversion of pseudocobalamin to cobalamin by hydrolyzing the nucleotide loop (Mok et al., 2020), which is then a substrate to other remodeling enzymes to finally yield cobalamin.

The salvage of vitamin B_{12} also depends on transporters that can capture the different intermediaries and cobamides to ensure the procurement for the prokaryote (Fig. 20.2).

The model for the transport of vitamin B_{12} is comprised of proteins: BtuB, a TonB-dependent receptor, as the outer membrane receptor; periplasmic protein BtuF, which transports vitamin B_{12} to the ABC complex on the inner membrane; and BtuCD, which finally introduces the cofactor into the cytoplasm (Pieńko et al., 2021). CbrT can also transport vitamin B_{12} at the expense of ATP hydrolysis in *Lactobacillus delbrueckii* with high specificity for vitamin B_{12} and for the precursor cobinamide (Santos et al., 2018). On the other hand, BtuM transports vitamin B_{12} into the cell from the periplasm but lacks any known assisting proteins or ATPase motifs and binds vitamin B_{12} covalently, unlike BtuCD (Rempel et al., 2018). Other transporters like Rv1819c from *Mycobacterium tuberculosis* can fit up to six vitamin B_{12} but lack any specific high-affinity binding sites; it is assumed that it is probably a more promiscuous transporter (Rempel et al., 2020). Recently, the transporter BacA from *Mycobacterium marinum*, highly similar to Rv1819c, has been described as capable of not only importing vitamin B_{12} but also exporting it at the expense of ATP hydrolysis, being the first described system for the export of vitamin B_{12} (Nijland et al., 2024). In addition to corrinoids, lower ligands can also be salvaged since not all prokaryotes can synthesize them de novo. Transporter CblT, in coordination with kinase CblS, can import and phosphorylate α-ribazole, which is the precursor of DMB (Gray and Escalante-Semerena, 2010).

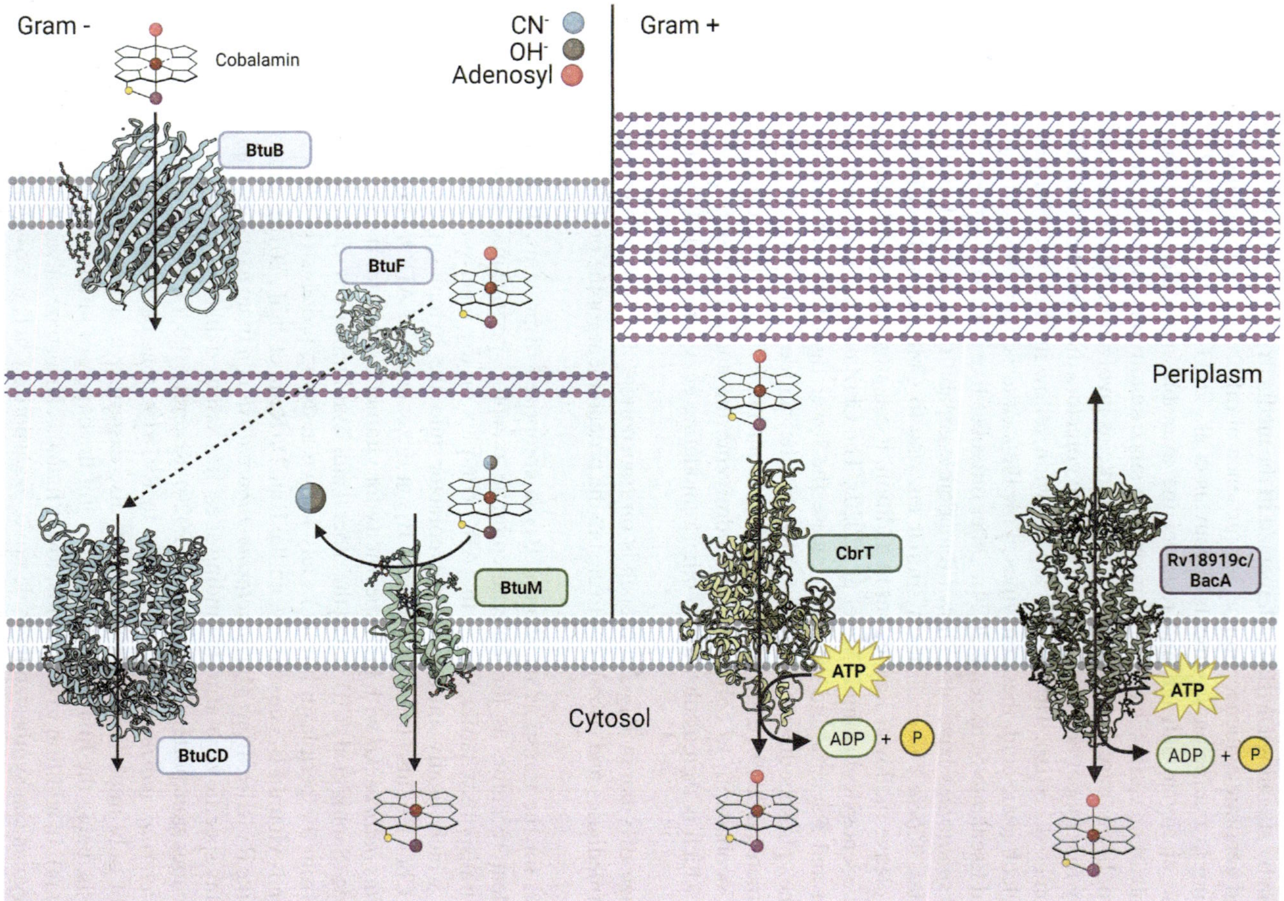

FIGURE 20.2 Cobalamin salvage transporters.

Transporters involved in the uptake of cobalamin for both Gram-negative (gram -) and Gram-positive (gram +) bacteria.

- Created in BioRender. Balsa, E. (2025) https://BioRender.com/ej2tydz

Genes involved in both the biosynthesis and salvage of vitamin B_{12} are also dependent on it for its regulation. S Adenosyl L methionine Uro III methyltransferase or CobA can be subject to negative feedback loops in response to different levels of vitamin B_{12} biosynthesis intermediaries like Uro III, although the inhibition threshold can vary between organisms (Blanche et al., 1989; Robin et al., 1991). Regardless, riboswitches have been reported as the main source of regulation. Generally, vitamin B_{12} binds to a DNA aptamer upstream of the genes, and it triggers the formation of the termination sequence, which inhibits transcription. It can also regulate translation of the mRNA. When vitamin B_{12} binds to the mRNA aptamer, it sequesters the ribosome binding site, blocking the start of translation. Transcription level regulation seems to be the most widespread for riboswitches involved with vitamin B_{12} synthesis (Nahvi et al., 2004).

In addition to enzymes that directly depend on vitamin B_{12} as a cofactor, there are operons and genes that depend on vitamin B_{12} for their expression (Romine et al., 2017). These riboswitches have ligand selectivity as well, where, depending on the microorganism, some forms of cobalamin might not be recognized by the aptamer. In *E. coli*, its riboswitch for the *btuB* gene only binds to adenosylcobalamin, while *P. megateirum metE* riboswitch can also bind purinylcobalamin in addition to adenosylcobalamin (Kennedy et al., 2022). Some of them can also have more promiscuous riboswitches like *mutA* from *Vesiculosa parvula*, which can bind not only cobalamin but also other cobamides with different lower ligands or without one (Johnson et al., 2012). It can even regulate gene expression in photosensing processes in bacteria like *Halomonas* (Romine et al., 2017). As many microorganisms can uptake different intermediaries of vitamin B_{12} in order to support growth depending on the availability of their environment (Sokolovskaya et al., 2020), having a wider range of selectivity in their riboswitches could be an evolutionary response to this environmental catalog of vitamin B_{12} intermediaries (Kennedy et al., 2022).

Environmental relevance of cobalamin

Cobalamin is an essential cofactor for many central metabolic processes, as previously discussed. Thus, its obtention plays a key role in the distribution of microorganisms in an ecosystem. For instance, it has been reported that over 50% of the algal species available in culture collections require this cofactor for their growth and depend on its obtention from an exogenous pool (Croft et al., 2005). Consequently, more attention has been paid to this vitamin, among others, as it would be a major factor controlling phenomena such as algal blooms (Cavari and Grossowicz, 1977).

The production of vitamin B_{12} is dependent on the presence of microorganisms capable of synthesizing it, as well as on the presence of Co. Vitamin B_{12} has been mainly studied in the context of aquatic environments, mainly marine, in order to study the symbiosis between algae and bacteria for the acquisition of this cofactor

(Bunbury et al., 2022). The concentrations of Co measured in the oceanic water are higher than those of vitamin B_{12}. The concentrations vary depending on the location, but in the case of Co, they range from 0.2 to 44 nM (Panzeca et al., 2009; Robertson, 1970), while the concentrations of vitamin B_{12} range from 0 to 158 pM (Kondo, 2023; Panzeca et al., 2009). In these ecosystems, it appears that cobalamin concentrations are influenced by anthropogenic activity since higher concentrations have been reported in areas affected, for instance, by sewage discharge (Panzeca et al., 2009). Another factor affecting cobalamin synthesis is the bioavailability of Co, which is related to its speciation and whether it is complexed with organic ligands (Panzeca et al., 2009; Saito et al., 2004). In these aquatic environments, the bacterial taxa involved in cobalamin production are members of the phylum Cyanobacteriota and of the classes Alphaproteobacteria, Gammaproteobacteria, and Bacteroidetes (Sañudo-Wilhelmy et al., 2014), as well as the archaeal phylum Thaumarchaeota (Doxey et al., 2015).

In contrast to aquatic environments, soils and terrestrial environments have been less studied in the context of this topic. Co concentrations are known to be higher in soils than in water, ranging from 0.4 to 29.9 μg Co/g of soil depending on the studied soil (Jörgensen and Joergensen, 2024). These higher concentrations are caused by human activity, mainly from domestic and industrial effluents (Mulligan et al., 2001). The concentrations of cobalamin reported in soils are higher than those reported in water samples, ranging from 0.08 to 14 ng/g of soil (Lu et al., 2020; Mozafar, 1994). Soils are known to harbor high microbial biomass (Khan et al., 2016). However, metagenomic studies of soils reported that less than 10% of the bacteria and archaea found in soil samples have the metabolic potential for the de novo synthesis of vitamin B_{12} (Lu et al., 2020). In fact, it has been reported that from a pool of 11,000 bacterial genomes, only 37% were predicted to be able to synthesize vitamin B_{12} de novo, and 86% of the studied bacteria encoded in their genomes for cobamide-dependent enzymes (Shelton et al., 2019). Thus, as in water environments, microorganisms will need to establish interactions with members of the surrounding communities in order to obtain this vitamin, leading to the so-called cobamide-dependent interactions (Sokolovskaya et al., 2020).

To measure it, vitamin B_{12} should first be obtained from the samples. When working with bacterial cultures, the cobalamin can be extracted by boiling the sample to release all the protein-bound vitamin B_{12} (Jörgensen and Joergensen, 2024). In the case of soil samples, it can be obtained by using an organic solvent extraction protocol (Heal et al., 2017). Finally, for water samples, it can be directly obtained from the sample, but it should be preconcentrated to quantify it. This can be achieved by using affinity columns through which the sample is pumped (Heal et al., 2014; Okbamichael and Sañudo-Wilhelmy, 2004).

There are two main ways in which cobalamin concentration can be determined: a biological and an analytical one. In the biological assay, a vitamin B_{12} dependent bacterial strain is incubated in the presence of the extracts obtained in the previous step, and then the growth of the strain is quantified over time. In parallel, the same strain is grown in the presence of known concentrations of cobalamin, and this information is used to determine the unknown concentration found in the extracts (Bunbury et al., 2022; Raux et al., 1996). In contrast, the analytical approach involves four main techniques.

1. Protein-binding assay. Labeled B_{12} is mixed with the sample that is going to be quantified. Then, this mixture is incubated in the presence of cobalamin-binding proteins, such as the intrinsic factor (IF). Finally, after the incubation, the concentration of the labeled B_{12} is quantified (Fedosov et al., 2023). Then, based on the amount of nonbound labeled B_{12}, the concentrations of the non-labeled B_{12} from the sample can be estimated.
2. Immunoassay. This assay is very similar to the previous one, but in this case, B_{12} interacts with antibodies specific for this vitamin rather than with proteins (Ispir et al., 2015).
3. Reversed-phase high-performance liquid chromatography (RP HPLC) or ultra HPLC (UHPLC) coupled to a mass spectrometer (MS). HPLC is an analytical technique that can be used to separate, identify, and quantify components from a mixture. The RP HPLC is a variation of this technique that allows the separation of polar and nonpolar molecules based on their hydrophobicity (Aguilar, 2004). The UHPLC is another variation of the method in which higher temperatures and pressures are applied, and the sample is pumped through a column that contains materials with a smaller particle size, allowing a better separation of the molecules in the sample (Taleuzzaman et al., 2015).
4. Raman microspectroscopy has also been used for the detection of cobamides inside microbial cells without the need to extract them first (Kanno et al., 2022). This is a technique in which the scattering of light (radiation) by the sample is studied. Each molecule has a characteristic Raman scattering, which can be used as a fingerprint for its identification (Kudelski, 2008).

Vitamin B_{12} in extreme environments and Río Tinto

Vitamin B_{12} can only be synthesized by a reduced set of prokaryotes, but its production is essential for the auxotrophic microorganisms, as it acts as a cofactor of many key metabolic activities. In this chapter, we have discussed this in the context of water and terrestrial environments in general, but this would also apply to extreme environments. However, to the best of our knowledge, the concentrations of vitamin B_{12} have never been studied in any extreme environment, probably due to the difficulty of working with this type of sample and also because the extreme conditions might affect vitamin B_{12}'s stability.

Although vitamin B_{12} has not been directly studied in extreme environments, the role that this cofactor plays or might play in the microorganisms that inhabit these ecosystems has been studied. For instance, two hyperthermophilic methanogenic archaea, *Methanopyrus kandleri* JCM9639, which can tolerate temperatures of up to 110°C (Kurr et al., 1991), and *Methanocaldococcus jannaschii* JCM 10,045, which has an optimal growth temperature of 85°C (Jones et al., 1983), have been reported to contain vitamin B_{12} by Raman spectroscopy analysis (Kanno et al., 2022). In these two cases, it is thought that cobamide would play a role in cell functions like DNA synthesis (Kanno et al., 2022).

Halomonas sp. HL-48 is a halophile isolated from Hot Lake, a hypersaline ecosystem found in Washington state (Khan et al., 2018). This microorganism has been reported to contain the genes necessary for vitamin B_{12} production, and it is thought that it would use this vitamin as a cofactor for ethanolamine synthesis and also as a metabolic regulator (Romine et al., 2017). Moreover, this strain has been described as one of the few microorganisms capable of vitamin B_{12} production in this ecosystem, so the current hypothesis is that it would also supply cobalamin to the auxotrophic microorganisms of the community (Romine et al., 2017).

Two isolates of the genus *Thioalkalivibrio* (*Thioalkalivibrio versutus* AL2 and *Thioalkalivibrio nitratis* ALJ2) isolated from soda lakes, which are characterized by having pH values ranging from 9.5 to 11, have been reported to overexpress genes related to vitamin B_{12} production in conditions of stress (Ahn et al., 2021). Upregulation of the cobalamin synthesis pathway has been reported in the presence of arsenite and at low temperature conditions (Ahn et al., 2019, 2021). Thus, vitamin B_{12} is most likely used to prevent further damage to the cell under stress conditions.

The role of vitamin B_{12} has also been studied in *Leptospirillum* sp. CF-1 is an acidophilic iron-oxidizing microorganism. In this case, the strain was exposed to oxidative stress in the presence and absence of vitamin B_{12}. These experiments showed that this exogenous cobamide has a protective effect against this stress, reduces the damage to biomolecules, and increases the growth and survival of the microbial cells. Moreover, this study also proved that when this strain is exposed to oxidative stress, there is an overexpression of genes related to cobalamin biosynthesis (Ferrer et al., 2016), similar to what happened in the case of the genus *Thioalkalivibrio*. Acidophiles usually live in acidic leachates, which tend to be enriched in heavy metals, especially Fe (Johnson et al., 2012).

Thus, cobalamin production would be very important, not only for the wide variety of enzymatic activities that require it, but most likely to allow microorganisms to reduce ambient stressors that might affect their viability, especially in ecosystems such as extreme environments.

Río Tinto, a river in the southwest of the Iberian Peninsula, draws its waters from the heart of the Iberian Pyrite Belt (IPB), which is one of the largest metallic sulfide deposits in the world. Known for its extremely low pH (around 2 and 3) and high concentrations of heavy metals such as Fe, Cd, Cu, and Mn, Río Tinto has been extensively studied for its extreme conditions. The river is often considered a Mars analogue, particularly for the Meridiani Planum region (Amils et al., 2014).

The ecosystem of Río Tinto hosts a rich microbial community, including both eukaryotes and prokaryotes. In the river's water column, three predominant prokaryotic genera (*Leptospirillum*, *Acidithiobacillus*, and *Acidiphilium*) are present, with *A. ferrooxidans* and *L. ferrooxidans* being the most abundant species (Amils, 2016). Although this river has been studied for more than 30 years, the presence and role of vitamin B_{12} in it have never been looked into. However, the role of this cofactor in this ecosystem can be indirectly determined by studying the genomes of the microorganisms found in it. The genomes can be analyzed to determine the ability of these microorganisms to synthesize this cofactor and also whether they encode for vitamin B_{12} dependent enzymes.

Currently, the only genome available from a microorganism isolated from Río Tinto is from *Acidiphilium* sp. PM (Martin-Uriz et al., 2014). Thus, the genomes of other acidophiles closely related to those found in the river have been used as an approximation to study the role of this cofactor. In particular, the genomes studied were those of *A. ferrooxidans* ATCC 23,270 and *L. ferrooxidans* C2-3. *Acidiphilium* sp. PM possesses the genes necessary for cobalamin synthesis, as identified by Shelton et al. (2019) (Fig. 20.3). In contrast, the type strains of *A. ferrooxidans* and *L. ferrooxidans* lack the proteins required for cobalamin synthesis (Fig. 20.3), so this could also be the case for the strains of these species found in the river. Moreover, both *Acidiphilium* sp. PM and *L. ferrooxidans* C2-3 encode in their genomes enzymes that need this vitamin as a cofactor. Therefore, we can conclude that they would need this vitamin for them to thrive in Río Tinto. In contrast, interestingly, *A. ferrooxidans* ATCC 23,270 does not encode any enzyme dependent on this cofactor. Therefore, although this microorganism cannot synthesize this vitamin, its metabolic functions seem to be independent of it, and it does not seem to require it for thriving in this ecosystem (Fig. 20.4).

In addition to these prokaryotes, cyanobacteria have been detected in biofilms in Río Tinto (Gómez et al., 2024). Cyanobacteria are capable of producing pseudocobalamin, a cobalamin analogue with adenine as the lower ligand, which can be converted into active cobalamin by other bacteria (Helliwell et al., 2016). Thus, these microorganisms could play a role as vitamin B_{12} producers in Río Tinto.

Eukaryotes in the river include Chlorophyta (e.g., *Chlamydomonas*, *Dunaliella*, and *Chlorella*), diatoms, filamentous algae like *Klebsormidium*, and flagellates like *Euglena* and *Ochromonas* (Amils, 2016). Several diatoms, such as *Thalassiosira pseudonana* (Bertrand et al., 2012), and green algae like *Chlamydomonas nivalis* (Kazamia et al., 2012), have been reported as auxotrophs for cobalamin. In fact, studies suggest that approximately 50% of cultured eukaryotic phytoplankton species require external vitamin B_{12} for growth, with 18 species involved in algal blooms also exhibiting this dependency (Bertrand and Allen, 2012; Tang et al., 2010). Therefore, these eukaryotes would need to establish interactions with other vitamin B_{12} producing microorganisms to be able to thrive in this ecosystem.

In conclusion, based on its genetic content, *Acidiphilium* sp. PM is most likely a cobalamin producer widespread throughout the river's water column. As such, this strain could sustain the vitamin B_{12} auxotrophic fractions of the algal and

		Tetrapyrrole precursor synthesis	Corrin ring synthesis	Nucleotide loop assembly
Deep subsurface of the IPB	*Aestuariimicrobium* sp. T2.26MG-19.2	●	○	◍
	Brevundimonas sp. T2.26MG-97	●	○	○
	Cellulomonas sp. T2.22MG-43	●	○	○
	Ciceribacter sp. T2.26MG-112.2	●	●	●
	Ciceribacter sp. T2.30D-1.1	●	●	●
	Citrobacter sp. T1.2D-1	●	●	●
	Desulfosporosinus sp. DEEP	○	●	●
	Lelliottia sp. T2.26D-8	●	○	○
	Microbacterium sp. T2.11-28	◍	○	○
	Niallia sp. T2.9-1	●	○	●
	Nocardioides sp. T2.26MG-1	●	●	◍
	Paenibacillus sp. T2.5-46A	●	○	◍
	Pleomorphomonas sp. T1.2MG-36	●	●	●
	Propionicimonas sp. T2.31MG-1	●	●	●
	Pseudomonas sp. T2.1D-1.1	●	●	●
	Shewanella sp. *T2.3D-1.1*	●	○	●
	Stutzerimonas sp. T2.31D-1	●	○	●
	Rhodoplanes sp. T2.26MG-98	●	●	●
	Tessaracoccus sp. T2.5-30	●	●	●
Río Tinto	*Acidithiobacillus ferrooxidans* ATCC 23270	●	○	○
	Leptospirillum ferrooxidans C2-3	●	◍	◍
	Acidiphilium sp. PM	●	●	●

Legend (%)

○ 0 < X < 0,33 ◍ 0,33 < X < 0,66 ● 0,66 < X < 1

FIGURE 20.3 Cobalamin synthesis protein distribution within the isolates' genomes.

The upper table contains a summary of the results found in the genomes of the isolates from the deep subsurface, while the lower table contains the results related to the river. The fill inside each circle corresponds to an interval of percentages corresponding to the number of proteins in each phase module for the synthesis of cobalamin, as indicated in the legend.

prokaryotic communities. Due to the presence of cyanobacteria, pseudocobalamin may also be present in the river, suggesting that the microbial community could harbor genes for converting it into active cobalamin. Not only would cobalamin sustain growth for deeply conserved enzymes for both prokaryotes and eukaryotes, but it could also reduce the oxidative stress generated by the presence of metals, which would be of high relevance in an extreme environment like Río Tinto, which contains high concentrations of them.

		Vitamin B12 synthesis	Salvage	B12 dependency
Deep subsurface of the IPB	*Aestuariimicrobium* sp. T2.26MG-19.2	Unlikely	Unlikely	Highly likely
	Brevundimonas sp. T2.26MG-97	Unlikely	Unlikely	Highly likely
	Cellulomonas sp. T2.22MG-43	Unlikely	Highly likely	Unlikely
	Ciceribacter sp. T2.26MG-112.2	Probably	Probably	Highly likely
	Ciceribacter sp. T2.30D-1.1	Probably	Probably	Highly likely
	Citrobacter sp. T1.2D-1	Highly likely	Highly likely	Highly likely
	Desulfosporosinus sp. DEEP	Highly likely	Unlikely	Highly likely
	Lelliottia sp. T2.26D-8	Unlikely	Highly likely	Highly likely
	Microbacterium sp. T2.11-28	Unlikely	Probably	Highly likely
	Niallia sp. T2.9-1	Unlikely	Probably	Highly likely
	Nocardioides sp. T2.26MG-1	Probably	Probably	Highly likely
	Paenibacillus sp. T2.5-46A	Unlikely	Unlikely	Highly likely
	Pleomorphomonas sp. T1.2MG-36	Probably	Probably	Highly likely
	Propionicimonas sp. T2.31MG-1	Probably	Unlikely	Highly likely
	Pseudomonas sp. T2.1D-1.1	Highly likely	Highly likely	Highly likely
	Shewanella sp. *T2.3D-1.1*	Unlikely	Highly likely	Highly likely
	Stutzerimonas sp. T2.31D-1	Unlikely	Probably	Highly likely
	Rhodoplanes sp. T2.26MG-98	Probably	Probably	Highly likely
	Tessaracoccus sp. T2.5-30	Probably	Unlikely	Highly likely
Río Tinto	*Acidithiobacillus ferrooxidans* ATCC 23270	Unlikely	Unlikely	Unlikely
	Leptospirillum ferrooxidans C2-3	Probably	Unlikely	Highly likely
	Acidiphilium sp. PM	Highly likely	Unlikely	Highly likely

Legend

○ Unlikely (white) — Probably (yellow) — Highly likely (green)

FIGURE 20.4 Vitamin B_{12} production and auxotrophy in the microorganisms related to Río Tinto and its subsurface.

Summary of cobalamin synthesis pathway completion, cobalamin transporters (salvage pathway), and the presence in the genomes of highly conserved enzymes that use this vitamin as a cofactor (epoxyqueuosine reductase, ribonucleotide reductase, methionine synthase, and methylmalonyl-CoA mutase). For cobalamin synthesis and salvage, white-filled circles stand for very unlikely, yellow-colored circles stand for probable, and green for highly likely, according to the threshold set by Shelton et al. (2019). For cobalamin auxotrophy, white-filled circles indicate the absence of auxotrophy, while the black-filled circles indicate auxotrophy.

Cobalamin in the deep subsurface of the Iberian Pyrite Belt

As discussed in other chapters of this book, the origin of the extreme conditions found in Río Tinto would be a bioreactor found in the deep subsurface of the IPB. Since vitamin B_{12} is now known to be essential for some of the microorganisms found in Río Tinto, that could also be the case in the deep subsurface.

Poly extreme environments like the deep subsurface remain a frontier for geomicrobiological studies due to technical difficulties to explore them. Microbes that inhabit such environments face conditions like high temperatures, low nutrient concentration and availability, high pressure, and low water activity (Bell et al., 2020; Escudero et al., 2018a). Unlike surface ecosystems that rely on photosynthesis, the deep subsurface operates through biogeochemical cycles that depend on chemosynthetic primary producers, hydrogen produced both geochemically and biologically (Sanz et al., 2021; Gregory et al., 2019), and electron acceptors necessary for anaerobic respiration in anoxic settings. Regardless of its inherent difficulties, some authors have tried to study its particular traits by either studying groundwater (Zhang et al., 2020; Daly et al., 2019) or its hard-rock matrix (Nuppunen-Puputti et al., 2022; Bell et al., 2020; Amils et al., 2023), showing that the deep subsurface contains a rich and diverse environment yet to be well understood. It has even been estimated that around 15% (68 Gt of C) of the global biomass is allocated in deep subsurface environments, being mostly composed of prokaryotes (Bar-On et al., 2018). Although other authors like Magnabosco et al. (2018) estimate that the continental deep subsurface holds between 4.2 and 12.6 Gt of C. Even though DNA extraction from these environments is extremely complicated, many studies have reported 16S-based community composition. Using the available data on deep terrestrial subsurface communities. Soares et al. (2023) revealed that phyla Bacillota and Pseudomonadota are predominant within the evaluated samples, with special mention to the order Burkholderiales and the genus *Pseudomonas* due to its widespread abundance in the analyzed samples.

Through several drilling projects and with a wide array of culture dependent and culture-independent techniques, the deep subsurface of the IPB has been sampled and probed for life and its signatures. Results point to an underground bioreactor that dissolves minerals and increases nutrient availability for the communities present in this environment. Over a span of 600 m of rock, biogeochemical cycles for nitrogen, carbon, iron, hydrogen, and sulfur have been documented (Fernández-Remolar et al., 2008; Puente-Sánchez et al., 2018; Amils et al., 2023).

Deep subsurface environments have the same requirements as other environments: a need for the essential building blocks of life, C, H, O, N, P, and S, but also micronutrients like Fe or vitamins like vitamin B_{12}. The deep subsurface of the IPB is no different in this regard. Recent comparative genomic analysis shows that bacteria isolated in this environment could depend on vitamin B_{12} for several enzymatic activities, like the methylmalonyl CoA mutase, have transporters that could import this cofactor, and have a putative primary producer that would also produce vitamin B_{12} (Mateos et al., 2023).

In the available pool of genomes, three have returned the necessary genes for the synthesis of vitamin B_{12}: *Desulfosporosinus* sp. DEEP, *Citrobacter* sp. T1.2D-1, and *Pseudomonas* sp. T2.1D-1.1 (Fig. 20.3). This is also particularly interesting since *Desulfosporosinus* sp. DEEP encodes the protein AcsCDE methyltransferase involved in the Wood Ljungdahl pathway for CO_2 fixation (Banerjee and Ragsdale, 2003). This isolate has been detected throughout the drilling core of the IPB from the 90 m to the 607 m below ground samples (Amils et al., 2023). Other genomes do have an almost complete set of vitamin B_{12} synthesis genes, like both *Ciceribacter* genomes (strains T2.26 MG-112.2 and T2.30D-1.1), *Rhodoplanes* sp. T2.26 MG-98, and *Pleomorphomonas* sp. T2.1D-1.1. Perhaps instead of synthesizing de novo, they could be remodeling available intermediaries into different cobamides or other tetrapyrrolic groups. Regardless, all genomes point to some form of auxotrophy for vitamin B_{12} in different enzymatic activities (Fig. 20.4). Remodeling enzymes like CbiB, CobA, CobU, CobS, and CobC are widespread within the studied microorganisms, although just five genomes return positive results for all of these enzymes. None of the available genomes has the complete set of proteins for the anaerobic synthesis of the lower ligand (BzABCDF). Interestingly enough, the BluB protein for the aerobic synthesis of the lower ligand from $FMNH_2$ has eight positive results out of the 18 assayed genomes. Noncanonical remodeling enzymes like CbiZ have also returned positive matches in *Niallia* sp. T2.9-1, *Paenibacillus* sp. T2-5-46A, and *Pleomorphomonas* T1.2 MG-36.

Cyanobacteria have been detected in the deep subsurface of the IPB both through Catalyzed Reporter Deposition Fluoresence in situ Hybridization (CARD FISH) and metagenomics (Puente-Sánchez et al., 2018; Amils et al., 2023). This is of special interest since Cyanobacteria have been reported to produce pseudocobalamin instead of adenosylcobalamin (Helliwell et al., 2016), which would amount to the presence of cobamide and increase the relevance of remodeling cobamide molecules to suit the needs of each prokaryote. Available annotation of the metagenomes from the IPBs' deep subsurface reveals multiple sequences involved in cobalamin synthesis associated with cyanobacteria. These include most of the tetrapyrrole precursor synthesis, as well as the coring synthesis and the nucleotide assembly proteins. Furthermore, cyanobacterial-associated sequences would hold all the necessary proteins, according to Shelton et al. (2019), threshold to synthesize cobalamin. Nevertheless, the resulting cofactor should include adenine as the lower ligand due to the overall absence of DMB synthesis genes in the analyzed genomes of cyanobacteria (Heal et al., 2017; Helliwell et al., 2016), although some cyanobacteria, like the genus *Trichodesmium*, encode the aerobic DMB synthesis protein, BluB, suggesting that they could produce DMB and therefore produce adenosylcobalamin (Walworth et al., 2018). In fact, a fraction of the sequences from this metagenome correspond to the genus *Trichodesmium* (Puente-Sánchez et al., 2018). Overall, these deep subsurface cyanobacteria would produce pseudocobalamin and perhaps a smaller fraction of adenosylcobalamin.

Bacteria like *Vibrio cholerae* and *Rhodobacter sphaeroides* have been shown to be capable of remodeling pseudocobalamin into cobalamin by removing the adenine lower ligand and introducing a DMB moiety (Gray and Escalante-Semerena, 2009; Ma et al., 2020). Given that CbiZ has been detected in our pool of genomes, it is possible that bacteria living in the deep subsurface could remodel pseudocobalamin into cobamide cofactors, like adenosylcobalamin. Additionally, methanogenic archaea have also been detected in the IPBs' deep subsurface (Sanz et al., 2021) and can produce vitamin B_{12} analogs like pseudocobalamin and factor III, which instead of DMB as the lower ligand, has a 5′ hydroxybenzimidazole molecule (Kannó et al., 2022). Both factor III and pseudocobalamin can facilitate the transfer of a methyl group to coenzyme M, which will then yield CH_4 (Muñoz-Velasco et al., 2018; Stupperich and Kräutler, 1988). Therefore, the deep subsurface would hold different pools of cobamides, which suit the different needs of its microbiome. Similar phenomena have also been observed in marine ecosystems where Thaumarchaeota and heterotrophic bacteria are the main source of cobalamin and cyanobacteria produce pseudocobalamin with different distribution patterns along the water column (Doxey et al., 2015; Heal et al., 2017).

The need for cobalamin can be derived from the predicted enzymes from the available pool of genomes from IPBs' deep subsurface. Cobalamin-dependent enzymes are also prevalent in the deep subsurface. Deeply rooted enzymatic activities like the synthesis of methionine, the synthesis of deoxyribonucleotides, and the conversion of epoxyqueosine to queosine in the synthesis of several amino acids (Miles et al., 2011) are present in almost all of the genomes. Although cobalamin-independent counterparts exist for these enzymes, some isolates lack them. For example, nine out of 18 genomes lack the MetE protein for the synthesis of methionine, and that does not require cobalamin. In the case of the nucleotide reductase, 15 genomes present the cobalamin-dependent enzyme NrdJZ/NrdZ; six of them do not have either the aerobic or anaerobic cobalamin-independent nucleotide reductase. The epoxyqueosine reductase cobalamin-dependent enzyme (QueG) has returned 11 positive results, while its B_{12}-independent counter (QueH) has returned none. Less conserved enzymes, such as 2 hydroxybutanoyl-CoA mutase and glycerol dehydratase, were detected in the sample, although no novel cobalamin-independent counterparts have been reported. With the exception of *Cellulomonas* sp. T2.22 MG-43, all of the isolates obtained from the IPBs' deep subsurface depend on cobalamin for at least one highly conserved enzymatic activity (Fig. 20.4).

In spite of the apparent need for this cofactor, most of the IPB deep subsurface genomes lack any known means of import of this cofactor, in spite of its apparent dependence (Mateos et al., 2023). Only *S. putrefaciens* T2.3D-1.1, *Pseudomonas* sp. T2.1D-1.1, *Citrobacter* sp. T1.2D-1, and *Lelliottia* sp. T2.23D-8 have the BtuB outer membrane transporter plus the ABC transporter BtuCDF, which allows the transport of cobalamin into the interior of the cell (Pieńko et al., 2021). Other less studied transporters like BtuM or Rv1819c from *Mycobacterium tuberculosis* have also returned one positive result each. Thus, these isolates should have alternate ways to obtain cobamides that have not been found. Additional proteins involved in the salvage of the lower ligand and its precursors (CblT) did not return any positive results in any of the available genomes of the IPBs' deep subsurface.

In the deep subsurface, microbial communities are tightly packed inside rock pores, forming biofilms where microbes most likely grow in a mutualistic fashion (Escudero et al., 2018). In such biofilms, it must be crucial to share different metabolites in order to survive the extreme limitation of nutrients in the deep subsurface. In the IPBs' deep subsurface, acetate, propionate, oxalate, H_2, CO_2, and CH_4 have been detected throughout 600 m of rock (Puente-Sánchez et al., 2018; Sanz et al., 2021), which could sustain growth for many different organisms, especially plausible cobamide producers like cyanobacteria, methanogenic archaea, or the *Desulfosporosinus* genus. In this complex and diverse environment where different pools of cobamides could coexist, the capability of remodeling could explain incomplete vitamin B_{12} pathways in the available genomes of isolates of the deep subsurface in order to obtain their needed cofactors in spite of structural diversity in the available precursors (Villa and Escalante-Semerena, 2024). Representatives from the genus *Shewanella* have been reported to connect periplasmic spaces through nanotubes and increase metabolic rates (Jiang et al., 2020). Such connections could even allow the exchange of some molecules (D'Souza et al., 2018), like cobamides, especially for those cells that might not have the necessary transporters.

Although it is not possible to differentiate between orthologs for O_2 dependent or independent pathways for B_{12} synthesis, recent studies have shown that small quantities of O_2 could be generated in environments that have been traditionally considered as anoxic. These O_2 anomalies have been measured in anoxic sediments, methane seeps in anoxic aquifers, or deep crystalline bedrock (Sweetman et al., 2024; Ruff et al., 2024). Some of the metabolisms that have been reported to generate O_2 include the dismutation of NO ($2NO \rightarrow N2 + O2$). NO can be generated through processes like denitrification, nitrate-dependent iron oxidation (NDFO), or dissimilatory nitrate reduction to ammonium (DNRA).

Denitrification is a biological process that consists of the reduction of nitrate (NO_3^-) to nitrogen gas (N_2). The reduction of NO_3^- generates NO_2^- which is then reduced to NO. NO can then be reduced to N_2O to finally yield N_2 after another reduction (Cutruzzolà, Frankenberg-Dinkel and O'Toole, 2016). Other prokaryotes can instead reduce NO_3^- to NH_4^+, and NO is produced due to the dismutation of HNO_2 in acid and anoxic conditions (Kern and Simon, 2009; Egas et al., 2024). Additionally, NDFO can also produce NO due to the reaction of Fe^{2+} with NO_2^- (Bryce et al., 2018). These processes are widespread in nature and have been reported in vitro and in silico in samples and isolates from the IPBs' deep subsurface (Amils et al., 2023; Gallego-Rodríguez et al., 2023). The deep subsurface of the IPB is rich in metallic sulfides, especially in pyrite (FeS_2). Its groundwater contains NO_3^-, with both neutral and acid pHs and Fe^{2+} and Fe^{3+}. Thus, creating a favorable environment for the generation of NO and its dismutation. Taking all of this into account, it is not beyond the realm of possibility that the existence of intermittent wisps of O_2 could not only sustain growth but also facilitate the aerobic synthesis of vitamin B_{12}. The absence of proteins involved in the anaerobic synthesis of the lower ligand, opposed to the presence of the BluB protein for its aerobic synthesis, could be a symptom of the presence of small quantities of O_2 in the IPBs' deep subsurface.

Overall, the IPBs' deep subsurface available genomes have on the genomic level clear indications of cobalamin auxotrophy (Fig. 20.4), especially for deeply conserved enzymes like the methionine synthase, ribonucleotide reductase, or epoxyqueuosine reductase. Additionally, we have identified possible cobalamin producers like *Desulfosporosinus* sp. DEEP, *Citrobacter* sp. T1.2D-1, and *Pseudomonas* sp. T2.1D-1.1, which could supply this essential cofactor. The presence of cyanobacteria and methanogenic archaea could hint at the existence of different cobamide pools. Thus, cobamide remodeling could also play a big role in the cofeeding of cobalamin in the biofilms of the IPBs' deep subsurface. Remodeling proteins are present in the available genomic sample, although just five genomes contain all of the necessary proteins to remodel precursors into adenosylcobalamin. The salvage of these cobamides depends on a series of transporters like BtuB or BtuM, which are indeed present on some of the IPBs' deep subsurface isolates' genomes. Still, for isolates like *Aestuariimicrobium* sp. T2.26 MG-19.2 or *Propionicimonas* sp. T2.31MG-1 There is no clear path for cobalamin or its precursors to get inside the cell (Fig. 20.4). There is still much to unravel about the role of important cofactors like cobalamin in extreme environments like the deep subsurface. Regardless, within the available genomes, there is a clear pattern of auxotrophy, with several possible producers. Thus, shining some light on the relevance that cobalamin might have in this extreme environment.

Concluding remarks

Cobalamin, commonly known as vitamin B_{12}, is a cofactor that has a deep impact in many different environments and across all domains of life. The IPB's Río Tinto and its deep subsurface do not seem to contradict this. From a genomic point of view, three isolates from the deep subsurface and one from the river could synthesize cobalamin. Additionally, most of the genomes collected from these environments seem to heavily depend on cobalamin for deeply conserved activities, except for one isolate from the deep subsurface and another one from the river. In the case of Río Tinto, although there is no information regarding the dependence of its eukaryotic communities on this cofactor, global studies on phytoplankton suggest that auxotrophy would be relevant as well. Moreover, the presence of cyanobacteria in both the deep subsurface and the basin of the river adds another layer of complexity since these microorganisms could produce pseudocobalamin instead of cobalamin. Regardless, the potential of cobalamin and pseudocobalamin remodeling would be low since most genomes lack the proteins required for this process. Nevertheless, the production of cobalamin appears especially relevant in extreme environments, where it not only sustains the growth of auxotrophic organisms, but also protects against different stressors such as high metal concentration and the associated oxidative stress.

AI disclosure

During the preparation of this work, the author(s) used ChatGPT in order to correct and modify grammar and sentences. After using this tool/service, the author(s) reviewed and edited the content as needed and take(s) full responsibility for the content of the publication.

References

Aguilar, M.I., 2004. Reversed-phase high-performance liquid chromatography. Methods Mol. Biol. 251, 9–22.

Ahn, A.-C., Jongepier, E., Merijn Schuurmans, J., Rijpstra, W.I.C., Sinninghe Damsté, J.S., Galinski, E.A., Roman, P., Sorokin, D., Muyzer, G., Mackelprang, R., 2021. Molecular and physiological adaptations to low temperature in Thioalkalivibrio strains isolated from soda lakes with different temperature regimes. mSystems 6 (2). https://doi.org/10.1128/msystems.01202-20.

Ahn, A.-C., Cavalca, L., Colombo, M., Merijn Schuurmans, J., Sorokin, D.Y., Muyzer, G., 2019. Transcriptomic analysis of two Thioalkalivibrio species under arsenite stress revealed a potential candidate gene for an alternative arsenite oxidation pathway. Front. Microbiol. 10. https://doi.org/10.3389/fmicb.2019.01514.

Amils, R., Escudero, C., Oggerin, M., Sánchez, F.P., Rodríguez, A.A., Remolar, D.F., Rodríguez, N., Villadangos, M.G., Sanz, J.L., Briones, C., Román, M.S.-, Gómez, F., Leandro, T., Moreno-Paz, M., Ballesteros, O.P.-, Molina, A., Tornos, F., Andrea, I.S.-, Timmis, K., Pieper, D.H., Parro, V., 2023. Coupled C, H, N, S and Fe biogeochemical cycles operating in the continental deep subsurface of the Iberian Pyrite Belt. Environ. Microbiol. 25 (2), 428–453. https://doi.org/10.1111/1462-2920.16291.

Amils, R., 2016. Lessons learned from thirty years of geomicrobiological studies of Río Tinto. Res. Microbiol. 167 (7), 539–545. https://doi.org/10.1016/j.resmic.2016.06.001.

Amils, R., Fernández-Remolar, D., Team, T.I., 2014. Río tinto: a geochemical and mineralogical terrestrial analogue of Mars. Life 4 (3), 511–534. https://doi.org/10.3390/life4030511.

Banerjee, R., Ragsdale, S.W., 2003. The many faces of vitamin B12: catalysis by cobalamin-dependent enzymes. Annu. Rev. Biochem. 72, 209–247. https://doi.org/10.1146/annurev.biochem.72.121801.161828.

Bar-On, Y.M., Phillips, R., Milo, R., 2018. The biomass distribution on Earth. Proc. Natl. Acad. Sci. 115 (25), 6506–6511. https://doi.org/10.1073/pnas.1711842115.

Beauvais, M., Schatt, P., Montiel, L., Logares, R., Galand, P.E., Bouget, F.Y., 2023. Functional redundancy of seasonal vitamin B12 biosynthesis pathways in coastal marine microbial communities. Environ. Microbiol. 25 (12), 3753–3770. https://doi.org/10.1111/1462-2920.16545.

Bell, E., Lamminmäki, T., Alneberg, J., Andersson, A.F., Qian, C., Xiong, W., Hettich, R.L., Frutschi, M., Bernier-Latmani, R., 2020. Active sulfur cycling in the terrestrial deep subsurface. ISME J. 14 (5), 1260–1272. https://doi.org/10.1038/s41396-020-0602-x.

Bertrand, E.M., Allen, A.E., Dupont, C.L., Norden-Krichmar, T.M., Bai, J., Valas, R.E., Saito, M.A., 2012. Influence of cobalamin scarcity on diatom molecular physiology and identification of a cobalamin acquisition protein. Proc. Natl. Acad. Sci. U. S. A 109 (26), E1762. https://doi.org/10.1073/pnas.1201731109.

Bertrand, E.M., Allen, A.E., 2012. Influence of vitamin B auxotrophy on nitrogen metabolism in eukaryotic phytoplankton. Front. Microbiol. 3. https://doi.org/10.3389/fmicb.2012.00375.

Blanche, F., Debussche, L., Thibaut, D., Crouzet, J., Cameron, B., 1989. Purification and characterization of S-adenosyl-L-methionine: uroporphyrinogen III methyltransferase from Pseudomonas denitrificans. J. Bacteriol. 171 (8), 4222–4231. https://doi.org/10.1128/jb.171.8.4222-4231.1989.

Borovok, I., Gorovitz, B., Yanku, M., Schreiber, R., Gust, B., Chater, K., Aharonowitz, Y., Cohen, G., 2004. Alternative oxygen-dependent and oxygen-independent ribonucleotide reductases in Streptomyces : cross-regulation and physiological role in response to oxygen limitation. Mol. Microbiol. 54 (4), 1022–1035. https://doi.org/10.1111/j.1365-2958.2004.04325.x.

Bryant, D.A., Hunter, C.N., Warren, M.J., 2020. Biosynthesis of the modified tetrapyrroles—the pigments of life. J. Biol. Chem. 295 (20), 6888–6925. https://doi.org/10.1074/jbc.REV120.006194.

Bryce, C., Blackwell, N., Schmidt, C., Otte, J., Huang, Y.-M., Kleindienst, S., Tomaszewski, E., Schad, M., Warter, V., Peng, C., Byrne, J.M., Kappler, A., 2018. Microbial anaerobic Fe(II) oxidation – ecology, mechanisms and environmental implications. Environ. Microbiol. 20 (10), 3462–3483. https://doi.org/10.1111/1462-2920.14328.

Bunbury, F., Deery, E., Sayer, A.P., Bhardwaj, V., Harrison, E.L., Warren, M.J., Smith, A.G., 2022. Exploring the onset of B12-based mutualisms using a recently evolved Chlamydomonas auxotroph and B12-producing bacteria. Environ. Microbiol. 24 (7), 3134–3147. https://doi.org/10.1111/1462-2920.16035.

Cavari, B., Grossowicz, N., 1977. Seasonal distribution of vitamin B12 in Lake Kinneret. Appl. Environ. Microbiol. 34 (2), 120–124. https://doi.org/10.1128/aem.34.2.120-124.1977.

Croft, M.T., Lawrence, A.D., Raux-Deery, E., Warren, M.J., Smith, A.G., 2005. Algae acquire vitamin B12 through a symbiotic relationship with bacteria. Nature 438 (7064), 90–93. https://doi.org/10.1038/nature04056.

Cutruzzolà, F., Frankenberg-Dinkel, N., O'Toole, G.A., 2016. Origin and impact of nitric oxide in Pseudomonas aeruginosa biofilms. J. Bacteriol. 198 (1), 55–65. https://doi.org/10.1128/jb.00371-15.

Daly, R.A., Roux, S., Borton, M.A., Morgan, D.M., Johnston, M.D., Booker, A.E., Hoyt, D. W., Meulia, T., Wolfe, R.A., Hanson, A.J., Mouser, P.J., Moore, J.D., Wunch, K., Sullivan, M.B., Wrighton, K.C., Wilkins, M.J., 2019. Viruses control dominant bacteria colonizing the terrestrial deep biosphere after hydraulic fracturing. Nat. Microbiol. 4 (2), 352–361. https://doi.org/10.1038/s41564-018-0312-6.

Deobald, D., Adrian, L., Schöne, C., Rother, M., Layer, G., 2018. Identification of a unique Radical SAM methyltransferase required for the sp3-C-methylation of an arginine residue of methyl-coenzyme M reductase. Sci. Rep. 8 (1). https://doi.org/10.1038/s41598-018-25716-x.

Doxey, A.C., Kurtz, D.A., Lynch, M.D.J., Sauder, L.A., Neufeld, J.D., 2015. Aquatic metagenomes implicate Thaumarchaeota in global cobalamin production. ISME J. 9 (2), 461–471. https://doi.org/10.1038/ismej.2014.142.

D'Souza, G., Shitut, S., Preussger, D., Yousif, G., Waschina, S., Kost, C., 2018. Ecology and evolution of metabolic cross-feeding interactions in bacteria. Nat. Prod. Rep. 35 (5), 455–488. https://doi.org/10.1039/c8np00009c.

Egas, R.A., Kurth, J.M., Boeren, S., Sousa, D.Z., Welte, C.U., Sánchez-Andrea, I., 2024. A novel mechanism for dissimilatory nitrate reduction to ammonium in Acididesulfobacillus acetoxydans. mSystems 9 (3). https://doi.org/10.1128/msystems.00967-23.

Escudero, C., Vera, M., Oggerin, M., Amils, R., 2018. Active microbial biofilms in deep poor porous continental subsurface rocks. Sci. Rep. 8 (1). https://doi.org/10.1038/s41598-018-19903-z.

Fang, H., Dong, L., Kang, J., Jiang, P., Sun, J., Zhang, D., 2018. Metabolic engineering of Escherichia coli for de novo biosynthesis of vitamin B12. Nat. Commun. 9 (1). https://doi.org/10.1038/s41467-018-07412-6.

Fang, Kang, Zhang, 2017. Microbial production of vitamin B12: a review and future perspectives. Microb. Cell Fact. 16.

Fedosov, S.N., Nexo, E., Heegaard, C.W., Goldin, J., Mason, J.B., 2023. Protein binding assays for an accurate differentiation of vitamin B12 from its inactive analogue. A study on edible cricket powder. Food Chem. X 19. https://doi.org/10.1016/j.fochx.2023.100824.

Fernández-Remolar, D.C., Prieto-Ballesteros, O., Rodríguez, N., Gómez, F., Amils, R., Gómez-Elvira, J., Stoker, C.R., 2008. Underground habitats in the Río tinto basin: a model for subsurface life habitats on mars. Astrobiology 8 (5), 1023–1047. https://doi.org/10.1089/ast.2006.0104.

Ferrer, A., Rivera, J., Zapata, C., Norambuena, J., Sandoval, Á., Chávez, R., Omar, O., Levicán, G., 2016. Cobalamin protection against oxidative stress in the acidophilic iron-oxidizing bacterium Leptospirillum group II CF-1. Front. Microbiol. 7 (MAY). https://doi.org/10.3389/fmicb.2016.00748.

Gagsteiger, J., Jahn, S., Heidinger, L., Gericke, L., Andexer, J.N., Friedrich, T., Loenarz, C., Layer, G., 2022. A cobalamin-dependent radical SAM enzyme catalyzes the unique Cα-methylation of glutamine in methyl-coenzyme M reductase. Angew. Chem., Int. Ed. 61 (32). https://doi.org/10.1002/anie.202204198.

Gallego-Rodríguez, V., Martínez-Bonilla, A., Rodríguez, N., Amils, R., 2023. Dark fermentation in the dark biosphere: the case of Citrobacter sp. T1.2D-12. Fermentation 9 (10), 887. https://doi.org/10.3390/fermentation9100887.

Gómez, F., Rodríguez, N., Rodríguez-Manfredi, J.A., Escudero, C., Carrasco-Ropero, I., Martínez, J.M., Ferrari, M., De Angelis, S., Frigeri, A., Fernández-Sampedro, M., Amils, R., 2024. Association of acidotolerant cyanobacteria to microbial mats below pH 1 in acidic mineral precipitates in Río Tinto River in Spain. Microorganisms 12 (4). https://doi.org/10.3390/microorganisms12040829.

Gray, M.J., Escalante-Semerena, J.C., 2010. A new pathway for the synthesis of α-ribazole-phosphate in Listeria innocua. Mol. Microbiol. 77 (6), 1429–1438. https://doi.org/10.1111/j.1365-2958.2010.07294.x.

Gray, M.J., Escalante-Semerena, J.C., 2009. The cobinamide amidohydrolase (cobyric acid-forming) CbiZ enzyme: a critical activity of the cobamide remodelling system of Rhodobacter sphaeroides. Mol. Microbiol. 74 (5), 1198–1210. https://doi.org/10.1111/j.1365-2958.2009.06928.x.

Gregory, S.P., Barnett, M.J., Field, L.P., Milodowski, A.E., 2019. Subsurface microbial hydrogen cycling: natural occurrence and implications for industry. Microorganisms 7 (2). https://doi.org/10.3390/microorganisms7020053.

Heal, K.R., Qin, W., Ribalet, F., Bertagnolli, A.D., Coyote-Maestas, W., Hmelo, L.R., Moffett, J.W., Devol, A.H., Virginia Armbrust, E., Stahl, D.A., Ingalls, A.E., 2017. Two distinct pools of B 12 analogs reveal community interdependencies in the ocean. Proc. Natl. Acad. Sci. 114 (2), 364–369. https://doi.org/10.1073/pnas.1608462114.

Heal, K.R., Carlson, L.T.r., Devol, A.H., Armbrust, E.V., Moffett, J.W., Stahl, D.A., Ingalls, A. E., 2014. Determination of four forms of vitamin B12 and other B vitamins in seawater by liquid chromatography/tandem mass spectrometry. Rapid communications in mass spectrometry : RCM (Rapid Commun. Mass Spectrom.) 28 (22), 2398–2404. https://doi.org/10.1002/rcm.7040.

Helliwell, K.E., Lawrence, A.D., Holzer, A., Kudahl, U.J., Sasso, S., Kräutler, B., Scanlan, D. J., Warren, M.J., Smith, A.G., 2016. Cyanobacteria and eukaryotic algae use different chemical variants of vitamin B12. Curr. Biol. 26 (8), 999–1008. https://doi.org/10.1016/j.cub.2016.02.041.

Ispir, E., Serdar, M.A., Ozgurtas, T., Gulbahar, O., Akln, K.O., Yesildal, F., Kurt, I., 2015. Comparison of four automated serum vitamin B12 assays. Clin. Chem. Lab. Med. 53 (8), 1205–1213. https://doi.org/10.1515/cclm-2014-0843.

Jeter, R.M., Olivera, B.M., Roth, J.R., 1984. Salmonella typhimurium synthesizes cobalamin (vitamin B12) de novo under anaerobic growth conditions. J. Bacteriol. 159 (1), 206–213. https://doi.org/10.1128/jb.159.1.206-213.1984.

Jiang, M., Xiong, Z., Chen, Y., 2020. Enhancement of denitrification performance with reduction of nitrite accumulation and N_2O emission by Shewanella oneidensis MR-1 in microbial denitrifying process. Water Res. 169, 115242. https://doi.org/10.1016/j.watres.2019.115242.

Johnson, J.E., Reyes, F.E., Polaski, J.T., Batey, R.T., 2012. B12 cofactors directly stabilize an mRNA regulatory switch. Nature 492 (7427), 133–137. https://doi.org/10.1038/nature11607.

Jones, W.J., Leigh, J.A., Mayer, F., Woese, C.R., Wolfe, R.S., 1983. Methanococcus jannaschii sp. nov., an extremely thermophilic methanogen from a submarine hydrothermal vent. Arch. Microbiol. 136 (4), 254–261. https://doi.org/10.1007/bf00425213.

Jörgensen, A.M., Joergensen, R.G., 2024. Soil contribution to the cobalamin (vitamin B12) supply of terrestrial organisms. Biol. Fertil. Soils 60 (5), 613–625. https://doi.org/10.1007/s00374-024-01828-7.

Kanno, N., Kato, S., Itoh, T., Ohkuma, M., Shigeto, S., 2022. Resonance Raman analysis of intracellular vitamin B12 analogs in methanogenic archaea. Analytical Science Advances 3 (5–6), 165–173. https://doi.org/10.1002/ansa.202100042.

Kazamia, E., Czesnick, H., Nguyen, T.T.V., Croft, M.T., Sherwood, E., Sasso, S., Hodson, S. J., Warren, M.J., Smith, A.G., 2012. Mutualistic interactions between vitamin B12-dependent algae and heterotrophic bacteria exhibit regulation. Environ. Microbiol. 14 (6), 1466–1476. https://doi.org/10.1111/j.1462-2920.2012.02733.x.

Kennedy, K.J., Widner, F.J., Sokolovskaya, O.M., Innocent, L.V., Procknow, R.R., Mok, K.C., Taga, M.E., 2022. Cobalamin riboswitches are broadly sensitive to corrinoid cofactors to enable an efficient gene regulatory strategy. bioRxiv. https://doi.org/10.1101/2022.02.20.481237.

Kern, M., Simon, J., 2009. Electron transport chains and bioenergetics of respiratory nitrogen metabolism in Wolinella succinogenes and other Epsilonproteobacteria. Biochim. Biophys. Acta Bioenerg. 1787 (6), 646–656. https://doi.org/10.1016/j.bbabio.2008.12.010.

Khan, K.S., Mack, R., Castillo, X., Kaiser, M., Joergensen, R.G., 2016. Microbial biomass, fungal and bacterial residues, and their relationships to the soil organic matter C/N/P/S ratios. Geoderma 271, 115–123. https://doi.org/10.1016/j.geoderma.2016.02.019.

Khan, N., Maezato, Y., McClure, R.S., Brislawn, C.J., Mobberley, J.M., Isern, N., Chrisler, W. B., Markillie, L.M., Barney, B.M., Song, H.S., Nelson, W.C., Bernstein, H.C., 2018. Phenotypic responses to interspecies competition and commensalism in a naturally-derived microbial co-culture. Sci. Rep. 8 (1). https://doi.org/10.1038/s41598-017-18630-1.

Kim, H.J., McCarty, R.M., Ogasawara, Y., Liu, Y.N., Mansoorabadi, S.O., Levieux, J., Liu, H. W., 2013. GenK-catalyzed C-6′ methylation in the biosynthesis of gentamicin: isolation and characterization of a cobalamin-dependent radical SAM enzyme. J. Am. Chem. Soc. 135 (22), 8093–8096. https://doi.org/10.1021/ja312641f.

Ko, Y., Ashok, S., Ainala, S.K., Sankaranarayanan, M., Chun, A.Y., Jung, G.Y., Park, S., 2014. Coenzyme B12 can be produced by engineered Escherichia coli under both anaerobic and aerobic conditions. Biotechnol. J. 9 (12), 1526–1535. https://doi.org/10.1002/biot.201400221.

Kondo, Y., 2023. Dynamics of vitamin B_{12} and its importance in the ecosystem in the ocean. Chikyukagaku (Geochemistry) 57, 179–189.

Kudelski, A., 2008. Analytical applications of Raman spectroscopy. Talanta 76 (1), 1–8. https://doi.org/10.1016/j.talanta.2008.02.042.

Kurr, M., Huber, R., König, H., Jannasch, H.W., Fricke, H., Trincone, A., Kristjansson, J.K., Stetter, K.O., 1991. Methanopyrus kandleri, gen. and sp. nov. represents a novel group of hyperthermophilic methanogens, growing at 110°C. Arch. Microbiol. 156 (4), 239–247. https://doi.org/10.1007/BF00262992.

Liao, R.Z., Chen, S.L., Siegbahn, P.E.M., 2016. Unraveling the mechanism and regioselectivity of the B12-dependent reductive dehalogenase PceA. Chem. Eur J. 22 (35), 12391–12399. https://doi.org/10.1002/chem.201601575.

Lu, X., Heal, K.R., Ingalls, A.E., Doxey, A.C., Neufeld, J.D., 2020. Metagenomic and chemical characterization of soil cobalamin production. ISME J. 14 (1), 53–66. https://doi.org/10.1038/s41396-019-0502-0.

Ma, A.T., Tyrell, B., Beld, J., 2020. Specificity of cobamide remodeling, uptake and utilization in Vibrio cholerae. Mol. Microbiol. 113 (1), 89–102. https://doi.org/10.1111/mmi.14402.

Magnabosco, C., Lin, L.-H., Dong, H., Bomberg, M., Ghiorse, W., Stan-Lotter, H., Pedersen, K., Kieft, T.L., van Heerden, E., Onstott, T.C., 2018. The biomass and biodiversity of the continental subsurface. Nat. Geosci. 11 (10), 707–717. https://doi.org/10.1038/s41561-018-0221-6.

Marous, D.R., Lloyd, E.P., Buller, A.R., Moshos, K.A., Grove, T.L., Blaszczyk, A.J., Booker, S.J., Townsend, C.A., 2015. Consecutive radical S-adenosylmethionine methylations form the ethyl side chain in thienamycin biosynthesis. Proc. Natl. Acad. Sci. U. S. A 112 (33), 10354–10358. https://doi.org/10.1073/pnas.1508615112.

Martin-Uriz, P.S., Mirete, S., Alcolea, P.J., Gomez, M.J., Amils, R., Gonzalez-Pastor, J.E., 2014. Nickel-resistance determinants in Acidiphilium sp. Pm identified by genome-wide functional screening. PLoS One 9 (4). https://doi.org/10.1371/journal.pone.0095041.

Mateos, G., Martínez-Bonilla, A., Martínez, J.M., Amils, R., 2023. Vitamin B12 auxotrophy in isolates from the deep subsurface of the Iberian Pyrite Belt. Genes 14 (7). https://doi.org/10.3390/genes14071339.

Miles, Z.D., McCarty, R.M., Molnar, G., Bandarian, V., 2011. Discovery of epoxyqueuosine (oQ) reductase reveals parallels between halorespiration and tRNA modification. Proc. Natl. Acad. Sci. 108 (18), 7368–7372. https://doi.org/10.1073/pnas.1018636108.

Mohammed, Y., Lee, B., Kang, Z., Du, G., 2014. Development of a two-step cultivation strategy for the production of vitamin B12 by Bacillus megaterium. Microb. Cell Fact. 13 (1). https://doi.org/10.1186/s12934-014-0102-7.

Mok, K.C., Sokolovskaya, O.M., Nicolas, A.M., Hallberg, Z.F., Deutschbauer, A., Carlson, H. K., Taga, M.E., 2020. Identification of a novel cobamide remodeling enzyme in the beneficial human gut bacterium Akkermansia muciniphila. mBio 11 (6), 1–18. https://doi.org/10.1128/mBio.02507-20.

Moore, S.J., Mayer, M.J., Biedendieck, R., Deery, E., Warren, M.J., 2014. Towards a cell factory for vitamin B12 production in Bacillus megaterium: bypassing of the cobalamin riboswitch control elements. New Biotechnology 31 (6), 553–561. https://doi.org/10.1016/j.nbt.2014.03.003.

Moore, S.J., Warren, M.J., 2012. The anaerobic biosynthesis of vitamin B12. Biochem. Soc. Trans. 40 (3), 581–586. https://doi.org/10.1042/BST20120066.

Mozafar, A., 1994. Enrichment of some B-vitamins in plants with application of organic fertilizers. Plant Soil 167 (2), 305–311. https://doi.org/10.1007/bf00007957.

Mulligan, C.N., Yong, R.N., Gibbs, B.F., 2001. Remediation technologies for metal-contaminated soils and groundwater: an evaluation. Eng. Geol. 60 (1–4), 193–207. https://doi.org/10.1016/s0013-7952(00)00101-0.

Muñoz-Velasco, I., García-Ferris, C., Hernandez-Morales, R., Lazcano, A., Peretó, J., Becerra, A., 2018. Methanogenesis on early stages of life: ancient but not primordial. Orig. Life Evol. Biosph. 48 (4), 407–420. https://doi.org/10.1007/s11084-018-9570-9.

Nahvi, A., Barrick, J.E., Breaker, R.R., 2004. Coenzyme B12 riboswitches are widespread genetic control elements in prokaryotes. Nucleic Acids Res. 32 (1), 143–150. https://doi.org/10.1093/nar/gkh167.

Nijland, M., Lefebvre, S.N., Thangaratnarajah, C., Slotboom, D.J., 2024. Bidirectional ATP-driven transport of cobalamin by the mycobacterial ABC transporter BacA. Nat. Commun. 15 (1). https://doi.org/10.1038/s41467-024-46917-1.

Nuppunen-Puputti, M., Kietäväinen, R., Raulio, M., Soro, A., Purkamo, L., Kukkonen, I., Bomberg, M., 2022. Epilithic microbial community functionality in deep oligotrophic continental bedrock. Front. Microbiol. 13. https://doi.org/10.3389/fmicb.2022.826048.

Okbamichael, M., Sañudo-Wilhelmy, S.A., 2004. A new method for the determination of Vitamin B12 in seawater. Anal. Chim. Acta 517 (1–2), 33–38. https://doi.org/10.1016/j.aca.2004.05.020.

Oren, A., Garrity, G.M., 2021. Valid publication of new names and new combinations effectively published outside the IJSEM. Int. J. Syst. Evol. Microbiol. 71 (5). https://doi.org/10.1099/ijsem.0.004773.

Panzeca, C., Beck, A.J., Tovar-Sanchez, A., Segovia-Zavala, J., Taylor, G.T., Gobler, C.J., Sañudo-Wilhelmy, S.A., 2009. Distributions of dissolved vitamin B12 and Co in coastal and open-ocean environments. Estuar. Coast Shelf Sci. 85 (2), 223–230. https://doi.org/10.1016/j.ecss.2009.08.016.

Parent, A., Guillot, A., Benjdia, A., Chartier, G., Leprince, J., Berteau, O., 2016. The B 12-radical SAM enzyme PoyC catalyzes valine C β-methylation during polytheonamide biosynthesis. J. Am. Chem. Soc. 138 (48), 15515–15518. https://doi.org/10.1021/jacs.6b06697.

Pieńko, T., Czarnecki, J., Równicki, M., Wojciechowska, M., Wierzba, A.J., Gryko, D., Bartosik, D., Trylska, J., 2021. Vitamin B12-peptide nucleic acids use the BtuB receptor to pass through the Escherichia coli outer membrane. Biophys. J. 120 (4), 725–737. https://doi.org/10.1016/j.bpj.2021.01.004.

Puente-Sánchez, F., Arce-Rodríguez, A., Oggerin, M., García-Villadangos, M., Moreno-Paz, M., Blanco, Y., Rodríguez, N., Bird, L., Lincoln, S.A., Tornos, F., Prieto-Ballesteros, O., Freeman, K.H., Pieper, D.H., Timmis, K.N., Amils, R., Parro, V., 2018. Viable cyanobacteria in the deep continental subsurface. Proc. Natl. Acad. Sci. 115 (42), 10702–10707. https://doi.org/10.1073/pnas.1808176115.

Raux, E., Lanois, A., Levillayer, F., Warren, M.J., Brody, E., Rambach, A., Thermes, C., 1996. Salmonella typhimurium cobalamin (vitamin B12) biosynthetic genes: functional studies in S. typhimurium and Escherichia coli. J. Bacteriol. 178 (3), 753–767. https://doi.org/10.1128/jb.178.3.753-767.1996.

Rempel, S., Gati, C., Nijland, M., Thangaratnarajah, C., Karyolaimos, A., de Gier, J.W., Guskov, A., Slotboom, D.J., 2020. A mycobacterial ABC transporter mediates the uptake of hydrophilic compounds. Nature 580 (7803), 409–412. https://doi.org/10.1038/s41586-020-2072-8.

Rempel, S., Colucci, E., de Gier, J.W., Guskov, A., Slotboom, D.J., 2018. Cysteine-mediated decyanation of vitamin B12 by the predicted membrane transporter BtuM. Nat. Commun. 9 (1). https://doi.org/10.1038/s41467-018-05441-9.

Robertson, D.E., 1970. The distribution of cobalt in Oceanic waters. Geochem. Cosmochim. Acta 34 (5), 553–567. https://doi.org/10.1016/0016-7037(70)90016-5.

Robin, C., Blanche, F., Cauchois, L., Cameron, B., Couder, M., Crouzet, J., 1991. Primary structure, expression in Escherichia coli, and properties of S-adenosyl-L-methionine:uroporphyrinogen III methyltransferase from Bacillus megaterium. J. Bacteriol. 173 (15), 4893–4896. https://doi.org/10.1128/jb.173.15.4893-4896.1991.

Rodionov, V., Yi, J., Kashina, A., Oladipo, A., Gross, S.P., 2003. Switching between microtubule- and actin-based transport systems in melanophores is controlled by cAMP levels. Curr. Biol. 13 (21), 1837–1847. https://doi.org/10.1016/j.cub.2003.10.027.

Romine, M.F., Rodionov, D.A., Maezato, Y., Anderson, L.N., Nandhikonda, P., Rodionova, I. A., Carre, A., Li, X., Xu, C., Clauss, T.R.W., Kim, Y.M., Metz, T.O., Wright, A.T., 2017a. Elucidation of roles for vitamin B12 in regulation of folate, ubiquinone, and methionine metabolism. Proc. Natl. Acad. Sci. U. S. A 114 (7), E1205. https://doi.org/10.1073/pnas.1612360114.

Romine, M.F., Rodionov, D.A., Maezato, Y., Osterman, A.L., Nelson, W.C., 2017b. Underlying mechanisms for syntrophic metabolism of essential enzyme cofactors in microbial communities. ISME J. 11 (6), 1434–1446. https://doi.org/10.1038/ismej.2017.2.

Ruff, S.E., Schwab, L., Vidal, E., Hemingway, J.D., Kraft, B., Murali, R., 2024. Widespread occurrence of dissolved oxygen anomalies, aerobic microbes, and oxygen-producing metabolic pathways in apparently anoxic environments. FEMS (Fed. Eur. Microbiol. Soc.) Microbiol. Ecol. 100 (11). https://doi.org/10.1093/femsec/fiae132.

Saito, M.A., Moffett, J.W., DiTullio, G.R., 2004. Cobalt and nickel in the Peru upwelling region: a major flux of labile cobalt utilized as a micronutrient. Glob. Biogeochem. Cycles 18 (4), 1–14. https://doi.org/10.1029/2003GB002216.

Santos, J.A., Rempel, S., Mous, S.T.M., Pereira, C.T., Beek, J.T., de Gier, J.W., Guskov, A., Slotboom, D.J., 2018. Functional and structural characterization of an ECF-type ABC transporter for vitamin B12. eLife 7. https://doi.org/10.7554/eLife.35828.

Sañudo-Wilhelmy, S.A., Gómez-Consarnau, L., Suffridge, C., Webb, E.A., 2014. The role of B vitamins in marine biogeochemistry. Ann. Rev. Mar. Sci 6, 339–367. https://doi.org/10.1146/annurev-marine-120710-100912.

Sanz, J.L., Rodriguez, N., Escudero, C., Carrizo, D., Amils, R., 2021. Biological production of H2, CH4 and CO2 in the deep subsurface of the Iberian Pyrite Belt. Environ. Microbiol. 23 (7), 3913–3922. https://doi.org/10.1111/1462-2920.15561.

Shelton, A.N., Seth, E.C., Mok, K.C., Han, A.W., Jackson, S.N., Haft, D.R., Taga, M.E., 2019. Uneven distribution of cobamide biosynthesis and dependence in bacteria predicted by comparative genomics. ISME J. 13 (3), 789–804. https://doi.org/10.1038/s41396-018-0304-9.

Soares, A., Edwards, A., An, D., Bagnoud, A., Bradley, J., Barnhart, E., Bomberg, M., Budwill, K., Caffrey, S.M., Fields, M., Gralnick, J., Kadnikov, V., Momper, L., Osburn, M., Mu, A., Moreau, J.W., Moser, D., Purkamo, L., Rassner, S.M., Sheik, C.S., Sherwood Lollar, B., Toner, B.M., Voordouw, G., Wouters, K., Mitchell, A.C., 2023. A global perspective on bacterial diversity in the terrestrial deep subsurface. Microbiology 169 (1). https://doi.org/10.1099/mic.0.001172.

Sokolovskaya, O.M., Shelton, A.N., Taga, M.E., 2020. Sharing vitamins: cobamides unveil microbial interactions. Science 369 (6499). https://doi.org/10.1126/science.aba0165.

Stupperich, E., Kräutler, B., 1988. Pseudo vitamin B12 or 5-hydroxybenzimidazolyl-cobamide are the corrinoids found in methanogenic bacteria. Arch. Microbiol. 149 (3), 268–271. https://doi.org/10.1007/BF00422016.

Sweetman, A.K., Smith, A.J., de Jonge, D.S.W., Hahn, T., Schroedl, P., Silverstein, M., Andrade, C., Edwards, R.L., Lough, A.J.M., Woulds, C., Homoky, W.B., Koschinsky, A., Fuchs, S., Kuhn, T., Geiger, F., Marlow, J.J., 2024. Evidence of dark oxygen production at the abyssal seafloor. Nat. Geosci. 17 (8), 737–739. https://doi.org/10.1038/s41561-024-01480-8.

Swithers, K.S., Petrus, A.K., Secinaro, M.A., Nesbø, C.L., Gogarten, J.P., Noll, K.M., Butzin, N.C., 2012. Vitamin B12 synthesis and salvage pathwayswere acquired by horizontal gene transfer to the Thermotogales. Genome Biol. Evol. 4 (8), 730–739. https://doi.org/10.1093/gbe/evs057.

Taleuzzaman, M., Ali, S., Gilani, S.J., Imam, S.S., Hafeez, A., 2015. Ultra performance liquid chromatography (UPLC)-a review. Austin J. Anal. Pharm. Chem. 2 (6).

Tang, Y.Z., Koch, F., Gobler, C.J., 2010. Most harmful algal bloom species are vitamin B1 and B12 auxotrophs. Proc. Natl. Acad. Sci. 107 (48), 20756–20761. https://doi.org/10.1073/pnas.1009566107.

Tripathi, A., Pandey, V.K., Panesar, P.S., Taufeeq, A., Mishra, H., Rustagi, S., Malik, S., Kovács, B., Suthar, T., Shaikh, A.M., 2024. Fermentative production of vitamin B 12 by Propionibacterium shermanii and Pseudomonas denitrificans and its promising health benefits: a review. Food Sci. Nutr. 12 (11), 8675–8691. https://doi.org/10.1002/fsn3.4428.

Villa, E.A., Escalante-Semerena, J.C., 2024. Corrinoid salvaging and cobamide remodeling in bacteria and archaea. J. Bacteriol. 206 (11). https://doi.org/10.1128/jb.00286-24.

Walworth, N.G., Lee, M.D., Suffridge, C., Qu, P., Fu, F.X., Saito, M.A., et al., 2018. Functional genomics and phylogenetic evidence suggest genus-wide cobalamin production by the globally distributed marine nitrogen fixer Trichodesmium. Front. Microbiol. 9.

Wood, J.M., Moura, I., Moura, J.J.G., Santos, M.H., Xavier, A.V., LeGall, J., Scandellari, M., 1982. Role of vitamin B 12 in methyl transfer for methane biosynthesis by Methanosarcina barkeri. Science 216 (4543), 303–305. https://doi.org/10.1126/science.7063887.

Yi, S., Seth, E.C., Men, Y.J., Stabler, S.P., Allen, R.H., Alvarez-Cohen, L., Taga, M.E., 2012. Versatility in corrinoid salvaging and remodeling pathways supports corrinoid-dependent metabolism in Dehalococcoides mccartyi. Appl. Environ. Microbiol. 78 (21), 7745–7752. https://doi.org/10.1128/AEM.02150-12.

Zhang, Y., Dekas, A.E., Hawkins, A.J., Parada, A.E., Gorbatenko, O., Li, K., Horne, R.N., 2020. Microbial community composition in deep-subsurface reservoir fluids reveals natural interwell connectivity. Water Resour. Res. 56 (2). https://doi.org/10.1029/2019WR025916.

Zheng, K., Rush, K.W., Date, S.S., Johs, A., Parks, J.M., Fleischhacker, A.S., Abernathy, M.J., Sarangi, R., Ragsdale, S.W., 2024. S-adenosyl-L-methionine is the unexpected methyl donor for the methylation of mercury by the membrane-associated HgcAB complex. Proc. Natl. Acad. Sci. U. S. A 121 (47). https://doi.org/10.1073/pnas.2408086121.

Zhou, J., Yu, X., Liu, J., Qin, W., He, Z., Stahl, D., Jiao, N., Zhou, J., Tu, Q., 2021. Vb12path for accurate metagenomic profiling of microbially driven cobalamin synthesis pathways. mSystems 6 (3). https://doi.org/10.1128/mSystems.00497-21.

Index

Note: Page numbers followed by "f" indicate figures and "t" indicate tables.

A

B

C

D

I

Q

R

S

T

9780443329487